FOR EXTRA PRACTICE

The following sections feature extra practice problems available when you register for MyMathLab.com. For your convenience, these exercises are also printed in the *Instructor's Resource Guide*. Please contact your instructor for assistance.

Elementary Algebra

CONCEPTS AND APPLICATIONS

Elementary Algebra

CONCEPTS AND APPLICATIONS

SIXTH EDITION

MARVIN L. BITTINGER
*Indiana University—
Purdue University at Indianapolis*

DAVID J. ELLENBOGEN
Community College of Vermont

Addison
Wesley

Boston • San Francisco • New York • London • Toronto • Sydney • Tokyo • Singapore
Madrid • Mexico City • Munich • Paris • Cape Town • Hong Kong • Montreal

Publisher	Jason A. Jordan
Acquisitions Editor	Jennifer Crum
Project Manager	Kari Heen
Assistant Editor	Greg Erb
Managing Editor	Ron Hampton
Production Supervisor	Kathleen A. Manley
Text Design	Geri Davis/The Davis Group, Inc.
Editorial and Production Services	Martha Morong/Quadrata, Inc.
Art Editor	Geri Davis/The Davis Group, Inc.
Marketing Manager	Dona Kenly
Illustrators	Network Graphics and Jim Bryant
Compositor	The Beacon Group
Cover Design	Dennis Schaefer
Cover Photograph	Stuart Westmorland/Index Stock Imagery
Prepress Supervisor	Caroline Fell
Print Buyer	Evelyn Beaton

Photo Credits
1, UPI **6,** The Image Bank **28,** UPI **69,** Christine D. Tuff **99,** Appalachian Trail Conference **118,** Christine D. Tuff **129,** © David Keaton, The Stock Market **164,** The Image Bank **170,** AP/Wide World Photos **182,** Mary Clay/Tom Stack & Associates **201,** Howard Sochurek, The Stock Market **273,** © Bob Daemmrich, The Stock Market **331,** © Bachmann, Stock Boston **376,** Todd Phillips/Third Coast Stock Source, Inc. **380,** American Honda **381,** Stuart Westmorland/Index Stock Imagery **393,** PhotoDisc **416,** Archive Photos **445,** PhotoDisc **470,** UPI **473,** PhotoDisc **491,** Ed Bock, The Stock Market **510,** The Toronto Star/B. Spremo **516,** Getty One

The product name "TI-83 Plus" and the likeness of the same product are used by permission of Texas Instruments.

Library of Congress Cataloging-in-Publication Data
Bittinger, Marvin L.
   Elementary algebra: concepts and applications / Marvin L. Bittinger, David J. Ellenbogen.—6th ed.
    p. cm.
   Includes index.
   ISBN 0-201-71965-7 (alk. paper)
   1. Algebra. I. Ellenbogen, David. II. Title.
QA152.2 .B5797 2001
512.9—dc21

00-068956

4 5 6 7 8 9 10—DOW—05 04 03 02

For my brother Tony,
with love

Contents

Preface

We are pleased to present the sixth edition of *Elementary Algebra: Concepts and Applications*. Each time we work on a new edition, it's a balancing act. On the one hand, we want to preserve the features, applications, and explanations that faculty have come to rely on and expect. On the other hand, we want to blend our own ideas for improvement with the many insights that we receive from faculty and students throughout North America. The result is a living document in which new features and applications are developed while successful features and popular applications from previous editions are updated and refined. Our goal, as always, is to present content that is easy to understand and has the depth required for success in this and future courses.

Appropriate for a one-term course in elementary algebra, this text is intended for those students who have a firm background in arithmetic. It is the first of three texts in an algebra series that also includes *Intermediate Algebra: Concepts and Applications*, Sixth Edition, by Bittinger/Ellenbogen and *Elementary and Intermediate Algebra: Concepts and Applications, A Combined Approach*, Third Edition, by Bittinger/Ellenbogen/Johnson.

Approach

Our goal, quite simply, is to help today's students both learn and retain mathematical concepts. To achieve this goal, we feel that we must prepare developmental-mathematics students for the transition from "skills-oriented" elementary and intermediate algebra courses to more "concept-oriented" college-level mathematics courses. This requires that we teach these same students critical thinking skills: to reason mathematically, to communicate mathematically, and to identify and solve mathematical problems. Following are some aspects of our approach that we use in this revision to help meet the challenges we all face teaching developmental mathematics.

Problem Solving

One distinguishing feature of our approach is our treatment of and emphasis on problem solving. We use problem solving and applications to motivate the material wherever possible, and we include real-life applications and problem-solving techniques throughout the text. Problem solving not only encourages students to think about how mathematics can be used, it helps to prepare them for more advanced material in future courses.

- In Chapter 2, we introduce the five-step process for solving problems: (1) Familiarize, (2) Translate, (3) Carry out, (4) Check, and (5) State the answer. These steps are then used consistently throughout the text whenever we encounter a problem-solving situation. Repeated use of this problem-solving strategy gives students a sense that they have a starting point for any type of problem they encounter, and frees them to focus on the mathematics necessary to successfully translate the problem situation. We often use estimation and carefully checked guesses to help with the Familiarize and Check steps (see pp. 103, 118, and 416). In this edition, we also use dimensional analysis as a quick check of certain problems (see pp. 162 and 164).

Applications

Interesting applications of mathematics help motivate both students and instructors. Solving applied problems gives students the opportunity to see their conceptual understanding put to use in a real way. In the Sixth Edition of *Elementary Algebra: Concepts and Applications*, not only have we increased the total number of applications and real-data problems overall, nearly 20 percent of our applications are new, and we have increased the number of source lines to better highlight the real-world data. As in the past, art is integrated into the applications and exercises to aid the student in visualizing the mathematics. (See pp. 140, 189, 323, and 414.)

Pedagogy

New! **Connecting the Concepts.**   To help students understand the "big picture," Connecting the Concepts subsections within each chapter (and highlighted in the table of contents) relate the concept at hand to previously learned and upcoming concepts. Because students may occasionally "lose sight of the forest because of the trees," we feel confident that this feature will help them keep better track of their bearings as they encounter new material. (See pp. 177, 311, 369, and 467.)

New! **Study Tips.**   Most plentiful in the first three chapters when students are still establishing their study habits, Study Tips are found in the margins and interspersed throughout the first six chapters. Our Study Tips range from how to approach assignments, to reminders of the various study aids that are available, to strategies for preparing for a final exam. (See pp. 75, 100, and 134.)

New! *Aha!* **Exercises.** Designated by $Aha!$, these exercises can be solved quickly if the student has the proper insight. The $Aha!$ designation is used the first time a new insight can be used on a particular type of exercise and indicates to the student that there is a simpler way to complete the exercise that requires less lengthy computation. It's then up to the student to find the simpler approach and, in subsequent exercises, to determine if and when that particular insight can be used again. Our hope is that the *Aha!* exercises will discourage rote learning and reward students who "look before they leap" into a problem. (See pp. 90, 189, 249, and 414.)

Technology Connections. Throughout each chapter, optional Technology Connection boxes help students use graphing calculator technology to better visualize a concept that they have just learned. To connect this feature to the exercise sets, certain exercises are marked with a graphing calculator icon and reinforce the use of this optional technology. (See pp. 80, 85, 112, 215, 223, 232, and 279.)

Skill Maintenance Exercises. Retaining mathematical skills is critical to a student's success in future courses. To this end, nearly every exercise set includes six to eight Skill Maintenance exercises that review skills and concepts from preceding chapters of the text. In this edition, not only have the Skill Maintenance exercises been increased by 50 percent, but they are now designed to provide extra practice with the specific skills needed for the next section of the text. (See pp. 182, 226, 297, and 414.)

Synthesis Exercises. Following the Skill Maintenance section, each exercise set ends with a group of Synthesis exercises designated by their own heading. These exercises offer opportunities for students to synthesize skills and concepts from earlier sections with the present material, and often provide students with deeper insights into the current topic. Synthesis exercises are generally more challenging than those in the main body of the exercise set and occasionally include *Aha!* exercises. (See pp. 76, 209, 243, and 368.)

Writing Exercises. In this edition, nearly every set of exercises includes at least four writing exercises. Two of these are more basic and appear just before the Skill Maintenance exercises. The other writing exercises are more challenging and appear as Synthesis exercises. All writing exercises are marked with ▨ and require answers that are one or more complete sentences. This type of problem has been found to aid in student comprehension, critical thinking, and conceptualization. Because some instructors may collect answers to writing exercises, and because more than one answer may be correct, answers to writing exercises are not listed at the back of the text. (See pp. 76, 140, 233, and 368.)

Collaborative Corners. In today's professional world, teamwork is essential. We continue to provide optional Collaborative Corner features throughout the text that require students to work in groups to explore and solve mathematical problems. On average, there are two to three Collaborative Corners per chapter, each one appearing after the appropriate exercise set. Additional Collaborative Corner activities and suggestions for directing collaborative learning appear in the *Printed Test Bank/Instructor's Resource Guide.* (See pp. 10, 124, 169, and 252.)

Cumulative Review. After Chapters 3, 6, and 9, we have included a Cumulative Review, which reviews skills and concepts from all preceding chapters of the text. (See pp. 199, 391, and 541.)

What's New in the Sixth Edition?

We have rewritten many key topics in response to user and reviewer feedback and have made significant improvements in design, art, pedagogy, and an expanded supplements package. Detailed information about the content changes is available in the form of a conversion guide. Please ask your local Addison-Wesley sales consultant for more information. Following is a list of the major changes in this edition.

New Design

You will see that the page dimension for this edition is larger, which allows for an open look and a typeface that is easier to read. In addition, we continue to pay close attention to the pedagogical use of color to make sure that it is used to present concepts in the clearest possible manner.

Content Changes

A variety of content changes have been made throughout the text. Some of the more significant changes are listed below.

- Chapter 3 now includes a gentle introduction to interpolation and extrapolation as well as a new section that stresses the connection between rate of change and graphing. Slope–intercept and point–slope form also now appear in this chapter.
- Chapter 6 has expanded material on rates and units. Section 6.8 of the Fifth Edition has been merged into Chapter 9 (see Section 9.4, below).
- Chapter 7 is now devoted to systems of equations and linear inequalities. The material is presented in a manner that allows instructors to cover this material earlier if they wish.
- Section 9.4 in Chapter 9 now covers a wide variety of formulas and provides practice in recognizing and solving all the various types of equations previously discussed.

Supplements for the Instructor

New! Annotated Instructor's Edition
(ISBN 0-201-65870-4)

The *Annotated Instructor's Edition* includes all the answers to the exercise sets, usually right on the page where the exercises appear, and Teaching Tips in the margins that give insights and classroom discussion suggestions that will be especially useful for new instructors. These handy answers and ready Teaching Tips will help both new and experienced instructors save classroom preparation time.

New! Web Site: www.MyMathLab.com

In addition to providing a wealth of resources for lecture-based courses, MyMathLab.com gives instructors a quick and easy way to create a complete on-line course based on *Elementary Algebra: Concepts and Applications*, Sixth Edition. MyMathLab.com is hosted nationally at no cost to instructors, students, or schools, and it provides access to an interactive learning environment where all content has been developed to directly enhance and reinforce the text. Using CourseCompass, a customized version of Blackboard, Inc.®, as the course-management platform, MyMathLab.com lets instructors administer preexisting tests and quizzes or create their own, and it provides detailed tracking of all student work as well as a wide array of communication tools for course participants. Within MyMathLab.com, students link directly from on-line pages of their text to supplementary resources such as tutorial software, interactive animations, and audio and video clips.

Printed Test Bank/ Instructor's Resource Guide (ISBN 0-201-73407-9)

The Instructor's Resource Guide portion of this supplement contains the following:

- Extra practice problems
- Black-line masters of grids and number lines for transparency masters or test preparation
- A videotape index and section cross-references to our tutorial software packages
- Additional collaborative learning activities and suggestions
- A syllabus conversion guide from the Fifth Edition to the Sixth Edition

The Printed Test Bank portion of this supplement contains the following:

- Six new alternate free-response test forms for each chapter, organized with the same topic order as the chapter tests in the main text. Each form includes synthesis questions, as appropriate, at the end of each test.
- Two new multiple-choice versions of each chapter test
- Eight new alternate test forms for the final examination: Alternate

Test Forms A, B, and C of the final examinations are organized by chapter and D, E, and F are organized by problem type.
- Answers to all tests

Instructor's Solutions Manual
(ISBN 0-201-73408-7)
The *Instructor's Solutions Manual* contains fully worked-out solutions to the odd-numbered exercises and brief solutions to the even-numbered exercises in the exercise sets.

Answer Book (ISBN 0-201-73796-5)
The *Answer Book* includes answers to all even-numbered and odd-numbered exercises.

TestGen-EQ/QuizMaster-EQ
(ISBN 0-201-73061-8)
Available on a dual-platform Windows/Macintosh CD-ROM, this fully networkable software enables instructors to build, edit, print, and administer tests using a computerized test bank of questions organized according to the contents of each chapter. Tests can be printed or saved for on-line testing via a network on the Web, and the software can generate a variety of grading reports for tests and quizzes.

InterAct Math Plus
(Windows ISBN 0-201-63555-0; Macintosh ISBN 0-201-64805-9)
This networkable software provides course management and on-line administration for Addison-Wesley's InterAct Math Tutorial Software (see "Supplements for the Student"). InterAct Math Plus enables instructors to create and administer on-line tests, summarize students' results, and monitor students' progress in the tutorial software, providing an invaluable teaching and tracking resource.

InterAct MathXL: www.mathxl.com
(12-month registration ISBN 0-201-71111-7, stand-alone)
The MathXL Web site provides diagnostic testing and tutorial help, all on-line using InterAct Math® tutorial software and TestGen-EQ testing software. Students can take chapter tests correlated to the text, receive individualized study plans based on those test results, work practice problems and receive tutorial instruction for areas in which they need improvement, and take further tests to gauge their progress. Instructors can customize tests and track all student test results, study plans, and practice work.

Supplements for the Student

New! Web Site: www.MyMathLab.com
Ideal for lecture-based, lab-based, and on-line courses, this state-of-the-art Web site provides students with a centralized point of access to the wide variety of on-line resources available with this text. The pages

of the actual book are loaded into MyMathLab.com, and as students work through a section of the on-line text, they can link directly from the pages to supplementary resources (such as tutorial software, interactive animations, and audio and video clips) that provide instruction, exploration, and practice beyond what is offered in the printed book. MyMathLab.com generates personalized study plans for students and allows instructors to track all student work on tutorials, quizzes, and tests. Complete course-management capabilities, including a host of communication tools for course participants, are provided to create a user-friendly and interactive on-line learning environment.

Student's Solutions Manual
(ISBN 0-201-65871-2)
The *Student's Solutions Manual* by Judith A. Penna contains completely worked-out solutions with step-by-step annotations for all the odd-numbered exercises in the text, with the exception of the Writing exercises. This manual also lists, without complete solutions, the answers for even-numbered text exercises.

InterAct Math® Tutorial CD-ROM
(ISBN 0-201-74623-9)
This interactive tutorial software provides algorithmically generated practice exercises that correlate at the objective level to the odd-numbered exercises in the text. Each practice exercise is accompanied by both an example and a guided solution designed to involve students in the solution process. For Windows users, selected problems also include a video clip that helps students visualize concepts. The software recognizes common student errors and provides appropriate feedback. Instructors can use InterAct Math Plus course management software to create, administer, and track on-line tests and monitor student performance during practice sessions.

InterAct MathXL www.mathxl.com
(12-month registration ISBN 0-201-71630-5, stand-alone)
The MathXL Web site provides diagnostic testing and tutorial help, all on-line, using InterAct Math® tutorial software and TestGen-EQ testing software. Students can take chapter tests correlated to the text, receive individualized study plans based on those test results, work practice problems and receive tutorial instruction for areas in which they need improvement, and take further tests to gauge their progress.

Videotapes (ISBN 0-201-74208-X)
Developed and produced especially for this text, the videotapes feature an engaging team of instructors, including the authors. These instructors present material and concepts by using examples and exercises from every section of the text in a format that stresses student interaction.

Digital Video Tutor
(ISBN 0-201-74642-5, stand-alone)
The videotapes for this text are now available on CD-ROM, making it easy and convenient for students to watch video segments from a

computer at home or on campus. The complete digitized video set, now affordable and portable for students, is ideal for distance learning or supplemental instruction.

AW Math Tutor Center

(ISBN 0-201-72170-8, stand-alone)

The AW Math Tutor Center is staffed by qualified mathematics instructors who provide students with tutoring on examples and odd-numbered exercises from the textbook. Tutoring is available via toll-free telephone, fax, or e-mail.

Acknowledgments

No book can be produced without a team of professionals who take pride in their work and are willing to put in long hours. Barbara Johnson, in particular, deserves special thanks for her work as development editor. Barbara's tireless devotion to all aspects of this project and her many fine suggestions have contributed immeasurably to the quality of this text. Laurie A. Hurley also deserves special thanks for her careful accuracy checks, well-thought-out suggestions, and uncanny eye for detail. Judy Penna's outstanding work in organizing and preparing the printed supplements and the indexes amounts to an inspection of the text that goes far beyond the call of duty and for which we will always be extremely grateful. Thanks to Tom Schicker for authoring the *Printed Test Bank*. Dawn Mulheron not only served as an accuracy checker, but was terrifically helpful in posting and double-checking the "catches" found by the checkers. Cassidy Ferraro and Deirdre Hurley provided enormous help, often in the face of great time pressure, as accuracy checkers. We are also indebted to Chris Burditt and Jann MacInnes for their many fine ideas that appear in our Collaborative Corners and Vince McGarry and Janet Wyatt for their recommendations for Teaching Tips featured in the *Annotated Instructor's Edition*.

Martha Morong, of Quadrata, Inc., provided editorial and production services of the highest quality imaginable—she is simply a joy to work with. Geri Davis, of the Davis Group, Inc., performed superb work as designer, art editor, and photo researcher, and always with a disposition that can brighten an otherwise gray day. Network Graphics generated the graphs, charts, and many of the illustrations. Not only are the people at Network reliable, but they clearly take pride in their work. The many hand-drawn illustrations appear thanks to Jim Bryant, a gifted artist with true mathematical sensibilities. Tom and Pam Hansen, of Copy Ship Fax Plus, consistently went the extra yard in providing the best in copying services.

Our team at Addison-Wesley deserves special thanks. Assistant Editor Greg Erb coordinated all the reviews, tracked down countless pieces of information, and managed many of the day-to-day details—always in a pleasant and reliable manner. Executive Project Manager Kari Heen expertly provided a steadying influence along with gentle prodding at just the right moments. Senior Acquisitions Editor Jenny Crum provided many fine suggestions along with unflagging support.

Senior Production Supervisor Kathy Manley exhibited patience when others would have shown frustration. Designer Dennis Schaefer's willingness to listen and then creatively respond resulted in a book that is beautiful to look at. Marketing Manager Dona Kenly skillfully kept us in touch with the needs of faculty; Executive Technology Producer Lorie Reilly provided us with the technological guidance so necessary for our many supplements; and Media Producer Tricia Mescall remains the steady hand responsible for our fine video series. Our publisher, Jason Jordan, deserves credit for assembling this fine team and remaining accessible to us on both a professional and personal level. To all of these people we owe a real debt of gratitude.

We also thank the students at the Community College of Vermont and the following professors for their thoughtful reviews and insightful comments.

Prerevision Diary Reviewers (Fifth Edition)
Rochelle Beatty, *Southwestern Oklahoma State University*
Heidi Howard, *Florida Community College—Jacksonville*

Manuscript Reviewers
Mazie Akana, *Leeward Community College*
Peter Arvanites, *Rockland Community College*
Rochelle Beatty, *Southwestern Oklahoma State University*
Joy Bjerke, *Reedley College*
Connie Buller, *Metropolitan Community College*
Michael Butler, *College of the Redwoods*
Wilton Clarke, *La Sierra University*
Beth Fraser, *Middlesex Community College*
Abel Gage, *Skagit Valley College*
Barbara Gardner, *Carroll Community College*
Lonnie Hass, *North Dakota State University*
Tony Julianelle, *University of Vermont*
Jennifer Laveglia, *Bellevue Community College*
William Livingston, *Missouri Southern State College*
Mitzi Logan, *Pitt Community College*
Jann MacInnes, *Florida Community College—Jacksonville–Kent*
Jean-Marie Magnier, *Springfield Technical Community College*
Mary Martin, *Consumnes River College*
Eric Matsuoka, *Leeward Community College*
Laurie McManus, *St. Louis Community College—Meramac*
Wayne Neidhardt, *Edmonds Community College*
Lymeda Singleton, *Abilene Christian University*
Gwen Terwilliger, *University of Toledo*
Virginia Urban, *Fashion Institute of Technology*

Finally, a special thank you to all those who so generously agreed to discuss their professional use of mathematics in our chapter openers. These dedicated people, none of whom we knew prior to writing this text, all share a desire to make math more meaningful to students. We cannot imagine a finer set of role models.

M.L.B.
D.J.E.

Feature Walkthrough

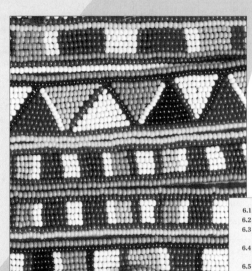

6
Rational Expressions and Equations

CHAPTER OPENERS
Each chapter opens with a list of the sections covered and a real-life application that includes a testimonial from a person in that field to show how integral mathematics is in problem solving. Real data are often used in these applications, as well as in many other exercises, and in "on the job" examples (like those students might find in the workplace) to increase student interest.

AN APPLICATION

In South Africa, the design of every woven handbag, or *gipatsi*, is created by repeating two or more patterns around the bag. If a weaver uses a four-strand, a six-strand, and an eight-strand pattern, what is the smallest number of strands needed in order for all three patterns to repeat a whole number of times?

This problem appears as Exercise 80 in Section 6.3.

Weavers use math when designing patterns, from calculating the size of the pattern and number of times it repeats to figuring the number of threads per inch and the weight of the thread. I also use math to convert between English and metric units when dyeing warps and woofs.

ESTELLE CARLSON
Handweaver and Designer
Los Angeles, California

E x a m p l e   4   Solve: $x^2 + 5x + 6 = 0$.

Solution   This equation differs from those so[...]
like terms to combine, and there is a squared t[...]
mial. Then we use the principle of zero produc[...]

$$x^2 + 5x + 6 = 0$$
$$(x + 2)(x + 3) = 0 \qquad \text{Fa[...]}$$
$$x + 2 = 0 \quad or \quad x + 3 = 0 \qquad \text{Us[...]}$$
$$x = -2 \quad or \qquad x = -3.$$

Check: For -2:

$$\frac{x^2 + 5x + 6 = 0}{(-2)^2 + 5(-2) + 6 \;?\; 0}$$
$$4 - 10 + 6$$
$$-6 + 6$$
$$0 \mid 0 \quad \text{TRUE}$$

For -3:

$$\frac{x^2 + 5x + 6 = 0}{(-3)^2 + 5(-3) + 6 \;?\; 0}$$
$$9 - 15 + 6$$
$$-6 + 6$$
$$0 \mid 0 \quad \text{TRUE}$$

The solutions are -2 and -3.

The principle of zero products is used even if the factoring consists of only removing a common factor.

Study Tip

Immediately after each quiz or test, write out a step-by-step solution to any questions you missed. Visit your professor during office hours or consult with a tutor for help with problems that are still giving you trouble. Misconceptions tend to resurface if they are not corrected as soon as possible.

for every horizontal distance of 100 ft, the [...]
of grade also occurs in skiing or snowboar[...]
very tame, but a 40% grade is considered s[...]

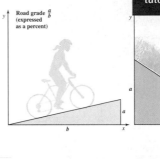

Road grade $\frac{a}{b}$ (expressed as a percent)

E x a m p l e   5   **Skiing.**   Among the steepest skiable terrain in North America, the Headwall on Mount Washington, in New Hampshire, drops 720 ft over a horizontal distance of 900 ft. Find the grade of the Headwall.

E x a m p l e   1

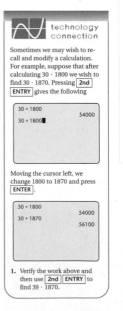

technology connection

Sometimes we may wish to re-call and modify a calculation. For example, suppose that after calculating $30 \cdot 1800$ we wish to find $30 \cdot 1870$. Pressing [2nd] [ENTRY] gives the following

```
30 * 1800
              54000
30 * 1800▮
```

Moving the cursor left, we change 1800 to 1870 and press [ENTER].

```
30 * 1800
              54000
30 * 1870
              56100
```

1. Verify the work above and then use [2nd] [ENTRY] to find $39 \cdot 1870$.

Furnace output.   Contractors in the Northeast use the formula $B = 30a$ to determine the minimum furnace output B, in British thermal units (Btu's), for a well-insulated house with a square feet of flooring (*Source*: U.S. Department of Energy). Determine the minimum furnace output for an 1800-ft^2 house that is well insulated.

Solution   We substitute 1800 for a and calculate B:

$$B = 30a = 30(1800) = 54,000.$$

The furnace should be able to provide at least 54,000 Btu's of h[...]

Solving for a Letter

Suppose that a contractor has an extra furnace and wants to de[...]
of the largest (well-insulated) house in which it can be used. Th[...]
substitute the amount of the furnace's output in Btu—say, 63,[...]
then solve for a:

$$63,000 = 30a \qquad \text{Replacing } B \text{ with } 63,000$$
$$2100 = a. \qquad \text{Dividing both sides by 30:}$$

$$\begin{array}{r} 2\,100 \\ 30\overline{)63,000} \\ 60 \\ \hline 3\,0 \\ 3\,0 \\ \hline 000 \end{array}$$

COLLABORATIVE CORNER FEATURE
A popular feature from the previous edition, optional Collaborative Corners are inserted throughout the text. Collaborative Corners give students the opportunity to work as a group to solve problems or to perform specially designed activities. There are two or three Collaborative Corners per chapter, each one appearing after the appropriate exercise set.

REAL-DATA APPLICATIONS
Applications have always been a strength of this text, and now the authors bring you even more of a good thing. This edition includes 20% new application and real-data problems, along with an increase in the total number of applications overall.

Visualizing Rates

Graphs allow us to visualize a rate of change. As a rule, the quantity listed in the numerator appears on the vertical axis and the quantity listed in the denominator appears on the horizontal axis.

Example 3

Communication. In 1998, there were approximately 69 million cellular telephone subscribers in the United States, and the figure was growing at a rate of about 14 million per year (*Source: Statistical Abstract of the United States*, 1999). Draw a linear graph to represent this information.

Solution To decide which labels to use on the axes, we note that the rate is given in millions of customers per year. Thus we list *Number of customers, in millions*, on the vertical axis and *Year* on the horizontal axis.

14 million per year is 14 million/yr; millions of customers is the vertical axis; year is the horizontal axis.

Next, we must decide on a scale for each axis that will allow us to plot the given information. If we count by 10's of millions on the vertical axis, we can easily reach 69 million without needing a terribly large graph. On the horizontal axis, we list several years, making certain that 1998 is included (see the figure on the left below).

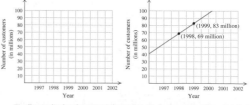

Finally, we display the given information. To do so, we plot the point that corresponds to (1998, 69 million). Then, to display the rate of growth, we move from that point to a second point that represents 14 million more users one year later. The coordinates of this point are (1998 + 1, 69 + 14 million), or (1999, 83 million). We draw a line passing through the two points, as shown in the figure on the right above.

CONNECTING THE CONCEPTS

In Chapters 1 and 2, we simplified expressions and solved equations. In Sections 3.1–3.3 of this chapter, we learned how a graph can be used to represent the solutions of an equation in two variables, like $y = 4 - x$ or $2x + 5y = 15$.

Graphs are used in many important applications. Sections 3.4 and 3.5 have shown us that the slope of a line can be used to represent the rate at which the quantity measured on the vertical axis changes with respect to the quantity measured on the horizontal axis.

In Section 3.6, we will return to the task of graphing equations. This time, however, our understanding of rates and slope will provide us with the tools necessary to develop shortcuts that will streamline our work.

NEW!

CONNECTING THE CONCEPTS
This feature highlights the importance of connecting concepts and invites students to pause and check that they understand the "big picture." This helps assure that students understand how concepts work together in several sections at once. For example, students are alerted to shifts made from solving equations to writing equivalent expressions. The pacing of this feature helps students increase their comprehension and maximize their retention of key concepts.

EXERCISES

NEW!

SKILL MAINTENANCE EXERCISES
As in the past, Skill Maintenance exercises appear in all exercise sets as a means of keeping past concepts fresh and previously covered skills sharp. Two changes to the Skill Maintenance exercises now improve this already popular feature: The number has been increased nearly 50% and they are now designed to provide extra practice with the specific skills needed for the very next section of the text.

SYNTHESIS EXERCISES
Synthesis exercises in this new edition guarantee an extensive and wide-ranging variety of problems in every exercise set. The Synthesis exercises allow students to combine concepts from more than one section and provide challenge for even the strongest students. Mixed in with these problems are occasional *Aha!* exercises (described above).

---

47. $27 + 12y + y^2$ **48.** $50 + 15x + x^2$

49. $t^2 - 0.3t - 0.10$ **50.** $y^2 - 0.2y - 0.08$

51. $p^2 + 3pq - 10q^2$ **52.** $a^2 - 2ab - 3b^2$

53. $m^2 + 5mn + 5n^2$ **54.** $x^2 - 11xy + 24y^2$

55. $s^2 - 2st - 15t^2$ **56.** $b^2 + 8bc - 20c^2$

57. $6a^{10} - 30a^9 - 84a^8$ **58.** $7x^9 - 28x^8 - 35x^7$

59. Marge factors $x^3 - 8x^2 + 15x$ as $(x^2 - 5x)(x - 3)$. Is she wrong? Why or why not? What advice would you offer?

60. Without multiplying $(x - 17)(x - 18)$, explain why it cannot possibly be a factorization of $x^2 + 35x + 306$.

SKILL MAINTENANCE

Solve.

61. $3x - 8 = 0$ **62.** $2x + 7 = 0$

Multiply.

63. $(x + 6)(3x + 4)$ **64.** $(7w + 6)^2$

65. In a recent year, 29,090 people were arrested for counterfeiting. This figure was down 1.2% from the year before. How many people were arrested the year before?

66. The first angle of a triangle is four times as large as the second. The measure of the third angle is 30° greater than that of the second. How large are the angles?

SYNTHESIS

67. When searching for a factorization, why do we list pairs of numbers with the correct *product* instead of pairs of numbers with the correct *sum*?

68. What is the advantage of writing out the prime factorization of c when factoring $x^2 + bx + c$ with a large value of c?

69. Find all integers b for which $a^2 + ba - 50$ can be factored.

70. Find all integers m for which $y^2 + my + 50$ can be factored.

Factor each of the following by first factoring out -1.

71. $30 + 7x - x^2$ **72.** $45 + 4x - x^2$

73. $24 - 10a - a^2$ **74.** $36 - 9a - a^2$

75. $84 - 8t - t^2$ **76.** $72 - 6t - t^2$

Factor completely.

77. $x^2 + \frac{1}{4}x - \frac{1}{8}$ **78.** $x^2 + \frac{1}{2}x - \frac{3}{16}$

79. $\frac{1}{3}a^3 - \frac{1}{3}a^2 - 2a$ **80.** $a^7 - \frac{25}{7}a^5 - \frac{30}{7}a^6$

81. $x^{2m} + 11x^m + 28$ **82.** $t^{2n} - 7t^n + 10$

Aha! **83.** $(a + 1)x^2 + (a + 1)3x + (a + 1)2$

84. $ax^2 - 5x^2 + 8ax - 40x - (a - 5)9$ (*Hint*: See Exercise 83.)

Find a polynomial in factored form for the shaded area in each figure. (Leave answers in terms of π.)

85.

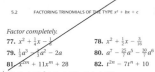

86.

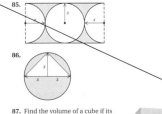

87. Find the volume of a cube if its surface area is $6x^2 + 36x + 54$ square meters.

88. A census taker asks a woman, "How many children do you have?"
"Three," she answers.
"What are their ages?"
She responds, "The product of their ages is 36. The sum of their ages is the house number next door."
The math-savvy census taker walks next door, reads the house number, appears puzzled, and returns to the woman, asking, "Is there something you forgot to tell me?"
"Oh yes," says the woman. "I'm sorry. The oldest child is at the park."
The census taker records the three ages, thanks the woman for her time, and leaves.
How old is each child? Explain how you reached this conclusion. (*Hint*: Consider factorizations.) (*Source*: Adapted from Harnadek, Anita, *Classroom Quickies*. Pacific Grove, CA: Critical Thinking Press and Software)

AHA! EXERCISES
In many exercise sets, students will see a new icon, $Aha!$. This icon indicates to students that there is a simpler way to complete the exercise without going through a lengthy computation. It's then up to the student to discover that simpler approach. The *Aha!* icon appears the first time a new insight can be used on a particular type of exercise. After that, it's up to the student to determine if and when that particular insight can be reused.

WRITING EXERCISES
Writing exercises, indicated by 🖉, provide opportunities for students to answer problems with one or more sentences. Often, these questions have more than one correct response and ask students to explain *why* a certain concept works as it does. In this new edition, two Writing exercises now precede the Skill Maintenance exercises, indicating that they are somewhat less challenging than those that follow the Skill Maintenance exercises. This allows for Writing exercises to be assigned to a wider cross section of the student body.

Elementary Algebra

CONCEPTS AND APPLICATIONS

1

Introduction to Algebraic Expressions

AN APPLICATION

The Viking 2 Lander spacecraft has determined that temperatures on Mars range from −125° Celsius (C) to 25°C (*Source*: The Lunar and Planetary Institute). Find the temperature range on Mars.

This problem appears as Example 11 in Section 1.6.

The math we use every day is mostly algebra and geometry. We determine the constraints, like weights and dimensions, and then design robotic equipment that works within those constraints. Using the math correctly makes all the difference in whether the equipment does its job or not.

SHANNON CROWELL
Development Engineer
Richland, Washington

*P*roblem solving is the focus of this text. Chapter 1 presents some preliminaries that are needed for the problem-solving approach that is developed in Chapter 2 and used throughout the rest of the book. These preliminaries include a review of arithmetic, a discussion of real numbers and their properties, and an examination of how real numbers are added, subtracted, multiplied, divided, and raised to powers.

Introduction to Algebra

1.1

Algebraic Expressions • Translating to Algebraic Expressions • Translating to Equations

This section introduces some of the basic concepts and expressions used in algebra. Solving real-world problems is an important part of algebra, so we will concentrate on the wordings and mathematical expressions that often arise in applications.

Algebraic Expressions

Probably the greatest difference between arithmetic and algebra is the use of *variables* in algebra. When a letter can represent a variety of different numbers, that letter is a **variable**. For example, if n represents the number of tickets sold for a Los Lobos concert, then n will vary, depending on factors like price and day of the week. Thus the number n is a variable. If every ticket costs $25, then a total of $25 \cdot n$ dollars will be paid for tickets. Note that $25 \cdot n$ means 25 *times* n. The number 25 is called a **constant** because it does not change.

Cost per Ticket (in dollars)	Number of Tickets Sold	Total Collected (in dollars)
25	n	$25 \cdot n$

The expression $25 \cdot n$ is a **variable expression** because its value varies with the choice of n. In this case, the total amount collected, $25 \cdot n$, will change with the number of tickets sold. In the following chart, we replace n with a variety of values and compute the total amount collected. In doing so, we are **evaluating the expression** $25 \cdot n$.

Cost per Ticket (in dollars), 25	Number of Tickets Sold, n	Total Collected (in dollars), $25 \cdot n$
25	400	$10,000
25	500	12,500
25	600	15,000

Variable expressions are examples of *algebraic expressions*. An **algebraic expression** consists of variables and/or numerals, often with operation signs and grouping symbols. Examples are

$$t + 97, \quad 5 \cdot x, \quad 3a - b, \quad 18 \div y, \quad \frac{9}{7}, \quad \text{and} \quad 4r(s + t).$$

Recall that a fraction bar is a division symbol: $\frac{9}{7}$, or 9/7, means $9 \div 7$. Similarly, multiplication can be written in several ways. For example, "5 times x" can be written as $5 \cdot x$, $5 \times x$, $5(x)$, or simply $5x$.

To evaluate an algebraic expression, we **substitute** a number for each variable in the expression. This replaces each variable with a number.

E x a m p l e   1

Evaluate each expression for the given values.

a) $x + y$ for $x = 37$ and $y = 28$
b) $5ab$ for $a = 2$ and $b = 3$

Solution

a) We substitute 37 for x and 28 for y and carry out the addition:

$$x + y = 37 + 28 = 65.$$

The number 65 is called the **value** of the expression.
b) We substitute 2 for a and 3 for b and multiply:

$$5ab = 5 \cdot 2 \cdot 3 = 10 \cdot 3 = 30.$$

E x a m p l e   2

The area A of a rectangle of length l and width w is given by the formula $A = lw$. Find the area when l is 17 in. and w is 10 in.

Solution We evaluate, using 17 in. for l and 10 in. for w, and carry out the multiplication:

$$\begin{aligned} A = lw &= (17 \text{ in.})(10 \text{ in.}) \\ &= (17)(10)(\text{in.})(\text{in.}) \\ &= 170 \text{ in}^2, \text{ or } 170 \text{ square inches.} \end{aligned}$$

Note that we use square units for area and $(\text{in.})(\text{in.}) = \text{in}^2$. Exponents like the 2 in the expression in^2 are discussed in detail in Section 1.8.

E x a m p l e   3

The area of a triangle with a base of length b and a height of length h is given by the formula $A = \frac{1}{2}bh$. Find the area when b is 8 m and h is 6.4 m.

Solution We substitute 8 m for b and 6.4 m for h and then multiply:

$$\begin{aligned} A = \tfrac{1}{2}bh &= \tfrac{1}{2}(8 \text{ m})(6.4 \text{ m}) \\ &= \tfrac{1}{2}(8)(6.4)(\text{m})(\text{m}) \\ &= 4(6.4) \text{ m}^2 \\ &= 25.6 \text{ m}^2, \text{ or } 25.6 \text{ square meters.} \end{aligned}$$

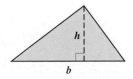

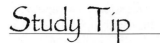

The examples in each section are designed to prepare you for success with the exercise set. Study the step-by-step solutions of the examples, noting that substitutions are highlighted in red. The time you spend studying the examples will save you valuable time when you do your assignment.

Translating to Algebraic Expressions

Before attempting to translate problems to equations, we need to be able to translate certain phrases to algebraic expressions.

Key Words

Addition (+)	Subtraction (−)	Multiplication (·)	Division (÷)
add	subtract	multiply	divide
sum of	difference of	product of	quotient of
plus	minus	times	divided by
more than	less than	twice	ratio of
increased by	decreased by	of	per

Example 4

Translate each phrase to an algebraic expression.

a) Four less than Jean's height, in inches
b) Eighteen more than a number
c) A week's pay, in dollars, divided by three

Solution   To help think through a translation, we sometimes begin with a specific number in place of a variable.

a) If the height were 60, then 4 less than 60 would mean $60 - 4$. If the height were 70, the translation would be $70 - 4$. If we use h to represent "Jean's height, in inches," the translation of "Four less than Jean's height, in inches" is $h - 4$.

b) If we knew the number to be 10, the translation would be $10 + 18$, or $18 + 10$. If we use t to represent "a number," the translation of "Eighteen more than a number" is

$$t + 18, \quad \text{or} \quad 18 + t.$$

c) We let w represent "a week's pay, in dollars." If the pay were $450, the translation would be $450 \div 3$, or $\frac{450}{3}$. Thus our translation of "a week's pay, in dollars, divided by three" is

$$w \div 3, \quad \text{or} \quad \frac{w}{3}.$$

Caution!   Because the order in which we subtract and divide affects the answer, answering $4 - h$ or $3 \div w$ in Examples 4(a) and 4(c) is incorrect.

Example 5   Translate each of the following.

a) Kate's age increased by five
b) Half of some number
c) Three more than twice a number
d) Six less than the product of two numbers
e) Seventy-six percent of the town's population

Solution

Phrase	*Algebraic Expression*
a) Kate's age increased by five	$a + 5$, or $5 + a$
b) Half of some number	$\frac{1}{2}t$, or $\frac{t}{2}$, or $t/2$, or $t \div 2$
c) Three more than twice a number	$2x + 3$
d) Six less than the product of two numbers	$mn - 6$
e) Seventy-six percent of the town's population	76% of p, or $0.76p$

Translating to Equations

The symbol = ("equals") indicates that the expressions on either side of the equals sign represent the same number. An **equation** is a number sentence with the verb =. Equations may be true, false, or neither true nor false.

Example 6   Determine whether each equation is true, false, or neither.

a) $8 \cdot 4 = 32$ b) $7 - 2 = 4$ c) $x + 6 = 13$

Solution

a) $8 \cdot 4 = 32$ The equation is *true*.
b) $7 - 2 = 4$ The equation is *false*.
c) $x + 6 = 13$ The equation is *neither* true nor false, because we do not know what number x represents.

> **Solution**
>
> A replacement or substitution that makes an equation true is called a *solution*. Some equations have more than one solution, and some have no solution. When all solutions have been found, we have *solved* the equation.

To determine whether a number is a solution, we evaluate all expressions in the equation. If the values on both sides of the equation are the same, the number is a solution.

E x a m p l e   7

Determine whether 7 is a solution of $x + 6 = 13$.

Solution

$$\begin{array}{c|c} x + 6 = 13 & \text{Writing the equation} \\ \hline 7 + 6 \ ? \ 13 & \text{Substituting 7 for } x \\ 13 \ | \ 13 & 13 = 13 \text{ is TRUE.} \end{array}$$

Since the left-hand and the right-hand sides are the same, 7 is a solution.

Although we do not study solving equations until Chapter 2, we can translate certain problem situations to equations now. The words "is the same as," "equal," "is," and "are" translate to "=."

E x a m p l e   8

Translate the following problem to an equation.

What number plus 478 is 1019?

Solution   We let y represent the unknown number. The translation then comes almost directly from the English sentence.

What number plus 478 is 1019?

$$y \qquad + \quad 478 \ = \ 1019$$

Note that "plus" translates to "+" and "is" translates to "=."

Sometimes it helps to reword a problem before translating.

E x a m p l e   9

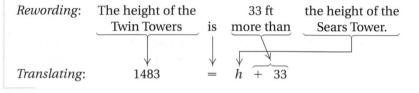

Translate the following problem to an equation.

The Petronas Twin Towers in Kuala Lumpur are the world's tallest buildings. At 1483 ft, they are 33 ft taller than the Sears Tower. (*Source*: *New York Times*) How tall is the Sears Tower?

Solution   We let h represent the height, in feet, of the Sears Tower. A rewording and translation follow:

Rewording: The height of the   33 ft   the height of the
                     Twin Towers   is   more than   Sears Tower.

Translating: 1483   = $h \ + \ 33$

technology connection

Technology Connections are activities that make use of the graphing calculator as a tool for using algebra. These activities use only basic features that are common to most graphing calculators. **(Henceforth in this text we will refer to all graphing utilities as graphers.)** In some cases, students may find the user's manual for their particular grapher helpful for exact keystrokes.

Although all graphers are not the same, most share the following characteristics.

Screen. The large screen can show graphs and tables as well as several operations at once. The screen has a different layout for different functions. Computations are performed in the **home screen**. On many calculators, the home screen is accessed by pressing 2nd QUIT . The **cursor** shows location on the screen, and the **contrast** determines how dark the characters appear.

Keypad. There are often options written above the keys as well as on them. To access those options, we press 2nd or ALPHA and then the key. Expressions are usually entered as they would appear in print. For example, to evaluate $3xy + x$ for $x = 65$ and $y = 92$, we press 3 × 65 × 92 + 65 and then ENTER or EXE .
The value of the expression, 18005, will appear at the right of the screen.

```
3*65*92+65
                    18005
■
```

Evaluate each of the following.

1. $27a - 18b$, for $a = 136$ and $b = 13$
2. $19xy - 9x + 13y$, for $x = 87$ and $y = 29$

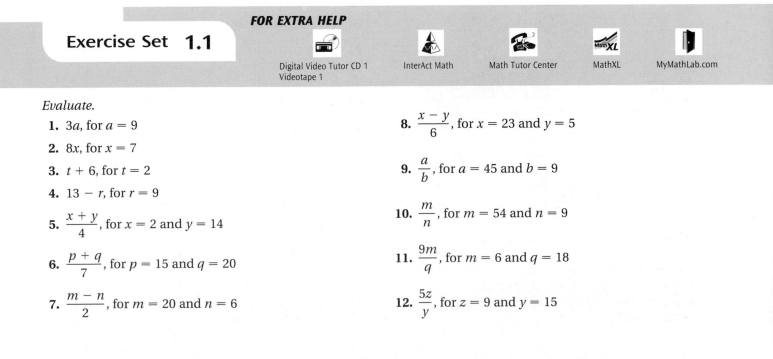

FOR EXTRA HELP

Digital Video Tutor CD 1
Videotape 1   InterAct Math   Math Tutor Center   MathXL   MyMathLab.com

Exercise Set 1.1

Evaluate.

1. $3a$, for $a = 9$

2. $8x$, for $x = 7$

3. $t + 6$, for $t = 2$

4. $13 - r$, for $r = 9$

5. $\dfrac{x + y}{4}$, for $x = 2$ and $y = 14$

6. $\dfrac{p + q}{7}$, for $p = 15$ and $q = 20$

7. $\dfrac{m - n}{2}$, for $m = 20$ and $n = 6$

8. $\dfrac{x - y}{6}$, for $x = 23$ and $y = 5$

9. $\dfrac{a}{b}$, for $a = 45$ and $b = 9$

10. $\dfrac{m}{n}$, for $m = 54$ and $n = 9$

11. $\dfrac{9m}{q}$, for $m = 6$ and $q = 18$

12. $\dfrac{5z}{y}$, for $z = 9$ and $y = 15$

Substitute to find the value of each expression.

13. *Hockey.*   The area of a rectangle with base b and height h is bh. A regulation hockey goal is 6 ft wide and 4 ft high. Find the area of the opening.

14. *Orbit time.*   A communications satellite orbiting 300 mi above the earth travels about 27,000 mi in one orbit. The time, in hours, for an orbit is

$$\frac{27,000}{v},$$

where v is the velocity, in miles per hour. How long will an orbit take at a velocity of 1125 mph?

15. *Zoology.*   A great white shark has triangular teeth. Each tooth measures about 5 cm across the base and has a height of 6 cm. Find the surface area of the front side of one such tooth. (See Example 3.)

16. *Work time.*   Enrico takes five times as long to do a job as Rosa does. Suppose t represents the time it takes Rosa to do the job. Then $5t$ represents the time it takes Enrico. How long does it take Enrico if Rosa takes **(a)** 30 sec? **(b)** 90 sec? **(c)** 2 min?

17. *Olympic softball.*   A softball player's batting average is h/a, where h is the number of hits and a is the number of "at bats." In the 2000 Summer Olympics, Crystl Bustos had 10 hits in 37 at bats. What was her batting average? Round to the nearest thousandth.

18. *Area of a parallelogram.*   The area of a parallelogram with base b and height h is bh. Find the area of the parallelogram when the height is 4 cm (centimeters) and the base is 6.5 cm.

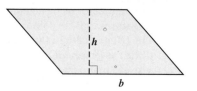

Translate to an algebraic expression.

19. 8 more than Jan's age

20. The product of 4 and a

21. 6 more than b

22. 7 more than Lou's weight

23. 9 less than c **24.** 4 less than d

25. 6 increased by q **26.** 11 increased by z

27. 9 times Phil's speed

28. c more than d

29. x less than y

30. 2 less than Lorrie's age

31. x divided by w

32. The quotient of two numbers

33. m subtracted from n

34. p subtracted from q

35. The sum of the box's length and height

36. The sum of d and f

37. The product of 9 and twice m

38. Paula's speed minus twice the wind speed

39. One quarter of some number

40. One third of the sum of two numbers

41. 64% of the women attending

42. 38% of a number

43. Lita had \$50 before paying x dollars for a pizza. How much remains?

44. Dino drove his pickup truck at 65 mph for t hours. How far did he go?

Determine whether the given number is a solution of the given equation.

45. 15; $x + 17 = 32$

46. 75; $y + 28 = 93$

47. 93; $a - 28 = 75$

48. 12; $8t = 96$

49. 63; $\dfrac{t}{7} = 9$

50. 52; $\dfrac{x}{8} = 6$

51. 3; $\dfrac{108}{x} = 36$

52. 7; $\dfrac{94}{y} = 12$

Translate each problem to an equation. Do not solve.

53. What number added to 73 is 201?

54. Seven times what number is 2303?

55. When 42 is multiplied by a number, the result is 2352. Find the number.

56. When 345 is added to a number, the result is 987. Find the number.

57. *Chess.*   A chess board has 64 squares. If you control 35 squares and your opponent controls the rest, how many does your opponent control?

58. *Hours worked.*   A carpenter charges $25 an hour. How many hours did she work if she billed a total of $53,400?

59. *Recycling.*   Currently, Americans recycle or compost 27% of all municipal solid waste. This is the same as recycling or composting 56 million tons. What is the total amount of waste generated?

60. *Travel to work.*   In the Northeast, the average commute to work is 24.5 min. The average commuting time in the West is 1.8 min less. How long is the average commute in the West?

To the student and the instructor: Writing exercises, denoted by ▯, should be answered using one or more English sentences. Because answers to many writing exercises will vary, solutions are not listed in the answers at the back of the book.

▯ 61. What is the difference between a variable, a variable expression, and an equation?

▯ 62. What does it mean to evaluate an algebraic expression?

SYNTHESIS

To the student and the instructor: Synthesis exercises *are designed to challenge students to extend the concepts or skills studied in each section. Many synthesis exercises will require the assimilation of skills and concepts from several sections.*

▯ 63. If the lengths of the sides of a square are doubled, is the area doubled? Why or why not?

▯ 64. Write a problem that translates to $1998 + t = 2006$.

65. Signs of Distinction charges $90 per square foot for handpainted signs. The town of Belmar commissioned a triangular sign with a base of 3 ft and a height of 2.5 ft. How much will the sign cost?

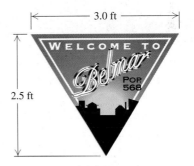

66. Find the area that is shaded.

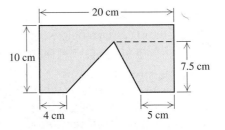

67. Evaluate $\dfrac{x - y}{3}$ when x is twice y and $x = 12$.

68. Evaluate $\dfrac{x + y}{2}$ when y is twice x and $x = 6$.

69. Evaluate $\dfrac{a + b}{4}$ when a is twice b and $a = 16$.

70. Evaluate $\dfrac{a - b}{3}$ when a is three times b and $a = 18$.

Answer each question with an algebraic expression.

71. If $w + 3$ is a whole number, what is the next whole number after it?

72. If $d + 2$ is an odd number, what is the preceding odd number?

Translate to an algebraic expression.

73. One third of one half of the product of two numbers

74. The perimeter of a rectangle with length l and width w (perimeter means distance around)

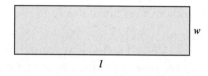

75. The perimeter of a square with side s (perimeter means distance around)

76. Ray's age 7 yr from now if he is 2 yr older than Monique and Monique is a years old

77. If the length of the height of a triangle is doubled, is its area also doubled? Why or why not?

CORNER

Teamwork

COLLABORATIVE

Focus: Group problem solving; working collaboratively

Time: 15 minutes

Group size: 2

Working and studying as a team often enables students to solve problems that are difficult to solve alone.

ACTIVITY

1. The left-hand column below contains the names of 12 colleges. A scrambled list of the names of their sports teams is on the right. As a group, match the names of the colleges to the teams.

1. University of Texas	**a.** Antelopes
2. Western State College of Colorado	**b.** Fighting Banana Slugs
3. University of North Carolina	**c.** Sea Warriors
4. University of Massachusetts	**d.** Gators
5. Hawaii Pacific University	**e.** Mountaineers
6. University of Nebraska	**f.** Sailfish
7. University of California, Santa Cruz	**g.** Longhorns
8. University of Southern Louisiana	**h.** Tarheels
9. Grand Canyon University	**i.** Seawolves
10. Palm Beach Atlantic College	**j.** Ragin' Cajuns
11. University of Alaska, Anchorage	**k.** Cornhuskers
12. University of Florida	**l.** Minutemen

2. After working for 5 min, confer with another group and reach mutual agreement.

3. Does the class agree on all 12 pairs?

4. Do you agree that group collaboration enhances our ability to solve problems?

The Commutative, Associative, and Distributive Laws

1.2

Equivalent Expressions • The Commutative Laws • The Associative Laws • The Distributive Law • The Distributive Law and Factoring

In order to solve equations, we must be able to manipulate algebraic expressions. The commutative, associative, and distributive laws discussed in this section enable us to write *equivalent expressions* that will simplify our work. Indeed, much of this text is devoted to finding equivalent expressions.

Equivalent Expressions

The expressions $4 + 4 + 4$, $3 \cdot 4$, and $4 \cdot 3$ all represent the same number, 12. Expressions that represent the same number are said to be **equivalent**. The equivalent expressions $t + 18$ and $18 + t$ were used on p. 4 when we translated "eighteen more than a number." To check that these expressions are equivalent, we make some choices for t:

$$\text{When } t = 3, \quad t + 18 = 3 + 18 \quad \text{and} \quad 18 + t = 18 + 3$$
$$= 21 \qquad\qquad\qquad = 21.$$

$$\text{When } t = 40, \quad t + 18 = 40 + 18 \quad \text{and} \quad 18 + t = 18 + 40$$
$$= 58 \qquad\qquad\qquad = 58.$$

The Commutative Laws

Recall that changing the order in addition or multiplication does not change the result. Equations like $3 + 78 = 78 + 3$ and $5 \cdot 14 = 14 \cdot 5$ illustrate this idea and show that addition and multiplication are **commutative**.

The Commutative Laws

For Addition. For any numbers a and b,

$$a + b = b + a.$$

(Changing the order of addition does not affect the answer.)

For Multiplication. For any numbers a and b,

$$ab = ba.$$

(Changing the order of multiplication does not affect the answer.)

Example 1

Use the commutative laws to write an expression equivalent to each of the following: **(a)** $y + 5$; **(b)** $9x$; **(c)** $7 + ab$.

Solution

a) $y + 5$ is equivalent to $5 + y$ by the commutative law of addition.

b) $9x$ is equivalent to $x9$ by the commutative law of multiplication.

c) $7 + ab$ is equivalent to $ab + 7$ by the commutative law of *addition*.

$7 + ab$ is also equivalent to $7 + ba$ by the commutative law of *multiplication*.

$7 + ab$ is also equivalent to $ba + 7$ by both commutative laws.

The Associative Laws

Parentheses are used to indicate groupings. We normally simplify within the parentheses first. For example,

$$3 + (8 + 4) = 3 + 12 \quad \text{and} \quad (3 + 8) + 4 = 11 + 4$$
$$= 15 \qquad\qquad\qquad = 15.$$

Similarly,

$$4 \cdot (2 \cdot 3) = 4 \cdot 6 \quad \text{and} \quad (4 \cdot 2) \cdot 3 = 8 \cdot 3$$
$$= 24 \qquad\qquad\qquad = 24.$$

Note that, so long as only addition or only multiplication appears in an expression, changing the grouping does not change the result. Equations such as $3 + (7 + 5) = (3 + 7) + 5$ and $4(5 \cdot 3) = (4 \cdot 5)3$ illustrate that addition and multiplication are **associative**.

> ### The Associative Laws
>
> *For Addition.* For any numbers a, b, and c,
>
> $$a + (b + c) = (a + b) + c.$$
>
> (Numbers can be grouped in any manner for addition.)
>
> *For Multiplication.* For any numbers a, b, and c,
>
> $$a \cdot (b \cdot c) = (a \cdot b) \cdot c.$$
>
> (Numbers can be grouped in any manner for multiplication.)

Example 2

Use an associative law to write an expression equivalent to each of the following: **(a)** $y + (z + 3)$; **(b)** $(8x)y$.

Solution

a) $y + (z + 3)$ is equivalent to $(y + z) + 3$ by the associative law of addition.

b) $(8x)y$ is equivalent to $8(xy)$ by the associative law of multiplication.

When only additions or only multiplications are involved, parentheses do not change the result. For that reason, we sometimes omit them altogether. Thus,

$$x + (y + 7) = x + y + 7, \quad \text{and} \quad l(wh) = lwh.$$

A sum such as $(5 + 1) + (3 + 5) + 9$ can be simplified by pairing numbers that add to 10. The associative and commutative laws allow us to do this:

$$(5 + 1) + (3 + 5) + 9 = 5 + 5 + 9 + 1 + 3$$
$$= 10 + 10 + 3 = 23.$$

E x a m p l e   3   Use the commutative and/or associative laws of addition to write at least two expressions equivalent to $(x + 5) + y$.

Solution

a) $(x + 5) + y = x + (5 + y)$ Using the associative law; $x + (5 + y)$ is one equivalent expression.

$\qquad\qquad\quad = x + (y + 5)$ Using the commutative law

b) $(x + 5) + y = y + (x + 5)$ Using the commutative law; $y + (x + 5)$ is one equivalent expression.

$\qquad\qquad\quad = y + (5 + x)$ Using the commutative law again

E x a m p l e   4   Use the commutative and/or associative laws of multiplication to rewrite $2(x3)$ as $6x$. Show and give reasons for each step.

Solution

$$2(x3) = 2(3x) \qquad \text{Using the commutative law}$$
$$= (2 \cdot 3)x \qquad \text{Using the associative law}$$
$$= 6x \qquad \text{Simplifying}$$

The Distributive Law

The *distributive law* is probably the single most important law for manipulating algebraic expressions. Unlike the commutative and associative laws, the distributive law uses multiplication together with addition.

You have already used the distributive law although you may not have realized it at the time. To illustrate, try to multiply $3 \cdot 21$ mentally. Many people find the product, 63, by thinking of 21 as $20 + 1$ and then multiplying 20 by 3 and 1 by 3. The sum of the two products, $60 + 3$, is 63. Note that if the 3 does not multiply both 20 and 1, the result will not be correct.

E x a m p l e   5

Compute in two ways: $4(7 + 2)$.

Solution

a) As in the discussion of $3(20 + 1)$ above, to compute $4(7 + 2)$, we can multiply both 7 and 2 by 4 and add the results:

$$4(7 + 2) = 4 \cdot 7 + 4 \cdot 2 \qquad \text{Multiplying both 7 and 2 by 4}$$
$$= 28 + 8 = 36. \qquad \text{Adding}$$

b) By first adding inside the parentheses, we get the same result in a different way:

$$4(7 + 2) = 4(9) \qquad \text{Adding; } 7 + 2 = 9$$
$$= 36. \qquad \text{Multiplying}$$

The Distributive Law

For any numbers a, b, and c,

$$a(b + c) = ab + ac.$$

(The product of a number and a sum can be written as the sum of two products.)

E x a m p l e   6

Multiply: $3(x + 2)$.

Solution Since $x + 2$ cannot be simplified unless a value for x is given, we use the distributive law:

$$3(x + 2) = 3 \cdot x + 3 \cdot 2 \qquad \text{Using the distributive law}$$
$$= 3x + 6. \qquad \text{Note that } 3 \cdot x \text{ is the same as } 3x.$$

The expression $3x + 6$ has two *terms*, $3x$ and 6. In general, a **term** is a number, a variable, or a product or quotient of numbers and/or variables. Thus, t, 29, $5ab$, and $2x/y$ are terms in $t + 29 + 5ab + 2x/y$. Note that terms are separated by plus signs.

E x a m p l e   7

List the terms in $7s + st + \dfrac{3}{t}$.

Solution Terms are separated by plus signs, so the terms in $7s + st + 3/t$ are $7s$, st, and $3/t$.

The distributive law can also be used when more than two terms are inside the parentheses.

Example 8

Multiply: $6(s + 2 + 5w)$.

Solution

$$6(s + 2 + 5w) = 6 \cdot s + 6 \cdot 2 + 6 \cdot 5w \qquad \text{Using the distributive law}$$
$$= 6s + 12 + (6 \cdot 5)w \qquad \text{Using the associative law for multiplication}$$
$$= 6s + 12 + 30w$$

Because of the commutative law of multiplication, the distributive law can be used on the "right": $(b + c)a = ba + ca$.

Example 9

Multiply: $(c + 4)5$.

Solution

$$(c + 4)5 = c \cdot 5 + 4 \cdot 5 \qquad \text{Using the distributive law on the right}$$
$$= 5c + 20$$

> **Caution!** To use the distributive law for removing parentheses, be sure to multiply *each* term inside the parentheses by the multiplier outside:
>
> $$a(b + c) \neq ab + c.$$

The Distributive Law and Factoring

If we use the distributive law in reverse, we have the basis of a process called **factoring**: $ab + ac = a(b + c)$. To **factor** an expression means to write an equivalent expression that is a product. The parts of the product are called **factors**. Note that "factor" can be used as either a verb or a noun.

Example 10

Use the distributive law to factor each of the following.

a) $3x + 3y$ **b)** $7x + 21y + 7$

Solution

a) By the distributive law,

$$3x + 3y = 3(x + y). \qquad \text{The } common \; factor \text{ is 3.}$$

b) $7x + 21y + 7 = 7 \cdot x + 7 \cdot 3y + 7 \cdot 1 \qquad \text{The common factor is 7.}$

$$= 7(x + 3y + 1) \qquad \text{Using the distributive law}$$

Be sure not to omit the 1 or the common factor, 7.

To check our factoring, we multiply to see if the original expression is obtained. For example, to check the **factorization** in Example 10(b), note that

$$7(x + 3y + 1) = 7x + 7 \cdot 3y + 7 \cdot 1$$
$$= 7x + 21y + 7.$$

Exercise Set 1.2

FOR EXTRA HELP

Digital Video Tutor CD 1   InterAct Math   Math Tutor Center   MathXL   MyMathLab.com
Videotape 1

Use the commutative law of addition to write an equivalent expression.

1. $7 + x$

2. $a + 2$

3. $ab + c$

4. $x + 3y$

5. $9x + 3y$

6. $3a + 7b$

7. $5(a + 1)$

8. $9(x + 5)$

Use the commutative law of multiplication to write an equivalent expression.

9. $2 \cdot a$

10. xy

11. st

12. $4x$

13. $5 + ab$

14. $x + 3y$

15. $5(a + 1)$

16. $9(x + 5)$

Use the associative law of addition to write an equivalent expression.

17. $(a + 5) + b$

18. $(5 + m) + r$

19. $r + (t + 7)$

20. $x + (2 + y)$

21. $(ab + c) + d$

22. $(m + np) + r$

Use the associative law of multiplication to write an equivalent expression.

23. $(8x)y$

24. $(9a)b$

25. $2(ab)$

26. $9(rp)$

27. $3[2(a + b)]$

28. $5[x(2 + y)]$

Use the commutative and/or associative laws to write two equivalent expressions. Answers may vary.

29. $r + (t + 6)$

30. $5 + (v + w)$

31. $(17a)b$

32. $x(3y)$

Use the commutative and/or associative laws to show why the expression on the left is equivalent to the expression on the right. Write a series of steps with labels, as in Example 4.

33. $(5 + x) + 2$ is equivalent to $x + 7$

34. $(2a)4$ is equivalent to $8a$

35. $(m3)7$ is equivalent to $21m$

36. $4 + (9 + x)$ is equivalent to $x + 13$

Multiply.

37. $4(a + 3)$

38. $3(x + 5)$

39. $6(1 + x)$

40. $6(v + 4)$

41. $3(x + 1)$

42. $9(x + 3)$

43. $8(3 + y)$

44. $7(s + 5)$

45. $9(2x + 6)$

46. $9(6m + 7)$

47. $5(r + 2 + 3t)$

48. $4(5x + 8 + 3p)$

49. $(a + b)2$

50. $(x + 2)7$

51. $(x + y + 2)5$

52. $(2 + a + b)6$

List the terms in each expression.

53. $x + xyz + 19$

54. $9 + 17a + abc$

55. $2a + \dfrac{a}{b} + 5b$

56. $3xy + 20 + \dfrac{4a}{b}$

Use the distributive law to factor each of the following. Check by multiplying.

57. $2a + 2b$

58. $5y + 5z$

59. $7 + 7y$

60. $13 + 13x$

61. $18x + 3$

62. $20a + 5$

63. $5x + 10 + 15y$

64. $3 + 27b + 6c$

65. $12x + 9$ **66.** $6x + 6$

67. $3a + 9b$ **68.** $5a + 15b$

69. $44x + 11y + 22z$ **70.** $14a + 56b + 7$

71. Is subtraction commutative? Why or why not?

72. Is division associative? Why or why not?

SKILL MAINTENANCE

To the student and the instructor: Skill maintenance exercises *review skills studied in earlier sections. Often these exercises are preparation for the next section. Answers at the back of the book indicate the section in which that material originally appeared.*

Translate to an algebraic expression.

73. Twice Kara's salary **74.** Half of m

SYNTHESIS

75. Are terms and factors the same thing? Why or why not?

76. Explain how the distributive, commutative, and associative laws can be used to show that $2(3x + 4y)$ is equivalent to $6x + 8y$.

Tell whether the expressions in each pairing are equivalent. Then explain why or why not.

77. $8 + 4(a + b)$ and $4(2 + a + b)$

78. $7 \div 3m$ and $m3 \div 7$

79. $(rt + st)5$ and $5t(r + s)$

80. $yax + ax$ and $xa(1 + y)$

81. $30y + x15$ and $5[2(x + 3y)]$

82. $[c(2 + 3b)]5$ and $10c + 15bc$

83. Evaluate the expressions $3(2 + x)$ and $6 + x$ for $x = 0$. Do your results indicate that $3(2 + x)$ and $6 + x$ are equivalent? Why or why not?

84. Factor $15x + 40$. Then evaluate both $15x + 40$ and the factorization for $x = 4$. Do your results *guarantee* that the factorization is correct? Why or why not? (*Hint*: See Exercise 83.)

CORNER

COLLABORATIVE

Mental Addition

Focus: Application of commutative and associative laws

Time: 10 minutes

Group size: 2–3

Legend has it that while still in grade school, the mathematician Carl Friedrich Gauss (1777–1855) was able to add the numbers from 1 to 100 mentally. Gauss did not add them sequentially, but rather paired 1 with 99, 2 with 98, and so on.

ACTIVITY

1. Use a method similar to Gauss's to simplify the following:

   $$1 + 2 + 3 + 4 + 5 + 6 + 7 + 8 + 9 + 10.$$

   One group member should add from left to right as a check.

2. Use Gauss's method to find the sum of the first 25 counting numbers:

   $$1 + 2 + 3 + \cdots + 23 + 24 + 25.$$

   Again, one student should add from left to right as a check.

3. How were the associative and commutative laws applied in parts (1) and (2) above?

4. Now use a similar approach involving both addition and division to find the sum of the first 10 counting numbers:

   $$1 + 2 + 3 + \cdots + 10$$
   $$+10 + 9 + 8 + \cdots + 1$$

5. Use the approach in step (4) to find the sum of the first 100 counting numbers. Are the associative and commutative laws applied in this method, too? How is the distributive law used in this approach?

Fraction Notation

1.3

Factors and Prime Factorizations • Fraction Notation •
Multiplication, Division, and Simplification • More Simplifying •
Addition and Subtraction

This section covers multiplication, addition, subtraction, and division with fractions. Although much of this may be review, note that fractional expressions that contain variables are also included.

Factors and Prime Factorizations

In order to be able to study addition and subtraction using fraction notation, we first review how *natural numbers* are factored. **Natural numbers** can be thought of as the counting numbers:

$$1, 2, 3, 4, 5, \ldots .^*$$

(The dots indicate that the established pattern continues without ending.) To factor a number, we simply express it as a product of two or more numbers.

Example 1

Write several factorizations of 12. Then list all factors of 12.

Solution The number 12 can be factored in several ways:

$$1 \cdot 12, \quad 2 \cdot 6, \quad 3 \cdot 4, \quad 2 \cdot 2 \cdot 3.$$

The factors of 12 are 1, 2, 3, 4, 6, and 12.

Some numbers have only two factors, the number itself and 1. Such numbers are called **prime**.

> ### Prime Number
>
> A *prime number* is a natural number that has exactly two different factors: the number itself and 1.

Example 2

Which of these numbers are prime? 7, 4, 1

Solution

7 is prime. It has exactly two different factors, 7 and 1.

4 is not prime. It has three different factors, 1, 2, and 4.

1 is not prime. It does not have two *different* factors.

---

*A similar collection of numbers, the **whole numbers**, includes 0: 0, 1, 2, 3, ...

If a natural number, other than 1, is not prime, we call it **composite**. Every composite number can be factored into a product of prime numbers. Such a factorization is called the **prime factorization** of that composite number.

E x a m p l e   3

Find the prime factorization of 36.

Solution   We first factor 36 in any way that we can. One way is like this:

$$36 = 4 \cdot 9.$$

The factors 4 and 9 are not prime, so we factor them:

$$36 = 4 \cdot 9$$
$$= 2 \cdot 2 \cdot 3 \cdot 3. \qquad \text{2 and 3 are both prime.}$$

The prime factorization of 36 is $2 \cdot 2 \cdot 3 \cdot 3$.

Fraction Notation

An example of **fraction notation** for a number is

$$\frac{2}{3}. \quad \begin{array}{l} \longleftarrow \text{Numerator} \\ \longleftarrow \text{Denominator} \end{array}$$

The top number is called the **numerator**, and the bottom number is called the **denominator**. When the numerator and the denominator are the same nonzero number, we have fraction notation for the number 1.

> **Fraction Notation for 1**
> For any number a, except 0,
> $$\frac{a}{a} = 1.$$
> (Any nonzero number divided by itself is 1.)

Multiplication, Division, and Simplification

Recall from arithmetic that fractions are multiplied as follows.

> **Multiplication of Fractions**
> For any two fractions a/b and c/d,
> $$\frac{a}{b} \cdot \frac{c}{d} = \frac{ac}{bd}.$$
> (The numerator of the product is the product of the two numerators. The denominator of the product is the product of the two denominators.)

E x a m p l e   4

Multiply: **(a)** $\dfrac{2}{3} \cdot \dfrac{7}{5}$; **(b)** $\dfrac{4}{x} \cdot \dfrac{8}{y}$.

Solution   We multiply numerators as well as denominators.

a) $\dfrac{2}{3} \cdot \dfrac{7}{5} = \dfrac{2 \cdot 7}{3 \cdot 5} = \dfrac{14}{15}$

b) $\dfrac{4}{x} \cdot \dfrac{8}{y} = \dfrac{4 \cdot 8}{x \cdot y} = \dfrac{32}{xy}$

Two numbers whose product is 1 are **reciprocals**, or **multiplicative inverses**, of each other. All numbers, except zero, have reciprocals. For example,

the reciprocal of $\frac{2}{3}$ is $\frac{3}{2}$ because $\frac{2}{3} \cdot \frac{3}{2} = \frac{6}{6} = 1$;

the reciprocal of 9 is $\frac{1}{9}$ because $9 \cdot \frac{1}{9} = \frac{9}{9} = 1$; and

the reciprocal of $\frac{1}{4}$ is 4 because $\frac{1}{4} \cdot 4 = 1$.

Reciprocals are used to rewrite division as multiplication.

Division of Fractions

To divide two fractions, multiply by the reciprocal of the divisor:

$$\frac{a}{b} \div \frac{c}{d} = \frac{a}{b} \cdot \frac{d}{c}.$$

E x a m p l e   5

Divide: $\dfrac{1}{2} \div \dfrac{3}{5}$.

Solution

$$\frac{1}{2} \div \frac{3}{5} = \frac{1}{2} \cdot \frac{5}{3} \qquad \frac{5}{3} \text{ is the reciprocal of } \frac{3}{5}$$

$$= \frac{5}{6}$$

When one of the fractions being multiplied is 1, multiplying yields an equivalent expression because of the *identity property of* 1. A similar property could be stated for division, but there is no need to do so here.

The Identity Property of 1

For any number a,

$$a \cdot 1 = a.$$

(Multiplying a number by 1 gives that same number.)

E x a m p l e 6

Multiply $\dfrac{4}{5} \cdot \dfrac{6}{6}$ to find an expression equivalent to $\dfrac{4}{5}$.

Solution   We have

$$\frac{4}{5} \cdot \frac{6}{6} = \frac{4 \cdot 6}{5 \cdot 6} = \frac{24}{30}.$$

Since $\frac{6}{6} = 1$, the expression $\frac{4}{5} \cdot \frac{6}{6}$ is equivalent to $\frac{4}{5} \cdot 1$, or simply $\frac{4}{5}$. Thus, $\frac{24}{30}$ is equivalent to $\frac{4}{5}$.

The steps of Example 6 are reversed by "removing a factor equal to 1"—in this case, $\frac{6}{6}$. By removing a factor that equals 1, we can *simplify* an expression like $\frac{24}{30}$ to an equivalent expression like $\frac{4}{5}$.

To simplify, we factor the numerator and the denominator, looking for the largest factor common to both. This is sometimes made easier by writing prime factorizations. After identifying common factors, we can express the fraction as a product of two fractions, one of which is in the form a/a.

E x a m p l e 7

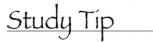

Study Tip

Take the time to include all the steps when working your homework problems. Doing so will help you organize your thinking and avoid computational errors. It will also give you complete, step-by-step solutions of the exercises that can be used when studying for quizzes and tests.

Simplify: **(a)** $\dfrac{15}{40}$;   **(b)** $\dfrac{36}{24}$.

Solution

a) Note that 5 is a factor of both 15 and 40:

$$\frac{15}{40} = \frac{3 \cdot 5}{8 \cdot 5} \qquad \text{Factoring the numerator and the denominator, using the common factor, 5}$$

$$= \frac{3}{8} \cdot \frac{5}{5} \qquad \text{Rewriting as a product of two fractions; } \frac{5}{5} = 1$$

$$= \frac{3}{8} \cdot 1 = \frac{3}{8}. \qquad \text{Using the identity property of 1 (removing a factor equal to 1)}$$

b) $\dfrac{36}{24} = \dfrac{2 \cdot 2 \cdot 3 \cdot 3}{2 \cdot 2 \cdot 2 \cdot 3} \qquad$ Writing the prime factorizations and identifying common factors; 12/12 could also be used.

$$= \frac{3}{2} \cdot \frac{2 \cdot 2 \cdot 3}{2 \cdot 2 \cdot 3} \qquad \text{Rewriting as a product of two fractions; } \frac{2 \cdot 2 \cdot 3}{2 \cdot 2 \cdot 3} = 1$$

$$= \frac{3}{2} \cdot 1 = \frac{3}{2} \qquad \text{Using the identity property of 1}$$

It is always wise to check your result to see if any common factors of the numerator and the denominator remain. (This will never happen if prime factorizations are used correctly.) If common factors remain, repeat the process by removing another factor equal to 1 to simplify your result.

More Simplifying

"Canceling" is a shortcut that you may have used for removing a factor equal to 1 when working with fraction notation. With *great* concern, we mention it as a possible way to speed up your work. Canceling can be used only when removing common factors in numerators and denominators. Canceling *cannot* be used in sums or differences. Our concern is that "canceling" be used with understanding. Example 7(b) might have been done faster as follows:

$$\frac{36}{24} = \frac{\cancel{2} \cdot \cancel{2} \cdot 3 \cdot \cancel{3}}{\cancel{2} \cdot \cancel{2} \cdot 2 \cdot \cancel{3}} = \frac{3}{2}, \quad \text{or} \quad \frac{36}{24} = \frac{3 \cdot \cancel{12}}{2 \cdot \cancel{12}} = \frac{3}{2}, \quad \text{or} \quad \frac{\overset{\overset{3}{\cancel{18}}}{\cancel{36}}}{\underset{\underset{2}{\cancel{12}}}{\cancel{24}}} = \frac{3}{2}.$$

Caution!   Unfortunately, canceling is often performed incorrectly:

$$\underbrace{\frac{\cancel{2} + 3}{\cancel{2}} = 3,}_{\downarrow} \qquad \underbrace{\frac{\cancel{4} - 1}{\cancel{4} - 2} = \frac{1}{2},}_{\downarrow} \qquad \underbrace{\frac{1\cancel{5}}{\cancel{5}4} = \frac{1}{4}.}_{\downarrow}$$

$$\text{Wrong!} \qquad\qquad \text{Wrong!} \qquad\qquad \text{Wrong!}$$

$$\frac{2 + 3}{2} = \frac{5}{2} \qquad \frac{4 - 1}{4 - 2} = \frac{3}{2} \qquad \frac{15}{54} = \frac{5 \cdot 3}{18 \cdot 3} = \frac{5}{18}$$

In each of these situations, the expressions canceled are *not* factors. Factors are parts of products. For example, in $2 \cdot 3$, the numbers 2 and 3 are factors, but in $2 + 3$, 2 and 3 are *not* factors. **If you can't factor, you can't cancel! If in doubt, don't cancel!**

Sometimes it is helpful to use 1 as a factor in the numerator or the denominator when simplifying.

Example 8

Simplify: $\dfrac{9}{72}$.

Solution

$$\frac{9}{72} = \frac{1 \cdot 9}{8 \cdot 9} \qquad \text{Factoring and using the identity property of 1 to write 9 as } 1 \cdot 9$$

$$= \frac{1 \cdot \cancel{9}}{8 \cdot \cancel{9}} = \frac{1}{8} \qquad \text{Simplifying by removing a factor equal to 1: } \frac{9}{9} = 1$$

Addition and Subtraction

When denominators are the same, fractions are added or subtracted by adding or subtracting numerators and keeping the same denominator.

Addition and Subtraction of Fractions

For any two fractions a/d and b/d,

$$\frac{a}{d} + \frac{b}{d} = \frac{a + b}{d} \quad \text{and} \quad \frac{a}{d} - \frac{b}{d} = \frac{a - b}{d}.$$

Example 9   Add and simplify: $\dfrac{4}{8} + \dfrac{5}{8}$.

Solution   The common denominator is 8. We add the numerators and keep the common denominator:

$$\dfrac{4}{8} + \dfrac{5}{8} = \dfrac{4+5}{8} = \dfrac{9}{8}$$ You can think of this as $4 \cdot \dfrac{1}{8} + 5 \cdot \dfrac{1}{8} = 9 \cdot \dfrac{1}{8}$, or $\dfrac{9}{8}$.

In arithmetic, we often write $1\frac{1}{8}$ rather than the "improper" fraction $\frac{9}{8}$. In algebra, $\frac{9}{8}$ is generally more useful and is quite "proper" for our purposes.

When denominators are different, we use the identity property of 1 and multiply to find a common denominator. Then we add, as in Example 9.

Example 10   Add or subtract as indicated: **(a)** $\dfrac{7}{8} + \dfrac{5}{12}$; **(b)** $\dfrac{9}{8} - \dfrac{4}{5}$.

Solution

a) The number 24 is divisible by both 8 and 12. We multiply both $\frac{7}{8}$ and $\frac{5}{12}$ by suitable forms of 1 to obtain two fractions with denominators of 24:

$$\dfrac{7}{8} + \dfrac{5}{12} = \dfrac{7}{8} \cdot \dfrac{3}{3} + \dfrac{5}{12} \cdot \dfrac{2}{2}$$ Multiplying by 1.
Since $8 \cdot 3 = 24$, we multiply $\frac{7}{8}$ by $\frac{3}{3}$.
Since $12 \cdot 2 = 24$, we multiply $\frac{5}{12}$ by $\frac{2}{2}$.

$$= \dfrac{21}{24} + \dfrac{10}{24}$$ Performing the multiplication

$$= \dfrac{31}{24}.$$ Adding fractions

b) $\dfrac{9}{8} - \dfrac{4}{5} = \dfrac{9}{8} \cdot \dfrac{5}{5} - \dfrac{4}{5} \cdot \dfrac{8}{8}$ Using 40 as a common denominator

$$= \dfrac{45}{40} - \dfrac{32}{40} = \dfrac{13}{40}$$ Subtracting fractions

After adding, subtracting, multiplying, or dividing, we may still need to simplify the answer.

Example 11   Perform the indicated operation and, if possible, simplify.

a) $\dfrac{7}{10} - \dfrac{1}{5}$ **b)** $8 \cdot \dfrac{5}{12}$ **c)** $\dfrac{\frac{5}{6}}{\frac{25}{9}}$

Some graphers can perform operations using fraction notation. Others may be able to convert answers given in decimal notation to fraction notation. Often this conversion is done using a command found in a **menu**, or a list of options that appears when a key is pressed. To select an item from a menu, we highlight its number and press ENTER or simply press the number of the item.

For example, to find fraction notation for $\frac{2}{15} + \frac{7}{12}$, we enter the expression as $2/15 + 7/12$. The answer is given in decimal notation. To convert this to fraction notation, we press MATH and select the Frac option. In this case, the notation Ans▶Frac shows that the grapher will convert .7166666667 to fraction notation.

```
2/15+7/12
                .7166666667
Ans▶Frac
                    43/60
```

We see that $\frac{2}{15} + \frac{7}{12} = \frac{43}{60}$.

Solution

a) $\frac{7}{10} - \frac{1}{5} = \frac{7}{10} - \frac{1}{5} \cdot \frac{2}{2}$ Using 10 as the common denominator

$\qquad = \frac{7}{10} - \frac{2}{10}$

$\qquad = \frac{5}{10} = \frac{1 \cdot \cancel{5}}{2 \cdot \cancel{5}} = \frac{1}{2}$ Removing a factor equal to 1: $\frac{5}{5} = 1$

b) $8 \cdot \frac{5}{12} = \frac{8 \cdot 5}{12}$ Multiplying numerators and denominators. Think of 8 as $\frac{8}{1}$.

$\qquad = \frac{2 \cdot 2 \cdot 2 \cdot 5}{2 \cdot 2 \cdot 3}$ Factoring; $\frac{4 \cdot 2 \cdot 5}{4 \cdot 3}$ can also be used.

$\qquad = \frac{\cancel{2} \cdot \cancel{2} \cdot 2 \cdot 5}{\cancel{2} \cdot \cancel{2} \cdot 3}$ Removing a factor equal to 1: $\frac{2 \cdot 2}{2 \cdot 2} = 1$

$\qquad = \frac{10}{3}$ Simplifying

c) $\dfrac{\frac{5}{6}}{\frac{25}{9}} = \frac{5}{6} \div \frac{25}{9}$ Rewriting horizontally. Remember that a fraction bar indicates division.

$\qquad = \frac{5}{6} \cdot \frac{9}{25}$ Multiplying by the reciprocal of $\frac{25}{9}$

$\qquad = \frac{5 \cdot 3 \cdot 3}{2 \cdot 3 \cdot 5 \cdot 5}$ Writing as one fraction and factoring

$\qquad = \frac{\cancel{5} \cdot \cancel{3} \cdot 3}{2 \cdot \cancel{3} \cdot \cancel{5} \cdot 5}$ Removing a factor equal to 1: $\frac{5 \cdot 3}{3 \cdot 5} = 1$

$\qquad = \frac{3}{10}$ Simplifying

Exercise Set **1.3**

FOR EXTRA HELP

Digital Video Tutor CD 1   InterAct Math   Math Tutor Center   MathXL   MyMathLab.com
Videotape 1

To the student and the instructor: *Beginning in this section, selected exercises are marked with the icon* Aha! *. These "Aha!" exercises can be answered most easily if the student pauses to inspect the exercise rather than proceed mechanically. This is done to discourage rote memorization. Some "Aha!" exercises are left unmarked,* *to encourage students to* always *pause before working a problem.*

Write at least two factorizations of each number. Then list all the factors of the number.

1. 50   **2.** 70   **3.** 42   **4.** 60

Find the prime factorization of each number. If the number is prime, state this.

5. 26   **6.** 15   **7.** 30

8. 55   **9.** 20   **10.** 50

11. 27   **12.** 98   **13.** 18

14. 54   **15.** 40   **16.** 56

17. 43   **18.** 120   **19.** 210

20. 79   **21.** 115   **22.** 143

Simplify.

23. $\dfrac{10}{14}$ **24.** $\dfrac{14}{21}$ **25.** $\dfrac{16}{56}$

26. $\dfrac{72}{27}$ **27.** $\dfrac{6}{48}$ **28.** $\dfrac{12}{70}$

29. $\dfrac{49}{7}$ **30.** $\dfrac{132}{11}$ **31.** $\dfrac{19}{76}$

32. $\dfrac{17}{51}$ **33.** $\dfrac{150}{25}$ **34.** $\dfrac{170}{34}$

35. $\dfrac{75}{80}$ **36.** $\dfrac{42}{50}$ **37.** $\dfrac{120}{82}$

38. $\dfrac{75}{45}$ **39.** $\dfrac{210}{98}$ **40.** $\dfrac{140}{350}$

Perform the indicated operation and, if possible, simplify.

41. $\dfrac{1}{2} \cdot \dfrac{3}{7}$ **42.** $\dfrac{11}{10} \cdot \dfrac{8}{5}$ **43.** $\dfrac{9}{2} \cdot \dfrac{3}{4}$

Aha! **44.** $\dfrac{11}{12} \cdot \dfrac{12}{11}$ **45.** $\dfrac{1}{8} + \dfrac{3}{8}$ **46.** $\dfrac{1}{2} + \dfrac{1}{8}$

47. $\dfrac{4}{9} + \dfrac{13}{18}$ **48.** $\dfrac{4}{5} + \dfrac{8}{15}$ **49.** $\dfrac{3}{a} \cdot \dfrac{b}{7}$

50. $\dfrac{x}{5} \cdot \dfrac{y}{z}$ **51.** $\dfrac{4}{a} + \dfrac{3}{a}$ **52.** $\dfrac{7}{a} - \dfrac{5}{a}$

53. $\dfrac{3}{10} + \dfrac{8}{15}$ **54.** $\dfrac{7}{8} + \dfrac{5}{12}$ **55.** $\dfrac{9}{7} - \dfrac{2}{7}$

56. $\dfrac{12}{5} - \dfrac{2}{5}$ **57.** $\dfrac{13}{18} - \dfrac{4}{9}$ **58.** $\dfrac{13}{15} - \dfrac{8}{45}$

Aha! **59.** $\dfrac{20}{30} - \dfrac{2}{3}$ **60.** $\dfrac{5}{7} - \dfrac{5}{21}$ **61.** $\dfrac{7}{6} \div \dfrac{3}{5}$

62. $\dfrac{7}{5} \div \dfrac{3}{4}$ **63.** $\dfrac{8}{9} \div \dfrac{4}{15}$ **64.** $\dfrac{9}{4} \div 9$

65. $12 \div \dfrac{3}{7}$ **66.** $\dfrac{1}{10} \div \dfrac{1}{5}$ *Aha!* **67.** $\dfrac{7}{13} \div \dfrac{7}{13}$

68. $\dfrac{17}{8} \div \dfrac{5}{6}$ **69.** $\dfrac{\frac{2}{5}}{\frac{5}{3}}$ **70.** $\dfrac{\frac{3}{8}}{\frac{1}{5}}$

71. $\dfrac{\frac{9}{1}}{\frac{1}{2}}$ **72.** $\dfrac{\frac{7}{3}}{5}$

📓 **73.** Under what circumstances would the sum of two fractions be easier to compute than the product of the same two fractions?

📓 **74.** Under what circumstances would the product of two fractions be easier to compute than the sum of the same two fractions?

SKILL MAINTENANCE

Use a commutative law to write an equivalent expression. There can be more than one correct answer.

75. $5(x + 3)$ **76.** $7 + (a + b)$

SYNTHESIS

📓 **77.** Bryce insists that $(2 + x)/8$ is equivalent to $(1 + x)/4$. What mistake do you think is being made and how could you demonstrate to Bryce that the two expressions are not equivalent?

📓 **78.** Use the word factor in two sentences—once as a noun and once as a verb.

79. *Packaging.*   Tritan Candies uses two sizes of boxes, 6 in. and 8 in. long. These are packed end to end in bigger cartons to be shipped. What is the shortest-length carton that will accommodate boxes of either size without any room left over? (Each carton must contain boxes of only one size; no mixing is allowed.)

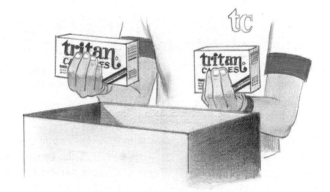

80. In the following table, the top number can be factored in such a way that the sum of the factors is the bottom number. For example, in the first column, 56 is factored as $7 \cdot 8$, since $7 + 8 = 15$, the bottom number. Find the missing numbers in each column.

Product	56	63	36	72	140	96	168
Factor	7						
Factor	8						
Sum	15	16	20	38	24	20	29

Simplify.

81. $\dfrac{16 \cdot 9 \cdot 4}{15 \cdot 8 \cdot 12}$

82. $\dfrac{9 \cdot 8xy}{2xy \cdot 36}$

83. $\dfrac{27pqrs}{9prst}$

84. $\dfrac{512}{192}$

85. $\dfrac{15 \cdot 4xy \cdot 9}{6 \cdot 25x \cdot 15y}$

86. $\dfrac{10x \cdot 12 \cdot 25y}{2 \cdot 30x \cdot 20y}$

87. $\dfrac{\frac{27ab}{15mn}}{\frac{18bc}{25np}}$

88. $\dfrac{\frac{45xyz}{24ab}}{\frac{30xz}{32ac}}$

Find the area of each figure.

89.

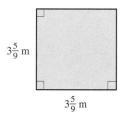

90.

91. Find the perimeter of a square with sides of length $3\frac{5}{9}$ m.

92. Find the perimeter of the rectangle in Exercise 89.

93. Make use of the properties and laws discussed in Sections 1.2 and 1.3 to explain why $x + y$ is equivalent to $(2y + 2x)/2$.

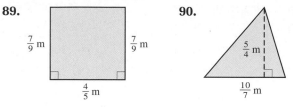

Positive and Negative Real Numbers

1.4

The Integers • The Rational Numbers • Real Numbers and Order • Absolute Value

A **set** is a collection of objects. The set containing 1, 3, and 7 is usually written $\{1, 3, 7\}$. In this section, we examine some important sets of numbers. More on sets can be found in Appendix A.

The Integers

Two sets of numbers were mentioned in Section 1.3. We represent these sets using dots on a number line.

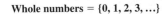

Natural numbers = {1, 2, 3, ...}

Whole numbers = {0, 1, 2, 3, ...}

To create the set of *integers,* we include all whole numbers, along with their *opposites.* To find the opposite of a number, we locate the number that is the same distance from 0 but on the other side of the number line. For example,

the opposite of 1 is negative 1, written -1;

and

the opposite of 3 is negative 3, written -3.

The **integers** consist of all whole numbers and their opposites.

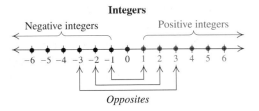

Opposites are discussed in more detail in Section 1.6. Note that, except for 0, opposites occur in pairs. Thus, 5 is the opposite of -5, just as -5 is the opposite of 5. Note that 0 acts as its own opposite.

> ### Set of Integers
> The set of integers = $\{..., -4, -3, -2, -1, 0, 1, 2, 3, 4, ...\}$.

Integers are associated with many real-world problems and situations.

Example 1

State which integer(s) corresponds to each situation.

a) In 1997, Tiger Woods set a U.S. Masters tournament record by finishing 18 strokes under par.
b) Death Valley is 280 ft below sea level.
c) Jaco's Bistro made $329 on Sunday, but lost $53 on Monday.

In 1997, Tiger Woods set a U.S. Masters tournament record by finishing 18 strokes under par.

Solution

a) The integer -18 corresponds to 18 under par.

b) The integer -280 corresponds to the situation (see the figure below). The elevation is -280 ft.

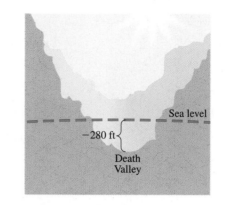

c) The integer 329 corresponds to making $329 on Sunday and -53 corresponds to losing $53 on Monday.

The Rational Numbers

Although numbers like $\frac{5}{9}$ are built out of integers, these numbers are not themselves integers. Another set, the **rational numbers**, contains fractions and decimals, as well as the integers. Some examples of rational numbers are

$$\frac{5}{9}, \quad -\frac{4}{7}, \quad 95, \quad -16, \quad 0, \quad \frac{-35}{8}, \quad 2.4, \quad -0.31.$$

In Section 1.7, we show that $-\frac{4}{7}$ can be written as $\frac{-4}{7}$ or $\frac{4}{-7}$. Indeed, every number listed above can be written as an integer over an integer. For example, 95 can be written as $\frac{95}{1}$ and 2.4 can be written as $\frac{24}{10}$. In this manner, any *ratio*nal number can be expressed as the *ratio* of two integers. Rather than attempt to list all rational numbers, we use this idea of ratio to describe the set as follows.

> ### Set of Rational Numbers
> The set of rational numbers $= \left\{ \dfrac{a}{b} \,\middle|\, a \text{ and } b \text{ are integers and } b \neq 0 \right\}.$
>
> This is read "the set of all numbers $\dfrac{a}{b}$, where a and b are integers and $b \neq 0$."

In Section 1.7, we explain why b cannot equal 0.

To *graph* a number is to mark its location on a number line.

E x a m p l e   2   Graph each of the following rational numbers.

a) $\frac{5}{2}$ **b)** -3.2 **c)** $\frac{11}{8}$

Solution

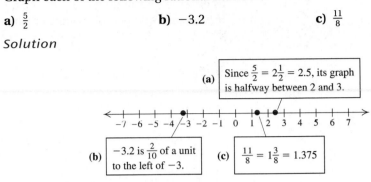

(a) Since $\frac{5}{2} = 2\frac{1}{2} = 2.5$, its graph is halfway between 2 and 3.

(b) -3.2 is $\frac{2}{10}$ of a unit to the left of -3.

(c) $\frac{11}{8} = 1\frac{3}{8} = 1.375$

Every rational number can be written as a fraction or a decimal.

E x a m p l e   3   Convert to decimal notation: $-\frac{5}{8}$.

Solution   We first find decimal notation for $\frac{5}{8}$. Since $\frac{5}{8}$ means $5 \div 8$, we divide.

$$
\begin{array}{r}
0.6\ 2\ 5 \\
8)\overline{5.0\ 0\ 0} \\
\underline{4\ 8\ 0\ 0} \\
2\ 0\ 0 \\
\underline{1\ 6\ 0} \\
4\ 0 \\
\underline{4\ 0} \\
0
\end{array}
$$

$\longleftarrow$ The remainder is 0.

Thus, $\frac{5}{8} = 0.625$, so $-\frac{5}{8} = -0.625$.

Because the division in Example 3 ends with the remainder 0, we consider -0.625 a **terminating decimal**. If we are "bringing down" zeros and a remainder reappears, we have a **repeating decimal**, as shown in the next example.

E x a m p l e   4   Convert to decimal notation: $\frac{7}{11}$.

Solution   We divide:

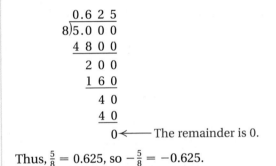

$$
\begin{array}{r}
0.6\ 3\ 6\ 3... \\
1\ 1)\overline{7.0\ 0\ 0\ 0} \\
\underline{6\ 6} \\
4\ 0 \\
\underline{3\ 3} \\
7\ 0 \\
\underline{6\ 6} \\
4\ 0
\end{array}
$$

4 reappears as a remainder.

We abbreviate repeating decimals by writing a bar over the repeating part—in this case, $0.\overline{63}$. Thus, $\frac{7}{11} = 0.\overline{63}$.

Although we do not prove it here, every rational number can be expressed as either a terminating or repeating decimal, and every terminating or repeating decimal can be expressed as a ratio of two integers.

Real Numbers and Order

Some numbers, when written in decimal form, neither terminate nor repeat. Such numbers are called **irrational numbers**.

What sort of numbers are irrational? One example is π (the Greek letter *pi*, read "pie"), which is used to find the area and circumference of a circle: $A = \pi r^2$ and $C = 2\pi r$.

Another irrational number, $\sqrt{2}$ (read "the square root of 2"), is the length of the diagonal of a square with sides of length 1. It is also the number that, when multiplied by itself, gives 2. No rational number can be multiplied by itself to get 2, although some approximations come close:

1.4 is an *approximation* of $\sqrt{2}$ because $(1.4)(1.4) = 1.96$;

1.41 is a better approximation because $(1.41)(1.41) = 1.9881$;

1.4142 is an even better approximation because $(1.4142)(1.4142) = 1.99996164$.

To approximate $\sqrt{2}$ on some calculators, simply press $\boxed{2}$ and then $\boxed{\sqrt{}}$. With other calculators, press $\boxed{\sqrt{}}$, $\boxed{2}$, and $\boxed{\text{ENTER}}$, or consult a manual.

E x a m p l e   5   Graph the real number $\sqrt{3}$ on a number line.

Solution   We use a calculator and approximate: $\sqrt{3} \approx 1.732$ ("$\approx$" means "approximately equals"). Then we locate this number on a number line.

The rational numbers and the irrational numbers together correspond to all the points on a number line and make up what is called the **real-number system**.

> ### Set of Real Numbers
> The set of real numbers = The set of all numbers corresponding to points on the number line.

The following figure shows the relationships among various kinds of numbers.

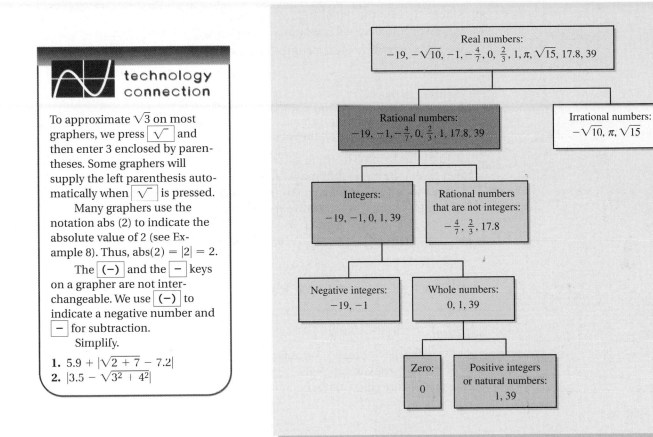

Real numbers are named in order on the number line, with larger numbers further to the right. For any two numbers, the one to the left is less than the one to the right. We use the symbol **<** to mean "**is less than.**" The sentence $-8 < 6$ means "-8 is less than 6." The symbol **>** means "**is greater than.**" The sentence $-3 > -7$ means "-3 is greater than -7."

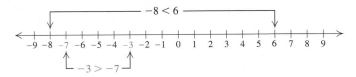

Example 6

Use either $<$ or $>$ for ▨ to write a true sentence.

a) 2 ▨ 9 **b)** -3.45 ▨ 1.32 **c)** 6 ▨ -12
d) -18 ▨ -5 **e)** $\frac{7}{11}$ ▨ $\frac{5}{8}$

Solution

a) Since 2 is to the left of 9 on a number line, we know that 2 is less than 9, so $2 < 9$.

b) Since -3.45 is to the left of 1.32, we have $-3.45 < 1.32$.

c) Since 6 is to the right of -12, we have $6 > -12$.

d) Since -18 is to the left of -5, we have $-18 < -5$.

e) We convert to decimal notation: $\frac{7}{11} = 0.\overline{63}$ and $\frac{5}{8} = 0.625$. Thus, $\frac{7}{11} > \frac{5}{8}$.
We also could have used a common denominator: $\frac{7}{11} = \frac{56}{88} > \frac{55}{88} = \frac{5}{8}$.

Sentences like "$a < -5$" and "$-3 > -8$" are **inequalities**. It is useful to remember that every inequality can be written in two ways. For example,

$$-3 > -8 \quad \text{has the same meaning as} \quad -8 < -3.$$

It may be helpful to think of an inequality sign as an "arrow" with the smaller side pointing to the smaller number.

Note that $a > 0$ means that a represents a positive real number and $a < 0$ means that a represents a negative real number.

Statements like $a \le b$ and $b \ge a$ are also inequalities. We read $a \le b$ as "a **is less than or equal to** b" and $a \ge b$ as "a **is greater than or equal to** b."

E x a m p l e   7

Classify each inequality as true or false.

a) $-3 \le 5$ **b)** $-3 \le -3$ **c)** $-5 \ge 4$

Solution

a) $-3 \le 5$ is *true* because $-3 < 5$ is true.

b) $-3 \le -3$ is *true* because $-3 = -3$ is true.

c) $-5 \ge 4$ is *false* since neither $-5 > 4$ nor $-5 = 4$ is true.

Absolute Value

There is a convenient terminology and notation for the distance a number is from 0 on a number line. It is called the **absolute value** of the number.

> ### Absolute Value
>
> We write $|a|$, read "the absolute value of a," to represent the number of units that a is from zero.

E x a m p l e   8

Find each absolute value: **(a)** $|-3|$; **(b)** $|7.2|$; **(c)** $|0|$.

Solution

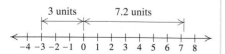

a) $|-3| = 3$ since -3 is 3 units from 0.

b) $|7.2| = 7.2$ since 7.2 is 7.2 units from 0.

c) $|0| = 0$ since 0 is 0 units from itself.

Distance is never negative, so numbers that are opposites have the same absolute value. If a number is nonnegative, its absolute value is the number itself. If a number is negative, its absolute value is its opposite.

Exercise Set 1.4

FOR EXTRA HELP

Digital Video Tutor CD 1
Videotape 1

InterAct Math

Math Tutor Center

MathXL

MyMathLab.com

Tell which real numbers correspond to each situation.

1. In Burlington, Vermont, the record low temperature for Washington's Birthday is 19° Fahrenheit (F) below zero. The record high for the date is 59°F above zero.

2. Karissa's golf score was 2 under par, while Alex's score was 5 over.

3. Using a NordicTrack exercise machine, Kit burned 150 calories. She then drank an isotonic drink containing 65 calories.

4. A painter earned $1200 one week and spent $800 the next.

5. The Dead Sea is 1286 feet below sea level, whereas Mt. Everest is 29,029 feet above sea level.

6. In bowling, the Jets are 34 pins behind the Strikers after one game. Describe the situation from the viewpoint of each team.

7. Janice deposited $750 in a savings account. Two weeks later, she withdrew $125.

8. In 1998, the world birthrate, per thousand, was 22. The death rate, per thousand, was 9. (*Source*: *Time Almanac 2000,* 1999)

9. During a video game, Cindy intercepted a missile worth 20 points, lost a starship worth 150 points, and captured a base worth 300 points.

10. Ignition occurs 10 seconds before liftoff. A spent fuel tank is detached 235 seconds after liftoff.

Graph each rational number on a number line.

11. $\frac{10}{3}$ **12.** $-\frac{17}{5}$ **13.** -4.3

14. 3.87 **15.** -2 **16.** 5

Find decimal notation.

17. $\frac{7}{8}$ **18.** $-\frac{1}{8}$ **19.** $-\frac{3}{4}$

20. $\frac{5}{6}$ **21.** $\frac{7}{6}$ **22.** $\frac{5}{12}$

23. $\frac{2}{3}$ **24.** $\frac{1}{4}$ **25.** $-\frac{1}{2}$

26. $-\frac{3}{8}$ Aha! **27.** $\frac{13}{100}$ **28.** $-\frac{7}{20}$

Write a true sentence using either $<$ or $>$.

29. $-8 \ \square \ 2$ **30.** $9 \ \square \ 0$

31. $7 \ \square \ 0$ **32.** $8 \ \square \ -8$

33. $-6 \ \square \ 6$ **34.** $0 \ \square \ -7$

35. $-8 \ \square \ -5$ **36.** $-4 \ \square \ -3$

37. $-5 \ \square \ -11$ **38.** $-3 \ \square \ -4$

39. $-12.5 \ \square \ -9.4$ **40.** $-10.3 \ \square \ -14.5$

41. $\frac{5}{12} \ \square \ \frac{11}{25}$ **42.** $-\frac{14}{17} \ \square \ -\frac{27}{35}$

For each of the following, write a second inequality with the same meaning.

43. $-7 > x$ **44.** $a > 9$

45. $-10 \leq y$ **46.** $12 \geq t$

Classify each inequality as true or false.

47. $-3 \geq -11$ **48.** $5 \leq -5$

49. $0 \geq 8$ **50.** $-5 \leq 7$

51. $-8 \leq -8$ **52.** $8 \geq 8$

Find each absolute value.

53. $|-23|$ **54.** $|-47|$ **55.** $|17|$

56. $|3.1|$ **57.** $|5.6|$ **58.** $\left|-\frac{2}{5}\right|$

59. $|329|$ **60.** $|-456|$ **61.** $\left|-\frac{9}{7}\right|$

62. $|8.02|$ **63.** $|0|$ **64.** $|-1.07|$

65. $|x|$, for $x = -8$ **66.** $|a|$, for $a = -5$

For Exercises 67–72, consider the following list:

$$-83, -4.7, 0, \tfrac{5}{9}, \pi, \sqrt{17}, 8.31, 62.$$

67. List all rational numbers.

68. List all natural numbers.

69. List all integers.

70. List all irrational numbers.

71. List all real numbers.

72. List all nonnegative integers.

73. Is every integer a rational number? Why or why not?

74. Is every integer a natural number? Why or why not?

SKILL MAINTENANCE

75. Evaluate $3xy$ for $x = 2$ and $y = 7$.

76. Use a commutative law to write an expression equivalent to $ab + 5$.

SYNTHESIS

77. Is the absolute value of a number always positive? Why or why not?

78. How many rational numbers are there between 0 and 1? Justify your answer.

79. Does "nonnegative" mean the same thing as "positive"? Why or why not?

List in order from least to greatest.

80. $13, -12, 5, -17$

81. $-23, 4, 0, -17$

82. $\tfrac{4}{5}, \tfrac{4}{3}, \tfrac{4}{8}, \tfrac{4}{6}, \tfrac{4}{9}, \tfrac{4}{2}, -\tfrac{4}{3}$

83. $-\tfrac{2}{3}, \tfrac{1}{2}, -\tfrac{3}{4}, -\tfrac{5}{6}, \tfrac{3}{8}, \tfrac{1}{6}$

Write a true sentence using either $<$, $>$, or $=$.

84. $|-5| \quad \blacksquare \quad |-2|$

85. $|4| \quad \blacksquare \quad |-7|$

86. $|-8| \quad \blacksquare \quad |8|$

87. $|23| \quad \blacksquare \quad |-23|$

88. $|-3| \quad \blacksquare \quad |5|$

89. $|-19| \quad \blacksquare \quad |-27|$

Solve. Consider only integer replacements.

90. $|x| = 7$

91. $|x| < 3$

92. $2 < |x| < 5$

Given that $0.3\overline{3} = \tfrac{1}{3}$ and $0.6\overline{6} = \tfrac{2}{3}$, express each of the following as a ratio of two integers.

93. $0.1\overline{1}$

94. $0.9\overline{9}$

95. $5.5\overline{5}$

96. $7.7\overline{7}$

To the student and instructor: *The calculator icon, ▥, is used to indicate those exercises designed to be solved with a calculator.*

97. When Helga's calculator gives a decimal value for $\sqrt{2}$ and that value is promptly squared, the result is 2. Yet when that same decimal approximation is entered by hand and then squared, the result is not exactly 2. Why do you suppose this is?

Addition of Real Numbers

1.5

Adding with a Number Line • Adding without a Number Line • Problem Solving • Combining Like Terms

We now consider addition of real numbers. To gain understanding, we will use a number line first. After observing the principles involved, we will develop rules that allow us to work more quickly without a number line.

Adding with a Number Line

To add $a + b$ on a number line, we start at a and move according to b.

a) If b is positive, we move to the right (the positive direction).
b) If b is negative, we move to the left (the negative direction).
c) If b is 0, we stay at a.

E x a m p l e   1

Add: $-4 + 9$.

Solution   To add on a number line, we locate the first number, -4, and then move 9 units to the right. Note that it requires 4 units to reach 0. The difference between 9 and 4 is where we finish.

$$-4 + 9 = 5$$

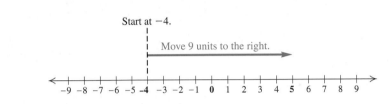

E x a m p l e   2

Add: $3 + (-5)$.

Solution   We locate the first number, 3, and then move 5 units to the left. Note that it requires 3 units to reach 0. The difference between 5 and 3 is 2, so we finish 2 units to the left of 0.

$$3 + (-5) = -2$$

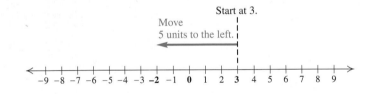

E x a m p l e   3

Add: $-4 + (-3)$.

Solution   After locating -4, we move 3 units to the left. We finish a total of 7 units to the left of 0.

$$-4 + (-3) = -7$$

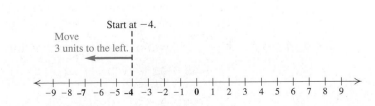

E x a m p l e   4   Add: $-5.2 + 0$.

Solution   We locate -5.2 and move 0 units. Thus we finish where we started, at -5.2.

$$-5.2 + 0 = -5.2$$

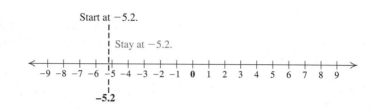

From Examples 1–4, we observe the following rules.

> ### Rules for Addition of Real Numbers
>
> 1. *Positive numbers*: Add as usual. The answer is positive.
> 2. *Negative numbers*: Add absolute values and make the answer negative (see Example 3).
> 3. *A positive number and a negative number*: Subtract the smaller absolute value from the greater absolute value. Then:
>   a) If the positive number has the greater absolute value, the answer is positive (see Example 1).
>   b) If the negative number has the greater absolute value, the answer is negative (see Example 2).
>   c) If the numbers have the same absolute value, the answer is 0.
> 4. *One number is zero*: The sum is the other number (see Example 4).

Rule 4 is known as the **identity property of 0**. It says that for any real number a, we have $a + 0 = a$.

Adding without a Number Line

The rules listed above can be used without drawing a number line.

E x a m p l e   5   Add without using a number line.

a) $-12 + (-7)$
b) $-1.4 + 8.5$
c) $-36 + 21$
d) $1.5 + (-1.5)$
e) $-\frac{7}{8} + 0$
f) $\frac{2}{3} + \left(-\frac{5}{8}\right)$

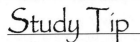

Solution

a) $-12 + (-7) = -19$ Two negatives. *Think*: Add the absolute values, 12 and 7, to get 19. Make the answer *negative*, -19.

b) $-1.4 + 8.5 = 7.1$ A negative and a positive. *Think*: The difference of absolute values is $8.5 - 1.4$, or 7.1. The positive number has the larger absolute value, so the answer is *positive*, 7.1.

c) $-36 + 21 = -15$ A negative and a positive. *Think*: The difference of absolute values is $36 - 21$, or 15. The negative number has the larger absolute value, so the answer is *negative*, -15.

d) $1.5 + (-1.5) = 0$ A negative and a positive. *Think*: Since the numbers are opposites, they have the same absolute value and the answer is 0.

e) $-\dfrac{7}{8} + 0 = -\dfrac{7}{8}$ One number is zero. The sum is the other number, $-\frac{7}{8}$.

f) $\dfrac{2}{3} + \left(-\dfrac{5}{8}\right) = \dfrac{16}{24} + \left(-\dfrac{15}{24}\right)$ This is similar to part (b) above.

$$= \dfrac{1}{24}$$

If we are adding several numbers, some positive and some negative, the commutative and associative laws allow us to add all the positives, then add all the negatives, and then add the results. Of course, we can also add from left to right, if we prefer.

E x a m p l e   6

Add: $15 + (-2) + 7 + 14 + (-5) + (-12)$.

Solution

$$15 + (-2) + 7 + 14 + (-5) + (-12)$$

$$= 15 + 7 + 14 + (-2) + (-5) + (-12) \qquad \text{Using the commutative law of addition}$$

$$= (15 + 7 + 14) + [(-2) + (-5) + (-12)] \qquad \text{Using the associative law of addition}$$

$$= 36 + (-19) \qquad \text{Adding the positives; adding the negatives}$$

$$= 17 \qquad \text{Adding a positive and a negative}$$

Problem Solving

Addition of real numbers occurs in many real-world applications.

E x a m p l e 7

Lake level. Lake Champlain straddles the border of New York and Vermont. From mid-November 1998 through mid-November 2000, the water level dropped $1\frac{1}{2}$ ft, rose $\frac{1}{4}$ ft, and dropped 1 ft (*Source: Burlington Free Press*, Nov. 21, 2000). By how much did the level change over the 2 yr?

Solution   The problem translates to a sum:

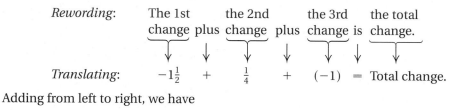

Rewording:	The 1st change	plus	the 2nd change	plus	the 3rd change	is	the total change.
Translating:	$-1\frac{1}{2}$	$+$	$\frac{1}{4}$	$+$	(-1)	$=$	Total change.

Adding from left to right, we have

$$-1\tfrac{1}{2} + \tfrac{1}{4} + (-1) = -1\tfrac{2}{4} + \tfrac{1}{4} + (-1) = -1\tfrac{1}{4} + (-1) = -2\tfrac{1}{4}.$$

The lake level dropped $2\frac{1}{4}$ ft between mid-November 1998 and mid-November 2000.

Combining Like Terms

When two terms have variable factors that are exactly the same, like $5ab$ and $7ab$, the terms are called **like**, or **similar**, **terms**.* The distributive law enables us to **combine**, or **collect**, **like terms**. The above rules for addition will again apply.

E x a m p l e 8

Combine like terms.

a) $-7x + 9x$ **b)** $2a + (-3b) + (-5a) + 9b$
c) $6 + y + (-3.5y) + 2$

Solution

a) $-7x + 9x = (-7 + 9)x$ Using the distributive law
    $\qquad\qquad\ = 2x$ Adding -7 and 9
b) $2a + (-3b) + (-5a) + 9b$
    $\quad = 2a + (-5a) + (-3b) + 9b$ Using the commutative law of addition
    $\quad = (2 + (-5))a + (-3 + 9)b$ Using the distributive law
    $\quad = -3a + 6b$ Adding

---

*Like terms are discussed in greater detail in Section 1.8.

c) $6 + y + (-3.5y) + 2 = y + (-3.5y) + 6 + 2$ Using the commutative law of addition

$$= (1 + (-3.5))y + 6 + 2$$ Using the distributive law

$$= -2.5y + 8$$ Adding

With practice we can leave out some steps, combining like terms mentally. Note that numbers like 6 and 2 in the expression $6 + y + (-3.5y) + 2$ are constants and are also considered to be like terms.

Exercise Set 1.5

FOR EXTRA HELP

Digital Video Tutor CD 1 Videotape 2   InterAct Math   Math Tutor Center   MathXL   MyMathLab.com

Add using a number line.

1. $4 + (-7)$ **2.** $2 + (-5)$

3. $-5 + 9$ **4.** $-3 + 8$

5. $8 + (-8)$ **6.** $6 + (-6)$

7. $-3 + (-5)$ **8.** $-4 + (-6)$

Add. Do not use a number line except as a check.

9. $-15 + 0$ **10.** $-6 + 0$

11. $0 + (-8)$ **12.** $0 + (-2)$

13. $12 + (-12)$ **14.** $17 + (-17)$

15. $-24 + (-17)$ **16.** $-17 + (-25)$

17. $-15 + 15$ **18.** $-18 + 18$

19. $18 + (-11)$ **20.** $8 + (-5)$

21. $10 + (-12)$ **22.** $9 + (-13)$

23. $-3 + 14$ **24.** $13 + (-6)$

25. $-14 + (-19)$ **26.** $11 + (-9)$

27. $19 + (-19)$ **28.** $-20 + (-6)$

29. $23 + (-5)$ **30.** $-15 + (-7)$

31. $-23 + (-9)$ **32.** $40 + (-8)$

33. $40 + (-40)$ **34.** $-25 + 25$

35. $85 + (-65)$ **36.** $63 + (-18)$

37. $-3.6 + 1.9$ **38.** $-6.5 + 4.7$

39. $-5.4 + (-3.7)$ **40.** $-3.8 + (-9.4)$

41. $\frac{-3}{5} + \frac{4}{5}$ **42.** $\frac{-2}{7} + \frac{3}{7}$

43. $\frac{-4}{7} + \frac{-2}{7}$ **44.** $\frac{-5}{9} + \frac{-2}{9}$

45. $-\frac{2}{5} + \frac{1}{3}$ **46.** $-\frac{4}{13} + \frac{1}{2}$

47. $\frac{-4}{9} + \frac{2}{3}$ **48.** $\frac{-1}{6} + \frac{1}{3}$

49. $35 + (-14) + (-19) + (-5)$

50. $28 + (-44) + 17 + 31 + (-94)$

Aha! **51.** $-4.9 + 8.5 + 4.9 + (-8.5)$

52. $24 + 3.1 + (-44) + (-8.2) + 63$

Solve. Write your answer as a complete sentence.

53. *Class size.* During the first two weeks of the semester, 5 students withdrew from Elisa's algebra class, 8 students were added to the class, and 6 students were dropped as "no-shows." By how many students did the original class size change?

54. *Telephone bills.* Maya's telephone bill for July was $82. She sent a check for $50 and then made $37 worth of calls in August. What was her new balance?

55. *Profits and losses.* The following table shows the profits and losses of Fax City over a 3-yr period.

Find the profit or loss after this period of time.

Year	Profit or loss
1998	−$26,500
1999	−$10,200
2000	+$32,400

56. *Yardage gained.* In a college football game, the quarterback attempted passes with the following results.

First try	13-yd gain
Second try	12-yd loss
Third try	21-yd gain

Find the total gain (or loss).

57. *Account balance.* Leah has $350 in a checking account. She writes a check for $530, makes a deposit of $75, and then writes a check for $90. What is the balance in the account?

58. *Credit card bills.* Lyle's credit card bill indicates that he owes $470. He sends a check to the credit card company for $45, charges another $160 in merchandise, and then pays off another $500 of his bill. What is Lyle's new balance?

59. *Stock growth.* In the course of one day, the value of a share of America Online rose $\frac{3}{16}$, dropped $\frac{1}{2}$, and then rose $\frac{1}{4}$. How much had the stock's value risen or fallen at the end of the day?

60. *Peak elevation.* The tallest mountain in the world, as measured from base to peak, is Mauna Kea in Hawaii. From a base 19,684 ft below sea level, it rises 33,480 ft. (*Source: The Guinness Book of Records,* 1999) What is the elevation of its peak?

Combine like terms.

61. $7a + 5a$

62. $3x + 8x$

63. $-3x + 12x$

64. $2m + (-7m)$

65. $5t + 8t$

66. $5a + 9a$

67. $7m + (-9m)$

68. $-4x + 4x$

69. $-5a + (-2a)$

70. $10n + (-17n)$

71. $-3 + 8x + 4 + (-10x)$

72. $8a + 5 + (-a) + (-3)$

Find the perimeter of each figure.

73.

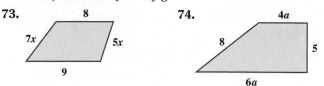

74.

75. **76.**

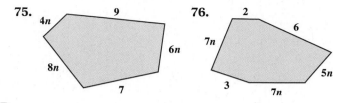

77. Explain in your own words why the sum of two negative numbers is negative.

78. Without performing the actual addition, explain why the sum of all integers from -10 to 10 is 0.

SKILL MAINTENANCE

79. Multiply: $7(3z + y + 2)$.

80. Divide and simplify: $\frac{7}{2} \div \frac{3}{8}$.

SYNTHESIS

81. Under what circumstances will the sum of one positive number and several negative numbers be positive?

82. Is it possible to add real numbers without knowing how to calculate $a - b$ with a and b both non-negative and $a \geq b$? Why or why not?

83. *Stock prices.* The value of EKB stock rose $2\frac{3}{8}$ and then dropped $3\frac{1}{4}$ before finishing at $64\frac{3}{8}$. What was the stock's original value?

84. *Sports card values.* The value of a sports card dropped $12 and then rose $17.50 before settling at $61. What was the original value of the card?

Find the missing term or terms.

85. $4x +$ _____ $+ (-9x) + (-2y) = -5x - 7y$

86. $-3a + 9b +$ _____ $+ 5a = 2a - 6b$

87. $3m + 2n +$ _____ $+ (-2m) = 2n + (-6m)$

88. _____ $+ 9x + (-4y) + x = 10x - 7y$

Aha! **89.** $7t + 23 +$ _____ $+$ _____ $= 0$

90. *Geometry.* The perimeter of a rectangle is $7x + 10$. If the length of the rectangle is 5, express the width in terms of x.

91. *Golfing.* After five rounds of golf, a golf pro was 3 under par twice, 2 over par once, 2 under par once, and 1 over par once. On average, how far above or below par was the golfer?

Subtraction of Real Numbers

1.6

Opposites and Additive Inverses • Subtraction • Problem Solving

In arithmetic, when a number b is subtracted from another number a, the difference, $a - b$, is the number that when added to b gives a. For example, $45 - 17 = 28$ because $28 + 17 = 45$. We will use this approach to develop an efficient way of finding the value of $a - b$ for any real numbers a and b. Before doing so, however, we must develop some terminology.

Opposites and Additive Inverses

Numbers such as 6 and -6 are *opposites,* or *additive inverses,* of each other. Whenever opposites are added, the result is 0; and whenever two numbers add to 0, those numbers are opposites.

E x a m p l e   1   Find the opposite of each number: **(a)** 34; **(b)** -8.3; **(c)** 0.

Solution

a) The opposite of 34 is -34: $34 + (-34) = 0$.
b) The opposite of -8.3 is 8.3: $-8.3 + 8.3 = 0$.
c) The opposite of 0 is 0: $0 + 0 = 0$.

To write the opposite, we use the symbol $-$, as follows.

> **Opposite**
>
> The *opposite,* or *additive inverse,* of a number a is written $-a$ (read "the opposite of a" or "the additive inverse of a").

Note that if we take a number, say 8, and find its opposite, -8, and then find the opposite of the result, we will have the original number, 8, again. Thus, for any number a,

$$-(-a) = a.$$

E x a m p l e   2   Find $-x$ and $-(-x)$ when $x = 16$.

Solution

If $x = 16$, then $-x = -16$.   The opposite of 16 is -16.
If $x = 16$, then $-(-x) = -(-16) = 16$.   The opposite of the opposite of 16 is 16.

E x a m p l e   3

Find $-x$ and $-(-x)$ when $x = -3$.

Solution

If $x = -3$, then $-x = -(-3) = 3$.   The opposite of -3 is 3.

If $x = -3$, then $-(-x) = -(-(-3)) = -(\ 3\ ) = -3$.

Note in Example 3 that an extra set of parentheses is used to show that we are substituting the negative number -3 for x. The notation $-\ -x$ is not used.

A symbol such as -8 is usually read "negative 8." It could be read "the additive inverse of 8," because the additive inverse of 8 is negative 8. It could also be read "the opposite of 8," because the opposite of 8 is -8.

A symbol like $-x$, which has a variable, should be read "the opposite of x" or "the additive inverse of x" and *not* "negative x," since to do so suggests that $-x$ represents a negative number. As we saw in Example 3, $-x$ can represent a positive number. This notation can be used to restate a result from Section 1.5 as *the law of opposites*:

> ### The Law of Opposites
>
> For any two numbers a and $-a$,
>
> $$a + (-a) = 0.$$
>
> (When opposites are added, their sum is 0.)

A negative number is said to have a "negative *sign*." A positive number is said to have a "positive *sign*." When we replace a number with its opposite, or additive inverse, we can say that we have "changed or reversed its sign."

E x a m p l e   4

Change the sign (find the opposite) of each number: **(a)** -3; **(b)** -10; **(c)** 14.

Solution

a) When we change the sign of -3, we obtain 3.

b) When we change the sign of -10, we obtain 10.

c) When we change the sign of 14, we obtain -14.

Subtraction

Opposites are helpful when subtraction involves negative numbers. To see why, look for a pattern in the following:

Subtracting		*Adding the Opposite*
$9 - 5 = 4$	since $4 + 5 = 9$	$9 + (-5) = 4$
$5 - 8 = -3$	since $-3 + 8 = 5$	$5 + (-8) = -3$
$-6 - 4 = -10$	since $-10 + 4 = -6$	$-6 + (-4) = -10$
$-7 - (-10) = 3$	since $3 + (-10) = -7$	$-7 + 10 = 3$
$-7 - (-2) = -5$	since $-5 + (-2) = -7$	$-7 + 2 = -5$

technology connection

On nearly all graphers, it is essential to distinguish between the key for negation and the key for subtraction. To enter a negative number, we use (−) and to subtract, we use − . This said, be careful not to rely on a calculator for computations that you will be expected to do by hand.

The matching results suggest that we can subtract by adding the opposite of the number being subtracted. This can always be done and often provides the easiest way to subtract real numbers.

Subtraction of Real Numbers

For any real numbers a and b,

$$a - b = a + (-b).$$

(To subtract, add the opposite, or additive inverse, of the number being subtracted.)

Example 5

Subtract each of the following and then check with addition.

a) $2 - 6$
b) $4 - (-9)$
c) $-4.2 - (-3.6)$

Solution

a) $2 - 6 = 2 + (-6) = -4$ The opposite of 6 is -6. We change the subtraction to addition and add the opposite. *Check*: $-4 + 6 = 2$.

b) $4 - (-9) = 4 + 9 = 13$ The opposite of -9 is 9. We change the subtraction to addition and add the opposite. *Check*: $13 + (-9) = 4$.

c) $-4.2 - (-3.6) = -4.2 + 3.6$ Adding the opposite of -3.6.
                                         Check: $-0.6 + (-3.6) = -4.2$.

$= -0.6$

The symbol "−" is read differently depending on where it appears. For example, $-5 - (-x)$ is read "negative five minus the opposite of x."

Example 6

Read each of the following and then subtract.

a) $3 - 5$
b) $-4.6 - (-9.8)$

Solution

a) $3 - 5$;   Read "three minus five"
    $3 - 5 = 3 + (-5) = -2$ Adding the opposite

b) $-4.6 - (-9.8)$;   Read "negative four point six minus negative nine point eight"
    $-4.6 - (-9.8) = -4.6 + 9.8 = 5.2$ Adding the opposite

E x a m p l e 7

Subtract $-\frac{3}{5}$ from $\frac{1}{5}$.

Solution   A common denominator exists. We subtract as follows:

$$\frac{1}{5} - \left(-\frac{3}{5}\right) = \frac{1}{5} + \frac{3}{5} \qquad \text{Adding the opposite}$$

$$= \frac{1+3}{5} = \frac{4}{5}.$$

Check: $\dfrac{4}{5} + \left(-\dfrac{3}{5}\right) = \dfrac{4}{5} + \dfrac{-3}{5} = \dfrac{4 + (-3)}{5} = \dfrac{1}{5}.$

E x a m p l e 8

Simplify: $8 - (-4) - 2 - (-5) + 3$.

Solution

$$8 - (-4) - 2 - (-5) + 3 = 8 + 4 + (-2) + 5 + 3 \qquad \begin{array}{l}\text{To subtract,} \\ \text{we add the} \\ \text{opposite.}\end{array}$$

$$= 18$$

Recall from Section 1.2 that the terms of an algebraic expression are separated by plus signs. This means that the terms of $5x - 7y - 9$ are $5x$, $-7y$, and -9, since $5x - 7y - 9 = 5x + (-7y) + (-9)$.

E x a m p l e 9

Identify the terms of $4 - 2ab + 7a - 9$.

Solution   We have

$$4 - 2ab + 7a - 9 = 4 + (-2ab) + 7a + (-9), \qquad \text{Rewriting as addition}$$

so the terms are 4, $-2ab$, $7a$, and -9.

E x a m p l e 10

Combine like terms.

a) $1 + 3x - 7x$ **b)** $-5a - 7b - 4a + 10b$
c) $4 - 3m - 9 + 7m$

Solution

a) $1 + 3x - 7x = 1 + 3x + (-7x)$ Adding the opposite

$\qquad\qquad\quad = 1 + (3 + (-7))x$ Using the distributive law.

$\qquad\qquad\quad = 1 + (-4)x$ Try to do this mentally.

$\qquad\qquad\quad = 1 - 4x$ Rewriting as subtraction to be more concise

b) $-5a - 7b - 4a + 10b = -5a + (-7b) + (-4a) + 10b$ Adding the opposite

$$= -5a + (-4a) + (-7b) + 10b$$ Using the commutative law of addition

$$= -9a + 3b$$ Combining like terms mentally

c) $4 - 3m - 9 + 7m = 4 + (-3m) + (-9) + 7m$ Rewriting as addition

$$= 4 + (-9) + (-3m) + 7m$$ Using the commutative law of addition

$$= -5 + 4m$$

Problem Solving

Subtraction is used to solve problems involving differences.

E x a m p l e   1 1

Changes in temperature.   The Viking 2 Lander spacecraft has determined that temperatures on Mars range from $-125°$ Celsius (C) to 25°C (*Source*: The Lunar and Planetary Institute). Find the temperature range on Mars.

Solution   It is helpful to make a drawing of the situation.

To find the difference between two temperatures, we always subtract the lower temperature from the higher temperature:

$$25 - (-125) = 25 + 125$$
$$= 150.$$

The temperature range on Mars is 150°C.

Exercise Set 1.6

FOR EXTRA HELP

Digital Video Tutor CD 1
Videotape 2

InterAct Math

Math Tutor Center

MathXL

MyMathLab.com

Find the opposite, or additive inverse.

1. 39   **2.** -17 **3.** -9

4. $\frac{7}{2}$ **5.** -3.14 **6.** 48.2

Find $-x$ when x is each of the following.

7. 23   **8.** -26 **9.** $-\frac{14}{3}$

10. $\frac{1}{328}$ **11.** 0.101 **12.** 0

Find $-(-x)$ when x is each of the following.

13. 72   **14.** 29

15. $-\frac{2}{5}$ **16.** -9.1

Change the sign. (Find the opposite.)

17. -1 **18.** -7

19. 7   **20.** 10

Write words for each of the following and then perform the subtraction.

21. $-3 - 5$ **22.** $-4 - 7$

23. $2 - (-9)$ **24.** $5 - (-8)$

25. $4 - 6$ **26.** $9 - 12$

27. $-5 - (-7)$ **28.** $-2 - (-5)$

Subtract.

29. $6 - 8$ **30.** $4 - 13$

31. $0 - 5$ **32.** $0 - 8$

33. $3 - 9$ **34.** $3 - 13$

35. $0 - 10$ **36.** $0 - 7$

37. $-9 - (-3)$ **38.** $-9 - (-5)$

Aha! **39.** $-8 - (-8)$ **40.** $-10 - (-10)$

41. $14 - 19$ **42.** $12 - 16$

43. $30 - 40$ **44.** $20 - 27$

45. $-7 - (-9)$ **46.** $-8 - (-3)$

47. $-9 - (-9)$ **48.** $-40 - (-40)$

49. $5 - 5$ **50.** $7 - 7$

51. $4 - (-4)$ **52.** $6 - (-6)$

53. $-7 - 4$ **54.** $-6 - 8$

55. $6 - (-10)$ **56.** $3 - (-12)$

57. $-14 - 2$ **58.** $-4 - 15$

59. $-4 - (-3)$ **60.** $-6 - (-5)$

61. $5 - (-6)$ **62.** $5 - (-12)$

63. $0 - 6$ **64.** $0 - 5$

65. $-3 - (-1)$ **66.** $-5 - (-2)$

67. $-9 - 16$ **68.** $-7 - 14$

69. $0 - (-1)$ **70.** $0 - (-5)$

71. $-9 - 0$ **72.** $-8 - 0$

73. $12 - (-5)$ **74.** $3 - (-7)$

75. $18 - 63$ **76.** $2 - 25$

77. $-18 - 63$ **78.** $-42 - 26$

79. $-45 - 4$ **80.** $-51 - 7$

81. $1.5 - 9.4$ **82.** $3.2 - 8.7$

83. $0.825 - 1$ **84.** $0.072 - 1$

85. $\frac{3}{7} - \frac{5}{7}$ **86.** $\frac{2}{11} - \frac{9}{11}$

87. $\frac{-2}{9} - \frac{5}{9}$ **88.** $\frac{-1}{5} - \frac{3}{5}$

89. $-\frac{2}{13} - \left(-\frac{5}{13}\right)$ **90.** $-\frac{4}{17} - \left(-\frac{9}{17}\right)$

Translate each phrase to mathematical language and simplify. See Example 11.

91. The difference between 3.8 and -5.2

92. The difference between -2.1 and -5.9

93. The difference between 114 and -79

94. The difference between 23 and -17

95. Subtract 37 from -21.

96. Subtract 19 from -7.

97. Subtract -25 from 9.

98. Subtract -31 from -5.

Simplify.

99. $25 - (-12) - 7 - (-2) + 9$

100. $22 - (-18) + 7 + (-42) - 27$

101. $-31 + (-28) - (-14) - 17$

102. $-43 - (-19) - (-21) + 25$

103. $-34 - 28 + (-33) - 44$

104. $39 + (-88) - 29 - (-83)$

Aha! **105.** $-93 + (-84) - (-93) - (-84)$

106. $84 + (-99) + 44 - (-18) - 43$

Identify the terms in each expression.

107. $-7x - 4y$

108. $7a - 9b$

109. $9 - 5t - 3st$

110. $-4 - 3x + 2xy$

Combine like terms.

111. $4x - 7x$

112. $3a - 14a$

113. $7a - 12a + 4$

114. $-9x - 13x + 7$

115. $-8n - 9 + n$

116. $-7 + 9n - 8$

117. $3x + 5 - 9x$

118. $2 + 3u - 7$

119. $2 - 6t - 9 - 2t$

120. $-5 + 3b - 7 - 5b$

121. $5y + (-3x) - 9x + 1 - 2y + 8$

122. $14 - (-5x) + 2z - (-32) + 4z - 2x$

123. $13x - (-2x) + 45 - (-21) - 7x$

124. $8x - (-2x) - 14 - (-5x) + 53 - 9x$

Solve.

125. *Record temperature drop.*   The greatest recorded temperature change in one day occurred in Browning, Montana, when the temperature fell from 44°F to −56°F (*Source: The Guinness Book of Records*, 1999). How much did the temperature drop?

126. *Loan repayment.*   Gisela owed Ramon $290. Ramon decides to "forgive" $125 of the debt. How much does Gisela owe?

127. *Elevation extremes.*   The lowest elevation in Asia, the Dead Sea, is 1312 ft below sea level. The highest elevation in Asia, Mount Everest, is 29,028 ft. (*Source: The World Almanac and Book of Facts* 2000) Find the difference in elevation.

128. *Elevation extremes.*   The elevation of Mount Whitney, the highest peak in California, is 14,776 ft more than the elevation of Death Valley, California (*Source: 1999 Information Please Almanac*). If Death Valley is 282 ft below sea level, find the elevation of Mount Whitney.

129. *Changes in elevation.*   The lowest point in Africa is Lake Assal, which is 156 m below sea level. The lowest point in South America is the Valdes Peninsula, which is 40 m below sea level. How much lower is Lake Assal than the Valdes Peninsula?

130. *Underwater elevation.*   The deepest point in the Pacific Ocean is the Marianas Trench, with a depth of 10,415 m. The deepest point in the Atlantic Ocean is the Puerto Rico Trench, with a depth of 8648 m. What is the difference in elevation of the two trenches?

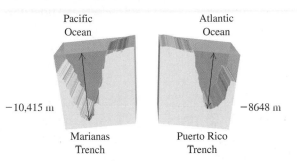

131. Jeremy insists that if you can *add* real numbers, then you can also *subtract* real numbers. Do you agree? Why or why not?

132. Are the expressions $-a + b$ and $a + (-b)$ opposites of each other? Why or why not?

SKILL MAINTENANCE

133. Find the area of a rectangle when the length is 36 ft and the width is 12 ft.

134. Find the prime factorization of 864.

SYNTHESIS

135. Why might it be advantageous to rewrite a long series of additions and subtractions as all additions?

136. If a and b are both negative, under what circumstances will $a - b$ be negative?

Tell whether each statement is true or false for all real numbers m and n. Use various replacements for m and n to support your answer.

137. If $m > n$, then $m - n > 0$.

138. If $m > n$, then $m + n > 0$.

139. If m and n are opposites, then $m - n = 0$.

140. If $m = -n$, then $m + n = 0$.

141. A gambler loses a wager and then loses "double or nothing" (meaning the gambler owes twice as much) twice more. After the three losses, the gambler's assets are −$20. Explain how much the gambler originally bet and how the $20 debt occurred.

142. If n is positive and m is negative, what is the sign of $n + (-m)$? Why?

1.7

Multiplication and Division of Real Numbers

Multiplication • Division

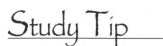
We now develop rules for multiplication and division of real numbers. Because multiplication and division are closely related, the rules are quite similar.

Multiplication

We already know how to multiply two nonnegative numbers. To see how to multiply a positive number and a negative number, consider the following pattern in which multiplication is regarded as repeated addition:

This number ⟶ $4(-5) = (-5) + (-5) + (-5) + (-5) = -20$ ← This number
decreases by $3(-5) = \qquad\quad (-5) + (-5) + (-5) = -15$ increases by
1 each time. $2(-5) = \qquad\qquad\qquad (-5) + (-5) = -10$ 5 each time.
                   $1(-5) = \qquad\qquad\qquad\qquad\quad (-5) = -5$
                   $0(-5) = \qquad\qquad\qquad\qquad\qquad\;\; 0 = \;\; 0$

This pattern illustrates that the product of a negative number and a positive number is negative.

> **The Product of a Negative Number and a Positive Number**
>
> To multiply a positive number and a negative number, multiply their absolute values. The answer is negative.

Example 1

Multiply: **(a)** $8(-5)$; **(b)** $-\frac{1}{3} \cdot \frac{5}{7}$.

Solution

a) $8(-5) = -40$ *Think*: $8 \cdot 5 = 40$; make the answer negative.

b) $-\frac{1}{3} \cdot \frac{5}{7} = -\frac{5}{21}$ *Think*: $\frac{1}{3} \cdot \frac{5}{7} = \frac{5}{21}$; make the answer negative.

The pattern developed above includes not just products of positive and negative numbers, but a product involving zero as well.

> **The Multiplicative Property of Zero**
>
> For any real number a,
>
> $$0 \cdot a = a \cdot 0 = 0.$$
>
> (The product of 0 and any real number is 0.)

Example 2

Multiply: $173(-452)0$.

Solution   We have

$$173(-452)0 = 173[(-452)0]$$ Using the associative law of multiplication

$$= 173[0]$$ Using the multiplicative property of zero

$$= 0.$$ Using the multiplicative property of zero again

Note that whenever 0 appears as a factor, the product will be 0.

We can extend the above pattern still further to examine the product of two negative numbers.

This number → $2(-5) =$ $(-5) + (-5) = -10$ ← This number
decreases by $1(-5) =$ $(-5) = -5$ increases by
1 each time. $0(-5) =$ $0 = 0$ 5 each time.
                $-1(-5) =$ $-(-5) = 5$
                $-2(-5) = -(-5) - (-5) = 10$

According to the pattern, the product of two negative numbers is positive.

> ### *The Product of Two Negative Numbers*
> To multiply two negative numbers, multiply their absolute values. The answer is positive.

Example 3

Multiply: **(a)** $(-6)(-8)$; **(b)** $(-1.2)(-3)$.

Solution

a) The absolute value of -6 is 6 and the absolute value of -8 is 8. Thus,

$$(-6)(-8) = 6 \cdot 8$$ Multiplying absolute values. The answer is positive.

$$= 48.$$

b) $(-1.2)(-3) = (1.2)(3)$ Multiplying absolute values. The answer is positive.

$$= 3.6$$ Try to go directly to this step.

When three or more numbers are multiplied, we can order and group the numbers as we please, because of the commutative and associative laws.

E x a m p l e   4

Multiply: **(a)** $-3(-2)(-5)$; **(b)** $-4(-6)(-1)(-2)$.

Solution

a) $-3(-2)(-5) = 6(-5)$ Multiplying the first two numbers. The product of two negatives is positive.

$\qquad\qquad\qquad = -30$ The product of a positive and a negative is negative.

b) $-4(-6)(-1)(-2) = 24 \cdot 2$ Multiplying the first two numbers and the last two numbers

$\qquad\qquad\qquad\qquad = 48$

We can see the following pattern in the results of Example 4.

The product of an even number of negative numbers is positive.

The product of an odd number of negative numbers is negative.

Division

Recall that $a \div b$, or $\frac{a}{b}$, is the number, if one exists, that when multiplied by b gives a. For example, to show that $10 \div 2$ is 5, we need only note that $5 \cdot 2 = 10$. Thus division can always be checked with multiplication.

E x a m p l e   5

Divide, if possible, and check your answer.

a) $14 \div (-7)$ **b)** $\dfrac{-32}{-4}$ **c)** $\dfrac{-10}{9}$ **d)** $\dfrac{-17}{0}$

Solution

a) $14 \div (-7) = -2$ We look for a number that when multiplied by -7 gives 14. That number is -2. *Check*: $(-2)(-7) = 14$.

b) $\dfrac{-32}{-4} = 8$ We look for a number that when multiplied by -4 gives -32. That number is 8. *Check*: $8(-4) = -32$.

c) $\dfrac{-10}{9} = -\dfrac{10}{9}$ We look for a number that when multiplied by 9 gives -10. That number is $-\frac{10}{9}$. *Check*: $-\frac{10}{9} \cdot 9 = -10$.

d) $\dfrac{-17}{0}$ is **undefined**.   We look for a number that when multiplied by 0 gives -17. There is no such number because the product of 0 and *any* number is 0, not -17.

The sign rules for division are the same as those for multiplication: The quotient of a positive number and a negative number is negative; the quotient of two negative numbers is positive.

> ### Rules for Multiplication and Division
>
> To multiply or divide two real numbers:
>
> **1.** Using the absolute values, multiply or divide, as indicated.
> **2.** If the signs are the same, the answer is positive.
> **3.** If the signs are different, the answer is negative.

Had Example 5(a) been written as $-14 \div 7$ or $-\frac{14}{7}$, rather than $14 \div (-7)$, the result would still have been -2. Thus from Examples 5(a)–5(c), we have the following:

$$\frac{-a}{b} = \frac{a}{-b} = -\frac{a}{b} \quad \text{and} \quad \frac{-a}{-b} = \frac{a}{b}.$$

E x a m p l e   6

Rewrite each of the following in two equivalent forms: **(a)** $\frac{5}{-2}$; **(b)** $-\frac{3}{10}$.

Solution   We use one of the properties just listed.

a) $\dfrac{5}{-2} = \dfrac{-5}{2}$ and $\dfrac{5}{-2} = -\dfrac{5}{2}$

b) $-\dfrac{3}{10} = \dfrac{-3}{10}$ and $-\dfrac{3}{10} = \dfrac{3}{-10}$

Since $\dfrac{-a}{b} = \dfrac{a}{-b} = -\dfrac{a}{b}$

When a fraction contains a negative sign, it may be helpful to rewrite (or simply visualize) the fraction in an equivalent form.

E x a m p l e   7

Perform the indicated operation: **(a)** $\left(-\frac{4}{5}\right)\left(\frac{-7}{3}\right)$; **(b)** $-\frac{2}{7} + \frac{9}{-7}$.

Solution

a) $\left(-\dfrac{4}{5}\right)\left(\dfrac{-7}{3}\right) = \left(-\dfrac{4}{5}\right)\left(-\dfrac{7}{3}\right)$ Rewriting $\dfrac{-7}{3}$ as $-\dfrac{7}{3}$

   $= \dfrac{28}{15}$ Try to go directly to this step.

b) Given a choice, we generally choose a positive denominator:

   $-\dfrac{2}{7} + \dfrac{9}{-7} = \dfrac{-2}{7} + \dfrac{-9}{7}$ Rewriting both fractions with a common denominator of 7

   $= \dfrac{-11}{7}$, or $-\dfrac{11}{7}$.

E x a m p l e   8

Find the reciprocal: **(a)** -27; **(b)** $\frac{-3}{4}$; **(c)** $-\frac{1}{5}$.

Solution

a) The reciprocal of -27 is $\frac{1}{-27}$. More often, this number is written as $-\frac{1}{27}$.

b) The reciprocal of $\frac{-3}{4}$ is $\frac{4}{-3}$, or, equivalently, $-\frac{4}{3}$.

c) The reciprocal of $-\frac{1}{5}$ is -5.

Recall that the opposite, or additive inverse, of a number is what we add to the number to get 0, whereas a reciprocal is what we multiply the number by to get 1. Compare the following.

Number	Opposite (Change the sign.)	Reciprocal (Invert but do not change the sign.)
$-\dfrac{3}{8}$	$\dfrac{3}{8}$	$-\dfrac{8}{3}$
19	-19	$\dfrac{1}{19}$
0	0	Undefined

$$\left(-\frac{3}{8}\right)\left(-\frac{8}{3}\right)=1$$

$$-\frac{3}{8}+\frac{3}{8}=0$$

When dividing with fraction notation, it is usually easier to multiply by a reciprocal. With decimal notation, it is usually easier to carry out division.

E x a m p l e   9   Divide: **(a)** $-\frac{2}{3}\div\left(-\frac{5}{4}\right)$; **(b)** $-\frac{3}{4}\div\frac{3}{10}$; **(c)** $27.9\div(-3)$.

Solution

a) $-\dfrac{2}{3}\div\left(-\dfrac{5}{4}\right)=-\dfrac{2}{3}\cdot\left(-\dfrac{4}{5}\right)=\dfrac{8}{15}$ Multiplying by the reciprocal

Be careful not to change the sign when taking a reciprocal!

b) $-\dfrac{3}{4}\div\dfrac{3}{10}=-\dfrac{3}{4}\cdot\left(\dfrac{10}{3}\right)=-\dfrac{30}{12}=-\dfrac{5}{2}\cdot\dfrac{6}{6}=-\dfrac{5}{2}$ Removing a factor equal to $1:\frac{6}{6}=1$

c) $27.9\div(-3)=\dfrac{27.9}{-3}=-9.3$ Dividing: $3\overline{)27.9}$, 9.3.
The answer is negative.

In Example 5(d), we explained why we cannot divide -17 by 0. This also explains why *no* nonzero number b can be divided by 0: Consider $b\div0$. Is there a number that when multiplied by 0 gives b? No, because the product of 0 and any number is 0, not b. We say that $b\div0$ is **undefined** for $b\neq0$. In the special case of $0\div0$, we look for a number r such that $0\div0=r$ and $r\cdot0=0$. But, $r\cdot0=0$ for *any* number r. For this reason, we say that $b\div0$ is undefined for any choice of b.*

Finally, note that $0\div7=0$ since $0\cdot7=0$. This can be written $0/7=0$.

E x a m p l e   1 0   Divide, if possible: **(a)** $\frac{0}{-2}$; **(b)** $\frac{5}{0}$.

Solution

a) $\dfrac{0}{-2}=0$ *Check:* $0(-2)=0$.

b) $\dfrac{5}{0}$ is undefined.

---

*Sometimes $0\div0$ is said to be *indeterminate*.

> **Division Involving Zero**
> For any real number a,
>
> $$\frac{a}{0} \text{ is undefined,}$$
>
> and for $a \neq 0$,
>
> $$\frac{0}{a} = 0.$$

Exercise Set 1.7

Multiply.

1. $-4 \cdot 9$

2. $-3 \cdot 7$

3. $-8 \cdot 7$

4. $-9 \cdot 2$

5. $8 \cdot (-3)$

6. $9 \cdot (-5)$

7. $-9 \cdot 8$

8. $-10 \cdot 3$

9. $-6 \cdot (-7)$

10. $-2 \cdot (-5)$

11. $-5 \cdot (-9)$

12. $-9 \cdot (-2)$

13. $17 \cdot (-10)$

14. $-12 \cdot (-10)$

15. $-12 \cdot 12$

16. $-13 \cdot (-15)$

17. $-25 \cdot (-48)$

18. $39 \cdot (-43)$

19. $-3.5 \cdot (-28)$

20. $97 \cdot (-2.1)$

21. $6 \cdot (-13)$

22. $7 \cdot (-9)$

23. $-7 \cdot (-3.1)$

24. $-4 \cdot (-3.2)$

25. $\frac{2}{3} \cdot \left(-\frac{3}{5}\right)$

26. $\frac{5}{7} \cdot \left(-\frac{2}{3}\right)$

27. $-\frac{3}{8} \cdot \left(-\frac{2}{9}\right)$

28. $-\frac{5}{8} \cdot \left(-\frac{2}{5}\right)$

29. $(-5.3)(2.1)$

30. $(-4.3)(9.5)$

31. $-\frac{5}{9} \cdot \frac{3}{4}$

32. $-\frac{8}{3} \cdot \frac{9}{4}$

33. $3 \cdot (-7) \cdot (-2) \cdot 6$

34. $9 \cdot (-2) \cdot (-6) \cdot 7$

Aha! **35.** $-27 \cdot (-34) \cdot 0$

36. $-43 \cdot (-74) \cdot 0$

37. $-\frac{1}{3} \cdot \frac{1}{4} \cdot \left(-\frac{3}{7}\right)$

38. $-\frac{1}{2} \cdot \frac{3}{5} \cdot \left(-\frac{2}{7}\right)$

39. $-2 \cdot (-5) \cdot (-3) \cdot (-5)$

40. $-3 \cdot (-5) \cdot (-2) \cdot (-1)$

41. $(-14) \cdot (-27) \cdot 0$

42. $7 \cdot (-6) \cdot 5 \cdot (-4) \cdot 3 \cdot (-2) \cdot 1 \cdot 0$

43. $(-8)(-9)(-10)$

44. $(-7)(-8)(-9)(-10)$

45. $(-6)(-7)(-8)(-9)(-10)$

46. $(-5)(-6)(-7)(-8)(-9)(-10)$

Divide, if possible, and check. If a quotient is undefined, state this.

47. $28 \div (-7)$

48. $\dfrac{24}{-3}$

49. $\dfrac{36}{-9}$

50. $26 \div (-13)$

51. $\dfrac{-16}{8}$

52. $-32 \div (-4)$

53. $\dfrac{-48}{-12}$

54. $-63 \div (-9)$

55. $\dfrac{-72}{9}$

56. $\dfrac{-50}{25}$

57. $-100 \div (-50)$

58. $\dfrac{-200}{8}$

59. $-108 \div 9$

60. $\dfrac{-64}{-7}$

61. $\dfrac{400}{-50}$

62. $-300 \div (-13)$

63. $\dfrac{28}{0}$

64. $\dfrac{0}{-5}$

65. $-4.8 \div 1.2$

66. $-3.9 \div 1.3$

67. $\dfrac{0}{-9}$

Aha! **68.** $\dfrac{(-4.9)(7.2)}{0}$

69. $0 \div 7$

70. $0 \div (-47)$

Write each number in two equivalent forms, as in Example 6.

71. $\dfrac{-8}{3}$

72. $\dfrac{-12}{7}$

73. $\dfrac{29}{-35}$

74. $\dfrac{9}{-14}$

75. $-\dfrac{7}{3}$

76. $-\dfrac{4}{15}$

77. $\dfrac{-x}{2}$

78. $\dfrac{9}{-a}$

Find the reciprocal of each number.

79. $\dfrac{4}{-5}$

80. $\dfrac{2}{-9}$

81. $-\dfrac{47}{13}$

82. $-\dfrac{31}{12}$

83. -10

84. 13

85. 4.3

86. -8.5

87. $\dfrac{-9}{4}$

88. $\dfrac{-6}{11}$

89. -1

90. $3/5$

Perform the indicated operation and, if possible, simplify. If a quotient is undefined, state this.

91. $\left(\dfrac{-7}{4}\right)\left(-\dfrac{3}{5}\right)$

92. $\left(-\dfrac{5}{6}\right)\left(\dfrac{-1}{3}\right)$

93. $\left(\dfrac{-6}{5}\right)\left(\dfrac{2}{-11}\right)$

94. $\left(\dfrac{7}{-2}\right)\left(\dfrac{-5}{6}\right)$

95. $\dfrac{-3}{8} + \dfrac{-5}{8}$

96. $\dfrac{-4}{5} + \dfrac{7}{5}$

Aha! **97.** $\left(\dfrac{-9}{5}\right)\left(\dfrac{5}{-9}\right)$

98. $\left(-\dfrac{2}{7}\right)\left(\dfrac{5}{-8}\right)$

99. $\left(-\dfrac{3}{11}\right) + \left(-\dfrac{6}{11}\right)$

100. $\left(-\dfrac{4}{7}\right) + \left(-\dfrac{2}{7}\right)$

101. $\dfrac{7}{8} \div \left(-\dfrac{1}{2}\right)$

102. $\dfrac{3}{4} \div \left(-\dfrac{2}{3}\right)$

103. $\dfrac{9}{5} \cdot \dfrac{-20}{3}$

104. $\dfrac{-5}{12} \cdot \dfrac{7}{15}$

105. $\left(-\dfrac{18}{7}\right) + \left(-\dfrac{3}{7}\right)$

106. $\left(-\dfrac{12}{5}\right) + \left(-\dfrac{3}{5}\right)$

Aha! **107.** $-\dfrac{5}{9} \div \left(-\dfrac{5}{9}\right)$

108. $-\dfrac{5}{4} \div \left(-\dfrac{3}{4}\right)$

109. $-44.1 \div (-6.3)$

110. $-6.6 \div 3.3$

111. $\dfrac{1}{9} - \dfrac{2}{9}$

112. $\dfrac{2}{7} - \dfrac{6}{7}$

113. $\dfrac{-3}{10} + \dfrac{2}{5}$

114. $\dfrac{-5}{9} + \dfrac{2}{3}$

115. $\dfrac{7}{10} \div \left(\dfrac{-3}{5}\right)$

116. $\left(\dfrac{-3}{5}\right) \div \dfrac{6}{15}$

117. $\dfrac{5}{7} - \dfrac{1}{-7}$

118. $\dfrac{4}{9} - \dfrac{1}{-9}$

119. $\dfrac{-4}{15} + \dfrac{2}{-3}$

120. $\dfrac{3}{-10} + \dfrac{-1}{5}$

121. Most calculators have a key, often appearing as $\boxed{1/x}$, for finding reciprocals. To use this key, enter a number and then press $\boxed{1/x}$ to find its reciprocal. What should happen if you enter a number and then press the reciprocal key twice? Why?

122. Multiplication can be regarded as repeated addition. Using this idea and a number line, explain why $3 \cdot (-5) = -15$.

SKILL MAINTENANCE

123. Simplify: $\dfrac{264}{468}$.

124. Combine like terms: $x + 12y + 11x - 14y - 9$.

SYNTHESIS

125. If two nonzero numbers are opposites of each other, are their reciprocals opposites of each other? Why or why not?

126. If two numbers are reciprocals of each other, are their opposites reciprocals of each other? Why or why not?

127. Show that the reciprocal of a sum is *not* the sum of the two reciprocals.

128. Which real numbers are their own reciprocals?

Tell whether each expression represents a positive number or a negative number when m and n are negative.

129. $\dfrac{m}{-n}$

130. $\dfrac{-n}{-m}$

131. $-m \cdot \left(\dfrac{-n}{m}\right)$

132. $-\left(\dfrac{n}{-m}\right)$

133. $(m + n) \cdot \dfrac{m}{n}$

134. $(-n - m)\dfrac{n}{m}$

135. What must be true of m and n if $-mn$ is to be (a) positive? (b) zero? (c) negative?

136. The following is a proof that a positive number times a negative number is negative. Provide a reason for each step. Assume that $a > 0$ and $b > 0$.

$$a(-b) + ab = a[-b + b]$$
$$= a(0)$$
$$= 0$$

Therefore, $a(-b)$ is the opposite of ab.

137. Is it true that for any numbers a and b, if a is larger than b, then the reciprocal of a is smaller than the reciprocal of b? Why or why not?

Exponential Notation and Order of Operations

1.8

Exponential Notation • Order of Operations • Simplifying and the Distributive Law • The Opposite of a Sum

Algebraic expressions often contain *exponential notation*. In this section, we learn how to use exponential notation as well as rules for the *order of operations*, in performing certain algebraic manipulations.

Exponential Notation

A product like $3 \cdot 3 \cdot 3 \cdot 3$, in which the factors are the same, is called a **power**. Powers occur often enough that a simpler notation called **exponential notation** is used. For

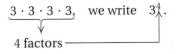

This is read "three to the fourth power," or simply, "three to the fourth." The number 4 is called an **exponent** and the number 3 a **base**.

Expressions like s^2 and s^3 are usually read "s squared" and "s cubed," respectively. This comes from the fact that a square with sides of length s has an area A given by $A = s^2$ and a cube with sides of length s has a volume V given by $V = s^3$.

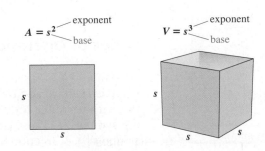

E x a m p l e   1   Write exponential notation for $10 \cdot 10 \cdot 10 \cdot 10 \cdot 10$.

Solution

Exponential notation is 10^5.   5 is the exponent.
10 is the base.

E x a m p l e   2   Evaluate: **(a)** 5^2; **(b)** $(-5)^3$; **(c)** $(2n)^3$.

Solution

a) $5^2 = 5 \cdot 5 = 25$ The second power indicates two factors of 5.

b) $(-5)^3 = (-5)(-5)(-5)$ The third power indicates three factors of -5.

$\qquad\quad = 25(-5)$ Using the associative law of multiplication

$\qquad\quad = -125$

c) $(2n)^3 = (2n)(2n)(2n)$ The third power indicates three factors of $2n$.

$\qquad\quad = 2 \cdot 2 \cdot 2 \cdot n \cdot n \cdot n$ Using the associative and commutative laws of multiplication

$\qquad\quad = 8n^3$

To determine what the exponent 1 will mean, look for a pattern in the following:

$$7 \cdot 7 \cdot 7 \cdot 7 = 7^4$$
$$7 \cdot 7 \cdot 7 = 7^3$$
$$7 \cdot 7 = 7^2$$
$$7 = 7^?$$

We divide by 7 each time.

The exponents decrease by 1 each time. To continue the pattern, we say that

$$7 = 7^1.$$

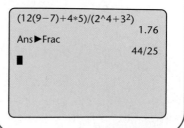

technology connection

On most graphers, grouping symbols, such as a fraction bar, must be replaced with parentheses. For example, to calculate

$$\frac{12(9 - 7) + 4 \cdot 5}{2^4 + 3^2},$$

we enter $(12(9 - 7) + 4 \cdot 5) \div$ $(2^4 + 3^2)$. To enter an exponential expression, we enter the base, press $\boxed{\wedge}$ and then enter the exponent. The $\boxed{x^2}$ key can be used to enter an exponent of 2. We can also convert to fraction notation if we wish.

```
(12(9−7)+4∗5)/(2^4+3²)
                    1.76
Ans▶Frac
                   44/25
■
```

> **Exponential Notation**
>
> For any natural number n,
>
> $$b^n \quad \text{means} \quad \overbrace{b \cdot b \cdot b \cdot b \cdots b}^{n \text{ factors}}.$$

Order of Operations

How should $4 + 2 \times 5$ be computed? If we multiply 2 by 5 and then add 4, the result is 14. If we add 2 and 4 first and then multiply by 5, the result is 30. Since these results differ, the order in which we perform operations matters. If grouping symbols such as parentheses ( ), brackets [ ], braces { }, absolute-value symbols | |, or fraction bars are used, they tell us what to do first. For example,

$$(4 + 2) \times 5 = 6 \times 5 = 30$$

and

$$4 + (2 \times 5) = 4 + 10 = 14.$$

In addition to grouping symbols, the following conventions exist for determining the order in which operations should be performed.

Rules for Order of Operations

1. Calculate within the innermost grouping symbols.
2. Simplify all exponential expressions.
3. Perform all multiplication and division, working from left to right.
4. Perform all addition and subtraction, working from left to right.

Thus the correct way to compute $4 + 2 \times 5$ is to first multiply 2 by 5 and then add 4. The result is 14.

E x a m p l e   3

Simplify: $15 - 2 \times 5 + 3$.

Solution   When no groupings or exponents appear, we always multiply or divide before adding or subtracting:

$$
\begin{aligned}
15 - 2 \times 5 + 3 &= 15 - 10 + 3 && \text{Multiplying} \\
&= 5 + 3 \\
&= 8.
\end{aligned}
$$

Subtracting and adding from left to right

Always calculate within parentheses first. When there are exponents and no parentheses, simplify powers before multiplying or dividing.

E x a m p l e   4

Simplify: **(a)** $(3 \cdot 4)^2$; **(b)** $3 \cdot 4^2$.

Solution

a) $(3 \cdot 4)^2 = (12)^2$ Working within parentheses first
$\qquad\qquad\;\; = 144$

b) $3 \cdot 4^2 = 3 \cdot 16$ Simplifying the power
$\qquad\quad\;\; = 48$ Multiplying

Note that $(3 \cdot 4)^2 \neq 3 \cdot 4^2$.

Caution!   Example 4 illustrates that, in general, $(ab)^2 \neq ab^2$.

E x a m p l e   5

Evaluate for $x = 5$: **(a)** $(-x)^2$; **(b)** $-x^2$.

Solution

a) $(-x)^2 = (-5)^2 = (-5)(-5) = 25$ We square the opposite of 5.

b) $-x^2 = -5^2 = -25$ We square 5 and then find the opposite.

Caution!   Example 5 illustrates that, in general, $(-x)^2 \neq -x^2$.

E x a m p l e   6

Evaluate $-15 \div 3(6 - a)^3$ for $a = 4$.

Solution

$$
\begin{aligned}
-15 \div 3(6 - a)^3 &= -15 \div 3(6 - 4)^3 && \text{Substituting 4 for } a \\
&= -15 \div 3(2)^3 && \text{Working within parentheses} \\
& && \text{first} \\
&= -15 \div 3 \cdot 8 && \text{Simplifying the exponential} \\
& && \text{expression} \\
&= -5 \cdot 8 && \text{Dividing and multiplying} \\
&= -40 && \text{from left to right}
\end{aligned}
$$

When combinations of grouping symbols are used, the rules still apply. We begin with the innermost grouping symbols and work to the outside.

E x a m p l e   7

Simplify: $8 \div 4 + 3[9 + 2(3 - 5)^3]$.

Solution

$$
\begin{aligned}
8 \div 4 + 3[9 + 2(3 - 5)^3] &= 8 \div 4 + 3[9 + 2(-2)^3] && \text{Doing the calculations} \\
& && \text{in the innermost} \\
& && \text{parentheses first} \\
&= 8 \div 4 + 3[9 + 2(-8)] && (-2)^3 = (-2)(-2)(-2) \\
& && \qquad\quad = -8 \\
&= 8 \div 4 + 3[9 + (-16)] \\
&= 8 \div 4 + 3[-7] && \text{Completing the calculations} \\
& && \text{within the brackets} \\
&= 2 + (-21) && \text{Multiplying and dividing from} \\
& && \text{left to right} \\
&= -19
\end{aligned}
$$

Example 8

Calculate: $\dfrac{12(9-7)+4\cdot 5}{3^4+2^3}$.

Solution An equivalent expression with brackets is

$$[12(9-7)+4\cdot 5]\div[3^4+2^3].$$

In effect, we need to simplify the numerator, simplify the denominator, and then divide the results:

$$\frac{12(9-7)+4\cdot 5}{3^4+2^3}=\frac{12(2)+4\cdot 5}{81+8}$$

$$=\frac{24+20}{89}=\frac{44}{89}.$$

Simplifying and the Distributive Law

Sometimes we cannot simplify within parentheses. When a sum or difference is within the parentheses, the distributive law often allows us to simplify the expression.

Example 9

Simplify: $5x-9+2(4x+5)$.

Solution

$$5x-9+2(4x+5)=5x-9+8x+10 \qquad \text{Using the distributive law}$$
$$=13x+1 \qquad \text{Combining like terms}$$

Now that exponents have been introduced, we can make our definition of *like* or *similar terms* more precise. **Like, or similar, terms** are either constant terms or terms containing the same variable(s) raised to the same power(s). Thus, 5 and -7, $19xy$ and $2yx$, and $4a^3b$ and a^3b are all pairs of like terms.

Example 10

Simplify: $7x^2+3(x^2+2x)-5x$.

Solution

$$7x^2+3(x^2+2x)-5x=7x^2+3x^2+6x-5x \qquad \text{Using the distributive law}$$
$$=10x^2+x \qquad \text{Combining like terms}$$

The Opposite of a Sum

When a number is multiplied by -1, the result is the opposite of that number. For example, $-1(7)=-7$ and $-1(-5)=5$.

> ### The Property of −1
>
> For any real number a,
>
> $$-1 \cdot a = -a.$$
>
> (Negative one times a is the opposite of a.)

When grouping symbols are preceded by a "−" symbol, we can multiply the grouping by −1 and use the distributive law. In this manner, we can find the *opposite*, or *additive inverse*, of a sum.

E x a m p l e   1 1

Write an expression equivalent to $-(3x + 2y + 4)$ without using parentheses.

Solution

$$
\begin{aligned}
-(3x + 2y + 4) &= -1(3x + 2y + 4) &&\text{Using the property of } -1 \\
&= -1(3x) + (-1)(2y) + (-1)4 &&\text{Using the distributive law} \\
&= -3x - 2y - 4 &&\text{Using the property of } -1
\end{aligned}
$$

Example 11 illustrates an important property of real numbers.

> ### The Opposite of a Sum
>
> For any real numbers a and b,
>
> $$-(a + b) = -a + (-b).$$
>
> (The opposite of a sum is the sum of the opposites.)

To remove parentheses from an expression like $-(x - 7y + 5)$, we can first rewrite the subtraction as addition:

$$
\begin{aligned}
-(x - 7y + 5) &= -(x + (-7y) + 5) &&\text{Rewriting as addition} \\
&= -x + 7y - 5. &&\text{Taking the opposite of a sum}
\end{aligned}
$$

This procedure is normally streamlined to one step in which we find the opposite by "removing parentheses and changing the sign of every term":

$$-(x - 7y + 5) = -x + 7y - 5.$$

E x a m p l e   1 2

Simplify: $3x - (4x + 2)$.

Solution

$$
\begin{aligned}
3x - (4x + 2) &= 3x + [-(4x + 2)] &&\text{Adding the opposite of } 4x + 2 \\
&= 3x + [-4x - 2] &&\text{Taking the opposite of } 4x + 2 \\
&= 3x + (-4x) + (-2) \\
&= 3x - 4x - 2 &&\text{Try to go directly to this step.} \\
&= -x - 2 &&\text{Combining like terms}
\end{aligned}
$$

In practice, the first three steps of Example 12 are usually skipped.

E x a m p l e   1 3

Simplify: $5t^2 - 2t - (4t^2 - 9t)$.

Solution

$$5t^2 - 2t - (4t^2 - 9t) = 5t^2 - 2t - 4t^2 + 9t \qquad \text{Removing parentheses and changing the sign of each term inside}$$

$$= t^2 + 7t \qquad \text{Combining like terms}$$

Expressions such as $7 - 3(x + 2)$ can be simplified as follows:

$$7 - 3(x + 2) = 7 + [-3(x + 2)] \qquad \text{Adding the opposite of } 3(x + 2)$$
$$= 7 + [-3x - 6] \qquad \text{Multiplying } x + 2 \text{ by } -3$$
$$= 7 - 3x - 6 \qquad \text{Try to go directly to this step.}$$
$$= 1 - 3x. \qquad \text{Combining like terms}$$

E x a m p l e   1 4

Simplify: **(a)** $3n - 2(4n - 5)$; **(b)** $7x^3 + 2 - [5(x^3 - 1) + 8]$.

Solution

a) $3n - 2(4n - 5) = 3n - 8n + 10 \qquad \text{Multiplying each term inside the parentheses by } -2$

$\qquad\qquad\qquad\quad = -5n + 10 \qquad \text{Combining like terms}$

b) $7x^3 + 2 - [5(x^3 - 1) + 8] = 7x^3 + 2 - [5x^3 - 5 + 8] \qquad \text{Removing parentheses}$

$\qquad\qquad\qquad\qquad = 7x^3 + 2 - [5x^3 + 3]$

$\qquad\qquad\qquad\qquad = 7x^3 + 2 - 5x^3 - 3 \qquad \text{Removing brackets}$

$\qquad\qquad\qquad\qquad = 2x^3 - 1 \qquad \text{Combining like terms}$

C O N N E C T I N G   T H E   C O N C E P T S

Algebra is a tool that can be used to solve problems. We have seen in this chapter that certain problems can be translated to algebraic expressions that can in turn be simplified (see Section 1.6). In Chapter 2, we will solve problems that require us to solve an equation.

As we progress through our study of algebra, it is important that we be able to distinguish between the two tasks of **simplifying an expression** and **solving an equation.** In Chapter 1, we did not solve equations, but we did simplify expressions. This enabled us to write *equivalent expressions* that were simpler than the given expression. In Chapter 2, we will continue to simplify expressions, but we will also begin to solve equations.

Exercise Set 1.8

FOR EXTRA HELP

Digital Video Tutor CD 1 Videotape 2   InterAct Math   Math Tutor Center   MathXL   MyMathLab.com

Write exponential notation.

1. $4 \cdot 4 \cdot 4$

2. $6 \cdot 6 \cdot 6 \cdot 6$

3. $x \cdot x \cdot x \cdot x \cdot x \cdot x \cdot x$

4. $y \cdot y \cdot y \cdot y \cdot y \cdot y$

5. $3t \cdot 3t \cdot 3t \cdot 3t \cdot 3t$

6. $5m \cdot 5m \cdot 5m \cdot 5m \cdot 5m$

Simplify.

7. 2^4

8. 5^3

9. $(-3)^2$

10. $(-7)^2$

11. -3^2

12. -7^2

13. 4^3

14. 9^1

15. $(-5)^4$

16. 5^4

17. 7^1

18. $(-1)^7$

19. $(3t)^4$

20. $(5t)^2$

21. $(-7x)^3$

22. $(-5x)^4$

23. $5 + 3 \cdot 7$

24. $3 - 4 \cdot 2$

25. $8 \cdot 7 + 6 \cdot 5$

26. $10 \cdot 5 + 1 \cdot 1$

27. $19 - 5 \cdot 3 + 3$

28. $14 - 2 \cdot 6 + 7$

29. $9 \div 3 + 16 \div 8$

30. $32 - 8 \div 4 - 2$

Aha! **31.** $84 \div 28 - 84 \div 28$

32. $18 - 6 \div 3 \cdot 2 + 7$

33. $4 - 8 \div 2 + 3^2$

34. $3(-10)^2 - 8 \div 2^2$

35. $9 - 3^2 \div 9(-1)$

36. $8 - (2 \cdot 3 - 9)$

37. $(8 - 2 \cdot 3) - 9$

38. $(8 - 2)(3 - 9)$

39. $(-24) \div (-3) \cdot \left(-\frac{1}{2}\right)$

40. $32 \div (-2)^2 \cdot 4$

41. $13(-10)^2 + 45 \div (-5)$

42. $5 \cdot 3^2 - 4^2 \cdot 2$

43. $2^4 + 2^3 - 10 \div (-1)^4$

44. $40 - 3^2 - 2^3 \div (-4)$

45. $5 + 3(2 - 9)^2$

46. $9 - (3 - 5)^3 - 4$

47. $[2 \cdot (5 - 8)]^2$

48. $3(5 - 7)^4 \div 4$

49. $\dfrac{7 + 2}{5^2 - 4^2}$

50. $\dfrac{5^2 - 3^2}{2 \cdot 6 - 4}$

51. $8(-7) + |6(-5)|$

52. $|10(-5)| + 1(-1)$

53. $\dfrac{(-2)^3 + 4^2}{3 - 5^2 + 3 \cdot 6}$

54. $\dfrac{7^2 - (-1)^5}{3 - 2 \cdot 3^2 + 5}$

55. $\dfrac{27 - 2 \cdot 3^2}{8 \div 2^2 - (-2)^2}$

56. $\dfrac{(-5)^2 - 4 \cdot 5}{3^2 + 4 \cdot 2(-1)^5}$

Evaluate.

57. $7 - 5x$, for $x = 3$

58. $1 + x^3$, for $x = -2$

59. $24 \div t^3$, for $t = -2$

60. $20 \div a \cdot 4$, for $a = 5$

61. $45 \div 3 \cdot a$, for $a = -1$

62. $50 \div 2 \cdot t$, for $t = -5$

63. $5x \div 15x^2$, for $x = 3$

64. $6a \div 12a^3$, for $a = 2$

Aha! **65.** $(12 \cdot 17) \div (17 \cdot 12)$

66. $-30 \div t(t + 4)^2$, for $t = -6$

67. $-x^2 - 5x$, for $x = -3$

68. $(-x)^2 - 5x$, for $x = -3$

69. $\dfrac{3a - 4a^2}{a^2 - 20}$, for $a = 5$

70. $\dfrac{a^3 - 4a}{a(a - 3)}$, for $a = -2$

Write an equivalent expression without using parentheses.

71. $-(9x + 1)$

72. $-(3x + 5)$

73. $-(7 - 2x)$

74. $-(6x - 7)$

75. $-(4a - 3b + 7c)$

76. $-(5x - 2y - 3z)$

77. $-(3x^2 + 5x - 1)$

78. $-(8x^3 - 6x + 5)$

Simplify.

79. $5x - (2x + 7)$

80. $7y - (2y + 9)$

81. $2a - (5a - 9)$

82. $11n - (3n - 7)$

83. $2x + 7x - (4x + 6)$

84. $3a + 2a - (4a + 7)$

85. $9t - 5r - 2(3r + 6t)$

86. $4m - 9n - 3(2m - n)$

87. $15x - y - 5(3x - 2y + 5z)$

88. $4a - b - 4(5a - 7b + 8c)$

89. $3x^2 + 7 - (2x^2 + 5)$

90. $5x^4 + 3x - (5x^4 + 3x)$

91. $5t^3 + t - 3(t + 2t^3)$

92. $8n^2 + n - 2(n + 3n^2)$

93. $12a^2 - 3ab + 5b^2 - 5(-5a^2 + 4ab - 6b^2)$

94. $-8a^2 + 5ab - 12b^2 - 6(2a^2 - 4ab - 10b^2)$

95. $-7t^3 - t^2 - 3(5t^3 - 3t)$

96. $9t^4 + 7t - 5(9t^3 - 2t)$

97. $5(2x - 7) - [4(2x - 3) + 2]$

98. $3(6x - 5) - [3(1 - 8x) + 5]$

99. Some students use the mnemonic device PEM-DAS to help remember the rules for the order of operations. Explain how this can be done.

100. Jake keys $18/2 \cdot 3$ into his calculator and expects the result to be 3. What mistake is he probably making?

SKILL MAINTENANCE

Translate to an algebraic expression.

101. Nine more than twice a number

102. Half of the sum of two numbers

SYNTHESIS

103. Write the sentence $(-x)^2 \neq -x^2$ in words. Explain why $(-x)^2$ and $-x^2$ are not equivalent.

104. Write the sentence $-|x| \neq -x$ in words. Explain why $-|x|$ and $-x$ are not equivalent.

Simplify.

105. $5t - \{7t - [4r - 3(t - 7)] + 6r\} - 4r$

106. $z - \{2z - [3z - (4z - 5z) - 6z] - 7z\} - 8z$

107. $\{x - [f - (f - x)] + [x - f]\} - 3x$

108. Is it true that for any real numbers a and b,
$$-(ab) = (-a)b = a(-b)?$$
Why or why not?

109. Is it true that for any real numbers a and b,
$$ab = (-a)(-b)?$$
Why or why not?

If $n > 0$, $m > 0$, and $n \neq m$, determine whether each of the following is true or false.

110. $-n + m = m - n$

111. $-n + m = -(n + m)$

112. $m - n = -(n - m)$

113. $n(-n - m) = -n^2 + nm$

114. $-m(n - m) = -(mn + m^2)$

115. $-m(-n + m) = m(n - m)$

116. $-n(-n - m) = n(n + m)$

Evaluate.

Aha! **117.** $[x + 3(2 - 5x) \div 7 + x](x - 3)$, for $x = 3$

Aha! **118.** $[x + 2 \div 3x] \div [x + 2 \div 3x]$, for $x = -7$

119. In Mexico, between 500 B.C. and 600 A.D., the Mayans represented numbers using powers of 20 and certain symbols. For example, the symbols

represent $4 \cdot 20^3 + 17 \cdot 20^2 + 10 \cdot 20^1 + 0 \cdot 20^0$. (*Source*: National Council of Teachers of Mathematics, 1906 Association Drive, Reston, VA 22091) Evaluate this number.

120. Examine the Mayan symbols and the numbers in Exercise 119. What numbers do

•,   ⚊,   and   ⬭

each represent?

COLLABORATIVE

CORNER

Select the Symbols

Focus: Order of operations
Time: 15 minutes
Group size: 2

One way to master the rules for the order of operations is to insert symbols within a display of numbers in order to obtain a predetermined result. For example, the display

$$1 \quad 2 \quad 3 \quad 4 \quad 5$$

can be used to obtain the result 21 as follows:

$$(1 + 2) \div 3 + 4 \cdot 5.$$

Note that without an understanding of the rules for the order of operations, solving a problem of this sort is impossible.

ACTIVITY

1. Each group should prepare an exercise similar to the example shown above. (Exponents are not allowed.) To do so, first select five single-digit numbers for display. Then insert operations and grouping symbols and calculate the result.
2. Pair with another group. Each group should give the other its result along with its five-number display, and challenge the other group to insert symbols that will make the display equal the result given.
3. Share with the entire class the various mathematical statements developed by each group.

Summary and Review 1

Key Terms

Variable, p. 2
Constant, p. 2
Variable expression, p. 2
Evaluate an expression, p. 2
Algebraic expression, p. 3
Substitute, p. 3
Value of an expression, p. 3
Equation, p. 5
Solution, p. 5
Equivalent expressions, p. 11
Term, p. 14
Factors, p. 15
Factorization, p. 16
Natural number, p. 18
Whole number, p. 18
Prime number, p. 18
Composite number, p. 19
Prime factorization, p. 19
Fraction notation, p. 19
Numerator, p. 19
Denominator, p. 19
Reciprocal, p. 20
Multiplicative inverse, p. 20
Simplifying, p. 21
Set, p. 26
Integer, p. 27
Rational number, p. 28
Terminating decimal, p. 29
Repeating decimal, p. 29
Irrational number, p. 30
Real-number system, p. 30
Less than, p. 31

Greater than, p. 31

Inequality, p. 32

Absolute value, p. 32

Combine like terms, p. 38

Opposite, p. 41

Additive inverse, p. 41

Undefined, p. 52

Indeterminate, p. 52

Power, p. 55

Exponential notation, p. 55

Exponent, p. 55

Base, p. 55

Like terms, p. 59

Important Properties and Formulas

Area of a rectangle: $A = lw$

Area of a triangle: $A = \frac{1}{2}bh$

Area of a parallelogram: $A = bh$

Commutative laws: $a + b = b + a; \quad ab = ba$

Associative laws: $a + (b + c) = (a + b) + c; \quad a(bc) = (ab)c$

Distributive law: $a(b + c) = ab + ac$

Identity property of 1: $1 \cdot a = a \cdot 1 = a$

Identity property of 0: $a + 0 = 0 + a = a$

Law of opposites: $a + (-a) = 0$

Multiplicative property of 0: $0 \cdot a = a \cdot 0 = 0$

Property of -1: $-1 \cdot a = -a$

Opposite of a sum: $-(a + b) = -a + (-b)$

Division involving 0: $\dfrac{0}{a} = 0; \quad \dfrac{a}{0}$ is undefined

$$\frac{-a}{b} = \frac{a}{-b} = -\frac{a}{b}, \qquad \frac{-a}{-b} = \frac{a}{b}$$

Rules for Order of Operations

1. Calculate within the innermost grouping symbols.
2. Simplify all exponential expressions.
3. Perform all multiplication and division, working from left to right.
4. Perform all addition and subtraction, working from left to right.

Review Exercises

Evaluate.

1. $5t$, for $t = 3$

2. $\dfrac{x - y}{3}$, for $x = 17$ and $y = 5$

3. $10 - y^2$, for $y = 4$

4. $-10 + a^2 \div (b + 1)$, for $a = 5$ and $b = 4$

Translate to an algebraic expression.

5. 7 less than z

6. The product of x and z

7. One more than the product of two numbers

8. Determine whether 35 is a solution of $x/5 = 8$.

9. Translate to an equation. Do not solve.

> In 1999, $4.6 billion worth of tea was sold wholesale in the United States. This was $2.8 billion more than the amount sold in 1990. How much tea was sold wholesale in 1990?

10. Use the commutative law of multiplication to write an expression equivalent to $2x + y$.

11. Use the associative law of addition to write an expression equivalent to $(2x + y) + z$.

12. Use the commutative and associative laws to write three expressions equivalent to $4(xy)$.

Multiply.

13. $6(3x + 5y)$

14. $8(5x + 3y + 2)$

Factor.

15. $21x + 15y$

16. $35x + 14 + 7y$

17. Find the prime factorization of 52.

Simplify.

18. $\dfrac{20}{48}$

19. $\dfrac{18}{8}$

Perform the indicated operation and, if possible, simplify.

20. $\dfrac{5}{12} + \dfrac{4}{9}$

21. $\dfrac{9}{16} \div 3$

22. $\dfrac{2}{3} - \dfrac{1}{15}$

23. $\dfrac{9}{10} \cdot \dfrac{16}{5}$

24. Tell which integers correspond to this situation: Renir has a debt of $45 and Raoul has $72 in his savings account.

25. Graph on a number line: $\frac{-1}{3}$.

26. Write an inequality with the same meaning as $-3 < x$.

27. Classify as true or false: $8 \geq 8$.

28. Classify as true or false: $0 \leq -1$.

29. Find decimal notation: $-\frac{7}{8}$.

30. Find the absolute value: $|-1|$.

31. Find $-(-x)$ when x is -7.

Simplify.

32. $4 + (-7)$

33. $-\frac{2}{3} + \frac{1}{12}$

34. $10 + (-9) + (-8) + 7$

35. $-3.8 + 5.1 + (-12) + (-4.3) + 10$

36. $-2 - (-7)$

37. $-\frac{9}{10} - \frac{1}{2}$

38. $-3.8 - 4.1$

39. $-9 \cdot (-6)$

40. $-2.7(3.4)$

41. $\frac{2}{3} \cdot \left(-\frac{3}{7}\right)$

42. $2 \cdot (-7) \cdot (-2) \cdot (-5)$

43. $35 \div (-5)$

44. $-5.1 \div 1.7$

45. $-\frac{3}{5} \div \left(-\frac{4}{5}\right)$

46. $|-3 \cdot 4 - 12 \cdot 2| - 8(-7)$

47. $|-12(-3) - 2^3 - (-9)(-10)|$

48. $120 - 6^2 \div 4 \cdot 8$

49. $(120 - 6^2) \div 4 \cdot 8$

50. $(120 - 6^2) \div (4 \cdot 8)$

51. $\dfrac{4(18 - 8) + 7 \cdot 9}{9^2 - 8^2}$

Combine like terms.

52. $11a + 2b + (-4a) + (-5b)$

53. $7x - 3y - 9x + 8y$

54. Find the opposite of -7.

55. Find the reciprocal of -7.

56. Write exponential notation for $2x \cdot 2x \cdot 2x \cdot 2x$.

57. Simplify: $(-5x)^3$.

Remove parentheses and simplify.

58. $2a - (5a - 9)$

59. $3(b + 7) - 5b$

60. $3[11x - 3(4x - 1)]$

61. $2[6(y - 4) + 7]$

62. $[8(x + 4) - 10] - [3(x - 2) + 4]$

63. Explain the difference between a constant and a variable.

64. Explain the difference between a term and a factor.

SYNTHESIS

65. Describe at least three ways in which the distributive law was used in this chapter.

66. Devise a rule for determining the sign of a negative quantity raised to a power.

67. Evaluate $a^{50} - 20a^{25}b^4 + 100b^8$ for $a = 1$ and $b = 2$.

68. If $0.090909\ldots = \frac{1}{11}$ and $0.181818\ldots = \frac{2}{11}$, what rational number is named by each of the following?

   a) $0.272727\ldots$ **b)** $0.909090\ldots$

Simplify.

69. $-\left|\frac{7}{8} - \left(-\frac{1}{2}\right) - \frac{3}{4}\right|$

70. $(|2.7 - 3| + 3^2 - |-3|) \div (-3)$

Match the phrase in the left column with the most appropriate choice from the right column.

___ **71.** A number is nonnegative.

___ **72.** The reciprocal of a sum

___ **73.** A number squared

___ **74.** The opposite of a sum

___ **75.** The opposite of an opposite is the original number.

___ **76.** The order in which numbers are added does not change the result.

___ **77.** A number is negative.

___ **78.** The absolute value of a product

___ **79.** A sum of a number and its reciprocal

___ **80.** The square of a sum

___ **81.** The absolute value of one number is less than the absolute value of another number.

A. a^2

B. $a + b = b + a$

C. $a < 0$

D. $a + \dfrac{1}{a}$

E. $|ab|$

F. $(a + b)^2$

G. $|a| < |b|$

H. $-(a + b)$

I. $a \geq 0$

J. $\dfrac{1}{a + b}$

K. $-(-a) = a$

Chapter Test 1

1. Evaluate $\dfrac{2x}{y}$ for $x = 10$ and $y = 5$.

2. Write an algebraic expression: Nine less than some number.

3. Find the area of a triangle when the height h is 30 ft and the base b is 16 ft.

4. Use the commutative law of addition to write an expression equivalent to $3p + q$.

5. Use the associative law of multiplication to write an expression equivalent to $x \cdot (4 \cdot y)$.

6. Determine whether 3 is a solution of $96 - a = 93$.

7. Translate to an equation. Do not solve.

On a hot summer day, Green River Electric met a demand of 2518 megawatts. This is only 282 megawatts less than its maximum production capability. What is the maximum capability of production?

Multiply.

8. $5(6 - x)$ **9.** $-5(y - 1)$

Factor.

10. $11 - 44x$ **11.** $7x + 21 + 14y$

12. Find the prime factorization of 300.

13. Simplify: $\frac{10}{35}$.

Write a true sentence using either < or >.

14. $-4 \quad \rule{0.8em}{0.8em} \quad 0$

15. $-3 \quad \rule{0.8em}{0.8em} \quad -8$

Find the absolute value.

16. $\left| \frac{9}{4} \right|$

17. $|-2.7|$

18. Find the opposite of $\frac{2}{3}$.

19. Find the reciprocal of $-\frac{4}{7}$.

20. Find $-x$ when x is -8.

21. Write an inequality with the same meaning as $x \leq -2$.

Compute and simplify.

22. $3.1 - (-4.7)$

23. $-8 + 4 + (-7) + 3$

24. $\frac{2}{5} + \frac{3}{8}$

25. $2 - (-8)$

26. $3.2 - 5.7$

27. $\frac{1}{8} - \left(-\frac{3}{4} \right)$

28. $4 \cdot (-12)$

29. $-\frac{1}{2} \cdot \left(-\frac{3}{8} \right)$

30. $-45 \div 5$

31. $-\frac{3}{5} \div \left(-\frac{4}{5} \right)$

32. $4.864 \div (-0.5)$

33. $-2(16) - |2(-8) - 5^3|$

34. $6 + 7 - 4 - (-3)$

35. $256 \div (-16) \div 4$

36. $2^3 - 10[4 - (-2 + 18)3]$

37. Combine like terms: $18y + 30a - 9a + 4y$.

38. Simplify: $(-2x)^4$.

Remove parentheses and simplify.

39. $5x - (3x - 7)$

40. $4(2a - 3b) + a - 7$

41. $4\{3[5(y - 3) + 9] + 2(y + 8)\}$

SYNTHESIS

42. Evaluate $\dfrac{5y - x}{4}$ when $x = 20$ and y is 4 less than x.

Simplify.

43. $\dfrac{13,800}{42,000}$

44. $|-27 - 3(4)| - |-36| + |-12|$

45. $a - \{3a - [4a - (2a - 4a)]\}$

2

Equations, Inequalities, and Problem Solving

AN APPLICATION

To cater a party, Curtis' Barbeque charges a $50 setup fee plus $15 per person. The cost of Hotel Pharmacy's end-of-season softball party cannot exceed $450. How many people can attend the party?

This problem appears as Example 1 in Section 2.7.

Most people might not associate math with barbeque, but in fact I use math in everything from ordering ingredients for my secret sauce to keeping the books for my business.

CURTIS TUFF
Chef and Owner,
Curtis' Barbeque
Putney, VT

*S*olving equations and inequalities is a recurring theme in much of mathematics. In this chapter, we will study some of the principles used to solve equations and inequalities. We will then use equations and inequalities to solve applied problems.

Solving Equations

2.1

Equations and Solutions • The Addition Principle • The Multiplication Principle

Solving equations is essential for problem solving in algebra. In this section, we study two of the most important principles used for this task.

Equations and Solutions

We have already seen that an equation is a number sentence stating that the expressions on either side of the equals sign represent the same number. Some equations, like $3 + 2 = 5$ or $2x + 6 = 2(x + 3)$, are *always* true and some, like $3 + 2 = 6$ or $x + 2 = x + 3$, are *never* true. In this text, we will concentrate on equations like $x + 6 = 13$ or $7x = 141$ that are *sometimes* true, depending on the replacement value for the variable.

> ### Solution of an Equation
>
> Any replacement for the variable that makes an equation true is called a *solution* of the equation. To *solve* an equation means to find all of its solutions.

To determine whether a number is a solution, we substitute that number for the variable throughout the equation. If the values on both sides of the equals sign are the same, then the number that was substituted is a solution.

Example 1

Determine whether 7 is a solution of $x + 6 = 13$.

Solution   We have

$$
\begin{array}{ll}
x + 6 = 13 & \text{Writing the equation} \\
\overline{7 + 6 \ ? \ 13} & \text{Substituting 7 for } x \\
13 \ | \ 13 \ \text{\scriptsize TRUE} & \text{Note that 7, not 13, is the solution.}
\end{array}
$$

Since the left-hand and the right-hand sides are the same, 7 is a solution.

E x a m p l e   2

Determine whether 19 is a solution of $7x = 141$.

Solution   We have

$$7x = 141 \qquad \text{Writing the equation}$$
$$7(19) \ ? \ 141 \qquad \text{Substituting 19 for } x$$
$$133 \ | \ 141 \quad \text{FALSE} \qquad \text{The statement } 133 = 141 \text{ is false.}$$

Since the left-hand and the right-hand sides differ, 19 is not a solution.

The Addition Principle

Consider the equation

$$x = 7.$$

We can easily see that the solution of this equation is 7. Replacing x with 7, we get

$$7 = 7, \quad \text{which is true.}$$

Now consider the equation

$$x + 6 = 13.$$

In Example 1, we found that the solution of $x + 6 = 13$ is also 7. Although the solution of $x = 7$ may seem more obvious, the equations $x + 6 = 13$ and $x = 7$ are **equivalent**.

> ### *Equivalent Equations*
> Equations with the same solutions are called *equivalent equations*.

There are principles that enable us to begin with one equation and end up with an equivalent equation, like $x = 7$, for which the solution is obvious. One such principle concerns addition. The equation $a = b$ says that a and b stand for the same number. Suppose this is true, and some number c is added to a. We get the same result if we add c to b, because a and b are the same number.

> ### *The Addition Principle*
> For any real numbers a, b, and c,
> $$a = b \quad \text{is equivalent to} \quad a + c = b + c.$$

To visualize the addition principle, consider a balance similar to one a jeweler might use. (See the figure on the following page.) When the two sides of the balance hold quantities of equal weight, the balance is level. If weight is added or removed, equally, on both sides, the balance will remain level.

$$a = b \qquad\qquad a + c = b + c$$

When using the addition principle, we often say that we "add the same number to both sides of an equation." We can also "subtract the same number from both sides," since subtraction can be regarded as the addition of an opposite.

E x a m p l e   3

Solve: $x + 5 = -7$.

Solution   We can add any number we like to both sides. Since -5 is the opposite, or additive inverse, of 5, we add -5 to each side:

$$x + 5 = -7$$

$x + 5 - 5 = -7 - 5$	Using the addition principle: adding -5 to both sides or subtracting 5 from both sides
$x + 0 = -12$	Simplifying; $x + 5 - 5 = x + 5 + (-5) = x + 0$
$x = -12.$	Using the identity property of 0

It is obvious that the solution of $x = -12$ is the number -12. To check the answer in the original equation, we substitute.

Check:

$$\begin{array}{c|c} \multicolumn{2}{c}{x + 5 = -7} \\ \hline -12 + 5\ ?\ -7 & \\ -7\ \big|\ -7 & \text{TRUE} \end{array}$$

The solution of the original equation is -12.

In Example 3, note that because we added the *opposite,* or *additive inverse,* of 5, the left side of the equation simplified to x plus the *additive identity,* 0, or simply x. These steps effectively replaced the 5 on the left with a 0. When solving $x + a = b$ for x, we simply add $-a$ to (or subtract a from) both sides.

E x a m p l e   4

Solve: $-6.5 = y - 8.4$.

Solution   The variable is on the right side this time. We can isolate y by adding 8.4 to each side:

$-6.5 = y - 8.4$	This can be regarded as $-6.5 = y + (-8.4)$.
$-6.5 + 8.4 = y - 8.4 + 8.4$	Using the addition principle: Adding 8.4 to both sides "eliminates" -8.4 on the right side.
$1.9 = y.$	$y - 8.4 + 8.4 = y + (-8.4) + 8.4$ $= y + 0 = y$

Check: $-6.5 = y - 8.4$

$-6.5 \; ? \; 1.9 - 8.4$

$-6.5 \; | \; -6.5$ TRUE

The solution is 1.9.

Note that the equations $a = b$ and $b = a$ have the same meaning. Thus, $-6.5 = y - 8.4$ could have been rewritten as $y - 8.4 = -6.5$.

Example 5

Solve: $-\frac{2}{3} + x = \frac{5}{2}$.

Solution   We have

$$-\frac{2}{3} + x = \frac{5}{2}$$

$$-\frac{2}{3} + x + \frac{2}{3} = \frac{5}{2} + \frac{2}{3} \qquad \text{Adding } \tfrac{2}{3} \text{ to both sides}$$

$$x = \frac{5}{2} + \frac{2}{3}$$

$$= \frac{5}{2} \cdot \frac{3}{3} + \frac{2}{3} \cdot \frac{2}{2} \qquad \text{Multiplying by 1 to obtain a common}$$
$$\text{denominator}$$

$$= \frac{15}{6} + \frac{4}{6}$$

$$= \frac{19}{6}.$$

The check is left to the student. The solution is $\frac{19}{6}$.

The Multiplication Principle

A second principle for solving equations concerns multiplying. Suppose a and b are equal. If a and b are multiplied by some number c, then ac and bc will also be equal.

> ### The Multiplication Principle
> For any real numbers a, b, and c, with $c \neq 0$,
>
> $$a = b \quad \text{is equivalent to} \quad a \cdot c = b \cdot c.$$

Example 6

Solve: $\frac{5}{4}x = 10$.

Solution   We can multiply both sides by any nonzero number we like. Since $\frac{4}{5}$ is the reciprocal of $\frac{5}{4}$, we multiply each side by $\frac{4}{5}$:

$$\frac{5}{4}x = 10$$

$$\frac{4}{5} \cdot \frac{5}{4}x = \frac{4}{5} \cdot 10 \qquad \text{Using the multiplication principle: Multiplying both}$$
$$\text{sides by } \tfrac{4}{5} \text{ "eliminates" the } \tfrac{5}{4} \text{ on the left.}$$

$$1 \cdot x = 8 \qquad \text{Simplifying}$$

$$x = 8. \qquad \text{Using the identity property of 1}$$

Check: $\dfrac{5}{4}x = 10$

$\dfrac{5}{4} \cdot 8 \;?\; 10$

$\phantom{\dfrac{5}{4} \cdot 8 }10 \;\Big|\; 10$ TRUE

The solution is 8.

In Example 6, to get x alone, we multiplied by the *reciprocal*, or *multiplicative inverse* of $\frac{5}{4}$. We then simplified the left-hand side to x times the *multiplicative identity*, 1, or simply x. These steps effectively replaced the $\frac{5}{4}$ on the left with 1.

Because division is the same as multiplying by a reciprocal, the multiplication principle also tells us that we can "divide both sides by the same nonzero number." That is,

$$\text{if } a = b, \text{ then }\quad \frac{1}{c} \cdot a = \frac{1}{c} \cdot b \quad \text{and} \quad \frac{a}{c} = \frac{b}{c} \qquad (\text{provided } c \neq 0).$$

In a product like $3x$, the multiplier 3 is called the **coefficient**. When the coefficient of the variable is an integer or a decimal, it is usually easiest to solve an equation by dividing on both sides. When the coefficient is in fraction notation, it is usually easier to multiply by the reciprocal.

Example 7   Solve: **(a)** $-4x = 92$;   **(b)** $12.6 = 3t$;   **(c)** $-x = 9$;   **(d)** $\dfrac{2y}{9} = \dfrac{8}{3}$.

Solution

a) $-4x = 92$

$\dfrac{-4x}{-4} = \dfrac{92}{-4}$ Using the multiplication principle: Dividing both sides by -4 is the same as multiplying by $-\frac{1}{4}$.

$1 \cdot x = -23$ Simplifying

$x = -23$ Using the identity property of 1

Check: $\dfrac{-4x = 92}{}$

$-4(-23) \;?\; 92$

$92 \;\Big|\; 92$ TRUE

The solution is -23.

b) $12.6 = 3t$

$\dfrac{12.6}{3} = \dfrac{3t}{3}$ Dividing both sides by 3 or multiplying both sides by $\frac{1}{3}$

$4.2 = 1t$

$4.2 = t$ Simplifying

Check: $\dfrac{12.6 = 3t}{}$

$12.6 \;?\; 3(4.2)$

$12.6 \;\Big|\; 12.6$ TRUE

The solution is 4.2.

c) To solve an equation like $-x = 9$, remember that when an expression is multiplied or divided by -1, its sign is changed. Here we multiply both sides by -1 to change the sign of $-x$:

$$-x = 9$$

$$(-1)(-x) = (-1)9 \qquad \text{Multiplying both sides by } -1 \text{ (Dividing by } -1 \text{ would also work)}$$

$$x = -9. \qquad \text{Note that } (-1)(-x) \text{ is the same as } (-1)(-1)x.$$

Check:
$$\frac{-x = 9}{-(-9) \; ? \; 9}$$
$$9 \mid 9 \quad \text{TRUE}$$

The solution is -9.

d) To solve an equation like $\frac{2y}{9} = \frac{8}{3}$, we rewrite the left-hand side as $\frac{2}{9} \cdot y$ and then use the multiplication principle:

$$\frac{2y}{9} = \frac{8}{3}$$

$$\frac{2}{9} \cdot y = \frac{8}{3} \qquad \text{Rewriting } \frac{2y}{9} \text{ as } \frac{2}{9} \cdot y$$

$$\frac{9}{2} \cdot \frac{2}{9} \cdot y = \frac{9}{2} \cdot \frac{8}{3} \qquad \text{Multiplying both sides by } \frac{9}{2}$$

$$1y = \frac{3 \cdot 3 \cdot 2 \cdot 4}{2 \cdot 3} \qquad \text{Removing a factor equal to 1: } \frac{3 \cdot 2}{2 \cdot 3} = 1$$

$$y = 12.$$

Check:
$$\frac{2y}{9} = \frac{8}{3}$$
$$\frac{2 \cdot 12}{9} \; ? \; \frac{8}{3}$$
$$\frac{24}{9}$$
$$\frac{8}{3} \mid \frac{8}{3} \quad \text{TRUE}$$

The solution is 12.

FOR EXTRA HELP

Exercise Set 2.1

Digital Video Tutor CD 1 Videotape 3   InterAct Math   Math Tutor Center   MathXL   MyMathLab.com

Solve using the addition principle. Don't forget to check!

1. $x + 8 = 23$

2. $x + 5 = 8$

3. $t + 9 = -4$

4. $y + 9 = 43$

5. $y + 7 = -3$

6. $t + 9 = -12$

7. $-5 = x + 8$

8. $-6 = y + 25$

9. $x - 9 = 6$

10. $x - 8 = 5$

11. $y - 6 = -14$

12. $x - 4 = -19$

13. $9 + t = 3$

14. $3 + t = 21$

15. $12 = -7 + y$

16. $15 = -9 + z$

17. $-5 + t = -9$

18. $-6 + y = -21$

19. $r + \frac{1}{3} = \frac{8}{3}$

20. $t + \frac{3}{8} = \frac{5}{8}$

21. $x + \frac{3}{5} = -\frac{7}{10}$

22. $x + \frac{2}{3} = -\frac{5}{6}$

23. $x - \frac{5}{6} = \frac{7}{8}$

24. $y - \frac{3}{4} = \frac{5}{6}$

25. $-\frac{1}{5} + z = -\frac{1}{4}$

26. $-\frac{1}{8} + y = -\frac{3}{4}$

27. $m + 3.9 = 5.4$

28. $y + 5.3 = 8.7$

29. $-9.7 = -4.7 + y$

30. $-7.8 = 2.8 + x$

Solve using the multiplication principle. Don't forget to check!

31. $5x = 80$

32. $3x = 39$

33. $9t = 36$

34. $6x = 72$

35. $84 = 7x$

36. $56 = 7t$

37. $-x = 23$

38. $100 = -x$

Aha! **39.** $-t = -8$

40. $-68 = -r$

41. $7x = -49$

42. $9x = -36$

43. $-12x = 72$

44. $-15x = 105$

45. $-3.4t = -20.4$

46. $-1.3a = -10.4$

47. $\dfrac{a}{4} = 13$

48. $\dfrac{y}{-8} = 11$

49. $\dfrac{3}{4}x = 27$

50. $\dfrac{4}{5}x = 16$

51. $\dfrac{-t}{5} = 9$

52. $\dfrac{-x}{6} = 9$

53. $\dfrac{2}{7} = \dfrac{x}{3}$

54. $\dfrac{1}{9} = \dfrac{z}{5}$

Aha! **55.** $-\dfrac{3}{5}r = -\dfrac{3}{5}$

56. $-\dfrac{2}{5}y = -\dfrac{4}{15}$

57. $\dfrac{-3r}{2} = -\dfrac{27}{4}$

58. $\dfrac{5x}{7} = -\dfrac{10}{14}$

Solve. The icon ▤ indicates an exercise designed to give practice using a calculator.

59. $4.5 + t = -3.1$

60. $\dfrac{3}{4}x = 18$

61. $-8.2x = 20.5$

62. $t - 7.4 = -12.9$

63. $12 = y + 29$

64. $96 = -\dfrac{3}{4}t$

65. $a - \dfrac{1}{6} = -\dfrac{2}{3}$

66. $-\dfrac{x}{7} = \dfrac{2}{9}$

67. $-24 = \dfrac{8x}{5}$

68. $\dfrac{1}{5} + y = -\dfrac{3}{10}$

69. $-\dfrac{4}{3}t = -16$

70. $\dfrac{17}{35} = -x$

▤ **71.** $-483.297 = -794.053 + t$

▤ **72.** $-0.2344x = 2028.732$

▤ **73.** When solving an equation, how do you determine what number to add, subtract, multiply, or divide by on both sides of that equation?

▤ **74.** What is the difference between equivalent expressions and equivalent equations?

SKILL MAINTENANCE

Simplify.

75. $9 - 2 \cdot 5^2 + 7$

76. $10 \div 2 \cdot 3^2 - 4$

77. $16 \div (2 - 3 \cdot 2) + 5$

78. $12 - 5 \cdot 2^3 + 4 \cdot 3$

SYNTHESIS

▤ **79.** To solve $-3.5 = 14t$, Anita adds 3.5 to both sides. Will this form an equivalent equation? Will it help solve the equation? Explain.

▤ **80.** Explain why it is not necessary to state a subtraction principle: For any real numbers a, b, and c, $a = b$ is equivalent to $a - c = b - c$.

Some equations, like $3 = 7$ or $x + 2 = x + 5$, have no solution and are called **contradictions**. *Other equations, like $7 = 7$ or $2x = 2x$, are true for all numbers and are called* **identities**. *Solve each of the following and if an identity or contradiction is found, state this.*

81. $2x = x + x$

82. $x + 5 + x = 2x$

83. $5x = 0$

84. $4x - x = 2x + x$

85. $x + 8 = 3 + x + 7$

86. $3x = 0$

Aha! **87.** $2|x| = -14$

88. $|3x| = 6$

Solve for x. Assume a, c, m ≠ 0.

89. $mx = 9.4m$

90. $x - 4 + a = a$

91. $\dfrac{7cx}{2a} = \dfrac{21}{a} \cdot c$

92. $5c + cx = 7c$

93. $5a = ax - 3a$

94. $|x| + 6 = 19$

95. If $x - 4720 = 1634$, find $x + 4720$.

96. Lydia makes a calculation and gets an answer of 22.5. On the last step, she multiplies by 0.3 when she should have divided by 0.3. What should the correct answer be?

97. Are the equations $x = 5$ and $x^2 = 25$ equivalent? Why or why not?

Using the Principles Together

2.2

Applying Both Principles • Combining Like Terms • Clearing Fractions and Decimals

CONNECTING THE CONCEPTS

We have stated that most of algebra involves either simplifying expressions (by writing equivalent expressions) or solving equations (by writing equivalent equations). In Section 2.1, we used the addition and multiplication principles to produce equivalent equations, like $x = 5$, from which the solution—in this case, 5—is obvious. Here in Section 2.2, we will find that more complicated equations can be solved by using both principles together and by using the commutative, associative, and distributive laws to write equivalent expressions.

An important strategy for solving a new problem is to find a way to make the new prob-

lem look like a problem we already know how to solve. This is precisely the approach taken in this section. You will find that the last steps of the examples in this section are nearly identical to the steps used for solving the examples of Section 2.1. What is new in this section appears in the early steps of each example.

Before reading this section, make sure that you thoroughly understand the material in Section 2.1. Without a solid grasp of how and when to use the addition and multiplication principles, the problems in this section will seem much more difficult than they really are.

Applying Both Principles

In the expression $5 + 3x$, the variable x is multiplied by 3 and then 5 is added. To reverse these steps, we first subtract 5 and then divide by 3. Thus, to solve $5 + 3x = 17$, we first subtract 5 from each side and then divide both sides by 3.

E x a m p l e   1

Solve: $5 + 3x = 17$.

Solution   We have

$$5 + 3x = 17$$

$$5 + 3x - 5 = 17 - 5$$ Using the addition principle: subtracting 5 from both sides (adding -5)

$$5 + (-5) + 3x = 12$$ Using a commutative law. Try to perform this step mentally.

First isolate the x-term. $3x = 12$ Simplifying

$$\frac{3x}{3} = \frac{12}{3}$$ Using the multiplication principle: dividing both sides by 3 $\left(\text{multiplying by } \frac{1}{3}\right)$

Then isolate x. $x = 4$.   Simplifying

Check:
$$\frac{5 + 3x = 17}{5 + 3 \cdot 4 \;?\; 17}$$ We use the rules for order of operations:
$$5 + 12 \mid$$ Find the product, $3 \cdot 4$, and then add.
$$17 \mid 17 \;\text{TRUE}$$

The solution is 4.

Multiplication by a negative number and subtraction are handled in much the same way.

E x a m p l e   2

Solve: $-5x - 6 = 16$.

Solution   In $-5x - 6$ we multiply first and then subtract. To reverse these steps, we first add 6 and then divide by -5.

$$-5x - 6 = 16$$

$$-5x - 6 + 6 = 16 + 6$$ Adding 6 to both sides

$$-5x = 22$$

$$\frac{-5x}{-5} = \frac{22}{-5}$$ Dividing both sides by -5

$$x = -\frac{22}{5}, \text{ or } -4\frac{2}{5}$$ Simplifying

Check:
$$\frac{-5x - 6 = 16}{-5\left(-\frac{22}{5}\right) - 6 \;?\; 16}$$
$$22 - 6 \mid$$
$$16 \mid 16 \;\text{TRUE}$$

The solution is $-\frac{22}{5}$.

Example 3

Solve: $45 - t = 13$.

Solution   We have

$$45 - t = 13$$
$$45 - t - 45 = 13 - 45 \qquad \text{Subtracting 45 from both sides}$$
$$\left.\begin{array}{l} 45 + (-t) + (-45) = 13 - 45 \\ 45 + (-45) + (-t) = 13 - 45 \end{array}\right\} \quad \text{Try to do these steps mentally.}$$
$$-t = -32 \qquad \text{Try to go directly to this step.}$$
$$(-1)(-t) = (-1)(-32) \qquad \begin{array}{l}\text{Multiplying both sides by } -1 \\ \text{(Dividing by } -1 \text{ would also work.)}\end{array}$$
$$t = 32.$$

Check:
$$\begin{array}{c} 45 - t = 13 \\ \hline 45 - 32 \ ? \ 13 \\ 13 \ \bigm| \ 13 \quad \text{TRUE} \end{array}$$

The solution is 32.

As our skills improve, many of the steps can be streamlined.

Example 4

Solve: $16.3 - 7.2y = -8.18$.

Solution   We have

$$16.3 - 7.2y = -8.18$$
$$-7.2y = -8.18 - 16.3 \qquad \begin{array}{l}\text{Subtracting 16.3 from both sides. We} \\ \text{write the subtraction of 16.3 on the} \\ \text{right side and remove 16.3 from the} \\ \text{left side.}\end{array}$$

$$-7.2y = -24.48$$
$$y = \frac{-24.48}{-7.2} \qquad \begin{array}{l}\text{Dividing both sides by } -7.2. \text{ We write} \\ \text{the division by } -7.2 \text{ on the right side} \\ \text{and remove the } -7.2 \text{ from the left side.}\end{array}$$

$$y = 3.4.$$

Check:
$$\begin{array}{c} 16.3 - 7.2y = -8.18 \\ \hline 16.3 - 7.2(3.4) \ ? \ -8.18 \\ 16.3 - 24.48 \ \bigm| \\ -8.18 \ \bigm| \ -8.18 \quad \text{TRUE} \end{array}$$

The solution is 3.4.

Combining Like Terms

If like terms appear on the same side of an equation, we combine them and then solve. Should like terms appear on both sides of an equation, we can use the addition principle to rewrite all like terms on one side.

Example 5

technology connection

Most graphers have a TABLE feature that enables the calculator to evaluate a variable expression for different choices of x. For example, to evaluate $6x + 5 - 7x$ for $x = 0, 1, 2, \ldots$, we first use Y= to enter $6x + 5 - 7x$ as y_1. We then use 2nd TBLSET to specify which x-values will be used. Using TblStart = 0, ΔTbl = 1, and selecting Auto twice, we can generate a table in which the value of $6x + 5 - 7x$ is listed for values of x starting at 0 and increasing by ones.

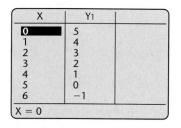

1. Create the above table on your grapher. Scroll up and down to extend the table.
2. Enter $10 - 4x + 7$ as y_2. Your table should now have three columns.
3. For what x-value is y_1 the same as y_2? Compare this with the solution of Example 5(c). Is this a reliable way to solve equations? Why or why not?

Solve.

a) $3x + 4x = -14$

b) $2x - 4 = -3x + 1$

c) $6x + 5 - 7x = 10 - 4x + 7$

d) $2 - 5(x + 5) = 3(x - 2) - 1$

Solution

a) $3x + 4x = -14$

$$7x = -14 \qquad \text{Combining like terms}$$

$$x = \frac{-14}{7} \qquad \text{Dividing both sides by 7}$$

$$x = -2$$

The check is left to the student. The solution is -2.

b) To solve $2x - 4 = -3x + 1$, we must first write only variable terms on one side and only constant terms on the other. This can be done by adding 4 to both sides, to get all constant terms on the right, and $3x$ to both sides, to get all variable terms on the left. We can add 4 first, or $3x$ first, or do both in one step.

$$2x - 4 = -3x + 1$$

> Isolate variable terms on one side and constant terms on the other side.

$$2x - 4 + 4 = -3x + 1 + 4 \qquad \text{Adding 4 to both sides}$$

$$2x = -3x + 5 \qquad \text{Simplifying}$$

$$2x + 3x = -3x + 3x + 5 \qquad \text{Adding } 3x \text{ to both sides}$$

$$5x = 5 \qquad \text{Combining like terms and simplifying}$$

$$x = \frac{5}{5} \qquad \text{Dividing both sides by 5}$$

$$x = 1 \qquad \text{Simplifying}$$

Check:

$$\frac{2x - 4 = -3x + 1}{2 \cdot 1 - 4 \;?\; -3 \cdot 1 + 1}$$

$$2 - 4 \;\Big|\; -3 + 1$$

$$-2 \;\Big|\; -2 \qquad \text{TRUE}$$

The solution is 1.

c) $6x + 5 - 7x = 10 - 4x + 7$

$$-x + 5 = 17 - 4x \qquad \text{Combining like terms on both sides}$$

$$-x + 5 + 4x = 17 - 4x + 4x \qquad \text{Adding } 4x \text{ to both sides}$$

$$5 + 3x = 17 \qquad \text{Simplifying. This is identical to Example 1.}$$

$$3x = 12 \qquad \text{Subtracting 5 from both sides and simplifying}$$

$$x = 4 \qquad \text{Dividing both sides by 3 and simplifying}$$

Check:

$$\frac{6x + 5 - 7x = 10 - 4x + 7}{6 \cdot 4 + 5 - 7 \cdot 4 \;?\; 10 - 4 \cdot 4 + 7}$$

$$24 + 5 - 28 \;\Big|\; 10 - 16 + 7$$

$$1 \;\Big|\; 1 \qquad \text{TRUE}$$

The solution is 4.

d) $2 - 5(x + 5) = 3(x - 2) - 1$

$\quad\quad 2 - 5x - 25 = 3x - 6 - 1$ Using the distributive law. This is now similar to part (c) above.

$\quad\quad -5x - 23 = 3x - 7$ Combining like terms on both sides

$\quad\quad -5x - 23 + 7 = 3x$

$\quad\quad -23 + 7 = 3x + 5x$ Adding 7 and $5x$ to both sides. This isolates the x-terms on one side and the constant terms on the other.

$\quad\quad\quad\quad -16 = 8x$ Simplifying

$\quad\quad\quad\quad -2 = x$ Dividing both sides by 8

The student can confirm that -2 checks and is the solution.

Clearing Fractions and Decimals

Equations are generally easier to solve when they do not contain fractions or decimals. The multiplication principle can be used to "clear" fractions or decimals, as shown here.

Clearing Fractions	Clearing Decimals
$\frac{1}{2}x + 5 = \frac{3}{4}$	$2.3x + 7 = 5.4$
$4(\frac{1}{2}x + 5) = 4 \cdot \frac{3}{4}$	$10(2.3x + 7) = 10 \cdot 5.4$
$2x + 20 = 3$	$23x + 70 = 54$

In each case, the resulting equation is equivalent to the original equation, but easier to solve.

The easiest way to clear an equation of fractions is to multiply *both sides* of the equation by the smallest, or *least*, common denominator.

Example 6

Solve: **(a)** $\frac{2}{3}x - \frac{1}{6} = 2x$; **(b)** $\frac{2}{5}(3x + 2) = 8$.

Solution

a) The number 6 is the least common denominator, so we multiply both sides by 6.

$$6\left(\frac{2}{3}x - \frac{1}{6}\right) = 6 \cdot 2x \quad \text{Multiplying both sides by 6}$$

$$6 \cdot \frac{2}{3}x - 6 \cdot \frac{1}{6} = 6 \cdot 2x$$

Caution! Be sure the distributive law is used to multiply *all* the terms by 6.

$$4x - 1 = 12x \quad \text{Simplifying. Note that the fractions are cleared.}$$

$$-1 = 8x \quad \text{Subtracting } 4x \text{ from both sides}$$

$$-\frac{1}{8} = x \quad \text{Dividing both sides by 8}$$

The number $-\frac{1}{8}$ checks and is the solution.

b) To solve $\frac{2}{5}(3x + 2) = 8$, we can multiply both sides by $\frac{5}{2}$ (or divide by $\frac{2}{5}$) to "undo" the multiplication by $\frac{2}{5}$ on the left side.

$$\frac{5}{2} \cdot \frac{2}{5}(3x + 2) = \frac{5}{2} \cdot 8 \qquad \text{Multiplying both sides by } \tfrac{5}{2}$$

$$3x + 2 = 20 \qquad \text{Simplifying; } \tfrac{5}{2} \cdot \tfrac{2}{5} = 1 \text{ and } \tfrac{5}{2} \cdot \tfrac{8}{1} = 20$$

$$3x = 18 \qquad \text{Subtracting 2 from both sides}$$

$$x = 6 \qquad \text{Dividing both sides by 3}$$

The student can confirm that 6 checks and is the solution.

To clear an equation of decimals, we count the greatest number of decimal places in any one number. If the greatest number of decimal places is 1, we multiply both sides by 10; if it is 2, we multiply by 100; and so on.

Example 7

Solve: $16.3 - 7.2y = -8.18$.

Solution The greatest number of decimal places in any one number is *two*. Multiplying by 100 will clear all decimals.

$$100(16.3 - 7.2y) = 100(-8.18) \qquad \text{Multiplying both sides by 100}$$

$$100(16.3) - 100(7.2y) = 100(-8.18) \qquad \text{Using the distributive law}$$

$$1630 - 720y = -818 \qquad \text{Simplifying}$$

$$-720y = -818 - 1630 \qquad \text{Subtracting 1630 from both sides}$$

$$-720y = -2448 \qquad \text{Combining like terms}$$

$$y = \frac{-2448}{-720} \qquad \text{Dividing both sides by } -720$$

$$y = 3.4$$

In Example 4, the same solution was found without clearing decimals. Finding the same answer two ways is a good check. The solution is 3.4.

An Equation-Solving Procedure

1. Use the multiplication principle to clear any fractions or decimals. (This is optional, but can ease computations.)
2. If necessary, use the distributive law to remove parentheses. Then combine like terms on each side.
3. Use the addition principle, as needed, to get all variable terms on one side and all constant terms on the other.
4. Combine like terms again, if necessary.
5. Multiply or divide to solve for the variable, using the multiplication principle.
6. Check all possible solutions in the original equation.

Exercise Set 2.2

Solve and check.

1. $5x + 3 = 38$

2. $3x + 6 = 30$

3. $8x + 4 = 68$

4. $6z + 3 = 57$

5. $7t - 8 = 27$

6. $6x - 3 = 15$

7. $3x - 9 = 33$

8. $5x - 9 = 41$

9. $8z + 2 = -54$

10. $4x + 3 = -21$

11. $-39 = 1 + 8x$

12. $-91 = 9t + 8$

13. $9 - 4x = 37$

14. $12 - 4x = 108$

15. $-7x - 24 = -129$

16. $-6z - 18 = -132$

17. $48 = 5x + 7x$

18. $4x + 5x = 45$

19. $27 - 6x = 99$

20. $32 - 7x = 11$

21. $4x + 3x = 42$

22. $6x + 19x = 100$

23. $-2a + 5a = 24$

24. $-4y - 7y = 33$

25. $-7y - 8y = -15$

26. $-10y - 2y = -48$

27. $10.2y - 7.3y = -58$

28. $3.4t - 1.2t = -44$

29. $x + \frac{1}{3}x = 8$

30. $x + \frac{1}{4}x = 10$

31. $9y - 35 = 4y$

32. $4x - 6 = 6x$

33. $6x - 5 = 7 + 2x$

34. $5y - 2 = 28 - y$

Aha! 35. $6x + 3 = 2x + 3$

36. $5y + 3 = 2y + 15$

37. $5 - 2x = 3x - 7x + 25$

38. $10 - 3x = 2x - 8x + 40$

39. $7 + 3x - 6 = 3x + 5 - x$

40. $5 + 4x - 7 = 4x - 2 - x$

41. $4y - 4 + y + 24 = 6y + 20 - 4y$

42. $5y - 10 + y = 7y + 18 - 5y$

Clear fractions or decimals, solve, and check.

43. $\frac{5}{4}x + \frac{1}{4}x = 2x + \frac{1}{2} + \frac{3}{4}x$

44. $\frac{7}{8}x - \frac{1}{4} + \frac{3}{4}x = \frac{1}{16} + x$

45. $\frac{2}{3} + \frac{1}{4}t = 6$

46. $-\frac{1}{2} + x = -\frac{5}{6} - \frac{1}{3}$

47. $\frac{2}{3} + 4t = 6t - \frac{2}{15}$

48. $\frac{1}{2} + 4m = 3m - \frac{5}{2}$

49. $\frac{1}{3}x + \frac{2}{5} = \frac{4}{15} + \frac{3}{5}x - \frac{2}{3}$

50. $1 - \frac{2}{3}y = \frac{9}{5} - \frac{1}{5}y + \frac{3}{5}$

51. $2.1x + 45.2 = 3.2 - 8.4x$

52. $0.91 - 0.2z = 1.23 - 0.6z$

53. $0.76 + 0.21t = 0.96t - 0.49$

54. $1.7t + 8 - 1.62t = 0.4t - 0.32 + 8$

55. $\frac{2}{5}x - \frac{3}{2}x = \frac{3}{4}x + 2$

56. $\frac{5}{16}y + \frac{3}{8}y = 2 + \frac{1}{4}y$

Solve and check.

57. $7(2a - 1) = 21$

58. $5(2t - 2) = 35$

59. $35 = 5(3x + 1)$

60. $9 = 3(5x - 2)$

61. $2(3 + 4m) - 6 = 48$

62. $3(5 + 3m) - 8 = 88$

63. $7r - (2r + 8) = 32$

64. $6b - (3b + 8) = 16$

65. $13 - 3(2x - 1) = 4$

66. $5(d + 4) = 7(d - 2)$

67. $3(t - 2) = 9(t + 2)$

68. $8(2t + 1) = 4(7t + 7)$

69. $7(5x - 2) = 6(6x - 1)$

70. $5(t + 3) + 9 = 3(t - 2) + 6$

71. $19 - (2x + 3) = 2(x + 3) + x$

72. $13 - (2c + 2) = 2(c + 2) + 3c$

73. $\frac{1}{4}(3t - 4) = 5$

74. $\frac{1}{3}(2x - 1) = 7$

75. $\frac{4}{3}(5x + 1) = 8$

76. $\frac{3}{4}(3t - 6) = 9$

77. $\frac{3}{2}(2x + 5) = -\frac{15}{2}$

78. $\frac{1}{6}\left(\frac{3}{4}x - 2\right) = -\frac{1}{5}$

79. $\frac{3}{4}\left(3x - \frac{1}{2}\right) - \frac{2}{3} = \frac{1}{3}$

80. $\frac{2}{3}\left(\frac{7}{8} - 4x\right) - \frac{5}{8} = \frac{3}{8}$

81. $0.7(3x + 6) = 1.1 - (x + 2)$

82. $0.9(2x + 8) = 20 - (x + 5)$

83. $a + (a - 3) = (a + 2) - (a + 1)$

84. $0.8 - 4(b - 1) = 0.2 + 3(4 - b)$

85. When an equation contains decimals, is it essential to clear the equation of decimals? Why or why not?

86. Why must the rules for the order of operations be understood before solving the equations in this section?

SKILL MAINTENANCE

Evaluate.

87. $3 - 5a$, for $a = 2$

88. $12 \div 4 \cdot t$, for $t = 5$

89. $7x - 2x$, for $x = -3$

90. $t(8 - 3t)$, for $t = -2$

SYNTHESIS

91. What procedure would you follow to solve an equation like $0.23x + \frac{17}{3} = -0.8 + \frac{3}{4}x$? Could your procedure be streamlined? If so, how?

92. Dave is determined to solve the equation $3x + 4 = -11$ by first using the multiplication principle to "eliminate" the 3. How should he proceed and why?

Solve. If an equation is an identity or a contradiction (see p. 76), state this.

93. $8.43x - 2.5(3.2 - 0.7x) = -3.455x + 9.04$

94. $0.008 + 9.62x - 42.8 = 0.944x + 0.0083 - x$

95. $-2[3(x - 2) + 4] = 4(5 - x) - 2x$

96. $0 = y - (-14) - (-3y)$

97. $3(x + 4) = 3(4 + x)$

98. $5(x - 7) = 3(x - 2) + 2x$

99. $2x(x + 5) - 3(x^2 + 2x - 1) = 9 - 5x - x^2$

100. $x(x - 4) = 3x(x + 1) - 2(x^2 + x - 5)$

101. $9 - 3x = 2(5 - 2x) - (1 - 5x)$

102. $2(7 - x) - 20 = 7x - 3(2 + 3x)$

Aha! **103.** $[7 - 2(8 \div (-2))]x = 0$

104. $\dfrac{x}{14} - \dfrac{5x + 2}{49} = \dfrac{3x - 4}{7}$

105. $\dfrac{5x + 3}{4} + \dfrac{25}{12} = \dfrac{5 + 2x}{3}$

CORNER

Step-by-Step Solutions

Focus: Solving linear equations

Time: 20 minutes

Group size: 3

In general, there is more than one correct sequence of steps for solving an equation. This makes it important that you write your steps clearly and logically so that others can follow your approach.

ACTIVITY

1. Each group member should select a different one of the following equations and, on a fresh sheet of paper, perform the first step of the solution.

$4 - 3(x - 3) = 7x + 6(2 - x)$

$5 - 7[x - 2(x - 6)] = 3x + 4(2x - 7) + 9$

$4x - 7[2 + 3(x - 5) + x] = 4 - 9(-3x - 19)$

2. Pass the papers around so that the second and third steps of each solution are performed by the other two group members. Before writing, make sure that the previous step is correct. If a mistake is discovered, return the problem to the person who made the mistake for repairs. Continue passing the problems around until all equations have been solved.

3. Each group should reach a consensus on what the three solutions are and then compare their answers to those of other groups.

2.3

Formulas

Evaluating Formulas • Solving for a Letter

Many applications of mathematics involve relationships among two or more quantities. An equation that represents such a relationship will use two or more letters and is known as a **formula**. Although most of the letters in this book represent variables, some—like c in $E = mc^2$ or π in $C = \pi d$—represent constants.

Evaluating Formulas

Example 1

Furnace output. Contractors in the Northeast use the formula $B = 30a$ to determine the minimum furnace output B, in British thermal units (Btu's), for a well-insulated house with a square feet of flooring (*Source*: U.S. Department of Energy). Determine the minimum furnace output for an 1800-ft^2 house that is well insulated.

Solution   We substitute 1800 for a and calculate B:

$$B - 30a = 30(1800) - 54,000.$$

The furnace should be able to provide at least 54,000 Btu's of heat.

Solving for a Letter

Suppose that a contractor has an extra furnace and wants to determine the size of the largest (well-insulated) house in which it can be used. The contractor can substitute the amount of the furnace's output in Btu's—say, 63,000—for B, and then solve for a:

$$63,000 = 30a \qquad \text{Replacing } B \text{ with } 63,000$$

$$2100 = a. \qquad \text{Dividing both sides by 30:}$$

```
      2 100
30)63,000
      60
       3 0
       3 0
        000
```

Were these calculations to be performed for a variety of furnaces, the contractor would find it easier to first solve $B = 30a$ for a, and *then* substitute values for B. This can be done in much the same way that we solved equations in Sections 2.1 and 2.2.

E x a m p l e   2

Solve for a: $B = 30a$.

Solution   We have

$$B = 30a \qquad \text{We want this letter alone.}$$

$$\frac{B}{30} = a. \qquad \text{Dividing both sides by 30}$$

The equation $a = B/30$ gives a quick, easy way to determine the floor area of the largest (well-insulated) house that a furnace with B Btu's could heat.

To see how the addition and multiplication principles apply to formulas, compare the following. In (A), we solve as usual; in (B), we do not simplify; and in (C), we *cannot* simplify since a, b, and c are unknown.

A. $5x + 2 = 12$

$\quad 5x = 12 - 2$

$\quad 5x = 10$

$\quad x = \dfrac{10}{5} = 2$

B. $5x + 2 = 12$

$\quad 5x = 12 - 2$

$\quad x = \dfrac{12 - 2}{5}$

C. $ax + b = c$

$\quad ax = c - b$

$\quad x = \dfrac{c - b}{a}$

E x a m p l e   3

Circumference of a circle.   The formula $C = 2\pi r$ gives the *circumference C* of a circle with radius r. Solve for r.

Solution   The **circumference** is the distance around a circle.

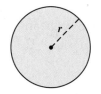

Given a radius r, we can use this equation to find a circle's circumference C. ———

Given a circle's circumference C, we can use this equation to find the radius r. ———

$$C = 2\pi r \qquad \text{We want this letter alone.}$$

$$\frac{C}{2\pi} = \frac{2\pi r}{2\pi} \qquad \text{Dividing both sides by } 2\pi$$

$$\frac{C}{2\pi} = r$$

E x a m p l e   4

Nutrition.   The number of calories K needed each day by a moderately active woman who weighs w pounds, is h inches tall, and is a years old, can be estimated using the formula

$$K = 917 + 6(w + h - a).*$$

Solve for h.

---

*Based on information from M. Parker (ed.), *She Does Math!* (Washington DC: Mathematical Association of America, 1995), p. 96.

Solution   We reverse the order in which the operations occur on the right side:

We want h alone.

$$K = 917 + 6(w + h - a)$$

$$K - 917 = 6(w + h - a) \quad \text{Subtracting 917 from both sides}$$

$$\frac{K - 917}{6} = w + h - a \quad \text{Dividing both sides by 6}$$

$$\frac{K - 917}{6} + a - w = h. \quad \text{Adding } a \text{ and subtracting } w \text{ on both sides}$$

This formula can be used to estimate a woman's height, if we know her age, weight, and caloric needs.

The above steps are similar to those used in Section 2.2 to solve equations. We use the addition and multiplication principles just as before. The main difference is the need to factor when combining like terms.

> **To Solve a Formula for a Given Letter**
> 1. If the letter for which you are solving appears in a fraction, use the multiplication principle to clear fractions.
> 2. Get all terms with the letter for which you are solving on one side of the equation and all other terms on the other side.
> 3. Combine like terms, if necessary. This may require factoring.
> 4. Multiply or divide to solve for the letter in question.

Example 5   Solve for x: $y = ax + bx - 4$.

Solution   We solve as follows:

$$y = ax + bx - 4 \quad \text{We want this letter alone.}$$

$$y + 4 = ax + bx \quad \text{Adding 4 to both sides}$$

$$y + 4 = x(a + b) \quad \text{Combining like terms by factoring out } x$$

$$\frac{y + 4}{a + b} = x. \quad \text{Dividing both sides by } a + b, \text{ or multiplying both sides by } 1/(a + b)$$

We can also write this as

$$x = \frac{y + 4}{a + b}.$$

> ***Caution!*** Had we performed the following steps in Example 5, we would *not* have solved for x:
>
> $$y = ax + bx - 4$$
> $$y - ax + 4 = bx \qquad \text{Subtracting } ax \text{ and adding 4 to both sides}$$
>
> Two occurrences of x
>
> $$\frac{y - ax + 4}{b} = x. \qquad \text{Dividing both sides by } b$$
>
> The mathematics of each step is correct, but since x occurs on both sides of the formula, *we have not solved the formula for x*. Remember that the letter being solved for should be alone on one side of the equation, with no occurrence of that letter on the other side!

Exercise Set 2.3

FOR EXTRA HELP

Digital Video Tutor CD 1
Videotape 3

InterAct Math

Math Tutor Center

MathXL

MyMathLab.com

1. *Distance from a storm.* The formula $M = \frac{1}{5}t$ can be used to determine how far M, in miles, you are from lightning when its thunder takes t seconds to reach your ears. If it takes 10 sec for the sound of thunder to reach you after you have seen the lightning, how far away is the storm?

2. *Electrical power.* The power rating P, in watts, of an electrical appliance is determined by

$$P = I \cdot V,$$

where I is the current, in amperes, and V is the voltage, measured in volts. If a kitchen requires 30 amps of current and the voltage in the house is 115 volts, what is the wattage of the kitchen?

3. *College enrollment.* At many colleges, the number of "full-time-equivalent" students f is given by

$$f = \frac{n}{15},$$

where n is the total number of credits for which students have enrolled in a given semester. Determine the number of full-time-equivalent students on a campus in which students registered for a total of 21,345 credits.

4. *Surface area of a cube.* The surface area A of a cube with side s is given by

$$A = 6s^2.$$

Find the surface area of a cube with sides of 3 in.

5. *Calorie density.* The calorie density D, in calories per ounce, of a food that contains c calories and weighs w ounces is given by

$$D = \frac{c}{w}.*$$

Eight ounces of fat-free milk contains 84 calories. Find the calorie density of fat-free milk.

6. *Wavelength of a musical note.* The wavelength w, in meters per cycle, of a musical note is given by

$$w = \frac{r}{f},$$

where r is the speed of the sound, in meters per second, and f is the frequency, in cycles per second. The speed of sound in air is 344 m/sec. What is the

---

Source: Nutrition Action Healthletter, March 2000, p. 9. Center for Science in the Public Interest, Suite 300; 1875 Connecticut Ave NW, Washington, D.C. 20008.

wavelength of a note whose frequency in air is 24 cycles per second?

7. *Absorption of ibuprofen.*   When 400 mg of the painkiller ibuprofen is swallowed, the number of milligrams n in the bloodstream t hours later (for $0 \leq t \leq 6$) is estimated by

$$n = 0.5t^4 + 3.45t^3 - 96.65t^2 + 347.7t.$$

How many milligrams of ibuprofen remain in the blood 1 hr after 400 mg has been swallowed?

8. *Size of a league schedule.*   When all n teams in a league play every other team twice, a total of N games are played, where

$$N = n^2 - n.$$

If a soccer league has 7 teams and all teams play each other twice, how many games are played?

Solve each formula for the indicated letter.

9. $A = bh,$ for b
(Area of parallelogram with base b and height h)

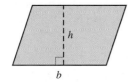

10. $A = bh,$ for h

11. $d = rt,$ for r
(A distance formula, where d is distance, r is speed, and t is time)

12. $d = rt,$ for t

13. $I = Prt,$ for P
(Simple-interest formula, where I is interest, P is principal, r is interest rate, and t is time)

14. $I = Prt,$ for t

15. $H = 65 - m,$ for m
(To determine the number of heating degree days H for a day with m degrees Fahrenheit as the average temperature)

16. $d = h - 64,$ for h
(To determine how many inches d above average an h-inch-tall woman is)

17. $P = 2l + 2w,$ for l
(Perimeter of a rectangle of length l and width w)

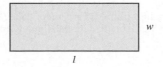

18. $P = 2l + 2w,$ for w

19. $A = \pi r^2,$ for π
(Area of a circle with radius r)

20. $A = \pi r^2,$ for r^2

21. $A = \frac{1}{2}bh,$ for h
(Area of a triangle with base b and height h)

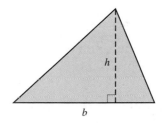

22. $A = \frac{1}{2}bh,$ for b

23. $E = mc^2,$ for m
(A relativity formula from physics)

24. $E = mc^2,$ for c^2

25. $Q = \dfrac{c + d}{2},$ for d

26. $Q = \dfrac{p - q}{2},$ for p

27. $A = \dfrac{a + b + c}{3},$ for b

28. $A = \dfrac{a + b + c}{3},$ for c

29. $M = \dfrac{A}{s},$ for A
(To compute the Mach number M for speed A and speed of sound s)

30. $P = \dfrac{ab}{c},$ for b

31. $A = at + bt,$ for t

32. $S = rx + sx,$ for x

33. *Area of a trapezoid.*   The formula

$$A = \tfrac{1}{2}ah + \tfrac{1}{2}bh$$

can be used to find the area A of a trapezoid with bases a and b and height h. Solve for h. (*Hint*: First clear fractions.)

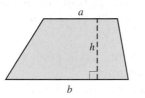

34. *Compounding interest.*   The formula

$$A = P + Prt$$

is used to find the amount A in an account when simple interest is added to an investment of P dollars (see Exercise 13). Solve for P.

35. *Chess rating.*   The formula

$$R = r + \frac{400(W - L)}{N}$$

is used to establish a chess player's rating R after that player has played N games, won W of them, and lost L of them. Here r is the average rating of the opponents (*Source*: The U.S. Chess Federation). Solve for L.

36. *Angle measure.*   The angle measure S, of a sector of a circle, is given by

$$S = \frac{360A}{\pi r^2},$$

where r is the radius, A is the area of the sector, and S is in degrees. Solve for r^2.

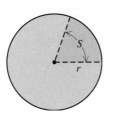

37. Naomi has a formula that allows her to convert Celsius temperatures to Fahrenheit temperatures. She needs a formula for converting Fahrenheit temperatures to Celsius temperatures. What advice can you give her?

38. Under what circumstances would it be useful to solve $d = rt$ for r? (See Exercise 11.)

SKILL MAINTENANCE

Multiply.

Aha! **39.** $0.79(38.4)0$

40. $(0.085)(108)$

Simplify.

41. $20 \div (-4) \cdot 2 - 3$

42. $5|8 - (2 - 7)|$

SYNTHESIS

43. The equations

$$P = 2l + 2w \quad \text{and} \quad w = \frac{P}{2} - l$$

are equivalent formulas involving the perimeter P, length l, and width w of a rectangle. Devise a problem for which the second of the two formulas would be more useful.

44. Describe a circumstance for which the answer to Exercise 34 would be useful.

45. The number of calories K needed each day by a moderately active man who weighs w kilograms, is h centimeters tall, and is a years old, can be determined by

$$K = 19.18w + 7h - 9.52a + 92.4.^*$$

If Janos is moderately active, weighs 82 kg, is 185 cm tall, and needs to consume 2627 calories a day, how old is he?

46. *Altitude and temperature.*   Air temperature drops about 1° Celsius (C) for each 100-m rise above ground level, up to 12 km (*Source*: *A Sourcebook of School Mathematics*, Mathematical Association of America, 1980). If the ground level temperature is t°C, find a formula for the temperature T at an elevation of h meters.

47. *Dosage size.*   Clark's rule for determining the size of a particular child's medicine dosage c is

$$c = \frac{w}{a} \cdot d,$$

where w is the child's weight, in pounds, and d is the usual adult dosage for an adult weighing a pounds. (*Source*: Olsen, June Looby, et al., *Medical Dosage Calculations*. Redwood City, CA: Addison Wesley, 1995). Solve for a.

---

*Based on information from M. Parker (ed.), *She Does Math!* (Washington DC: Mathematical Association of America, 1995), p. 96.

48. *Weight of a fish.*   An ancient fisherman's formula for estimating the weight of a fish is
$$w = \frac{lg^2}{800},$$
where w is the weight, in pounds, l is the length, in inches, and g is the girth (distance around the midsection), in inches. Estimate the girth of a 700-lb yellow tuna that is 8 ft long.

Solve each formula for the given letter.

49. $\dfrac{y}{z} \div \dfrac{z}{t} = 1$, for y

50. $ac = bc + d$, for c

51. $qt = r(s + t)$, for t

52. $3a = c - a(b + d)$, for a

53. *Furnace output.*   The formula
$$B = 50a$$
is used in New England to estimate the minimum furnace output B, in Btu's, for an old, poorly insulated house with a square feet of flooring. Find an equation for determining the number of Btu's saved by insulating an old house. (*Hint*: See Example 1.)

54. Revise the formula in Example 4 so that a woman's weight in kilograms (2.2046 lb = 1 kg) and her height in centimeters (0.3937 in. = 1 cm) are used.

55. Revise the formula in Exercise 45 so that a man's weight in pounds (2.2046 lb = 1 kg) and his height in inches (0.3937 in. = 1 cm) are used.

2.4

Applications with Percent

Converting Between Percent Notation and Decimal Notation •
Solving Percent Problems

Recently Middlesex Toy and Hobby installed a new cash register and the sales clerks inadvertently set up the machine to print out "totals" on each receipt without separating each transaction into "merchandise" and the five-percent "sales tax." For tax purposes, the shop needs a formula for separating each total into the amount spent on merchandise and the amount spent on tax. Before developing such a formula, we need to review the basics of percent problems.

Converting Between Percent Notation and Decimal Notation

Nutritionists recommend that no more than 30% of the calories in a person's diet come from fat. This means that of every 100 calories consumed, no more than 30 should come from fat. Thus, 30% is a ratio of 30 to 100.

Calories consumed

Calories from fat
30%

The percent symbol % means "per hundred." We can regard the percent symbol as part of a name for a number. For example,

$$30\% \quad \text{is defined to mean} \quad \frac{30}{100}, \quad \text{or} \quad 30 \times \frac{1}{100}, \quad \text{or} \quad 30 \times 0.01.$$

Percent Notation

$$n\% \quad \text{means} \quad \frac{n}{100}, \quad \text{or} \quad n \times \frac{1}{100}, \quad \text{or} \quad n \times 0.01.$$

Example 1

Convert to decimal notation: **(a)** 78%;   **(b)** 1.3%.

Solution

a) $78\% = 78 \times 0.01$ Replacing % with $\times 0.01$
    $= 0.78$

b) $1.3\% = 1.3 \times 0.01$ Replacing % with $\times 0.01$
    $= 0.013$

As shown above, multiplication by 0.01 simply moves the decimal point two places to the left.

To convert from percent notation to decimal notation, move the decimal point two places to the left and drop the percent symbol.

Example 2

Convert 43.67% to decimal notation.

Solution

43.67%   0.43.67 $43.67\% = 0.4367$

Move the decimal point two places to the left.

The procedure used in Examples 1 and 2 can be reversed:

$$0.38 = 38 \times 0.01$$
$$= 38\%. \qquad \text{Replacing} \times 0.01 \text{ with } \%$$

To convert from decimal notation to percent notation, move the decimal point two places to the right and write a percent symbol.

Example 3

Convert to percent notation: **(a)** 1.27;   **(b)** $\frac{1}{4}$;   **(c)** 0.3.

Solution

a) We first move the decimal point
    two places to the right: 1.27.

    and then write a % symbol: 127%   This is the same as multiply-
                                                    ing 1.27 by 100 and writing %

b) Note that $\frac{1}{4} = 0.25$. We move the decimal point two places to the right:

0.25.

and then write a % symbol:

25%   Multiplying by 100 and writing %

c) We first move the decimal point two places to the right (recall that $0.3 = 0.30$):

0.30.

and then write a % symbol:

30%   Multiplying by 100 and writing %

Solving Percent Problems

To solve problems involving percents, we translate to mathematical language and then solve an equation.

Example 4

What is 11% of 49?

Solution

Translate: What is 11% of 49?

$a = 0.11 \cdot 49$ "of" means multiply; 11% = 0.11

$a = 5.39$

A way of checking answers is by estimating as follows:

$$11\% \times 49 \approx 10\% \times 50$$
$$= 0.10 \times 50 = 5.$$

Since 5 is close to 5.39, our answer is reasonable.

Thus, 5.39 is 11% of 49. The answer is 5.39.

Example 5

3 is 16 percent of what?

Solution

Translate: 3 is 16 percent of what?

$3 = 0.16 \cdot y$

$\dfrac{3}{0.16} = y$ Dividing both sides by 0.16

$18.75 = y$

Thus, 3 is 16 percent of 18.75. The answer is 18.75.

E x a m p l e 6

What percent of $50 is $16?

Solution

Translate: What percent of $50 is $16?

$$n \cdot 50 = 16$$

$$n = \frac{16}{50} \qquad \text{Dividing both sides by 50}$$

$$n = 0.32 = 32\% \qquad \text{Converting to percent notation}$$

Thus, 32% of $50 is $16. The answer is 32%.

Examples 4–6 represent the three basic types of percent problems.

E x a m p l e 7

Retail sales.   Recently, receipts from Middlesex Toy and Hobby indicated the total amount paid (including tax), but not the price of the merchandise. Given that the sales tax was 5%, find the following.

a) The cost of the merchandise when the total read $31.50
b) A formula for the cost of the merchandise c when the total reads T dollars

Solution

a) When tax is added to the cost of an item, the customer actually pays more than 100% of the item's price. When sales tax is 5%, the total paid is 105% of the price of the merchandise. Thus if $c =$ the cost of the merchandise, we have

$31.50 is 105% of c

$$31.50 = 1.05 \cdot c$$

$$\frac{31.50}{1.05} = c \qquad \text{Dividing both sides by 1.05}$$

$$30 = c. \qquad \text{Simplifying}$$

The merchandise cost $30 before tax.
b) When the total is T dollars, we modify the approach used in part (a):

$$T = 1.05c$$

$$\frac{T}{1.05} = c. \qquad \text{Dividing both sides by 1.05}$$

As a check, note that when T is $31.50, we have $31.50 ÷ 1.05 = $30. Since this matches the result of part (a), our formula is probably correct.

The formula $c = T/1.05$ can be used to find the cost of the merchandise when the total T is known and the sales tax is 5%.

Find decimal notation.

1. 82%   **2.** 49%   **3.** 9%   **4.** 91.3%

5. 43.7%   **6.** 2%   **7.** 0.46%   **8.** 4.8%

Find percent notation.

9. 0.29   **10.** 0.78   **11.** 0.998

12. 0.358   **13.** 1.92   **14.** 1.39

15. 2.1   **16.** 9.2   **17.** 0.0068

18. 0.0095   **19.** $\frac{3}{8}$ **20.** $\frac{3}{4}$

21. $\frac{7}{25}$ **22.** $\frac{4}{5}$ **23.** $\frac{2}{3}$ **24.** $\frac{5}{6}$

Solve.

25. What percent of 68 is 17?

26. What percent of 150 is 39?

27. What percent of 125 is 30?

28. What percent of 300 is 57?

29. 14 is 30% of what number?

30. 54 is 24% of what number?

31. 0.3 is 12% of what number?

32. 7 is 175% of what number?

33. What number is 35% of 240?

34. What number is 1% of one million?

35. What percent of 60 is 75?

Aha! **36.** What percent of 70 is 70?

37. What is 2% of 40?

38. What is 40% of 2?

Aha! **39.** 25 is what percent of 50?

40. 8 is 2% of what number?

41. *Student loans.* To finance her community college education, Sarah takes out a Stafford loan for $3500. After a year, Sarah decides to pay off the interest, which is 8% of $3500. How much will she pay?

42. *Student loans.* Paul takes out a subsidized federal Stafford loan for $2400. After a year, Paul decides to pay off the interest, which is 7% of $2400. How much will he pay?

43. *Votes for president.* In 2000, Al Gore received 48.62 million votes. This accounted for 48.36% of all votes cast. How many people voted in the 2000 presidential election?

44. *Lotteries and education.* In 1997, $6.2 billion of state lottery money was used for education. This accounted for 52% of all lottery proceeds for the year (*Source*: *Statistical Abstract of the United States*, 1999). What was the total amount of lottery proceeds in 1997? (Round to the nearest tenth of a billion.)

45. *Infant health.* In a study of 300 pregnant women with "poor" diets, 8% had babies in good or excellent health. How many women in this group had babies in good or excellent health?

46. *Infant health.* In a study of 300 pregnant women with "good-to-excellent" diets, 95% had babies in good or excellent health. How many women in this group had babies in good or excellent health?

47. *Nut consumption.* Each American consumes, on average, 2.25 lb of tree nuts each year (*Source*: *USA Today*, 2/17/00). Of this amount, 25% is almonds. How many pounds of almonds does the average American consume each year?

48. *Junk mail.* The U.S. Postal Service reports that we open and read 78% of the junk mail that we receive. A business sends out 9500 advertising brochures. How many of them can the business expect to be opened and read?

49. *Left-handed bowlers.* It has been determined by sociologists that 17% of the population is left-handed. Each week 160 bowlers enter a tournament conducted by the Professional Bowlers Association. How many would you expect to be left-handed? (Round to the nearest one.)

50. *Kissing and colds.* In a medical study, it was determined that if 800 people kiss someone else who has a cold, only 56 will actually catch the cold. What percent is this?

51. On a test of 88 items, a student got 76 correct. What percent were correct?

52. A baseball player had 13 hits in 25 times at bat. What percent were hits?

53. A bill at Officeland totaled $37.80. How much did the merchandise cost if the sales tax is 5%?

54. Doreen's checkbook shows that she wrote a check for $987 for building materials. What was the price of the materials if the sales tax is 5%?

55. *Deducting sales tax.*   A tax-exempt school group received a bill of $157.41 for educational software. The bill incorrectly included sales tax of 6%. How much should the school group pay?

56. *Deducting sales tax.*   A tax-exempt charity received a bill of $145.90 for a sump pump. The bill incorrectly included sales tax of 5%. How much does the charity owe?

57. *Cost of self-employment.*   Because of additional taxes and fewer benefits, it has been estimated that a self-employed person must earn 20% more than a non–self-employed person performing the same task(s). If Roy earns $15 an hour working for Village Copy, how much would he need to earn on his own for a comparable income?

58. Refer to Exercise 57. Clara earns $12 an hour working for Round Edge stairbuilders. How much would Clara need to earn on her own for a comparable income?

59. *Calorie content.*   Pepperidge Farm Light Style 7 Grain Bread® has 140 calories in a 3-slice serving. This is 15% less than the number of calories in a serving of regular bread. How many calories are in a serving of regular bread?

60. *Fat content.*   Peek Freans Shortbread Reduced Fat Cookies® contain 35 calories of fat in each serving. This is 40% less than the fat content in the leading imported shortbread cookie. How many calories of fat are in a serving of the leading shortbread cookie?

61. Campus Bookbuyers pays $30 for a book and sells it for $60. Is this a 100% markup or a 50% markup? Explain.

62. If Julian leaves a $12 tip for a $90 dinner, is he being generous, stingy, or neither? Explain.

SKILL MAINTENANCE

Translate to an algebraic expression.

63. 5 more than some number

64. 4 less than Tino's weight

65. The product of 8 and twice *a*

66. 1 more than the product of two numbers

SYNTHESIS

67. Does the following advertisement provide a convincing argument that summertime is when most burglaries occur? Why or why not?

When you go away, the burglars will stay.

FBI statistics show that over 26% of home burglaries occur between Memorial Day and Labor Day.*

For just $479, SND Security Systems provides the peace of mind you deserve, 24 hours a day, 365 days a year.

Call SND today; *THIS DEAL IS A STEAL.*

SND Summer Sale!

Now Only **$479**

SND Security SYSTEMS

*1993 FBI Uniform Crime Report

68. Erin is returning a tent that she bought during a 25%-off storewide sale that has ended. She is offered store credit for 125% of what she paid (not to be used on sale items). Is this fair to Erin? Why or why not?

69. The community of Bardville has 1332 left-handed females. If 48% of the community is female and 15% of all females are left-handed, how many people are in the community?

70. Rollie's Music charges $11.99 for a compact disc. Sound Warp charges $13.99 but you have a coupon for $2 off. In both cases, a 7% sales tax is charged on the *regular* price. How much does the disc cost at each store?

71. The new price of a car is 25% higher than the old price. The old price is what percent lower than the new price?

72. Claude pays 26% of his pretax earnings in taxes. What percentage of his *post*-tax earnings is this?

73. *U.S. birth rate.* There were 3.88 million births in 1997 and 3.94 million births in 1998 (*Source: Burlington Free Press,* page 2A, March 29, 2000). By what percentage did the number of births increase?

Aha! **74.** Would it be better to receive a 5% raise and then an 8% raise or the other way around? Why?

75. Herb is in the 30% tax bracket. This means that 30¢ of each dollar earned goes to taxes. Which would cost him the least: contributing $50 that is tax-deductible or contributing $40 that is not tax-deductible? Explain.

CORNER

Sales and Discounts

Focus: Applications and models using percent

Time: 15 minutes

Group size: 3

Materials: Calculators are optional.

Often a store will reduce the price of an item by a fixed percentage. When the sale ends, the items are returned to their original prices. Suppose a department store reduces all sporting goods 20%, all clothing 25%, and all electronics 10%.

ACTIVITY

1. Each group member should select one of the following items: a $50 basketball, an $80 jacket, or a $200 portable sound system. Fill in the first three columns of the first three rows of the chart below.

2. Apply the appropriate discount and determine the sale price of your item. Fill in the fourth column of the chart.

3. Next, find a multiplier that can be used to convert the sale price back to the original price and fill in the remaining column of the chart. Does this multiplier depend on the price of the item?

4. Working as a group, compare the results of part (3) for all three items. Then develop a formula for a multiplier that will restore a sale price to its original price, p, after a discount r has been applied. Complete the fourth row of the table and check that your formula will duplicate the results of part (3).

5. Use the formula from part (4) to find the multiplier that a store would use to return an item to its original price after a "30% off" sale expires. Fill in the last line on the chart.

6. Inspect the last column of your chart. How can these multipliers be used to determine the percentage by which a sale price is increased when a sale ends?

Original Price, p	Discount, r	$1 - r$	Sale Price	Multiplier to convert back to p
p	r	$1 - r$		
	.30			

Problem Solving

2.5

Five Steps for Problem Solving • Applying the Five Steps

Probably the most important use of algebra is as a tool for problem solving. In this section, we develop a problem-solving approach that will be used throughout the remainder of the text.

Five Steps for Problem Solving

In Section 2.4, we solved a problem in which Middlesex Toy and Hobby needed a formula. To solve the problem, we *familiarized* ourselves with percent notation so that we could then *translate* the problem into an equation. At the end of the section, we *solved* the equation, *checked* the solution, and *stated* the answer.

> ### Five Steps for Problem Solving in Algebra
> 1. *Familiarize* yourself with the problem.
> 2. *Translate* to mathematical language. (This often means writing an equation.)
> 3. *Carry out* some mathematical manipulation. (This often means *solving* an equation.)
> 4. *Check* your possible answer in the original problem.
> 5. *State* the answer clearly.

Of the five steps, the most important is probably the first one: becoming familiar with the problem. Here are some hints for familiarization.

> ### To Become Familiar with a Problem
> 1. Read the problem carefully. Try to visualize the problem.
> 2. Reread the problem, perhaps aloud. Make sure you understand all important words.
> 3. List the information given and the question(s) to be answered. Choose a variable (or variables) to represent the unknown and specify what the variable represents. For example, let L = length in centimeters, d = distance in miles, and so on.
> 4. Look for similarities between the problem and other problems you have already solved.
> 5. Find more information. Look up a formula in a book, at a library, or on-line. Consult a reference librarian or an expert in the field.

(continued)

6. Make a table that uses all the information you have available. Look for patterns that may help in the translation.
7. Make a drawing and label it with known and unknown information, using specific units if given.
8. Think of a possible answer and check the guess. Observe the manner in which the guess is checked.

Applying the Five Steps

E x a m p l e   1

Hiking.   In 1998, at age 79, Earl Shaffer became the oldest person to hike all 2100 miles of the Appalachian Trail—from Springer Mountain, Georgia, to Mount Katahdin, Maine (*Source*: Appalachian Trail Conference). At one point, Shaffer stood atop Big Walker Mountain, Virginia, which is three times as far from the northern end as from the southern end. How far was Shaffer from each end of the trail?

Solution

1. **Familiarize.**   It may be helpful to make a drawing.

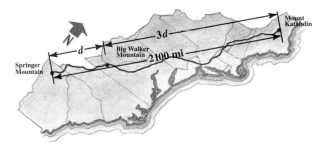

To gain some familiarity, let's suppose that Shaffer stood 600 mi from Springer Mountain. Three times 600 mi is 1800 mi. Since 600 mi + 1800 mi = 2400 mi and 2400 mi > 2100 mi, we see that our guess is too large. Rather than guess again, we let

   d = the distance, in miles, to the southern end,

and

   $3d$ = the distance, in miles, to the northern end.

(We could also let x = the distance to the northern end and $\frac{1}{3}x$ = the distance to the southern end.)

2. **Translate.**   From the drawing, we see that the lengths of the two parts of the trail must add up to 2100 mi. This leads to our translation.

Rewording: $\underbrace{\text{Distance to southern end}}$ plus $\underbrace{\text{distance to northern end}}$ is 2100 mi

Translating: d $+$ $3d$ $=$ 2100

3. **Carry out.**   We solve the equation:

$$d + 3d = 2100$$
$$4d = 2100 \qquad \text{Combining like terms}$$
$$d = 525. \qquad \text{Dividing both sides by 4}$$

4. **Check.**   As expected, d is less than 600 mi. If $d = 525$ mi, then $3d =$ 1575 mi. Since 525 mi + 1575 mi = 2100 mi, we have a check.

5. **State.**   Atop Big Walker Mountain, Shaffer stood 525 mi from Springer Mountain and 1575 mi from Mount Katahdin.

E x a m p l e   2

Page numbers.   The sum of two consecutive page numbers is 305. Find the page numbers.

Solution

1. **Familiarize.**   If the meaning of the word consecutive is unclear, we should consult a dictionary or someone who might know. Consecutive numbers are integers that are one unit apart. Thus, 18 and 19 are consecutive numbers, as are -24 and -23. Let's "guess and check": If the first page number is 40, the next would be 41. Since $40 + 41 = 81$ and $81 < 305$, our guess is much too small. Suppose the first page number is 130. The next page would then be 131. Since $130 + 131 = 261$ and $261 < 305$, our guess is still a bit too small. We could continue guessing, but algebra offers a more direct approach. Let's have

    $$x = \text{the first page number}$$

    and, since the two numbers must be one unit apart,

    $$x + 1 = \text{the next page number.}$$

2. **Translate.**   We reword the problem and translate as follows.

    Rewording: First page number   plus   next page number   is   305

    Translating: $x \qquad + \qquad (x + 1) \qquad = \qquad 305$

3. **Carry out.**   We solve the equation:

    $$x + (x + 1) = 305$$
    $$2x + 1 = 305 \qquad \text{Using an associative law and combining like terms}$$
    $$2x = 304 \qquad \text{Subtracting 1 from both sides}$$
    $$x = 152. \qquad \text{Dividing both sides by 2}$$

    If x is 152, then $x + 1$ is 153.

4. **Check.**   Our possible answers are 152 and 153. These are consecutive integers and their sum is 305, so the answers check in the original problem.

5. **State.**   The page numbers are 152 and 153.

Do not be surprised if your success rate drops some as you work through the exercises in this section. *This is normal.* Your success rate will increase as you gain experience with these types of problems and use some of the study tips already listed.

Example 3

Taxi rates.   In Bermuda, a taxi ride costs $4.80 plus $1.68 for each mile traveled. Debbie and Alex have budgeted $18 for a taxi ride (excluding tip). How far can they travel on their $18 budget?

Solution

1. **Familiarize.**   Suppose the taxi takes Debbie and Alex 5 mi. Such a ride would cost $4.80 + 5($1.68), or $4.80 + $8.40 = $13.20. Since 13.20 < 18, we know that the actual answer exceeds 5 mi. Rather than guess again, we let

   s = the distance, in miles, driven by the taxi for $18.

2. **Translate.**   We rephrase the problem and translate.

   Rewording: The initial charge plus the mileage charge is $18.

   Translating: $4.80   + s($1.68)   = $18

3. **Carry out.**   We solve as follows:

$$4.80 + 1.68s = 18$$
$$1.68s = 13.20 \qquad \text{Subtracting 4.80 from both sides}$$
$$s = \frac{13.20}{1.68} \qquad \text{Dividing both sides by 1.68}$$
$$s \approx 7.8. \qquad \text{Simplifying}$$

   In the work above, the symbol $\approx$ means *is approximately equal to.* We use it here because we rounded off our result.

4. **Check.**   A 7.8-mi taxi ride would cost $4.80 + 7.8($1.68), or $17.90. Since we rounded down when finding s, the check will not be precise.

5. **State.**   Debbie and Alex can take a 7.8-mi taxi ride and stay within their budget.

   The division in Example 3 would normally be rounded *up* to 7.9. In the circumstances of the problem, however, this would lead to a taxi fare slightly in excess of $18. When solving problems, be careful whenever you round off to round in the appropriate direction.

E x a m p l e 4

Gardening.   A rectangular community garden is to be enclosed with 92 m of fencing. In order to allow for compost storage, the garden must be 4 m longer than it is wide. Determine the dimensions of the garden.

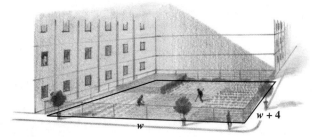

Solution

1. **Familiarize.**   Recall that the perimeter of a rectangle is twice the length plus twice the width. Suppose the garden were 30 m wide. The length would then be 30 + 4 m, or 34 m, and the perimeter would be 2 · 30 m + 2 · 34 m, or 128 m. This shows that for the perimeter to be 92 m, the width must be less than 30 m. Instead of guessing again, we let w = the width of the garden, in meters. Since the garden is "4 m longer than it is wide," we let $w + 4$ = the length of the garden, in meters.

2. **Translate.**   To translate, we use $w + 4$ as the length and 92 as the perimeter.

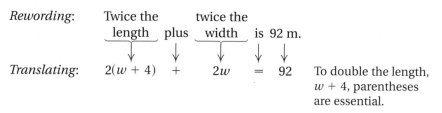

Rewording: Twice the length   plus   twice the width   is 92 m.

Translating: $2(w + 4)$ + $2w$ =   92   To double the length, $w + 4$, parentheses are essential.

3. **Carry out.**   We solve the equation:

$$2(w + 4) + 2w = 92$$
$$2w + 8 + 2w = 92 \qquad \text{Using the distributive law}$$
$$4w + 8 = 92 \qquad \text{Combining like terms}$$
$$4w = 84$$
$$w = 21.$$

The dimensions appear to be w = 21 m and l, or $w + 4$, = 25 m.

4. **Check.**   If the width is 21 m and the length 25 m, then the garden is 4 m longer than it is wide. The perimeter is 2(25 m) + 2(21 m), or 92 m, and since 92 m of fencing is available, we have a check.

5. **State.**   The garden should be 21 m wide and 25 m long.

> **Caution!**   Always be sure to answer the original problem completely. For instance, in Example 1 we need to find *two* numbers: the distances from *each* end of the trail to the hiker. Similarly, in Example 2 we needed to find two page numbers and in Example 4 we needed to find two dimensions, not just the width.

E x a m p l e   5

Selling a home.   The McCanns are planning to sell their home. If they want to be left with $117,500 after paying 6% of the selling price to a realtor as a commission, for how much must they sell the house?

Solution

1. **Familiarize.**   Suppose the McCanns sell the house for $120,000. A 6% commission can be determined by finding 6% of $120,000:

   6% of $120,000 = 0.06($120,000) = $7200.

   Subtracting this commission from $120,000 would leave the McCanns with

   $120,000 − $7200 = $112,800.

   This shows that in order for the McCanns to clear $117,500, the house must sell for more than $120,000. To determine what the sale price must be, we could check more guesses. Instead, we let x = the selling price, in dollars. With a 6% commission, the realtor would receive $0.06x$.

2. **Translate.**   We reword the problem and translate as follows.

   Rewording: Selling price less commission is amount remaining.

   Translating: x − $0.06x$ =   117,500

3. **Carry out.** We solve the equation:

$$x - 0.06x = 117{,}500$$

$$1x - 0.06x = 117{,}500$$

$$0.94x = 117{,}500 \qquad \text{Combining like terms. Had we noted that after the commission has been paid, 94\% remains, we could have begun with this equation.}$$

$$x = \frac{117{,}500}{0.94} \qquad \text{Dividing both sides by 0.94}$$

$$x = 125{,}000.$$

4. **Check.** To check, we first find 6% of $125,000:

$$6\% \text{ of } \$125{,}000 = 0.06(\$125{,}000) = \$7500. \qquad \text{This is the commission.}$$

Next, we subtract the commission to find the remaining amount:

$$\$125{,}000 - \$7500 = \$117{,}500.$$

Since, after the commission, the McCanns are left with $117,500, our answer checks. Note that the $125,000 sale price is greater than $120,000, as predicted in the *Familiarize* step.

5. **State.** To be left with $117,500, the McCanns must sell the house for $125,000.

E x a m p l e   6

Angles in a triangle. The second angle of a triangle is 20° greater than the first. The third angle is twice as large as the first. How large are the angles?

Solution

1. **Familiarize.** We make a drawing. In this case, the measure of the first angle is x, the measure of the second angle is $x + 20$, and the measure of the third angle is $2x$.

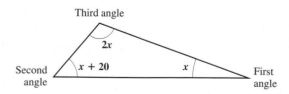

2. **Translate.** To translate, we need to recall that the sum of the measures of the angles in a triangle is 180°.

Rewording: Measure of first angle + measure of second angle + measure of third angle is 180°

Translating: $x + (x + 20) + 2x = 180$

3. **Carry out.**   We solve:

$$x + (x + 20) + 2x = 180$$
$$4x + 20 = 180$$
$$4x = 160$$
$$x = 40.$$

The measures for the angles appear to be:

First angle: $x = 40°$,
Second angle: $x + 20 = 40 + 20 = 60°$,
Third angle: $2x = 2(40) = 80°$.

4. **Check.**   Consider 40°, 60°, and 80°. The second angle is 20° greater than the first, the third is twice the first, and the sum is 180°. These numbers check.

5. **State.**   The measures of the angles are 40°, 60°, and 80°.

We close this section with some tips to aid you in problem solving.

> ### Problem-Solving Tips
>
> 1. The more problems you solve, the more your skills will improve.
> 2. Look for patterns when solving problems. Each time you study an example in a text, you may observe a pattern for problems that you will encounter later in the exercise sets or in other practical situations.
> 3. When translating in mathematics, consider the dimensions of the variables and constants in the equation. The variables that represent length should all be in the same unit, those that represent money should all be in dollars or all in cents, and so on.
> 4. Make sure that units appear in the answer whenever appropriate and that you have completely answered the original problem.

Exercise Set 2.5

Solve. Even though you might find the answer quickly in some other way, practice using the five-step problem-solving process.

1. Three less than twice a number is 19. What is the number?

2. Two fewer than ten times a number is 78. What is the number?

3. Five times the sum of 3 and some number is 70. What is the number?

4. Twice the sum of 4 and some number is 34. What is the number?

5. *Price of sneakers.*   Amy paid $63.75 for a pair of New Balance 903 running shoes during a 15%-off sale. What was the regular price?

6. *Price of a CD player.*   Doug paid $72 for a shockproof portable CD player during a 20%-off sale. What was the regular price?

7. *Price of a textbook.*   Evelyn paid $89.25, including 5% tax, for her biology textbook. How much did the book itself cost?

8. *Price of a printer.*   Jake paid $100.70, including 6% tax, for a color printer. How much did the printer itself cost?

9. *Running.*   In 1997, Yiannis Kouros of Australia set the record for the greatest distance run in 24 hr by running 188 mi (*Source: Guinness World Records 2000 Millennium Edition*). After 8 hr, he was approximately twice as far from the finish line as he was from the start. How far had he run?

10. *Sled-dog racing.*   The Iditarod sled-dog race extends for 1049 mi from Anchorage to Nome. If a musher is twice as far from Anchorage as from Nome, how many miles has the musher traveled?

11. The sum of three consecutive page numbers is 60. Find the numbers.

12. The sum of three consecutive page numbers is 99. Find the numbers.

13. The sum of two consecutive odd numbers is 60. Find the numbers. (*Hint*: Odd numbers, like even numbers, are separated by two units.)

14. The sum of two consecutive odd integers is 108. What are the integers?

15. The sum of two consecutive even integers is 126. What are the integers?

16. The sum of two consecutive even numbers is 50. Find the numbers.

17. *Oldest groom.*   The world's oldest groom was 19 yr older than his bride (*Source: Guinness World Records 2000 Millennium Edition*). Together, their ages totaled 187 yr. How old were the bride and the groom?

18. *Oldest divorcees.*   In the world's oldest divorcing couple, the woman was 6 yr younger than the man (*Source: Guinness World Records 2000 Millennium Edition*). Together, their ages totaled 188 yr. How old were the man and the woman?

19. *Angles of a triangle.*   The second angle of a triangle is three times as large as the first. The third angle is 30° more than the first. Find the measure of each angle.

20. *Angles of a triangle.*   The second angle of a triangle is four times as large as the first. The third angle is 45° less than the sum of the other two angles. Find the measure of each angle.

21. *Angles of a triangle.*   The second angle of a triangle is three times as large as the first. The third angle is 10° more than the sum of the other two angles. Find the measure of the third angle.

22. *Angles of a triangle.*   The second angle of a triangle is four times as large as the first. The third angle is 5° more than the sum of the other two angles. Find the measure of the second angle.

23. *Page numbers.*   The sum of the page numbers on the facing pages of a book is 385. What are the page numbers?

24. *Page numbers.*   The sum of the page numbers on the facing pages of a book is 281. What are the page numbers?

25. *Perimeter of a triangle.*   The perimeter of a triangle is 195 mm. If the lengths of the sides are consecutive odd integers, find the length of each side.

26. *Hancock Building dimensions.*   The top of the John Hancock Building in Chicago is a rectangle whose length is 60 ft more than the width. The perimeter is 520 ft. Find the width and the length of the rectangle. Find the area of the rectangle.

27. *Dimensions of a state.*   The perimeter of the state of Wyoming is 1280 mi. The width is 90 mi less than the length. Find the width and the length.

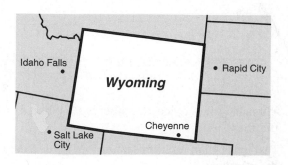

28. *Copier paper.*   The perimeter of standard-size copier paper is 99 cm. The width is 6.3 cm less than the length. Find the length and the width.

29. *Stock prices.*   Sarah's investment in America Online stock grew 28% to $448. How much did she invest?

30. *Savings interest.*   Sharon invested money in a savings account at a rate of 6% simple interest. After 1 yr, she has $6996 in the account. How much did Sharon originally invest?

31. *Credit cards.*   The balance in Will's Mastercard® account grew 2%, to $870, in onc month. What was his balance at the beginning of the month?

32. *Loan interest.*   Alvin borrowcd moncy from a cousin at a rate of 10% simple interest. After 1 yr, $7194 paid off the loan. How much did Alvin borrow?

33. *Taxi fares.*   In Beniford, taxis charge $3 plus 75¢ per mile for an airport pickup. How far from the airport can Courtney travel for $12?

34. *Taxi fares.*   In Cranston, taxis charge $4 plus 90¢ per mile for an airport pickup. How far from the airport can Ralph travel for $17.50?

35. *Truck rentals.*   Truck-Rite Rentals rents trucks at a daily rate of $49.95 plus 39¢ per mile. Concert Productions has budgeted $100 for renting a truck to haul equipment to an upcoming concert. How far can they travel in one day and stay within their budget?

36. *Truck rentals.*   Fine Line Trucks rents an 18-ft truck for $42 plus 35¢ per mile. Judy needs a truck for one day to deliver a shipment of plants. How far can she drive and stay within a budget of $70?

37. *Complementary angles.*   The sum of the measures of two *complementary* angles is 90°. If one angle measures 15° more than twice the measure of its complement, find the measure of each angle.

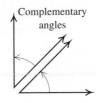

Complementary angles

38. *Supplementary angles.*   The sum of the measures of two *supplementary* angles is 180°. If one angle measures 45° less than twice the measure of its supplement, find the measure of each angle.

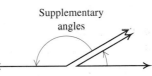

Supplementary angles

39. *Cricket chirps and temperature.*   The equation $T = \frac{1}{4}N + 40$ can be used to determine the temperature T, in degrees Fahrenheit, given the number of times N a cricket chirps per minute. Determine the number of chirps per minute for a temperature of 80°F.

40. *Race time.*   The equation $R = -0.028t + 20.8$ can be used to predict the world record in the 200-m dash, where R is the record in seconds and t is the number of years since 1920. In what year will the record be 18.0 sec?

41. Sean claims he can solve most of the problems in this section by guessing. Is there anything wrong with this approach? Why or why not?

42. When solving Exercise 17, Beth used a to represent the bride's age and Ben used a to represent the groom's age. Is one of these approaches preferable to the other? Why or why not?

SKILL MAINTENANCE

Write a true sentence using either < or >.

43. −9 ▨ 5

44. 1 ▨ 3

45. −4 ▨ 7

46. −9 ▨ −12

SYNTHESIS

47. Write a problem for a classmate to solve. Devise it so that the problem can be translated to the equation $x + (x + 2) + (x + 4) = 375$.

48. Write a problem for a classmate to solve. Devise it so that the solution is "Audrey can drive the rental truck for 50 mi without exceeding her budget."

49. *Discounted dinners.* Kate's "Dining Card" entitles her to $10 off the price of a meal after a 15% tip has been added to the cost of the meal. If, after the discount, the bill is $32.55, how much did the meal originally cost?

50. *Test scores.* Pam scored 78 on a test that had 4 fill-ins worth 7 points each and 24 multiple-choice questions worth 3 points each. She had one fill-in wrong. How many multiple-choice questions did Pam get right?

51. *Gettysburg Address.* Abraham Lincoln's 1863 Gettysburg Address refers to the year 1776 as "four *score* and seven years ago." Determine what a score is.

52. One number is 25% of another. The larger number is 12 more than the smaller. What are the numbers?

53. *Perimeter of a rectangle.* The width of a rectangle is three fourths of the length. The perimeter of the rectangle becomes 50 cm when the length and the width are each increased by 2 cm. Find the length and the width.

54. *Angles in a quadrilateral.* The measures of the angles in a quadrilateral are consecutive odd numbers. Find the measure of each angle.

55. *Angles in a pentagon.* The measures of the angles in a pentagon are consecutive even numbers. Find the measure of each angle.

56. *Sharing fruit.* Apples are collected in a basket for six people. One third, one fourth, one eighth, and one fifth of the apples are given to four people, respectively. The fifth person gets ten apples, and one apple remains for the sixth person. Find the original number of apples in the basket.

57. *Discounts.* In exchange for opening a new credit account, Filene's Department Stores® subtracts 10% from all purchases made the day the account is established. Julio is opening an account and has a coupon for which he receives 10% off the first day's reduced price of a camera. If Julio's final price is $77.75, what was the price of the camera before the two discounts?

58. *Winning percentage.* In a basketball league, the Falcons won 15 of their first 20 games. In order to win 60% of the total number of games, how many more games will they have to play, assuming they win only half of the remaining games?

59. *Music-club purchases.* During a recent sale, BMG Music Service® charged $8.49 for the first CD ordered and $3.99 for all others. For shipping and handling, BMG charged $2.47 for the first CD, $2.28 for the second CD, and $1.99 for all others. The total cost of a shipment (excluding tax) was $65.07. How many CD's were in the shipment?

60. *Test scores.* Ella has an average score of 82 on three tests. Her average score on the first two tests is 85. What was the score on the third test?

61. *Taxi fares.* In New York City, a taxi ride costs $2 plus 30¢ per $\frac{1}{5}$ mile and 20¢ per minute stopped in traffic. Due to traffic, Glenda's taxi took 20 min to complete what is usually a 10-min drive. If she is charged $13 for the ride, how far did Glenda travel?

62. A school purchases a piano and must choose between paying $2000 at the time of purchase or $2150 at the end of one year. Which option should the school select and why?

63. Annette claims the following problem has no solution: "The sum of the page numbers on facing pages is 191. Find the page numbers." Is she correct? Why or why not?

64. The perimeter of a rectangle is 101.74 cm. If the length is 4.25 cm longer than the width, find the dimensions of the rectangle.

65. The second side of a triangle is 3.25 cm longer than the first side. The third side is 4.35 cm longer than the second side. If the perimeter of the triangle is 26.87 cm, find the length of each side.

Aha!

Solving Inequalities

2.6

Solutions of Inequalities • Graphs of Inequalities •
Solving Inequalities Using the Addition Principle •
Solving Inequalities Using the Multiplication Principle •
Using the Principles Together

Many real-world situations translate to *inequalities*. For example, a student might need to register for *at least* 12 credits; an elevator might be designed to hold *at most* 2000 pounds; a tax credit might be allowable for families with incomes of *less than* $25,000; and so on. Before solving applications of this type, we must adapt our equation-solving principles to the solving of inequalities.

Solutions of Inequalities

Recall from Section 1.4 that an inequality is a number sentence containing $>$ (is greater than), $<$ (is less than), $\geq$ (is greater than or equal to), or $\leq$ (is less than or equal to). Inequalities like

$$-7 > x, \qquad t < 5, \qquad 5x - 2 \geq 9, \quad \text{and} \quad -3y + 8 \leq -7$$

are true for some replacements of the variable and false for others.

Example 1   Determine whether the given number is a solution of $x < 2$: **(a)** -3; **(b)** 2.

Solution

a) Since $-3 < 2$ is true, -3 is a solution.
b) Since $2 < 2$ is false, 2 is not a solution.

Example 2   Determine whether the given number is a solution of $y \geq 6$: **(a)** 6; **(b)** -4.

Solution

a) Since $6 \geq 6$ is true, 6 is a solution.
b) Since $-4 \geq 6$ is false, -4 is not a solution.

Graphs of Inequalities

Because the solutions of inequalities like $x < 2$ are too numerous to list, it is helpful to make a drawing that represents all the solutions. The **graph** of an inequality is such a drawing. Graphs of inequalities in one variable can be drawn on a number line by shading all points that are solutions. Open dots are used to indicate endpoints that are *not* solutions and closed dots indicate endpoints that *are* solutions.

E x a m p l e   3   Graph each inequality: **(a)** $x < 2$; **(b)** $y \geq -3$; **(c)** $-2 < x \leq 3$.

Solution

a) The solutions of $x < 2$ are those numbers less than 2. They are shown on the graph by shading all points to the left of 2. The open dot at 2 and the shading to its left indicates that 2 is *not* part of the graph, but numbers like 1.2 and 1.99 are.

b) The solutions of $y \geq -3$ are shown on the number line by shading the point for -3 and all points to the right of -3. The closed dot at -3 indicates that -3 *is* part of the graph.

c) The inequality $-2 < x \leq 3$ is read "-2 is less than x *and* x is less than or equal to 3," or "x is greater than -2 *and* less than or equal to 3." To be a solution of $-2 < x \leq 3$, a number must be a solution of both $-2 < x$ *and* $x \leq 3$. The number 1 is a solution, as are -0.5, 1.9, and 3. The open dot indicates that -2 is *not* a solution, whereas the closed dot indicates that 3 *is* a solution. The other solutions are shaded.

Solving Inequalities Using the Addition Principle

Consider a balance similar to one that appears in Section 2.1. When one side of the balance holds more weight than the other, the balance tips in that direction. If equal amounts of weight are then added to or subtracted from both sides of the balance, the balance remains tipped in the same direction.

The balance illustrates the idea that when a number, such as 2, is added to (or subtracted from) both sides of a true inequality, such as $3 < 7$, we get another true inequality:

$$3 + 2 < 7 + 2, \quad \text{or} \quad 5 < 9.$$

Similarly, if we add -4 to both sides of $x + 4 < 10$, we get an *equivalent* inequality:

$$x + 4 + (-4) < 10 + (-4), \quad \text{or} \quad x < 6.$$

We say that $x + 4 < 10$ and $x < 6$ are **equivalent**, which means that both inequalities have the same solution set.

> **The Addition Principle for Inequalities**
> For any real numbers a, b, and c:
>
> $$a < b \text{ is equivalent to } a + c < b + c;$$
> $$a \leq b \text{ is equivalent to } a + c \leq b + c;$$
> $$a > b \text{ is equivalent to } a + c > b + c;$$
> $$a \geq b \text{ is equivalent to } a + c \geq b + c.$$

As with equations, our goal is to isolate the variable on one side.

E x a m p l e   4

Solve $x + 2 > 8$ and then graph the solution.

Solution   We use the addition principle, subtracting 2 from both sides:

$$x + 2 - 2 > 8 - 2 \qquad \text{Subtracting 2 from, or adding } -2 \text{ to, both sides}$$
$$x > 6.$$

From the inequality $x > 6$, we can determine the solutions easily. Any number greater than 6 makes $x > 6$ true and is a solution of that inequality as well as the inequality $x + 2 > 8$. The graph is as follows:

Because most inequalities have an infinite number of solutions, we cannot possibly check them all. A partial check can be made using one of the possible solutions. For this example, we can substitute any number greater than 6—say, 6.1—into the original inequality:

$$\frac{x + 2 > 8}{6.1 + 2 \ ? \ 8}$$
$$8.1 \ | \ 8 \quad \text{TRUE} \qquad 8.1 > 8 \text{ is a true statement.}$$

Since $8.1 > 8$ is true, 6.1 is a solution. Any number greater than 6 is a solution.

Although the inequality $x > 6$ is easy to solve (we merely replace x with numbers greater than 6), it is worth noting that $x > 6$ is an *inequality*, not a *solution*. In fact, the solutions of $x > 6$ are numbers. To describe the set of all solutions, we will use **set-builder notation** to write the *solution set* of Example 4 as

$$\{x \mid x > 6\}.$$

This notation is read

"The set of all x such that x is greater than 6."

Thus a number is in $\{x \mid x > 6\}$ if that number is greater than 6. From now on, solutions of inequalities will be written using set-builder notation.

E x a m p l e 5

technology connection

As a partial check of Example 5, we can let $y_1 = 3x - 1$ and $y_2 = 2x - 5$. By scrolling up or down, you can note that for $x \le -4$, we have $y_1 \le y_2$.

X	Y₁	Y₂
−5	−16	−15
−4	−13	−13
−3	−10	−11
−2	−7	−9
−1	−4	−7
0	−1	−5
1	2	−3

X = −5

Solve $3x - 1 \le 2x - 5$ and then graph the solution.

Solution We have

$$3x - 1 \le 2x - 5$$
$$3x - 1 + 1 \le 2x - 5 + 1 \qquad \text{Adding 1 to both sides}$$
$$3x \le 2x - 4 \qquad \text{Simplifying}$$
$$3x - 2x \le 2x - 4 - 2x \qquad \text{Subtracting } 2x \text{ from both sides}$$
$$x \le -4. \qquad \text{Simplifying}$$

The graph is as follows:

Any number less than or equal to -4 is a solution, so the solution set is $\{x \mid x \le -4\}$.

Solving Inequalities Using the Multiplication Principle

There is a multiplication principle for inequalities similar to that for equations, but it must be modified when multiplying both sides by a negative number. Consider the true inequality

$$3 < 7.$$

If we multiply both sides by a *positive* number, say 2, we get another true inequality:

$$3 \cdot 2 < 7 \cdot 2, \quad \text{or} \quad 6 < 14. \qquad \text{TRUE}$$

If we multiply both sides by a negative number, say -2, we get a *false* inequality:

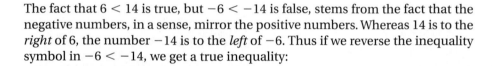

$$3 \cdot (-2) < 7 \cdot (-2), \quad \text{or} \quad -6 < -14. \qquad \text{FALSE}$$

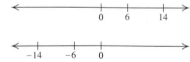

The fact that $6 < 14$ is true, but $-6 < -14$ is false, stems from the fact that the negative numbers, in a sense, mirror the positive numbers. Whereas 14 is to the *right* of 6, the number -14 is to the *left* of -6. Thus if we reverse the inequality symbol in $-6 < -14$, we get a true inequality:

$$-6 > -14. \qquad \text{TRUE}$$

> ### The Multiplication Principle for Inequalities
>
> For any real numbers a and b, and for any *positive* number c:
>
> $$a < b \quad \text{is equivalent to} \quad ac < bc, \quad \text{and}$$
> $$a > b \quad \text{is equivalent to} \quad ac > bc.$$
>
> For any real numbers a and b, and for any *negative* number c:
>
> $$a < b \quad \text{is equivalent to} \quad ac > bc, \quad \text{and}$$
> $$a > b \quad \text{is equivalent to} \quad ac < bc.$$
>
> Similar statements hold for $\leq$ and $\geq$.

E x a m p l e   6

Solve and graph each inequality: **(a)** $\frac{1}{4}x < 7$; **(b)** $-2y < 18$.

Solution

a) $\frac{1}{4}x < 7$

$\quad 4 \cdot \frac{1}{4}x < 4 \cdot 7$ Multiplying both sides by 4, the reciprocal of $\frac{1}{4}$

⎿————— The symbol stays the same, since 4 is positive.

$\quad\quad x < 28$ Simplifying

The solution set is $\{x \mid x < 28\}$. The graph is as follows:

b) $-2y < 18$

$\quad \dfrac{-2y}{-2} > \dfrac{18}{-2}$ Multiplying both sides by $-\frac{1}{2}$, or dividing both sides by -2

⎿————— At this step, we reverse the inequality, because $-\frac{1}{2}$ is negative.

$\quad\quad y > -9$ Simplifying

As a partial check, we substitute a number greater than -9, say -8, into the original inequality:

$$\frac{-2y < 18}{-2(-8) \ ? \ 18}$$
$$16 \ \mid \ 18 \ \text{TRUE} \quad\quad 16 < 18 \text{ is a true statement.}$$

The solution set is $\{y \mid y > -9\}$. The graph is as follows:

Using the Principles Together

We use the addition and multiplication principles together to solve inequalities much as we did when solving equations.

E x a m p l e 7

Solve: **(a)** $6 - 5y > 7$; **(b)** $2x - 9 \le 7x + 1$.

Solution

a)
$$6 - 5y > 7$$
$$-6 + 6 - 5y > -6 + 7 \qquad \text{Adding } -6 \text{ to both sides}$$
$$-5y > 1 \qquad \text{Simplifying}$$
$$-\tfrac{1}{5} \cdot (-5y) < -\tfrac{1}{5} \cdot 1 \qquad \text{Multiplying both sides by } -\tfrac{1}{5}, \text{ or dividing both sides by } -5$$

Remember to reverse the inequality symbol!

$$y < -\tfrac{1}{5} \qquad \text{Simplifying}$$

As a check, we substitute a number smaller than $-\tfrac{1}{5}$, say -1, into the original inequality:

$$\frac{6 - 5y > 7}{6 - 5(-1)\ ?\ 7}$$
$$6 - (-5)$$
$$11\ \Big|\ 7 \quad \text{TRUE} \qquad 11 > 7 \text{ is a true statement.}$$

The solution set is $\left\{ y \mid y < -\tfrac{1}{5} \right\}$. We show the graph in the margin for reference.

b)
$$2x - 9 \le 7x + 1$$
$$2x - 9 - 1 \le 7x + 1 - 1 \qquad \text{Subtracting 1 from both sides}$$
$$2x - 10 \le 7x \qquad \text{Simplifying}$$
$$2x - 10 - 2x \le 7x - 2x \qquad \text{Subtracting } 2x \text{ from both sides}$$
$$-10 \le 5x \qquad \text{Simplifying}$$
$$\frac{-10}{5} \le \frac{5x}{5} \qquad \text{Dividing both sides by 5}$$
$$-2 \le x \qquad \text{Simplifying}$$

The solution set is $\{x \mid -2 \le x\}$, or $\{x \mid x \ge -2\}$.

All of the equation-solving techniques used in Sections 2.1 and 2.2 can be used with inequalities provided we remember to reverse the inequality symbol when multiplying or dividing both sides by a negative number.

E x a m p l e 8

Solve: **(a)** $16.3 - 7.2p \leq -8.18$; **(b)** $3(x - 2) - 1 < 2 - 5(x + 6)$.

Solution

a) The greatest number of decimal places in any one number is *two*. Multiplying both sides by 100 will clear decimals. Then we proceed as before.

$$16.3 - 7.2p \leq -8.18$$

$100(16.3 - 7.2p) \leq 100(-8.18)$	Multiplying both sides by 100
$100(16.3) - 100(7.2p) \leq 100(-8.18)$	Using the distributive law
$1630 - 720p \leq -818$	Simplifying
$-720p \leq -818 - 1630$	Subtracting 1630 from both sides
$-720p \leq -2448$	Simplifying
$p \geq \dfrac{-2448}{-720}$	Dividing both sides by -720

Remember to reverse the symbol.

$$p \geq 3.4$$

The solution set is $\{\, p \,|\, p \geq 3.4 \,\}$.

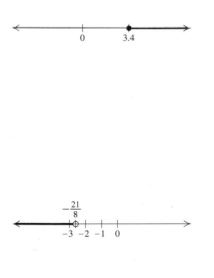

b) $3(x - 2) - 1 < 2 - 5(x + 6)$

$3x - 6 - 1 < 2 - 5x - 30$	Using the distributive law to remove parentheses
$3x - 7 < -5x - 28$	Simplifying
$3x + 5x < -28 + 7$	Adding $5x$ and also 7 to both sides. This isolates the x-terms on one side.
$8x < -21$	Simplifying
$x < -\dfrac{21}{8}$	Dividing both sides by 8

The solution set is $\left\{ x \,|\, x < -\frac{21}{8} \right\}$.

Exercise Set 2.6

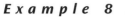
Determine whether each number is a solution of the given inequality.

1. $x > -2$

   a) 5   **b)** 0   **c)** -1.9

   d) -7.3 **e)** 1.6

2. $y < 5$

   a) 0   **b)** 5   **c)** 4.99

   d) -13 **e)** $7\frac{1}{4}$

3. $x \geq 6$

   a) -6 **b)** 0 **c)** 6

   d) 6.01 **e)** $-3\frac{1}{2}$

4. $x \leq 10$

   a) 4 **b)** -10 **c)** 0

   d) 10.2 **e)** -4.7

Graph on a number line.

5. $x \leq 7$ **6.** $y < 2$

7. $t > -2$ **8.** $y > 4$

9. $1 \leq m$ **10.** $0 \leq t$

11. $-3 < x \leq 5$ **12.** $-5 \leq x < 2$

13. $0 < x < 3$ **14.** $-5 \leq x \leq 0$

Describe each graph using set-builder notation.

15.

16.

17.

18.

19.

20.

21.

22.

Solve using the addition principle. Graph and write set-builder notation for the answers.

23. $y + 2 > 9$ **24.** $y + 6 > 9$

25. $x + 8 \leq -10$ **26.** $x + 9 \leq -12$

27. $x - 3 < 7$ **28.** $x - 3 < 14$

29. $5 \leq t + 8$ **30.** $4 \leq t + 9$

31. $y - 7 > -12$ **32.** $y - 10 > -16$

33. $2x + 4 \leq x + 9$ **34.** $2x + 4 \leq x + 1$

Solve using the addition principle. Write the answers in set-builder notation.

35. $5x - 6 \geq 4x - 1$ **36.** $3x - 9 \geq 2x + 11$

37. $y + \frac{1}{3} \leq \frac{5}{6}$ **38.** $x + \frac{1}{4} \leq \frac{1}{2}$

39. $t - \frac{1}{8} > \frac{1}{2}$ **40.** $y - \frac{1}{3} > \frac{1}{4}$

41. $-9x + 17 > 17 - 8x$ **42.** $-8n + 12 > 12 - 7n$

Aha! **43.** $-23 < -t$ **44.** $19 < -x$

Solve using the multiplication principle. Graph and write set-builder notation for the answers.

45. $5x < 35$ **46.** $8x \geq 32$

47. $9y \leq 81$ **48.** $350 > 10t$

49. $-7x < 13$ **50.** $8y < 17$

51. $-24 > 8t$ **52.** $-16x < -64$

Solve using the multiplication principle. Write the answers in set-builder notation.

53. $7y \geq -2$ **54.** $5x > -3$

55. $-2y \leq \frac{1}{5}$ **56.** $-2x \geq \frac{1}{5}$

57. $-\frac{8}{5} > -2x$ **58.** $-\frac{5}{8} < -10y$

Solve using the addition and multiplication principles.

59. $7 + 3x < 34$ **60.** $5 + 4y < 37$

61. $6 + 5y \geq 26$ **62.** $7 + 8x \geq 71$

63. $4t - 5 \leq 23$ **64.** $5y - 9 \leq 21$

65. $13x - 7 < -46$ **66.** $8y - 4 < -52$

67. $16 < 4 - 3y$ **68.** $22 < 6 - 8x$

69. $39 > 3 - 9x$ **70.** $40 > 5 - 7y$

71. $5 - 6y > 25$ **72.** $8 - 2y > 14$

73. $-3 < 8x + 7 - 7x$ **74.** $-5 < 9x + 8 - 8x$

75. $6 - 4y > 4 - 3y$ **76.** $7 - 8y > 5 - 7y$

77. $7 - 9y \leq 4 - 8y$ **78.** $6 - 13y \leq 4 - 12y$

79. $33 - 12x < 4x + 97$

80. $27 - 11x > 14x - 18$

81. $2.1x + 43.2 > 1.2 - 8.4x$

82. $0.96y - 0.79 \leq 0.21y + 0.46$

83. $0.7n - 15 + n \geq 2n - 8 - 0.4n$

84. $1.7t + 8 - 1.62t < 0.4t - 0.32 + 8$

85. $\frac{x}{3} - 4 \leq 1$

86. $\frac{2}{3} - \frac{x}{5} < \frac{4}{15}$

87. $3 < 5 - \dfrac{t}{7}$

88. $2 > 9 - \dfrac{x}{5}$

89. $4(2y - 3) < 36$

90. $3(2y - 3) > 21$

91. $3(t - 2) \geq 9(t + 2)$

92. $8(2t + 1) > 4(7t + 7)$

93. $3(r - 6) + 2 < 4(r + 2) - 21$

94. $5(t + 3) + 9 > 3(t - 2) + 6$

95. $\dfrac{2}{3}(2x - 1) \geq 10$

96. $\dfrac{4}{5}(3x + 4) \leq 20$

97. $\dfrac{3}{4}\left(3x - \dfrac{1}{2}\right) - \dfrac{2}{3} < \dfrac{1}{3}$

98. $\dfrac{2}{3}\left(\dfrac{7}{8} - 4x\right) - \dfrac{5}{8} < \dfrac{3}{8}$

99. Are the inequalities $x > -3$ and $3 > -x$ equivalent? Why or why not?

100. Are the inequalities $t > -7$ and $7 < -t$ equivalent? Why or why not?

SKILL MAINTENANCE

Translate to an algebraic expression.

101. The sum of 3 and some number

102. Twice the sum of two numbers

103. Three less than twice a number

104. Five more than twice a number

SYNTHESIS

105. Explain in your own words why it is necessary to reverse the inequality symbol when multiplying both sides of an inequality by a negative number.

106. Explain how it is possible for the graph of an inequality to consist of just one number. (*Hint*: See Example 3c.)

Solve.

107. $6[4 - 2(6 + 3t)] > 5[3(7 - t) - 4(8 + 2t)] - 20$

108. $27 - 4[2(4x - 3) + 7] \geq 2[4 - 2(3 - x)] - 3$

Solve for x.

109. $-(x + 5) \geq 4a - 5$

110. $\dfrac{1}{2}(2x + 2b) > \dfrac{1}{3}(21 + 3b)$

111. $y < ax + b$ (Assume $a > 0$.)

112. $y < ax + b$ (Assume $a < 0$.)

113. Determine whether each number is a solution of the inequality $|x| < 3$.

a) 3.2 b) -2 c) -3
d) -2.9 e) 3 f) 1.7

114. Graph the solutions of $|x| < 3$ on a number line.

Aha! **115.** Determine the solution set of $|x| > -3$.

116. Determine the solution set of $|x| < 0$.

Solving Applications with Inequalities

2.7

Translating to Inequalities • Solving Problems

The five steps for problem solving can be used for problems involving inequalities.

Translating to Inequalities

Before solving problems that involve inequalities, we list some important phrases to look for. Sample translations are listed as well.

Study Tip

Make an effort to do your homework as soon as possible after each class. Make this part of your routine, choosing a time and a place where you can focus with a minimum of interruptions.

Important Words	Sample Sentence	Translation
is at least is at most	Bill is at least 21 years old. At most 5 students dropped the course.	$b \geq 21$ $n \leq 5$
cannot exceed	To qualify, earnings cannot exceed $12,000.	$r \leq 12{,}000$
must exceed is less than is more than is between	The speed must exceed 15 mph. Tucker's weight is less than 50 lb. Boston is more than 200 miles away. The film is between 90 and 100 minutes long.	$s > 15$ $w < 50$ $d > 200$ $90 < t < 100$
no more than no less than	Bing weighs no more than 90 lb. Valerie scored no less than 8.3.	$w \leq 90$ $s \geq 8.3$

Solving Problems

Example 1

Catering costs.   To cater a party, Curtis' Barbeque charges a $50 setup fee plus $15 per person. The cost of Hotel Pharmacy's end-of-season softball party cannot exceed $450. How many people can attend the party?

Solution

1. **Familiarize.**   Suppose that 20 people were to attend the party. The cost would then be $50 + $15 · 20, or $350. This shows that more than 20 people could attend without exceeding $450. Instead of making another guess, we let n represent the number of people in attendance.

2. **Translate.**   The cost of the party will be $50 for the setup fee plus $15 times the number of people attending. We can reword as follows:

   Rewording: The setup fee plus the cost of the meals cannot exceed $450.

   Translating: 50 $+$ $15 \cdot n$ $\leq$ 450

3. **Carry out.**   We solve for n:

   $$50 + 15n \leq 450$$
   $$15n \leq 400 \qquad \text{Subtracting 50 from both sides}$$
   $$n \leq \frac{400}{15} \qquad \text{Dividing both sides by 15}$$
   $$n \leq 26\frac{2}{3}. \qquad \text{Simplifying}$$

4. **Check.**   Although the solution set of the inequality is all numbers less than or equal to $26\frac{2}{3}$, since n represents the number of people in attendance, we round *down* to 26. If 26 people attend, the cost will be $50 + $15 · 26, or $440, and if 27 attend, the cost will exceed $450.

5. **State.**   At most 26 people can attend the party.

> **Caution!** Solutions of problems should always be checked using the original wording of the problem. In some cases, answers might need to be whole numbers or integers or rounded off in a particular direction.

E x a m p l e   2

Nutrition. The U.S. Department of Health and Human Services and the Department of Agriculture recommend that for a typical 2000-calorie daily diet, no more than 65 g of fat be consumed. In the first three days of a four-day vacation, Phil consumed 70 g, 62 g, and 80 g of fat. Determine (in terms of an inequality) how many grams of fat Phil can consume on the fourth day if he is to average no more than 65 g of fat per day.

Solution

1. **Familiarize.** Suppose Phil consumed 64 g of fat on the fourth day. His daily average for the vacation would then be

$$\frac{70\text{ g} + 62\text{ g} + 80\text{ g} + 64\text{ g}}{4} = 69\text{ g}.$$

This shows that Phil cannot consume 64 g of fat on the fourth day, if he is to average no more than 65 g of fat per day. Let's have x represent the number of grams of fat that Phil consumes on the fourth day.

2. **Translate.** We reword the problem and translate as follows:

Rewording: The average consumption of fat — should be no more than — 65 g.

Translating: $\dfrac{70 + 62 + 80 + x}{4}$ $\leq$ 65

3. **Carry out.** Because of the fraction, it is convenient to use the multiplication principle first:

$$\frac{70 + 62 + 80 + x}{4} \leq 65$$

$$4\left(\frac{70 + 62 + 80 + x}{4}\right) \leq 4 \cdot 65 \qquad \text{Multiplying both sides by 4}$$

$$70 + 62 + 80 + x \leq 260$$

$$212 + x \leq 260 \qquad \text{Simplifying}$$

$$x \leq 48. \qquad \text{Subtracting 212 from both sides}$$

4. **Check.** As a partial check, we show that Phil can consume 48 g of fat on the fourth day and not exceed a 65-g average for the four days:

$$\frac{70 + 62 + 80 + 48}{4} = \frac{260}{4} = 65.$$

5. **State.** Phil's average fat intake for the vacation will not exceed 65 g per day if he consumes no more than 48 g of fat on the fourth day.

Exercise Set 2.7

Translate to an inequality.

1. A number is at least 7.

2. A number is greater than or equal to 5.

3. The baby weighs more than 2 kilograms (kg).

4. Between 75 and 100 people attended the concert.

5. The speed of the train was between 90 and 110 mph.

6. At least 400,000 people attended the Million Man March.

7. At most 1,200,000 people attended the Million Man March.

8. The amount of acid is not to exceed 40 liters (L).

9. The cost of gasoline is no less than $1.50 per gallon.

10. The temperature is at most $-2°C$.

Use an inequality and the five-step process to solve each problem.

11. *Blueprints.* To make copies of blueprints, Vantage Reprographics charges a $5 setup fee plus $4 per copy. Myra can spend no more than $65 for the copying. What numbers of copies will allow her to stay within budget?

12. *Banquet costs.* The women's volleyball team can spend at most $450 for its awards banquet at a local restaurant. If the restaurant charges a $40 setup fee plus $16 per person, at most how many can attend?

13. *Truck rentals.* Ridem rents trucks at a daily rate of $42.95 plus $0.46 per mile. The Letsons want a one-day truck rental, but must stay within a budget of $200. What mileages will allow them to stay within budget? Round to the nearest tenth of a mile.

14. *Phone costs.* Simon claims that it costs him at least $3.00 every time he calls an overseas customer. If his typical call costs 75¢ plus 45¢ for each minute, how long do his calls typically last?

15. *Parking costs.* Laura is certain that every time she parks in the municipal garage it costs her at least $2.20. If the garage charges 45¢ plus 25¢ for each half hour, for how long is Laura's car generally parked?

16. *Furnace repairs.* RJ's Plumbing and Heating charges $25 plus $30 per hour for emergency service. Gary remembers being billed over $100 for an emergency call. How long was RJ's there?

17. *College tuition.* Angelica's financial aid stipulates that her tuition not exceed $1000. If her local community college charges a $35 registration fee plus $375 per course, what is the greatest number of courses for which Angelica can register?

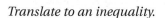

Tuition, Fees, and Refunds

Fees and Refunds Policy
Fees are due at the time of registration.

Tuition	
Registration	$375 per course
Late Registration	$35
Additional Drop/Add	$25
Late Payment	$25
	1% of past due balance

Credit Card Policy
Students ... wish to use a credit card to pay their ... re responsible for supplying valid ... ation on the Registration Form. If ... t, the registration is not complete ...

... 's Office will inform the student that the ... as been declined. In order to complete the ... tion process, the student must then make ... ative arrangements for payment. If the declined ... ge results in a late payment, a late fee may be ...

Tuition Refunds
The following schedule is used to determine the portion of tuition that will be refunded, depending upon the date that a student withdraws from a course or from the college.

Prior to the start of classes	100%
Before the second class meeting	90%
Before the fourth class meeting	25%
After the fourth class meeting	0%

18. *Van rentals.* Atlas rents a cargo van at a daily rate of $44.95 plus $0.39 per mile. A business has budgeted $250 for a one-day van rental. What mileages will allow the business to stay within budget? (Round to the nearest tenth of a mile.)

19. *Grade average.* Nadia is taking a literature course in which four tests are given. To get a B, a student must average at least 80 on the four tests. Nadia scored 82, 76, and 78 on the first three tests. What scores on the last test will earn her at least a B?

20. *Quiz average.* Rod's quiz grades are 73, 75, 89, and 91. What scores on a fifth quiz will make his average quiz grade at least 85?

21. *Nutrition.* Following the guidelines of the Food and Drug Administration, Dale tries to eat at least 5 servings of fruits or vegetables each day. For the first six days of one week, she had 4, 6, 7, 4, 6, and 4 servings. How many servings of fruits or

vegetables should Dale eat on Saturday, in order to average at least 5 servings per day for the week?

22. *College course load.*   To remain on financial aid, Millie needs to complete an average of at least 7 credits per quarter each year. In the first three quarters of 2001, Millie completed 5, 7, and 8 credits. How many credits of course work must Millie complete in the fourth quarter if she is to remain on financial aid?

23. *Music lessons.*   Band members at Colchester Middle School are expected to average at least 20 min of practice time per day. One week Monroe practiced 15 min, 28 min, 30 min, 0 min, 15 min, and 25 min. How long must he practice on the seventh day if he is to meet expectations?

24. *Electrician visits.*   Dot's Electric made 17 customer calls last week and 22 calls this week. How many calls must be made next week in order to maintain an average of at least 20 for the three-week period?

25. *Perimeter of a rectangle.*   The width of a rectangle is fixed at 8 ft. What lengths will make the perimeter at least 200 ft? at most 200 ft?

26. *Perimeter of a triangle.*   One side of a triangle is 2 cm shorter than the base. The other side is 3 cm longer than the base. What lengths of the base will allow the perimeter to be greater than 19 cm?

27. *Perimeter of a pool.*   The perimeter of a rectangular swimming pool is not to exceed 70 ft. The length

is to be twice the width. What widths will meet these conditions?

28. *Volunteer work.*   George and Joan do volunteer work at a hospital. Joan worked 3 more hr than George, and together they worked more than 27 hr. What possible numbers of hours did each work?

29. *Cost of road service.*   Rick's Automotive charges $50 plus $15 for each (15-min) unit of time when making a road call. Twin City Repair charges $70 plus $10 for each unit of time. Under what circumstances would it be more economical for a motorist to call Rick's?

30. *Cost of clothes.*   Angelo is shopping for a new pair of jeans and two sweaters of the same kind. He is determined to spend no more than $120.00 for the clothes. He buys jeans for $21.95. How much can Angelo spend for each sweater?

31. *Area of a rectangle.*   The width of a rectangle is fixed at 4 cm. For what lengths will the area be less than 86 cm²?

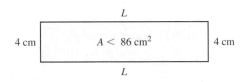

32. *Area of a rectangle.*   The width of a rectangle is fixed at 16 yd. For what lengths will the area be at least 264 yd²?

33. *Insurance-covered repairs.*   Most insurance companies will replace a vehicle if an estimated repair exceeds 80% of the "blue-book" value of the vehicle. Michelle's insurance company paid $8500 for repairs to her Subaru after an accident. What can be concluded about the blue-book value of the car?

34. *Insurance-covered repairs.*   Following an accident, Jeff's Ford pickup was replaced by his insurance company because the damage was so extensive. Before the damage, the blue-book value of the truck was $21,000. How much would it have cost to repair the truck? (See Exercise 33.)

35. *Body temperature.*   A person is considered to be feverish when his or her temperature is higher than 98.6°F. The formula $F = \frac{9}{5}C + 32$ can be used to convert Celsius temperatures C to Fahrenheit temperatures F. For which Celsius temperatures is a person considered feverish?

36. *Melting butter.*   Butter stays solid at Fahrenheit temperatures below 88°. Use the formula in Exercise 35 to determine those Celsius temperatures for which butter stays solid.

37. *Fat content in foods.*   Reduced Fat Skippy® peanut butter contains 12 g of fat per serving. In order for a food to be labeled "reduced fat," it must have at least 25% less fat than the regular item. What can you conclude about the number of grams of fat in a serving of the regular Skippy peanut butter?

38. *Fat content in foods.*   Reduced Fat Chips Ahoy!® cookies contain 5 g of fat per serving. What can you conclude about the number of grams of fat in regular Chips Ahoy! cookies (see Exercise 37)?

39. *Well drilling.*   All Seasons Well Drilling offers two plans. Under the "pay-as-you-go" plan, they charge $500 plus $8 a foot for a well of any depth. Under their "guaranteed-water" plan, they charge a flat fee of $4000 for a well that is guaranteed to provide adequate water for a household. For what depths would it save a customer money to use the pay-as-you-go plan?

40. *Running.*   In the course of a week, Tony runs 6 mi, 3 mi, 5 mi, 5 mi, 4 mi, and 4 mi. How far should his next run be if he is to average at least 5 mi per day?

41. *Track records.*   The formula $R = -0.012t + 20.8$, where R is in seconds, can be used to predict the world record in the 200-m dash t years after 1920. For what years will the world record be less than 19.8 sec?

42. *Track records.*   The formula $R = -0.0084t + 3.85$, where R is in minutes, can be used to predict the world record in the 1500-m run t years after 1930. For what years will the world record be less than 3.22 min?

43. *Pond depth.*   On July 1, Garrett's Pond was 25 ft deep. Since that date, the water level has dropped $\frac{2}{3}$ ft per week. For what dates will the water level not exceed 21 ft?

44. *Weight gain.*   A 9-lb puppy is gaining weight at a rate of $\frac{3}{4}$ lb per week. When will the puppy's weight exceed $22\frac{1}{2}$ lb?

45. *Area of a triangular flag.*   As part of an outdoor education course, Wanda needs to make a bright-colored triangular flag with an area of at least 3 ft². What heights can the triangle be if the base is $1\frac{1}{2}$ ft?

46. *Area of a triangular sign.*   Zoning laws in Harrington prohibit displaying signs with areas exceeding 12 ft². If Flo's Marina is ordering a triangular sign with an 8-ft base, how tall can the sign be?

47. *Toll charges.*   The equation $y = 0.027x + 0.19$ can be used to determine the approximate cost y, in dollars, of driving x miles on the Indiana toll road. For what mileages x will the cost be at most $6?

48. *Price of a movie ticket.*   The average price of a movie ticket can be estimated by the equation $P = 0.1522Y - 298.592$, where Y is the year and P is the average price, in dollars. The price is lower than what might be expected due to senior-citizen discounts, children's prices, and special volume discounts. For what years will the average price of a movie ticket be at least $6? (Include the year in which the $6 ticket first occurs.)

49. If f represents Fran's age and t represents Todd's age, write a sentence that would translate to $t + 3 < f$.

50. Explain how the meanings of "Five more than a number" and "Five is more than a number" differ.

SKILL MAINTENANCE

Simplify.

51. $\dfrac{9 - 5}{6 - 4}$

52. $\dfrac{8 - 5}{12 - 6}$

53. $\dfrac{8 - (-2)}{1 - 4}$

54. $\dfrac{7 - 9}{4 - (-6)}$

SYNTHESIS

55. Write a problem for a classmate to solve. Devise the problem so the answer is "The Rothmans can drive 90 mi without exceeding their truck rental budget."

56. Write a problem for a classmate to solve. Devise the problem so the answer is "At most 18 passengers can go on the boat." Design the problem so that at least one number in the solution must be rounded down.

57. *Parking fees.* Mack's Parking Garage charges $4.00 for the first hour and $2.50 for each additional hour. For how long has a car been parked when the charge exceeds $16.50?

58. *Ski wax.* Green ski wax works best between 5° and 15° Fahrenheit. Determine those Celsius temperatures for which green ski wax works best. (See Exercise 35.)

Aha! **59.** The area of a square can be no more than 64 cm². What lengths of a side will allow this?

Aha! **60.** The sum of two consecutive odd integers is less than 100. What is the largest pair of such integers?

61. *Nutritional standards.* In order for a food to be labeled "lowfat," it must have fewer than 3 g of fat per serving. Reduced fat Tortilla Pops® contain 60% less fat than regular nacho cheese tortilla chips, but still cannot be labeled lowfat. What can you conclude about the fat content of a serving of nacho cheese tortilla chips?

62. *Parking fees.* When asked how much the parking charge is for a certain car (see Exercise 57), Mack replies "between 14 and 24 dollars." For how long has the car been parked?

63. Alice's Books allows customers to select one free book for every 10 books purchased. The price of that book cannot exceed the average cost of the 10 books. Neoma has bought 9 books that average $12 per book. How much should her tenth book cost if she wants to select a $15 book for free?

64. After 9 quizzes, Blythe's average is 84. Is it possible for Blythe to improve her average by two points with the next quiz? Why or why not?

65. Arnold and Diaz Booksellers offers a preferred-customer card for $25. The card entitles a customer to a 10% discount on all purchases for a period of one year. Under what circumstances would an individual save money by purchasing a card?

CORNER

Calling Plans

Focus: Problem solving and inequalities

Time: 20 minutes

Group size: 4

Materials: Calculators

A recent ad for "Five Line" is represented below.

	Rate per minute	Monthly fee	Distance	Restrictions
MCI 5¢ Everyday	25¢ in the day, 5¢ evening and weekends	$1.95	State-to-state	Minimum monthly bill of $5.00
Five Line	5¢	None	State-to-state AND in-state	Each completed call costs a minimum of 50¢.
AT&T Seven Sense	7¢	$5.95 ($4.95 with internet billing)	State-to-state	
Sprint Nickel Nights	10¢ in the day, 5¢ in the evening	$5.95	State-to-state	

ACTIVITY

The following table lists one month of Kate's calls. There are 34 calls for a total of 173 minutes.

Length of call (in minutes)	Number of calls of given length
1	13
2	6
3	5
4	3
5	2
9	1
16	1
18	1
33	1
35	1

1. Assume all calls are state-to-state calls. Each group member should choose a different one of the four plans and compute the bill (not including taxes and other fees) for each of the following situations.
   a) All 34 calls are evening (off-peak) calls.
   b) All 34 calls are daytime (peak) calls.
2. Is there a best plan for evening phone use? Is there a best plan for daytime phone use?
3. For each plan, describe the type of caller, if one exists, for whom that plan works best.
4. How can inequalities be used to compare these plans?

Summary and Review 2

Key Terms

Important Properties and Formulas

Solving Equations

Addition principle:

For any real numbers a, b, and c,

$a = b$ is equivalent to $a + c = b + c$.

Multiplication principle:

For any real numbers a, b, and c, with $c \neq 0$,

$a = b$ is equivalent to $a \cdot c = b \cdot c$.

Solving Inequalities

Addition principle:

For any real numbers a, b, and c,

$a < b$ is equivalent to $a + c < b + c$;
$a > b$ is equivalent to $a + c > b + c$.

Multiplication principle:

For any real numbers a and b and any *positive* number c,

$a < b$ is equivalent to $ac < bc$;
$a > b$ is equivalent to $ac > bc$.

For any real numbers a and b and any *negative* number c,

$a < b$ is equivalent to $ac > bc$;
$a > b$ is equivalent to $ac < bc$.

Similar statements hold for $\leq$ and $\geq$.

An Equation-Solving Procedure

1. Use the multiplication principle to clear any fractions or decimals. (This is optional, but can ease computations.)
2. If necessary, use the distributive law to remove parentheses. Then combine like terms on each side.
3. Use the addition principle, as needed, to get all variable terms on one side and all constant terms on the other.
4. Combine like terms again, if necessary.
5. Multiply or divide to solve for the variable, using the multiplication principle.
6. Check all possible solutions in the original equation.

To Solve a Formula for a Given Letter

1. If the letter for which you are solving appears in a fraction, use the multiplication principle to clear fractions.
2. Get all terms with the letter for which you are solving on one side of the equation and all other terms on the other side.
3. Combine like terms, if necessary. This may require factoring.
4. Multiply or divide to solve for the letter in question.

Percent Notation

$n\%$ means $\dfrac{n}{100}$, or $n \times \dfrac{1}{100}$, or $n \times 0.01$.

Five Steps for Problem Solving in Algebra

1. *Familiarize* yourself with the problem.
2. *Translate* to mathematical language. (This often means writing an equation.)
3. *Carry out* some mathematical manipulation. (This often means *solving* an equation.)
4. *Check* your possible answer in the original problem.
5. *State* the answer clearly.

Review Exercises

Solve.

1. $x + 9 = -16$

2. $-8x = -56$

3. $-\dfrac{x}{4} = 17$

4. $n - 7 = -6$

5. $15x = -60$

6. $x - 0.1 = 1.01$

7. $-\frac{2}{3} + x = -\frac{1}{6}$

8. $\frac{4}{5}y = -\frac{3}{16}$

9. $5z + 3 = 41$

10. $5 - x = 13$

11. $5t + 9 = 3t - 1$

12. $7x - 6 = 25x$

13. $\frac{1}{4}x - \frac{5}{8} = \frac{3}{8}$

14. $14y = 23y - 17 - 10$

15. $0.22y - 0.6 = 0.12y + 3 - 0.8y$

16. $\frac{1}{4}x - \frac{1}{8}x = 3 - \frac{1}{16}x$

17. $3(x + 5) = 36$

18. $4(5x - 7) = -56$

19. $8(x - 2) = 5(x + 4)$

20. $-5x + 3(x + 8) = 16$

Solve each formula for the given letter.

21. $C = \pi d$, for d

22. $V = \dfrac{1}{3}Bh$, for B

23. $A = \dfrac{a + b}{2}$, for a

24. Find decimal notation: 0.9%.

25. Find percent notation: $\frac{11}{25}$.

26. What percent of 60 is 12?

27. 42 is 30% of what number?

Determine whether the given number is a solution of the inequality $x \le 4$.

28. -3

29. 7

30. 4

Graph on a number line.

31. $5x - 6 < 2x + 3$

32. $-2 < x \le 5$

33. $y > 0$

Solve. Write the answers in set-builder notation.

34. $t + \frac{2}{3} \ge \frac{1}{6}$

35. $9x \ge 63$

36. $2 + 6y > 20$

37. $7 - 3y \ge 27 + 2y$

38. $3x + 5 < 2x - 6$

39. $-4y < 28$

40. $3 - 4x < 27$

41. $4 - 8x < 13 + 3x$

42. $-3y \ge -36$

43. $-4x \le \frac{1}{3}$

Solve.

44. An ink jet printer sold for $139 in July. This was $28 less than the cost in February. Find the cost in February.

45. A can of powdered infant formula makes 120 oz of formula. How many 6-oz bottles of formula will the can make?

46. A 12-ft "two by four" is cut into two pieces. One piece is 2 ft longer than the other. How long are the pieces?

47. About 52% of all charitable contributions are made to religious organizations. In 1997, $75 billion was given to religious organizations (*Source: Statistical Abstract of the United States*, 1999). How much was given to charities in general?

48. The sum of two consecutive odd integers is 116. Find the integers.

49. The perimeter of a rectangle is 56 cm. The width is 6 cm less than the length. Find the width and the length.

50. After a 25% reduction, a picnic table is on sale for $120. What was the regular price?

51. In 1997, the average male with a bachelor's degree or above earned about $66,000. This was 57% more than the average woman with a comparable background. How much did the average woman with a comparable background earn?

52. The measure of the second angle of a triangle is 50° more than that of the first. The measure of the third angle is 10° less than twice the first. Find the measures of the angles.

53. Jason has budgeted an average of $95 a month for entertainment. For the first five months of the year, he has spent $98, $89, $110, $85, and $83. How much can Jason spend in the sixth month without exceeding his average budget?

54. The length of a rectangle is 43 cm. For what widths is the perimeter greater than 120 cm?

55. How does the multiplication principle for equations differ from the multiplication principle for inequalities?

SYNTHESIS

56. Explain how checking the solutions of an equation differs from checking the solutions of an inequality.

57. The combined length of the Nile and Amazon Rivers is 13,108 km (*Source: Statistical Abstract of the United States, 1999*). If the Amazon were 234 km longer, it would be as long as the Nile. Find the length of each river.

58. Consumer experts advise us never to pay the sticker price for a car. A rule of thumb is to pay the sticker price minus 20% of the sticker price, plus $200. A car is purchased for $15,080 using the rule. What was the sticker price?

Solve.

59. $2|n| + 4 = 50$

60. $|3n| = 60$

61. $y = 2a - ab + 3$, for a

Chapter Test 2

Solve.

1. $x + 8 = 17$

2. $t - 3 = 12$

3. $3x = -18$

4. $-\frac{4}{7}x = -28$

5. $3t + 7 = 2t - 5$

6. $\frac{1}{2}x - \frac{3}{5} = \frac{2}{5}$

7. $8 - y = 16$

8. $-\frac{2}{5} + x = -\frac{3}{4}$

9. $3(x + 2) = 27$

10. $-3x + 6(x + 4) = 9$

11. $\frac{5}{6}(3x + 1) = 20$

Solve. Write the answers in set-builder notation.

12. $x + 6 > 1$

13. $14x + 9 > 13x - 4$

14. $\frac{1}{3}x < \frac{7}{8}$

15. $-2y \geq 26$

16. $4y \leq -32$

17. $-5x \geq \frac{1}{4}$

18. $4 - 6x > 40$

19. $5 - 9x \geq 19 + 5x$

Solve each formula for the given letter.

20. $A = 2\pi rh$, for r

21. $w = \dfrac{P + l}{2}$, for l

22. Find decimal notation: 230%.

23. Find percent notation: 0.054.

24. What number is 32% of 50?

25. What percent of 75 is 33?

Graph on a number line.

26. $y < 4$

27. $-2 \leq x \leq 2$

Solve.

28. The perimeter of a rectangle is 36 cm. The length is 4 cm greater than the width. Find the width and the length.

29. Kari is taking a 240-mi bicycle trip through Vermont. She has three times as many miles to go as she has already ridden. How many miles has she biked so far?

30. The perimeter of a triangle is 249 mm. If the sides are consecutive odd integers, find the length of each side.

31. Wholesale tea sales in 1999 were 255% of what they were in 1990 (*Source: Burlington Free Press, 2/14/00*). In 1990, there was $1.8 billion in wholesale tea sales. How much tea was sold wholesale in 1999?

32. Find all numbers for which six times the number is greater than the number plus 30.

33. The width of a rectangular ballfield is 96 yd. Find all possible lengths so that the perimeter of the ballfield will be at least 540 yd.

SYNTHESIS

Solve.

34. $c = \dfrac{2cd}{a - d}$, for d

35. $3|w| - 8 = 37$

36. A movie theater had a certain number of tickets to give away. Five people got the tickets. The first got one third of the tickets, the second got one fourth of the tickets, and the third got one fifth of the tickets. The fourth person got eight tickets, and there were five tickets left for the fifth person. Find the total number of tickets given away.

3

Introduction to Graphing

AN APPLICATION

As part of an ill-fated expedition to climb Mt. Everest, the world's tallest peak, author Jon Krakauer departed "The Balcony," elevation 27,600 ft, at 7:00 A.M. Krakauer reached the summit, elevation 29,028 ft, at 1:25 P.M. (*Source*: Krakauer, Jon, *Into Thin Air, the Illustrated Edition*. New York: Random House, 1998) Determine Krakauer's average rate of ascent, in feet per minute and in minutes per foot.

This problem appears as Exercise 11 in Section 3.4.

*I*n expeditions, we are always converting between feet and meters. Also, when we are working in exotic markets with parallel markets, we must often calculate the percentage difference between, say, a 65-rupee exchange rate and an 80-rupee exchange rate.

ALAN BURGESS
Mountain Expedition Guide
Salt Lake City, Utah

W*e now begin our study of graphing. First we will examine graphs as they commonly appear in newspapers or magazines and develop some terminology. Following that, we will graph certain equations and study the connection between rate and slope. We will also learn how graphs can be used as a problem-solving tool for certain applications.*

Our work in this chapter centers on solving equations that contain two variables.

Reading Graphs, Plotting Points, and Estimating Values

3.1

Problem Solving with Graphs • Points and Ordered Pairs • Estimations and Predictions

Today's print and electronic media make almost constant use of graphs. This can be attributed to the widespread availability of graphing software and the large quantity of information that a graph can display. In this section, we consider problem solving with bar graphs, line graphs, and circle graphs. Then we examine graphs that use a coordinate system.

Problem Solving with Graphs

A *bar graph* is a convenient way of showing comparisons. In every bar graph, certain categories, such as body weight in the example below, are paired with certain numbers.

Example 1

Driving under the influence. A blood-alcohol level of 0.08% or higher makes driving illegal in the United States. This bar graph shows how many drinks a person of a certain weight would need to consume in 1 hr to achieve a blood-alcohol level of 0.08% (*Source*: Adapted from soberup.com and vsa.vassar.edu/~source/drugs/alcohol.html). Note that a 12-oz beer, a 5-oz glass of wine, or a cocktail containing $1\frac{1}{2}$ oz of distilled liquor all count as one drink.

a) Approximately how many drinks would a 200-lb person have consumed if he or she had a blood-alcohol level of 0.08%?

b) What can be concluded about the weight of someone who can consume 3 drinks in an hour without reaching a blood-alcohol level of 0.08%?

Solution

a) We go to the top of the bar that is above the body weight 200 lb. Then we move horizontally from the top of the bar to the vertical scale listing numbers of drinks. It appears that approximately 4 drinks will give a 200-lb person a blood-alcohol level of 0.08%.

b) By moving up the vertical scale to the number 3, and then moving horizontally, we see that the first bar to reach a height of 3 corresponds to a weight of 140 lb. Thus an individual should weigh over 140 lb if he or she wishes to consume 3 drinks in an hour without exceeding a blood-alcohol level of 0.08%.

Circle graphs, or *pie charts*, are often used to show what percent of the whole each particular item in a group represents.

E x a m p l e   2

Color preference.   The circle graph below shows the favorite colors of Americans and the percentage that prefers each color (*Source*: Vitaly Komar and Alex Melamid: The Most Wanted Paintings on the Web). Because of rounding, the total is slightly less than 100%. There are approximately 272 million Americans, and three quarters of them live in cities. How many urban Americans choose red as their favorite color?

What is Your Favorite Color?

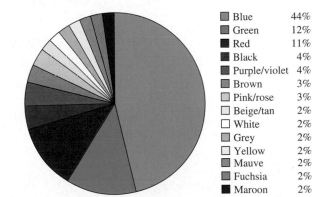

Blue	44%
Green	12%
Red	11%
Black	4%
Purple/violet	4%
Brown	3%
Pink/rose	3%
Beige/tan	2%
White	2%
Grey	2%
Yellow	2%
Mauve	2%
Fuchsia	2%
Maroon	2%

Solution

1. Familiarize.   The problem involves percents, so if we were unsure of how to solve percent problems, we might review Section 2.4. We are told that three quarters of all Americans live in cities. Since there are about 272 million Americans, this amounts to

$$\frac{3}{4} \cdot 272, \quad \text{or 204 million urban Americans.}$$

The chart indicates that 11% of the U.S. population prefers red. We let r = the number of urban Americans who select red as their favorite color.

For the purpose of this problem, we will assume that the color preferences of urban Americans are typical of all Americans.

2. **Translate.**   We reword and translate the problem as follows:

   Rewording: What is 11% of 204 million?

   Translating: r = 11% · 204,000,000

3. **Carry out.**   We solve the equation:

   $r = 0.11 · 204,000,000 = 22,440,000.$

4. **Check.**   The check is left to the student.

5. **State.**   About 22,440,000 urban Americans choose red as their favorite color.

E x a m p l e   3

Exercise and pulse rate.   The following line graph shows the relationship between a person's resting pulse rate and months of regular exercise.*

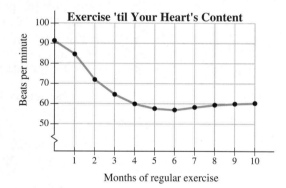

Exercise 'til Your Heart's Content

Beats per minute

Months of regular exercise

a) How many months of regular exercise are required to lower the pulse rate as much as possible?
b) How many months of regular exercise are needed to achieve a pulse rate of 65 beats per minute?

Solution

a) The lowest point on the graph occurs above the number 6. Thus, after 6 months of regular exercise, the pulse rate is lowered as much as possible.
b) We locate 65 on the vertical scale and then move right until the line is reached. At that point, we move down to the horizontal scale and read the information we are seeking.

---

*Data from *Body Clock* by Dr. Martin Hughes (New York: Facts on File, Inc.), p. 60.

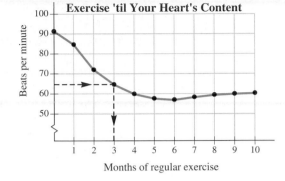

The pulse rate is 65 beats per minute after 3 months of regular exercise.

Points and Ordered Pairs

The line graph in Example 3 contains a collection of points. Each point pairs up a number of months of exercise with a pulse rate. To create such a graph, we **graph**, or **plot**, pairs of numbers on a plane. This is done using two perpendicular number lines called **axes** (pronounced "ak-sēz"; singular, **axis**). The point at which the axes cross is called the **origin**. Arrows on the axes indicate the positive directions.

Consider the pair $(3, 4)$. The numbers in such a pair are called **coordinates**. The **first coordinate** in this case is 3 and the **second coordinate** is 4. To plot $(3, 4)$, we start at the origin, move horizontally to the 3, move up vertically 4 units, and then make a "dot." Thus, $(3, 4)$ is located above 3 on the first axis and to the right of 4 on the second axis.

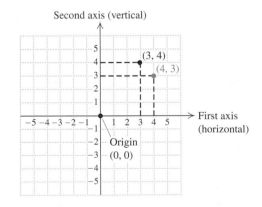

The point $(4, 3)$ is also plotted in the figure above. Note that $(3, 4)$ and $(4, 3)$ are different points. For this reason, coordinate pairs are called **ordered pairs**—the order in which the numbers appear is important.

E x a m p l e   4

Plot the point $(-3, 4)$.

Solution   The first number, -3, is negative. Starting at the origin, we move 3 units in the negative horizontal direction (3 units to the left). The second number, 4, is positive, so we move 4 units in the positive vertical direction (up). The point $(-3, 4)$ is above -3 on the first axis and to the left of 4 on the second axis.

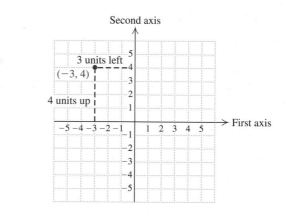

To find the coordinates of a point, we see how far to the right or left of the origin the point is and how far above or below the origin it is. Note that the coordinates of the origin itself are $(0, 0)$.

E x a m p l e   5

Find the coordinates of points A, B, C, D, E, F, and G.

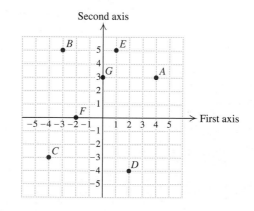

Solution   Point A is 4 units to the right of the origin and 3 units above the origin. Its coordinates are $(4, 3)$. The coordinates of the other points are as follows:

B: $(-3, 5)$; C: $(-4, -3)$; D: $(2, -4)$;

E: $(1, 5)$; F: $(-2, 0)$; G: $(0, 3)$.

The horizontal and vertical axes divide the plane into four regions, or **quadrants**, as indicated by Roman numerals in the following figure. In region I (the *first quadrant*), both coordinates of any point are positive. In region II (the *second quadrant*), the first coordinate is negative and the second is positive. In region III (the *third quadrant*), both coordinates are negative. In region IV (the *fourth quadrant*), the first coordinate is positive and the second is negative.

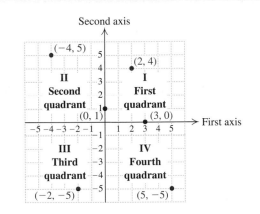

Note that the point $(-4, 5)$ is in the second quadrant and the point $(5, -5)$ is in the fourth quadrant. The points $(3, 0)$ and $(0, 1)$ are on the axes and are not considered to be in any quadrant.

Estimations and Predictions

It is possible to use line graphs to estimate real-life quantities that are not already known. To do so, we approximate the coordinates of an unknown point by using two points with known coordinates. When the unknown point is located *between* the two points, this process is called **interpolation**.* Sometimes a graph passing through the known points is *extended* to predict future values. Making predictions in this manner is called **extrapolation**.*

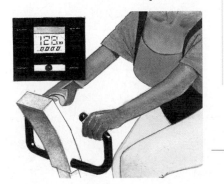

E x a m p l e   6

Aerobic exercise.   A person's target heart rate is the number of beats per minute that bring the most aerobic benefit to his or her heart. The target heart rate for a 20-year-old is 150 beats per minute and for a 60-year-old, 120 beats per minute.

a) Estimate the target heart rate for a 35-year-old.
b) Predict what the target heart rate would be for a 75-year-old.

*Both interpolation and extrapolation can be performed using more than two known points and using curves other than lines.

Solution

a) We first draw a horizontal axis for "Age" and a vertical axis for "Target heart rate" on a piece of graph paper. Next, we number the axes, using a scale that will permit us to view both the given and the desired data. The given information allows us to then plot (20, 150) and (60, 120).

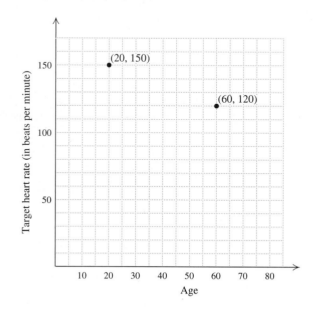

Next, we draw a line segment connecting the points. To estimate the target heart rate for a 35-year-old, we locate 35 on the horizontal axis. From there we move vertically to a point on the line and then left to the other axis. We see then that the target heart rate for a 35-year-old is approximately 140 beats per minute.

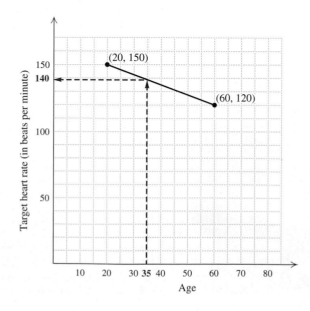

b) To predict the target heart rate for a 75-year-old, we extend the line segment on the graph until it is above 75 on the horizontal axis. Next, we proceed as in part (a) above, moving vertically to the line and then to the vertical axis. We predict that the target heart rate for a 75-year-old is approximately 110 beats per minute.

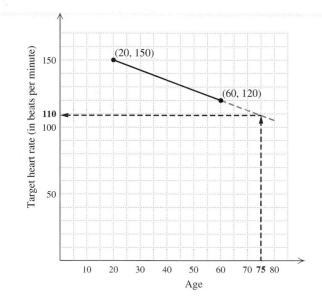

Exercise Set 3.1

FOR EXTRA HELP

Digital Video Tutor CD 2
Videotape 5

InterAct Math

Math Tutor Center

MathXL

MyMathLab.com

Blood alcohol level.   *Use the bar graph in Example 1 to answer Exercises 1–4.*

1. Approximately how many drinks would a 100-lb person have consumed in 1 hr to reach a blood-alcohol level of 0.08%?

2. Approximately how many drinks would a 160-lb person have consumed in 1 hr to reach a blood-alcohol level of 0.08%?

3. What can you conclude about the weight of someone who has consumed 4 drinks in 1 hr without reaching a blood-alcohol level of 0.08%?

4. What can you conclude about the weight of someone who has consumed 5 drinks in 1 hr without reaching a blood-alcohol level of 0.08%?

Favorite color.   *Use the information in Example 2 to answer Exercises 5–8.*

5. About one third of all Americans live in the South. How many Southerners choose brown as their favorite color?

6. About 50% of all Americans are at least 35 years old. How many Americans who are 35 or older choose purple/violet as their favorite color?

7. About one eighth of all Americans are senior citizens. How many senior citizens choose black as their favorite color?

8. About one fourth of all Americans are minors (under 18 years old). How many minors choose yellow as their favorite color?

Sorting solid waste.   *Use the following pie chart to answer Exercises 9–12.*

Sorting Solid Waste

Paper and cardboard 38.6%

Yard waste 12.8%

Food waste 10.1%

Metals 7.7%

Glass 5.5%

Plastics 9.9%

Wood 5.3%

Other 10.1%

Source: Statistical Abstract of the United States, 1999

9. In 1998, Americans generated 210 million tons of waste. How much of the waste was plastic?

10. In 2000, the average American generated 4.4 lb of waste per day. How much of that was paper and cardboard?

11. Americans are recycling about 26% of all glass that is in the waste stream. How much glass did Americans recycle in 1998? (See Exercise 9.)

12. Americans are recycling about 5% of all plastic waste. How much plastic does the average American recycle each day? (Use the information in Exercise 10.)

Recorded music.   *Use the following line graphs to answer Exercises 13–18. The graphs show the percentage of all recordings sold that were CD or cassette. (Source: Recording Industry Association of America, Washington, D.C.)*

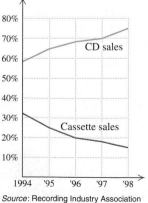

80%
70%
60%
50%
40%
30%
20%
10%

CD sales

Cassette sales

1994   '95   '96   '97   '98

Source: Recording Industry Association of America, Washington, D.C.

13. Approximately what percent of the recordings sold in 1997 were CDs?

14. Approximately what percent of the recordings sold in 1996 were cassettes?

15. In what year were approximately 25% of the recordings sold as cassettes?

16. In what year were approximately 75% of the recordings sold as CDs?

17. In what year did sales of CDs increase the most?

18. In what year did sales of cassettes change the least?

Plot each group of points.

19. $(1, 2), (-2, 3), (4, -1), (-5, -3), (4, 0), (0, -2)$

20. $(-2, -4), (4, -3), (5, 4), (-1, 0), (-4, 4), (0, 5)$

21. $(4, 4), (-2, 4), (5, -3), (-5, -5), (0, 4), (0, -4),$ $(3, 0), (-4, 0)$

22. $(2, 5), (-1, 3), (3, -2), (-2, -4), (0, 4), (0, -5),$ $(5, 0), (-5, 0)$

In Exercises 23–26, find the coordinates of points A, B, C, D, and E.

23.

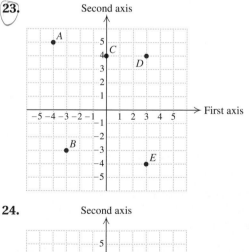

Second axis

First axis

24.

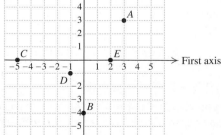

Second axis

First axis

25.

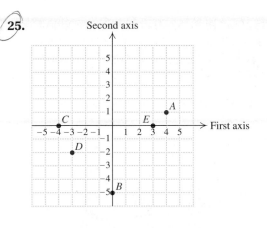

26.

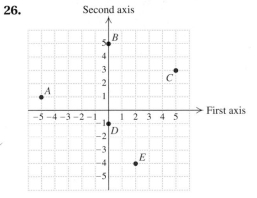

In which quadrant is each point located?

27. $(7, -2)$

28. $(-1, -4)$

29. $(-4, -3)$

30. $(1, -5)$

31. $(2, 1)$

32. $(-4, 6)$

33. $(-4.9, 8.3)$

34. $(7.5, 2.9)$

35. In which quadrants are the first coordinates positive?

36. In which quadrants are the second coordinates negative?

37. In which quadrants do both coordinates have the same sign?

38. In which quadrants do the first and second coordinates have opposite signs?

39. *Birth rate among teenagers.* The birth rate among teenagers, measured in births per 1000 females age 15–19, fell steadily from 62.1 in 1991 to 49.6 in 1999 (*Source*: National Center for Health Statistics; 1999 figure is preliminary).

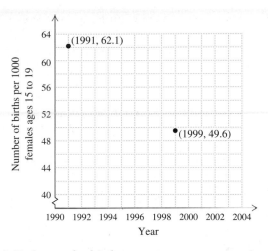

a) Estimate the birth rate among teenagers in 1995.

b) Predict the birth rate among teenagers in 2003.

40. *Food-stamp program participation.* Due to changing rules and a booming economy, monthly participation in the U.S. food-stamp program dropped from approximately 27.5 million people in 1994 to approximately 20 million in 1999 (*Source*: U.S. Department of Agriculture).

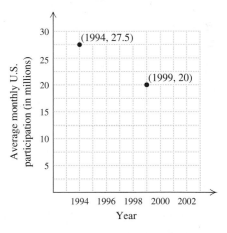

a) Estimate the number of participants in 1997.

b) Predict the number of participants in 2003.

In Exercises 41–46, display the given information in a graph similar to those used in Exercises 39 and 40. Use the horizontal axis to represent time. Then answer parts (a) and (b).

41. *Cigarette smoking.* The percentage of people age 26 to 34 who smoke has dropped from 45.7% in 1985 to 32.5% in 1998 (*Source*: *The World Almanac and Book of Facts 2000*, 1999).

a) Approximate the percentage of people age 26–34 who smoked in 1990.

b) Predict the percentage of people age 26–34 who will smoke in 2003.

42. *Cigarette smoking.* The percentage of people age 18–25 who smoke has changed from 47.4% in 1985 to 41.8% in 1998 (*Source: The World Almanac and Book of Facts 2000*, 1999).

a) Approximate the percentage of people age 18–25 who smoked in 1990.

b) Predict the percentage of people age 18–25 who will smoke in 2003.

43. *College enrollment.* U.S. college enrollment has grown from approximately 60.3 million in 1990 to 68.3 million in 2000 (*Source: The World Almanac and Book of Facts 2000*, 1999).

a) Approximate the U.S. college enrollment for 1996.

b) Predict the U.S. college enrollment for 2005.

44. *High school enrollment.* U.S. high school enrollment has changed from approximately 12.5 million in 1990 to 14.9 million in 2000 (*Source: The World Almanac and Book of Facts 2000*, 1999).

a) Approximate the U.S. high school enrollment for 1996.

b) Predict the U.S. high school enrollment for 2005.

45. *Aging baby boomers.* The number of U.S. residents over the age of 65 was approximately 31 million in 1990 and 34.4 million in 2000. (*Source: Statistical Abstract of the United States*).

Aha! **a)** Estimate the number of U.S. residents over the age of 65 in 1995.

b) Predict the number of U.S. residents over the age of 65 in 2010.

46. *Urban population.* The percentage of the U.S. population that resides in metropolitan areas increased from about 78% in 1980 to about 80% in 1996.

a) Estimate the percentage of the U.S. population residing in metropolitan areas in 1992.

b) Predict the percentage of the U.S. population residing in metropolitan areas in 2008.

47. What do all of the points on the vertical axis of a graph have in common?

48. The following graph was included in a mailing sent by Agway® to their oil customers in 2000. What information is missing from the graph and why is the graph misleading?

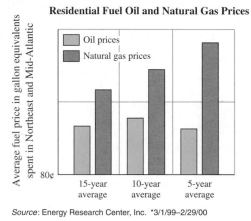

Source: Energy Research Center, Inc. *3/1/99–2/29/00

SKILL MAINTENANCE

Simplify.

49. $4 \cdot 3 - 6 \cdot 5$

50. $5(-2) + 3(-7)$

51. $-\frac{1}{2}(-6) + 3$

52. $-\frac{2}{3}(-12) - 7$

Solve for y.

53. $3x - 2y = 6$

54. $7x - 4y = 14$

SYNTHESIS

55. Describe what the result would be if the first and second coordinates of every point in the following graph of an arrow were interchanged.

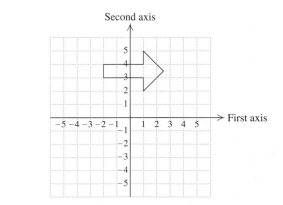

56. What advantage(s) does the use of a line graph have over that of a bar graph?

57. In which quadrant(s) could a point be located if its coordinates are reciprocals of each other?

58. In which quadrant(s) could a point be located if the point's second coordinate is $-\frac{2}{3}$ times the value of the first coordinate?

59. The points $(-1, 1)$, $(4, 1)$, and $(4, -5)$ are three vertices of a rectangle. Find the coordinates of the fourth vertex.

60. The pairs $(-2, -3)$, $(-1, 2)$, and $(4, -3)$ can serve as three (of four) vertices for three different parallelograms. Find the fourth vertex of each parallelogram.

61. Graph eight points such that the sum of the coordinates in each pair is 7.

62. Graph eight points such that the first coordinate minus the second coordinate is 1.

63. Find the perimeter of a rectangle if three of its vertices are $(5, -2)$, $(-3, -2)$, and $(-3, 3)$.

64. Find the area of a triangle whose vertices have coordinates $(0, 9)$, $(0, -4)$, and $(5, -4)$.

Coordinates on the globe. *Coordinates can also be used to describe the location on a sphere: 0° latitude is the equator and 0° longitude is a line from the North Pole to the South Pole through France and Algeria. In the figure shown here, hurricane Clara is at a point about 260 mi northwest of Bermuda near latitude 36.0° North, longitude 69.0° West.*

65. Approximate the latitude and the longitude of Bermuda.

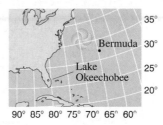

66. Approximate the latitude and the longitude of Lake Okeechobee.

67. In the *Star Trek* science-fiction series, a three-dimensional coordinate system is used to locate objects in space. If the center of a planet is used as the origin, how many "quadrants" will exist? Why? If possible, sketch a three-dimensional coordinate system and label each "quadrant."

68. The graph accompanying Example 3 flattens out. Why do you think this occurs?

C O R N E R
You Sank My Battleship!

COLLABORATIVE

Focus: Graphing points

Time: 15–25 minutes

Group size: 3–5

Materials: Graph paper

In the game Battleship®, a player places a miniature ship on a grid that only that player can see. An opponent guesses at coordinates that might "hit" the "hidden" ship. The following activity is similar to this game.

ACTIVITY

1. Using only integers from -10 to 10 (inclusive), one group member should secretly record the coordinates of a point on a slip of paper. (This point is the hidden "battleship.")
2. The other group members can then ask up to 10 "yes/no" questions in an effort to determine the coordinates of the secret point. Be sure to phrase each question mathematically (for example, "Is the x-coordinate negative?")
3. The group member who selected the point should answer each question. On the basis of the answer given, another group member should cross out the points no longer under consideration. All group members should check that this is done correctly.
4. If the hidden point has not been determined after 10 questions have been answered, the secret coordinates should be revealed to all group members.
5. Repeat parts (1)–(4) until each group member has had the opportunity to select the hidden point and answer questions.

Graphing Linear Equations

3.2

Solutions of Equations • Graphing Linear Equations

We have seen how bar, line, and circle graphs can represent information. Now we begin to learn how graphs can be used to represent solutions of equations.

Solutions of Equations

When an equation contains two variables, solutions must be ordered pairs in which each number in the pair replaces a letter in the equation. Unless stated otherwise, the first number in each pair replaces the variable that occurs first alphabetically.

E x a m p l e 1

Determine whether each of the following pairs is a solution of $4b - 3a = 22$: **(a)** $(2, 7)$; **(b)** $(1, 6)$.

Solution

a) We substitute 2 for a and 7 for b (alphabetical order of variables):

$$\frac{4b - 3a = 22}{4 \cdot 7 - 3 \cdot 2 \; ? \; 22}$$
$$28 - 6$$
$$22 \;\big|\; 22 \quad \text{TRUE}$$

Since $22 = 22$ is *true*, the pair $(2, 7)$ *is* a solution.

b) In this case, we replace a with 1 and b with 6:

$$\frac{4b - 3a = 22}{4 \cdot 6 - 3 \cdot 1 \; ? \; 22}$$
$$24 - 3$$
$$21 \;\big|\; 22 \quad \text{FALSE}$$

Since $21 = 22$ is *false*, the pair $(1, 6)$ is *not* a solution.

E x a m p l e 2

Show that the pairs $(3, 7)$, $(0, 1)$, and $(-3, -5)$ are solutions of $y = 2x + 1$. Then graph the three points to determine another pair that is a solution.

Solution   To show that a pair is a solution, we substitute, replacing x with the first coordinate and y with the second coordinate of each pair:

$$\frac{y = 2x + 1}{7 \; ? \; 2 \cdot 3 + 1}$$
$$6 + 1$$
$$7 \;\big|\; 7 \quad \text{TRUE}$$

$$\frac{y = 2x + 1}{1 \; ? \; 2 \cdot 0 + 1}$$
$$0 + 1$$
$$1 \;\big|\; 1 \quad \text{TRUE}$$

$$\frac{y = 2x + 1}{-5 \; ? \; 2(-3) + 1}$$
$$-6 + 1$$
$$-5 \;\big|\; -5 \quad \text{TRUE}$$

In each of the three cases, the substitution results in a true equation. Thus the pairs $(3, 7)$, $(0, 1)$, and $(-3, -5)$ are all solutions. We graph them below, labeling the "first" axis x and the "second" axis y. Note that the three points appear to "line up." Will other points that line up with these points also represent solutions of $y = 2x + 1$? To find out, we use a ruler and sketch a line passing through $(-3, -5)$, $(0, 1)$ and $(3, 7)$.

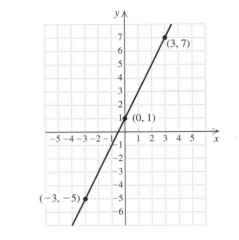

The line appears to pass through $(2, 5)$. Let's check if this pair is a solution of $y = 2x + 1$:

$$
\begin{array}{c|l}
\multicolumn{2}{l}{y = 2x + 1} \\
\hline
5 \;\; ? & 2 \cdot 2 + 1 \\
 & 4 + 1 \\
5 & 5 \qquad\qquad \text{TRUE}
\end{array}
$$

We see that $(2, 5)$ *is* a solution. You should perform a similar check for at least one other point that appears to be on the line.

Example 2 leads us to suspect that *any* point on the line passing through $(3, 7)$, $(0, 1)$, and $(-3, -5)$ represents a solution of $y = 2x + 1$. In fact, every solution of $y = 2x + 1$ is represented by a point on this line and every point on this line represents a solution. The line is called the **graph** of the equation.

Graphing Linear Equations

Equations like $y = 2x + 1$ or $4b + 3a = 22$ are said to be **linear** because the graph of each equation is a line. In general, any equation that can be written in the form $y = mx + b$ or $Ax + By = C$ (where m, b, A, B, and C are constants and A and B are not both 0) is linear.

To *graph* an equation is to make a drawing that represents its solutions. Linear equations can be graphed as follows.

> **To Graph a Linear Equation**
>
> 1. Select a value for one coordinate and calculate the corresponding value of the other coordinate. Form an ordered pair. This pair is one solution of the equation.
> 2. Repeat step (1) to find a second ordered pair. A third ordered pair can be used as a check.
> 3. Plot the ordered pairs and draw a straight line passing through the points. The line represents all solutions of the equation.

E x a m p l e   3

Graph: $y = -3x + 1$.

Solution   We select a convenient value for x, compute y, and form an ordered pair. Then we repeat the process for other choices of x.

If $x = 2$,   then $y = -3 \cdot 2 + 1 = -5$,   and $(2, -5)$ is a solution.

If $x = 0$,   then $y = -3 \cdot 0 + 1 = 1$,   and $(0, 1)$ is a solution.

If $x = -1$,   then $y = -3(-1) + 1 = 4$,   and $(-1, 4)$ is a solution.

Results are often listed in a table, as shown below. The points corresponding to each pair are then plotted.

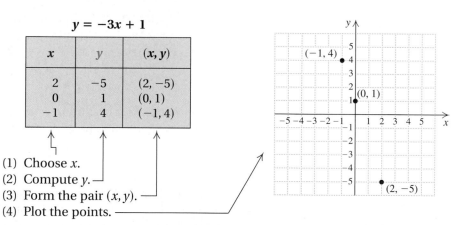

$y = -3x + 1$

x	y	(x, y)
2	-5	$(2, -5)$
0	1	$(0, 1)$
-1	4	$(-1, 4)$

(1) Choose x.
(2) Compute y.
(3) Form the pair (x, y).
(4) Plot the points.

Note that all three points line up. If they didn't, we would know that we had made a mistake, because the equation is linear. When only two points are plotted, an error is more difficult to detect.

Finally, we use a ruler or other straight-edge to draw a line. Every point on the line represents a solution of $y = -3x + 1$.

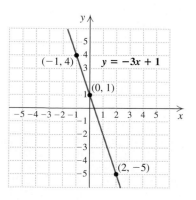

E x a m p l e   4

Graph: $y = 2x - 3$.

Solution   We select some x-values and compute y-values.

If $x = 4$,   then $y = 2 \cdot 4 - 3 = 5$,   and $(4, 5)$ is a solution.

If $x = 1$,   then $y = 2 \cdot 1 - 3 = -1$,   and $(1, -1)$ is a solution.

If $x = 0$,   then $y = 2 \cdot 0 - 3 = -3$,   and $(0, -3)$ is a solution.

$y = 2x - 3$

x	y	(x, y)
4	5	$(4, 5)$
1	−1	$(1, -1)$
0	−3	$(0, -3)$

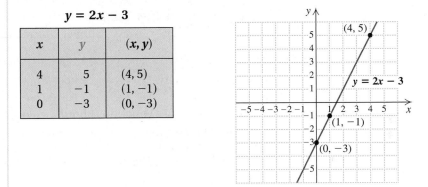

E x a m p l e   5

Graph: $4x + 2y = 12$.

Solution   To form ordered pairs, we can replace either variable with a number and then calculate the other coordinate:

If $y = 0$, we have $4x + 2 \cdot 0 = 12$
$$4x = 12$$
$$x = 3,$$

so $(3, 0)$ is a solution.

If $x = 0$, we have $4 \cdot 0 + 2y = 12$
$$2y = 12$$
$$y = 6,$$

so $(0, 6)$ is a solution.

If $y = 2$, we have $4x + 2 \cdot 2 = 12$
$$4x + 4 = 12$$
$$4x = 8$$
$$x = 2,$$

so $(2, 2)$ is a solution.

$4x + 2y = 12$

x	y	(x, y)
3	0	$(3, 0)$
0	6	$(0, 6)$
2	2	$(2, 2)$

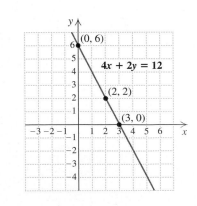

Note that in Examples 3 and 4 the variable y is isolated on one side of the equation. This generally simplifies calculations, so it is important to be able to solve for y before graphing.

Example 6

Graph $3y = 2x$ by first solving for y.

Solution To isolate y, we divide both sides by 3, or multiply both sides by $\frac{1}{3}$:

$$3y = 2x$$

$$\tfrac{1}{3} \cdot 3y = \tfrac{1}{3} \cdot 2x \qquad \text{Using the multiplication principle to multiply both sides by } \tfrac{1}{3}$$

$$\left. \begin{aligned} 1y &= \tfrac{2}{3} \cdot x \\ y &= \tfrac{2}{3}x. \end{aligned} \right\} \quad \text{Simplifying}$$

Because all the equations above are equivalent, we can use $y = \frac{2}{3}x$ to draw the graph of $3y = 2x$.

To graph $y = \frac{2}{3}x$, we can select x-values that are multiples of 3. This will allow us to avoid fractions when the corresponding y-values are computed.

$$\left. \begin{aligned} &\text{If } x = 3, && \text{then } y = \tfrac{2}{3} \cdot 3 = 2. \\ &\text{If } x = -3, && \text{then } y = \tfrac{2}{3}(-3) = -2. \\ &\text{If } x = 6, && \text{then } y = \tfrac{2}{3} \cdot 6 = 4. \end{aligned} \right\} \quad \begin{array}{l} \text{Note that when multiples of 3 are} \\ \text{substituted for } x, \text{ the } y\text{-coordinates} \\ \text{are not fractions.} \end{array}$$

The following table lists these solutions. Next, we plot the points and see that they form a line. Finally, we draw and label the line.

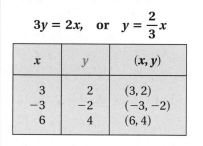

$3y = 2x$,   or $y = \dfrac{2}{3}x$

x	y	(x, y)
3	2	$(3, 2)$
-3	-2	$(-3, -2)$
6	4	$(6, 4)$

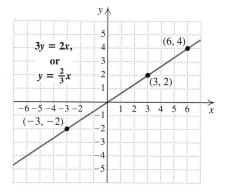

Example 7

Graph $x + 5y = -10$ by first solving for y.

Solution We have

$$x + 5y = -10$$

$$5y = -x - 10 \qquad \text{Adding } -x \text{ to both sides}$$

$$y = \tfrac{1}{5}(-x - 10) \qquad \text{Multiplying both sides by } \tfrac{1}{5}$$

$$y = -\tfrac{1}{5}x - 2. \qquad \text{Using the distributive law}$$

> ***Caution!*** It is very important to multiply *both* $-x$ and -10 by $\frac{1}{5}$.

Thus, $x + 5y = -10$ is equivalent to $y = -\frac{1}{5}x - 2$. If we choose x-values that are multiples of 5, we can avoid fractions when calculating the corresponding y-values.

If $x = 5$,   then $y = -\frac{1}{5} \cdot 5 - 2 = -1 - 2 = -3$.

If $x = 0$,   then $y = -\frac{1}{5} \cdot 0 - 2 = 0 - 2 = -2$.

If $x = -5$,   then $y = -\frac{1}{5}(-5) - 2 = 1 - 2 = -1$.

$x + 5y = -10$, or $y = -\dfrac{1}{5}x - 2$

x	y	(x, y)
5	−3	(5, −3)
0	−2	(0, −2)
−5	−1	(−5, −1)

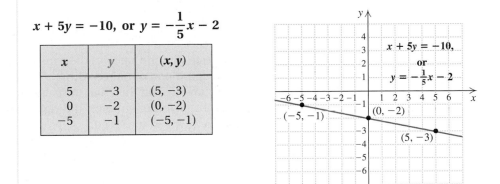

Linear equations appear in many real-life situations.

Example 8

Value of an office machine.   The value, in thousands of dollars, of Dupligraphix's color copier t years after purchase is given by $v = -\frac{1}{2}t + 3$. Graph the equation and then use the graph to estimate the value of the copier $2\frac{1}{2}$ yr after the date of purchase.

Solution   We graph $v = -\frac{1}{2}t + 3$ by selecting values for t and then calculating the associated value v. Since time cannot be negative in this case, we select nonnegative values for t.

If $t = 0$,   then $v = -\frac{1}{2} \cdot 0 + 3 = 3$.

If $t = 4$,   then $v = -\frac{1}{2} \cdot 4 + 3 = 1$.

If $t = 8$,   then $v = -\frac{1}{2} \cdot 8 + 3 = -1$.

t	v
0	3
4	1
8	−1

We label the axes and plot the points (see the figure on the left at the top of the next page). The points line up, so our calculations are probably correct. However, including any points below the horizontal axis seems unrealistic since the value of the copier cannot be negative. Thus, when drawing the graph, we end the solid line at the horizontal axis.

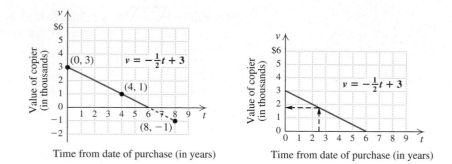

To estimate the value of the copier after $2\frac{1}{2}$ yr, we need to determine what second coordinate is paired with $2\frac{1}{2}$. To do this, we locate the point on the line that is above $2\frac{1}{2}$ and then find the value on the vertical axis that corresponds to that point (see the figure on the right above). It appears that after $2\frac{1}{2}$ yr, the copier is worth about $1800.

Caution!   When the coordinates of a point are read from a graph, as in Example 8, values should not be considered exact.

Many equations in two variables have graphs that are not straight lines. Three such graphs are shown below. As before, each graph represents the solutions of the given equation. Graphing calculators are especially helpful when drawing these *nonlinear* graphs. Nonlinear graphs are studied in Chapter 9 and in more advanced courses.

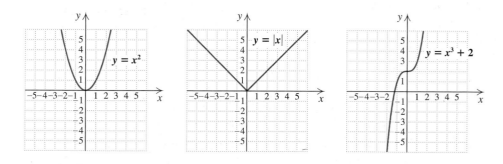

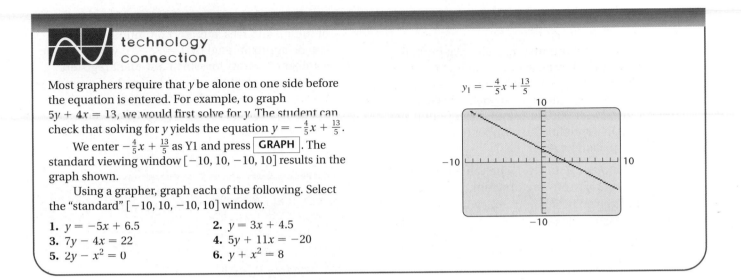

technology connection

Most graphers require that y be alone on one side before the equation is entered. For example, to graph $5y + 4x = 13$, we would first solve for y. The student can check that solving for y yields the equation $y = -\frac{4}{5}x + \frac{13}{5}$.

We enter $-\frac{4}{5}x + \frac{13}{5}$ as Y1 and press $\boxed{\textbf{GRAPH}}$. The standard viewing window $[-10, 10, -10, 10]$ results in the graph shown.

Using a grapher, graph each of the following. Select the "standard" $[-10, 10, -10, 10]$ window.

1. $y = -5x + 6.5$ **2.** $y = 3x + 4.5$
3. $7y - 4x = 22$ **4.** $5y + 11x = -20$
5. $2y - x^2 = 0$ **6.** $y + x^2 = 8$

$y_1 = -\frac{4}{5}x + \frac{13}{5}$

Exercise Set 3.2

FOR EXTRA HELP

Digital Video Tutor CD 2 Videotape 5   InterAct Math   Math Tutor Center   MathXL   MyMathLab.com

Determine whether each equation has the given ordered pair as a solution.

1. $y = 7x + 1$; $(0, 2)$

2. $y = 2x + 3$; $(0, 3)$

3. $3y + 2x = 12$; $(4, 2)$

4. $5x - 3y = 15$; $(0, 5)$

5. $4a - 3b = 11$; $(2, -1)$

6. $3q - 2p = -8$; $(1, -2)$

In Exercises 7–14, an equation and two ordered pairs are given. Show that each pair is a solution of the equation. Then graph the two pairs to determine another solution. Answers may vary.

7. $y = x - 2$; $(3, 1), (-2, -4)$

8. $y = x + 3$; $(-1, 2), (4, 7)$

9. $y = \frac{1}{2}x + 3$; $(4, 5), (-2, 2)$

10. $y = \frac{1}{2}x - 1$; $(6, 2), (0, -1)$

11. $y + 3x = 7$; $(2, 1), (4, -5)$

12. $2y + x = 5$; $(-1, 3), (7, -1)$

13. $4x - 2y = 10$; $(0, -5), (4, 3)$

14. $6x - 3y = 3$; $(1, 1), (-1, -3)$

Graph each equation.

15. $y = x - 1$ **16.** $y = x + 1$

17. $y = x$ **18.** $y = -x$

19. $y = \frac{1}{2}x$ **20.** $y = \frac{1}{3}x$

21. $y = x + 2$ **22.** $y = x + 3$

23. $y = 3x - 2$ **24.** $y = 2x + 2$

25. $y = \frac{1}{2}x + 1$ **26.** $y = \frac{1}{3}x - 4$

27. $x + y = -5$ **28.** $x + y = 4$

29. $y = \frac{5}{3}x - 2$ **30.** $y = \frac{5}{2}x + 3$

31. $x + 2y = 8$ **32.** $x + 2y = -6$

33. $y = \frac{3}{2}x + 1$ **34.** $y = -\frac{2}{3}x + 4$

35. $6x - 3y = 9$ **36.** $8x - 4y = 12$

37. $8y + 2x = -4$ **38.** $6y + 2x = 8$

Solve by graphing. Label all axes, and show where each solution is located on the graph.

39. *Bottled water.* The number of gallons of bottled water w consumed by an average American in one year is given by $w = \frac{1}{2}t + 5$, where t is the number of years since 1990 (based on an article in *New York Times Magazine*, 8/30/98). Graph the equation and use the graph to predict the number of gallons consumed per person in 2004.

40. *Value of computer software.* The value v of a shopkeeper's inventory software program, in hundreds of dollars, is given by $v = -\frac{3}{4}t + 6$, where t is the number of years since the shopkeeper first bought the program. Graph the equation and use the graph to estimate what the program is worth 4 yr after it was first purchased.

41. *Increasing life expectancy.* A smoker is 15 times more likely to die from lung cancer than a nonsmoker. An ex-smoker who stopped smoking t years ago is w times more likely to die from lung cancer than a nonsmoker, where

$$t + w = 15.^*$$

Graph the equation and use the graph to estimate how much more likely it is for Sandy to die from lung cancer than Polly, if Polly never smoked and Sandy quit $2\frac{1}{2}$ years ago.

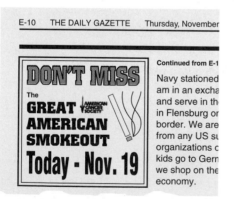

42. *Price of printing.* The price p, in cents, of a photocopied and bound lab manual is given by $p = \frac{7}{2}n + 20$, where n is the number of pages in the manual. Graph the equation and use the graph to estimate the cost of a 25-page manual. (*Hint:* Count by 5's on both axes.)

---

*Source: Data from *Body Clock* by Dr. Martin Hughes, p. 60. New York: Facts on File, Inc.

43. *Cost of college.* The cost T, in hundreds of dollars, of tuition and fees at many community colleges can be approximated by $T = \frac{6}{5}c + 1$, where c is the number of credits for which a student registers (based on information provided by the Community College of Vermont). Graph the equation and use the graph to estimate the cost of tuition and fees when a student registers for 4 three-credit courses.

44. *Cost of college.* The cost C, in thousands of dollars, of a year at a private four-year college (all expenses) can be approximated by $C = \frac{5}{4}t + 21$, where t is the number of years since 1995 (based on information in *Statistical Abstract of the United States*, 1998). Graph the equation and use the graph to estimate the cost of a year at a private four-year college in 2005.

45. *Coffee consumption.* The number of gallons of coffee n consumed each year by the average U.S. consumer can be approximated by $n = \frac{5}{2}d + 20$, where d is the number of years since 1994 (based on information in *Statistical Abstract of the United States*, 1998). Graph the equation and use the graph to estimate what the average coffee consumption was in 2000.

46. *Record temperature drop.* On January 22, 1943, the temperature T, in degrees Fahrenheit, in Spearfish, South Dakota, could be approximated by $T = -2m + 54$, where m is the number of minutes since 9:00 A.M. that morning (*Source*: 2000 *Information Please Almanac*). Graph the equation and use the graph to estimate the temperature at 9:15 A.M.

47. The equations $3x + 4y = 8$ and $y = -\frac{3}{4}x + 2$ are equivalent. Which equation would be easier to graph and why?

48. Suppose that a linear equation is graphed by plotting three points and that the three points line up with each other. Does this *guarantee* that the equation is being correctly graphed? Why or why not?

SKILL MAINTENANCE

Solve and check.

49. $5x + 3 \cdot 0 = 12$

50. $2x - 5 \cdot 0 = 9$

51. $7 \cdot 0 - 4y = 10$

Solve.

52. $pq + p = w$, for p

53. $Ax + By = C$, for y

54. $A = \dfrac{T + Q}{2}$, for Q

SYNTHESIS

55. Janice consistently makes the mistake of plotting the x-coordinate of an ordered pair using the y-axis, and the y-coordinate using the x-axis. How will Janice's incorrect graph compare with the appropriate graph?

56. Explain how the graph in Example 8 can be used to determine when the value of the color copier has dropped to $1500.

57. *Bicycling.* Long Beach Island in New Jersey is a long, narrow, flat island. For exercise, Lauren routinely bikes to the northern tip of the island and back. Because of the steady wind, she uses one gear going north and another for her return. Lauren's bike has 14 gears and the sum of the two gears used on her ride is always 18. Write and graph an equation that represents the different pairings of gears that Lauren uses. Note that there are no fractional gears on a bicycle.

In Exercises 58–61, try to find an equation for the graph shown.

58.

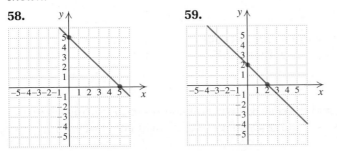

59.

60.

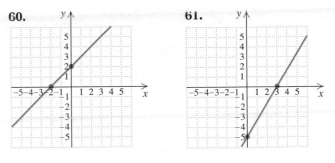

61.

62. Translate to an equation:

   d dimes and n nickels total $1.75.

Then graph the equation and use the graph to determine three different combinations of dimes and nickels that total $1.75 (see also Exercise 75).

63. Translate to an equation:

   d $25 dinners and l $5 lunches total $225.

Then graph the equation and use the graph to determine three different combinations of lunches and dinners that total $225 (see also Exercise 75).

Use the suggested x-values $-3, -2, -1, 0, 1, 2,$ *and* 3 *to graph each equation.*

64. $y = |x|$

Aha! **65.** $y = -|x|$

Aha! **66.** $y = |x| - 2$

67. $y = -|x| + 2$

68. $y = |x| + 3$

For Exercises 69–74, use a grapher to graph the equation. Use a $[-10, 10, -10, 10]$ *window.*

69. $y = -2.8x + 3.5$

70. $y = 4.5x + 2.1$

71. $y = 2.8x - 3.5$

72. $y = -4.5x - 2.1$

73. $y = x^2 + 4x + 1$

74. $y = -x^2 + 4x - 7$

75. Study the graph of Exercise 62 or 63. Does *every* point on the graph represent a solution of the associated problem? Why or why not?

COLLABORATIVE

CORNER

Follow the Bouncing Ball

Focus: Graphing and problem solving

Time: 25 minutes

Group size: 3

Materials: Each group will need a rubber ball, graph paper, and a tape measure.

Does a rubber ball always rebound a fixed percentage of the height from which it is dropped? The following activity attempts to answer this. Please be sure to read all steps before beginning.

ACTIVITY

1. One group member should hold a rubber ball 5 ft above the floor. A second group member should measure this height for accuracy. The ball should then be dropped and caught at the peak of its bounce. The second group member should measure this rebound height and the third group member should record the measurement.

2. Repeat part (1) two more times from the same height. Then find the average of the three rebound heights and use that number to form the ordered pair (original height, rebound height).

3. Repeat parts (1) and (2) four more times at heights of 4 ft, 3 ft, 2 ft, and 1 ft to find five ordered pairs. Graph the five pairs on graph paper and draw a straight line starting at (0, 0) that comes as close as possible to all six points.

4. Use the graph in part (3) to predict the rebound height for a ball dropped from a height of 7 ft. Then perform the drop and check your prediction.

5. Repeat part (4), dropping the ball from a "new" height of the group's choice.

6. Does it appear that a rubber ball will always rebound a fixed percentage of its original height? Why or why not?

7. Compare results from other class groups. What conclusions can you draw?

Graphing and Intercepts

3.3

Intercepts • Using Intercepts to Graph • Graphing Horizontal or Vertical Lines

Unless a line is horizontal or vertical, it will cross both axes. Knowing where the axes are crossed gives us another way of graphing linear equations.

Intercepts

In Example 5 of Section 3.2, we graphed $4x + 2y = 12$ by plotting the points $(3, 0)$, $(0, 6)$, and $(2, 2)$ and then drawing the line.

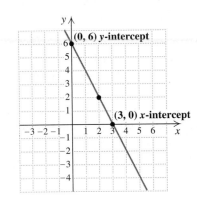

The point at which a graph crosses the y-axis is called the **y-intercept**. In the figure above, the y-intercept is $(0, 6)$. The x-coordinate of a y-intercept is always 0.

The point at which a graph crosses the x-axis is called the **x-intercept**. In the figure above, the x-intercept is $(3, 0)$. The y-coordinate of an x-intercept is always 0.

It is possible for the graph of a curve to have more than one y-intercept or more than one x-intercept.

E x a m p l e 1   For the graph shown below, **(a)** give the coordinates of any x-intercepts and **(b)** give the coordinates of any y-intercepts.

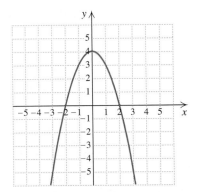

Solution

a) The x-intercepts are the points at which the graph crosses the x-axis. For the graph shown, the x-intercepts are $(-2, 0)$ and $(2, 0)$.

b) The y-intercept is the point at which the graph crosses the y-axis. For the graph shown, the y-intercept is $(0, 4)$.

Using Intercepts to Graph

It is important to know how to locate a graph's intercepts from the equation being graphed.

> ### To Find Intercepts
>
> To find the y-intercept(s) of an equation's graph, replace x with 0 and solve for y.
>
> To find the x-intercept(s) of an equation's graph, replace y with 0 and solve for x.

E x a m p l e   2   Find the y-intercept and the x-intercept of the graph of $2x + 4y = 20$.

Solution   To find the y-intercept, we let $x = 0$ and solve for y:

$$2 \cdot 0 + 4y = 20 \qquad \text{Replacing } x \text{ with } 0$$
$$4y = 20$$
$$y = 5.$$

Thus the y-intercept is $(0, 5)$.
   To find the x-intercept, we let $y = 0$ and solve for x:

$$2x + 4 \cdot 0 = 20 \qquad \text{Replacing } y \text{ with } 0$$
$$2x = 20$$
$$x = 10.$$

Thus the x-intercept is $(10, 0)$.

As we saw in Example 5 of Section 3.2, intercepts can be used to graph a linear equation.

E x a m p l e   3   Graph $2x + 4y = 20$ using intercepts.

Solution   In Example 2, we showed that the y-intercept is $(0, 5)$ and the x-intercept is $(10, 0)$. Before drawing a line, we plot a third point as a check. We substitute any convenient value for x and solve for y.
   If we let $x = 5$, then

$$2 \cdot 5 + 4y = 20 \qquad \text{Substituting 5 for } x$$
$$10 + 4y = 20$$
$$4y = 10 \qquad \text{Subtracting 10 from both sides}$$
$$y = \tfrac{10}{4}, \text{ or } 2\tfrac{1}{2}. \qquad \text{Solving for } y$$

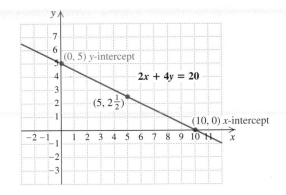

The point $\left(5, 2\frac{1}{2}\right)$ appears to line up with the intercepts, so our work is probably correct. To finish, we draw and label the line.

Note that when we solved for the y-intercept, we simplified $2x + 4y = 20$ to $4y = 20$. Thus, to find the y-intercept, we can momentarily ignore the x-term and solve the remaining equation.

In a similar manner, when we solved for the x-intercept, we simplified $2x + 4y = 20$ to $2x = 20$. Thus, to find the x-intercept, we can momentarily ignore the y-term and then solve this remaining equation.

Example 4

Graph $3x - 2y = 6$ using intercepts.

Solution   To find the y-intercept, we let $x = 0$. This amounts to temporarily ignoring the x-term and then solving:

$$-2y = 6 \qquad \text{For } x = 0, \text{ we have } 3 \cdot 0 - 2y, \text{ or simply } -2y.$$
$$y = -3.$$

The y-intercept is $(0, -3)$.
    To find the x-intercept, we let $y = 0$. This amounts to temporarily disregarding the y-term and then solving:

$$3x = 6 \qquad \text{For } y = 0, \text{ we have } 3x - 2 \cdot 0, \text{ or simply } 3x.$$
$$x = 2.$$

The x-intercept is $(2, 0)$.
    To find a third point, we replace x with 4 and solve for y:

$$3 \cdot 4 - 2y = 6 \qquad \text{Numbers other than 4 can be used for } x.$$
$$12 - 2y = 6$$
$$-2y = -6$$
$$y = 3. \qquad \text{This means that } (4, 3) \text{ is on the graph.}$$

The point $(4, 3)$ appears to line up with the intercepts, so we draw the graph.

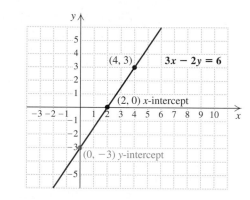

 technology connection

When an equation has been entered into a grapher, we may not be able to see both intercepts. For example, if $y = -0.8x + 17$ is graphed in the $[-10, 10, -10, 10]$ window, neither intercept is visible.

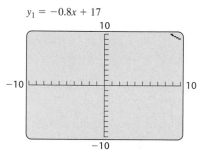

To better view the intercepts, we can change the window dimensions or we can zoom out. The ZOOM feature

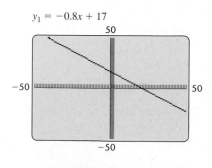

allows us to reduce or magnify a graph or a portion of a graph. Before zooming, the ZOOM *factors* must be set in the memory of the ZOOM key. If we zoom out with factors set at 5, both intercepts are visible but the axes are heavily drawn, as shown in the preceding figure.

This suggests that the *scales* of the axes should be changed. To do this, we use the WINDOW menu and set Xscl to 5 and Yscl to 5. The resulting graph has tick marks 5 units apart and clearly shows both intercepts. Other choices for Xscl and Yscl can also be made.

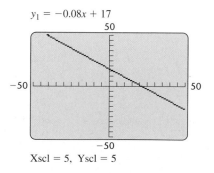

Xscl = 5, Yscl = 5

Graph each equation so that both intercepts can be easily viewed. Zoom or adjust the window settings so that tick marks can be clearly seen on both axes.

1. $y = -0.72x - 15$
2. $y - 2.13x = 27$
3. $5x + 6y = 84$
4. $2x - 7y = 150$
5. $19x - 17y = 200$
6. $6x + 5y = 159$

Graphing Horizontal or Vertical Lines

The equations graphed in Examples 3 and 4 are both in the form $Ax + By = C$. We have already stated that any equation in the form $Ax + By = C$ is linear, provided A and B are not both zero. What if A or B (but not both) is zero? We will find that when A is zero, there is no x-term and the graph is a horizontal line. We will also find that when B is zero, there is no y-term and the graph is a vertical line.

Example 5

Study Tip

Don't be hesitant to ask questions in class at appropriate times. Most instructors welcome questions and encourage students to ask them. Other students in your class probably have the same questions you do.

Graph: $y = 3$.

Solution We can regard the equation $y = 3$ as $0 \cdot x + y = 3$. No matter what number we choose for x, we find that y must be 3 if the equation is to be solved. Consider the following table.

y = 3

Choose any number for x. →

x	y	(x, y)
-2	3	$(-2, 3)$
0	3	$(0, 3)$
4	3	$(4, 3)$

y must be 3.

All pairs will have 3 as the y-coordinate.

When we plot the ordered pairs $(-2, 3)$, $(0, 3)$, and $(4, 3)$ and connect the points, we obtain a horizontal line. Any ordered pair $(x, 3)$ is a solution, so the line is parallel to the x-axis with y-intercept $(0, 3)$.

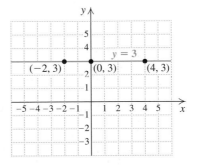

Example 6

Graph: $x = -4$.

Solution We can regard the equation $x = -4$ as $x + 0 \cdot y = -4$. We make up a table with all -4's in the x-column.

x = −4

x must be -4. →

x	y	(x, y)
-4	-5	$(-4, -5)$
-4	1	$(-4, 1)$
-4	3	$(-4, 3)$

Choose any number for y.

All pairs will have -4 as the x-coordinate.

When we plot the ordered pairs $(-4, -5)$, $(-4, 1)$, and $(-4, 3)$ and connect them, we obtain a vertical line. Any ordered pair $(-4, y)$ is a solution. The line is parallel to the y-axis with x-intercept $(-4, 0)$.

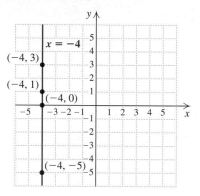

Linear Equations in One Variable

The graph of $y = b$ is a horizontal line, with y-intercept $(0, b)$.

The graph of $x = a$ is a vertical line, with x-intercept $(a, 0)$.

E x a m p l e   7

Write an equation for each graph.

a)

b)

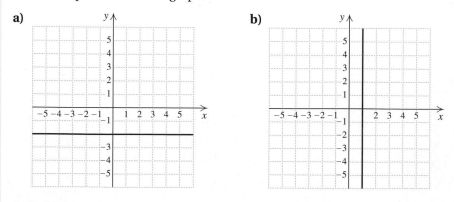

Solution

a) Note that every point on the horizontal line passing through $(0, -2)$ has -2 as the y-coordinate. Thus the equation of the line is $y = -2$.

b) Note that every point on the vertical line passing through $(1, 0)$ has 1 as the x-coordinate. Thus the equation of the line is $x = 1$.

Exercise Set 3.3

For Exercises 1–8, find **(a)** *the coordinates of the y-intercept and* **(b)** *the coordinates of all x-intercepts.*

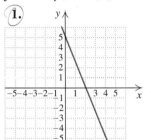

1.

2.

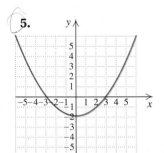

3.

4.

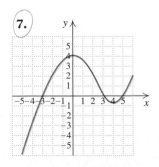

5.

6.

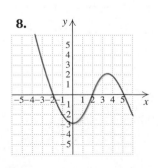

7.

8.

For Exercises 9–16, find **(a)** *the coordinates of any y-intercept and* **(b)** *the coordinates of any x-intercept. Do not graph.*

9. $5x + 3y = 15$

10. $5x + 2y = 20$

11. $7x - 2y = 28$

12. $4x - 3y = 24$

13. $-4x + 3y = 10$

14. $-2x + 3y = 7$

Aha! **15.** $y = 9$

16. $x = 8$

Find the intercepts. Then graph.

17. $x + 2y = 6$

18. $3x + 2y = 12$

19. $6x + 9y = 36$

20. $x + 3y = 6$

21. $-x + 3y = 9$

22. $-x + 2y = 8$

23. $2x - y = 8$

24. $3x + y = 9$

25. $y = -3x + 6$

26. $y = 2x - 6$

27. $5x - 10 = 5y$

28. $3x - 9 = 3y$

29. $2x - 5y = 10$

30. $2x - 3y = 6$

31. $6x + 2y = 12$

32. $4x + 5y = 20$

33. $x - 1 = y$

34. $3x + 2y = 8$

35. $2x - 6y = 18$

36. $2x + 7y = 6$

37. $4x - 3y = 12$

38. $3x - y = 2$

39. $-3x = 6y - 2$

40. $y = -3 - 3x$

41. $3 = 2x - 5y$

42. $-4x = 8y - 5$

43. $x + 2y = 0$

44. $y - 3x = 0$

Graph.

45. $y = 5$

46. $y = 2$

47. $x = 4$

48. $x = 6$

49. $y = -2$

50. $y = -4$

51. $x = -1$

52. $x = -6$

53. $y = 7$

54. $y = 1$

55. $x = 1$

56. $x = -2$

57. $y = 0$

58. $y = \frac{3}{2}$

59. $x = -\frac{5}{2}$

60. $x = 0$

Aha! **61.** $-5y = 15$

62. $12x = -36$

63. $35 + 7y = 0$

64. $-3x - 24 = 0$

Write an equation for each graph.

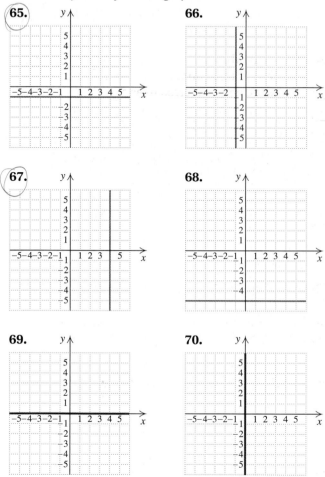

65.

66.

67.

68.

69.

70.

71. Explain in your own words why the graph of $y = 8$ is a horizontal line.

72. Explain in your own words why the graph of $x = -4$ is a vertical line.

SKILL MAINTENANCE

Translate to an algebraic expression.

73. 7 less than d

74. 5 more than w

75. The sum of 2 and a number

76. The product of 3 and a number

77. Twice the sum of two numbers

78. Half of the sum of two numbers

SYNTHESIS

79. Describe what the graph of $x + y = C$ will look like for any choice of C.

80. If the graph of a linear equation has one point that is both the x- and the y-intercepts, what is that point? Why?

81. Write an equation for the x-axis.

82. Write an equation of the line parallel to the x-axis and passing through $(3, 5)$.

83. Write an equation of the line parallel to the y-axis and passing through $(-2, 7)$.

84. Find the coordinates of the point of intersection of the graphs of $y = x$ and $y = 6$.

85. Find the coordinates of the point of intersection of the graphs of the equations $x = -3$ and $y = x$.

86. Write an equation of the line shown in Exercise 1.

87. Write an equation of the line shown in Exercise 4.

88. Find the value of C such that the graph of $3x + C = 5y$ has an x-intercept of $(-4, 0)$.

89. Find the value of C such that the graph of $4x = C - 3y$ has a y-intercept of $(0, -8)$.

90. For A and B nonzero, the graphs of $Ax + D = C$ and $By + D = C$ will be parallel to an axis. Explain why.

In Exercises 91–96, find the intercepts of each equation algebraically. Then adjust the window and scale so that the intercepts can be checked graphically with no further window adjustments.

91. $3x + 2y = 50$

92. $2x - 7y = 80$

93. $y = 0.2x - 9$

94. $y = 1.3x - 15$

95. $25x - 20y = 1$

96. $50x + 25y = 1$

3.4

Rates

Rates of Change • Visualizing Rates

Rates of Change

Because graphs make use of two axes, they allow us to visualize how two quantities change with respect to each other. A number accompanied by units is used to represent this type of change and is referred to as a *rate*.

> ### Rate
> A *rate* is a ratio that indicates how two quantities change with respect to each other.

Rates occur often in everyday life:

A town that grows by 3400 residents over a period of 2 yr has a *growth rate* of $\frac{3400}{2}$, or 1700, residents per year.

A person running 150 m in 20 sec is moving at a *rate* of $\frac{150}{20}$, or 7.5, m/sec (meters per second).

A class of 25 students pays a total of $93.75 to visit a museum. The *rate* is $\frac{\$93.75}{25}$, or $3.75, per student.

> ***Caution!*** To calculate a rate, it is important to keep track of the units being used.

Example 1

On January 3, Nell rented a Ford Focus with a full tank of gas and 9312 mi on the odometer. On January 7, she returned the car with 9630 mi on the odometer.* If the rental agency charged Nell $108 for the rental and needed 12 gal of gas to fill up the gas tank, find the following rates:

a) The car's rate of gas consumption, in miles per gallon
b) The average cost of the rental, in dollars per day
c) The car's rate of travel, in miles per day

---

*For all rental problems, assume that the pickup time was later in the day than the return time so that no late fees were applied.

Solution

a) The rate of gas consumption, in miles per gallon, is found by dividing the number of miles traveled by the number of gallons used for that amount of driving:

$$\text{Rate, in miles per gallon} = \frac{9630 \text{ mi} - 9312 \text{ mi}}{12 \text{ gal}} \qquad \text{The word "per"}$$
$$\text{indicates}$$
$$\text{division.}$$
$$= \frac{318 \text{ mi}}{12 \text{ gal}}$$
$$= 26.5 \text{ mi/gal} \qquad \text{Dividing}$$
$$= 26.5 \text{ miles per gallon.}$$

b) The average cost of the rental, in dollars per day, is found by dividing the cost of the rental by the number of days:

$$\text{Rate, in dollars per day} = \frac{108 \text{ dollars}}{4 \text{ days}} \qquad \text{From January 3 to January}$$
$$7 \text{ is } 7 - 3 = 4 \text{ days.}$$
$$= 27 \text{ dollars/day}$$
$$= \$27 \text{ per day.}$$

c) The car's rate of travel, in miles per day, is found by dividing the number of miles traveled by the number of days:

$$\text{Rate, in miles per day} = \frac{318 \text{ mi}}{4 \text{ days}} \qquad 9630 \text{ mi} - 9312 \text{ mi} = 318 \text{ mi}$$
$$\text{From January 3 to January 7}$$
$$\text{is } 7 - 3 = 4 \text{ days.}$$
$$= 79.5 \text{ mi/day}$$
$$= 79.5 \text{ mi per day.}$$

Many problems involve a rate of travel, or *speed*. The **speed** of an object is found by dividing the distance traveled by the time required to travel that distance.

Example 2

Transportation. The Atlantic City Express is a bus that makes regular trips between New York City and Atlantic City, New Jersey. At 6:00 P.M., the bus is at mileage marker 70 on the Garden State Parkway, and at 8:00 P.M. it is at marker 200. Find the average speed of the bus.

Solution Speed is the distance traveled divided by the time spent traveling:

$$\text{Bus speed} = \frac{\text{Distance traveled}}{\text{Time spent traveling}}$$
$$= \frac{\text{Change in mileage}}{\text{Change in time}}$$
$$= \frac{130 \text{ mi}}{2 \text{ hr}} \qquad 200 \text{ mi} - 70 \text{ mi} = 130 \text{ mi;}$$
$$8:00 \text{ P.M.} - 6:00 \text{ P.M.} = 2 \text{ hr}$$
$$= 65 \frac{\text{mi}}{\text{hr}}$$
$$= 65 \text{ miles per hour} \qquad \text{This } \textit{average} \text{ speed does not}$$
$$\text{indicate by how much the bus}$$
$$\text{speed may vary along the route.}$$

Visualizing Rates

Graphs allow us to visualize a rate of change. As a rule, the quantity listed in the numerator appears on the vertical axis and the quantity listed in the denominator appears on the horizontal axis.

E x a m p l e   3

Communication.   In 1998, there were approximately 69 million cellular telephone subscribers in the United States, and the figure was growing at a rate of about 14 million per year (*Source*: *Statistical Abstract of the United States*, 1999). Draw a linear graph to represent this information.

Solution   To decide which labels to use on the axes, we note that the rate is given in millions of customers per year. Thus we list *Number of customers, in millions*, on the vertical axis and *Year* on the horizontal axis.

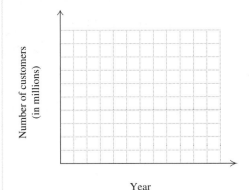

14 million per year is 14 million/yr; millions of customers is the vertical axis; year is the horizontal axis.

Next, we must decide on a scale for each axis that will allow us to plot the given information. If we count by 10's of millions on the vertical axis, we can easily reach 69 million without needing a terribly large graph. On the horizontal axis, we list several years, making certain that 1998 is included (see the figure on the left below).

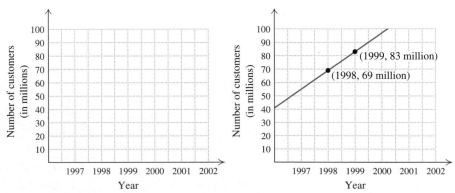

Finally, we display the given information. To do so, we plot the point that corresponds to (1998, 69 million). Then, to display the rate of growth, we move from that point to a second point that represents 14 million more users one year later. The coordinates of this point are (1998 + 1, 69 + 14 million), or (1999, 83 million). We draw a line passing through the two points, as shown in the figure on the right above.

E x a m p l e   4

Haircutting.   Gary's Barber Shop has a graph displaying data from a recent day of work.

a) What rate can be determined from the graph?
b) What is that rate?

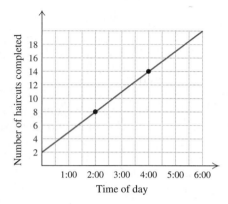

Solution

a) Because the vertical axis shows the number of haircuts completed and the horizontal axis lists the time in hour-long increments, we can find the rate *Number of haircuts per hour.*

b) The points (2:00, 8 haircuts) and (4:00, 14 haircuts) are both on the graph. This tells us that in the 2 hr between 2:00 and 4:00, there were $14 - 8 = 6$ haircuts completed. Thus the rate is

$$\frac{14 \text{ haircuts} - 8 \text{ haircuts}}{4{:}00 - 2{:}00} = \frac{6 \text{ haircuts}}{2 \text{ hours}}$$

$$= 3 \text{ haircuts per hour.}$$

FOR EXTRA HELP

Exercise Set 3.4

Digital Video Tutor CD 2
Videotape 5

InterAct Math

Math Tutor Center

MathXL

MyMathLab.com

Solve. For Exercises 1–8, round answers to the nearest cent. On Exercises 1 and 2, assume that the pickup time was later in the day than the return time so that no late fees were applied.

1. *Van rentals.*   Late on July 1, Frank rented a Dodge Caravan with a full tank of gas and 13,741 mi on the odometer. On July 4, he returned the van with 14,014 mi on the odometer. The rental agency charged Frank $118 for the rental and needed 13 gal of gas to fill up the tank.

a) Find the van's rate of gas consumption, in miles per gallon.
b) Find the average cost of the rental, in dollars per day.
c) Find the rate of travel, in miles per day.
d) Find the rental rate, in cents per mile.

2. *Car rentals.*   On February 10, Maggie rented a Chevy Blazer with a full tank of gas and 13,091 mi on the odometer. On February 12, she returned the vehicle with 13,322 mi on the odometer. The rental agency charged $92 for the rental and needed 14 gal of gas to fill the tank.

   a) Find the Blazer's rate of gas consumption, in miles per gallon.

   b) Find the average cost of the rental, in dollars per day.

   c) Find the rate of travel, in miles per day.

   d) Find the rental rate, in cents per mile.

3. *Bicycle rentals.*   At 2:00, Denise rented a mountain bike from The Slick Rock Cyclery. She returned the bike at 5:00, after cycling 18 mi. Denise paid $10.50 for the rental.

   a) Find Denise's average speed, in miles per hour.

   b) Find the rental rate, in dollars per hour.

   c) Find the rental rate, in dollars per mile.

4. *Bicycle rentals.*   At 9:00, Blair rented a mountain bike from The Bike Rack. He returned the bicycle at 11:00, after cycling 14 mi. Blair paid $12 for the rental.

   a) Find Blair's average speed, in miles per hour.

   b) Find the rental rate, in dollars per hour.

   c) Find the rental rate, in dollars per mile.

5. *Temporary help.*   A typist from Jobsite Services, Inc., reports to AKA International for work at 9:00 A.M. and leaves at 5:00 P.M. after having typed from the end of page 12 to the end of page 48 of a prospectus. AKA International pays $128 for the typist's services.

   a) Find the rate of pay, in dollars per hour.

   b) Find the average typing rate, in number of pages per hour.

   c) Find the rate of pay, in dollars per page.

6. *Temporary help.*   A typist for Kelly Services reports to 3E's Properties for work at 10:00 A.M. and leaves at 6:00 P.M. after having typed from the end of page 8 to the end of page 50 of a proposal. 3E's pays $120 for the typist's services.

   a) Find the rate of pay, in dollars per hour.

   b) Find the average typing rate, in number of pages per hour.

   c) Find the rate of pay, in dollars per page.

7. *Two-year-college tuition.*   The average tuition at a public two-year college was $1239 in 1996 and $1318 in 1998 (*Source: Statistical Abstract of the United States,* 1999). Find the rate at which tuition was increasing.

8. *Four-year-college tuition.*   The average tuition at a public four-year college was $2977 in 1995 and $3489 in 1998 (*Source: Statistical Abstract of the United States,* 1999, p. 199). Find the rate at which tuition was increasing.

9. *Elevators.*   At 2:38, Serge entered an elevator on the 34th floor of the Regency Hotel. At 2:40, he stepped off at the 5th floor.

   a) Find the elevator's average rate of travel, in number of floors per minute.

   b) Find the elevator's average rate of travel, in seconds per floor.

10. *Snow removal.*   By 1:00 P.M., Erin had already shoveled 2 driveways, and by 6:00 P.M., the number was up to 7.

   a) Find Erin's shoveling rate, in number of driveways per hour.

   b) Find Erin's shoveling rate, in hours per driveway.

11. *Mountaineering.*   As part of an ill-fated expedition to climb Mt. Everest in 1996, author Jon Krakauer departed "The Balcony," elevation 27,600 ft, at 7:00 A.M. and reached the summit, elevation 29,028 ft, at 1:25 P.M. (*Source:* Krakauer, Jon, *Into Thin Air, the Illustrated Edition.* New York: Random House, 1998)

   a) Find Krakauer's average rate of ascent, in feet per minute.

   b) Find Krakauer's average rate of ascent, in minutes per foot.

12. *Mountaineering.*   The fastest ascent of Mt. Everest was accomplished by Kaji Sherpa on October 17, 1998. Kaji Sherpa climbed from base camp, elevation 17,552 ft, to the summit, elevation 29,028 ft, in 20 hr 24 min (*Source: Guinness Book of World Records* 2000, Millenium Edition).

   a) Find Kaji Sherpa's rate of ascent, in feet per minute.

   b) Find Kaji Sherpa's rate of ascent, in minutes per foot.

In Exercises 13–20, draw a linear graph to represent the given information. Be sure to label and number the axes appropriately (see Example 3).

13. *Law enforcement.* In 1996, there were approximately 37 million crimes reported in the United States, and the figure was dropping at a rate of about 2.5 million per year (*Source: Statistical Abstract of the United States,* 1999).

14. *Fire fighting.* In 1996, there were approximately 2 million fires in the United States, and the figure was dropping at a rate of about 0.2 million per year (*Source: Statistical Abstract of the United States,* 1999).

15. *Train travel.* At 3:00 P.M., the Boston–Washington Metroliner had traveled 230 mi and was cruising at a rate of 90 miles per hour.

16. *Plane travel.* At 4:00 P.M., the Seattle–Los Angeles shuttle had traveled 400 mi and was cruising at a rate of 300 miles per hour.

17. *Wages.* By 2:00 P.M., Diane had earned $50. She continued earning money at a rate of $15 per hour.

18. *Wages.* By 3:00 P.M., Arnie had earned $70. He continued earning money at a rate of $12 per hour.

19. *Telephone bills.* Roberta's phone bill was already $7.50 when she made a call for which she was charged at a rate of $0.10 per minute.

20. *Telephone bills.* At 3:00 P.M., Larry's phone bill was $6.50 and increasing at a rate of 7¢ per minute.

In Exercises 21–30, use the graph provided to calculate a rate of change in which the units of the horizontal axis are used in the denominator.

21. *Hairdresser.* Eve's Custom Cuts has a graph displaying data from a recent day of work. At what rate does Eve work?

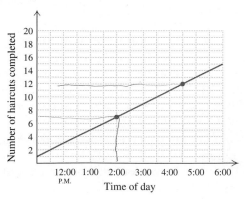

22. *Manicures.* The following graph shows data from a recent day's work at the O'Hara School of Cosmetology. At what rate do they work?

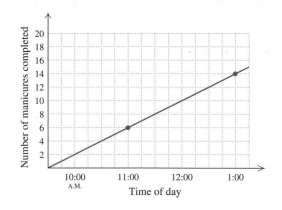

23. *Train travel.* The following graph shows data from a recent train ride from Chicago to St. Louis. At what rate did the train travel?

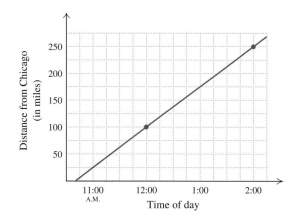

24. *Train travel.* The following graph shows data from a recent train ride from Denver to Kansas City. At what rate did the train travel?

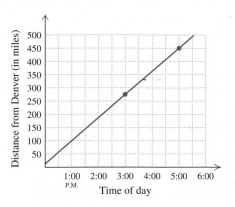

25. *Cost of a telephone call.*   The following graph shows data from a recent AT&T phone call between Burlington, VT, and Austin, TX. At what rate was the customer being billed?

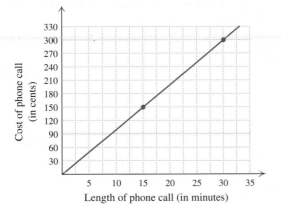

Length of phone call (in minutes)

26. *Cost of a telephone call.*   The following graph shows data from a recent MCI phone call between San Francisco, CA, and Pittsburgh, PA. At what rate was the customer being billed?

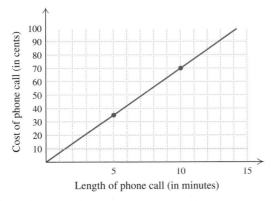

Length of phone call (in minutes)

27. *Depreciation of an office machine.*   Data regarding the value of a particular color copier is represented in the following graph. At what rate is the value changing?

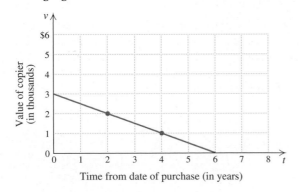

Time from date of purchase (in years)

28. *NASA spending.*   Data regarding the amount spent on the National Aeronautics and Space Administration (NASA) is represented in the following graph (based on information in the *Statistical Abstract of the United States*, 1999). At what rate is the amount spent on NASA changing?

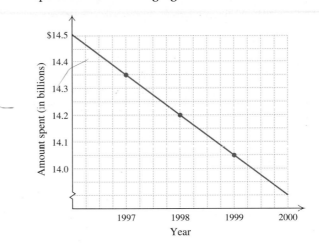

Year

29. *Gas mileage.*   The following graph shows data for a Honda Odyssey driven on interstate highways. At what rate was the vehicle consuming gas?

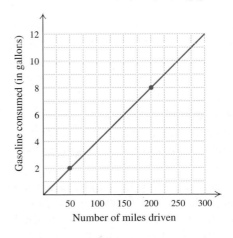

Number of miles driven

30. *Gas mileage.*   The following graph shows data for a Ford Explorer driven on city streets. At what rate was the vehicle consuming gas?

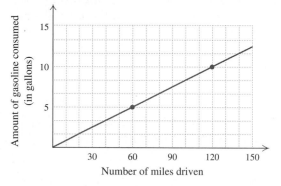

31. What does a negative rate of travel indicate? Explain.

32. Explain how to convert from kilometers per hour to meters per second.

SKILL MAINTENANCE

33. $-2 - (-7)$

34. $-9 - (-3)$

35. $\dfrac{5 - (-4)}{-2 - 7}$

36. $\dfrac{8 - (-4)}{2 - 11}$

37. $\dfrac{-4 - 8}{7 - (-2)}$

38. $\dfrac{-5 - 3}{6 - (-4)}$

SYNTHESIS

39. Write an exercise similar to Exercises 21–30 for a classmate to solve. Design the problem so that the solution is "The plane was traveling at a rate of 300 miles per hour."

40. Write an exercise similar to Exercises 1–12 for a classmate to solve. Design the problem so that the solution is "The motorcycle's rate of gas consumption was 65 miles per gallon."

41. *Aviation.*   A Boeing 737 climbs from sea level to a cruising altitude of 31,500 ft at a rate of 6300 ft/min. After cruising for 3 min, the jet is forced to land, descending at a rate of 3500 ft/min. Represent the flight with a graph in which altitude is measured on the vertical axis and time on the horizontal axis.

42. *Wages with commissions.*   Each salesperson at Mike's Bikes is paid $140 a week plus 13% of all sales up to $2000, and then 20% on any sales in excess of $2000. Draw a graph in which sales are measured on the horizontal axis and wages on the vertical axis. Then use the graph to estimate the wages paid when a salesperson sells $2700 in merchandise in one week.

43. *Gas mileage.*   Suppose that a Honda motorcycle goes twice as far as a Honda Odyssey on the same amount of gas (see Exercise 29). Draw a graph that reflects this information.

44. *Taxi fares.*   The driver of a New York City Yellow Cab recently charged $2 plus 50¢ for each fifth of a mile traveled. Draw a graph that could be used to determine the cost of a fare.

45. *Navigation.*   In 3 sec, Penny walks 24 ft, to the bow (front) of a tugboat. The boat is cruising at a rate of 5 feet per second. What is Penny's rate of travel with respect to land?

46. *Aviation.*   Tim's F-16 jet is moving forward at a deck speed of 95 mph aboard an aircraft carrier that is traveling 39 mph in the same direction. How fast is the jet traveling, in minutes per mile, with respect to the sea?

47. *Running.*   Annette ran from the 4-km mark to the 7-km mark of a 10-km race in 15.5 min. At this rate, how long would it take Annette to run a 5-mi race?

48. *Running.*   Jerod ran from the 2-mi marker to the finish line of a 5-mi race in 25 min. At this rate, how long would it take Jerod to run a 10-km race?

49. Marcy picks apples twice as fast as Ryan. By 4:30, Ryan had already picked 4 bushels of apples. Fifty minutes later, his total reached $5\frac{1}{2}$ bushels. Find Marcy's picking rate. Give your answer in number of bushels per hour.

50. By 3:00, Catanya and Chad had already made 46 candles. Forty minutes later, the total reached 64 candles. Find the rate at which Catanya and Chad made candles. Give your answer as a number of candles per hour.

CORNER

Determining Depreciation Rates

COLLABORATIVE

Focus: Modeling, graphing, and rates

Time: 30 minutes

Group size: 3

Materials: Graph paper and straightedges

From the minute a new car is driven out of the dealership, it *depreciates,* or drops in value with the passing of time. The N.A.D.A.® Official Used Car Guide (often called the "blue book" despite its orange cover) is a monthly listing of the trade-in values of used cars. The data below are taken from the N.A.D.A. guides for June and December of 2000.

Car	Trade-in Value in June 2000	Trade-in Value in December 2000
1998 Chevrolet Camaro convertible V6	$16,200	$14,900
1998 Ford Mustang convertible, 2 door, GT	$16,425	$15,325
1998 VW Passat GLS Turbo, 4 cylinder	$15,975	$14,850

ACTIVITY

1. Each group member should select a different one of the cars listed in the table above as his or her own. Assuming that the values are dropping linearly, each student should draw a line representing the trade-in value of his or her car. Draw all three lines on the same graph. Let the horizontal axis represent the time, in months, since June 2000, and let the vertical axis represent the trade-in value of each car. Decide as a group how many months or dollars each square should represent. Make the drawings as neat as possible.
2. At what *rate* is each car depreciating and how are the different rates illustrated in the graph of part (1)?
3. If one of the three cars had to be sold in June 2001, which one would your group sell and why? Compare answers with other groups.

Slope

3.5

Rate and Slope • Horizontal and Vertical Lines •
Applications

In Section 3.4, we introduced *rate* as a method of measuring how two quantities change with respect to each other. In this section, we will discuss how rate can be related to the slope of a line.

Rate and Slope

Suppose that a car manufacturer operates two plants: one in Michigan and one in Pennsylvania. Knowing that the Michigan plant produces 3 cars every 2 hours and the Pennsylvania plant produces 5 cars every 4 hours, we can set up tables listing the number of cars produced after various amounts of time.

Michigan Plant	
Hours Elapsed	**Cars Produced**
0	0
2	3
4	6
6	9
8	12

Pennsylvania Plant	
Hours Elapsed	**Cars Produced**
0	0
4	5
8	10
12	15
16	20

By comparing the number of cars produced at each plant over a specified period of time, we can compare the two production rates. For example, the Michigan plant produces 3 cars every 2 hours, so its *rate* is $3 \div 2 = 1\frac{1}{2}$, or $\frac{3}{2}$ cars per hour. Since the Pennsylvania plant produces 5 cars every 4 hours, its rate is $5 \div 4 = 1\frac{1}{4}$, or $\frac{5}{4}$ cars per hour.

Let's now graph the pairs of numbers listed in the tables, using the horizontal axis for time and the vertical axis for the number of cars produced. Note that the rate in the Michigan plant is slightly greater so its graph is slightly steeper.

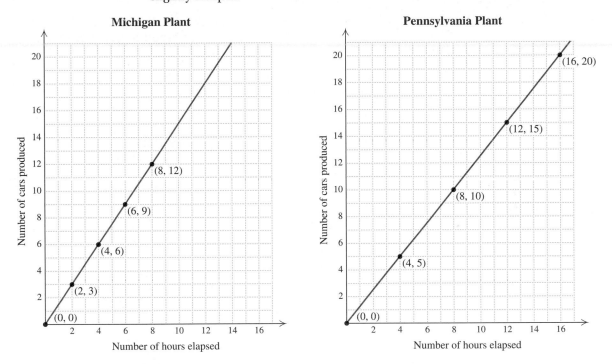

The rates $\frac{3}{2}$ and $\frac{5}{4}$ can also be found using the coordinates of any two points that are on the line. For example, we can use the points $(6, 9)$ and $(8, 12)$ to find the production rate for the Michigan plant. To do so, remember that these coordinates tell us that after 6 hr, 9 cars have been produced, and after 8 hr, 12 cars have been produced. In the 2 hr between the 6-hr and 8-hr points, $12 - 9$, or 3 cars were produced. Thus,

$$\text{Michigan production rate} = \frac{\text{change in number of cars produced}}{\text{corresponding change in time}}$$

$$= \frac{12 - 9 \text{ cars}}{8 - 6 \text{ hr}}$$

$$= \frac{3 \text{ cars}}{2 \text{ hr}} = \frac{3}{2} \text{ cars per hour.}$$

Because the line is straight, the same rate is found using *any* pair of points on the line. For example, using $(0, 0)$ and $(4, 6)$, we have

$$\text{Michigan production rate} = \frac{6 - 0 \text{ cars}}{4 - 0 \text{ hr}} = \frac{6 \text{ cars}}{4 \text{ hr}} = \frac{3}{2} \text{ cars per hour.}$$

Note that the rate is always the vertical change divided by the associated horizontal change.

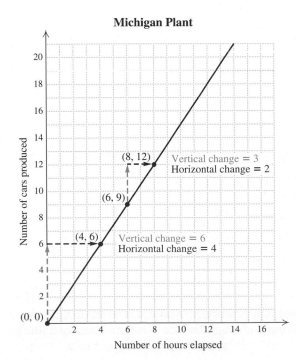

E x a m p l e   1   Use the graph of car production at the Pennsylvania plant to find the rate of production.

Solution   We can use any two points on the line, such as $(12, 15)$ and $(16, 20)$:

$$\frac{\text{Pennsylvania}}{\text{production rate}} = \frac{\text{change in number of cars produced}}{\text{corresponding change in time}}$$

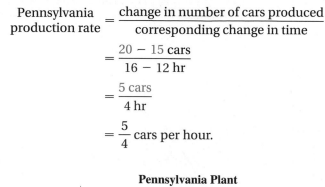

$$= \frac{20 - 15 \text{ cars}}{16 - 12 \text{ hr}}$$

$$= \frac{5 \text{ cars}}{4 \text{ hr}}$$

$$= \frac{5}{4} \text{ cars per hour.}$$

Pennsylvania Plant

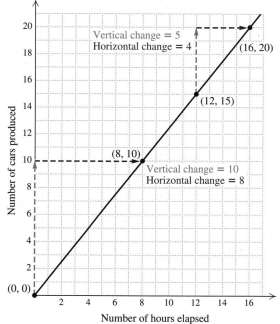

As a check, we can use another pair of points, like $(0, 0)$ and $(8, 10)$:

$$\frac{\text{Pennsylvania}}{\text{production rate}} = \frac{10 - 0 \text{ cars}}{8 - 0 \text{ hr}}$$

$$= \frac{10 \text{ cars}}{8 \text{ hr}}$$

$$= \frac{5}{4} \text{ cars per hour.}$$

When the axes of a graph are simply labeled x and y, it is useful to know the ratio of vertical change to horizontal change. This ratio is a measure of a line's slant, or **slope**, and is the rate at which y is changing with respect to x.

Consider a line passing through $(2, 3)$ and $(6, 5)$, as shown below. We find the ratio of vertical change, or *rise*, to horizontal change, or *run*, as follows:

$$\text{Ratio of vertical change} \atop \text{to horizontal change} = \frac{\text{change in } y}{\text{change in } x} = \frac{\text{rise}}{\text{run}}$$

$$= \frac{5 - 3}{6 - 2}$$

$$= \frac{2}{4}, \text{ or } \frac{1}{2}.$$

Note that these calculations can be performed without viewing a graph.

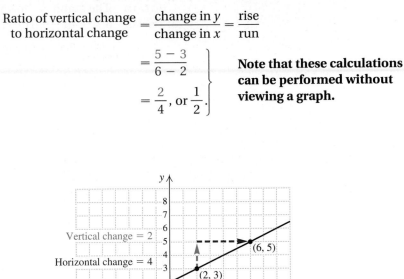

Thus the y-coordinates of points on this line increase at a rate of 2 units for every 4-unit increase in x, 1 unit for every 2-unit increase in x, or $\frac{1}{2}$ unit for every 1-unit increase in x. The slope of the line is $\frac{1}{2}$.

Slope

The *slope* of the line containing points (x_1, y_1) and (x_2, y_2) is given by

$$m = \frac{\text{change in } y}{\text{change in } x} = \frac{\text{rise}}{\text{run}} = \frac{y_2 - y_1}{x_2 - x_1}.$$

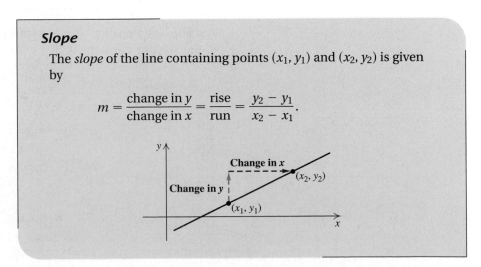

E x a m p l e   2

Graph the line containing the points $(-4, 3)$ and $(2, -6)$ and find the slope.

Solution The graph is shown below. From $(-4, 3)$ to $(2, -6)$, the change in y, or rise, is $-6 - 3$, or -9. The change in x, or run, is $2 - (-4)$, or 6. Thus,

$$\text{Slope} = \frac{\text{change in } y}{\text{change in } x}$$

$$= \frac{\text{rise}}{\text{run}}$$

$$= \frac{-6 - 3}{2 - (-4)}$$

$$= \frac{-9}{6}$$

$$= -\frac{9}{6}, \text{ or } -\frac{3}{2}.$$

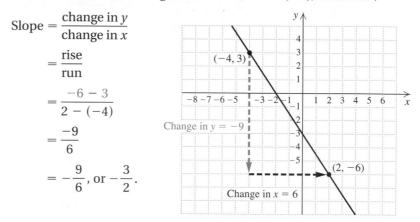

Caution! When we use the formula

$$m = \frac{y_2 - y_1}{x_2 - x_1},$$

it makes no difference which point is considered (x_1, y_1). What matters is that we subtract the y-coordinates in the same order that we subtract the x-coordinates.

To illustrate, we reverse *both* of the subtractions in Example 2. The slope is still $-\frac{3}{2}$:

$$\text{Slope} = \frac{\text{change in } y}{\text{change in } x} = \frac{3 - (-6)}{-4 - 2} = \frac{9}{-6} = -\frac{3}{2}.$$

If a line has a positive slope, it slants up from left to right. The larger the slope, the steeper the slant. A line with negative slope slants down from left to right.

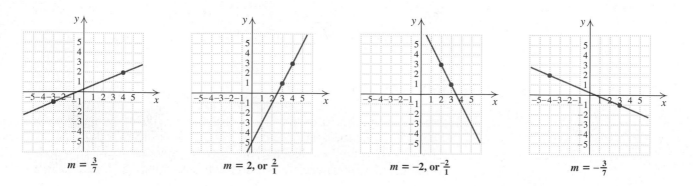

$m = \frac{3}{7}$ $m = 2$, or $\frac{2}{1}$ $m = -2$, or $\frac{-2}{1}$ $m = -\frac{3}{7}$

Horizontal and Vertical Lines

What about the slope of a horizontal or a vertical line?

Example **3**

Find the slope of the line $y = 4$.

Solution   Consider the points $(2, 4)$ and $(-3, 4)$, which are on the line. The change in y, or the rise, is $4 - 4$, or 0. The change in x, or the run, is $-3 - 2$, or -5. Thus,

$$m = \frac{4 - 4}{-3 - 2}$$

$$= \frac{0}{-5}$$

$$= 0$$

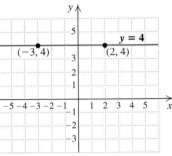

Any two points on a horizontal line have the same y-coordinate. Thus the change in y is 0, so the slope is 0.

A horizontal line has slope 0.

Example **4**

Find the slope of the line $x = -3$.

Solution   Consider the points $(-3, 4)$ and $(-3, -2)$, which are on the line. The change in y, or the rise, is $-2 - 4$, or -6. The change in x, or the run, is $-3 - (-3)$, or 0. Thus,

$$m = \frac{-2 - 4}{-3 - (-3)}$$

$$= \frac{-6}{0} \quad \text{(undefined)}$$

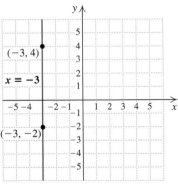

Since division by 0 is not defined, the slope of this line is not defined. The answer to a problem of this type is "The slope of this line is undefined."

The slope of a vertical line is undefined.

Applications

We have seen that slope has many real-world applications, ranging from car speed to production rate. Some applications use slope to measure steepness. For example, numbers like 2%, 3%, and 6% are often used to represent the **grade** of a road, a measure of a road's steepness. That is, a 3% grade means that

Study Tip

Immediately after each quiz or test, write out a step-by-step solution to any questions you missed. Visit your professor during office hours or consult with a tutor for help with problems that are still giving you trouble. Misconceptions tend to resurface if they are not corrected as soon as possible.

for every horizontal distance of 100 ft, the road rises or drops 3 ft. The concept of grade also occurs in skiing or snowboarding, where a 4% grade is considered very tame, but a 40% grade is considered steep.

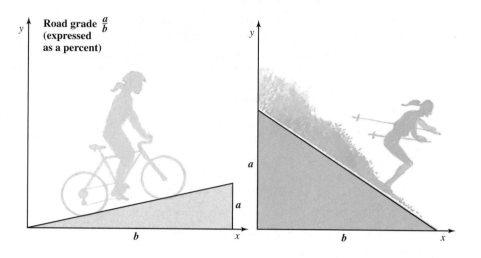

Example 5

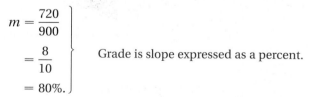

Skiing.   Among the steepest skiable terrain in North America, the Headwall on Mount Washington, in New Hampshire, drops 720 ft over a horizontal distance of 900 ft. Find the grade of the Headwall.

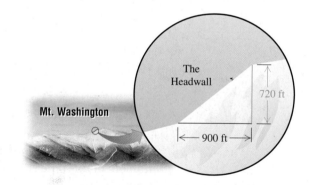

Solution   The grade of the Headwall is its slope, expressed as a percent:

$$m = \frac{720}{900}$$

$$= \frac{8}{10}$$

$$= 80\%.$$

Grade is slope expressed as a percent.

Carpenters use slope when designing stairs, ramps, or roof pitches. Another application occurs in the engineering of a dam—the force or strength of a river depends on how much the river drops over a specified distance.

CONNECTING THE CONCEPTS

In Chapters 1 and 2, we simplified expressions and solved equations. In Sections 3.1–3.3 of this chapter, we learned how a graph can be used to represent the solutions of an equation in two variables, like $y = 4 - x$ or $2x + 5y = 15$.

Graphs are used in many important applications. Sections 3.4 and 3.5 have shown us that the slope of a line can be used to represent the rate at which the quantity measured on the vertical axis changes with respect to the quantity measured on the horizontal axis.

In Section 3.6, we will return to the task of graphing equations. This time, however, our understanding of rates and slope will provide us with the tools necessary to develop shortcuts that will streamline our work.

Exercise Set 3.5

FOR EXTRA HELP

Digital Video Tutor CD 2
Videotape 6   InterAct Math   Math Tutor Center   MathXL   MyMathLab.com

1. Find the rate at which a runner burns calories.

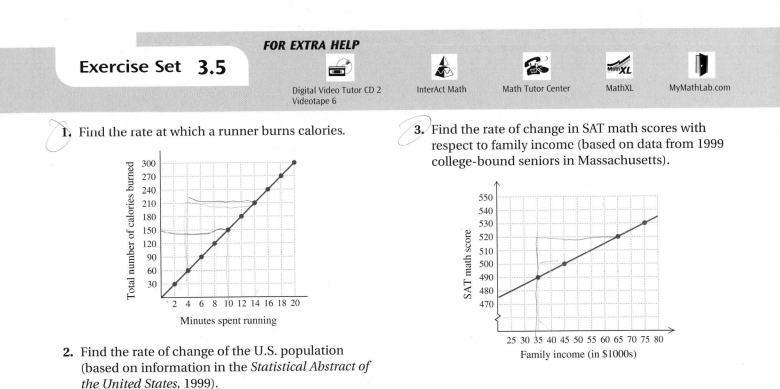

2. Find the rate of change of the U.S. population (based on information in the *Statistical Abstract of the United States,* 1999).

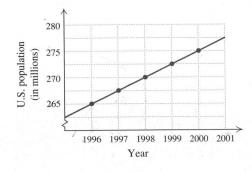

3. Find the rate of change in SAT math scores with respect to family income (based on data from 1999 college-bound seniors in Massachusetts).

4. Find the rate of change in SAT verbal scores with respect to family income (based on data from 1999 college-bound seniors in Massachusetts).

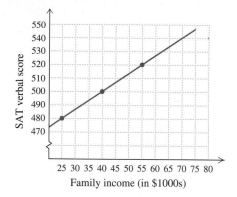

Family income (in $1000s)

5. Find the rate of change in the percent of the U.S. budget spent on defense (based on data from the U.S. Office of Management and Budget).

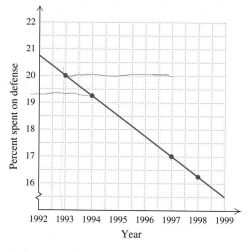

6. Find the rate of change in the percent of U.S. workers that are unemployed (*Source*: U.S. Bureau of Labor Statistics).

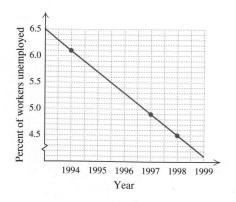

Find the slope, if it is defined, of each line. If the slope is undefined, state this.

7.

8.

9.

10.

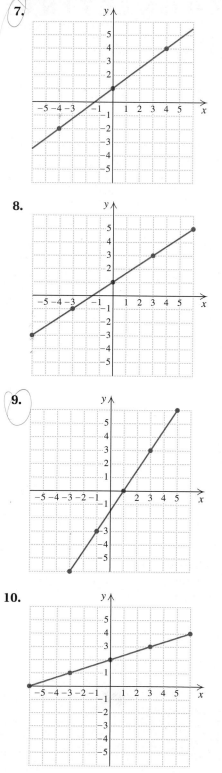

11.

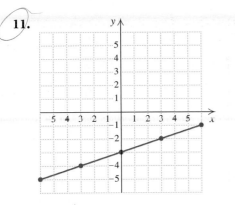

15.

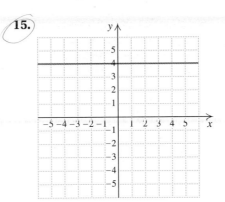

12.

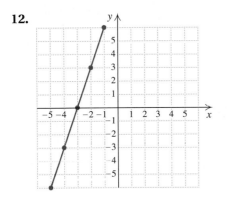

16.

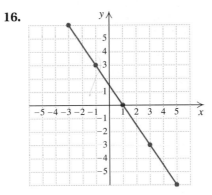

13.

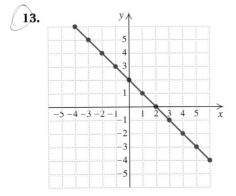

17.

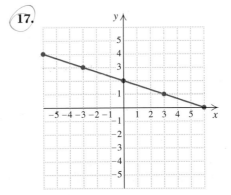

14.

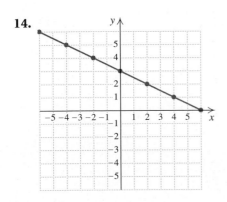

18.

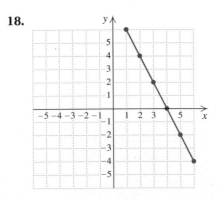

19.

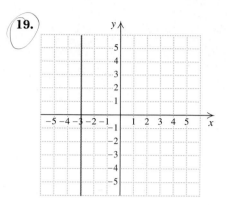

20.

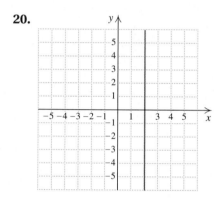

21.

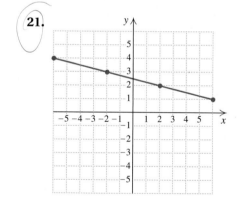

22.

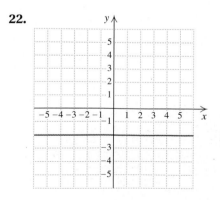

23.

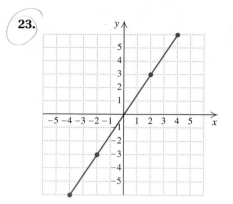

24.

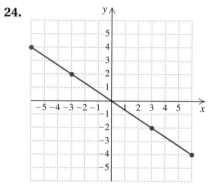

25.

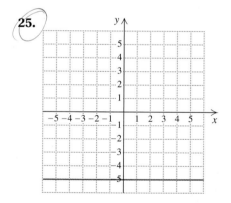

26.

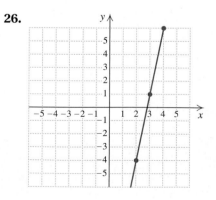

27.

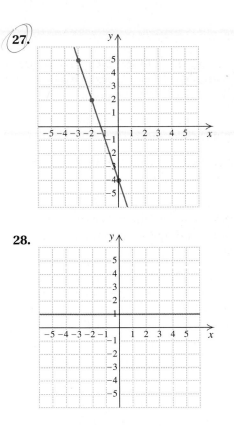

28.

Find the slope of the line containing each given pair of points. If the slope is undefined, state this.

29. (1, 2) and (5, 8)

30. (2, 1) and (6, 9)

31. (−2, 4) and (3, 0)

32. (−4, 2) and (2, −3)

33. (−4, 0) and (5, 7)

34. (3, 0) and (6, 2)

35. (0, 8) and (−3, 10)

36. (0, 9) and (4, 7)

37. (−2, 3) and (−6, 5)

38. (−2, 4) and (6, −7)

Aha! **39.** $\left(-2, \frac{1}{2}\right)$ and $\left(-5, \frac{1}{2}\right)$

40. (−5, −1) and (2, 3)

41. (3, 4) and (9, −7)

42. (−10, 3) and (−10, 4)

43. (6, −4) and (6, 5)

44. (5, −2) and (−4, −2)

Find the slope of each line. If the slope is undefined, state this.

45. $x = -3$ **46.** $x = -4$

47. $y = 4$ **48.** $y = 17$

49. $x = 9$ **50.** $x = 6$

51. $y = -9$ **52.** $y = -4$

53. *Surveying.* Tucked between two ski areas, Vermont Route 108 rises 106 m over a horizontal distance of 1325 m. What is the grade of the road?

54. *Navigation.* Capital Rapids drops 54 ft vertically over a horizontal distance of 1080 ft. What is the slope of the rapids?

55. *Architecture.* To meet federal standards, a wheelchair ramp cannot rise more than 1 ft over a horizontal distance of 12 ft. Express this slope as a grade.

56. *Engineering.* At one point, Yellowstone's Beartooth Highway rises 315 ft over a horizontal distance of 4500 ft. Find the grade of the road.

57. *Carpentry.* Find the slope (or pitch) of the roof.

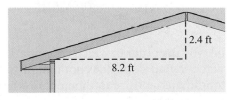

58. *Exercise.* Find the slope (or grade) of the treadmill.

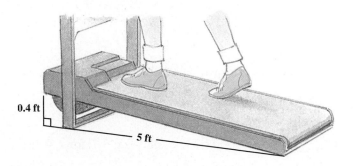

59. *Surveying.*   From a base elevation of 9600 ft, Longs Peak, Colorado, rises to a summit elevation of 14,255 ft over a horizontal distance of 15,840 ft. Find the grade of Longs Peak.

60. *Construction.*   Public buildings regularly include steps with 7-in. risers and 11-in. treads. Find the grade of such a stairway.

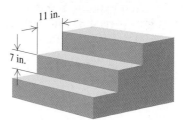

61. Explain why the order in which coordinates are subtracted to find slope does not matter so long as *y*-coordinates and *x*-coordinates are subtracted in the same order.

62. If one line has a slope of −3 and another has a slope of 2, which line is steeper? Why?

SKILL MAINTENANCE

Solve.

63. $ax + by = c$, for y

64. $rx - mn = p$, for r

65. $ax - by = c$, for y

66. $rs + nt = q$, for t

Evaluate.

67. $\frac{2}{3}x - 5$, for $x = 12$

68. $\frac{3}{5}x - 7$, for $x = 15$

SYNTHESIS

69. The points $(-4, -3)$, $(1, 4)$, $(4, 2)$, and $(-1, -5)$ are vertices of a quadrilateral. Use slopes to explain why the quadrilateral is a parallelogram.

70. Can the points $(-4, 0)$, $(-1, 5)$, $(6, 2)$, and $(2, -3)$ be vertices of a parallelogram? Why or why not?

71. A line passes through $(4, -7)$ and never enters the first quadrant. What numbers could the line have for its slope?

72. A line passes through $(2, 5)$ and never enters the second quadrant. What numbers could the line have for its slope?

73. *Architecture.*   Architects often use the equation $x + y = 18$ to determine the height y, in inches, of the riser of a step when the tread is x inches wide. Express the slope of stairs designed with this equation without using the variable y.

In Exercises 74 and 75, the slope of each line is $-\frac{2}{3}$, but the numbering on one axis is missing. How many units should each tick mark on that unnumbered axis represent?

74.

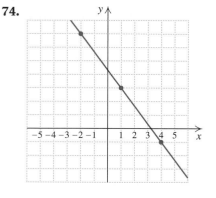

75.

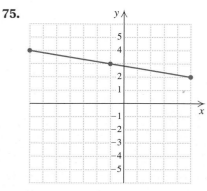

Slope–Intercept Form

3.6

Using the *y*-intercept and the Slope to Graph a Line • Equations in Slope–Intercept Form • Graphing and Slope–Intercept Form

If we know the slope and the *y*-intercept of a line, it is possible to graph the line. In this section, we will discover that a line's slope and *y*-intercept can be determined directly from the line's equation, provided the equation is written in a certain form.

Using the *y*-intercept and the Slope to Graph a Line

Let's modify the car production situation that first appeared in Section 3.5. Suppose that as a new workshift begins, 4 cars have already been produced. At the Michigan plant, 3 cars were being produced every 2 hours, a rate of $\frac{3}{2}$ cars per hour. If this rate remains the same regardless of how many cars have already been produced, the table and graph shown here can be made.

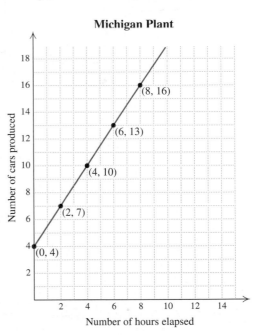

Michigan Plant

Michigan Plant	
Hours Elapsed	**Cars Produced**
0	4
2	7
4	10
6	13
8	16

To confirm that the production rate is still $\frac{3}{2}$, we calculate the slope. Recall that

$$\text{Slope} = \frac{\text{change in } y}{\text{change in } x} = \frac{\text{rise}}{\text{run}} = \frac{y_2 - y_1}{x_2 - x_1},$$

where (x_1, y_1) and (x_2, y_2) are any two points on the graphed line. Here we select $(0, 4)$ and $(2, 7)$:

$$\text{Slope} = \frac{\text{change in } y}{\text{change in } x} = \frac{7 - 4}{2 - 0} = \frac{3}{2}.$$

Knowing that the slope is $\frac{3}{2}$, we could have drawn the graph by plotting $(0, 4)$ and from there moving *up* 3 units and *to the right* 2 units. This would have located the point $(2, 7)$. Using $(0, 4)$ and $(2, 7)$, we can then draw the line. This is the method used in the next example.

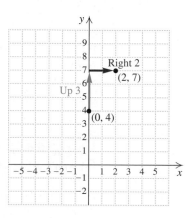

E x a m p l e   1

Draw a line that has slope $\frac{1}{4}$ and y-intercept $(0, 2)$.

Solution   We plot $(0, 2)$ and from there move *up* 1 unit and *to the right* 4 units. This locates the point $(4, 3)$. We plot $(4, 3)$ and draw a line passing through $(0, 2)$ and $(4, 3)$, as shown on the right below.

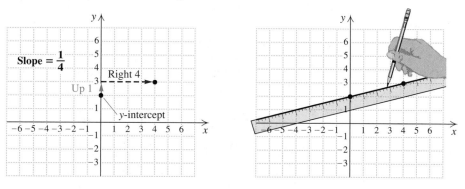

Equations in Slope–Intercept Form

It is not difficult to find a line's slope and y-intercept from its equation. Recall from Section 3.3 that to find the y-intercept of an equation's graph, we replace x with 0 and solve the resulting equation for y. For example, to find the y-intercept of the graph of $y = 2x + 3$, we replace x with 0 and solve as follows:

$$y = 2x + 3$$
$$= 2 \cdot 0 + 3 = 0 + 3 = 3. \qquad \text{The } y\text{-intercept is } (0, 3).$$

The y-intercept of the graph of $y = 2x + 3$ is $(0, 3)$. It can be similarly shown that the graph of $y = mx + b$ has the y-intercept $(0, b)$.

To calculate the slope of the graph of $y = 2x + 3$, we need two ordered pairs that are solutions of the equation. The y-intercept $(0, 3)$ is one pair; a second pair, $(1, 5)$, can be found by substituting 1 for x. We then have

$$\text{Slope} = \frac{\text{change in } y}{\text{change in } x} = \frac{5 - 3}{1 - 0} = \frac{2}{1} = 2.$$

Note that the slope, 2, is also the x-coefficient in $y = 2x + 3$. It can be similarly shown that the graph of any equation of the form $y = mx + b$ has slope m (see Exercise 77).

The Slope–Intercept Equation

The equation $y = mx + b$ is called the *slope–intercept equation*. The equation represents a line of slope m with y-intercept $(0, b)$.

The equation of any nonvertical line can be written in this form.

Example 2

Find the slope and the y-intercept of each line.

a) $y = \frac{4}{5}x - 8$ **b)** $2x + y = 5$ **c)** $3x + 4y = 7$

Solution

a) We rewrite $y = \frac{4}{5}x - 8$ as $y = \frac{4}{5}x + (-8)$. Now we simply read the slope and the y-intercept from the equation:

$$y = \frac{4}{5}x + (-8).$$

The slope is $\frac{4}{5}$. The y-intercept is $(0, -8)$.

b) We first solve for y to find an equivalent equation in the form $y = mx + b$:

$$2x + y = 5$$
$$y = -2x + 5. \qquad \text{Adding } -2x \text{ to both sides}$$

The slope is -2. The y-intercept is $(0, 5)$.

c) We rewrite the equation in the form $y = mx + b$:

$$3x + 4y = 7$$
$$4y = -3x + 7 \qquad \text{Adding } -3x \text{ to both sides}$$
$$y = \tfrac{1}{4}(-3x + 7) \qquad \text{Multiplying both sides by } \tfrac{1}{4}$$
$$y = -\tfrac{3}{4}x + \tfrac{7}{4}. \qquad \text{Using the distributive law}$$

The slope is $-\frac{3}{4}$, or $\frac{-3}{4}$, or $\frac{3}{-4}$. The y-intercept is $\left(0, \frac{7}{4}\right)$.

Example 3

A line has slope $-\frac{12}{5}$ and y-intercept $(0, 11)$. Find an equation of the line.

Solution   We use the slope–intercept equation, substituting $-\frac{12}{5}$ for m and 11 for b:

$$y = mx + b = -\frac{12}{5}x + 11.$$

The desired equation is $y = -\frac{12}{5}x + 11$.

Example 4

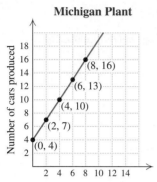

Michigan Plant

Determine an equation for the graph of car production shown at the beginning of this section.

Solution   To write an equation for a line, we can use slope–intercept form, provided the slope and the y-intercept are known. Using the coordinates of two points, we already found that the slope, or rate of production, is $\frac{3}{2}$. Since $(0, 4)$ is given, we know the y-intercept as well. The desired equation is

$$y = \frac{3}{2}x + 4, \qquad \text{Using } \frac{3}{2} \text{ for } m \text{ and 4 for } b$$

where y is the number of cars produced after x hours.

Graphing and Slope–Intercept Form

In Example 1, we drew a graph, knowing only the slope and the y-intercept. In Example 2, we determined the slope and the y-intercept of a line by examining its equation. We now combine the two procedures to develop a quick way to graph a linear equation.

Example 5

Graph: **(a)** $y = \frac{3}{4}x + 5$; **(b)** $2x + 3y = 3$.

Solution

a) From the equation $y = \frac{3}{4}x + 5$, we see that the slope of the graph is $\frac{3}{4}$ and the y-intercept is $(0, 5)$. We plot $(0, 5)$ and then consider the slope, $\frac{3}{4}$. Starting at $(0, 5)$, we plot a second point by moving *up* 3 units (since the numerator is *positive* and corresponds to the change in y) and *to the right* 4 units (since the denominator is *positive* and corresponds to the change in x). We reach a new point, $(4, 8)$.

We can also rewrite the slope as $\frac{-3}{-4}$. We again start at the y-intercept, $(0, 5)$, but move *down* 3 units (since the numerator is *negative* and corresponds to the change in y) and *to the left* 4 units (since the denominator is *negative* and corresponds to the change in x). We reach another point, $(-4, 2)$. Once two or three points have been plotted, the line representing all solutions of $y = \frac{3}{4}x + 5$ can be drawn.

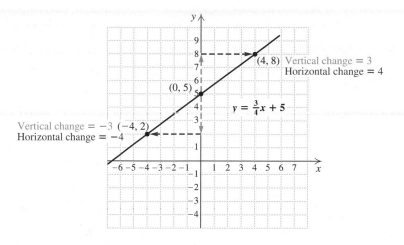

b) To graph $2x + 3y = 3$, we first rewrite it in slope–intercept form:

$$2x + 3y = 3$$
$$3y = -2x + 3 \qquad \text{Adding } -2x \text{ to both sides}$$
$$y = \tfrac{1}{3}(-2x + 3) \qquad \text{Multiplying both sides by } \tfrac{1}{3}$$
$$y = -\tfrac{2}{3}x + 1. \qquad \text{Using the distributive law}$$

To graph $y - -\frac{2}{3}x + 1$, we first plot the y-intercept, $(0, 1)$. We can think of the slope as $\frac{-2}{3}$. Starting at $(0, 1)$ and using the slope, we find a second point by moving *down* 2 units (since the numerator is *negative*) and *to the right* 3 units (since the denominator is *positive*). We plot the new point, $(3, -1)$. In a similar manner, we can move from the point $(3, -1)$ to locate a third point, $(6, -3)$. The line can then be drawn.

Since $-\frac{2}{3} = \frac{2}{-3}$, an alternative approach is to again plot $(0, 1)$, but this time move *up* 2 units (since the numerator is *positive*) and *to the left* 3 units (since the denominator is *negative*). This leads to another point on the graph, $(-3, 3)$.

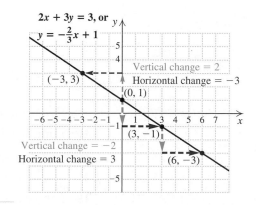

It is important to be able to use both $\frac{2}{-3}$ and $\frac{-2}{3}$ to draw the graph.

Slope–intercept form allows us to quickly determine the slope of a line by simply inspecting its equation. This can be especially helpful when attempting to decide whether two lines are parallel.

Example 6

technology connection

Using a standard $[-10, 10, -10, 10]$ window, graph the equations $y_1 = \frac{2}{3}x + 1$, $y_2 = \frac{3}{8}x + 1$, $y_3 = \frac{2}{3}x + 5$, and $y_4 = \frac{3}{8}x + 5$. If you can, use your grapher in the MODE that graphs equations *simultaneously.* Once all lines have been drawn, try to decide which equation corresponds to each line. After matching equations with lines, you can check your matches by using TRACE and the up and down arrow keys to move from one line to the next. The number of the equation will appear in a corner of the screen.

1. Graph $y_1 = -\frac{3}{4}x - 2$, $y_2 = -\frac{1}{5}x - 2$, $y_3 = -\frac{3}{4}x - 5$, and $y_4 = -\frac{1}{5}x - 5$ using the SIMULTANEOUS mode. Then match each line with the corresponding equation. Check using TRACE.

Determine whether the graphs of $y = -3x + 4$ and $6x + 2y = -10$ are parallel.

Solution Recall that *parallel* lines extend indefinitely without intersecting. Thus, when two lines have the same slope but different y-intercepts, they are parallel.

One of the two equations given,

$$y = -3x + 4,$$

represents a line with slope -3 and y-intercept $(0, 4)$. To find the slope of the other line, we need to rewrite

$$6x + 2y = -10$$

in slope–intercept form:

$$6x + 2y = -10$$
$$2y = -6x - 10 \qquad \text{Adding } -6x \text{ to both sides}$$
$$y = -3x - 5. \qquad \text{The slope is } -3 \text{ and the } y\text{-intercept is } (0, -5).$$

Since both lines have slope -3 but different y-intercepts, the graphs are parallel. There is no need for us to actually graph either equation.

FOR EXTRA HELP

Exercise Set **3.6**

Digital Video Tutor CD 2
Videotape 6

InterAct Math

Math Tutor Center

MathXL

MyMathLab.com

Draw a line that has the given slope and y-intercept.

1. Slope $\frac{2}{5}$; y-intercept $(0, 1)$

2. Slope $\frac{3}{5}$; y-intercept $(0, -1)$

3. Slope $\frac{5}{3}$; y-intercept $(0, -2)$

4. Slope $\frac{5}{2}$; y-intercept $(0, 1)$

5. Slope $-\frac{3}{4}$; y-intercept $(0, 5)$

6. Slope $-\frac{4}{5}$; y-intercept $(0, 6)$

7. Slope 2; y-intercept $(0, -4)$

8. Slope -2; y-intercept $(0, -3)$

9. Slope -3; y-intercept $(0, 2)$

10. Slope 3; y-intercept $(0, 4)$

Find the slope and the y-intercept of each line.

11. $y = \frac{3}{7}x + 5$

12. $y = -\frac{3}{8}x + 6$

13. $y = -\frac{5}{6}x + 2$

14. $y = \frac{7}{2}x + 4$

15. $y = \frac{9}{4}x - 7$

16. $y = \frac{2}{9}x - 1$

17. $y = -\frac{2}{5}x$

18. $y = \frac{4}{3}x$

19. $-2x + y = 4$

20. $-5x + y = 5$

21. $3x - 4y = 12$

22. $3x - 2y = 18$

23. $x - 5y = -8$

24. $x - 6y = 9$

Aha! **25.** $y = 4$

26. $y - 3 = 5$

Find the slope–intercept equation for the line with the indicated slope and y-intercept.

27. Slope 3; y-intercept $(0, 7)$

28. Slope -4; y-intercept $(0, -2)$

29. Slope $\frac{7}{8}$; y-intercept $(0, -1)$

30. Slope $\frac{5}{7}$; y-intercept $(0, 4)$

31. Slope $-\frac{5}{3}$; y-intercept $(0, -8)$

32. Slope $\frac{3}{4}$; y-intercept $(0, 23)$

Aha! **33.** Slope 0; y-intercept $(0, 3)$

34. Slope 7; y-intercept $(0, 0)$

Graph.

35. $y = \frac{3}{5}x + 2$

36. $y = -\frac{3}{5}x - 1$

37. $y = -\frac{3}{5}x + 1$

38. $y = \frac{3}{5}x - 2$

39. $y = \frac{5}{3}x + 3$

40. $y = \frac{5}{3}x - 2$

41. $y = -\frac{3}{2}x - 2$

42. $y = -\frac{4}{3}x + 3$

43. $2x + y = 1$

44. $3x + y = 2$

45. $3x - y = 4$

46. $2x - y = 5$

47. $2x + 3y = 9$

48. $4x + 5y = 15$

49. $x - 4y = 12$

50. $x + 5y = 20$

Solve.

51. *Cost of water.*   Freda and Phil are keeping track of their water bills. One month, while they were away on vacation, they used no water and were billed $9. The next month they used 70,000 gal and their bill was for $16. Let y represent the size of a monthly bill and x the amount of water used (in 10,000-gal units). Find and graph an equation of the form $y = mx + b$. Then determine the rate that they pay in dollars per 10,000 gallons.

52. *Cost of cable TV.*   Allegra and Larry are keeping track of how much their cable TV service costs. When service began, they paid $50 for installation. After one month, their total costs had risen to $70, and after two months they had paid a total of $90 for their cable service. Let y represent the amount paid for x months of service. Find and graph an equation of the form $y = mx + b$. Then determine their monthly rate.

53. *Refrigerator size.*   Kitchen designers recommend that a refrigerator be selected on the basis of the number of people in the household. For 1–2 people, a 16 ft³ model is suggested. For each additional person, an additional 1.5 ft³ is recommended. If x is the number of residents in excess of 2, find the slope–intercept equation for the recommended size of a refrigerator.

54. *Telephone service.*   In a recent promotion, AT&T charged a monthly fee of $4.95 plus 7¢ for each minute of long-distance phone calls. If x is the number of minutes of long-distance calls, find the slope–intercept equation for the monthly bill.

Determine whether each pair of equations represents parallel lines.

55. $y = \frac{2}{3}x + 7$,
$y = \frac{2}{3}x - 5$

56. $y = -\frac{5}{4}x + 1$,
$y = \frac{5}{4}x + 3$

57. $y = 2x - 5$,
$4x + 2y = 9$

58. $y = -3x + 1$,
$6x + 2y = 8$

59. $3x + 4y = 8$,
$7 - 12y = 9x$

60. $3x = 5y - 2$,
$10y = 4 - 6x$

61. Can a horizontal line be graphed using the method of Example 5? Why or why not?

62. Can a vertical line be graphed using the method of Example 5? Why or why not?

SKILL MAINTENANCE

Solve.

63. $y - k = m(x - h)$, for y

64. $y - 9 = -2(x + 4)$, for y

Simplify.

65. $-5 - (-7)$

66. $7 - (-9)$

67. $-3 - 6$

68. $-2 - 8$

SYNTHESIS

69. Explain how it is possible for an incorrect graph to be drawn, even after plotting three points that line up.

70. Which would you prefer, and why: graphing an equation of the form $y = mx + b$ or graphing an equation of the form $Ax + By = C$?

Two lines are perpendicular if either the product of their slopes is -1, or one line is vertical and the other horizontal. For Exercises 71–76, determine whether each pair of equations represents perpendicular lines.

71. $3y = 5x - 3$,
$3x + 5y = 10$

72. $y + 3x = 10$,
$2x - 6y = 18$

73. $3x + 5y = 10$,
$15x + 9y = 18$

74. $10 - 4y = 7x$,
$7y + 21 = 4x$

75. $x = 5$,
$y = \frac{1}{2}$

76. $y = -2x$,
$x = \frac{1}{2}$

77. Show that the slope of the line given by $y = mx + b$ is m. (*Hint*: Substitute both 0 and 1 for x to find two pairs of coordinates. Then use the formula, Slope = change in y/change in x.)

78. Write an equation of the line with the same slope as the line given by $5x + 2y = 8$ and the same y-intercept as the line given by $3x - 7y = 10$.

79. Write an equation of the line parallel to the line given by $2x - 6y = 10$ and having the same y-intercept as the line given by $9x + 6y = 18$.

80. Find an equation of the line parallel to the line given by $3x - 2y = 8$ and having the same y-intercept as the line given by $2y + 3x = -4$.

81. Find an equation of the line perpendicular to the line given by $2x + 5y = 6$ (see Exercises 71–76) that passes through $(2, 6)$. (*Hint*: Draw a graph.)

82. *Aerobic exercise.* The formula $T = -\frac{3}{4}a + 165$ can be used to determine the *target heart rate*, in beats per minute, for a person, a years old, participating in aerobic exercise. Graph the equation and interpret the significance of its slope.

CORNER

Draw the Graph and Match the Math

COLLABORATIVE

Focus: Slope–intercept form

Time: 15–20 minutes

Group size: 3

Materials: Graph paper and straightedges

It is important not only to be able to graph equations written in slope–intercept form, but to be able to match a linear graph with an appropriate equation.

ACTIVITY

1. Each group member should select a different one of the following sets of equations:

A. $y = \frac{4}{3}x - 5$,
$y = \frac{4}{3}x + 2$,
$y = \frac{1}{2}x - 5$,
$y = \frac{1}{2}x + 2$;

B. $y = \frac{2}{5}x + 1$,
$y = \frac{3}{4}x + 1$,
$y = \frac{2}{5}x - 1$,
$y = \frac{3}{4}x - 1$;

C. $y = \frac{3}{5}x + 2$,
$y = \frac{3}{5}x - 2$,
$y = \frac{4}{3}x + 2$,
$y = \frac{4}{3}x - 2$,

2. Working independently, each group member should graph the four equations he or she

has selected. Do not label the graphs with their corresponding equations, but instead list the four equations (in any random order) across the top of the graph paper.

3. After all group members have completed part (2), the sheets should be passed, clockwise, to the person on the left. This person should then attempt to match each of the four equations listed at the top of the graph paper with the appropriate graph below. If no graph appears to be appropriate, discuss the relevant equation with the group member who drew the graphs. If necessary, turn to the third group member for help in identifying any incorrect graphs.

4. Once all four equations and graphs have been matched, share your answers with the rest of the group. Make sure everyone agrees on all of the matches.

3.7

Point–Slope Form

Writing Equations in Point–Slope Form • Graphing and Point–Slope Form

There are many applications in which a slope—or a rate of change—and an ordered pair are known. When the ordered pair is the y-intercept, an equation in slope–intercept form can be easily produced. When the ordered pair represents a point other than the y-intercept, a different form, known as *point–slope form*, is more convenient.

Writing Equations in Point–Slope Form

Consider a line with slope 2 passing through the point $(4, 1)$, as shown in the figure. In order for a point (x, y) to be on the line, the coordinates x and y must be solutions of the slope equation

$$\frac{y - 1}{x - 4} = 2.$$

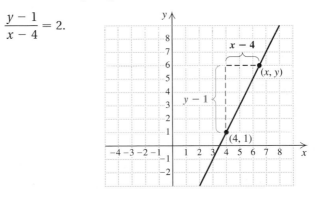

Take a moment to examine this equation. Pairs like $(5, 3)$ and $(3, -1)$ are solutions, since

$$\frac{3 - 1}{5 - 4} = 2 \quad \text{and} \quad \frac{-1 - 1}{3 - 4} = 2.$$

Note, however, that $(4, 1)$ is not itself a solution of the equation:

$$\frac{1 - 1}{4 - 4} \neq 2.$$

To avoid this difficulty, we can use the multiplication principle:

$$(x - 4) \cdot \frac{y - 1}{x - 4} = 2(x - 4) \qquad \text{Multiplying both sides by } x - 4$$

$$y - 1 = 2(x - 4). \qquad \text{Removing a factor equal to 1: } \frac{x - 4}{x - 4} = 1$$

This is considered **point–slope form** for the line shown above. A point–slope equation can be written any time a line's slope and a point on the line are known.

> ### The Point–Slope Equation
>
> The equation $y - y_1 = m(x - x_1)$ is called the *point–slope equation* for the line with slope m that contains the point (x_1, y_1).

Point–slope form is especially useful in more advanced mathematics courses, where problems similar to the following often arise.

E x a m p l e 1

Write a point–slope equation for the line with slope $\frac{1}{5}$ that contains the point $(7, 2)$.

Solution   We substitute $\frac{1}{5}$ for m, 7 for x_1, and 2 for y_1:

$$y - y_1 = m(x - x_1) \qquad \text{Using the point–slope equation}$$
$$y - 2 = \tfrac{1}{5}(x - 7) \qquad \text{Substituting}$$

E x a m p l e 2

Write a point–slope equation for the line with slope $-\frac{4}{3}$ that contains the point $(1, -6)$.

Solution   We substitute $-\frac{4}{3}$ for m, 1 for x_1, and -6 for y_1:

$$y - y_1 = m(x - x_1) \qquad \text{Using the point–slope equation}$$
$$y - (-6) = -\tfrac{4}{3}(x - 1). \qquad \text{Substituting}$$

E x a m p l e 3

Write the slope–intercept equation for the line with slope 3 that contains the point $(1, 9)$.

Solution   There are two parts to this solution. First, we write an equation in point–slope form:

$$y - y_1 = m(x - x_1)$$
$$y - 9 = 3(x - 1). \qquad \text{Substituting}$$

Next, we find an equivalent equation of the form $y = mx + b$:

$$y - 9 = 3(x - 1)$$
$$y - 9 = 3x - 3 \qquad \text{Using the distributive law}$$
$$y = 3x + 6. \qquad \text{Adding 9 to both sides to get slope–intercept form}$$

Graphing and Point–Slope Form

When we know a line's slope and a point that is on the line, we can draw the graph, much as we did in Section 3.6.

E x a m p l e   4

Graph the line with slope 2 that passes through $(-3, 1)$.

Solution   We plot $(-3, 1)$, move *up* 2 and *to the right* 1 $\left(\text{since } 2 = \frac{2}{1}\right)$, and draw the line.

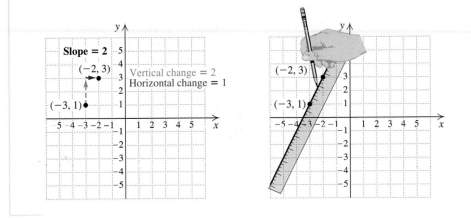

E x a m p l e   5

Graph: $y - 2 = 3(x - 4)$.

Solution   Since $y - 2 = 3(x - 4)$ is in point–slope form, we know that the line has slope 3, or $\frac{3}{1}$, and passes through the point $(4, 2)$. We plot $(4, 2)$ and then find a second point by moving *up* 3 units and *to the right* 1 unit. The line can then be drawn, as shown below.

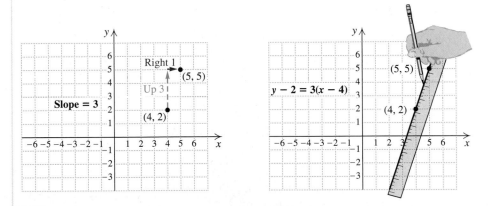

E x a m p l e   6

Graph: $y + 4 = -\frac{5}{2}(x + 3)$.

Solution   Once we have written the equation in point–slope form, $y - y_1 = m(x - x_1)$, we can proceed much as we did in Example 5. To find an equivalent equation in point–slope form, we subtract opposites instead of adding:

$$y + 4 = -\frac{5}{2}(x + 3)$$
$$y - (-4) = -\frac{5}{2}(x - (-3)).$$ Subtracting a negative instead of adding a positive. This is now in point–slope form.

From this last equation, $y - (-4) = -\frac{5}{2}(x - (-3))$, we see that the line passes through $(-3, -4)$ and has slope $-\frac{5}{2}$, or $\frac{5}{-2}$.

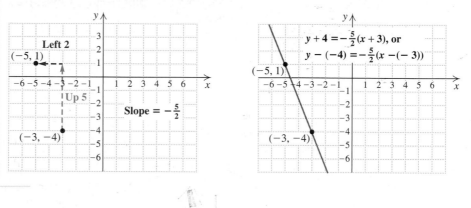

Exercise Set 3.7

Write a point–slope equation for the line with the given slope that contains the given point.

1. $m = 6;\ (2, 7)$ **2.** $m = 4;\ (3, 5)$

3. $m = \frac{3}{5};\ (9, 2)$ **4.** $m = \frac{2}{3};\ (4, 1)$

5. $m = -4;\ (3, 1)$ **6.** $m = -5;\ (6, 2)$

7. $m = \frac{3}{2};\ (5, -4)$ **8.** $m = \frac{4}{3};\ (7, -1)$

9. $m = \frac{5}{4};\ (-2, 6)$ **10.** $m = \frac{7}{2};\ (-3, 4)$

11. $m = -2;\ (-4, -1)$ **12.** $m = -3;\ (-2, -5)$

13. $m = 1;\ (-2, 8)$ **14.** $m = -1;\ (-3, 6)$

Write the slope–intercept equation for the line with the given slope that contains the given point.

15. $m = 2;\ (5, 7)$ **16.** $m = 3;\ (6, 2)$

17. $m = \frac{7}{4};\ (4, -2)$ **18.** $m = \frac{8}{3};\ (3, -4)$

19. $m = -3;\ (1, -5)$ **20.** $m = -2;\ (3, -1)$

21. $m = -4;\ (-2, -1)$ **22.** $m = -5;\ (-1, -4)$

23. $m = \frac{2}{3};\ (6, 5)$ **24.** $m = \frac{3}{2};\ (4, 7)$

25. $m = -\frac{5}{6};\ (3, 2)$ **26.** $m = -\frac{3}{4};\ (2, 5)$

27. Graph the line with slope $\frac{4}{3}$ that passes through the point $(1, 2)$.

28. Graph the line with slope $\frac{2}{5}$ that passes through the point $(3, 4)$.

29. Graph the line with slope $-\frac{3}{4}$ that passes through the point $(2, 5)$.

30. Graph the line with slope $-\frac{3}{2}$ that passes through the point $(1, 4)$.

Graph.

31. $y - 2 = \frac{1}{2}(x - 1)$ **32.** $y - 5 = \frac{1}{3}(x - 2)$

33. $y - 1 = -\frac{1}{2}(x - 3)$ **34.** $y - 1 = -\frac{1}{4}(x - 3)$

35. $y + 2 = \frac{1}{2}(x - 3)$ **36.** $y - 1 = \frac{1}{3}(x + 5)$

37. $y + 4 = 3(x + 1)$ **38.** $y + 3 = 2(x + 1)$

39. $y - 4 = -2(x + 1)$ **40.** $y + 3 = -1(x - 4)$

41. $y + 3 = -(x + 2)$ **42.** $y + 4 = 3(x + 2)$

43. $y + 1 = -\frac{3}{5}(x + 2)$ **44.** $y + 2 = -\frac{2}{3}(x + 1)$

45. $y - 1 = -\frac{7}{2}(x + 5)$ **46.** $y - 3 = -\frac{7}{4}(x + 1)$

47. Can equations for horizontal or vertical lines be written in point–slope form? Why or why not?

48. Describe a situation in which it is easier to graph the equation of a line in point–slope form rather than slope–intercept form.

SKILL MAINTENANCE

Simplify.

49. $(-5)^3$

50. $(-2)^6$

51. $3 \cdot 2^4 - 5 \cdot 2^3$

52. $5 \cdot 3^2 - 7 \cdot 3$

53. $(-2)^3(-3)^2$

54. $(5 - 7)^2(3 - 2 \cdot 2)$

SYNTHESIS

55. Describe a procedure that can be used to write the slope–intercept equation for any nonvertical line passing through two given points.

56. Any nonvertical line has many equations in point–slope form, but only one in slope–intercept form. Why is this?

Graph.

Aha! **57.** $y - 3 = 0(x - 52)$

58. $y + 4 = 0(x + 93)$

Write two different point–slope equations for the line passing through each pair of points.

59. $(1, 2)$ and $(3, 7)$

60. $(3, 1)$ and $(7, 3)$

61. $(-1, 2)$ and $(3, 8)$

62. $(-3, 1)$ and $(4, 3)$

63. $(-3, 8)$ and $(1, -2)$

64. $(-2, 7)$ and $(4, -3)$

Write the slope–intercept equation for each line shown.

65.

66.

67.

68.

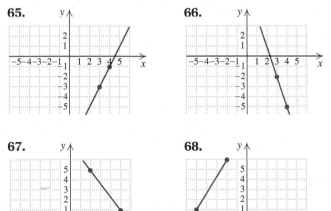

Write the slope–intercept equation for the line containing the given pair of points.

69. $(1, 5)$ and $(4, 2)$

70. $(3, 7)$ and $(4, 8)$

71. $(-3, 1)$ and $(3, 5)$

72. $(-2, 3)$ and $(2, 5)$

73. $(5, 0)$ and $(0, -2)$

74. $(-2, 0)$ and $(0, 3)$

75. $(-2, -4)$ and $(2, -1)$

76. $(-3, 5)$ and $(-1, -3)$

77. Write a point–slope equation of the line passing through $(-4, 7)$ that is parallel to the line given by $2x + 3y = 11$.

78. Write a point–slope equation of the line passing through $(3, -1)$ that is parallel to the line given by $4x - 5y = 9$.

Aha! **79.** Write an equation of the line parallel to the line given by $y = 3 - 4x$ that passes through $(0, 7)$.

80. Write the slope–intercept equation of the line that has the same y-intercept as the line $x - 3y = 6$ and contains the point $(5, -1)$.

81. Write the slope–intercept equation of the line that contains the point $(-1, 5)$ and is parallel to the line passing through $(2, 7)$ and $(-1, -3)$.

82. Write the slope–intercept equation of the line that has x-intercept $(-2, 0)$ and is parallel to $4x - 8y = 12$.

83. Why is slope–intercept form more useful than point–slope form when using a grapher? How can point–slope form be modified so that it is more easily used with graphers?

Summary and Review 3

Key Terms

Bar graph, p. 130
Circle graph, p. 131
Pie chart, p. 131
Graph, plot, p. 133
Axis (plural, axes), p. 133
Origin, p. 133
Coordinates, p. 133
Ordered pairs, p. 133
Quadrants, p. 135
Interpolation, p. 135
Extrapolation, p. 135
Graph of equation, p. 143

Linear equation, p. 143
Nonlinear graph, p. 148
y-intercept, p. 153
x-intercept, p. 153
Rate, p. 161
Speed, p. 162
Slope, p. 172
Grade, p. 175
Slope–intercept form, p. 185
Parallel, p. 188
Point–slope form, p. 192

Important Properties and Formulas

To Graph a Linear Equation

1. Select a value for one coordinate and calculate the corresponding value of the other coordinate. Form an ordered pair. This pair is one solution of the equation.
2. Repeat step (1) to find at least one other ordered pair.
3. Plot the ordered pairs and draw a straight line passing through the points. The line represents all solutions of the equation.

To Find Intercepts

To find a y-intercept, let $x = 0$ and solve for y.

To find an x-intercept, let $y = 0$ and solve for x.

$$\text{Slope} = m = \frac{\text{change in } y}{\text{change in } x} = \frac{\text{rise}}{\text{run}} = \frac{y_2 - y_1}{x_2 - x_1}$$

Horizontal line: Slope is 0.
Vertical line: Slope is undefined.
Parallel lines: Slopes equal or both lines are vertical

Slope–intercept equation: $y = mx + b$
Point–slope equation: $y - y_1 = m(x - x_1)$

Review Exercises

The following circle graph shows a breakdown of the charities to which Americans donated in a recent year (Source: Giving USA 1998/American Association of Fund-Raising Counsel Trust for Philanthropy). Use the graph for Exercises 1 and 2.

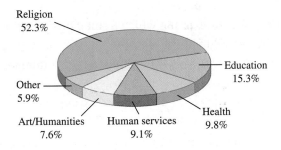

Religion 52.3%

Education 15.3%

Other 5.9%

Art/Humanities 7.6%

Human services 9.1%

Health 9.8%

1. The citizens of Ferrisburg are typical of Americans in general with regard to charitable contributions. If the citizens donated a total of $2 million to charities, how much was given to education?

2. About 5% of the Sophrins' income of $70,000 goes to charity. If their giving habits are typical, approximate the amount that they contribute to human services.

Plot each point.

3. $(2, -3)$ **4.** $(1, 0)$ **5.** $(2, 4)$

In which quadrant is each point located?

6. $(5, -10)$ **7.** $(-16.5, -20.3)$ **8.** $(-14, 7)$

Find the coordinates of each point in the figure.

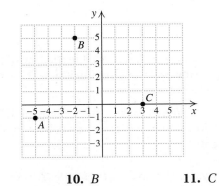

9. A **10.** B **11.** C

Determine whether the equation $y = 2x - 5$ has each ordered pair as a solution.

12. $(-3, 1)$ **13.** $(3, 1)$

14. Show that the ordered pairs $(0, -3)$ and $(2, 1)$ are solutions of the equation $2x - y = 3$. Then use the graph of the two points to determine another solution. Answers may vary.

Graph.

15. $y = x - 5$ **16.** $y = -\frac{1}{4}x$

17. $y = -x + 4$ **18.** $4x + y = 3$

19. $4x + 5 = 3$ **20.** $5x - 2y = 10$

21. *Meal service.*   At 8:30 A.M., the Colchester Boy Scouts had served 45 people at their annual pancake breakfast. By 9:15, the total served had reached 65.

   a) Find the Boy Scouts' serving rate, in number of meals per minute.

   b) Find the Boy Scouts' serving rate, in minutes per meal.

22. *U.S. population.*   The following graph shows data for the size of the U.S. population (*Source: Statistical Abstract of the United States, 1999*). At what rate has the population been growing?

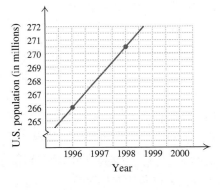

Find the slope of each line.

23.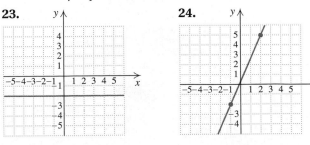

24.

25.

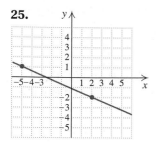

Find the slope of the line containing the given pair of points.

26. $(6, 8)$ and $(-2, -4)$

27. $(5, 1)$ and $(-1, 1)$

28. $(-3, 0)$ and $(-3, 5)$

29. $(-8.3, 4.6)$ and $(-9.9, 1.4)$

30. A road drops 369.6 ft vertically over a horizontal distance of 5280 ft. What is the grade of the road?

31. Find the x-intercept and the y-intercept of the line given by $3x + 2y = 18$.

32. Find the slope and the y-intercept of the line given by $2x + 4y = 20$.

33. Write the slope–intercept equation of the line with slope $-\frac{3}{4}$ and y-intercept $(0, 6)$.

34. Write a point–slope equation for the line with slope $-\frac{1}{2}$ that contains the point $(3, 6)$.

35. Write the slope–intercept equation for the line with slope 4 that contains the point $(-3, -7)$.

Graph.

36. $y = \frac{2}{3}x - 5$

37. $2x + y = 4$

38. $y = 6$

39. $x = -2$

40. $y + 2 = -\frac{1}{2}(x - 3)$

SYNTHESIS

41. Describe two ways in which a small business might make use of graphs.

42. Explain why the first coordinate of the y-intercept is always 0.

43. Find the value of m in $y = mx + 3$ such that $(-2, 5)$ is on the graph.

44. Find the value of b in $y = -5x + b$ such that $(3, 4)$ is on the graph.

45. Find the area and the perimeter of a rectangle for which $(-2, 2)$, $(7, 2)$, and $(7, -3)$ are three of the vertices.

46. Find three solutions of $y = 4 - |x|$.

Chapter Test 3

Use of tax dollars.   *The following pie chart shows how federal income tax dollars are spent.*

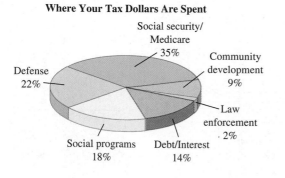

Where Your Tax Dollars Are Spent

Social security/Medicare 35%
Community development 9%
Defense 22%
Law enforcement 2%
Social programs 18%
Debt/Interest 14%

1. Debbie pays 18% of her taxable income of $31,200 in taxes. How much of her income will go to law enforcement?

2. Larry pays 16% of his taxable income of $27,000 in taxes. How much of his income will go to social programs?

In which quadrant is each point located?

3. $\left(-\frac{1}{2}, 7\right)$

4. $(-5, -6)$

Find the coordinates of each point in the figure.

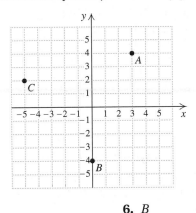

5. *A*

6. *B*

7. *C*

Graph.

8. $y = 2x - 1$

9. $2x - 4y = -8$

10. $y + 1 = 6$

11. $y = \frac{3}{4}x$

12. $y = 7$

13. $2x - y = 3$

14. $x = -1$

Find the x- and y-intercepts. Do not graph.

15. $5x - 3y = 30$

16. $x = 10 - 4y$

Find the slope of the line containing each pair of points.

17. $(4, -1)$ and $(6, 8)$

18. $(-3, -5)$ and $(9, 2)$

19. *Running.* Ted reached the 3-km mark of a race at 2:15 P.M. and the 6-km mark at 2:24 P.M. What is his running rate?

20. Find the slope and the *y*-intercept of the line given by $y - 3x = 7$.

Graph.

21. $y = \frac{1}{4}x - 2$

22. $y + 4 = -\frac{1}{2}(x - 3)$

23. Write a point–slope equation for the line of slope -3 that contains the point $(6, 8)$.

SYNTHESIS

24. Write an equation of the line that is parallel to the graph of $2x - 5y = 6$ and has the same *y*-intercept as the graph of $3x + y = 9$.

25. A diagonal of a square connects the points $(-3, -1)$ and $(2, 4)$. Find the area and the perimeter of the square.

Cumulative Review 1–3

1. Evaluate $\dfrac{x}{2y}$ for $x = 60$ and $y = 2$.

2. Multiply: $3(4x - 5y + 7)$.

3. Factor: $15x - 9y + 3$.

4. Find the prime factorization of 42.

5. Find decimal notation: $\frac{9}{20}$.

6. Find the absolute value: $|-4|$.

7. Find the opposite of $-\frac{1}{4}$.

8. Find the reciprocal of $-\frac{1}{4}$.

9. Combine like terms: $2x - 5y + (-3x) + 4y$.

10. Find decimal notation: 78.5%.

Simplify.

11. $\frac{3}{5} - \frac{5}{12}$

12. $3.4 + (-0.8)$

13. $(-2)(-1.4)(2.6)$

14. $\frac{3}{8} \div \left(-\frac{9}{10}\right)$

15. $2 - [32 \div (4 + 2^2)]$

16. $-5 + 16 \div 2 \cdot 4$

17. $y - (3y + 7)$

18. $3(x - 1) - 2[x - (2x + 7)]$

Solve.

19. $1.5 = 2.7 + x$

20. $\frac{2}{7}x = -6$

21. $5x - 9 = 36$

22. $\frac{2}{3} = \frac{-m}{10}$

23. $5.4 - 1.9x = 0.8x$

24. $x - \frac{7}{8} = \frac{3}{4}$

25. $2(2 - 3x) = 3(5x + 7)$

26. $\frac{1}{4}x - \frac{2}{3} = \frac{3}{4} + \frac{1}{3}x$

27. $y + 5 - 3y = 5y - 9$

28. $x - 28 < 20 - 2x$

29. $2(x + 2) \geq 5(2x + 3)$

30. Solve $A = 2\pi rh + \pi r^2$ for h.

31. In which quadrant is the point $(3, -1)$ located?

32. Graph on a number line: $-1 < x \leq 2$.

Graph.

33. $y = -2$

34. $2x + 5y = 10$

35. $y = -2x + 1$

36. $y = \frac{2}{3}x$

Find the coordinates of the x- and y-intercepts. Do not graph.

37. $2x - 7y = 21$

38. $y = 4x + 5$

Solve.

39. *Donating blood.* Each year 8 million Americans donate blood. This is 5% of those healthy enough to do so (*Source*: *Indianapolis Star,* 10/6/96). How many Americans are eligible to donate blood?

40. *Blood types.* There are 117 million Americans with either O-positive or O-negative blood. Those with O-positive blood outnumber those with O-negative blood by 85.8 million. How many Americans have O-negative blood?

41. Tina paid $126 for a cordless drill, including a 5% sales tax. How much did the drill itself cost?

42. A 143-m wire is cut into three pieces. The second is 3 m longer than the first. The third is four fifths as long as the first. How long is each piece?

43. Cory's contract stipulates that he cannot work more than 40 hr per week. For the first 4 days of one week, he worked 7, 10, 9, and 6 hr. How many hours can he work the fifth day and not violate his contract?

44. *Fund raising.* The graph at the top of the next column shows data from a recent road race held to raise money for Amnesty International. At what rate was money raised?

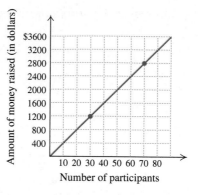

45. Find the slope of the line containing the points $(-4, 1)$ and $(2, -1)$.

46. Write an equation of the line with slope $\frac{2}{7}$ and y-intercept $(0, -4)$.

47. Find the slope and the y-intercept of the line given by $2x + 6y = 18$.

Graph.

48. $y = \frac{4}{3}x - 2$

49. $2x + 3y = -12$

50. Write a point–slope equation of the line with slope $-\frac{3}{8}$ that contains the point $(-6, 4)$.

SYNTHESIS

51. *Yearly earnings.* Paula's salary at the end of a year is $26,780. This reflects a 4% salary increase that preceded a 3% cost-of-living adjustment during the year. What was her salary at the beginning of the year?

Solve. If no solution exists, state this.

52. $4|x| - 13 = 3$

53. $4(x + 2) = 9(x - 2) + 16$

54. $2(x + 3) + 4 = 0$

55. $\frac{2 + 5x}{4} = \frac{11}{28} + \frac{8x + 3}{7}$

56. $5(7 + x) = (x + 6)5$

57. Solve $p = \frac{2}{m + Q}$ for Q.

58. The points $(-3, 0)$, $(0, 7)$, $(3, 0)$, and $(0, -7)$ are vertices of a parallelogram. Find four equations of lines that intersect to form the parallelogram.

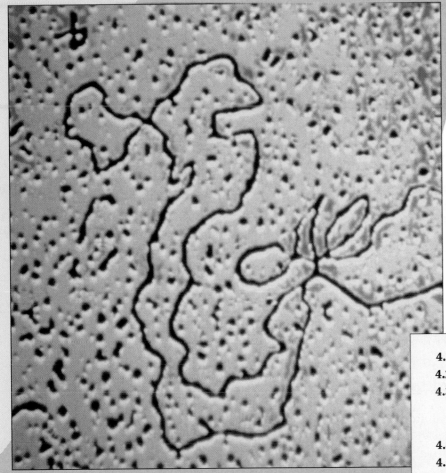

4
Polynomials

AN APPLICATION

A strand of DNA (deoxyribonucleic acid) is about 1.5 m long and 1.3×10^{-10} cm wide (*Source*: Human Genome Project Information). How many times longer is DNA than it is wide?

This problem appears as Exercise 153 in Section 4.8.

A s a biochemist, I cannot conduct an experiment without using math. For example, to investigate the effect of enzyme-to-protein ratios, I must calculate how much protein I have, on the basis of the concentration and volume, and then calculate the total amount of enzyme to add.

KIRA L. FORD
Biochemist
Indianapolis, Indiana

*O*ur work in Chapter 3 concentrated on using graphs to represent solutions of equations in two variables. Here in Chapter 4 we will focus on finding equivalent expressions, not on solving equations.

Algebraic expressions such as $16t^2$, $5a^2 - 3ab$, and $3x^2 - 7x + 5$ are called polynomials. Polynomials occur frequently in applications and appear in most branches of mathematics. Thus learning to add, subtract, multiply, and divide polynomials is an important part of most courses in elementary algebra and is the focus of this chapter.

Exponents and Their Properties

4.1

Multiplying Powers with Like Bases • Dividing Powers with Like Bases • Zero as an Exponent • Raising a Power to a Power • Raising a Product or a Quotient to a Power

In Section 4.2, we begin our study of polynomials. Before doing so, however, we must develop some rules for manipulating exponents.

Multiplying Powers with Like Bases

Recall from Section 1.8 that an expression like a^3 means $a \cdot a \cdot a$. We can use this fact to find the product of two expressions that have the same base:

$$a^3 \cdot a^2 = (a \cdot a \cdot a)(a \cdot a)$$ There are three factors in a^3; two factors in a^2.

$$= a \cdot a \cdot a \cdot a \cdot a$$ Using an associative law

$$= a^5.$$

Note that the exponent in a^5 is the sum of the exponents in $a^3 \cdot a^2$. That is, $3 + 2 = 5$. Similarly,

$$b^4 \cdot b^3 = (b \cdot b \cdot b \cdot b)(b \cdot b \cdot b)$$

$$= b^7, \quad \text{where } 4 + 3 = 7.$$

Adding the exponents gives the correct result.

The Product Rule

For any number a and any positive integers m and n,

$$a^m \cdot a^n = a^{m+n}.$$

(To multiply powers with the same base, keep the base and add the exponents.)

E x a m p l e   1

Multiply and simplify each of the following. (Here "simplify" means express the product as one base to a power whenever possible.)

a) $x^2 \cdot x^9$
b) $5 \cdot 5^8 \cdot 5^3$
c) $(r + s)^7(r + s)^6$
d) $(a^3b^2)(a^3b^5)$

Solution

a) $x^2 \cdot x^9 = x^{2+9}$ Adding exponents: $a^m \cdot a^n = a^{m+n}$
         $= x^{11}$

b) $5 \cdot 5^8 \cdot 5^3 = 5^1 \cdot 5^8 \cdot 5^3$ Recall that $x^1 = x$ for any number x.
         $= 5^{1+8+3}$ Adding exponents
         $= 5^{12}$ —————| *Caution!* $5^{12} \neq 5 \cdot 12$. |

c) $(r + s)^7(r + s)^6 = (r + s)^{7+6}$ The base here is $r + s$.
         $= (r + s)^{13}$ ———| *Caution!* $(r + s)^{13} \neq r^{13} + s^{13}$. |

d) $(a^3b^2)(a^3b^5) = a^3b^2a^3b^5$ Using an associative law
         $= a^3a^3b^2b^5$ Using a commutative law
         $= a^6b^7$ Adding exponents

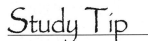

Study Tip

When you feel confident in your command of a topic, don't hesitate to help classmates experiencing trouble. Your understanding and retention of a concept will deepen when you explain it to someone else and your classmate will appreciate your help.

Dividing Powers with Like Bases

Recall that any expression that is divided or multiplied by 1 is unchanged. This, together with the fact that anything (besides 0) divided by itself is 1, can lead to a rule for division:

$$\frac{a^5}{a^2} = \frac{a \cdot a \cdot a \cdot a \cdot a}{a \cdot a}$$

$$= \frac{a \cdot a \cdot a}{1} \cdot \frac{a \cdot a}{a \cdot a}$$

$$= \frac{a \cdot a \cdot a}{1} \cdot 1$$

$$= a \cdot a \cdot a = a^3.$$

Note that the exponent in a^3 is the difference of the exponents in a^5/a^2. Similarly,

$$\frac{x^4}{x^3} = \frac{x \cdot x \cdot x \cdot x}{x \cdot x \cdot x} = \frac{x}{1} \cdot \frac{x \cdot x \cdot x}{x \cdot x \cdot x} = \frac{x}{1} \cdot 1 = x^1, \quad \text{or } x.$$

Subtracting the exponents gives the correct result.

> ### The Quotient Rule
>
> For any nonzero number a and any positive integers m and n for which $m > n$,
>
> $$\frac{a^m}{a^n} = a^{m-n}.$$
>
> (To divide powers with the same base, subtract the exponent of the denominator from the exponent of the numerator.)

E x a m p l e   2   Divide and simplify. (Here "simplify" means express the quotient as one base to a power whenever possible.)

a) $\dfrac{x^8}{x^2}$ **b)** $\dfrac{7^9}{7^4}$ **c)** $\dfrac{(5a)^{12}}{(5a)^4}$ **d)** $\dfrac{p^5 q^7}{p^2 q}$

Solution

a) $\dfrac{x^8}{x^2} = x^{8-2}$ Subtracting exponents: $\dfrac{a^m}{a^n} = a^{m-n}$

     $= x^6$

b) $\dfrac{7^9}{7^4} = 7^{9-4}$

     $= 7^5$

c) $\dfrac{(5a)^{12}}{(5a)^4} = (5a)^{12-4} = (5a)^8$ The base here is $5a$.

d) $\dfrac{p^5 q^7}{p^2 q} = \dfrac{p^5}{p^2} \cdot \dfrac{q^7}{q^1} = p^{5-2} \cdot q^{7-1} = p^3 q^6$ Using the quotient rule twice

Zero as an Exponent

The quotient rule can be used to help determine what 0 should mean when it appears as an exponent. Consider a^4/a^4, where a is nonzero. Since the numerator and the denominator are the same,

$$\frac{a^4}{a^4} = 1.$$

On the other hand, using the quotient rule would give us

$$\frac{a^4}{a^4} = a^{4-4} = a^0. \qquad \text{Subtracting exponents}$$

Since $a^0 = a^4/a^4 = 1$, this suggests that $a^0 = 1$ for any nonzero value of a.

> **The Exponent Zero**
> For any real number a, $a \neq 0$,
> $$a^0 = 1.$$
> (Any nonzero number raised to the 0 power is 1.)

Note that in the above box, 0^0 is not defined. For this text, we will assume that expressions like a^m do not represent 0^0.

Example 3

Simplify: **(a)** 1948^0; **(b)** $(-9)^0$; **(c)** $(3x)^0$; **(d)** $(-1)9^0$.

Solution

a) $1948^0 = 1$ Any nonzero number raised to the 0 power is 1.

b) $(-9)^0 = 1$ Any nonzero number raised to the 0 power is 1.
                      The base here is -9.

c) $(3x)^0 = 1$, for any $x \neq 0$.   The parentheses indicate that the base is $3x$.

d) We have

$$(-1)9^0 = (-1)1 = -1.$$ The base here is 9.

Recall that, unless there are calculations within parentheses, exponents are calculated before multiplication. Since multiplying by -1 is the same as finding the opposite, the expression $(-1)9^0$ could have been written as -9^0. Note that although $-9^0 = -1$, part (b) shows that $(-9)^0 = 1$.

> **Caution!** $-9^0 \neq (-9)^0$, and, in general, $-a^n \neq (-a)^n$.

Raising a Power to a Power

Consider an expression like $(7^2)^4$:

$$(7^2)^4 = (7^2)(7^2)(7^2)(7^2)$$ There are four factors of 7^2.

$$= (7 \cdot 7)(7 \cdot 7)(7 \cdot 7)(7 \cdot 7)$$ We could also use the product rule.

$$= 7 \cdot 7 \cdot 7 \cdot 7 \cdot 7 \cdot 7 \cdot 7 \cdot 7$$ Using an associative law

$$= 7^8.$$

Note that the exponent in 7^8 is the product of the exponents in $(7^2)^4$. Similarly,

$$(y^5)^3 = y^5 \cdot y^5 \cdot y^5$$ There are three factors of y^5.

$$= (y \cdot y \cdot y \cdot y \cdot y)(y \cdot y \cdot y \cdot y \cdot y)(y \cdot y \cdot y \cdot y \cdot y)$$

$$= y^{15}.$$

Once again, we get the same result if we multiply exponents:

$$(y^5)^3 = y^{5 \cdot 3} = y^{15}.$$

> ### The Power Rule
>
> For any number a and any whole numbers m and n,
>
> $$(a^m)^n = a^{mn}.$$
>
> (To raise a power to a power, multiply the exponents and leave the base unchanged.)

Remember that for this text we assume that 0^0 is not considered.

E x a m p l e   4   Simplify: **(a)** $(m^2)^5$; **(b)** $(3^5)^4$.

Solution

a) $(m^2)^5 = m^{2 \cdot 5}$ Multiplying exponents: $(a^m)^n = a^{mn}$

   $= m^{10}$

b) $(3^5)^4 = 3^{5 \cdot 4}$

   $= 3^{20}$

Raising a Product or a Quotient to a Power

When an expression inside parentheses is raised to a power, the inside expression is the base. Let's compare $2a^3$ and $(2a)^3$:

$$2a^3 = 2 \cdot a \cdot a \cdot a; \qquad \text{The base is } a.$$

$$(2a)^3 = (2a)(2a)(2a) \qquad \text{The base is } 2a.$$

$$= (2 \cdot 2 \cdot 2)(a \cdot a \cdot a) \qquad \text{Using an associative and a commutative law}$$

$$= 2^3 a^3$$

$$= 8a^3.$$

We see that $2a^3$ and $(2a)^3$ are *not* equivalent. Note too that $(2a)^3$ can be simplified by cubing each factor. This leads to the following rule for raising a product to a power.

> ### Raising a Product to a Power
>
> For any numbers a and b and any whole number n,
>
> $$(ab)^n = a^n b^n.$$
>
> (To raise a product to a power, raise each factor to that power.)

E x a m p l e   5

Simplify: **(a)** $(4a)^3$; **(b)** $(-5x^4)^2$; **(c)** $(a^7b)^2(a^3b^4)$.

Solution

a) $(4a)^3 = 4^3a^3 = 64a^3$ Raising each factor to the third power and simplifying

b) $(-5x^4)^2 = (-5)^2(x^4)^2$ Raising each factor to the second power. Parentheses are important here.

$\qquad\qquad = 25x^8$ Simplifying $(-5)^2$ and using the product rule

c) $(a^7b)^2(a^3b^4) = (a^7)^2b^2a^3b^4$ Raising a product to a power

$\qquad\qquad\quad = a^{14}b^2a^3b^4$ Multiplying exponents

$\qquad\qquad\quad = a^{17}b^6$ Adding exponents

Caution!   The rule $(ab)^n = a^nb^n$ applies only to *products* raised to a power, not to sums or differences. For example, $(3 + 4)^2 \neq 3^2 + 4^2$ since $7^2 \neq 9 + 16$.

There is a similar rule for raising a quotient to a power.

Raising a Quotient to a Power

For any numbers a and b, $b \neq 0$, and any whole number n,

$$\left(\frac{a}{b}\right)^n = \frac{a^n}{b^n}.$$

(To raise a quotient to a power, raise the numerator to the power and divide by the denominator to the power.)

E x a m p l e   6

Simplify: **(a)** $\left(\dfrac{x}{5}\right)^2$; **(b)** $\left(\dfrac{5}{a^4}\right)^3$; **(c)** $\left(\dfrac{3a^4}{b^3}\right)^2$.

Solution

a) $\left(\dfrac{x}{5}\right)^2 = \dfrac{x^2}{5^2} = \dfrac{x^2}{25}$ Squaring the numerator and the denominator

b) $\left(\dfrac{5}{a^4}\right)^3 = \dfrac{5^3}{(a^4)^3}$ Raising a quotient to a power

$\qquad\qquad = \dfrac{125}{a^{4\cdot 3}} = \dfrac{125}{a^{12}}$ Using the power rule and simplifying

c) $\left(\dfrac{3a^4}{b^3}\right)^2 = \dfrac{(3a^4)^2}{(b^3)^2}$ Raising a quotient to a power

$\qquad\qquad = \dfrac{3^2(a^4)^2}{b^{3\cdot 2}} = \dfrac{9a^8}{b^6}$ Raising a product to a power and using the power rule

In the following summary of definitions and rules, we assume that no denominators are 0 and 0^0 is not considered.

Definitions and Properties of Exponents

For any whole numbers m and n,

1 as an exponent:	$a^1 = a$
0 as an exponent:	$a^0 = 1$
The Product Rule:	$a^m \cdot a^n = a^{m+n}$
The Quotient Rule:	$\dfrac{a^m}{a^n} = a^{m-n}$
The Power Rule:	$(a^m)^n = a^{mn}$
Raising a product to a power:	$(ab)^n = a^n b^n$
Raising a quotient to a power:	$\left(\dfrac{a}{b}\right)^n = \dfrac{a^n}{b^n}$

Exercise Set 4.1

FOR EXTRA HELP

Digital Video Tutor CD 2 Videotape 7   InterAct Math   Math Tutor Center   MathXL   MyMathLab.com

Simplify. Assume that no denominator is zero and 0^0 is not considered.

1. $r^4 \cdot r^6$

2. $8^4 \cdot 8^3$

3. $9^5 \cdot 9^3$

4. $n^3 \cdot n^{20}$

5. $a^6 \cdot a$

6. $y^7 \cdot y^9$

7. $5^7 \cdot 5^8$

8. $t^0 \cdot t^{16}$

9. $(3y)^4(3y)^8$

10. $(2t)^8(2t)^{17}$

11. $(5t)(5t)^6$

12. $(8x)^0(8x)^1$

13. $(a^2b^7)(a^3b^2)$

14. $(m-3)^4(m-3)^5$

15. $(x+1)^5(x+1)^7$

16. $(a^8b^3)(a^4b)$

17. $r^3 \cdot r^7 \cdot r^0$

18. $s^4 \cdot s^5 \cdot s^2$

19. $(xy^4)(xy)^3$

20. $(a^3b)(ab)^4$

21. $\dfrac{7^5}{7^2}$

22. $\dfrac{4^7}{4^3}$

23. $\dfrac{x^{15}}{x^3}$

24. $\dfrac{a^{10}}{a^2}$

25. $\dfrac{t^5}{t}$

26. $\dfrac{x^7}{x}$

27. $\dfrac{(5a)^7}{(5a)^6}$

28. $\dfrac{(3m)^9}{(3m)^8}$

Aha! **29.** $\dfrac{(x+y)^8}{(x+y)^8}$

30. $\dfrac{(a-b)^4}{(a-b)^3}$

31. $\dfrac{18m^5}{6m^2}$

32. $\dfrac{30n^7}{6n^3}$

33. $\dfrac{a^9b^7}{a^2b}$

34. $\dfrac{r^{10}s^7}{r^2s}$

35. $\dfrac{m^9n^8}{m^0n^4}$

36. $\dfrac{a^{10}b^{12}}{a^2b^0}$

Simplify.

37. x^0 when $x = 13$

38. y^0 when $y = 38$

39. $5x^0$ when $x = -4$

40. $7m^0$ when $m = 1.7$

41. $8^0 + 5^0$

42. $(8+5)^0$

43. $(-3)^1 - (-3)^0$

44. $(-4)^0 - (-4)^1$

Simplify. Assume that no denominator is zero and 0^0 is not considered.

45. $(x^4)^7$

46. $(a^3)^8$

47. $(5^8)^2$

48. $(2^5)^3$

49. $(m^7)^5$

50. $(n^9)^2$

51. $(t^{20})^4$

52. $(t^3)^9$

53. $(7x)^2$

54. $(5a)^2$

55. $(-2a)^3$

56. $(-3x)^3$

57. $(4m^3)^2$

58. $(5n^4)^2$

59. $(a^2b)^7$

60. $(xy^4)^9$

61. $(x^3y)^2(x^2y^5)$

62. $(a^4b^6)(a^2b)^5$

63. $(2x^5)^3(3x^4)$

64. $(5x^3)^2(2x^7)$

65. $\left(\dfrac{a}{4}\right)^3$

66. $\left(\dfrac{3}{x}\right)^4$

67. $\left(\dfrac{7}{5a}\right)^2$

68. $\left(\dfrac{5x}{2}\right)^3$

69. $\left(\dfrac{a^4}{b^3}\right)^5$

70. $\left(\dfrac{x^5}{y^2}\right)^7$

71. $\left(\dfrac{y^3}{2}\right)^2$

72. $\left(\dfrac{a^5}{2}\right)^3$

73. $\left(\dfrac{x^2y}{z^3}\right)^4$

74. $\left(\dfrac{x^3}{y^2z}\right)^5$

75. $\left(\dfrac{a^3}{-2b^5}\right)^4$

76. $\left(\dfrac{x^5}{-3y^3}\right)^4$

77. $\left(\dfrac{5x^7y}{2z^4}\right)^3$

78. $\left(\dfrac{4a^2b}{3c^7}\right)^3$

Aha! **79.** $\left(\dfrac{4x^3y^5}{3z^7}\right)^0$

80. $\left(\dfrac{5a^7}{2b^5c}\right)^0$

81. Explain in your own words why $-5^2 \neq (-5)^2$.

82. Under what circumstances should exponents be added?

SKILL MAINTENANCE

Factor.

83. $3s - 3r + 3t$

84. $-7x + 7y - 7z$

Combine like terms.

85. $9x + 2y - x - 2y$

86. $5a - 7b - 8a + b$

Use the commutative law of addition to write an equivalent expression.

87. $3x + 2y$

88. $2xy + 5z$

SYNTHESIS

89. Under what conditions does a^n represent a negative number? Why?

90. Using the quotient rule, explain why 9^0 is 1.

91. Suppose that the width of a square is three times the width of a second square (see the figure at the top of the next column). How do the areas of the squares compare? Why?

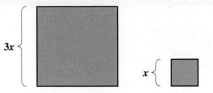

92. Suppose that the width of a cube is twice the width of a second cube. How do the volumes of the cubes compare? Why?

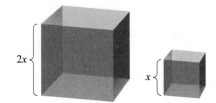

Find a value of the variable that shows that the two expressions are not *equivalent. Answers may vary.*

93. $(a + 5)^2$; $a^2 + 5^2$

94. $3x^2$; $(3x)^2$

95. $\dfrac{a + 7}{7}$; a

96. $\dfrac{t^6}{t^2}$; t^3

Simplify.

97. $a^{10k} \div a^{2k}$

98. $y^{4x} \cdot y^{2x}$

99. $\dfrac{\left(\frac{1}{2}\right)^3\left(\frac{2}{3}\right)^4}{\left(\frac{5}{6}\right)^3}$

100. $\dfrac{x^{5t}(x^t)^2}{(x^{3t})^2}$

101. Solve for x:
$$\frac{t^{26}}{t^x} = t^x.$$

Replace ▬ with $>$, $<$, or $=$ to write a true sentence.

102. 3^5 ▬ 3^4

103. 4^2 ▬ 4^3

104. 4^3 ▬ 5^3

105. 4^3 ▬ 3^4

106. 9^7 ▬ 3^{13}

107. 25^8 ▬ 125^5

In computer science, 1 K of memory refers to 1 kilobyte, or 1×10^3 bytes, of memory. This is really an approximation of 1×2^{10} bytes (since computer memory uses powers of 2). Use the fact that $10^3 \approx 2^{10}$ to estimate each of the following powers of 2. Then compute the power of 2 with a calculator and find the difference between the exact value and the approximation.

108. 2^{14}

109. 2^{22}

110. 2^{26}

111. 2^{31}

112. Dana's research project requires 56 K of memory. How many bytes is this?

113. The cash register at Justin's shop has 64 K of memory. How many bytes is this?

Polynomials

4.2

Terms • Types of Polynomials • Degree and Coefficients •
Combining Like Terms • Evaluating Polynomials and Applications

We now examine an important algebraic expression known as a *polynomial*.
Certain polynomials have appeared earlier in this text so you already have
some experience working with them.

Terms

At this point, we have seen a variety of algebraic expressions like

$$3a^2b^4, \qquad 2l + 2w, \quad \text{and} \quad 5x^2 + x - 2.$$

Of these, $3a^2b^4$, $2l$, $2w$, $5x^2$, x, and -2 are examples of *terms*. A **term** can be a
number (like -2), a variable (like x), or a product of numbers and/or variables,
which may be raised to powers (like $3a^2b^4$, $2l$, $2w$, or $5x^2$).*

Types of Polynomials

A term that is a product of constants and/or variables is called a **monomial**.† All
the terms listed above are monomials. Other examples of monomials are

$$7, \qquad t, \qquad 23x^2y, \quad \text{and} \quad \tfrac{3}{7}a^5.$$

A **polynomial** is a monomial or a sum of monomials. The following are examples of polynomials:

$$4x + 7, \quad \tfrac{2}{3}t^2, \quad 6a + 7, \quad -5n^2 + n - 1, \quad 42r^5, \quad x, \quad \text{and} \quad 0.$$

The following algebraic expressions are *not* polynomials:

$$\textbf{(1)} \ \frac{x + 3}{x - 4}, \qquad \textbf{(2)} \ 5x^3 - 2x^2 + \frac{1}{x}, \qquad \textbf{(3)} \ \frac{1}{x^3 - 2}.$$

Expressions (1) and (3) are not polynomials because they represent quotients,
not sums. Expression (2) is not a polynomial because $1/x$ is not a monomial.

When a polynomial is written as a sum of monomials, each monomial is
called a *term of the polynomial.*

---

*Later in this text, expressions like $5x^{3/2}$ and $2a^{-7}b$ will be discussed. Such expressions are also
considered terms.

†Note that a term, but not a monomial, can include division by a variable.

E x a m p l e   1

Identify the terms of the polynomial $3t^4 - 5t^6 - 4t + 2$.

Solution   The terms are $3t^4$, $-5t^6$, $-4t$, and 2. We can see this by rewriting all subtractions as additions of opposites:

$$3t^4 - 5t^6 - 4t + 2 = 3t^4 + (-5t^6) + (-4t) + 2.$$

These are the terms of the polynomial.

A polynomial that is composed of two terms is called a **binomial**, whereas those composed of three terms are called **trinomials**. Polynomials with four or more terms have no special name.

Monomials	Binomials	Trinomials	No Special Name
$4x^2$	$2x + 4$	$3t^3 + 4t + 7$	$4x^3 - 5x^2 + xy - 8$
9	$3a^5 + 6bc$	$6x^7 - 8z^2 + 4$	$z^5 + 2z^4 - z^3 + 7z + 3$
$-7a^{19}b^5$	$-9x^7 - 6$	$4x^2 - 6x - \frac{1}{2}$	$4x^6 - 3x^5 + x^4 - x^3 + 2x - 1$

Degree and Coefficients

The **degree of a term** is the number of variable factors in that term. Thus the degree of $7t^2$ is 2 because $7t^2$ has two variable factors: $7t^2 = 7 \cdot t \cdot t$.

E x a m p l e   2

Determine the degree of each term: **(a)** $8x^4$; **(b)** $3x$; **(c)** 7.

Solution

a) The degree of $8x^4$ is 4. x^4 represents 4 variable factors: $x \cdot x \cdot x \cdot x$.
b) The degree of $3x$ is 1.   There is 1 variable factor.
c) The degree of 7 is 0.   There is no variable factor.

The part of a term that is a constant factor is the **coefficient** of that term. Thus the coefficient of $3x$ is 3, and the coefficient for the term 7 is simply 7.

E x a m p l e   3

Identify the coefficient of each term in the polynomial

$$4x^3 - 7x^2y + x - 8.$$

Solution

The coefficient of $4x^3$ is 4.

The coefficient of $-7x^2y$ is -7.

The coefficient of the third term is 1, since $x = 1x$.

The coefficient of -8 is simply -8.

The **leading term** of a polynomial is the term of highest degree. Its coefficient is called the **leading coefficient** and its degree is referred to as the **degree of the polynomial**. To see how this terminology is used, consider the polynomial

$$3x^2 - 8x^3 + 5x^4 + 7x - 6.$$

The *terms* are $3x^2,\quad -8x^3,\quad 5x^4,\quad 7x,\quad$ and $\quad -6.$

The *coefficients* are $3,\quad -8,\quad 5,\quad 7,\quad$ and $\quad -6.$

The *degree of each term* is $2,\quad 3,\quad 4,\quad 1,\quad$ and $\quad 0.$

The *leading term* is $5x^4$ and the *leading coefficient* is 5.

The *degree of the polynomial* is 4.

Combining Like Terms

Recall from Section 1.8 that *like*, or *similar*, *terms* are either constant terms or terms containing the same variable(s) raised to the same power(s). To simplify certain polynomials, we can often *combine*, or *collect*, like terms.

E x a m p l e   4   Identify the like terms in $4x^3 + 5x - 7x^2 + 2x^3 + x^2$.

Solution

Like terms: $4x^3$ and $2x^3$ Same variable and exponent

Like terms: $-7x^2$ and x^2 Same variable and exponent

E x a m p l e   5   Combine like terms.

a) $2x^3 - 6x^3$ **b)** $5x^2 + 7 + 2x^4 + 4x^2 - 11 - 2x^4$

c) $7a^3 - 5a^2 + 9a^3 + a^2$ **d)** $\frac{2}{3}x^4 - x^3 - \frac{1}{6}x^4 + \frac{2}{5}x^3 - \frac{3}{10}x^3$

Solution

a) $2x^3 - 6x^3 = (2 - 6)x^3$ Using the distributive law

$\qquad\qquad\quad = -4x^3$

b) $5x^2 + 7 + 2x^4 + 4x^2 - 11 - 2x^4 = 5x^2 + 4x^2 + 2x^4 - 2x^4 + 7 - 11$

$\qquad = (5 + 4)x^2 + (2 - 2)x^4 + (7 - 11)$ ⎫

$\qquad = 9x^2 + 0x^4 + (-4)$ ⎬ These steps are often done mentally.

$\qquad = 9x^2 - 4$ ⎭

c) $7a^3 - 5a^2 + 9a^3 + a^2 = 7a^3 - 5a^2 + 9a^3 + 1a^2$ When a variable to a power appears without a coefficient, we can write in 1.

$\qquad\qquad\qquad = 16a^3 - 4a^2$

d) $\frac{2}{3}x^4 - x^3 - \frac{1}{6}x^4 + \frac{2}{5}x^3 - \frac{3}{10}x^3 = \left(\frac{2}{3} - \frac{1}{6}\right)x^4 + \left(-1 + \frac{2}{5} - \frac{3}{10}\right)x^3$

$\qquad\qquad\qquad = \left(\frac{4}{6} - \frac{1}{6}\right)x^4 + \left(-\frac{10}{10} + \frac{4}{10} - \frac{3}{10}\right)x^3$

$\qquad\qquad\qquad = \frac{3}{6}x^4 - \frac{9}{10}x^3$

$\qquad\qquad\qquad = \frac{1}{2}x^4 - \frac{9}{10}x^3$

Note in Example 5 that the solutions are written so that the term of highest degree appears first, followed by the term of next highest degree, and so on. This is known as **descending order** and is the form in which answers will normally appear.

Evaluating Polynomials and Applications

When each variable in a polynomial is replaced with a number, the polynomial then represents a number, or *value*, that can be calculated using the rules for order of operations.

E x a m p l e   6

Evaluate $-x^2 + 3x + 9$ for $x = -2$.

Solution For $x = -2$, we have

$$-x^2 + 3x + 9 = -(-2)^2 + 3(-2) + 9 \qquad \text{The negative sign in front of } x^2 \text{ remains.}$$

$$= -4 + (-6) + 9$$
$$= -10 + 9 = -1.$$

E x a m p l e   7

Games in a sports league.   In a sports league of n teams in which each team plays every other team twice, the total number of games to be played is given by the polynomial

$$n^2 - n.$$

A girl's soccer league has 10 teams. How many games are played if each team plays every other team twice?

Solution   We evaluate the polynomial for $n = 10$:

$$n^2 - n = 10^2 - 10$$
$$= 100 - 10$$
$$= 90.$$

The league plays 90 games.

E x a m p l e   8

Medical dosage.   The concentration, in parts per million, of a certain antibiotic in the bloodstream after t hours is given by the polynomial

$$-0.05t^2 + 2t + 2.$$

Find the concentration after 2 hr.

Solution   To find the concentration after 2 hr, we evaluate the polynomial for $t = 2$:

$$-0.05t^2 + 2t + 2 = -0.05(2)^2 + 2(2) + 2$$
$$= -0.05(4) + 2(2) + 2$$
$$= -0.2 + 4 + 2$$
$$= 5.8.$$

The concentration after 2 hr is 5.8 parts per million.

Sometimes, a graph can be used to estimate the value of a polynomial visually.

E x a m p l e   9

Medical dosage.   In the following graph, the polynomial from Example 8 has been graphed by evaluating it for several choices of *t*. Use the graph to estimate the concentration *c* of antibiotic in the bloodstream after 14 hr.

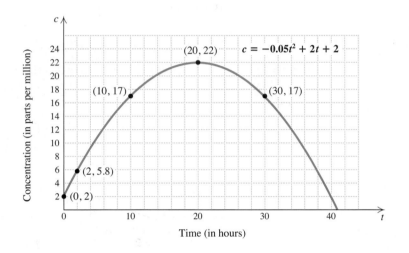

Solution   To estimate the concentration after 14 hr, we locate 14 on the horizontal axis. From there, we move vertically until we meet the curve at some point. From that point, we move horizontally to the *c*-axis.

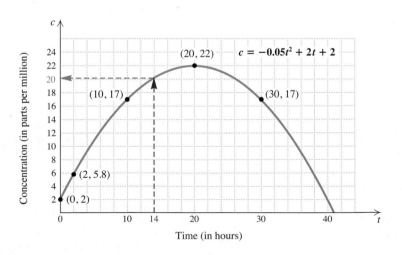

After 14 hr, the concentration of antibiotic in the bloodstream is about 20 parts per million. (For $t = 14$, the value of $-0.05t^2 + 2t + 2$ is approximately 20.)

technology connection

One way to evaluate a polynomial is to use the TRACE key. For example, to evaluate $-0.05x^2 + 2x + 2$ in Example 9, we can use TRACE and then enter the x-value in which we are interested (in this case, 14). The value of the polynomial appears as y, and the cursor automatically appears at $(14, 20.2)$. The Value option of the CALC menu works in a similar way.

1. Use TRACE or CALC Value to find the value of

$$-0.05x^2 + 2x + 2$$

for $x = 25$.

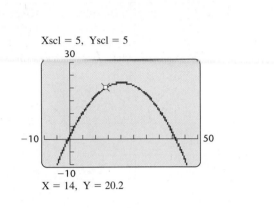

Xscl = 5, Yscl = 5

X = 14, Y = 20.2

FOR EXTRA HELP

Exercise Set 4.2

Digital Video Tutor CD 2
Videotape 7

InterAct Math

Math Tutor Center

MathXL

MyMathLab.com

Identify the terms of each polynomial.

1. $7x^4 + x^3 - 5x + 8$

2. $5a^3 + 4a^2 - a - 7$

3. $-t^4 + 7t^3 - 3t^2 + 6$

4. $n^5 - 4n^3 + 2n - 8$

Determine the coefficient and the degree of each term in each polynomial.

5. $4x^5 + 7x$

6. $9a^3 - 4a^2$

7. $9t^2 - 3t + 4$

8. $7x^4 + 5x - 3$

9. $7a^4 + 9a + a^3$

10. $6t^5 - 3t^2 - t$

11. $x^4 - x^3 + 4x - 3$

12. $3a^4 - a^3 + a - 9$

For each of the following polynomials, (a) list the degree of each term; (b) determine the leading term and the leading coefficient; and (c) determine the degree of the polynomial.

13. $2a^3 + 7a^5 + a^2$

14. $5x - 9x^2 + 3x^6$

15. $2t + 3 + 4t^2$

16. $3a^2 - 7 + 2a^4$

17. $9x^4 + x^2 + x^7 + 4$

18. $8 + 6x^2 - 3x - x^5$

19. $9a - a^4 + 3 + 2a^3$

20. $-x + 2x^5 - 5x^2 + x^6$

21. Complete the following table for the polynomial
$$7x^2 + 8x^5 - 4x^3 + 6 - \tfrac{1}{2}x^4.$$

Term	Coefficient	Degree of the Term	Degree of the Polynomial
		5	
$-\tfrac{1}{2}x^4$			
	-4		
		2	
	6		

22. Complete the following table for the polynomial
$$-3x^4 + 6x^3 - 2x^2 + 8x + 7.$$

Term	Coefficient	Degree of the Term	Degree of the Polynomial
	-3		
$6x^3$			
		2	
		1	
	7		

Classify each polynomial as a monomial, binomial, trinomial, or none of these.

23. $x^2 - 23x + 17$

24. $-9x^2$

25. $x^3 - 7x^2 + 2x - 4$

26. $t^3 + 4$

27. $8t^2 + 5t$

28. $4x^2 + 12x + 9$

29. 17

30. $2x^4 - 7x^3 + x^2 + x - 6$

Combine like terms. Write all answers in descending order.

31. $7x^2 + 3x + 4x^2$

32. $5a + 7a^2 + 3a$

33. $3a^4 - 2a + 2a + a^4$

34. $9b^5 + 3b^2 - 2b^5 - 3b^2$

35. $2x^2 - 6x + 3x + 4x^2$

36. $3x^4 - 7x + x^4 - 2x$

37. $9x^3 + 2x - 4x^3 + 5 - 3x$

38. $6x^2 + 2x^4 - 2x^2 - x^4 - 4x^2$

39. $10x^2 + 2x^3 - 3x^3 - 4x^2 - 6x^2 - x^4$

40. $8x^5 - x^4 + 2x^5 + 5x^4 - 4x^4 - x^6$

41. $\frac{1}{5}x^4 + 7 - 2x^2 + 3 - \frac{2}{15}x^4 + 2x^2$

42. $\frac{1}{6}x^3 + 3x^2 - \frac{1}{3}x^3 + 7 + x^2 - 10$

43. $5.9x^2 - 2.1x + 6 + 3.4x - 2.5x^2 - 0.5$

44. $7.4x^3 - 4.9x + 2.9 - 3.5x - 4.3 + 1.9x^3$

45. $6t - 9t^3 + 8t^4 + 4t + 2t^4 + 7t - 3t^3$

46. $5b^2 - 3b + 7b^2 - 4b^3 + 4b - 9b^2 + 10b^3$

Evaluate each polynomial for $x = 3$.

47. $-7x + 5$

48. $-5x + 9$

49. $2x^2 - 3x + 7$

50. $4x^2 - 6x + 9$

Evaluate each polynomial for $x = -2$.

51. $5x + 7$

52. $7 - 3x$

53. $x^2 - 3x + 1$

54. $5x - 9 + x^2$

55. $-3x^3 + 7x^2 - 4x - 5$

56. $-2x^3 - 4x^2 + 3x + 1$

Memorizing words. Participants in a psychology experiment were able to memorize an average of M words in t minutes, where $M = -0.001t^3 + 0.1t^2$. Use the following graph for Exercises 57–62.

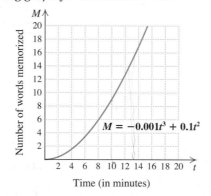

57. Estimate the number of words memorized after 10 min.

58. Estimate the number of words memorized after 14 min.

59. Find the approximate value of M for $t = 8$.

60. Find the approximate value of M for $t = 12$.

61. Estimate the value of M when t is 13.

62. Estimate the value of M when t is 7.

63. *Skydiving.* During the first 13 sec of a jump, the number of feet that a skydiver falls in t seconds is approximated by the polynomial
$$11.12t^2.$$

Approximately how far has a skydiver fallen 10 sec after jumping from a plane?

64. *Skydiving.*   For jumps that exceed 13 sec, the polynomial $173t - 369$ can be used to approximate the distance, in feet, that a skydiver has fallen in t seconds. Approximately how far has a skydiver fallen 20 sec after jumping from a plane?

Daily accidents.   *The average number of accidents per day involving drivers of age r can be approximated by the polynomial*

$$0.4r^2 - 40r + 1039.$$

65. Evaluate the polynomial for $r = 18$ to find the daily number of accidents involving 18-year-old drivers.

66. Evaluate the polynomial for $r = 20$ to find the daily number of accidents involving 20-year-old drivers.

Total revenue.   *Gigabytes Electronics is selling a new type of computer monitor.* Total revenue *is the total amount of money taken in. The firm estimates that for the monitor's first year, revenue from the sale of x monitors is*

$$250x - 0.5x^2 \text{ dollars.}$$

67. What is the total revenue from the sale of 40 monitors?

68. What is the total revenue from the sale of 60 monitors?

Total cost.   *Gigabytes Electronics estimates that the total cost of producing x monitors is given by*

$$4000 + 0.6x^2 \text{ dollars.}$$

69. What is the total cost of producing 200 monitors?

70. What is the total cost of producing 300 monitors?

Circumference.   *The circumference of a circle of radius r is given by the polynomial $2\pi r$, where π is an irrational number. For an approximation of π, use 3.14.*

71. Find the circumference of a circle with radius 10 cm.

72. Find the circumference of a circle with radius 5 ft.

Area of a circle.   *The area of a circle of radius r is given by the polynomial πr^2.*

73. Find the area of a circle with radius 7 m.

74. Find the area of a circle with radius 6 ft.

75. Explain how it is possible for a term to not be a monomial.

76. Is it possible to evaluate polynomials without understanding the rules for order of operations? Why or why not?

SKILL MAINTENANCE

Simplify.

77. $-19 + 24$

78. $5 - 14$

Factor.

79. $5x + 15$

80. $7a - 21$

81. A family spent $2011 to drive a car one year, during which the car was driven 14,800 mi. The family spent $972 for insurance and $114 for registration and oil. The only other cost was for gasoline. How much did gasoline cost per mile?

82. The sum of the page numbers on the facing pages of a book is 549. What are the page numbers?

SYNTHESIS

83. Suppose that the coefficients of a polynomial are all integers and the polynomial is evaluated for some integer. Must the value of the polynomial then also be an integer? Why or why not?

84. Is it easier to evaluate a polynomial before or after like terms have been combined? Why?

85. Construct a polynomial in x (meaning that x is the variable) of degree 5 with four terms and coefficients that are integers.

86. Construct a trinomial in y of degree 4 with coefficients that are rational numbers.

87. What is the degree of $(5m^5)^2$?

88. Construct three like terms of degree 4.

Simplify.

89. $\frac{9}{2}x^8 + \frac{1}{9}x^2 + \frac{1}{2}x^9 + \frac{9}{2}x + \frac{9}{2}x^9 + \frac{8}{9}x^2 + \frac{1}{2}x - \frac{1}{2}x^8$

90. $(3x^2)^3 + 4x^2 \cdot 4x^4 - x^4(2x)^2 + ((2x)^2)^3 - 100x^2(x^2)^2$

91. A polynomial in x has degree 3. The coefficient of x^2 is 3 less than the coefficient of x^3. The coefficient of x is three times the coefficient of x^2. The remaining constant is 2 more than the coefficient of x^3. The sum of the coefficients is -4. Find the polynomial.

92. *Path of the Olympic arrow.*   The Olympic flame at the 1992 Summer Olympics was lit by a flaming arrow. As the arrow moved d meters horizontally from the archer, its height h, in meters, was approximated by the polynomial

$$-0.0064d^2 + 0.8d + 2.$$

Complete the table for the choices of d given. Then plot the points and draw a graph representing the path of the arrow.

d	$-0.0064d^2 + 0.8d + 2$
0	
30	
60	
90	
120	

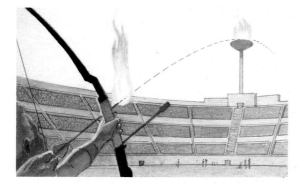

Semester averages.   Professor Kopecki calculates a student's average for her course using

$$A = 0.3q + 0.4t + 0.2f + 0.1h,$$

with q, t, f, and h representing a student's quiz average, test average, final exam score, and homework average, respectively. In Exercises 93 and 94, find the given student's course average rounded to the nearest tenth.

93. Mary Lou: quizzes: 60, 85, 72, 91; final exam: 84; tests: 89, 93, 90; homework: 88

94. Nigel: quizzes: 95, 99, 72, 79; final exam: 91; tests: 68, 76, 92; homework: 86

95. *Daily accidents.*   The average number of accidents per day involving drivers of age r can be approximated by the polynomial

$$0.4r^2 - 40r + 1039.$$

For what age is the number of daily accidents smallest?

In Exercises 96 and 97, complete the table for the given choices of t. Then plot the points and connect them with a smooth curve representing the graph of the polynomial.

96.

t	$-t^2 + 10t - 18$
3	
4	
5	
6	
7	

97.

t	$-t^2 + 6t - 4$
1	
2	
3	
4	
5	

Addition and Subtraction of Polynomials

4.3

Addition of Polynomials • Opposites of Polynomials • Subtraction of Polynomials • Problem Solving

Addition of Polynomials

To add two polynomials, we write a plus sign between them and combine like terms.

E x a m p l e   1

Add.

a) $(-5x^3 + 6x - 1) + (4x^3 + 3x^2 + 2)$

b) $\left(\frac{2}{3}x^4 + 3x^2 - 7x + \frac{1}{2}\right) + \left(-\frac{1}{3}x^4 + 5x^3 - 3x^2 + 3x - \frac{1}{2}\right)$

Solution

a) $(-5x^3 + 6x - 1) + (4x^3 + 3x^2 + 2)$

$= (-5 + 4)x^3 + 3x^2 + 6x + (-1 + 2)$ Combining like terms; using the distributive law

$= -x^3 + 3x^2 + 6x + 1$ Note that $-1x^3 = -x^3$.

b) $\left(\frac{2}{3}x^4 + 3x^2 - 7x + \frac{1}{2}\right) + \left(-\frac{1}{3}x^4 + 5x^3 - 3x^2 + 3x - \frac{1}{2}\right)$

$= \left(\frac{2}{3} - \frac{1}{3}\right)x^4 + 5x^3 + (3 - 3)x^2 + (-7 + 3)x + \left(\frac{1}{2} - \frac{1}{2}\right)$ Combining like terms

$= \frac{1}{3}x^4 + 5x^3 - 4x$

After some practice, polynomial addition is often performed mentally.

E x a m p l e   2

Add: $(2 - 3x + x^2) + (-5 + 7x - 3x^2 + x^3)$

Solution   We have

$(2 - 3x + x^2) + (-5 + 7x - 3x^2 + x^3)$

$= (2 - 5) + (-3 + 7)x + (1 - 3)x^2 + x^3$ You might do this step mentally.

$= -3 + 4x - 2x^2 + x^3$.   Then you would write only this.

The polynomials in the last example are written with the terms arranged according to degree, from least to greatest. Such an arrangement is called *ascending order*. As a rule, answers are written in ascending order when the polynomials being added are given in ascending order. When the polynomials being added are given in descending order, the answer is written in descending order.

We can also add polynomials by writing like terms in columns. Sometimes this makes like terms easier to see.

E x a m p l e   3

Add: $9x^5 - 2x^3 + 6x^2 + 3$ and $5x^4 - 7x^2 + 6$ and $3x^6 - 5x^5 + x^2 + 5$.

Solution   We arrange the polynomials with like terms in columns.

$$\begin{array}{l}
9x^5 \quad\quad - 2x^3 + 6x^2 + \ 3 \\
\quad\quad 5x^4 \quad\quad - 7x^2 + \ 6 \quad\text{We leave spaces for missing terms.} \\
\underline{3x^6 - 5x^5 \quad\quad\quad\quad + 1x^2 + \ 5} \quad\text{Writing } x^2 \text{ as } 1x^2 \\
3x^6 + 4x^5 + 5x^4 - 2x^3 \quad\quad + 14 \quad\text{Adding}
\end{array}$$

The answer is $3x^6 + 4x^5 + 5x^4 - 2x^3 + 14$.

CONNECTING THE CONCEPTS

In many ways, polynomials are to algebra as numbers are to arithmetic. Like numbers, polynomials can be added, subtracted, multiplied, and divided. In Sections 4.3–4.7, we examine how these operations are performed. Thus we are again learning how to write equivalent expressions, not how to solve equations.

Note that equations like those in Examples 1–3 are written to show how one expression can be rewritten in an equivalent form. This is very different from solving an equation.

There is much to learn about manipulating polynomials. Not until Section 5.6 will we return to the task of solving equations.

Opposites of Polynomials

In Section 1.8, we used the property of -1 to show that the opposite of a sum is the sum of the opposites. This idea can be extended.

> ### The Opposite of a Polynomial
>
> To find an equivalent polynomial for the *opposite*, or *additive inverse*, of a polynomial, change the sign of every term. This is the same as multiplying the polynomial by -1.

E x a m p l e   4   Write the opposite of $4x^5 - 7x^3 - 8x + \frac{5}{6}$ in two different forms.

Solution

i) $-\left(4x^5 - 7x^3 - 8x + \frac{5}{6}\right)$

ii) $-4x^5 + 7x^3 + 8x - \frac{5}{6}$ Changing the sign of every term

Thus, $-\left(4x^5 - 7x^3 - 8x + \frac{5}{6}\right)$ is equivalent to $-4x^5 + 7x^3 + 8x - \frac{5}{6}$. Both expressions represent the opposite of $4x^5 - 7x^3 - 8x + \frac{5}{6}$.

E x a m p l e   5   Simplify: $-(-7x^4 - \frac{5}{9}x^3 + 8x^2 - x + 67)$.

Solution

$$-\left(-7x^4 - \frac{5}{9}x^3 + 8x^2 - x + 67\right) = 7x^4 + \frac{5}{9}x^3 - 8x^2 + x - 67$$

Subtraction of Polynomials

We can now subtract one polynomial from another by adding the opposite of the polynomial being subtracted.

Example 6

Subtract: $(9x^5 + x^3 - 2x^2 + 4) - (-2x^5 + x^4 - 4x^3 - 3x^2)$.

Solution

$$(9x^5 + x^3 - 2x^2 + 4) - (-2x^5 + x^4 - 4x^3 - 3x^2)$$

$$= 9x^5 + x^3 - 2x^2 + 4 + 2x^5 - x^4 + 4x^3 + 3x^2 \qquad \text{Adding the} \\ \text{opposite}$$

$$= 11x^5 - x^4 + 5x^3 + x^2 + 4 \qquad \text{Combining like terms}$$

Example 7

Subtract: $(7x^5 + x^3 - 9x) - (3x^5 - 4x^3 + 5)$.

Solution

$$(7x^5 + x^3 - 9x) - (3x^5 - 4x^3 + 5)$$

$$= 7x^5 + x^3 - 9x + (-3x^5) + 4x^3 - 5 \qquad \text{Adding the opposite}$$

$$= 7x^5 + x^3 - 9x - 3x^5 + 4x^3 - 5 \qquad \text{Try to go directly to} \\ \text{this step.}$$

$$= 4x^5 + 5x^3 - 9x - 5 \qquad \text{Combining like terms}$$

To subtract using columns, we first replace the coefficients in the polynomial being subtracted with their opposites. We then add as before.

Example 8

Write in columns and subtract: $(5x^2 - 3x + 6) - (9x^2 - 5x - 3)$.

Solution

i) $\quad \begin{array}{r} 5x^2 - 3x + 6 \\ -(9x^2 - 5x - 3) \\ \hline \end{array}$ $\qquad$ Writing similar terms in columns

ii) $\quad \begin{array}{r} 5x^2 - 3x + 6 \\ -9x^2 + 5x + 3 \\ \hline \end{array}$ $\qquad$ Changing signs and removing parentheses

iii) $\quad \begin{array}{r} 5x^2 - 3x + 6 \\ -9x^2 + 5x + 3 \\ \hline -4x^2 + 2x + 9 \end{array}$ $\qquad$ Adding

If you can do so without error, you can arrange the polynomials in columns, mentally find the opposite of each term being subtracted, and write the answer.

Example 9

Write in columns and subtract: $(x^3 + x^2 + 2x - 12) - (-2x^3 + x^2 - 3x)$.

Solution We have

$$\begin{array}{r} x^3 + x^2 + 2x - 12 \\ -(-2x^3 + x^2 - 3x \qquad) \\ \hline 3x^3 \qquad\quad + 5x - 12. \end{array} \qquad \text{Leaving space for the missing term}$$

Problem Solving

Example   10

Find a polynomial for the sum of the areas of rectangles A, B, C, and D.

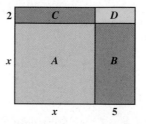

Solution

1. **Familiarize.**   Recall that the area of a rectangle is the product of its length and width.

2. **Translate.**   We translate the problem to mathematical language. The sum of the areas is a sum of products. We find each product and then add:

Area of A plus   area of B plus   area of C plus   area of D

$$x \cdot x \qquad + \qquad 5x \qquad + \qquad 2x \qquad + \qquad 2 \cdot 5.$$

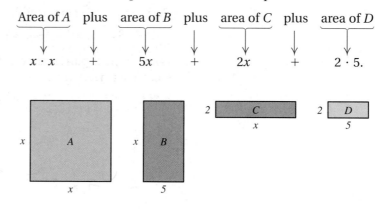

3. **Carry out.**   We combine like terms:

$$x^2 + 5x + 2x + 10 = x^2 + 7x + 10.$$

4. **Check.**   A partial check is to replace x with a number, say 3. Then we evaluate $x^2 + 7x + 10$ and compare that result with an alternative calculation:

$$3^2 + 7 \cdot 3 + 10 = 9 + 21 + 10 = 40.$$

When we substitute 3 for x and calculate the total area by regarding the figure as one large rectangle, we should also get 40:

$$\text{Total area} = (x + 5)(x + 2) = (3 + 5)(3 + 2) = 8 \cdot 5 = 40.$$

Our check is only partial, since it is possible for an incorrect answer to equal 40 when evaluated for $x = 3$. This would be unlikely, especially if a second choice of x, say $x = 5$, also checks. We leave that check to the student.

5. **State.**   A polynomial for the sum of the areas is $x^2 + 7x + 10$.

E x a m p l e   1 1   An 8-ft by 8-ft shed is placed on a square lawn x ft on a side. Find a polynomial for the remaining area.

Solution

1. **Familiarize.**   We make a drawing of the situation as follows.

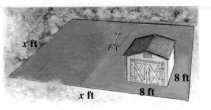

2. **Translate.**   We reword the problem and translate as follows.

   Rewording: $\underbrace{\text{Area of lawn}}$ − $\underbrace{\text{area of shed}}$ = $\underbrace{\text{area left over}}$

   Translating: x ft $\cdot$ x ft   −   8 ft $\cdot$ 8 ft   = Area left over

3. **Carry out.**   We carry out the manipulation by multiplying:

   x^2 ft^2 − 64 ft^2 = Area left over.

4. **Check.**   As a partial check, note that the units in the answer are square feet (ft^2), a measure of area, as expected.

5. **State.**   The remaining area in the yard is $(x^2 − 64)$ ft^2.

---

 technology connection

To check polynomial addition or subtraction, we can let y_1 = the expression before the addition or subtraction has been performed and y_2 = the simplified sum or difference. If the addition or subtraction is correct, y_1 will equal y_2 and $y_2 − y_1$ will be 0. We enter $y_2 − y_1$ as y_3, using the VARS key. Below is a check of Example 7 in which

$$y_1 = (7x^5 + x^3 − 9x) − (3x^5 − 4x^3 + 5),$$
$$y_2 = 4x^5 + 5x^3 − 9x − 5,$$

and

$$y_3 = y_2 − y_1.$$

We graph only y_3. If indeed y_1 and y_2 are equivalent, then y_3 should equal 0. This means its graph should coincide with the x-axis. The TRACE or TABLE features can confirm that y_3 is always 0, or we can select y_3 to be drawn bold at the $\boxed{\text{Y=}}$ window.

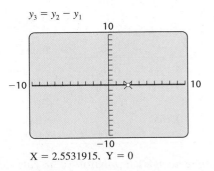

$y_3 = y_2 − y_1$

X = 2.5531915. Y = 0

1. Use a grapher to check Examples 1, 2, and 6.

Exercise Set 4.3

FOR EXTRA HELP

Digital Video Tutor CD 2
Videotape 7

InterAct Math   Math Tutor Center   MathXL   MyMathLab.com

Add.

1. $(2x + 3) + (-7x + 6)$

2. $(5x + 1) + (-9x + 4)$

3. $(-6x + 2) + (x^2 + x - 3)$

4. $(x^2 - 5x + 4) + (8x - 9)$

5. $(7t^2 - 3t + 6) + (2t^2 + 8t - 9)$

6. $(9a^2 + 4a - 5) + (6a^2 - 3a - 1)$

7. $(2m^3 - 4m^2 + m - 7) + (4m^3 + 7m^2 - 4m - 2)$

8. $(5n^3 - n^2 + 4n - 3) + (2n^3 - 4n^2 + 3n - 4)$

9. $(3 + 6a + 7a^2 + 8a^3) + (4 + 7a - a^2 + 6a^3)$

10. $(7 + 4t - 5t^2 + 6t^3) + (2 + t + 6t^2 - 4t^3)$

11. $(9x^8 - 7x^4 + 2x^2 + 5) + (8x^7 + 4x^4 - 2x)$

12. $(4x^5 - 6x^3 - 9x + 1) + (6x^3 + 9x^2 + 9x)$

13. $\left(\frac{1}{4}x^4 + \frac{2}{3}x^3 + \frac{5}{8}x^2 + 7\right) + \left(-\frac{3}{4}x^4 + \frac{3}{8}x^2 - 7\right)$

14. $\left(\frac{1}{3}x^9 + \frac{1}{5}x^5 - \frac{1}{2}x^2 + 7\right) + \left(-\frac{1}{5}x^9 + \frac{1}{4}x^4 - \frac{3}{5}x^5\right)$

15. $(5.3t^2 - 6.4t - 9.1) + (4.2t^3 - 1.8t^2 + 7.3)$

16. $(4.9a^3 + 3.2a^2 - 5.1a) + (2.1a^2 - 3.7a + 4.6)$

17. $\begin{array}{r} -3x^4 + 6x^2 + 2x - 1 \\ -\ 3x^2 + 2x + 1 \\ \hline \end{array}$

18. $\begin{array}{r} -4x^3 + 8x^2 + 3x - 2 \\ -\ 4x^2 + 3x + 2 \\ \hline \end{array}$

19. $\begin{array}{r} 0.15x^4 + 0.10x^3 - 0.9x^2 \\ -\ 0.01x^3 + 0.01x^2 + x \\ 1.25x^4 \qquad\quad + 0.11x^2 \qquad + 0.01 \\ 0.27x^3 \qquad\qquad\quad + 0.99 \\ -0.35x^4 \qquad\qquad + 15x^2 \quad - 0.03 \\ \hline \end{array}$

20. $\begin{array}{r} 0.05x^4 + 0.12x^3 - 0.5x^2 \\ -\ 0.02x^3 + 0.02x^2 + 2x \\ 1.5x^4 \qquad\quad + 0.01x^2 \qquad + 0.15 \\ 0.25x^3 \qquad\qquad\quad + 0.85 \\ -0.25x^4 \qquad\qquad + 10x^2 \quad - 0.04 \\ \hline \end{array}$

Write the opposite of each polynomial in two different forms, as in Example 4.

21. $-t^3 + 4t^2 - 9$

22. $-4x^3 - 5x^2 + 2x$

23. $12x^4 - 3x^3 + 3$

24. $5a^3 + 2a - 17$

Simplify.

25. $-(8x - 9)$

26. $-(-6x + 5)$

27. $-(3a^4 - 5a^2 + 9)$

28. $-(-6a^3 + 2a^2 - 7)$

29. $-\left(-4x^4 + 6x^2 + \frac{3}{4}x - 8\right)$

30. $-(-5x^4 + 4x^3 - x^2 + 0.9)$

Subtract.

31. $(7x + 4) - (-2x + 1)$

32. $(5x + 6) - (-2x + 4)$

33. $(-5t + 4) - (t^2 + 2t - 1)$

34. $(a^2 - 5a + 2) - (3a^2 + 2a - 4)$

35. $(6x^4 + 3x^3 - 1) - (4x^2 - 3x + 3)$

36. $(-4x^2 + 2x) - (3x^3 - 5x^2 + 3)$

37. $(1.2x^3 + 4.5x^2 - 3.8x) - (-3.4x^3 - 4.7x^2 + 23)$

38. $(0.5x^4 - 0.6x^2 + 0.7) - (2.3x^4 + 1.8x - 3.9)$

Aha! 39. $(7x^3 - 2x^2 + 6) - (7x^3 - 2x^2 + 6)$

40. $(8x^5 + 3x^4 + x - 1) - (8x^5 + 3x^4 - 1)$

41. $(6 + 5a + 3a^2 - a^3) - (2 + 3a - 4a^2 + 2a^3)$

42. $(7 + t - 5t^2 + 2t^3) - (1 + 2t - 4t^2 + 5t^3)$

43. $\left(\frac{5}{8}x^3 - \frac{1}{4}x - \frac{1}{3}\right) - \left(-\frac{1}{8}x^3 + \frac{1}{4}x - \frac{1}{3}\right)$

44. $\left(\frac{1}{5}x^3 + 2x^2 - \frac{3}{10}\right) - \left(-\frac{2}{5}x^3 + 2x^2 + \frac{7}{1000}\right)$

45. $(0.07t^3 - 0.03t^2 + 0.01t) - (0.02t^3 + 0.04t^2 - 1)$

46. $(0.9a^3 + 0.2a - 5) - (0.7a^4 - 0.3a - 0.1)$

47. $\begin{array}{r} x^2 + 5x + 6 \\ -(x^2 + 2x + 1) \\ \hline \end{array}$

48. $\begin{array}{r} x^3 + 3x^2 + 1 \\ -(x^3 +\ x^2 - 5) \\ \hline \end{array}$

49. $\begin{array}{r} 5x^4 + 6x^3 - 9x^2 \\ -(-6x^4 - 6x^3 +\ x^2) \\ \hline \end{array}$

50. $5x^4 - 2x^3 + 6x^2$
$\underline{-(7x^4 + 6x^3 + 7x^2)}$

51. Solve.
a) Find a polynomial for the sum of the areas of the rectangles shown in the figure.
b) Find the sum of the areas when $x = 5$ and $x = 7$.

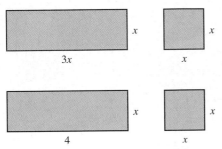

52. Solve.
a) Find a polynomial for the sum of the areas of the circles shown in the figure.
b) Find the sum of the areas when $r = 5$ and $r = 11.3$.

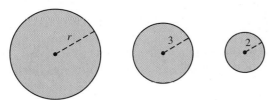

Find a polynomial for the perimeter of each figure in Exercises 53 and 54.

53.

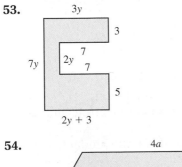

54.

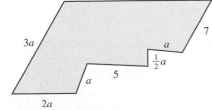

Find two algebraic expressions for the area of each figure. First, regard the figure as one large rectangle,

and then regard the figure as a sum of four smaller rectangles.

55.

56.

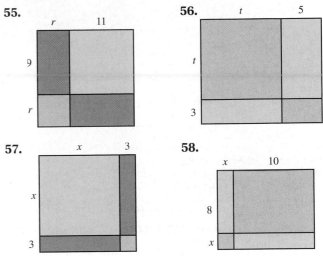

57.

58.

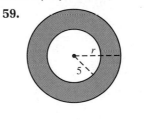

Find a polynomial for the shaded area of each figure.

59.

60.

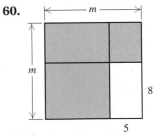

61.

62.

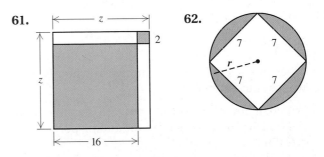

63. Find $(y - 2)^2$ by subtracting the white areas from y^2.

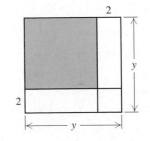

64. Find $(10 - 2x)^2$ by subtracting the white areas from 10^2.

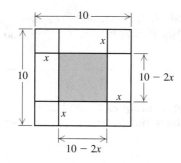

65. Is the sum of two trinomials always a trinomial? Why or why not?

66. What advice would you offer to a student who is successful at adding, but not subtracting, polynomials?

SKILL MAINTENANCE

Simplify.

67. $5(4 + 3) - 5 \cdot 4 - 5 \cdot 3$

68. $7(2 + 6) - 7 \cdot 2 - 7 \cdot 6$

69. $2(5t + 7) + 3t$

70. $3(4t - 5) + 2t$

Solve.

71. $2(x + 3) > 5(x - 3) + 7$

72. $7(x - 8) \le 4(x - 5)$

SYNTHESIS

73. What can be concluded about two polynomials whose sum is zero?

74. Which, if any, of the commutative, associative, and distributive laws are needed for adding polynomials? Why?

Simplify.

75. $(6t^2 - 7t) + (3t^2 - 4t + 5) - (9t - 6)$

76. $(3x^2 - 4x + 6) - (-2x^2 + 4) + (-5x - 3)$

77. $(-8y^2 - 4) - (3y + 6) - (2y^2 - y)$

78. $(5x^3 - 4x^2 + 6) - (2x^3 + x^2 - x) + (x^3 - x)$

79. $(-y^4 - 7y^3 + y^2) + (-2y^4 + 5y - 2) - (-6y^3 + y^2)$

80. $(-4 + x^2 + 2x^3) - (-6 - x + 3x^3) - (-x^2 - 5x^3)$

81. $(345.099x^3 - 6.178x) - (94.508x^3 - 8.99x)$

Find a polynomial for the surface area of the right rectangular solid.

82.   **83.**

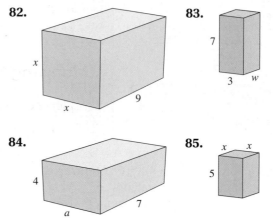

84.   **85.**

86. *Total profit.* Hadley Electronics is marketing a new kind of stereo. Total revenue is the total amount of money taken in. The firm determines that when it sells x stereos, its total revenue is given by

$$R = 280x - 0.4x^2.$$

Total cost is the total cost of producing x stereos. Hadley Electronics determines that the total cost of producing x stereos is given by

$$C = 5000 + 0.6x^2.$$

The total profit P is

(Total Revenue) − (Total Cost) = $R - C$.

a) Find a polynomial for total profit.

b) What is the total profit on the production and sale of 75 stereos?

c) What is the total profit on the production and sale of 100 stereos?

87. Does replacing each occurrence of the variable x in $4x^7 - 6x^3 + 2x$ with its opposite result in the opposite of the polynomial? Why or why not?

4.4

Multiplying Monomials • Multiplying a Monomial and a Polynomial • Multiplying Any Two Polynomials • Checking by Evaluating

We now multiply polynomials using techniques based largely on the distributive, associative, and commutative laws and the rules for exponents.

Multiplying Monomials

Consider $(3x)(4x)$. We multiply as follows:

$$
\begin{aligned}
(3x)(4x) &= 3 \cdot x \cdot 4 \cdot x && \text{Using an associative law}\\
&= 3 \cdot 4 \cdot x \cdot x && \text{Using a commutative law}\\
&= (3 \cdot 4) \cdot x \cdot x && \text{Using an associative law}\\
&= 12x^2.
\end{aligned}
$$

> ### To Multiply Monomials
>
> To find an equivalent expression for the product of two monomials, multiply the coefficients and then multiply the variables using the product rule for exponents.

E x a m p l e   1

Multiply: **(a)** $(5x)(6x)$; **(b)** $(3a)(-a)$; **(c)** $(-7x^5)(4x^3)$.

Solution

a) $(5x)(6x) = (5 \cdot 6)(x \cdot x)$ Multiplying the coefficients; multiplying the variables

$= 30x^2$ Simplifying

b) $(3a)(-a) = (3a)(-1a)$ Writing $-a$ as $-1a$ can ease calculations.

$= (3)(-1)(a \cdot a)$ Using an associative and a commutative law

$= -3a^2$

c) $(-7x^5)(4x^3) = (-7 \cdot 4)(x^5 \cdot x^3)$

$\left. \begin{aligned} &= -28x^{5+3}\\ &= -28x^8 \end{aligned} \right\}$ Using the product rule for exponents

After some practice, you can try writing only the answer.

Multiplying a Monomial and a Polynomial

To find an equivalent expression for the product of a monomial, such as $2x$, and a polynomial, such as $5x + 3$, we use the distributive law.

E x a m p l e   2

Multiply: **(a)** x and $x + 3$; **(b)** $5x(2x^2 - 3x + 4)$.

Solution

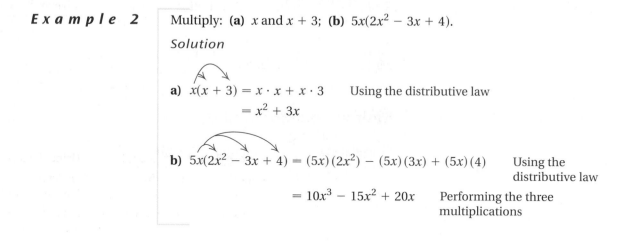

a) $x(x + 3) = x \cdot x + x \cdot 3$ Using the distributive law
$$= x^2 + 3x$$

b) $5x(2x^2 - 3x + 4) = (5x)(2x^2) - (5x)(3x) + (5x)(4)$ Using the distributive law
$$= 10x^3 - 15x^2 + 20x$$ Performing the three multiplications

The product in Example 2(a) can be visualized as the area of a rectangle with width x and length $x + 3$.

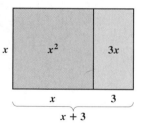

Note that the total area can be expressed as $x(x + 3)$ or, by adding the two smaller areas, $x^2 + 3x$.

The Product of a Monomial and a Polynomial

To multiply a monomial and a polynomial, multiply each term of the polynomial by the monomial.

Try to do this mentally, when possible.

E x a m p l e   3

Multiply: $2x^2(x^3 - 7x^2 + 10x - 4)$.

Solution

Think: $2x^2 \cdot x^3 - 2x^2 \cdot 7x^2 + 2x^2 \cdot 10x - 2x^2 \cdot 4$

$$2x^2(x^3 - 7x^2 + 10x - 4) = 2x^5 - 14x^4 + 20x^3 - 8x^2$$

Multiplying Any Two Polynomials

Before considering the product of *any* two polynomials, let's look at the product of two binomials.

To find an equivalent expression for the product of two binomials, we again begin by using the distributive law. This time, however, it is a *binomial* rather than a monomial that is being distributed.

E x a m p l e   4   Multiply each of the following.

a) $x + 5$ and $x + 4$ **b)** $4x - 3$ and $x - 2$

Solution

a) $(x + 5)\ (x + 4) = (x + 5)\ x + (x + 5)\ 4$ Using the distributive law

$= x(x + 5) + 4(x + 5)$ Using the commutative law for multiplication

$= x \cdot x + x \cdot 5 + 4 \cdot x + 4 \cdot 5$ Using the distributive law (twice)

$= x^2 + 5x + 4x + 20$ Multiplying the monomials

$= x^2 + 9x + 20$ Combining like terms

b) $(4x - 3)\ (x - 2) = (4x - 3)\ x - (4x - 3)\ 2$ Using the distributive law

$= x(4x - 3) - 2(4x - 3)$ Using the commutative law for multiplication. This step is often omitted.

$= x \cdot 4x - x \cdot 3 - 2 \cdot 4x - 2(-3)$ Using the distributive law (twice)

$= 4x^2 - 3x - 8x + 6$ Multiplying the monomials

$= 4x^2 - 11x + 6$ Combining like terms

To visualize the product in Example 4(a), consider a rectangle of length $x + 5$ and width $x + 4$.

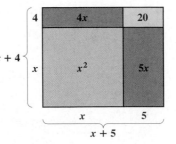

The total area can be expressed as $(x + 5)(x + 4)$ or, by adding the four smaller areas, $x^2 + 5x + 4x + 20$.

Let's consider the product of a binomial and a trinomial. Again we make repeated use of the distributive law.

E x a m p l e   5

Multiply: $(x^2 + 2x - 3)(x + 4)$.

Solution

$(x^2 + 2x - 3) \ (x + 4)$

$= (x^2 + 2x - 3) \ x + \ (x^2 + 2x - 3) \ 4$ Using the distributive law

$= x(x^2 + 2x - 3) + 4(x^2 + 2x - 3)$ Using the commutative law

$= x \cdot x^2 + x \cdot 2x - x \cdot 3 + 4 \cdot x^2 + 4 \cdot 2x - 4 \cdot 3$ Using the distributive law (twice)

$= x^3 + 2x^2 - 3x + 4x^2 + 8x - 12$ Multiplying the monomials

$= x^3 + 6x^2 + 5x - 12$ Combining like terms

Perhaps you have discovered the following in the preceding examples.

The Product of Two Polynomials

To multiply two polynomials P and Q, select one of the polynomials, say P. Then multiply each term of P by every term of Q and combine like terms.

To use columns for long multiplication, multiply each term in the top row by every term in the bottom row. We write like terms in columns, and then add the results. Such multiplication is like multiplying with whole numbers:

$$
\begin{array}{r}
3\ \ 2\ \ 1 \\
\times\ \ \ \ 1\ \ 2 \\
\hline
6\ \ 4\ \ 2 \\
3\ \ 2\ \ 1\ \ \ \\
\hline
3\ \ 8\ \ 5\ \ 2
\end{array}
\qquad
\begin{array}{r}
300 + 20 + 1 \\
\times\ \ \ \ \ \ \ \ \ \ 10 + 2 \\
\hline
600 + 40 + 2 \\
3000 + 200 + 10\ \ \ \ \ \ \ \ \ \\
\hline
3000 + 800 + 50 + 2
\end{array}
$$

Multiplying the top row by 2
Multiplying the bottom row by 10
Adding

E x a m p l e   6

Multiply: $(4x^3 - 2x^2 + 3x)(x^2 + 2x)$.

Solution

$$
\begin{array}{r}
4x^3 - 2x^2 + 3x \\
x^2 + 2x \\
\hline
8x^4 - 4x^3 + 6x^2 \\
4x^5 - 2x^4 + 3x^3\ \ \ \ \ \ \ \ \ \ \ \\
\hline
4x^5 + 6x^4 -\ \ x^3 + 6x^2
\end{array}
$$

Multiplying the top row by $2x$
Multiplying the top row by x^2
Combining like terms

———— Line up like terms in columns.

If a term is missing, it helps to leave space for it so that like terms can be easily aligned.

E x a m p l e   7

Multiply: $(-2x^2 - 3)(5x^3 - 3x + 4)$.

Solution

$$
\begin{array}{r}
5x^3 \qquad -3x + 4 \\
-2x^2 \qquad -3 \\
\hline
-15x^3 \qquad + 9x - 12 \\
-10x^5 + 6x^3 - 8x^2 \\
\hline
-10x^5 - 9x^3 - 8x^2 + 9x - 12
\end{array}
$$

Multiplying by -3
Multiplying by $-2x^2$
Combining like terms

With practice, some steps can be skipped. Sometimes we multiply horizontally, while still aligning like terms.

E x a m p l e   8

Multiply: $(2x^3 + 3x^2 - 4x + 6)(3x + 5)$.

Solution

$$
(2x^3 + 3x^2 - 4x + 6)(3x + 5) = \overbrace{6x^4 + 9x^3 - 12x^2 + 18x}^{\text{Multiplying by } 3x}
$$
$$
+ \underbrace{10x^3 + 15x^2 - 20x + 30}_{\text{Multiplying by } 5}
$$
$$
= 6x^4 + 19x^3 + 3x^2 - 2x + 30
$$

Checking by Evaluating

How can we be certain that our multiplication (or addition or subtraction) of polynomials is correct? One check is to simply review our calculations. A different type of check, used in Example 10 of Section 4.3, makes use of the fact that equivalent expressions have the same value when evaluated for the same replacement. Thus a quick, partial, check of Example 8 can be made by selecting a convenient replacement for x (say, 1) and comparing the values of the expressions $(2x^3 + 3x^2 - 4x + 6)(3x + 5)$ and $6x^4 + 19x^3 + 3x^2 - 2x + 30$:

$$
(2x^3 + 3x^2 - 4x + 6)(3x + 5) = (2 \cdot 1^3 + 3 \cdot 1^2 - 4 \cdot 1 + 6)(3 \cdot 1 + 5)
$$
$$
= (2 + 3 - 4 + 6)(3 + 5)
$$
$$
= 7 \cdot 8 = 56;
$$

$$
6x^4 + 19x^3 + 3x^2 - 2x + 30 = 6 \cdot 1^4 + 19 \cdot 1^3 + 3 \cdot 1^2 - 2 \cdot 1 + 30
$$
$$
= 6 + 19 + 3 - 2 + 30
$$
$$
= 28 - 2 + 30 = 56.
$$

Since the value of both expressions is 56, the multiplication in Example 8 is very likely correct.

It is possible, by chance, for two expressions that are not equivalent to share the same value when evaluated. For this reason, checking by evaluating is only a partial check. Consult your instructor for the checking approach that he or she prefers.

technology
connection

Tables can also be used to check polynomial multiplication. To illustrate, we can check Example 8 by entering $y_1 = (2x^3 + 3x^2 - 4x + 6)(3x + 5)$ and $y_2 = 6x^4 + 19x^3 + 3x^2 - 2x + 30$.

When $\boxed{\text{TABLE}}$ is then pressed, we are shown two columns of values—one for y_1 and one for y_2. If our multiplication was correct, the columns of values will match.

X	Y₁	Y₂
-3	36	36
-2	-10	-10
-1	22	22
0	30	30
1	56	56
2	286	286
3	1050	1050

X = -3

1. Form a table and scroll up and down to check Example 8.
2. Check Example 8 using the method discussed in Section 4.3: Let

$$y_1 = (2x^3 + 3x^2 - 4x + 6)(3x + 5),$$

$$y_2 = 6x^4 + 19x^3 + 3x^2 - 2x + 30,$$

and

$$y_3 = y_2 - y_1.$$

Then check that y_3 is always 0.

Exercise Set 4.4

FOR EXTRA HELP

Digital Video Tutor CD 3
Videotape 7

InterAct Math

Math Tutor Center

MathXL

MyMathLab.com

Multiply.

1. $(5x^4)6$

2. $(4x^3)7$

3. $(-x^2)(-x)$

4. $(-x^3)(x^4)$

5. $(-x^5)(x^3)$

6. $(-x^6)(-x^2)$

7. $(7t^5)(4t^3)$

8. $(10a^2)(3a^2)$

9. $(-0.1x^6)(0.2x^4)$

10. $(0.3x^3)(-0.4x^6)$

11. $\left(-\frac{1}{5}x^3\right)\left(-\frac{1}{3}x\right)$

12. $\left(-\frac{1}{4}x^4\right)\left(\frac{1}{5}x^8\right)$

13. $19t^2 \cdot 0$

14. $(-5n^3)(-1)$

15. $7x^2(-2x^3)(2x^6)$

16. $(-4y^5)(6y^2)(-3y^3)$

17. $3x(-x + 5)$

18. $2x(4x - 6)$

19. $4x(x + 1)$

20. $3x(x + 2)$

21. $(a + 9)3a$

22. $(a - 7)4a$

23. $x^2(x^3 + 1)$

24. $-2x^3(x^2 - 1)$

25. $3x(2x^2 - 6x + 1)$

26. $-4x(2x^3 - 6x^2 - 5x + 1)$

27. $5t^2(3t + 6)$

28. $7t^2(2t + 1)$

29. $-6x^2(x^2 + x)$

30. $-4x^2(x^2 - x)$

31. $\frac{2}{3}a^4(6a^5 - 12a^3 - \frac{5}{8})$

32. $\frac{3}{4}t^5(8t^6 - 12t^4 + \frac{12}{7})$

33. $(x + 6)(x + 3)$

34. $(x + 5)(x + 2)$

35. $(x + 5)(x - 2)$

36. $(x + 6)(x - 2)$

37. $(a - 6)(a - 7)$

38. $(a - 4)(a - 8)$

39. $(x + 3)(x - 3)$

40. $(x + 6)(x - 6)$

41. $(5 - x)(5 - 2x)$

42. $(3 + x)(6 + 2x)$

43. $\left(t + \frac{3}{2}\right)\left(t + \frac{4}{3}\right)$

44. $\left(a - \frac{2}{5}\right)\left(a + \frac{5}{2}\right)$

45. $\left(\frac{1}{4}a + 2\right)\left(\frac{3}{4}a - 1\right)$

46. $\left(\frac{2}{5}t - 1\right)\left(\frac{3}{5}t + 1\right)$

Draw and label rectangles similar to those following Examples 2 and 4 to illustrate each product.

47. $x(x + 5)$

48. $x(x + 2)$

49. $(x + 1)(x + 2)$

50. $(x + 3)(x + 1)$

51. $(x + 5)(x + 3)$

52. $(x + 4)(x + 6)$

53. $(3x + 2)(3x + 2)$

54. $(5x + 3)(5x + 3)$

Multiply and check.

55. $(x^2 - x + 5)(x + 1)$

56. $(x^2 + x - 7)(x + 2)$

57. $(2a + 5)(a^2 - 3a + 2)$

58. $(3t + 4)(t^2 - 5t + 1)$

59. $(y^2 - 7)(2y^3 + y + 1)$

60. $(a^2 + 4)(5a^3 - 3a - 1)$

61. $(5x^3 - 7x^2 + 1)(x - 3x^2)$

62. $(4x^3 - 5x - 3)(1 + 2x^2)$

63. $(x^2 - 3x + 2)(x^2 + x + 1)$

64. $(x^2 + 5x - 1)(x^2 - x + 3)$

65. $(2t^2 - 5t - 4)(3t^2 - t + 1)$

66. $(5t^2 - t + 1)(2t^2 + t - 3)$

67. $(x + 1)(x^3 + 7x^2 + 5x + 4)$

68. $(x + 2)(x^3 + 5x^2 + 9x + 3)$

69. $\left(x - \frac{1}{2}\right)\left(2x^3 - 4x^2 + 3x - \frac{2}{5}\right)$

70. $\left(x + \frac{1}{3}\right)\left(6x^3 - 12x^2 - 5x + \frac{1}{2}\right)$

71. Is it possible to understand polynomial multiplication without understanding the distributive law? Why or why not?

72. The polynomials

$$(a + b + c + d) \quad \text{and} \quad (r + s + m + p)$$

are multiplied. Without performing the multiplication, determine how many terms the product will contain. Provide a justification for your answer.

SKILL MAINTENANCE

Simplify.

73. $5 - 3 \cdot 2 + 7$

74. $4 + 6 \cdot 5 - 3$

75. $(8 - 2)(8 + 2) + 2^2 - 8^2$

76. $(7 - 3)(7 + 3) + 3^2 - 7^2$

SYNTHESIS

77. Under what conditions will the product of two binomials be a trinomial?

78. How can the following figure be used to show that $(x + 3)^2 \neq x^2 + 9$?

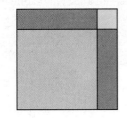

Find a polynomial for the shaded area of each figure.

79.

80.

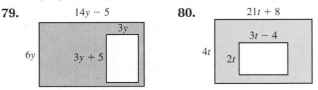

For each figure, determine what the missing number must be in order for the figure to have the given area.

81. Area is $x^2 + 7x + 10$

82. Area is $x^2 + 8x + 15$

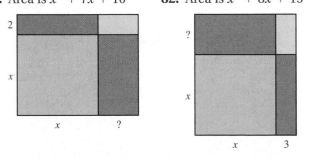

83. A box with a square bottom is to be made from a 12-in.-square piece of cardboard. Squares with side x are cut out of the corners and the sides are folded up. Find the polynomials for the volume and the outside surface area of the box.

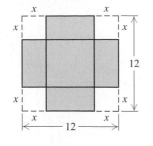

84. An open wooden box is a cube with side x cm. The box, including its bottom, is made of wood that is 1 cm thick. Find a polynomial for the interior volume of the cube.

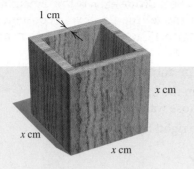

1 cm

x cm

x cm

x cm

85. Find a polynomial for the volume of the solid shown below.

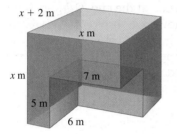

$x + 2$ m

x m

x m

7 m

5 m

6 m

86. A side of a cube is $(x + 2)$ cm long. Find a polynomial for the volume of the cube.

87. A rectangular garden is twice as long as it is wide and is surrounded by a sidewalk that is 4 ft wide (see the figure below). The area of the sidewalk is 256 ft^2. Find the dimensions of the garden.

← 4 ft →

Compute and simplify.

88. $(x + 3)(x + 6) + (x + 3)(x + 6)$

Aha! **89.** $(x - 2)(x - 7) - (x - 7)(x - 2)$

90. $(x + 5)^2 - (x - 3)^2$

Aha! **91.** Extend the pattern and simplify

$$(x - a)(x - b)(x - c)(x - d) \cdots (x - z).$$

92. Use a grapher to check your answers to Exercises 21, 41, and 61. Use graphs, tables, or both, as directed by your instructor.

CORNER

Slick Tricks with Algebra

Focus: Polynomial multiplication
Time: 15 minutes
Group size: 2

Consider the following dialogue.

Jinny: Cal, let me do a number trick with you. Think of a number between 1 and 7. I'll have you perform some manipulations to this number, you'll tell me the result, and I'll tell you your number.

Cal: OK. I've thought of a number.

Jinny: Good. Write it down so I can't see it. Now double it, and then subtract x from the result.

Cal: Hey, this is algebra!

Jinny: I know. Now square your binomial. After you're through squaring, subtract x^2.

Cal: How did you know I had an x^2? I *thought* this was rigged!

Jinny: It is. Now divide each of the remaining terms by 4 and tell me either your constant term or your x-term. I'll tell you the other term and the number you chose.

Cal: OK. The constant term is 16.

Jinny: Then the other term is $-4x$ and the number you chose was 4.

Cal: You're right! How did you do it?

ACTIVITY

1. Each group member should follow Jinny's instructions. Then determine how Jinny determined Cal's number and the other term.
2. Suppose that, at the end, Cal told Jinny the x-term. How would Jinny have determined Cal's number and the other term?
3. Would Jinny's "trick" work with *any* real number? Why do you think she specified numbers between 1 and 7?

Special Products

4.5

Products of Two Binomials • Multiplying Sums and Differences of Two Terms • Squaring Binomials • Multiplications of Various Types

Certain products of two binomials occur so often that it is helpful to be able to compute them quickly. In this section, we develop methods for computing "special" products more quickly than we were able to in Section 4.4.

Products of Two Binomials

In Section 4.4, we found the product $(x + 5)(x + 4)$ by using the distributive law a total of three times (see p. 229). Note that each term in $x + 5$ is multiplied by each term in $x + 4$. To shorten our work, we can go right to this step:

$$(x + 5)(x + 4) = x \cdot x + x \cdot 4 + 5 \cdot x + 5 \cdot 4$$
$$= x^2 + 4x + 5x + 20$$
$$= x^2 + 9x + 20.$$

Note that the product $x \cdot x$ is found by multiplying the *First* terms of each binomial, $x \cdot 4$ is found by multiplying the *Outer* terms of the two binomials, $5 \cdot x$ is the product of the *Inner* terms of the two binomials, and $5 \cdot 4$ is the product of the *Last* terms of each binomial:

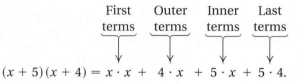

$$(x + 5)(x + 4) = x \cdot x + 4 \cdot x + 5 \cdot x + 5 \cdot 4.$$

To remember this shortcut for multiplying, we use the initials **FOIL**.

The FOIL Method

To multiply two binomials, $A + B$ and $C + D$, multiply the First terms AC, the Outer terms AD, the Inner terms BC, and then the Last terms BD. Then combine like terms, if possible.

$$(A + B)(C + D) = AC + AD + BC + BD$$

1. Multiply First terms: AC.
2. Multiply Outer terms: AD.
3. Multiply Inner terms: BC.
4. Multiply Last terms: BD.

$$\downarrow$$
FOIL

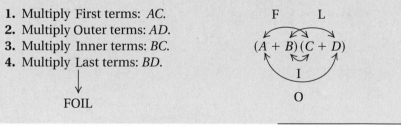

Because addition is commutative, the individual multiplications can be performed in any order. Both FLOI and FIOL yield the same result as FOIL, but FOIL is most easily remembered and most widely used.

Example 1

Multiply: $(x + 8)(x^2 + 5)$.

Solution

$$\overset{\text{F} \qquad \text{O} \quad \text{I} \qquad \text{L}}{(x + 8)(x^2 + 5) = x^3 + 5x + 8x^2 + 40} \qquad \text{There are no like terms.}$$
$$= x^3 + 8x^2 + 5x + 40 \qquad \text{Writing in descending order}$$

After multiplying, remember to combine any like terms.

E x a m p l e   2

Multiply.

a) $(x + 7)(x + 4)$
b) $(y + 3)(y - 2)$
c) $(4t^3 + 5t)(3t^2 - 2)$
d) $(3 - 4x)(7 - 5x^3)$

Solution

a) $(x + 7)(x + 4) = x^2 + 4x + 7x + 28$ Using FOIL
$= x^2 + 11x + 28$ Combining like terms

b) $(y + 3)(y - 2) - y^2 - 2y + 3y - 6$
$= y^2 + y - 6$

c) $(4t^3 + 5t)(3t^2 - 2) = 12t^5 - 8t^3 + 15t^3 - 10t$ Remember to add exponents when multiplying terms with the same base.

$= 12t^5 + 7t^3 - 10t$

d) $(3 - 4x)(7 - 5x^3) = 21 - 15x^3 - 28x + 20x^4$

$= 21 - 28x - 15x^3 + 20x^4$ Because the original binomials are in *ascending* order, we write the answer that way.

Multiplying Sums and Differences of Two Terms

Consider the product of the sum and difference of the same two terms, such as

$(x + 5)(x - 5)$.

Since this is the product of two binomials, we can use FOIL. In doing so, we find that the "outer" and "inner" products are opposites:

a) $(x + 5)(x - 5) = x^2 - 5x + 5x - 25$
$= x^2 - 25;$

b) $(3a - 2)(3a + 2) = 9a^2 + 6a - 6a - 4$ The "outer" and "inner" terms "drop out." Their sum is zero.
$= 9a^2 - 4;$

c) $\left(x^3 + \frac{2}{7}\right)\left(x^3 - \frac{2}{7}\right) = x^6 - \frac{2}{7}x^3 + \frac{2}{7}x^3 - \frac{4}{49}$
$= x^6 - \frac{4}{49}.$

Because opposites always add to zero, for products like $(x + 5)(x - 5)$ we can use a shortcut that is faster than FOIL.

> ### *The Product of a Sum and Difference*
>
> The product of the sum and difference of the same two terms is the square of the first term minus the square of the second term:
>
> $$(A + B)(A - B) = \underbrace{A^2 - B^2}.$$
>
> This is called a *difference of squares*.

Example 3

Multiply.

a) $(x + 4)(x - 4)$
b) $(5 + 2w)(5 - 2w)$
c) $(3a^4 - 5)(3a^4 + 5)$

Solution

$$(A + B)(A - B) = A^2 - B^2 \qquad \text{Saying the words can help:}$$

a) $(x + 4)(x - 4) = x^2 - 4^2$ "The square of the first term, x^2, minus the square of the second, 4^2"

$$= x^2 - 16 \qquad \text{Simplifying}$$

b) $(5 + 2w)(5 - 2w) = 5^2 - (2w)^2$

$$= 25 - 4w^2 \qquad \text{Squaring both 5 and } 2w$$

c) $(3a^4 - 5)(3a^4 + 5) = (3a^4)^2 - 5^2$

$$= 9a^8 - 25 \qquad \begin{array}{l}\text{Using the rules for exponents.}\\ \text{Remember to multiply exponents}\\ \text{when raising a power to a power.}\end{array}$$

Squaring Binomials

Consider the square of a binomial, such as $(x + 3)^2$. This can be expressed as $(x + 3)(x + 3)$. Since this is the product of two binomials, we can use FOIL. But again, this product occurs so often that a faster method has been developed. Look for a pattern in the following:

a) $(x + 3)^2 = (x + 3)(x + 3)$

$$= x^2 + 3x + 3x + 9$$
$$= x^2 + 6x + 9;$$

b) $(5 - 3p)^2 = (5 - 3p)(5 - 3p)$

$$= 25 - 15p - 15p + 9p^2$$
$$= 25 - 30p + 9p^2;$$

c) $(a^3 - 7)^2 = (a^3 - 7)(a^3 - 7)$

$$= a^6 - 7a^3 - 7a^3 + 49$$
$$= a^6 - 14a^3 + 49.$$

Perhaps you noticed that in each product the "outer" and "inner" products are identical. The other two terms, the "first" and "last" products, are squares.

> ### The Square of a Binomial
>
> The square of a binomial is the square of the first term, plus twice the product of the two terms, plus the square of the last term:
>
> $$(A + B)^2 = A^2 + 2AB + B^2;$$
> $$(A - B)^2 = A^2 - 2AB + B^2.$$
>
> These are called *perfect-square trinomials.**

E x a m p l e   4

Multiply: **(a)** $(x + 7)^2$; **(b)** $(t - 5)^2$; **(c)** $(3a + 0.4)^2$; **(d)** $(5x - 3x^4)^2$.

Solution

$$(A + B)^2 = A^2 + 2 \cdot A \cdot B + B^2$$

Saying the words can help:

a) $(x + 7)^2 = x^2 + 2 \cdot x \cdot 7 + 7^2$

"The square of the first term, x^2, plus twice the product of the terms, $2 \cdot 7x$, plus the square of the second term, 7^2"

$$= x^2 + 14x + 49$$

b) $(t - 5)^2 = t^2 - 2 \cdot t \cdot 5 + 5^2$
$$= t^2 - 10t + 25$$

c) $(3a + 0.4)^2 = (3a)^2 + 2 \cdot 3a \cdot 0.4 + 0.4^2$
$$= 9a^2 + 2.4a + 0.16$$

d) $(5x - 3x^4)^2 = (5x)^2 - 2 \cdot 5x \cdot 3x^4 + (3x^4)^2$
$$= 25x^2 - 30x^5 + 9x^8$$ Using the rules for exponents

Caution!   Although the square of a product is the product of the squares, the square of a sum is *not* the sum of the squares. That is, $(AB)^2 = A^2B^2$, but

The term $2AB$ is missing.

$$(A + B)^2 \neq A^2 + B^2.$$

To confirm this inequality, note that

$$(7 + 5)^2 = 12^2 = 144,$$

whereas

$$7^2 + 5^2 = 49 + 25 = 74, \quad \text{and} \quad 74 \neq 144.$$

---

*In some books, these are called *trinomial squares*.

Geometrically, $(A + B)^2$ can be viewed as the area of a square with sides of length $A + B$:

$$(A + B)(A + B) = (A + B)^2.$$

This is equal to the sum of the areas of the four smaller regions:

$$A^2 + AB + AB + B^2 = A^2 + 2AB + B^2.$$

Thus,

$$(A + B)^2 = A^2 + 2AB + B^2.$$

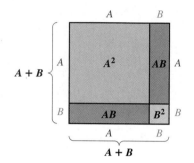

Multiplications of Various Types

Recognizing patterns often helps when new problems are encountered. To simplify a new multiplication problem, always examine what type of product it is so that the best method for finding that product can be used. To do this, ask yourself questions similar to the following.

> ### Multiplying Two Polynomials
>
> **1.** Is the multiplication the product of a monomial and a polynomial? If so, multiply each term of the polynomial by the monomial.
>
> **2.** Is the multiplication the product of two binomials? If so:
>
>   **a)** Is it the product of the sum and difference of the *same* two terms? If so, use the pattern
>
> $$(A + B)(A - B) = A^2 - B^2.$$
>
>   **b)** Is the product the square of a binomial? If so, use the pattern
>
> $$(A + B)(A + B) = (A + B)^2 = A^2 + 2AB + B^2,$$
>
> or
>
> $$(A - B)(A - B) = (A - B)^2 = A^2 - 2AB + B^2.$$
>
>   **c)** If neither (a) nor (b) applies, use FOIL.
>
> **3.** Is the multiplication the product of two polynomials other than those above? If so, multiply each term of one by every term of the other. Use columns if you wish.

E x a m p l e   5

Multiply.

a) $(x + 3)(x - 3)$ **b)** $(t + 7)(t - 5)$ **c)** $(x + 7)(x + 7)$

d) $2x^3(9x^2 + x - 7)$ **e)** $(p + 3)(p^2 + 2p - 1)$ **f)** $\left(3x + \frac{1}{4}\right)^2$

Solution

a) $(x + 3)(x - 3) = x^2 - 9$ This is the product of the sum and difference of the same two terms.

b) $(t + 7)(t - 5) = t^2 - 5t + 7t - 35$ Using FOIL
$$= t^2 + 2t - 35$$

c) $(x + 7)(x + 7) = x^2 + 14x + 49$ This is the square of a binomial.

d) $2x^3(9x^2 + x - 7) = 18x^5 + 2x^4 - 14x^3$ Multiplying each term of the trinomial by the monomial

e)
$$
\begin{array}{r}
p^2 + 2p - 1 \\
p + 3 \\
\hline
3p^2 + 6p - 3 \\
p^3 + 2p^2 - p \\
\hline
p^3 + 5p^2 + 5p - 3
\end{array}
$$
Using columns to multiply a binomial and a trinomial

Multiplying by 3
Multiplying by p

f) $\left(3x + \frac{1}{4}\right)^2 = 9x^2 + 2(3x)\left(\frac{1}{4}\right) + \frac{1}{16}$ Squaring a binomial
$$= 9x^2 + \frac{3}{2}x + \frac{1}{16}$$

Exercise Set **4.5**

Multiply.

1. $(x + 4)(x^2 + 3)$ **2.** $(x^2 - 3)(x - 1)$

3. $(x^3 + 6)(x + 2)$ **4.** $(x^4 + 2)(x + 12)$

5. $(y + 2)(y - 3)$ **6.** $(a + 2)(a + 2)$

7. $(3x + 2)(3x + 5)$ **8.** $(4x + 1)(2x + 7)$

9. $(5x - 6)(x + 2)$ **10.** $(t - 9)(t + 9)$

11. $(1 + 3t)(2 - 3t)$ **12.** $(7 - a)(2 + 3a)$

13. $(2x - 7)(x - 1)$ **14.** $(2x - 1)(3x + 1)$

15. $\left(p - \frac{1}{4}\right)\left(p + \frac{1}{4}\right)$ **16.** $\left(q + \frac{3}{4}\right)\left(q + \frac{3}{4}\right)$

17. $(x - 0.1)(x + 0.1)$ **18.** $(x + 0.3)(x - 0.4)$

19. $(2x^2 + 6)(x + 1)$ **20.** $(2x^2 + 3)(2x - 1)$

21. $(-2x + 1)(x + 6)$ **22.** $(-x + 4)(2x - 5)$

23. $(a + 9)(a + 9)$ **24.** $(2y + 7)(2y + 7)$

25. $(1 + 3t)(1 - 5t)$ **26.** $(1 + 2t)(1 - 3t^2)$

27. $(x^2 + 3)(x^3 - 1)$ **28.** $(x^4 - 3)(2x + 1)$

29. $(3x^2 - 2)(x^4 - 2)$ **30.** $(x^{10} + 3)(x^{10} - 3)$

31. $(2t^3 + 5)(2t^3 + 3)$ **32.** $(5t^2 + 1)(2t^2 + 3)$

33. $(8x^3 + 5)(x^2 + 2)$ **34.** $(4 - 2x)(5 - 2x^2)$

35. $(4x^2 + 3)(x - 3)$ **36.** $(7x - 2)(2x - 7)$

Multiply. Try to recognize what type of product each multiplication is before multiplying.

37. $(x + 8)(x - 8)$ **38.** $(x + 1)(x - 1)$

39. $(2x + 1)(2x - 1)$

40. $(x^2 + 1)(x^2 - 1)$

41. $(5m - 2)(5m + 2)$

42. $(3x^4 + 2)(3x^4 - 2)$

43. $(2x^2 + 3)(2x^2 - 3)$

44. $(6x^5 - 5)(6x^5 + 5)$

45. $(3x^4 - 1)(3x^4 + 1)$

46. $(t^2 - 0.2)(t^2 + 0.2)$

47. $(x^4 + 7)(x^4 - 7)$

48. $(t^3 + 4)(t^3 - 4)$

49. $\left(t - \frac{3}{4}\right)\left(t + \frac{3}{4}\right)$

50. $\left(m - \frac{2}{3}\right)\left(m + \frac{2}{3}\right)$

51. $(x + 2)^2$

52. $(2x - 1)^2$

53. $(3x^5 + 1)^2$

54. $(4x^3 + 1)^2$

55. $\left(a - \frac{2}{5}\right)^2$

56. $\left(t - \frac{1}{5}\right)^2$

57. $(t^3 + 3)^2$

58. $(a^4 + 2)^2$

59. $(2 - 3x^4)^2$

60. $(5 - 2t^3)^2$

61. $(5 + 6t^2)^2$

62. $(3p^2 - p)^2$

63. $(7x - 0.3)^2$

64. $(4a - 0.6)^2$

65. $5a^3(2a^2 - 1)$

66. $9x^3(2x^2 - 5)$

67. $(a - 3)(a^2 + 2a - 4)$

68. $(x^2 - 5)(x^2 + x - 1)$

69. $(3 - 2x^3)^2$

70. $(x - 4x^3)^2$

71. $4x(x^2 + 6x - 3)$

72. $8x(-x^5 + 6x^2 + 9)$

73. $(-t^3 + 1)^2$

74. $(-x^2 + 1)^2$

75. $3t^2(5t^3 - t^2 + t)$

76. $-5x^3(x^2 + 8x - 9)$

77. $(6x^4 - 3)^2$

78. $(8a^3 + 5)^2$

79. $(3x + 2)(4x^2 + 5)$

80. $(2x^2 - 7)(3x^2 + 9)$

81. $(5 - 6x^4)^2$

82. $(3 - 4t^5)^2$

83. $(a + 1)(a^2 - a + 1)$

84. $(x - 5)(x^2 + 5x + 25)$

Find the total area of all shaded rectangles.

85.

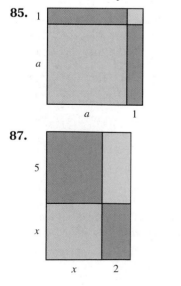

86.

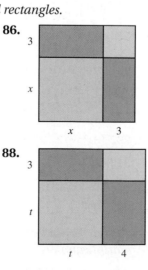

87.

88.

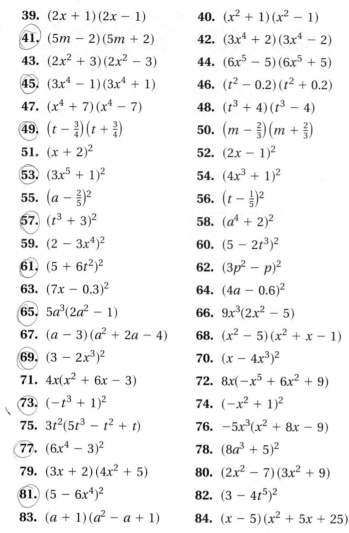

89.

90.

91.

92.

93.

94.

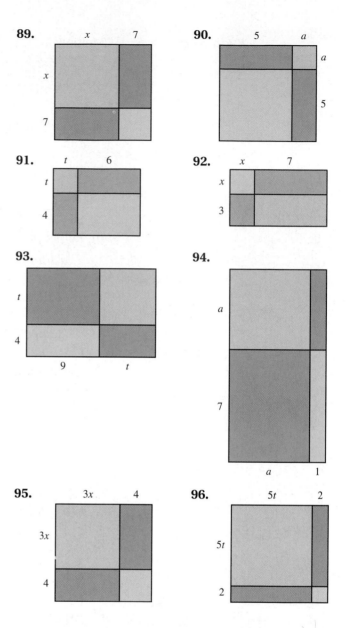

95.

96.

Draw and label rectangles similar to those in Exercises 85–96 to illustrate each of the following.

97. $(x + 5)^2$

98. $(x + 8)^2$

99. $(t + 9)^2$

100. $(a + 12)^2$

101. $(3 + x)^2$

102. $(7 + t)^2$

103. Blair feels that since he can find the product of any two binomials using FOIL, he needn't study the other special products. What advice would you give him?

104. Under what conditions is the product of two binomials a binomial?

SKILL MAINTENANCE

105. *Energy use.*   In an apartment, lamps, an air conditioner, and a television set are all operating at the same time. The lamps take 10 times as many watts as the television set, and the air conditioner takes 40 times as many watts as the television set. The total wattage used in the apartment is 2550 watts. How many watts are used by each appliance?

106. In what quadrant is the point $(-3, 4)$ located?

Solve.

107. $5xy = 8$, for y

108. $3ab = c$, for a

109. $ax - b = c$, for x

110. $st + r = u$, for t

SYNTHESIS

111. Anais claims that by writing $19 \cdot 21$ as $(20 - 1)(20 + 1)$, she can find the product mentally. How is this possible?

112. The product $(A + B)^2$ can be regarded as the sum of the areas of four regions (as shown following Example 4). How might one visually represent $(A + B)^3$? Why?

Multiply.

Aha! **113.** $(4x^2 + 9)(2x + 3)(2x - 3)$

114. $(9a^2 + 1)(3a - 1)(3a + 1)$

Aha! **115.** $(3t - 2)^2(3t + 2)^2$

116. $(5a + 1)^2(5a - 1)^2$

117. $(t^3 - 1)^4(t^3 + 1)^4$

118. $(32.41x + 5.37)^2$

Calculate as the difference of squares.

119. 18×22 [*Hint*: $(20 - 2)(20 + 2)$.]

120. 93×107

Solve.

121. $(x + 2)(x - 5) = (x + 1)(x - 3)$

122. $(2x + 5)(x - 4) = (x + 5)(2x - 4)$

The height of a box is 1 more than its length l, and the length is 1 more than its width w. Find a polynomial for the volume V in terms of the following.

123. The length l

124. The width w

Find a polynomial for the total shaded area in each figure.

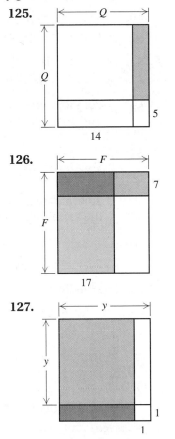

125.

126.

127.

128. Find three consecutive integers for which the sum of the squares is 65 more than three times the square of the smallest integer.

129. Use a grapher and the method developed on p. 232 to check your answers to Exercises 17, 47, and 83.

Polynomials in Several Variables

4.6

Evaluating Polynomials • Like Terms and Degree •
Addition and Subtraction • Multiplication

Thus far, the polynomials that we have studied have had only one variable. Polynomials such as

$$5x + x^2y - 3y + 7, \qquad 9ab^2c - 2a^3b^2 + 8a^2b^3 + 15, \quad \text{and}$$
$$4m^2 - 9n^2$$

contain two or more variables. In this section, we will add, subtract, multiply, and evaluate such **polynomials in several variables.**

Evaluating Polynomials

To evaluate a polynomial in two or more variables, we substitute numbers for the variables. Then we compute, using the rules for order of operations.

E x a m p l e   1

Evaluate the polynomial $4 + 3x + xy^2 + 8x^3y^3$ for $x = -2$ and $y = 5$.

Solution   We substitute -2 for x and 5 for y:

$$4 + 3x + xy^2 + 8x^3y^3 = 4 + 3(-2) + (-2) \cdot 5^2 + 8(-2)^3 \cdot 5^3$$
$$= 4 - 6 - 50 - 8000 = -8052.$$

E x a m p l e   2

Surface area of a right circular cylinder.   The surface area of a right circular cylinder is given by the polynomial

$$2\pi rh + 2\pi r^2,$$

where h is the height and r is the radius of the base. A 12-oz can has a height of 4.7 in. and a radius of 1.2 in. Approximate its surface area.

Solution   We evaluate the polynomial for $h = 4.7$ in. and $r = 1.2$ in. If 3.14 is used to approximate π, we have

$$2\pi rh + 2\pi r^2 \approx 2(3.14)(1.2 \text{ in.})(4.7 \text{ in.}) + 2(3.14)(1.2 \text{ in.})^2$$
$$\approx 2(3.14)(1.2 \text{ in.})(4.7 \text{ in.}) + 2(3.14)(1.44 \text{ in}^2)$$
$$\approx 35.4192 \text{ in}^2 + 9.0432 \text{ in}^2 \approx 44.4624 \text{ in}^2.$$

If the π key of a calculator is used, we have

$$2\pi rh + 2\pi r^2 \approx 2(3.141592654)(1.2 \text{ in.})(4.7 \text{ in.}) + 2(3.141592654)(1.2 \text{ in.})^2$$
$$\approx 44.48495197 \text{ in}^2.$$

Note that the unit in the answer (square inches) is a unit of area. The surface area is about 44.5 in^2 (square inches).

Like Terms and Degree

Recall that the degree of a term is the number of variable factors in the term. For example, the degree of $5x^2$ is 2 because there are two variable factors in $5 \cdot x \cdot x$. Similarly, the degree of $5a^2b^4$ is 6 because there are 6 variable factors in $5 \cdot a \cdot a \cdot b \cdot b \cdot b \cdot b$. Note that 6 can be found by adding the exponents 2 and 4.

As we learned in Section 4.2, the degree of a polynomial is the degree of the term of highest degree.

E x a m p l e 3   Identify the coefficient and the degree of each term and the degree of the polynomial

$$9x^2y^3 - 14xy^2z^3 + xy + 4y + 5x^2 + 7.$$

Solution

Term	Coefficient	Degree	Degree of the Polynomial
$9x^2y^3$	9	5	
$-14xy^2z^3$	-14	6	6
xy	1	2	
$4y$	4	1	
$5x^2$	5	2	
7	7	0	

Note in Example 3 that although both xy and $5x^2$ have degree 2, they are *not* like terms. *Like*, or *similar*, *terms* either have exactly the same variables with exactly the same exponents or are constants. For example,

$$8a^4b^7 \text{ and } 5b^7a^4 \text{ are like terms}$$

and

$$-17 \text{ and } 3 \text{ are like terms,}$$

but

$$-2x^2y \text{ and } 9xy^2 \text{ are } not \text{ like terms.}$$

As always, combining like terms is based on the distributive law.

E x a m p l e 4   Combine like terms.

a) $9x^2y + 3xy^2 - 5x^2y - xy^2$
b) $7ab - 5ab^2 + 3ab^2 + 6a^3 + 9ab - 11a^3 + b - 1$

Solution

a) $9x^2y + 3xy^2 - 5x^2y - xy^2 = (9 - 5)x^2y + (3 - 1)xy^2$

$\qquad\qquad\qquad\qquad\qquad = 4x^2y + 2xy^2$ Try to go directly to this step.

b) $7ab - 5ab^2 + 3ab^2 + 6a^3 + 9ab - 11a^3 + b - 1$

$\qquad = -2ab^2 + 16ab - 5a^3 + b - 1$

Addition and Subtraction

The procedure used for adding polynomials in one variable is used to add polynomials in several variables.

Example 5

Add.

a) $(-5x^3 + 3y - 5y^2) + (8x^3 + 4x^2 + 7y^2)$
b) $(5ab^2 - 4a^2b + 5a^3 + 2) + (3ab^2 - 2a^2b + 3a^3b - 5)$

Solution

a) $(-5x^3 + 3y - 5y^2) + (8x^3 + 4x^2 + 7y^2)$

$\qquad = (-5 + 8)x^3 + 4x^2 + 3y + (-5 + 7)y^2$ Try to do this step mentally.

$\qquad = 3x^3 + 4x^2 + 3y + 2y^2$

b) $(5ab^2 - 4a^2b + 5a^3 + 2) + (3ab^2 - 2a^2b + 3a^3b - 5)$

$\qquad = 8ab^2 - 6a^2b + 5a^3 + 3a^3b - 3$

When subtracting a polynomial, remember to find the opposite of each term in that polynomial and then add.

Example 6

Subtract: $(4x^2y + x^3y^2 + 3x^2y^3 + 6y) - (4x^2y - 6x^3y^2 + x^2y^2 - 5y)$.

Solution

$(4x^2y + x^3y^2 + 3x^2y^3 + 6y) - (4x^2y - 6x^3y^2 + x^2y^2 - 5y)$

$\qquad = 4x^2y + x^3y^2 + 3x^2y^3 + 6y - 4x^2y + 6x^3y^2 - x^2y^2 + 5y$

$\qquad = 7x^3y^2 + 3x^2y^3 - x^2y^2 + 11y$ Combining like terms

Multiplication

To multiply polynomials in several variables, multiply each term of one polynomial by every term of the other, just as we did in Sections 4.4 and 4.5.

E x a m p l e   7

Multiply: $(3x^2y - 2xy + 3y)(xy + 2y)$.

Solution

$$
\begin{array}{r}
3x^2y - 2xy + 3y \\
xy + 2y \\
\hline
6x^2y^2 - 4xy^2 + 6y^2 \\
3x^3y^2 - 2x^2y^2 + 3xy^2 \\
\hline
3x^3y^2 + 4x^2y^2 - \ xy^2 + 6y^2
\end{array}
$$

Multiplying by $2y$
Multiplying by xy
Adding

The special products discussed in Section 4.5 can speed up our work.

E x a m p l e   8

Multiply.

a) $(p + 5q)(2p - 3q)$

b) $(3x + 2y)^2$

c) $(a^3 - 7a^2b)^2$

d) $(3x^2y + 2y)(3x^2y - 2y)$

e) $(-2x^3y^2 + 5t)(2x^3y^2 + 5t)$

f) $(2x + 3 - 2y)(2x + 3 + 2y)$

Solution

$$
\begin{array}{cccc}
\text{F} & \text{O} & \text{I} & \text{L}
\end{array}
$$

a) $(p + 5q)(2p - 3q) = 2p^2 - 3pq + 10pq - 15q^2$

$\qquad\qquad\qquad\quad = 2p^2 + 7pq - 15q^2$ Combining like terms

$$(A + B)^2 = A^2 + 2 \cdot A \cdot B + B^2$$

b) $(3x + 2y)^2 = (3x)^2 + 2(3x)(2y) + (2y)^2$ Squaring a binomial

$\qquad\qquad\quad = 9x^2 + 12xy + 4y^2$

$$(A - B)^2 = A^2 - 2 \cdot A \cdot B + B^2$$

c) $(a^3 - 7a^2b)^2 = (a^3)^2 - 2(a^3)(7a^2b) + (7a^2b)^2$ Squaring a binomial

$\qquad\qquad\qquad = a^6 - 14a^5b + 49a^4b^2$ Using the rules for exponents

$$(A + B)(A - B) = A^2 - B^2$$

d) $(3x^2y + 2y)(3x^2y - 2y) = (3x^2y)^2 - (2y)^2$ Recognizing the pattern

$\qquad\qquad\qquad\qquad\quad = 9x^4y^2 - 4y^2$ Using the rules for exponents

e) $(-2x^3y^2 + 5t)(2x^3y^2 + 5t) = (5t - 2x^3y^2)(5t + 2x^3y^2)$ Using the commutative law for addition twice

$\qquad\qquad\qquad\qquad\qquad\qquad = (5t)^2 - (2x^3y^2)^2$ Multiplying the sum and the difference of the same two terms

$\qquad\qquad\qquad\qquad\qquad\qquad = 25t^2 - 4x^6y^4$

$$(\quad A \quad - \quad B)(\quad A \quad + \quad B) = \quad A^2 \quad - \quad B^2$$

f) $(\ 2x + 3\ - 2y)(\ 2x + 3\ + 2y) = (\ 2x + 3\ )^2 - (2y)^2$ Multiplying a sum and a difference

$$= 4x^2 + 12x + 9 - 4y^2$$ Squaring a binomial

Note that in Example 8 we recognized patterns that might have eluded some students, particularly in parts (e) and (f). In part (e), we could have used FOIL, and in part (f), we could have used long multiplication, but doing so would have been slower. By carefully inspecting a problem before "jumping in," we can often save ourselves considerable work.

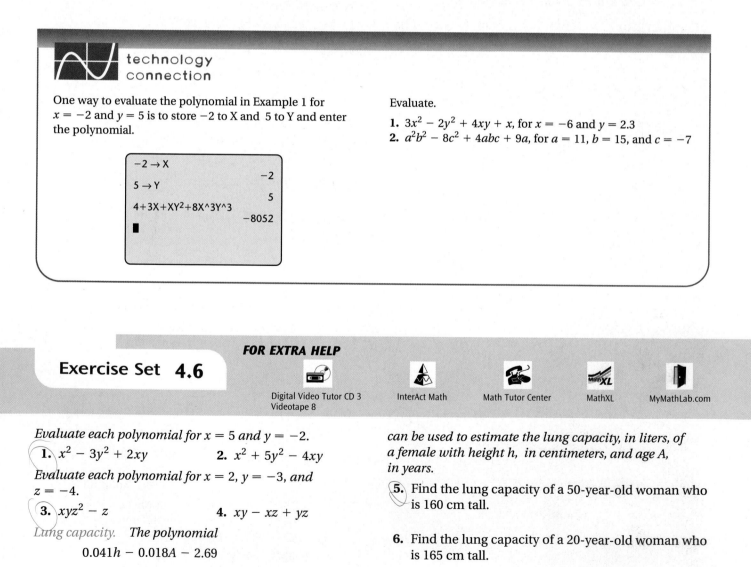

technology connection

One way to evaluate the polynomial in Example 1 for $x = -2$ and $y = 5$ is to store -2 to X and 5 to Y and enter the polynomial.

```
-2 → X
                              -2
5 → Y
                               5
4+3X+XY²+8X^3Y^3
                            -8052
■
```

Evaluate.

1. $3x^2 - 2y^2 + 4xy + x$, for $x = -6$ and $y = 2.3$
2. $a^2b^2 - 8c^2 + 4abc + 9a$, for $a = 11$, $b = 15$, and $c = -7$

Exercise Set 4.6

Evaluate each polynomial for $x = 5$ and $y = -2$.

1. $x^2 - 3y^2 + 2xy$ **2.** $x^2 + 5y^2 - 4xy$

Evaluate each polynomial for $x = 2$, $y = -3$, and $z = -4$.

3. $xyz^2 - z$ **4.** $xy - xz + yz$

Lung capacity.   *The polynomial*

$$0.041h - 0.018A - 2.69$$

can be used to estimate the lung capacity, in liters, of a female with height h, in centimeters, and age A, in years.

5. Find the lung capacity of a 50-year-old woman who is 160 cm tall.

6. Find the lung capacity of a 20-year-old woman who is 165 cm tall.

Altitude of a launched object.   *The altitude of an object, in meters, is given by the polynomial*

$$h + vt - 4.9t^2,$$

where h is the height, in meters, at which the launch occurs, v is the initial upward speed (or velocity), in meters per second, and t is the number of seconds for which the object is airborne.

7. A model rocket is launched from atop the Eiffel Tower in Paris, 300 m above the ground. If the initial upward speed is 40 meters per second (m/s), how high above the ground will the rocket be 2 sec after having been launched?

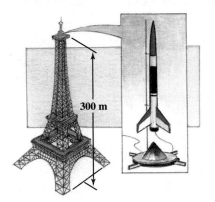

300 m

8. A golf ball is thrown upward with an initial speed of 30 m/s by a golfer atop the Washington Monument, 160 m above the ground. How high above the ground will the ball be after 3 sec?

Surface area of a silo.   *A silo is a structure that is shaped like a right circular cylinder with a half sphere on top. The surface area of a silo of height h and radius r (including the area of the base) is given by the polynomial $2\pi rh + \pi r^2$.*

9. A container of tennis balls is silo-shaped, with a height of $7\frac{1}{2}$ in. and a radius of $1\frac{1}{4}$ in. Find the surface area of the container. Use 3.14 for π.

r

PRO BOUND

h

10. A $1\frac{1}{2}$-oz bottle of roll-on deodorant has a height of 4 in. and a radius of $\frac{3}{4}$ in. Find the surface area of the bottle if the bottle is shaped like a silo. Use 3.14 for π.

Identify the coefficient and the degree of each term of each polynomial. Then find the degree of each polynomial.

11. $x^3y - 2xy + 3x^2 - 5$

12. $xy^2 - y^2 + 9x^2y + 7$

13. $17x^2y^3 - 3x^3yz - 7$

14. $6 - xy + 8x^2y^2 - y^5$

Combine like terms.

15. $7a + b - 4a - 3b$

16. $8r + s - 5r - 4s$

17. $3x^2y - 2xy^2 + x^2 + 5x$

18. $m^3 + 2m^2n - 3m^2 + 3mn^2$

19. $2u^2v - 3uv^2 + 6u^2v - 2uv^2 + 7u^2$

20. $3x^2 + 6xy + 3y^2 - 5x^2 - 10xy$

21. $5a^2c - 2ab^2 + a^2b - 3ab^2 + a^2c - 2ab^2$

22. $3s^2t + r^2t - 4st^2 - s^2t + 3st^2 - 7r^2t$

Add or subtract, as indicated.

23. $(4x^2 - xy + y^2) + (-x^2 - 3xy + 2y^2)$

24. $(2r^3 + 3rs - 5s^2) - (5r^3 + rs + 4s^2)$

25. $(3a^4 - 5ab + 6ab^2) - (9a^4 + 3ab - ab^2)$

26. $(2r^2t - 5rt + rt^2) - (7r^2t + rt - 5rt^2)$

Aha! **27.** $(5r^2 - 4rt + t^2) + (-6r^2 - 5rt - t^2) + (-5r^2 + 4rt - t^2)$

28. $(2x^2 - 3xy + y^2) + (-4x^2 - 6xy - y^2) + (4x^2 + 6xy + y^2)$

29. $(x^3 - y^3) - (-2x^3 + x^2y - xy^2 + 2y^3)$

30. $(a^3 + b^3) - (-5a^3 + 2a^2b - ab^2 + 3b^3)$

31. $(2y^4x^2 - 5y^3x) + (5y^4x^2 - y^3x) + (3y^4x^2 - 2y^3x)$

32. $(5a^2b + 7ab) + (9a^2b - 5ab) + (a^2b - 6ab)$

33. Subtract $7x + 3y$ from the sum of $4x + 5y$ and $-5x + 6y$.

34. Subtract $5a + 2b$ from the sum of $2a + b$ and $3a - 4b$.

Multiply.

35. $(3z - u)(2z + 3u)$ **36.** $(5x + y)(2x - 3y)$

37. $(xy + 7)(xy - 4)$ **38.** $(ab + 3)(ab - 5)$

39. $(2a - b)(2a + b)$ **40.** $(a - 3b)(a + 3b)$

41. $(5rt - 2)(3rt + 1)$ **42.** $(3xy - 1)(4xy + 2)$

43. $(m^3n + 8)(m^3n - 6)$ **44.** $(3 - c^2d^2)(4 + c^2d^2)$

45. $(6x - 2y)(5x - 3y)$ **46.** $(7a - 6b)(5a + 4b)$

47. $(pq + 0.2)(0.4pq - 0.1)$

48. $(ab - 0.6)(0.2ab + 0.3)$

49. $(x + h)^2$ **50.** $(r + t)^2$

51. $(4a + 5b)^2$ **52.** $(3x + 2y)^2$

53. $(c^2 - d)(c^2 + d)$ **54.** $(p^3 - 5q)(p^3 + 5q)$

55. $(ab + cd^2)(ab - cd^2)$ **56.** $(xy + pq)(xy - pq)$

Aha! **57.** $(a + b - c)(a + b + c)$

58. $(x + y + z)(x + y - z)$

59. $[a + b + c][a - (b + c)]$

60. $(a + b + c)(a - b - c)$

Find the total area of each shaded area.

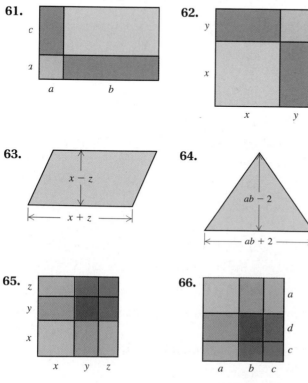

61.
c
a
a b

62.
y
x
x y

63.
$x - z$
$x + z$

64.
$ab - 2$
$ab + 2$

65.
z
y
x
x y z

66.
a
d
c
a b c

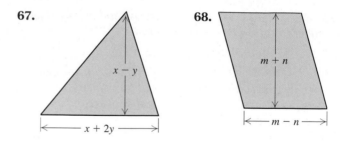

67.
$x - y$
$x + 2y$

68.
$m + n$
$m - n$

Draw and label rectangles similar to those in Exercises 61, 62, 65, and 66 to illustrate each product.

69. $(r + s)(u + v)$ **70.** $(m + r)(n + v)$

71. $(a + b + c)(a + d + f)$ **72.** $(r + s + t)^2$

73. Is it possible for a polynomial in 4 variables to have a degree less than 4? Why or why not?

74. A fourth-degree polynomial is multiplied by a third-degree polynomial. What is the degree of the product? Explain your reasoning.

SKILL MAINTENANCE

Simplify.

75. $5 + \dfrac{7 + 4 + 2 \cdot 5}{3}$ **76.** $9 - \dfrac{2 + 6 \cdot 3 + 4}{6}$

77. $(4 + 3 \cdot 5 + 8) \div 3 \cdot 3$

78. $(5 + 2 \cdot 7 + 5) \div 2 \cdot 3$

79. $[3 \cdot 5 - 4 \cdot 2 + 7(-3)] \div (-2)$

80. $(7 - 3 \cdot 9 - 2 \cdot 5) \div (-6)$

SYNTHESIS

81. Can the sum of two trinomials in several variables be a binomial in one variable? Why or why not?

82. Can the sum of two trinomials in several variables be a trinomial in one variable? Why or why not?

Find a polynomial for the shaded area. (Leave results in terms of π where appropriate.)

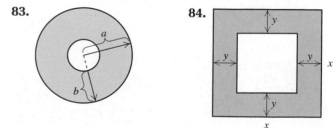

83.
a
b

84.
y
y y x
y
x

85.

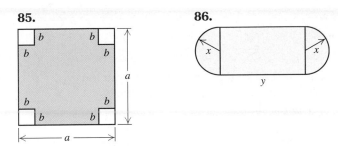

86.

87. Find the shaded area in this figure using each of the approaches given below. Then check that both answers match.

a) Find the shaded area by subtracting the area of the unshaded square from the total area of the figure.

b) Find the shaded area by adding the areas of the three shaded rectangles.

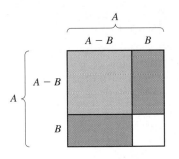

Find a polynomial for the surface area of each solid object shown. (Leave results in terms of π.)

88.

89.

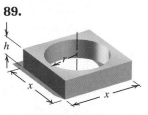

90. The observatory at Danville University is shaped like a silo that is 40 ft high and 30 ft wide (see Exercise 9). The Heavenly Bodies Astronomy Club is to paint the exterior of the observatory using paint that covers 250 ft² per gallon. How many gallons should they purchase? Explain your reasoning.

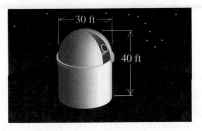

91. Multiply: $(x + a)(x - b)(x - a)(x + b)$.

92. *Interest compounded annually.* An amount of money P that is invested at the yearly interest rate r grows to the amount

$$P(1 + r)^t$$

after t years. Find a polynomial that can be used to determine the amount to which P will grow after 2 yr.

93. *Yearly depreciation.* An investment P that drops in value at the yearly rate r drops in value to

$$P(1 - r)^t$$

after t years. Find a polynomial that can be used to determine the value to which P has dropped after 2 yr.

94. Suppose that $10,400 is invested at 8.5% compounded annually. How much is in the account at the end of 5 yr? (See Exercise 92.)

95. A $90,000 investment in computer hardware is depreciating at a yearly rate of 12.5%. How much is the investment worth after 4 yr? (See Exercise 93.)

CORNER

Finding the Magic Number

Focus: Evaluating polynomials in several variables

Time: 15–25 minutes

Group size: 3

Materials: A coin for each person

When a team nears the end of its schedule in first place, fans begin to discuss the team's "magic number." A team's magic number is the combined number of wins by that team and losses by the second-place team that guarantee the leading team a first-place finish. For example, if the Cubs' magic number is 3 over the Reds, any combination of Cubs wins and Reds losses that totals 3 will guarantee a first-place finish for the Cubs, regardless of how subsequent games are decided. A team's magic number is computed using the polynomial

$$G - P - L + 1,$$

where G is the length of the season, in games, P is the number of games that the leading team has played, and L is the total number of games that the second-place team has lost minus the total number of games that the leading team has lost.

ACTIVITY

1. The standings below are from a fictitious baseball league. Together, the group should calculate the Jaguars' magic number with respect to the Catamounts as well as the Jaguars' magic number with respect to the Wildcats. (Assume that the schedule is 162 games long.)

	W	L
Jaguars	92	64
Catamounts	90	66
Wildcats	89	66

2. Each group member should play the role of one of the teams. To simulate each team's remaining games, coin tosses will be performed. If a group member correctly predicts the side (heads or tails) that comes up, the coin toss represents a win for that team. Should the other side appear, the toss represents a loss. Assume that these games are against other (unlisted) teams in the league. Each group member should perform three coin tosses and then update the standings.

3. Recalculate the two magic numbers, using the updated standings from part (2).

4. Slowly—one coin toss at a time—play out the remainder of the season. Record all wins and losses, update the standings, and recalculate the magic numbers each time all three group members have completed a round of coin tosses.

5. Examine the work in part (4) and explain why a magic number of 0 indicates that a team has been eliminated from contention.

4.7

Division of Polynomials

Dividing by a Monomial • Dividing by a Binomial

In this section, we study division of polynomials. We will find that polynomial division is similar to division in arithmetic.

Dividing by a Monomial

We first consider division by a monomial. When dividing a monomial by a monomial, we use the quotient rule of Section 4.1 to subtract exponents when bases are the same. For example,

$$\frac{15x^{10}}{3x^4} = 5x^{10-4}$$
$$= 5x^6$$

Caution! The coefficients are divided but the exponents are subtracted.

and

$$\frac{42a^2b^5}{-3ab^2} = \frac{42}{-3}a^{2-1}b^{5-2}$$
$$= -14ab^3.$$

To divide a polynomial by a monomial, we note that since

$$\frac{A}{C} + \frac{B}{C} = \frac{A+B}{C},$$

it follows that

$$\frac{A+B}{C} = \frac{A}{C} + \frac{B}{C}. \quad \text{Switching the left and right sides of the equation}$$

This is actually how we perform divisions like $86 \div 2$: Although we might simply write

$$\frac{86}{2} = 43,$$

we are really saying

$$\frac{80+6}{2} = \frac{80}{2} + \frac{6}{2} = 40 + 3.$$

Similarly, to divide a polynomial by a monomial, we divide each term by the monomial:

$$\frac{80x^5 + 6x^3}{2x^2} = \frac{80x^5}{2x^2} + \frac{6x^3}{2x^2}$$
$$= \frac{80}{2}x^{5-2} + \frac{6}{2}x^{3-2} \quad \text{Dividing coefficients and subtracting exponents}$$
$$= 40x^3 + 3x.$$

Study Tip

It is never too soon to begin reviewing for the final examination. Take a few minutes each week to review important problems, formulas, and properties. There is also at least one Connecting the Concepts in each chapter. Spend time reviewing the information in this special feature.

E x a m p l e 1

Divide $x^4 + 15x^3 - 6x^2$ by $3x$.

Solution We have

$$\frac{x^4 + 15x^3 - 6x^2}{3x} = \frac{x^4}{3x} + \frac{15x^3}{3x} - \frac{6x^2}{3x}$$

$$= \frac{1}{3}x^{4-1} + \frac{15}{3}x^{3-1} - \frac{6}{3}x^{2-1} \qquad \text{Dividing coeffi-}$$
$$\text{cients and sub-}$$
$$\text{tracting exponents}$$

$$= \frac{1}{3}x^3 + 5x^2 - 2x. \qquad \text{This is the quotient.}$$

To check, we multiply our answer by $3x$, using the distributive law:

$$3x\left(\frac{1}{3}x^3 + 5x^2 - 2x\right) = 3x \cdot \frac{1}{3}x^3 + 3x \cdot 5x^2 - 3x \cdot 2x$$
$$= x^4 + 15x^3 - 6x^2.$$

This is the polynomial that was being divided, so our answer, $\frac{1}{3}x^3 + 5x^2 - 2x$, checks.

E x a m p l e 2

Divide and check: $(10a^5b^4 - 2a^3b^2 + 6a^2b) \div 2a^2b$.

Solution We have

$$\frac{10a^5b^4 - 2a^3b^2 + 6a^2b}{2a^2b} = \frac{10a^5b^4}{2a^2b} - \frac{2a^3b^2}{2a^2b} + \frac{6a^2b}{2a^2b}$$

$$= \frac{10}{2}a^{5-2}b^{4-1} - \frac{2}{2}a^{3-2}b^{2-1} + \frac{6}{2} \qquad \text{Dividing}$$
$$\text{coefficients}$$
$$\text{and subtracting}$$
$$\text{exponents}$$

$$= 5a^3b^3 - ab + 3.$$

Check: $2a^2b(5a^3b^3 - ab + 3) = 2a^2b \cdot 5a^3b^3 - 2a^2b \cdot ab + 2a^2b \cdot 3$
$$= 10a^5b^4 - 2a^3b^2 + 6a^2b$$

Our answer, $5a^3b^3 - ab + 3$, checks.

Dividing by a Binomial

For divisors with more than one term, we use long division, much as we do in arithmetic. Polynomials are written in descending order and any missing terms are written in, using 0 for the coefficients.

E x a m p l e   3

Divide $x^2 + 5x + 6$ by $x + 3$.

Solution   We have

Divide the first term, x^2, by the first term in the divisor: $x^2/x = x$. Ignore the term 3 for the moment.

$$
\begin{array}{r}
x \phantom{{}+ 5x + 6} \\
x + 3 \overline{)\, x^2 + 5x + 6} \\
-(x^2 + 3x) \phantom{{}+ 6} \\
\hline
2x \phantom{{}+ 6}
\end{array}
$$

Multiply $x + 3$ by x.

Subtract by mentally changing signs and adding: $x^2 + 5x - (x^2 + 3x) = 2x$.

Now we "bring down" the next term—in this case, 6.

$$
\begin{array}{r}
x + 2 \\
x + 3 \overline{)\, x^2 + 5x + 6} \\
-(x^2 + 3x) \\
\hline
2x + 6 \\
-(2x + 6) \\
\hline
0
\end{array}
$$

Divide $2x$ by x: $2x/x = 2$.

Multiply 2 by the divisor, $x + 3$.

Subtract: $(2x + 6) - (2x + 6) = 0$.

The quotient is $x + 2$. The remainder is 0, expressed as R 0. A remainder of 0 is generally not listed in an answer.

Check: To check, we multiply the quotient by the divisor and add the remainder, if any, to see if we get the dividend:

Divisor	Quotient		Remainder		Dividend
$(x + 3)$	$(x + 2)$	$+$	0	$=$	$x^2 + 5x + 6.$

Our answer, $x + 2$, checks.

E x a m p l e   4

Divide: $(2x^2 + 5x - 1) \div (2x - 1)$.

Solution   We have

Divide the first term by the first term: $2x^2/(2x) = x$.

$$
\begin{array}{r}
x \phantom{{}+ 5x - 1} \\
2x - 1 \overline{)\, 2x^2 + 5x - 1} \\
-(2x^2 - x) \phantom{{}- 1} \\
\hline
6x \phantom{{}- 1}
\end{array}
$$

Multiply $2x - 1$ by x.

Subtract by mentally changing signs and adding: $2x^2 + 5x - (2x^2 - x) = 6x$.

Now, we bring down the next term of the dividend, -1.

$$
\begin{array}{r}
x + 3 \\
2x - 1 \overline{)\, 2x^2 + 5x - 1} \\
-(2x^2 - x) \\
\hline
6x - 1 \\
-(6x - 3) \\
\hline
2
\end{array}
$$

Divide $6x$ by $2x$: $6x/(2x) = 3$.

Multiply 3 by the divisor, $2x - 1$.

Note that $-1 - (-3) = -1 + 3 = 2$.

The answer is $x + 3$ with R 2.

Another way to write $x + 3 \text{ R } 2$ is as

$$\text{Quotient} \quad \underbrace{x + 3}_{} + \underbrace{\frac{2}{2x - 1}}_{}. \qquad \begin{matrix} \leftarrow \text{Remainder} \\ \\ \leftarrow \text{Divisor} \end{matrix}$$

(This is the way answers will be given at the back of the book.)

Check: To check, we multiply the quotient by the divisor and add the remainder:

$$(2x - 1)(x + 3) + 2 = 2x^2 + 5x - 3 + 2$$
$$= 2x^2 + 5x - 1. \qquad \text{Our answer checks.}$$

Our division procedure ends when the degree of the remainder is less than that of the divisor. Check that this was indeed the case in Example 4.

E x a m p l e   5   Divide each of the following.

a) $(x^3 + 1) \div (x + 1)$
b) $(x^4 - 3x^2 + 2x - 3) \div (x^2 - 5)$

Solution

a)
$$\begin{array}{r} x^2 - x + 1 \\ x + 1\overline{)x^3 + 0x^2 + 0x + 1} \quad \leftarrow \text{Fill in the missing terms.} \\ \underline{-(x^3 + x^2)} \\ -x^2 + 0x \quad \leftarrow \text{Subtracting } x^3 + x^2 \text{ from } x^3 + 0x^2 \text{ and} \\ \underline{-(-x^2 - x)} \text{bringing down the } 0x \\ x + 1 \quad \leftarrow \text{Subtracting } -x^2 - x \text{ from } -x^2 + 0x \text{ and} \\ \underline{-(x + 1)} \text{bringing down the } 1 \\ 0 \end{array}$$

The answer is $x^2 - x + 1$.

Check: $(x + 1)(x^2 - x + 1) = x^3 - x^2 + x + x^2 - x + 1$
$$= x^3 + 1.$$

b)
$$\begin{array}{r} x^2 + 2 \\ x^2 - 5\overline{)x^4 + 0x^3 - 3x^2 + 2x - 3} \\ \underline{-(x^4 - 5x^2)} \\ 2x^2 + 2x - 3 \\ \underline{-(2x^2 - 10)} \\ 2x + 7 \quad \leftarrow \end{array}$$

Writing in the missing term

Subtracting $x^4 - 5x^2$ from $x^4 - 3x^2$ and bringing down $2x - 3$

Subtracting $2x^2 - 10$ from $2x^2 + 2x - 3$

Since the remainder, $2x + 7$, is of lower degree than the divisor, the division process stops. The answer is $x^2 + 2$, with R $2x + 7$, or

$$x^2 + 2 + \frac{2x + 7}{x^2 - 5}.$$

Check: $(x^2 - 5)(x^2 + 2) + 2x + 7 = x^4 + 2x^2 - 5x^2 - 10 + 2x + 7$
$$= x^4 - 3x^2 + 2x - 3.$$

Exercise Set 4.7

FOR EXTRA HELP

Digital Video Tutor CD 3
Videotape 8

InterAct Math

Math Tutor Center

MathXL

MyMathLab.com

Divide and check.

1. $\dfrac{40x^5 - 16x}{8}$

2. $\dfrac{12a^4 - 3a^2}{6}$

3. $\dfrac{u - 2u^2 + u^7}{u}$

4. $\dfrac{50x^5 - 7x^4 + x^2}{x}$

5. $(15t^3 - 24t^2 + 6t) \div (3t)$

6. $(20t^3 - 15t^2 + 30t) \div (5t)$

7. $(25x^6 - 20x^4 - 5x^2) \div (-5x^2)$

8. $(16x^6 + 32x^5 - 8x^2) \div (-8x^2)$

9. $(24t^5 - 40t^4 + 6t^3) \div (4t^3)$

10. $(18t^6 - 27t^5 - 3t^3) \div (9t^3)$

11. $\dfrac{6x^2 - 10x + 1}{2}$

12. $\dfrac{9x^2 + 3x - 2}{3}$

13. $\dfrac{4x^3 + 6x^2 + 4x}{2x^2}$

14. $\dfrac{10x^4 + 15x^3 + 5x}{5x^2}$

15. $\dfrac{9r^2s^2 + 3r^2s - 6rs^2}{-3rs}$

16. $\dfrac{4x^4y - 8x^6y^2 + 12x^8y^6}{4x^4y}$

17. $(x^2 + 4x - 12) \div (x - 2)$

18. $(x^2 - 6x + 8) \div (x - 4)$

19. $(t^2 - 10t - 20) \div (t - 5)$

20. $(t^2 + 8t - 15) \div (t + 4)$

21. $(2x^2 + 11x - 5) \div (x + 6)$

22. $(3x^2 - 2x - 13) \div (x - 2)$

23. $\dfrac{a^3 + 8}{a + 2}$

24. $\dfrac{t^3 + 27}{t + 3}$

25. $\dfrac{t^2 - 15}{t - 4}$

26. $\dfrac{a^2 - 23}{a - 5}$

27. $(3x^2 + 11x - 4) \div (3x - 1)$

28. $(10x^2 + 13x - 3) \div (5x - 1)$

29. $(6a^2 + 17a + 8) \div (2a + 5)$

30. $(10a^2 + 19a + 9) \div (2a + 3)$

31. $\dfrac{2t^3 - 9t^2 + 11t - 3}{2t - 3}$

32. $\dfrac{8t^3 - 22t^2 - 5t + 12}{4t + 3}$

33. $(t^3 - t^2 + t - 1) \div (t + 1)$

34. $(x^3 - x^2 + x - 1) \div (x - 1)$

35. $(t^4 + 4t^2 + 3t - 6) \div (t^2 + 5)$

36. $(t^4 - 2t^2 + 4t - 5) \div (t^2 - 3)$

37. $(4x^4 - 4x^2 - x - 3) \div (2x^2 - 3)$

38. $(6x^4 - 3x^2 + x - 4) \div (2x^2 + 1)$

39. How is the distributive law used when dividing a polynomial by a binomial?

40. On an assignment, Emmy Lou *incorrectly* writes

$$\dfrac{12x^3 - 6x}{3x} = 4x^2 - 6x.$$

What mistake do you think she is making and how might you convince her that a mistake has been made?

SKILL MAINTENANCE

Simplify.

41. $-4 + (-13)$

42. $-8 + (-15)$

43. $-9 - (-7)$

44. $-2 - (-7)$

45. The perimeter of a rectangle is 640 ft. The length is 15 ft greater than the width. Find the length of the rectangle.

46. Solve: $3(2x - 1) = 7x - 5$.

47. Graph: $3x - 2y = 12$.

48. Plot the points $(4, -1)$, $(0, 5)$, $(-2, 3)$, and $(-3, 0)$.

SYNTHESIS

49. Explain how the quotient of two binomials can have more than two terms.

50. Explain how to form a trinomial for which division by $x - 5$ results in a remainder of 7.

Divide.

51. $(10x^{9k} - 32x^{6k} + 28x^{3k}) \div (2x^{3k})$

52. $(45a^{8k} + 30a^{6k} - 60a^{4k}) \div (3a^{2k})$

53. $(6t^{3h} + 13t^{2h} - 4t^h - 15) \div (2t^h + 3)$

54. $(x^4 + a^2) \div (x + a)$

55. $(5a^3 + 8a^2 - 23a - 1) \div (5a^2 - 7a - 2)$

56. $(15y^3 - 30y + 7 - 19y^2) \div (3y^2 - 2 - 5y)$

57. Divide the sum of $4x^5 - 14x^3 - x^2 + 3$ and $2x^5 + 3x^4 + x^3 - 3x^2 + 5x$ by $3x^3 - 2x - 1$.

58. Divide $5x^7 - 3x^4 + 2x^2 - 10x + 2$ by the sum of $(x - 3)^2$ and $5x - 8$.

If the remainder is 0 when one polynomial is divided by another, the divisor is a factor *of the dividend. Find the value(s) of c for which x − 1 is a factor of each polynomial.*

59. $x^2 - 4x + c$

60. $2x^2 - 3cx - 8$

61. $c^2x^2 + 2cx + 1$

Negative Exponents and Scientific Notation

4.8

Negative Integers as Exponents • Scientific Notation • Multiplying and Dividing Using Scientific Notation

We now attach a meaning to negative exponents. Once we understand both positive and negative exponents, we can study a method of writing numbers known as *scientific notation*.

Negative Integers as Exponents

Let's define negative exponents so that the rules that apply to whole-number exponents will hold for all integer exponents. To do so, consider a^{-5} and the rule for adding exponents:

$$a^{-5} = a^{-5} \cdot 1 \qquad \text{Using the identity property of 1}$$

$$= \frac{a^{-5}}{1} \cdot \frac{a^5}{a^5} \qquad \text{Writing 1 as } \frac{a^5}{a^5} \text{ and } a^{-5} \text{ as } \frac{a^{-5}}{1}$$

$$= \frac{a^{-5+5}}{a^5} \qquad \text{Adding exponents}$$

$$= \frac{1}{a^5}. \qquad -5 + 5 = 0 \text{ and } a^0 = 1$$

This leads to our definition of negative exponents.

> **Negative Exponents**
> For any real number a that is nonzero and any integer n,
>
> $$a^{-n} = \frac{1}{a^n}.$$
>
> (The numbers a^{-n} and a^n are reciprocals of each other.)

E x a m p l e   1   Express using positive exponents and, if possible, simplify.

a) m^{-3} **b)** 4^{-2} **c)** $(-3)^{-2}$ **d)** ab^{-1}

Solution

a) $m^{-3} = \dfrac{1}{m^3}$ m^{-3} is the reciprocal of m^3.

b) $4^{-2} = \dfrac{1}{4^2} = \dfrac{1}{16}$ 4^{-2} is the reciprocal of 4^2.
Note that $4^{-2} \neq 4(-2)$.

c) $(-3)^{-2} = \dfrac{1}{(-3)^2} = \dfrac{1}{(-3)(-3)} = \dfrac{1}{9}$ $(-3)^{-2}$ is the reciprocal of $(-3)^2$.
Note that $(-3)^{-2} \neq -\dfrac{1}{3^2}$.

d) $ab^{-1} = a\left(\dfrac{1}{b^1}\right) = a\left(\dfrac{1}{b}\right) = \dfrac{a}{b}$ b^{-1} is the reciprocal of b^1.

> **Caution!**   A negative exponent does not, in itself, indicate that an expression is negative. As shown in Example 1,
>
> $$4^{-2} \neq 4(-2) \quad \text{and} \quad (-3)^{-2} \neq -\frac{1}{3^2}.$$

The following is another way to illustrate why negative exponents are defined as they are.

On this side, we divide by 5 at each step.		On this side, the exponents decrease by 1.
	$125 = 5^3$	
	$25 = 5^2$	
	$5 = 5^1$	
	$1 = 5^0$	
	$\dfrac{1}{5} = 5^?$	
	$\dfrac{1}{25} = 5^?$	

To continue the pattern, it follows that

$$\frac{1}{5} = \frac{1}{5^1} = 5^{-1}, \qquad \frac{1}{25} = \frac{1}{5^2} = 5^{-2}, \quad \text{and, in general,} \quad \frac{1}{a^n} = a^{-n}.$$

E x a m p l e 2

Express $\dfrac{1}{x^7}$ using negative exponents.

Solution   We know that $\dfrac{1}{a^n} = a^{-n}$. Thus,

$$\frac{1}{x^7} = x^{-7}.$$

The rules for powers still hold when exponents are negative.

E x a m p l e 3

Simplify. Do not use negative exponents in the answer.

a) $t^5 \cdot t^{-2}$ **b)** $(y^{-5})^{-7}$ **c)** $(5x^2y^{-3})^4$

d) $\dfrac{x^{-4}}{x^{-5}}$ **e)** $\dfrac{1}{t^{-5}}$ **f)** $\dfrac{s^{-3}}{t^{-5}}$

Solution

a) $t^5 \cdot t^{-2} = t^{5+(-2)} = t^3$ Adding exponents

b) $(y^{-5})^{-7} = y^{(-5)(-7)}$ Multiplying exponents
$$= y^{35}$$

c) $(5x^2y^{-3})^4 = 5^4(x^2)^4(y^{-3})^4$ Raising each factor to the fourth power
$$= 625x^8y^{-12} = \frac{625x^8}{y^{12}}$$

d) $\dfrac{x^{-4}}{x^{-5}} = x^{-4-(-5)} = x^1 = x$ We subtract exponents even if the exponent in the denominator is negative.

e) Since $\dfrac{1}{a^n} = a^{-n}$, we have

$$\frac{1}{t^{-5}} = t^{-(-5)} = t^5.$$

f) $\dfrac{s^{-3}}{t^{-5}} = s^{-3} \cdot \dfrac{1}{t^{-5}}$

$$= \frac{1}{s^3} \cdot t^5 = \frac{t^5}{s^3}$$ Using the result from part (e) above

The result from Example 3(f) can be generalized.

Factors and Negative Exponents

For any nonzero real numbers a and b and any integers m and n,

$$\frac{a^{-n}}{b^{-m}} = \frac{b^m}{a^n}.$$

(A factor can be moved to the other side of the fraction bar if the sign of the exponent is changed.)

Example 4   Simplify: $\dfrac{5x^{-7}}{y^2 z^{-4}}$.

Solution   We can move the factors x^{-7} and z^{-4} to the other side of the fraction if we change the sign of each exponent:

$$\frac{5x^{-7}}{y^2 z^{-4}} = \frac{5z^4}{y^2 x^7}, \quad \text{or} \quad \frac{5z^4}{x^7 y^2}.$$

Another way to change the sign of the exponent is to take the reciprocal of the base. To understand why this is true, note that

$$\left(\frac{s}{t}\right)^{-5} = \frac{s^{-5}}{t^{-5}} = \frac{t^5}{s^5} = \left(\frac{t}{s}\right)^5.$$

This often provides the easiest way to simplify an expression containing a negative exponent.

> **Reciprocals and Negative Exponents**
>
> For any nonzero real numbers a and b and any integer n,
>
> $$\left(\frac{a}{b}\right)^{-n} = \left(\frac{b}{a}\right)^{n}.$$
>
> (Any base to a power is equal to the reciprocal of the base raised to the opposite power.)

Example 5   Simplify: $\left(\dfrac{x^4}{2y}\right)^{-3}$.

Solution

$$\left(\frac{x^4}{2y}\right)^{-3} = \left(\frac{2y}{x^4}\right)^{3} \qquad \text{Taking the reciprocal of the base and changing the sign of the exponent}$$

$$= \frac{(2y)^3}{(x^4)^3} \qquad \text{Raising a quotient to a power by raising both the numerator and denominator to the power}$$

$$= \frac{2^3 y^3}{x^{12}} \qquad \text{Raising a product to a power; using the power rule in the denominator}$$

$$= \frac{8y^3}{x^{12}} \qquad \text{Cubing 2}$$

Definitions and Properties of Exponents

The following summary assumes that no denominators are 0 and that 0^0 is not considered. For any integers m and n,

1 as an exponent:	$a^1 = a$
0 as an exponent:	$a^0 = 1$
Negative exponents:	$a^{-n} = \dfrac{1}{a^n},$
	$\dfrac{a^{-n}}{b^{-m}} = \dfrac{b^m}{a^n},$
	$\left(\dfrac{a}{b}\right)^{-n} = \left(\dfrac{b}{a}\right)^n$
The Product Rule:	$a^m \cdot a^n = a^{m+n}$
The Quotient Rule:	$\dfrac{a^m}{a^n} = a^{m-n}$
The Power Rule:	$(a^m)^n = a^{mn}$
Raising a product to a power:	$(ab)^n = a^n b^n$
Raising a quotient to a power:	$\left(\dfrac{a}{b}\right)^n = \dfrac{a^n}{b^n}$

Scientific Notation

When we are working with the very large or very small numbers that frequently occur in science, **scientific notation** provides a useful way of writing numbers. The following are examples of scientific notation.

The mass of the earth:

6.0×10^{24} kilograms (kg) = 6,000,000,000,000,000,000,000,000 kg

The mass of a hydrogen atom:

1.7×10^{-24} g = 0.0000000000000000000000017 g

Scientific Notation

Scientific notation for a number is an expression of the type

$$N \times 10^m,$$

where N is at least 1 but less than 10 ($1 \leq N < 10$), N is expressed in decimal notation, and m is an integer.

Converting from scientific to decimal notation involves multiplying by a power of 10. Consider the following.

Scientific Notation $N \times 10^m$	Multiplication	Decimal Notation
4.52×10^2	4.52×100	452.
4.52×10^1	4.52×10	45.2
4.52×10^0	4.52×1	4.52
4.52×10^{-1}	4.52×0.1	0.452
4.52×10^{-2}	4.52×0.01	0.0452

Note that when m, the power of 10, is positive, the decimal point moves right m places in decimal notation. When m is negative, the decimal point moves left $|m|$ places. We generally try to perform this multiplication mentally.

Example 6

Convert to decimal notation: **(a)** 7.893×10^5; **(b)** 4.7×10^{-8}.

Solution

a) Since the exponent is positive, the decimal point moves to the right:

7.89300.
5 places

$7.893 \times 10^5 = 789{,}300$ The decimal point moves 5 places to the right.

b) Since the exponent is negative, the decimal point moves to the left:

0.00000004.7
8 places

$4.7 \times 10^{-8} = 0.000000047$ The decimal point moves 8 places to the left.

To convert from decimal to scientific notation, this procedure is reversed.

Example 7

Write in scientific notation: **(a)** 83,000; **(b)** 0.0327.

Solution

a) We need to find m such that $83{,}000 = 8.3 \times 10^m$. To change 8.3 to 83,000 requires moving the decimal point 4 places to the right. This can be accomplished by multiplying by 10^4. Thus,

$$83{,}000 = 8.3 \times 10^4.$$ This is scientific notation.

b) We need to find m such that $0.0327 = 3.27 \times 10^m$. To change 3.27 to 0.0327 requires moving the decimal point 2 places to the left. This can be accomplished by multiplying by 10^{-2}. Thus,

$$0.0327 = 3.27 \times 10^{-2}.$$ This is scientific notation.

Conversions to and from scientific notation are often made mentally. Remember that positive exponents are used when representing large numbers and negative exponents are used when representing numbers between 0 and 1.

Multiplying and Dividing Using Scientific Notation

Products and quotients of numbers written in scientific notation are found using the rules for exponents.

Example 8

Simplify.

a) $(1.8 \times 10^9) \cdot (2.3 \times 10^{-4})$ **b)** $(3.41 \times 10^5) \div (1.1 \times 10^{-3})$

Solution

a) $(1.8 \times 10^9) \cdot (2.3 \times 10^{-4})$

$\qquad = 1.8 \times 2.3 \times 10^9 \times 10^{-4}$ Using the associative and commutative laws

$\qquad = 4.14 \times 10^{9+(-4)}$ Adding exponents

$\qquad = 4.14 \times 10^5$

b) $(3.41 \times 10^5) \div (1.1 \times 10^{-3})$

$\qquad = \dfrac{3.41 \times 10^5}{1.1 \times 10^{-3}}$

$\qquad = \dfrac{3.41}{1.1} \times \dfrac{10^5}{10^{-3}}$

$\qquad = 3.1 \times 10^{5-(-3)}$ Subtracting exponents

$\qquad = 3.1 \times 10^8$

When a problem is stated using scientific notation, we normally use scientific notation for the answer.

Example 9

Simplify.

a) $(3.1 \times 10^5) \cdot (4.5 \times 10^{-3})$ **b)** $(7.2 \times 10^{-7}) \div (8.0 \times 10^6)$

Solution

a) We have

$$(3.1 \times 10^5) \cdot (4.5 \times 10^{-3}) = 3.1 \times 4.5 \times 10^5 \times 10^{-3}$$
$$= 13.95 \times 10^2.$$

Our answer is not yet in scientific notation because 13.95 is not between 1 and 10. We convert to scientific notation as follows:

$$13.95 \times 10^2 = 1.395 \times 10^1 \times 10^2 \qquad \text{Substituting } 1.395 \times 10^1 \text{ for } 13.95$$
$$= 1.395 \times 10^3. \qquad \text{Adding exponents}$$

b) $(7.2 \times 10^{-7}) \div (8.0 \times 10^6) = \dfrac{7.2 \times 10^{-7}}{8.0 \times 10^6} = \dfrac{7.2}{8.0} \times \dfrac{10^{-7}}{10^6}$

$\qquad\qquad\qquad\qquad\qquad = 0.9 \times 10^{-13}$

$\qquad\qquad\qquad\qquad\qquad = 9.0 \times 10^{-1} \times 10^{-13}$ Substituting 9.0×10^{-1} for 0.9

$\qquad\qquad\qquad\qquad\qquad = 9.0 \times 10^{-14}$ Adding exponents

technology connection

A key labeled $\boxed{\wedge}$ or $\boxed{\text{EE}}$ is often used to enter scientific notation into a calculator. Sometimes this is a secondary function, meaning that another key—often labeled $\boxed{\text{SHIFT}}$ or $\boxed{\text{2nd}}$—must be pressed first.

To check Example 8(a), we press

$1.8\ \boxed{\text{EE}}\ 9\ \boxed{\times}\ 2.3\ \boxed{\text{EE}}\ \boxed{(-)}\ 4.$

When we then press $\boxed{=}$ or $\boxed{\text{ENTER}}$, the result 4.14E5 appears. This represents 4.14×10^5. On some calculators,

the mode SCI must be used in order to display scientific notation.

Calculate each of the following.

1. $(3.8 \times 10^9) \cdot (4.5 \times 10^7)$
2. $(2.9 \times 10^{-8}) \div (5.4 \times 10^6)$
3. $(9.2 \times 10^7) \div (2.5 \times 10^{-9})$

FOR EXTRA HELP

Exercise Set 4.8

Digital Video Tutor CD 3
Videotape 8   InterAct Math   Math Tutor Center   MathXL   MyMathLab.com

Express using positive exponents. Then, if possible, simplify.

1. 5^{-2}
2. 2^{-4}
3. 10^{-4}
4. 5^{-3}
5. $(-2)^{-6}$
6. $(-3)^{-4}$
7. x^{-8}
8. t^{-5}
9. xy^{-2}
10. $a^{-3}b$
11. $r^{-5}t$
12. xy^{-9}
13. $\dfrac{1}{t^{-7}}$
14. $\dfrac{1}{z^{-9}}$
15. $\dfrac{1}{h^{-8}}$
16. $\dfrac{1}{a^{-12}}$
17. 7^{-1}
18. 3^{-1}
19. $\left(\dfrac{2}{5}\right)^{-2}$
20. $\left(\dfrac{3}{4}\right)^{-2}$
21. $\left(\dfrac{a}{2}\right)^{-3}$
22. $\left(\dfrac{x}{3}\right)^{-4}$
23. $\left(\dfrac{s}{t}\right)^{-7}$
24. $\left(\dfrac{r}{v}\right)^{-5}$

Express using negative exponents.

25. $\dfrac{1}{7^2}$
26. $\dfrac{1}{5^2}$
27. $\dfrac{1}{t^6}$
28. $\dfrac{1}{y^2}$
29. $\dfrac{1}{a^4}$
30. $\dfrac{1}{t^5}$
31. $\dfrac{1}{p^8}$
32. $\dfrac{1}{m^{12}}$
33. $\dfrac{1}{5}$
34. $\dfrac{1}{8}$
35. $\dfrac{1}{t}$
36. $\dfrac{1}{m}$

Simplify. Do not use negative exponents in the answer.

37. $2^{-5} \cdot 2^8$
38. $5^{-8} \cdot 5^9$
39. $x^{-2} \cdot x^{-7}$
40. $x^{-2} \cdot x^{-9}$
41. $t^{-3} \cdot t$
42. $y^{-5} \cdot y$
43. $(a^{-2})^9$
44. $(x^{-5})^6$
45. $(t^{-3})^{-6}$
46. $(a^{-4})^{-7}$
47. $(t^4)^{-3}$
48. $(t^5)^{-2}$

49. $(x^{-2})^{-4}$ **50.** $(t^{-6})^{-5}$ **51.** $(ab)^{-3}$

52. $(xy)^{-6}$ **53.** $(mn)^{-7}$ **54.** $(ab)^{-9}$

55. $(3x^{-4})^2$ **56.** $(2a^{-5})^3$ **57.** $(5r^{-4}t^3)^2$

58. $(4x^5y^{-6})^3$ **59.** $\dfrac{t^7}{t^{-3}}$ **60.** $\dfrac{x^7}{x^{-2}}$

61. $\dfrac{y^{-7}}{y^{-3}}$ **62.** $\dfrac{z^{-6}}{z^{-2}}$ **63.** $\dfrac{y^{-4}}{y^{-9}}$

64. $\dfrac{a^{-6}}{a^{-10}}$ **65.** $\dfrac{x^6}{x}$ **66.** $\dfrac{x}{x^{-1}}$

67. $\dfrac{a^{-7}}{b^{-9}}$ **68.** $\dfrac{x^{-6}}{y^{-10}}$ Aha! **69.** $\dfrac{t^{-7}}{t^{-7}}$

70. $\dfrac{a^{-5}}{b^{-7}}$ **71.** $\dfrac{3x^{-5}}{y^{-6}z^{-2}}$ **72.** $\dfrac{4a^{-6}}{b^{-5}c^{-7}}$

73. $\dfrac{3t^4}{s^{-2}u^{-4}}$ **74.** $\dfrac{5x^{-8}}{y^{-3}z^2}$ **75.** $(x^4y^5)^{-3}$

76. $(t^5x^3)^{-4}$ **77.** $(x^{-6}y^{-2})^{-4}$ **78.** $(x^{-2}y^{-7})^{-5}$

79. $(a^{-5}b^7c^{-2})(a^{-3}b^{-2}c^6)$

80. $(x^3y^{-4}z^{-5})(x^{-4}y^{-2}z^9)$

81. $\left(\dfrac{a^4}{3}\right)^{-2}$ **82.** $\left(\dfrac{y^2}{2}\right)^{-2}$ **83.** $\left(\dfrac{7}{x^{-3}}\right)^2$

84. $\left(\dfrac{3}{a^{-2}}\right)^3$ **85.** $\left(\dfrac{m^{-1}}{n^{-4}}\right)^3$ **86.** $\left(\dfrac{x^2y}{z^{-5}}\right)^3$

87. $\left(\dfrac{2a^2}{3b^4}\right)^{-3}$ **88.** $\left(\dfrac{a^2b}{cd^3}\right)^{-5}$ Aha! **89.** $\left(\dfrac{5x^{-2}}{3y^{-2}z}\right)^0$

90. $\left(\dfrac{4a^3b^{-2}}{5c^{-3}}\right)^1$

Convert to decimal notation.

91. 7.12×10^4 **92.** 8.92×10^2

93. 8.92×10^{-3} **94.** 7.26×10^{-4}

95. 9.04×10^8 **96.** 1.35×10^7

97. 2.764×10^{-10} **98.** 9.043×10^{-3}

99. 4.209×10^7 **100.** 5.029×10^8

Convert to scientific notation.

101. 490,000   **102.** 71,500

103. 0.00583   **104.** 0.0814

105. 78,000,000,000   **106.** 3,700,000,000,000

107. 907,000,000,000,000,000

108. 168,000,000,000,000

109. 0.000000527

110. 0.00000000648

111. 0.000000018

112. 0.00000000002

113. 1,094,000,000,000,000

114. 1,030,200,000,000,000,000

Multiply or divide, and write scientific notation for the result.

115. $(4 \times 10^7)(2 \times 10^5)$

116. $(1.9 \times 10^8)(3.4 \times 10^{-3})$

117. $(3.8 \times 10^9)(6.5 \times 10^{-2})$

118. $(7.1 \times 10^{-7})(8.6 \times 10^{-5})$

119. $(8.7 \times 10^{-12})(4.5 \times 10^{-5})$

120. $(4.7 \times 10^5)(6.2 \times 10^{-12})$

121. $\dfrac{8.5 \times 10^8}{3.4 \times 10^{-5}}$

122. $\dfrac{5.6 \times 10^{-2}}{2.5 \times 10^5}$

123. $(3.0 \times 10^6) \div (6.0 \times 10^9)$

124. $(1.5 \times 10^{-3}) \div (1.6 \times 10^{-6})$

125. $\dfrac{7.5 \times 10^{-9}}{2.5 \times 10^{12}}$

126. $\dfrac{4.0 \times 10^{-3}}{8.0 \times 10^{20}}$

127. Without performing actual computations, explain why 3^{-29} is smaller than 2^{-29}.

128. What is it about scientific notation that makes it so useful?

SKILL MAINTENANCE

Simplify.

129. $(3 - 8)(9 - 12)$ **130.** $(2 - 9)^2$

131. $7 \cdot 2 + 8^2$ **132.** $5 \cdot 6 - 3 \cdot 2 \cdot 4$

133. Plot the points $(-3, 2)$, $(4, -1)$, $(5, 3)$, and $(-5, -2)$.

134. Solve $cx + bt = r$ for t.

SYNTHESIS

135. Explain what requirements must be met in order for x^{-n} to represent a negative integer.

136. Explain why scientific notation cannot be used without an understanding of the rules for exponents.

137. Simplify:
$$\frac{4.2 \times 10^8[(2.5 \times 10^{-5}) \div (5.0 \times 10^{-9})]}{3.0 \times 10^{-12}}.$$

138. Write the reciprocal of 1.25×10^{-6} in scientific notation.

139. Write the reciprocal of 2.5×10^9 in scientific notation.

140. Write $8^{-3} \cdot 32 \div 16^2$ as a power of 2.

141. Write $81^3 \cdot 27 \div 9^2$ as a power of 3.

Simplify.

142. $(7^{-12})^2 \cdot 7^{25}$

Aha! **143.** $\dfrac{125^{-4}(25^2)^4}{125}$

144. $\dfrac{27^{-2}(81^2)^3}{9^8}$

145. Determine whether each of the following is true for all pairs of integers m and n and all positive numbers x and y.
a) $x^m \cdot y^n = (xy)^{mn}$
b) $x^m \cdot y^m = (xy)^{2m}$
c) $(x - y)^m = x^m - y^m$

Simplify.

146. $\dfrac{7.4 \times 10^{29}}{(5.4 \times 10^{-6})(2.8 \times 10^8)}$

147. $\dfrac{5.8 \times 10^{17}}{(4.0 \times 10^{-13})(2.3 \times 10^4)}$

148. $\dfrac{(7.8 \times 10^7)(8.4 \times 10^{23})}{2.1 \times 10^{-12}}$

149. $\dfrac{(2.5 \times 10^{-8})(6.1 \times 10^{-11})}{1.28 \times 10^{-3}}$

Write scientific notation for each answer.

150. *Household income.* In 1997, there were about 102.5 million households in the United States. The average income of these households (before taxes) was about $49,700 (*Source: Statistical Abstract of the United States, 1999*). Find the total income generated by these households.

151. *Computers.* A gigabyte is a measure of a computer's storage capacity. One gigabyte holds about one billion bytes of information. If a firm's computer network contains 2500 gigabytes of memory, how many bytes are in the network?

152. *River discharge.* The average discharge at the mouth of the Amazon River is 4,200,000 cubic feet per second. How much water is discharged from the Amazon River in 1 yr?

153. *Biology.* A strand of DNA (deoxyribonucleic acid) is about 1.5 m long and 1.3×10^{-10} cm wide (*Source*: Human Genome Project Information). How many times longer is DNA than it is wide?

154. *Water contamination.* In the United States, 200 million gal of used motor oil is improperly disposed of each year. One gallon of used oil can contaminate one million gallons of drinking water (*Source: The Macmillan Visual Almanac*). How many gallons of drinking water can 200 million gallons of oil contaminate?

Summary and Review 4

Key Terms

Polynomial, p. 210
Term, p. 210
Monomial, p. 210
Binomial, p. 211
Trinomial, p. 211
Degree of a term, p. 211
Coefficient, p. 211
Leading term, p. 212
Leading coefficient, p. 212
Degree of a polynomial, p. 212

Descending order, p. 213
Ascending order, p. 219
Opposite of a polynomial, p. 220
FOIL, p. 236
Difference of squares, p. 238
Perfect-square trinomial, p. 239
Polynomial in several variables, p. 244
Like terms, p. 245
Scientific notation, p. 258

Important Properties and Formulas

Definitions and Properties of Exponents

Assuming that no denominator is 0 and that 0^0 is not considered, for any integers m and n,

1 as an exponent: $\qquad$ $a^1 = a$

0 as an exponent: $\qquad$ $a^0 = 1$

Negative exponents: $\qquad$ $a^{-n} = \dfrac{1}{a^n},$

$\qquad\qquad\qquad\qquad\quad$ $\dfrac{a^{-n}}{b^{-m}} = \dfrac{b^m}{a^n},$

$\qquad\qquad\qquad\qquad\quad$ $\left(\dfrac{a}{b}\right)^{-n} = \left(\dfrac{b}{a}\right)^n$

The Product Rule: $\qquad$ $a^m \cdot a^n = a^{m+n}$

The Quotient Rule: $\qquad$ $\dfrac{a^m}{a^n} = a^{m-n}$

The Power Rule: $\qquad$ $(a^m)^n = a^{mn}$

Raising a product to a power: $\quad$ $(ab)^n = a^n b^n$

Raising a quotient to a power: $\quad$ $\left(\dfrac{a}{b}\right)^n = \dfrac{a^n}{b^n}$

Special Products of Polynomials

$(A + B)(A - B) = A^2 - B^2$
$(A + B)(A + B) = A^2 + 2AB + B^2$
$(A - B)(A - B) = A^2 - 2AB + B^2$

Scientific notation: $\quad N \times 10^m$, where $1 \le N < 10$ and m is an integer

Review Exercises

Simplify.

1. $y^7 \cdot y^3 \cdot y$

2. $(3x)^5 \cdot (3x)^9$

3. $t^8 \cdot t^0$

4. $\dfrac{4^5}{4^2}$

5. $\dfrac{(a+b)^4}{(a+b)^4}$

6. $\left(\dfrac{3t^4}{2s^3}\right)^2$

7. $(-2xy^2)^3$

8. $(2x^3)(-3x)^2$

9. $(a^2b)(ab)^5$

Identify the terms of each polynomial.

10. $3x^2 + 6x + \frac{1}{2}$

11. $-4y^5 + 7y^2 - 3y - 2$

List the coefficients of the terms in each polynomial.

12. $7x^2 - x + 7$

13. $4x^3 + 6x^2 - 5x + \frac{5}{3}$

For each polynomial, (a) list the degree of each term; (b) determine the leading term and the leading coefficient; and (c) determine the degree of the polynomial.

14. $4t^2 + 6 + 15t^5$

15. $-2x^5 + x^4 - 3x^2 + x$

Classify each polynomial as a monomial, a binomial, a trinomial, or none of these.

16. $4x^3 - 1$

17. $4 - 9t^3 - 7t^4 + 10t^2$

18. $7y^2$

Combine like terms and write in descending order.

19. $5x - x^2 + 4x$

20. $\frac{3}{4}x^3 + 4x^2 - x^3 + 7$

21. $-2x^4 + 16 + 2x^4 + 9 - 3x^5$

22. $3x^2 - 2x + 3 - 5x^2 - 1 - x$

23. $-x + \frac{1}{2} + 14x^4 - 7x^2 - 1 - 4x^4$

Evaluate each polynomial for $x = -1$.

24. $7x - 10$

25. $x^2 - 3x + 6$

Add or subtract.

26. $(3x^4 - x^3 + x - 4) + (x^5 + 7x^3 - 3x - 5)$

27. $(3x^4 - 5x^3 + 3x^2) + (4x^5 + 4x^3) + (-5x^5 - 5x^2)$

28. $(5x^2 - 4x + 1) - (3x^2 + 7)$

29. $(3x^5 - 4x^4 + 2x^2 + 3) - (2x^5 - 4x^4 + 3x^3 + 4x^2 - 5)$

30. $\begin{aligned} -\tfrac{3}{4}x^4 + \tfrac{1}{2}x^3 \qquad\qquad\quad + \tfrac{7}{8} \\ -\tfrac{1}{4}x^3 - \; x^2 - \tfrac{7}{4}x \\ +\tfrac{3}{2}x^4 \qquad\quad + \tfrac{2}{3}x^2 \qquad\quad - \tfrac{1}{2} \end{aligned}$

31. $\begin{aligned} 2x^5 \qquad\quad -\; x^3 \qquad\quad + x + 3 \\ -(3x^5 - x^4 + 4x^3 + 2x^2 - x + 3) \end{aligned}$

32. The length of a rectangle is 3 m greater than its width.

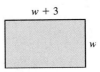

$w + 3$

w

  a) Find a polynomial for the perimeter.
  b) Find a polynomial for the area.

Multiply.

33. $3x(-4x^2)$

34. $(7x + 1)^2$

35. $(a - 7)(a + 4)$

36. $(m + 5)(m - 5)$

37. $(4x^2 - 5x + 1)(3x - 2)$

38. $(x - 9)^2$

39. $3t^2(5t^3 - 2t^2 + 4t)$

40. $(a - 7)(a + 7)$

41. $(x - 0.3)(x - 0.75)$

42. $(x^4 - 2x + 3)(x^3 + x - 1)$

43. $(3x - 5)^2$

44. $(2t^2 + 3)(t^2 - 7)$

45. $\left(a - \frac{1}{2}\right)\left(a + \frac{2}{3}\right)$

46. $(3x^2 + 4)(3x^2 - 4)$

47. $(2 - x)(2 + x)$

48. $(2x + 3y)(x - 5y)$

49. Evaluate $2 - 5xy + y^2 - 4xy^3 + x^6$ for $x = -1$ and $y = 2$.

Identify the coefficient and the degree of each term of each polynomial. Then find the degree of each polynomial.

50. $x^5y - 7xy + 9x^2 - 8$

51. $x^2y^5z^9 - y^{40} + x^{13}z^{10}$

Combine like terms.

52. $y + w - 2y + 8w - 5$

53. $6m^3 + 3m^2n + 4mn^2 + m^2n - 5mn^2$

Add or subtract.

54. $(5x^2 - 7xy + y^2) + (-6x^2 - 3xy - y^2)$

55. $(6x^3y^2 - 4x^2y - 6x) - (-5x^3y^2 + 4x^2y + 6x^2 - 6)$

Multiply.

56. $(p - q)(p^2 + pq + q^2)$ **57.** $\left(3a^4 - \frac{1}{3}b^3\right)^2$

58. Find a polynomial for the shaded area.

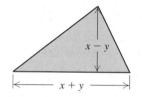

Divide.

59. $(10x^3 - x^2 + 6x) \div 2x$

60. $(6x^3 - 5x^2 - 13x + 13) \div (2x + 3)$

61. $\dfrac{t^4 + t^3 + 2t^2 - t - 3}{t + 1}$

62. Express using a positive exponent: m^{-7}.

63. Express using a negative exponent: $\dfrac{1}{t^8}$.

Simplify.

64. $7^2 \cdot 7^{-4}$ **65.** $\dfrac{a^{-5}b}{a^8b^8}$ **66.** $(x^3)^{-4}$

67. $(2x^{-3}y)^{-2}$ **68.** $\left(\dfrac{2x}{y}\right)^{-3}$

69. Convert to decimal notation: 8.3×10^6.

70. Convert to scientific notation: 0.0000328.

Multiply or divide and write scientific notation for the result.

71. $(3.8 \times 10^4)(5.5 \times 10^{-1})$

72. $\dfrac{1.28 \times 10^{-8}}{2.5 \times 10^{-4}}$

73. *Blood donors.* Every 4–6 weeks, one of the authors of this text donates 1.14×10^6 cubic millimeters (two pints) of blood platelets to the American Red Cross. In one cubic millimeter of blood, there are about 2×10^5 platelets. Approximate the number of platelets in a typical donation by this author.

SYNTHESIS

74. Explain why $5x^3$ and $(5x)^3$ are not equivalent expressions.

75. If two polynomials of degree n are added, is the sum also of degree n? Why or why not?

76. How many terms are there in each of the following?
   a) $(x - a)(x - b) + (x - a)(x - b)$
   b) $(x + a)(x - b) + (x - a)(x + b)$

77. Combine like terms:
   $$-3x^5 \cdot 3x^3 - x^6(2x)^2 + (3x^4)^2 + (2x^2)^4 - 40x^2(x^3)^2.$$

78. A polynomial has degree 4. The x^2-term is missing. The coefficient of x^4 is 2 times the coefficient of x^3. The coefficient of x is 3 less than the coefficient of x^4. The remaining coefficient is 7 less than the coefficient of x. The sum of the coefficients is 15. Find the polynomial.

Aha! **79.** Multiply: $[(x - 5) - 4x^3][(x - 5) + 4x^3]$.

80. Solve: $(x - 7)(x + 10) = (x - 4)(x - 6)$.

Chapter Test 4

Simplify.

1. $t^2 \cdot t^5 \cdot t$

2. $(x + 3)^5(x + 3)^6$

3. $\dfrac{3^5}{3^2}$

4. $\dfrac{(2x)^5}{(2x)^5}$

5. $(x^3)^2$

6. $(-3y^2)^3$

7. $(3x^2)(-2x^5)^3$

8. $(a^3b^2)(ab)^3$

9. Classify the polynomial as a monomial, a binomial, a trinomial, or none of these:
$$6t^2 - 9t.$$

10. Identify the coefficient of each term of the polynomial:
$$\tfrac{1}{3}x^5 - x + 7.$$

11. Determine the degree of each term, the leading term and the leading coefficient, and the degree of the polynomial:
$$2t^3 - t + 7t^5 + 4.$$

12. Evaluate the polynomial $x^2 + 5x - 1$ for $x = -2$.

Combine like terms and write in descending order.

13. $4a^2 - 6 + a^2$

14. $y^2 - 3y - y + \tfrac{3}{4}y^2$

15. $3 - x^2 + 2x^3 + 5x^2 - 6x - 2x + x^5$

Add or subtract.

16. $(3x^5 + 5x^3 - 5x^2 - 3) + (x^5 + x^4 - 3x^2 + 2x - 4)$

17. $\left(x^4 + \tfrac{2}{3}x + 5\right) + \left(4x^4 + 5x^2 + \tfrac{1}{3}x\right)$

18. $(2x^4 + x^3 - 8x^2 - 6x - 3) - (6x^4 - 8x^2 + 2x)$

19. $(x^3 - 0.4x^2 - 12) - (x^5 - 0.3x^3 + 0.4x^2 - 9)$

Multiply.

20. $-3x^2(4x^2 - 3x - 5)$

21. $\left(x - \tfrac{1}{3}\right)^2$

22. $(5t - 7)(5t + 7)$

23. $(3b + 5)(b - 3)$

24. $(x^6 - 4)(x^8 + 4)$

25. $(8 - y)(6 + 5y)$

26. $(2x + 1)(3x^2 - 5x - 3)$

27. $(8a + 3)^2$

28. Combine like terms:
$$x^3y - y^3 + xy^3 + 8 - 6x^3y - x^2y^2 + 11.$$

29. Subtract:
$$(8a^2b^2 - ab + b^3) - (-6ab^2 - 7ab - ab^3 + 5b^3).$$

30. Multiply: $(3x^5 - 4y^5)(3x^5 + 4y^5)$.

Divide.

31. $(12x^4 + 9x^3 - 15x^2) \div 3x^2$

32. $(6x^3 - 8x^2 - 14x + 13) \div (3x + 2)$

33. Express using a positive exponent: 5^{-3}.

34. Express using a negative exponent: $\dfrac{1}{y^8}$.

Simplify.

35. $t^{-4} \cdot t^{-2}$

36. $\dfrac{x^3y^2}{x^8y^{-3}}$

37. $(2a^3b^{-1})^{-4}$

38. $\left(\dfrac{ab}{c}\right)^{-3}$

39. Convert to scientific notation: 3,900,000,000.

40. Convert to decimal notation: 5×10^{-8}.

Multiply or divide and write scientific notation for the result.

41. $\dfrac{5.6 \times 10^6}{3.2 \times 10^{-11}}$

42. $(2.4 \times 10^5)(5.4 \times 10^{16})$

43. A CD-ROM can contain about 600 million pieces of information. How many sound files, each needing 40,000 pieces of information, can a CD-ROM hold? Write scientific notation for the answer.

SYNTHESIS

44. The height of a box is 1 less than its length, and the length is 2 more than its width. Express the volume in terms of the length.

45. Solve: $x^2 + (x - 7)(x + 4) = 2(x - 6)^2$.

5
Polynomials and Factoring

AN APPLICATION

An outdoor-education ropes course includes a cable that slopes downward from a height of 37 ft to a height of 30 ft. The trees that the cable connects are 24 ft apart. How long is the cable?

This problem appears as Exercise 17 in Section 5.7.

*O*utdoor education is a part of many of today's physical education programs. Most students probably never think about it, but proper design of a ropes course requires an understanding of both math and physics.

BETH LOGSDON
Physical Education Teacher
Olathe, Kansas

n Chapter 1, we learned that factoring is multiplying reversed. To factor a polynomial is to find an equivalent expression that is a product. In Sections 5.1–5.5, we factor to find equivalent expressions. In Sections 5.6 and 5.7, we use factoring to solve equations, many of which arise from real-world problems. Factoring polynomials requires a solid command of the multiplication methods studied in Chapter 4.

Introduction to Factoring

5.1

Factoring Monomials • Factoring When Terms Have a Common Factor • Factoring by Grouping • Checking by Evaluating

Just as a number like 15 can be factored as $3 \cdot 5$, a polynomial like $x^2 + 7x$ can be factored as $x(x + 7)$. In both cases, we ask ourselves, "What was multiplied to obtain the given result?" The situation is much like a popular television game show in which an "answer" is given and participants must find a "question" to which the answer corresponds.

Factoring

To *factor* a polynomial is to find an equivalent expression that is a product.

Factoring Monomials

To factor a monomial, we find two monomials whose product is equivalent to the original monomial. For example, $20x^2$ can be factored as $2 \cdot 10x^2$, $4x \cdot 5x$, or $10x \cdot 2x$, as well as several other ways.

Example 1

Find three factorizations of $15x^3$.

Solution

a) $15x^3 = (3 \cdot 5)(x \cdot x^2)$ Thinking of how 15 and x^3 factor
$= (3x)(5x^2)$ The factors here are $3x$ and $5x^2$.

b) $15x^3 = (3 \cdot 5)(x^2 \cdot x)$
$= (3x^2)(5x)$ The factors here are $3x^2$ and $5x$.

c) $15x^3 = ((-5)(-3))x^3$
$= (-5)(-3x^3)$ The factors here are -5 and $-3x^3$.

Recall from Section 1.2 that the word "factor" can be a verb or a noun, depending on the context in which it appears.

Factoring When Terms Have a Common Factor

To multiply a polynomial of two or more terms by a monomial, we multiply each term by the monomial, using the distributive law $a(b + c) = ab + ac$. To factor, we do the reverse. We rewrite a polynomial as a product, using the distributive law with the sides switched: $ab + ac = a(b + c)$. Consider the following:

Multiply

$$3x(x^2 + 2x - 4)$$
$$= 3x \cdot x^2 + 3x \cdot 2x - 3x \cdot 4$$
$$= 3x^3 + 6x^2 - 12x;$$

Factor

$$3x^3 + 6x^2 - 12x$$
$$= 3x \cdot x^2 + 3x \cdot 2x - 3x \cdot 4$$
$$= 3x(x^2 + 2x - 4).$$

In the factorization on the right, note that since $3x$ appears as a factor of $3x^3$, $6x^2$, and $-12x$, it is a *common factor* for all the terms of the trinomial $3x^3 + 6x^2 - 12x$.

To factor a polynomial with two or more terms, always try to first find a factor common to all terms. In some cases, there may not be a common factor (other than 1). If a common factor *does* exist, we generally use the common factor with the largest possible coefficient and the largest possible exponent. Such a factor is called the *largest*, or *greatest*, *common factor*.

E x a m p l e   2

Factor: $5x^2 + 15$.

Solution   We have

$$5x^2 + 15 = 5 \cdot x^2 + 5 \cdot 3 \qquad \text{Factoring each term}$$
$$= 5(x^2 + 3). \qquad \text{Factoring out the common factor, 5}$$

To check, we multiply: $5(x^2 + 3) = 5 \cdot x^2 + 5 \cdot 3 = 5x^2 + 15$. Since $5x^2 + 15$ is the original polynomial, the factorization $5(x^2 + 3)$ checks.

Caution! $5 \cdot x^2 + 5 \cdot 3$ is a factorization of the *terms* of $5x^2 + 15$, but not of the polynomial itself. The factorization of $5x^2 + 15$ is $5(x^2 + 3)$.

When asked to factor a polynomial in which all terms contain the same variable raised to various powers, we factor out the largest power possible.

E x a m p l e   3

Factor: $24x^3 + 30x^2$.

Solution   The largest factor common to 24 and 30 is 6. The largest power of x common to x^3 and x^2 is x^2. (To see this, think of x^3 as $x^2 \cdot x$.) Thus the largest common factor of $24x^3$ and $30x^2$ is $6x^2$. We factor as follows:

$$24x^3 + 30x^2 = 6x^2 \cdot 4x + 6x^2 \cdot 5 \qquad \text{Factoring each term}$$
$$= 6x^2(4x + 5). \qquad \text{Factoring out } 6x^2$$

Check: $6x^2(4x + 5) = 6x^2 \cdot 4x + 6x^2 \cdot 5 = 24x^3 + 30x^2$, as expected. The factorization $6x^2(4x + 5)$ checks.

Suppose in Example 3 that you did not recognize the *largest* common factor, and removed only part of it, as follows:

$$24x^3 + 30x^2 = 2x^2 \cdot 12x + 2x^2 \cdot 15 \qquad 2x^2 \text{ is a common factor.}$$
$$= 2x^2(12x + 15). \qquad 12x + 15 \text{ itself has a common factor.}$$

Note that $12x + 15$ still has a common factor, 3. To find the largest common factor, continue factoring out common factors, as follows, until no more exist:

$$= 2x^2[3(4x + 5)] \qquad \text{Factoring } 12x + 15. \text{ Remember to rewrite the first common factor, } 2x^2.$$
$$= 6x^2(4x + 5). \qquad \text{Using an associative law}$$

Since $4x + 5$ cannot be factored any further, we say that we have factored *completely*.

E x a m p l e   4

Factor: $15x^5 - 12x^4 + 27x^3 - 3x^2$.

Solution   We have

$$15x^5 - 12x^4 + 27x^3 - 3x^2$$
$$= 3x^2 \cdot 5x^3 - 3x^2 \cdot 4x^2 + 3x^2 \cdot 9x - 3x^2 \cdot 1 \qquad \text{Try to do this mentally.}$$
$$= 3x^2(5x^3 - 4x^2 + 9x - 1). \qquad \text{Factoring out } 3x^2$$

Caution!   Don't forget the term -1. The check below shows why it is essential.

Since $5x^3 - 4x^2 + 9x - 1$ has no common factor, we are finished, except for a check:

$$3x^2(5x^3 - 4x^2 + 9x - 1) = 15x^5 - 12x^4 + 27x^3 - 3x^2. \qquad \text{Our factorization checks.}$$

The factorization is $3x^2(5x^3 - 4x^2 + 9x - 1)$.

If you spot the largest common factor without writing out a factorization of each term, you can write the answer in one step.

E x a m p l e   5

Factor: **(a)** $8m^3 - 16m$; **(b)** $14p^2y^3 - 8py^2 + 2py$.

Solution

a) $8m^3 - 16m = 8m(m^2 - 2)$
b) $14p^2y^3 - 8py^2 + 2py = 2py(7py^2 - 4y + 1)$ \} Determine the largest common factor by inspection; then carefully fill in the parentheses.

The checks are left to the student.

> ### Tips for Factoring
> 1. Factor out the largest common factor, if one exists.
> 2. Factoring can always be checked by multiplying. Multiplication should yield the original polynomial.

Factoring by Grouping

Sometimes algebraic expressions contain a common factor with two or more terms.

E x a m p l e   6

Factor: $x^2(x + 1) + 2(x + 1)$.

Solution   The binomial $x + 1$ is a factor of both $x^2(x + 1)$ and $2(x + 1)$. Thus, $x + 1$ is a common factor:

$$x^2(x + 1) + 2(x + 1) = (x + 1)x^2 + (x + 1)2 \qquad \text{Using a commutative law twice}$$

$$= (x + 1)(x^2 + 2). \qquad \text{Factoring out the common factor, } x + 1$$

To check, we could simply reverse the above steps.
    The factorization is $(x + 1)(x^2 + 2)$.

In Example 6, the common binomial factor was clearly visible. How do we find such a factor in a polynomial like $5x^3 - x^2 + 15x - 3$? Although there is no factor, other than 1, common to all four terms, $5x^3 - x^2$ and $15x - 3$ can be grouped and factored separately:

$$5x^3 - x^2 = x^2(5x - 1) \quad \text{and} \quad 15x - 3 = 3(5x - 1).$$

Note that $5x^3 - x^2$ and $15x - 3$ share a common factor of $5x - 1$. This means that the original polynomial, $5x^3 - x^2 + 15x - 3$, can be factored:

$$5x^3 - x^2 + 15x - 3 = (5x^3 - x^2) + (15x - 3)$$

Using an associative law. This is generally done mentally.

$$= x^2(5x - 1) + 3(5x - 1)$$

Factoring each binomial

$$= (5x - 1)(x^2 + 3).$$

Factoring out the common factor, $5x - 1$

If a polynomial can be split into groups of terms and the groups share a common factor, then the original polynomial can be factored. This method, known as **factoring by grouping**, can be tried on any polynomial with four or more terms.

Example 7

Factor by grouping.

a) $2x^3 + 8x^2 + x + 4$
b) $8x^4 + 6x - 28x^3 - 21$

Solution

a) $2x^3 + 8x^2 + x + 4 = 2x^2(x + 4) + 1(x + 4)$

Factoring $2x^3 + 8x^2$ to find a common binomial factor. Writing the 1 helps with the next step.

$$= (x + 4)(2x^2 + 1)$$

Factoring out the common factor, $x + 4$. The 1 is essential in the factor $2x^2 + 1$.

Check: $(x + 4)(2x^2 + 1) = x \cdot 2x^2 + x \cdot 1 + 4 \cdot 2x^2 + 4 \cdot 1$ Using FOIL
$$= 2x^3 + x + 8x^2 + 4$$
$$= 2x^3 + 8x^2 + x + 4$$ Using a commutative law

The factorization is $(x + 4)(2x^2 + 1)$.

b) We can factor $8x^4 + 6x$ as $2x(4x^3 + 3)$ and $-28x^3 - 21$ as $7(-4x^3 - 3)$ or $-7(4x^3 + 3)$. We use $-7(4x^3 + 3)$ because it shares a common factor, $4x^3 + 3$, with $2x(4x^3 + 3)$:

$$8x^4 + 6x - 28x^3 - 21 = 2x(4x^3 + 3) - 7(4x^3 + 3)$$

Factoring two binomials. Using -7 gives a common binomial factor.

$$= (4x^3 + 3)(2x - 7)$$

Factoring out the common factor, $4x^3 + 3$

Check: $(4x^3 + 3)(2x - 7) = 8x^4 - 28x^3 + 6x - 21$
$$= 8x^4 + 6x - 28x^3 - 21$$ This is the original polynomial.

The factorization is $(4x^3 + 3)(2x - 7)$.

Although factoring by grouping can be useful, some polynomials, like $x^3 + x^2 + 2x - 2$, cannot be factored this way. Factoring polynomials of this type is beyond the scope of this text.

Checking by Evaluating

We have seen that one way to check a factorization is to multiply. A second type of check, discussed toward the end of Section 4.4, uses the fact that equivalent expressions have the same value when evaluated for the same replacement. Thus a quick, partial check of Example 7(a) can be made by using a convenient replacement for x (say, 1) and evaluating both $2x^3 + 8x^2 + x + 4$ and $(x + 4)(2x^2 + 1)$:

$$2 \cdot 1^3 + 8 \cdot 1^2 + 1 + 4 = 2 + 8 + 1 + 4$$
$$= 15;$$
$$(1 + 4)(2 \cdot 1^2 + 1) - 5 \cdot 3$$
$$= 15.$$

Since the value of both expressions is the same, the factorization is probably correct.

Keep in mind that it is possible, by chance, for two expressions that are not equivalent to share the same value when evaluated. Because of this, unless several values are used (at least one more than the degree of the polynomial, it turns out), evaluating offers only a partial check. Consult with your instructor before making extensive use of this type of check.

---

technology connection

We saw in the Technology Connection on p. 232 that a Table of values can be used to check that two expressions are equal. Thus to check Example 7(a), we let $y_1 = 2x^3 + 8x^2 + x + 4$ and $y_2 = (x + 4)(2x^2 + 1)$:

ΔTBL = 1

X	Y1	Y2
0	4	4
1	15	15
2	54	54
3	133	133
4	264	264
5	459	459
6	730	730
X = 0		

No matter how far up or down we scroll, $y_1 = y_2$. Thus Example 7(a) is correct.

1. Use a Table to check Example 7(b).

Exercise Set **5.1**

Find three factorizations for each monomial. Answers may vary.

1. $10x^3$

2. $6x^3$

3. $-15a^4$

4. $-8t^5$

5. $26x^5$

6. $25x^4$

Factor. Remember to use the largest common factor and to check by multiplying.

7. $x^2 + 8x$

8. $x^2 + 6x$

9. $10t^2 - 5t$

10. $5a^2 - 15a$

11. $x^3 + 6x^2$

12. $4x^4 + x^2$

13. $8x^4 - 24x^2$

14. $5x^5 + 10x^3$

15. $2x^2 + 2x - 8$

16. $6x^2 + 3x - 15$

17. $7a^6 - 10a^4 - 14a^2$

18. $10t^5 - 15t^4 + 9t^3$

19. $2x^8 + 4x^6 - 8x^4 + 10x^2$

20. $5x^4 - 15x^3 - 25x - 10$

21. $x^5y^5 + x^4y^3 + x^3y^3 - x^2y^2$

22. $x^9y^6 - x^7y^5 + x^4y^4 + x^3y^3$

23. $5a^3b^4 + 10a^2b^3 - 15a^3b^2$

24. $21r^5t^4 - 14r^4t^6 + 21r^3t^6$

Factor.

25. $y(y - 2) + 7(y - 2)$

26. $b(b + 5) + 3(b + 5)$

27. $x^2(x + 3) - 7(x + 3)$

28. $3z^2(2z + 9) + (2z + 9)$

29. $y^2(y + 8) + (y + 8)$

30. $x^2(x - 7) - 3(x - 7)$

Factor by grouping, if possible, and check.

31. $x^3 + 3x^2 + 4x + 12$

32. $6z^3 + 3z^2 + 2z + 1$

33. $3a^3 + 9a^2 + 2a + 6$

34. $3a^3 + 2a^2 + 6a + 4$

35. $9x^3 - 12x^2 + 3x - 4$

36. $10x^3 - 25x^2 + 4x - 10$

37. $4t^3 - 20t^2 + 3t - 15$

38. $6a^3 - 8a^2 + 9a - 12$

39. $7x^3 + 2x^2 - 14x - 4$

40. $5x^3 + 4x^2 - 10x - 8$

41. $6a^3 - 7a^2 + 6a - 7$

42. $7t^3 - 5t^2 + 7t - 5$

43. $x^3 + 8x^2 - 3x - 24$

44. $x^3 + 7x^2 - 2x - 14$

45. $2x^3 + 12x^2 - 5x - 30$

46. $3x^3 + 15x^2 - 5x - 25$

47. $w^3 - 7w^2 + 4w - 28$

48. $p^3 + p^2 - 3p + 10$

49. $x^3 - x^2 - 2x + 5$

50. $y^3 + 8y^2 - 2y - 16$

51. $2x^3 - 8x^2 - 9x + 36$

52. $20g^3 - 4g^2 - 25g + 5$

53. In answering a factoring problem, Taylor says the largest common factor is $-5x^2$ and Natasha says the largest common factor is $5x^2$. Can they both be correct? Why or why not?

54. Write a two-sentence paragraph in which the word "factor" is used at least once as a noun and once as a verb.

SKILL MAINTENANCE

Simplify.

55. $(x + 3)(x + 5)$

56. $(x + 2)(x + 7)$

57. $(a - 7)(a + 3)$

58. $(a + 5)(a - 8)$

59. $(2x + 5)(3x - 4)$

60. $(3t + 2)(4t - 7)$

61. $(3t - 5)^2$

62. $(2t - 9)^2$

SYNTHESIS

63. Marlene recognizes that evaluating provides only a partial check of her factoring. Because of this, she often performs a second check with a different replacement value. Is this a good idea? Why or why not?

64. Josh says that for Exercises 1–52 there is no need to print answers at the back of the book. Is he correct in saying this? Why or why not?

Factor, if possible.

65. $4x^5 + 6x^3 + 6x^2 + 9$

66. $x^6 + x^4 + x^2 + 1$

67. $x^{12} + x^7 + x^5 + 1$

68. $x^3 + x^2 - 2x + 2$

Aha! **69.** $5x^5 - 5x^4 + x^3 - x^2 + 3x - 3$

Aha! **70.** $ax^2 + 2ax + 3a + x^2 + 2x + 3$

71. Write a polynomial of degree 7 for which $3x^2y^3$ is the largest common factor. Answers may vary.

Factoring Trinomials of the Type $x^2 + bx + c$

5.2

Constant Term Positive • Constant Term Negative

We now learn how to factor trinomials like

$$x^2 + 5x + 4 \quad \text{or} \quad x^2 + 3x - 10,$$

for which no common factor exists and the leading coefficient is 1. As preparation for the factoring that follows, compare the following multiplications:

$$
\begin{array}{cccc}
\text{F} & \text{O} & \text{I} & \text{L} \\
\downarrow & \downarrow & \downarrow & \downarrow
\end{array}
$$

$$
\begin{aligned}
(x + 2)(x + 5) &= x^2 + 5x + 2x + 2 \cdot 5 \\
&= x^2 + \quad 7x \quad + \quad 10;
\end{aligned}
$$

$$
\begin{aligned}
(x - 2)(x - 5) &= x^2 - 5x - 2x + \\
&= x^2 - \quad 7x \quad + \quad 10;
\end{aligned}
$$

$$
\begin{aligned}
(x + 3)(x - 7) &= x^2 - 7x + 3x + 3(-7) \\
&= x^2 - \quad 4x \quad - \quad 21;
\end{aligned}
$$

$$
\begin{aligned}
(x - 3)(x + 7) &= x^2 + 7x - 3x + (-3)7 \\
&= x^2 + \quad 4x \quad - \quad 21.
\end{aligned}
$$

Note that for all four products:

- The product of the two binomials is a trinomial.
- The coefficient of x in the trinomial is the sum of the constant terms in the binomials.
- The constant term in the trinomial is the product of the constant terms in the binomials.

These observations lead to a method for factoring certain trinomials. The first type we consider has a positive constant term, just as in the first two multiplications above.

Constant Term Positive

To factor a polynomial like $x^2 + 7x + 10$, we think of FOIL in reverse. The x^2 resulted from x times x, which suggests that the first term of each binomial factor is x. Next, we look for numbers p and q such that

$$x^2 + 7x + 10 = (x + p)(x + q).$$

To get the middle term and the last term of the trinomial, we need two numbers p and q whose product is 10 and whose sum is 7. Those numbers are 2 and 5. Thus the factorization is

$$(x + 2)(x + 5). \qquad \textit{Check: } (x + 2)(x + 5) = x^2 + 5x + 2x + 10$$
$$= x^2 + 7x + 10$$

E x a m p l e   1

Factor: $x^2 + 5x + 6$.

Solution   Think of FOIL in reverse. The first term of each factor is x:

$$(x +\quad)(x +\quad).$$

To complete the factorization, we need a constant term for each of these binomial factors. The constants must have a product of 6 and a sum of 5. We list some pairs of numbers that multiply to 6.

Pairs of Factors of 6	Sums of Factors
1, 6	7
2, 3	5 ←
−1, −6	−7
−2, −3	−5

The numbers we seek are 2 and 3.

Since

$$2 \cdot 3 = 6 \quad \text{and} \quad 2 + 3 = 5,$$

the factorization of $x^2 + 5x + 6$ is $(x + 2)(x + 3)$. To check, we simply multiply the two binomials.

Check: $(x + 2)(x + 3) = x^2 + 3x + 2x + 6$
$$= x^2 + 5x + 6. \quad \text{The product is the original polynomial.}$$

Note that since 5 and 6 are both positive, when factoring $x^2 + 5x + 6$ we need not consider negative factors of 6. Note too that changing the signs of the factors changes only the sign of the sum.

At the beginning of this section, we considered the multiplication $(x - 2)(x - 5)$. For this product, the resulting trinomial, $x^2 - 7x + 10$, has a positive constant term but a negative coefficient of x. This is because the *product* of two negative numbers is always positive, whereas the *sum* of two negative numbers is always negative.

To Factor $x^2 + bx + c$ When c Is Positive

When the constant term of a trinomial is positive, look for two numbers with the same sign. The sign is that of the middle term:

$$x^2 - 7x + 10 = (x - 2)(x - 5);$$

$$x^2 + 7x + 10 = (x + 2)(x + 5).$$

E x a m p l e *2*   Factor: $y^2 - 8y + 12$.

Solution Since the constant term is positive and the coefficient of the middle term is negative, we look for a factorization of 12 in which both factors are negative. Their sum must be -8.

Pairs of Factors of 12	Sums of Factors
$-1, -12$	-13
$-2, -6$	-8
$-3, -4$	-7

We need a sum of -8.
The numbers we need are -2 and -6.

The factorization of $y^2 - 8y + 12$ is $(y - 2)(y - 6)$. The check is left to the student.

Constant Term Negative

As we saw in two of the multiplications earlier in this section, the product of two binomials can have a negative constant term:

$$(x + 3)(x - 7) = x^2 - 4x - 21$$

and

$$(x - 3)(x + 7) = x^2 + 4x - 21.$$

Note that when the signs of the constants in the binomials are reversed, only the sign of the middle term in the product changes.

E x a m p l e *3*   Factor: $x^2 - 8x - 20$.

Solution The constant term, -20, must be expressed as the product of a negative number and a positive number. Since the sum of these two numbers must be negative (specifically, -8), the negative number must have the greater absolute value.

Pairs of Factors of -20	Sums of Factors
$1, -20$	-19
$2, -10$	-8
$4, -5$	-1
$5, -4$	1
$10, -2$	8
$20, -1$	19

The numbers we need are 2 and -10.

Because these sums are all positive, for this problem all of the corresponding pairs can be disregarded. Note that in all three pairs, the positive number has the greater absolute value.

The numbers that we are looking for are 2 and -10.

Check: $(x + 2)(x - 10) = x^2 - 10x + 2x - 20$
$$= x^2 - 8x - 20.$$

The factorization of $x^2 - 8x - 20$ is $(x + 2)(x - 10)$.

> ### To Factor $x^2 + bx + c$ When c Is Negative
>
> When the constant term of a trinomial is negative, look for two numbers whose product is negative. One must be positive and the other negative:
>
> $$x^2 - 4x - 21 = (x + 3)(x - 7);$$
>
> $$x^2 + 4x - 21 = (x - 3)(x + 7).$$
>
> Select the two numbers so that the number with the larger absolute value has the same sign as b, the coefficient of the middle term.

Example 4

Factor: $t^2 - 24 + 5t$.

Solution It helps to first write the trinomial in descending order: $t^2 + 5t - 24$. The factorization of the constant term, -24, must have one factor positive and one factor negative. The sum must be 5, so the positive factor must have the larger absolute value. Thus we consider only pairs of factors in which the positive factor has the larger absolute value.

Pairs of Factors of -24	Sums of Factors
$-1, 24$	23
$-2, 12$	10
$-3,\ 8$	5 ← The numbers we need are -3 and 8.
$-4,\ 6$	2

The factorization is $(t - 3)(t + 8)$. The check is left to the student.

Polynomials in two or more variables, such as $a^2 + 4ab - 21b^2$, are factored in a similar manner.

Example 5

Factor: $a^2 + 4ab - 21b^2$.

Solution It may help to write the trinomial in the equivalent form
$$a^2 + 4ba - 21b^2.$$

This way we think of $-21b^2$ as the "constant" term and $4b$ as the "coefficient" of the middle term. Then we try to express $-21b^2$ as a product of two factors whose sum is $4b$. Those factors are $-3b$ and $7b$.

Check: $(a - 3b)(a + 7b) = a^2 + 7ab - 3ba - 21b^2$
$$= a^2 + 4ab - 21b^2.$$

The factorization is $(a - 3b)(a + 7b)$.

Example 6

Factor: $x^2 - x + 5$.

Solution   Since 5 has very few factors, we can easily check all possibilities.

Pairs of Factors of 5	Sums of Factors
5, 1	6
−5, −1	−6

Since there are no factors whose sum is -1, the polynomial is *not* factorable into binomials.

In this text, a polynomial like $x^2 - x + 5$ that cannot be factored further is said to be **prime**. In more advanced courses, polynomials like $x^2 - x + 5$ can be factored and are not considered prime.

Often factoring requires two or more steps. In general, when told to factor, we should *factor completely*. This means that the final factorization should not contain any factors that can be factored further.

Example 7

Factor: $2x^3 - 20x^2 + 50x$.

Solution   *Always* look first for a common factor. This time there is one, $2x$, which we factor out first:
$$2x^3 - 20x^2 + 50x = 2x(x^2 - 10x + 25).$$

Now consider $x^2 - 10x + 25$. Since the constant term is positive and the coefficient of the middle term is negative, we look for a factorization of 25 in which both factors are negative. Their sum must be -10.

Pairs of Factors of 25	Sums of Factors	
−25, −1	−26	
−5, −5	−10 ←	The numbers we need are −5 and −5.

The factorization of $x^2 - 10x + 25$ is $(x - 5)(x - 5)$, or $(x - 5)^2$.

> ***Caution!***   When factoring involves more than one step, be careful to write out the *entire* factorization.

Check: $2x(x-5)(x-5) = 2x[x^2 - 10x + 25]$ Multiplying binomials
$$= 2x^3 - 20x^2 + 50x.$$ Using the distributive law

The factorization of $2x^3 - 20x^2 + 50x$ is $2x(x-5)(x-5)$, or $2x(x-5)^2$.

Once any common factors have been factored out, the following summary can be used to factor $x^2 + bx + c$.

> **To Factor $x^2 + bx + c$**
>
> 1. Find a pair of factors that have c as their product and b as their sum.
>
>   a) If c is positive, its factors will have the same sign as b.
>   b) If c is negative, one factor will be positive and the other will be negative. Select the factors such that the factor with the larger absolute value is the factor with the same sign as b.
>
> 2. Check by multiplying.

FOR EXTRA HELP

Exercise Set 5.2

Digital Video Tutor CD 3   InterAct Math   Math Tutor Center   MathXL   MyMathLab.com
Videotape 9

Factor completely. Remember that you can check by multiplying. If a polynomial is prime, state this.

1. $x^2 + 6x + 5$ **2.** $x^2 + 7x + 6$

3. $x^2 + 7x + 10$ **4.** $x^2 + 7x + 12$

5. $y^2 + 11y + 28$ **6.** $x^2 - 6x + 9$

7. $a^2 + 11a + 30$ **8.** $x^2 + 9x + 14$

9. $x^2 - 5x + 4$ **10.** $b^2 + 5b + 4$

11. $z^2 - 8z + 7$ **12.** $a^2 - 4a - 12$

13. $x^2 - 8x + 15$ **14.** $d^2 - 7d + 10$

15. $y^2 - 11y + 10$ **16.** $x^2 - 2x - 15$

17. $x^2 + x - 42$ **18.** $x^2 + 2x - 15$

19. $2x^2 - 14x - 36$ **20.** $3y^2 - 9y - 84$

21. $x^3 - 6x^2 - 16x$ **22.** $x^3 - x^2 - 42x$

23. $y^2 + 4y - 45$ **24.** $x^2 + 7x - 60$

25. $-2x - 99 + x^2$ **26.** $x^2 - 72 + 6x$

27. $c^4 + c^3 - 56c^2$ **28.** $5b^2 + 25b - 120$

29. $2a^2 - 4a - 70$ **30.** $x^5 - x^4 - 2x^3$

31. $x^2 + x + 1$ **32.** $x^2 + 2x + 3$

33. $7 - 2p + p^2$ **34.** $11 - 3w + w^2$

35. $x^2 + 20x + 100$ **36.** $x^2 + 20x + 99$

37. $3x^3 - 63x^2 - 300x$ **38.** $2x^3 - 40x^2 + 192x$

39. $x^2 - 21x - 72$ **40.** $4x^2 + 40x + 100$

41. $x^2 - 25x + 144$ **42.** $y^2 - 21y + 108$

43. $a^4 + a^3 - 132a^2$ **44.** $a^6 + 9a^5 - 90a^4$

45. $x^2 - \frac{2}{5}x + \frac{1}{25}$ **46.** $t^2 + \frac{2}{3}t + \frac{1}{9}$

47. $27 + 12y + y^2$

48. $50 + 15x + x^2$

49. $t^2 - 0.3t - 0.10$

50. $y^2 - 0.2y - 0.08$

51. $p^2 + 3pq - 10q^2$

52. $a^2 - 2ab - 3b^2$

53. $m^2 + 5mn + 5n^2$

54. $x^2 - 11xy + 24y^2$

55. $s^2 - 2st - 15t^2$

56. $b^2 + 8bc - 20c^2$

57. $6a^{10} - 30a^9 - 84a^8$

58. $7x^9 - 28x^8 - 35x^7$

59. Marge factors $x^3 - 8x^2 + 15x$ as $(x^2 - 5x)(x - 3)$. Is she wrong? Why or why not? What advice would you offer?

60. Without multiplying $(x - 17)(x - 18)$, explain why it cannot possibly be a factorization of $x^2 + 35x + 306$.

SKILL MAINTENANCE

Solve.

61. $3x - 8 = 0$

62. $2x + 7 = 0$

Multiply.

63. $(x + 6)(3x + 4)$

64. $(7w + 6)^2$

65. In a recent year, 29,090 people were arrested for counterfeiting. This figure was down 1.2% from the year before. How many people were arrested the year before?

66. The first angle of a triangle is four times as large as the second. The measure of the third angle is 30° greater than that of the second. How large are the angles?

SYNTHESIS

67. When searching for a factorization, why do we list pairs of numbers with the correct *product* instead of pairs of numbers with the correct *sum*?

68. What is the advantage of writing out the prime factorization of c when factoring $x^2 + bx + c$ with a large value of c?

69. Find all integers b for which $a^2 + ba - 50$ can be factored.

70. Find all integers m for which $y^2 + my + 50$ can be factored.

Factor each of the following by first factoring out -1.

71. $30 + 7x - x^2$

72. $45 + 4x - x^2$

73. $24 - 10a - a^2$

74. $36 - 9a - a^2$

75. $84 - 8t - t^2$

76. $72 - 6t - t^2$

Factor completely.

77. $x^2 + \frac{1}{4}x - \frac{1}{8}$

78. $x^2 + \frac{1}{2}x - \frac{3}{16}$

79. $\frac{1}{3}a^3 - \frac{1}{3}a^2 - 2a$

80. $a^7 - \frac{25}{7}a^5 - \frac{30}{7}a^6$

81. $x^{2m} + 11x^m + 28$

82. $t^{2n} - 7t^n + 10$

Aha! **83.** $(a + 1)x^2 + (a + 1)3x + (a + 1)2$

84. $ax^2 - 5x^2 + 8ax - 40x - (a - 5)9$
(*Hint*: See Exercise 83.)

Find a polynomial in factored form for the shaded area in each figure. (Leave answers in terms of π.)

85.

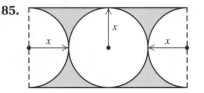

86.

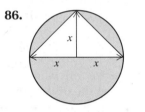

87. Find the volume of a cube if its surface area is $6x^2 + 36x + 54$ square meters.

88. A census taker asks a woman, "How many children do you have?"

"Three," she answers.

"What are their ages?"

She responds, "The product of their ages is 36. The sum of their ages is the house number next door."

The math-savvy census taker walks next door, reads the house number, appears puzzled, and returns to the woman, asking, "Is there something you forgot to tell me?"

"Oh yes," says the woman. "I'm sorry. The oldest child is at the park."

The census taker records the three ages, thanks the woman for her time, and leaves.

How old is each child? Explain how you reached this conclusion. (*Hint*: Consider factorizations.) (*Source*: Adapted from Harnadek, Anita, *Classroom Quickies*. Pacific Grove, CA: Critical Thinking Press and Software)

CORNER

Visualizing Factoring

Focus: Visualizing factoring

Time: 20–30 minutes

Group size: 3

Materials: Graph paper and scissors

The product $(x + 2)(x + 3)$ can be regarded as the area of a rectangle with width $x + 2$ and length $x + 3$. Similarly, factoring a polynomial like $x^2 + 5x + 6$ can be thought of as determining the length and the width of a rectangle that has area $x^2 + 5x + 6$. This is the approach used below.

ACTIVITY

1. **a)** To factor $x^2 + 11x + 10$ geometrically, the group needs to cut out shapes like those below to represent x^2, $11x$, and 10. This can be done by either tracing the figures below or by selecting a value for x, say 4, and using the squares on the graph paper to cut out the following:

    x^2:   Using the value selected for x, cut out a square that is x units on each side.

    $11x$:   Using the value selected for x, cut out a rectangle that is 1 unit wide and x units long. Repeat this to form 11 such strips.

    10:   Cut out two rectangles with whole-number dimensions and an area of 10. One should be 2 units by 5 units and the other 1 unit by 10 units.

    b) The group, working together, should then attempt to use one of the two rectangles with area 10, along with all of the other shapes, to piece together one large rectangle. Only one of the rectangles with area 10 will work.

    c) From the large rectangle formed in part (b), use the length and the width to determine the factorization of $x^2 + 11x + 10$. Where do the dimensions of the rectangle representing 10 appear in the factorization?

2. Repeat step (1) above, but this time use the other rectangle with area 10, and use only 7 of the 11 strips, along with the x^2-shape. Piece together the shapes to form one large rectangle. What factorization do the dimensions of this rectangle suggest?

3. Cut out rectangles with area 12 and use the above approach to factor $x^2 + 8x + 12$. What dimensions should be used for the rectangle with area 12?

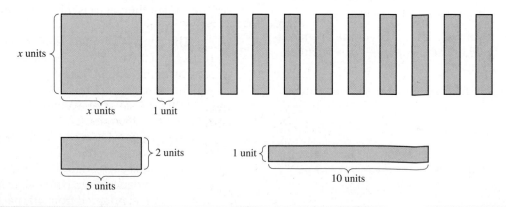

Factoring Trinomials of the Type $ax^2 + bx + c$

5.3

Factoring with FOIL • The Grouping Method

In Section 5.2, we learned a FOIL-based method for factoring trinomials of the type $x^2 + bx + c$. Now we learn to factor trinomials in which the leading, or x^2, coefficient is not 1. First we will use another FOIL-based method and then we will use an alternative method that involves factoring by grouping. Use the method that you prefer or the one selected by your instructor.

Factoring with FOIL

Before factoring trinomials of the type $ax^2 + bx + c$, consider the following:

$$\begin{array}{cccc} \text{F} & \text{O} & \text{I} & \text{L} \end{array}$$
$$(2x + 5)(3x + 4) = 6x^2 + 8x + 15x + 20$$
$$= 6x^2 + 23x + 20.$$

To factor $6x^2 + 23x + 20$, we reverse the multiplication above and look for two binomials whose product is this trinomial. The product of the First terms must be $6x^2$. The product of the Outer terms plus the product of the Inner terms must be $23x$. The product of the Last terms must be 20. This leads us to

$$(2x + 5)(3x + 4).$$ This factorization is verified above.

How can such a factorization be found without first seeing the corresponding multiplication? Our first approach relies on trial and error and FOIL.

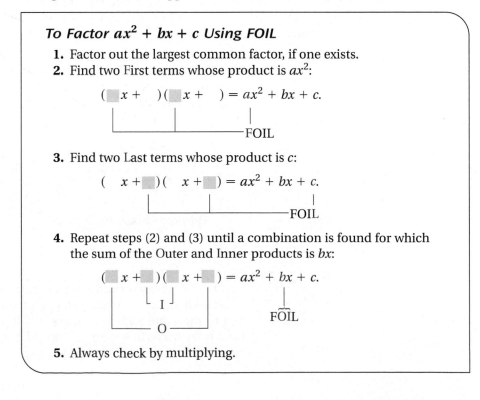

To Factor $ax^2 + bx + c$ Using FOIL

1. Factor out the largest common factor, if one exists.
2. Find two First terms whose product is ax^2:

$$(\ \square x + \ )(\ \square x + \ ) = ax^2 + bx + c.$$

$\vdash$ FOIL

3. Find two Last terms whose product is c:

$$(\ x + \square)(\ x + \square) = ax^2 + bx + c.$$

$\vdash$ FOIL

4. Repeat steps (2) and (3) until a combination is found for which the sum of the Outer and Inner products is bx:

$$(\ \square x + \square)(\ \square x + \square) = ax^2 + bx + c.$$

I
O
$\overline{\text{FOIL}}$

5. Always check by multiplying.

E x a m p l e   1

Factor: $3x^2 - 10x - 8$.

Solution

1. First, check for a common factor. In this case, there is none (other than 1 or -1).

2. Find two **First** terms whose product is $3x^2$.

   The only possibilities for the **First** terms are $3x$ and x, so any factorization must be of the form

   $$(3x + \quad)(x + \quad).$$

3. Find two **Last** terms whose product is -8.

   Possible factorizations of -8 are

   $$(-8) \cdot 1, \quad 8 \cdot (-1), \quad (-2) \cdot 4, \quad \text{and} \quad 2 \cdot (-4).$$

   Since the First terms are not identical, we must also consider

   $$1 \cdot (-8), \quad (-1) \cdot 8, \quad 4 \cdot (-2), \quad \text{and} \quad (-4) \cdot 2.$$

4., 5. Inspect the **Outer** and **Inner** products resulting from steps (2) and (3). Look for a combination in which the sum of the products is the middle term, $-10x$. This may take several tries.

Trial	*Product*	
$(3x - 8)(x + 1)$	$3x^2 + 3x - 8x - 8$	
	$= 3x^2 - 5x - 8$	Wrong middle term
$(3x + 8)(x - 1)$	$3x^2 - 3x + 8x - 8$	
	$= 3x^2 + 5x - 8$	Wrong middle term
$(3x - 2)(x + 4)$	$3x^2 + 12x - 2x - 8$	
	$= 3x^2 + 10x - 8$	Wrong middle term
$(3x + 2)(x - 4)$	$3x^2 - 12x + 2x - 8$	
	$= 3x^2 - 10x - 8$	Correct middle term!
$(3x + 1)(x - 8)$	$3x^2 - 24x + x - 8$	
	$= 3x^2 - 23x - 8$	Wrong middle term
$(3x - 1)(x + 8)$	$3x^2 + 24x - x - 8$	
	$= 3x^2 + 23x - 8$	Wrong middle term
$(3x + 4)(x - 2)$	$3x^2 - 6x + 4x - 8$	
	$= 3x^2 - 2x - 8$	Wrong middle term
$(3x - 4)(x + 2)$	$3x^2 + 6x - 4x - 8$	
	$= 3x^2 + 2x - 8$	Wrong middle term

The correct factorization is $(3x + 2)(x - 4)$.

Two observations can be made from Example 1. First, we listed all possible trials even though we usually stop after finding the correct factorization. We did this to show that each trial differs only in the middle term of the product. Second, note that as in Section 5.2, only the sign of the middle term changes when the signs in the binomials are reversed.

Example 2

Factor: $10x^2 + 37x + 7$.

Solution

1. There is no common factor (other than 1 or -1).

2. Because $10x^2$ factors as $10x \cdot x$ or $5x \cdot 2x$, we have two possibilities:

$$(10x + \quad)(x + \quad) \quad \text{or} \quad (5x + \quad)(2x + \quad).$$

3. There are two pairs of factors of 7 and each can be listed two ways:

$$1, 7 \qquad -1, -7$$

and

$$7, 1 \qquad -7, -1.$$

4., 5. From steps (2) and (3), we see that there are 8 possibilities for factorizations. Look for **O**uter and **I**nner products for which the sum is the middle term. Because all coefficients in $10x^2 + 37x + 7$ are positive, we need consider only positive factors of 7.

Trial	Product	
$(10x + 1)(x + 7)$	$10x^2 + 70x + 1x + 7$	
	$= 10x^2 + 71x + 7$	Wrong middle term
$(10x + 7)(x + 1)$	$10x^2 + 10x + 7x + 7$	
	$= 10x^2 + 17x + 7$	Wrong middle term
$(5x + 7)(2x + 1)$	$10x^2 + 5x + 14x + 7$	
	$= 10x^2 + 19x + 7$	Wrong middle term
$(5x + 1)(2x + 7)$	$10x^2 + 35x + 2x + 7$	
	$= 10x^2 + 37x + 7$	Correct middle term!

The correct factorization is $(5x + 1)(2x + 7)$.

Example 3

Factor: $24x^2 - 76x + 40$.

Solution

1. First, we factor out the largest common factor, 4:

$$4(6x^2 - 19x + 10).$$

2. Next, we factor $6x^2 - 19x + 10$. Since $6x^2$ can be factored as $3x \cdot 2x$ or $6x \cdot x$, we have two possibilities:

$$(3x + \quad)(2x + \quad) \quad \text{or} \quad (6x + \quad)(x + \quad).$$

3. There are four pairs of factors of 10 and each can be listed two ways:

$$10, 1 \qquad -10, -1 \qquad 5, 2 \qquad -5, -2$$

and

$$1, 10 \qquad -1, -10 \qquad 2, 5 \qquad -2, -5.$$

4., 5. The two possibilities from step (2) and the eight possibilities from step (3) give $2 \cdot 8$, or 16 possibilities for factorizations. We look for **O**uter and **I**nner products resulting from steps (2) and (3) for which the sum is the middle term, $-19x$. Since the sign of the middle term is negative, but the sign of the last term, 10, is positive, the two factors of 10 must both be negative. This means only four pairings from step (3) need be considered. We first try these factors with $(3x +)(2x +)$. If none gives the correct factorization of $6x^2 - 19x + 10$, then we will consider $(6x +)(x +)$.

Trial	*Product*	
$(3x - 10)(2x - 1)$	$6x^2 - 3x - 20x + 10$	
	$= 6x^2 - 23x + 10$	Wrong middle term
$(3x - 1)(2x - 10)$	$6x^2 - 30x - 2x + 10$	
	$= 6x^2 - 32x + 10$	Wrong middle term
$(3x - 5)(2x - 2)$	$6x^2 - 6x - 10x + 10$	
	$= 6x^2 - 16x + 10$	Wrong middle term
$(3x - 2)(2x - 5)$	$6x^2 - 15x - 4x + 10$	
	$= 6x^2 - 19x + 10$	Correct middle term!

Since we have a correct factorization, we need not consider

$$(6x +)(x +).$$

Look again at the possibility $(3x - 5)(2x - 2)$. Without multiplying, we can reject such a possibility. To see why, note that

$$(3x - 5)(2x - 2) = (3x - 5)2(x - 1).$$

The expression $2x - 2$ has a common factor, 2. But we removed the *largest* common factor in step (1). If $2x - 2$ were one of the factors, then 2 would be *another* common factor in addition to the original 4. Thus, $(2x - 2)$ cannot be part of the factorization of $6x^2 - 19x + 10$. Similar reasoning can be used to reject $(3x - 1)(2x - 10)$ as a possible factorization.

Once the largest common factor is factored out, none of the remaining factors can have a common factor.

The factorization of $6x^2 - 19x + 10$ is $(3x - 2)(2x - 5)$, but do not forget the common factor! The factorization of $24x^2 - 76x + 40$ is

$$4(3x - 2)(2x - 5).$$

Tips for Factoring $ax^2 + bx + c$

To factor $ax^2 + bx + c\, (a > 0)$:

- Always factor out the largest common factor, if one exists.
- Once the largest common factor has been factored out of the original trinomial, no binomial factor can contain a common factor (other than 1 or −1).
- If c is positive, then the signs in both binomial factors must match the sign of b.

(continued)

- Reversing the signs in the binomials reverses the sign of the middle term of their product.
- Organize your work so that you can keep track of which possibilities have or have not been checked.
- Always check by multiplying.

E x a m p l e 4

Factor: $10x + 8 - 3x^2$.

Solution An important problem-solving strategy is to find a way to make new problems look like problems we already know how to solve. The factoring tips above apply only to trinomials of the form $ax^2 + bx + c$, with $a > 0$. This leads us to rewrite $10x + 8 - 3x^2$ in descending order:

$$10x + 8 - 3x^2 = -3x^2 + 10x + 8.$$ Writing in descending order

Although $-3x^2 + 10x + 8$ looks similar to the trinomials we have factored, the tips above require a positive leading coefficient. This can be attained by factoring out -1:

$$-3x^2 + 10x + 8 = -1(3x^2 - 10x - 8)$$ Factoring out -1 changes the signs of the coefficients.

$$= -1(3x + 2)(x - 4).$$ Using the result from Example 1

The factorization of $10x + 8 - 3x^2$ is $-1(3x + 2)(x - 4)$.

E x a m p l e 5

Factor: $6p^2 - 13pq - 28q^2$.

Solution Since no common factor exists, we examine the first term, $6p^2$. There are two possibilities:

$$(2p + \quad)(3p + \quad) \quad \text{or} \quad (6p + \quad)(p + \quad).$$

The last term, $-28q^2$, has the following pairs of factors:

$$28q, -q \qquad 14q, -2q \qquad 7q, -4q,$$

and

$$-28q, \quad q \qquad -14q, \quad 2q \qquad -7q, \quad 4q,$$

as well as each of the pairings reversed.

Some trials, like $(2p + 28q)(3p - q)$ and $(2p + 14q)(3p - 2q)$, cannot be correct because both $(2p + 28q)$ and $(2p + 14q)$ contain a common factor, 2. We try $(2p + 7q)(3p - 4q)$:

$$(2p + 7q)(3p - 4q) = 6p^2 - 8pq + 21pq - 28q^2$$
$$= 6p^2 + 13pq - 28q^2.$$

Our trial is incorrect, but only because of the sign of the middle term. To correctly factor $6p^2 - 13pq - 28q^2$, we simply change the signs in the binomials:

$$(2p - 7q)(3p + 4q) = 6p^2 + 8pq - 21pq - 28q^2$$
$$= 6p^2 - 13pq - 28q^2.$$

The correct factorization is $(2p - 7q)(3p + 4q)$.

The Grouping Method

Another method of factoring trinomials of the type $ax^2 + bx + c$ is known as the *grouping method*. The grouping method relies on rewriting $ax^2 + bx + c$ in the form $ax^2 + px + qx + c$ and then factoring by grouping. To develop this method, consider the following*:

$$(2x + 5)(3x + 4) = 2x \cdot 3x + 2x \cdot 4 + 5 \cdot 3x + 5 \cdot 4 \qquad \text{Using FOIL}$$
$$= 2 \cdot 3 \cdot x^2 + 2 \cdot 4x + 5 \cdot 3x + 5 \cdot 4$$
$$= \underset{\underset{a}{\uparrow\downarrow}}{2 \cdot 3 \cdot x^2} + \underset{\underset{b}{\uparrow\downarrow}}{(2 \cdot 4 + 5 \cdot 3)x} + \underset{\underset{c}{\uparrow\downarrow}}{5 \cdot 4}$$
$$= \quad 6x^2 \quad + \quad 23x \quad + \quad 20.$$

Note that reversing these steps shows that $6x^2 + 23x + 20$ can be rewritten as $6x^2 + 8x + 15x + 20$ and then factored by grouping. Note that the numbers that add to b (in this case, $2 \cdot 4$ and $5 \cdot 3$), also multiply to ac (in this case, $2 \cdot 3 \cdot 5 \cdot 4$).

To Factor $ax^2 + bx + c$, Using the Grouping Method

1. Factor out the largest common factor, if one exists.
2. Multiply the leading coefficient a and the constant c.
3. Find a pair of factors of ac whose sum is b.
4. Rewrite the middle term, bx, as a sum or difference using the factors found in step (3).
5. Factor by grouping.
6. Always check by multiplying.

Example 6

Factor: $3x^2 - 10x - 8$.

Solution

1. First, we note that there is no common factor (other than 1 or -1).
2. We multiply the leading coefficient, 3, and the constant, -8:

$$3(-8) = -24.$$

---

*This discussion was inspired by a lecture given by Irene Doo at Austin Community College.

3. We next look for a factorization of -24 in which the sum of the factors is the coefficient of the middle term, -10.

Pairs of Factors of -24	Sums of Factors
1, -24	-23
-1, 24	23
2, -12	-10 ←
-2, 12	10
3, -8	-5
-3, 8	5
4, -6	-2
-4, 6	2

$2 + (-12) = -10$

We normally stop listing pairs of factors once we have found the one we are after.

4. Next, we express the middle term as a sum or difference using the factors found in step (3):

$$-10x = 2x - 12x.$$

5. We now factor by grouping as follows:

$$3x^2 - 10x - 8 = 3x^2 + 2x - 12x - 8$$ Substituting $2x - 12x$ for $-10x$. We could also use $-12x + 2x$.

$$= x(3x + 2) - 4(3x + 2)$$ Factoring by grouping; see Section 5.1

$$= (3x + 2)(x - 4).$$ Factoring out the common factor, $3x + 2$

6. *Check:* $(3x + 2)(x - 4) = 3x^2 - 10x - 8.$

The factorization of $3x^2 - 10x - 8$ is $(3x + 2)(x - 4)$.

Example 7

Factor: $8x^3 + 22x^2 - 6x$.

Solution

1. We factor out the largest common factor, $2x$:

$$8x^3 + 22x^2 - 6x = 2x(4x^2 + 11x - 3).$$

2. To factor $4x^2 + 11x - 3$ by grouping, we multiply the leading coefficient, 4, and the constant term, -3:

$$4(-3) = -12.$$

3. We next look for factors of -12 that add to 11.

Pairs of Factors of -12	Sums of Factors
1, -12	-11
-1, 12	11 ←
.	.
.	.
.	.

Since $-1 + 12 = 11$, there is no need to list other pairs of factors.

4. We then rewrite the $11x$ in $4x^2 + 11x - 3$ using

$$11x = -1x + 12x, \quad \text{or} \quad 11x = 12x - 1x.$$

5. Next, we factor by grouping:

$$4x^2 + 11x - 3 = 4x^2 + 12x - 1x - 3 \qquad \text{Rewriting the middle term; } -1x + 12x \text{ could also be used.}$$

$$= 4x(x + 3) - 1(x + 3) \qquad \text{Factoring by grouping. Removing } -1 \text{ reveals the common factor, } x + 3.$$

$$= (x + 3)(4x - 1). \qquad \text{Factoring out the common factor}$$

6. The factorization of $4x^2 + 11x - 3$ is $(x + 3)(4x - 1)$. But don't forget the common factor, $2x$. The factorization of the original trinomial is

$$2x(x + 3)(4x - 1).$$

Exercise Set 5.3

FOR EXTRA HELP

Digital Video Tutor CD 3 Videotape 9   InterAct Math   Math Tutor Center   MathXL   MyMathLab.com

Factor completely. If a polynomial is prime, state this.

1. $2x^2 + 7x - 4$

2. $3x^2 + x - 4$

3. $3t^2 + 4t - 15$

4. $5t^2 + t - 18$

5. $6x^2 - 23x + 7$

6. $6x^2 - 13x + 6$

7. $7x^2 + 15x + 2$

8. $3x^2 + 4x + 1$

9. $9a^2 - 6a - 8$

10. $4a^2 - 4a - 15$

11. $3x^2 - 5x - 2$

12. $15x^2 - 19x - 10$

Aha! **13.** $12t^2 - 6t - 6$

14. $18t^2 + 36t - 32$

15. $18t^2 + 3t - 10$

16. $2t^2 + 5t + 2$

17. $15x^2 + 19x + 6$

18. $12x^2 - 31x + 20$

19. $35x^2 + 34x + 8$

20. $28x^2 + 38x - 6$

21. $4 + 6t^2 - 13t$

22. $9 + 8t^2 - 18t$

23. $25x^2 + 40x + 16$

24. $49t^2 + 42t + 9$

25. $16a^2 + 78a + 27$

26. $24x^2 + 47x - 2$

27. $18t^2 + 24t - 10$

28. $35x^2 - 57x - 44$

29. $2x^2 - 15 - x$

30. $2t^2 - 19 - 6t$

31. $6x^2 + 33x + 15$

32. $12x^2 + 28x - 24$

33. $20x^2 - 25x + 5$

34. $30x^2 - 24x - 54$

35. $12x^2 + 68x - 24$

36. $6x^2 + 21x + 15$

37. $4x + 1 + 3x^2$

38. $-9 + 18x^2 + 21x$

Factor. Use factoring by grouping even though it would seem reasonable to first combine like terms.

39. $y^2 + 4y - 2y - 8$

40. $x^2 + 5x - 2x - 10$

41. $8t^2 - 6t - 28t + 21$

42. $35t^2 - 40t + 21t - 24$

43. $6x^2 + 4x + 9x + 6$

44. $3x^2 - 2x + 3x - 2$

45. $2t^2 + 6t - t - 3$

46. $5t^2 + 10t - t - 2$

47. $3a^2 - 12a - a + 4$

48. $2a^2 - 10a - a + 5$

Factor completely. If a polynomial is prime, state this.

49. $9t^2 + 14t + 5$

50. $16t^2 + 23t + 7$

51. $16x^2 + 32x + 7$

52. $9x^2 + 18x + 5$

53. $10a^2 + 25a - 15$

54. $10a^2 - 3a - 18$

55. $2x^2 + 6x - 14$

56. $14x^2 - 35x + 14$

57. $18x^3 + 21x^2 - 9x$

58. $6x^3 - 4x^2 - 10x$

59. $89x + 64 + 25x^2$

60. $47 - 42y + 9y^2$

61. $168x^3 + 45x^2 + 3x$

62. $144x^5 - 168x^4 + 48x^3$

63. $14t^4 - 19t^3 - 3t^2$

64. $70a^4 - 68a^3 + 16a^2$

65. $3x + 45x^2 - 18$

66. $2x + 24x^2 - 40$

67. $9a^2 + 18ab + 8b^2$

68. $3p^2 - 16pq - 12q^2$

69. $35p^2 + 34pq + 8q^2$

70. $10s^2 + 4st - 6t^2$

71. $18x^2 - 6xy - 24y^2$

72. $30a^2 + 87ab + 30b^2$

73. $24a^2 - 34ab + 12b^2$

74. $15a^2 - 5ab - 20b^2$

75. $35x^2 + 34x^3 + 8x^4$

76. $19x^3 - 3x^2 + 14x^4$

77. $18u^7 + 8u^6 + 9u^8$

78. $40a^8 + 16a^7 + 25a^9$

79. Asked to factor $2x^2 - 18x + 36$, Amy *incorrectly* answers

$$2x^2 - 18x + 36 = 2(x^2 + 9x + 18)$$
$$= 2(x + 3)(x + 6).$$

If this were a 10-point quiz question, how many points would you take off? Why?

80. Asked to factor $4x^2 + 28x + 48$, Herb *incorrectly* answers

$$4x^2 + 28x + 48 = (2x + 6)(2x + 8)$$
$$= 2(x + 3)(x + 4).$$

If this were a 10-point quiz question, how many points would you take off? Why?

SKILL MAINTENANCE

81. The earth is a sphere (or ball) that is about 40,000 km in circumference. Find the radius of the earth, in kilometers and in miles. Use 3.14 for π. (*Hint:* 1 km ≈ 0.62 mi.)

82. The second angle of a triangle is 10° less than twice the first. The third angle is 15° more than four times the first. Find the measure of the second angle.

Multiply.

83. $(3x + 1)^2$

84. $(5x - 2)^2$

85. $(4t - 5)^2$

86. $(7a + 1)^2$

87. $(5x - 2)(5x + 2)$

88. $(2x - 3)(2x + 3)$

89. $(2t + 7)(2t - 7)$

90. $(4a + 7)(4a - 7)$

SYNTHESIS

91. Which one of the six tips listed after Example 3 do you find most helpful? Which one do you find least helpful? Explain how you made these determinations.

92. For the trinomial $ax^2 + bx + c$, suppose that a is the product of three different prime factors and c is the product of another two prime factors. How many possible factorizations (like those in Example 1) exist? Explain how you determined your answer.

Factor. If a polynomial is prime, state this.

93. $9a^2b^2 - 15ab - 2$

94. $18x^2y^2 - 3xy - 10$

95. $8x^2y^3 + 10xy^2 + 2y$

96. $9a^2b^3 + 25ab^2 + 16$

97. $9t^{10} + 12t^5 + 4$

98. $16t^{10} - 8t^5 + 1$

99. $-15x^{2m} + 26x^m - 8$

100. $20x^{2n} + 16x^n + 3$

101. $a^{2n+1} - 2a^{n+1} + a$

102. $3a^{6n} - 2a^{3n} - 1$

103. $3(a + 1)^{n+1}(a + 3)^2 - 5(a + 1)^n(a + 3)^3$

104. $7(t - 3)^{2n} + 5(t - 3)^n - 2$

<table>
<tr><td>

Factoring Perfect-Square Trinomials and Differences of Squares

</td><td>

5.4

Recognizing Perfect-Square Trinomials • Factoring Perfect-Square Trinomials • Recognizing Differences of Squares • Factoring Differences of Squares • Factoring Completely

</td></tr>
</table>

In Section 4.5, we studied some shortcuts for finding the products of certain binomials. Reversing these procedures provides shortcuts for factoring certain polynomials.

Recognizing Perfect-Square Trinomials

Some trinomials are squares of binomials. For example, $x^2 + 10x + 25$ is the square of the binomial $x + 5$. To see this, we can calculate $(x + 5)^2$. It is $x^2 + 2 \cdot x \cdot 5 + 5^2$, or $x^2 + 10x + 25$. A trinomial that is the square of a binomial is called a **perfect-square trinomial**.

In Section 4.5, we considered squaring binomials as a special-product rule:

$$(A + B)^2 = A^2 + 2AB + B^2;$$
$$(A - B)^2 = A^2 - 2AB + B^2.$$

Reading the right sides first, we see that these equations can be used to factor perfect-square trinomials. Note that in order for a trinomial to be the square of a binomial, it must have the following:

1. Two terms, A^2 and B^2, must be squares, such as

   $4, \quad x^2, \quad 81m^2, \quad 16t^2.$

2. There must be no minus sign before A^2 or B^2.
3. The remaining term is either $2 \cdot A \cdot B$ or $-2 \cdot A \cdot B$, where A and B are the square roots of A^2 and B^2.

E x a m p l e   1

Determine whether each of the following is a perfect-square trinomial.

a) $x^2 + 6x + 9$ **b)** $t^2 - 8t - 9$ **c)** $16x^2 + 49 - 56x$

Solution

a) To see if $x^2 + 6x + 9$ is a perfect-square trinomial, note that:

   1. Two terms, x^2 and 9, are squares.
   2. There is no minus sign before x^2 or 9.
   3. The remaining term, $6x$, is $2 \cdot x \cdot 3$, where x and 3 are the square roots of x^2 and 9.

   Thus, $x^2 + 6x + 9$ *is* a perfect-square trinomial.

b) To see if $t^2 - 8t - 9$ is a perfect-square trinomial, note that:

   1. Two terms, t^2 and 9, are squares. But:
   2. Since 9 is being subtracted, $t^2 - 8t - 9$ *is not* a perfect-square trinomial.

c) To see if $16x^2 + 49 - 56x$ is a perfect-square trinomial, it helps to first write it in descending order:

$$16x^2 - 56x + 49.$$

Next, note that:

1. Two terms, $16x^2$ and 49, are squares.
2. There is no minus sign before $16x^2$ or 49.
3. Twice the product of the square roots, $2 \cdot 4x \cdot 7$, is $56x$, the opposite of the remaining term, $-56x$.

Thus, $16x^2 + 49 - 56x$ *is* a perfect-square trinomial.

Factoring Perfect-Square Trinomials

Either of the factoring methods from Section 5.3 can be used to factor perfect-square trinomials, but a faster method is to recognize the following patterns.

> ### Factoring a Perfect-Square Trinomial
> $$A^2 + 2AB + B^2 = (A + B)^2; \qquad A^2 - 2AB + B^2 = (A - B)^2$$

Each factorization uses the square roots of the squared terms and the sign of the remaining term.

E x a m p l e 2

Factor: **(a)** $x^2 + 6x + 9$; **(b)** $x^2 + 49 - 14x$; **(c)** $16x^2 - 40x + 25$.

Solution

a) $x^2 + 6x + 9 = x^2 + 2 \cdot x \cdot 3 + 3^2 = (x + 3)^2$ The sign of the middle term is positive.

$$A^2 + 2 \ A \ B + B^2 = (A + B)^2$$

b) $x^2 + 49 - 14x = x^2 - 14x + 49$ Using a commutative law to write in descending order

$$= x^2 - 2 \cdot x \cdot 7 + 7^2 = (x - 7)^2$$

$$A^2 - 2 \ A \ B + B^2 = (A - B)^2$$

c) $16x^2 - 40x + 25 = (4x)^2 - 2 \cdot 4x \cdot 5 + 5^2 = (4x - 5)^2$

$$A^2 \ - 2 \ A \ B + B^2 = (A - B)^2$$

With practice, it is possible to spot perfect-square trinomials as they occur and factor them quickly.

E x a m p l e   3

Factor: $4p^2 - 12pq + 9q^2$.

Solution   We have

$$4p^2 - 12pq + 9q^2 = (2p)^2 - 2(2p)(3q) + (3q)^2 \qquad \text{Recognizing the perfect-square trinomial}$$

$$= (2p - 3q)^2. \qquad \text{The sign of the middle term is negative.}$$

Check: $(2p - 3q)(2p - 3q) = 4p^2 - 12pq + 9q^2$.

The factorization is $(2p - 3q)^2$.

E x a m p l e   4

Factor: $75m^3 + 60m^2 + 12m$.

Solution   *Always* look first for a common factor. This time there is one, $3m$:

$$75m^3 + 60m^2 + 12m = 3m[25m^2 + 20m + 4] \qquad \text{Factoring out the largest common factor}$$

$$= 3m[(5m)^2 + 2(5m)(2) + 2^2] \qquad \text{Recognizing the perfect-square trinomial. Try to do this mentally.}$$

$$= 3m(5m + 2)^2.$$

Check: $3m(5m + 2)^2 = 3m(5m + 2)(5m + 2)$

$$= 3m(25m^2 + 20m + 4)$$

$$= 75m^3 + 60m^2 + 12m.$$

The factorization is $3m(5m + 2)^2$.

Recognizing Differences of Squares

Some binomials represent the difference of two squares. For example, the binomial $16x^2 - 9$ is a difference of two expressions, $16x^2$ and 9, that are squares. To see this, note that $16x^2 = (4x)^2$ and $9 = 3^2$.

Any expression, like $16x^2 - 9$, that can be written in the form $A^2 - B^2$ is called a **difference of squares**. Note that for a binomial to be a difference of squares, it must have the following.

1. There must be two expressions, both squares, such as

$$25x^2, \qquad 9, \qquad 4x^2y^2, \qquad 1, \qquad x^6, \qquad 49y^8.$$

2. The terms in the binomial must have different signs.

Note that in order for a term to be a square, its coefficient must be a perfect square and the power(s) of the variable(s) must be even.

Example 5

Determine whether each of the following is a difference of squares.

a) $9x^2 - 64$ **b)** $25 - t^3$ **c)** $-4x^{10} + 36$

Solution

a) To see if $9x^2 - 64$ is a difference of squares, note that:

1. The first expression is a square: $9x^2 = (3x)^2$.
   The second expression is a square: $64 = 8^2$.
2. The terms have different signs.

Thus, $9x^2 - 64$ is a difference of squares, $(3x)^2 - 8^2$.

b) To see if $25 - t^3$ is a difference of squares, note that:

1. The expression t^3 is not a square.

Thus, $25 - t^3$ is not a difference of squares.

c) To see if $-4x^{10} + 36$ is a difference of squares, note that:

1. The expressions $4x^{10}$ and 36 are squares: $4x^{10} = (2x^5)^2$ and $36 = 6^2$.
2. The terms have different signs.

Thus, $-4x^{10} + 36$ is a difference of squares, $6^2 - (2x^5)^2$.

Factoring Differences of Squares

To factor a difference of squares, we reverse a pattern from Section 4.5:

> **Factoring a Difference of Squares**
> $A^2 - B^2 = (A + B)(A - B).$

Once we have identified the expressions that are playing the roles of A and B, the factorization can be written directly.

Example 6

Factor: **(a)** $x^2 - 4$; **(b)** $m^2 - 9p^2$; **(c)** $9 - 16t^{10}$; **(d)** $50x^2 - 8x^8$.

Solution

a) $x^2 - 4 = x^2 - 2^2 = (x + 2)(x - 2)$

$$A^2 - B^2 = (A + B)(A - B)$$

b) $m^2 - 9p^2 = m^2 - (3p)^2 = (m + 3p)(m - 3p)$

$$A^2 - B^2 = (A + B)(A - B)$$

c) $9 - 16t^{10} = 3^2 - (4t^5)^2$ Using the rules for powers

$\hspace{3.5cm} A^2 - B^2$

$\hspace{2cm} = (3 + 4t^5)(3 - 4t^5)$ Try to go directly to this step.

$\hspace{2cm} (A + B)(A - B)$

d) *Always* check first for a common factor. This time there is one, $2x^2$:

$$50x^2 - 8x^8 = 2x^2(25 - 4x^6) \qquad \text{Factoring out the common factor}$$

$$= 2x^2[5^2 - (2x^3)^2] \qquad \text{Recognizing } A^2 - B^2. \text{ Try to do this mentally.}$$

$$= 2x^2(5 + 2x^3)(5 - 2x^3). \qquad \text{Factoring the difference of squares}$$

Check: $\quad 2x^2(5 + 2x^3)(5 - 2x^3) = 2x^2(25 - 4x^6)$
$$= 50x^2 - 8x^8.$$

The factorization is $2x^2(5 + 2x^3)(5 - 2x^3)$.

Caution!   Note in Example 6 that a difference of squares is *not* the square of the difference; that is,

$$A^2 - B^2 \neq (A - B)^2. \qquad \text{To see this, note that}$$
$$(A - B)^2 = A^2 - 2AB + B^2.$$

Factoring Completely

Sometimes, as in Examples 4 and 6(d), a *complete* factorization requires two or more steps. In general, a factorization is complete when no factor can be factored further.

E x a m p l e 7

Factor: $p^4 - 16$.

Solution   We have

$$p^4 - 16 = (p^2)^2 - 4^2 \qquad \text{Recognizing } A^2 - B^2$$

$$= (p^2 + 4)(p^2 - 4) \qquad \text{Factoring a difference of squares}$$

$$= (p^2 + 4)(p + 2)(p - 2). \qquad \text{Factoring further. The factor } p^2 - 4 \text{ is itself a difference of squares.}$$

Check: $\quad (p^2 + 4)(p + 2)(p - 2) = (p^2 + 4)(p^2 - 4)$
$$= p^4 - 16.$$

The factorization is $(p^2 + 4)(p + 2)(p - 2)$.

Note in Example 7 that the factor $p^2 + 4$ is a *sum* of squares that cannot be factored further.

Caution!   Apart from possibly removing a common factor, you cannot factor a sum of squares. In particular,

$$A^2 + B^2 \neq (A + B)^2.$$

Consider $25x^2 + 100$. Here a sum of squares has a common factor, 25. Factoring, we get $25(x^2 + 4)$, where $x^2 + 4$ is prime.

As you proceed through the exercises, these suggestions may prove helpful.

Tips for Factoring

1. Always look first for a common factor! If there is one, factor it out.
2. Be alert for perfect-square trinomials and differences of squares. Once recognized, they can be factored without trial and error.
3. Always factor completely.
4. Check by multiplying.

Exercise Set 5.4

FOR EXTRA HELP

Digital Video Tutor CD 3 Videotape 9   InterAct Math   Math Tutor Center   MathXL   MyMathLab.com

Determine whether each of the following is a perfect-square trinomial.

1. $x^2 - 18x + 81$

2. $x^2 - 16x + 64$

3. $x^2 + 16x - 64$

4. $x^2 - 14x - 49$

5. $x^2 - 3x + 9$

6. $x^2 + 2x + 4$

7. $9x^2 - 36x + 24$

8. $36x^2 - 24x + 16$

Factor completely. Remember to look first for a common factor and to check by multiplying. If a polynomial is prime, state this.

9. $x^2 - 16x + 64$

10. $x^2 - 14x + 49$

11. $x^2 + 14x + 49$

12. $x^2 + 16x + 64$

13. $3x^2 - 6x + 3$

14. $5x^2 - 10x + 5$

15. $4 + 4x + x^2$

16. $4 + x^2 - 4x$

17. $18x^2 - 12x + 2$

18. $25x^2 + 10x + 1$

19. $49 + 56y + 16y^2$

20. $120m + 75 + 48m^2$

21. $x^5 - 18x^4 + 81x^3$

22. $2x^2 - 40x + 200$

23. $2x^3 - 4x^2 + 2x$

24. $x^3 + 24x^2 + 144x$

25. $20x^2 + 100x + 125$

26. $12x^2 + 36x + 27$

27. $49 - 42x + 9x^2$

28. $64 - 112x + 49x^2$

29. $16x^2 + 24x + 9$

30. $2a^2 + 28a + 98$

31. $2 + 20x + 50x^2$

32. $9x^2 + 30x + 25$

33. $4p^2 + 12pq + 9q^2$

34. $25m^2 + 20mn + 4n^2$

35. $a^2 - 12ab + 49b^2$

36. $x^2 - 7xy + 9y^2$

37. $64m^2 + 16mn + n^2$

38 $81p^2 - 18pq + q^2$

39. $32s^2 - 80st + 50t^2$

40. $36a^2 + 96ab + 64b^2$

Determine whether each of the following is a difference of squares.

41. $x^2 - 100$ **42.** $x^2 - 36$ **43.** $x^2 + 36$

44. $x^2 + 4$ **45.** $9t^2 - 32$ **46.** $x^2 - 50y^2$

47. $-25 + 4t^2$ **48.** $-1 + 49t^2$

Factor completely. Remember to look first for a common factor. If a polynomial is prime, state this.

49. $y^2 - 4$ **50.** $x^2 - 36$

51. $p^2 - 9$ **52.** $q^2 + 1$

53. $-49 + t^2$ **54.** $-64 + m^2$

55. $6a^2 - 54$ **56.** $x^2 - 8x + 16$

57. $49x^2 - 14x + 1$ **58.** $3t^2 - 12$

59. $200 - 2t^2$ **60.** $98 - 8w^2$

61. $80a^2 - 45$ **62.** $25x^2 - 4$

63. $5t^2 - 80$ **64.** $4t^2 - 64$

65. $8x^2 - 98$ **66.** $24x^2 - 54$

67. $36x - 49x^3$ **68.** $16x - 81x^3$

69. $49a^4 - 20$ **70.** $25a^4 - 9$

71. $t^4 - 1$ **72.** $x^4 - 16$

73. $3x^3 - 24x^2 + 48x$ **74.** $2a^4 - 36a^3 + 162a^2$

75. $48t^2 - 27$ **76.** $125t^2 - 45$

77. $a^8 - 2a^7 + a^6$ **78.** $x^8 - 8x^7 + 16x^6$

79. $7a^2 - 7b^2$ **80.** $6p^2 - 6q^2$

81. $25x^2 - 4y^2$ **82.** $16a^2 - 9b^2$

83. $1 - a^4b^4$ **84.** $75 - 3m^4n^4$

85. $18t^2 - 8s^2$ **86.** $49x^2 - 16y^2$

87. Explain in your own words how to determine whether a polynomial is a difference of squares.

88. Explain in your own words how to determine if a polynomial is a perfect-square trinomial.

SKILL MAINTENANCE

89. About 5 L of oxygen can be dissolved in 100 L of water at 0°C. This is 1.6 times the amount that can be dissolved in the same volume of water at 20°C. How much oxygen can be dissolved at 20°C?

90. Bonnie is taking an astronomy course. To get an A, she must average at least 90 after four exams. Bonnie scored 96, 98, and 89 on the first three

tests. Determine (in terms of an inequality) what scores on the last test will earn her an A.

Simplify.

91. $(x^3y^5)(x^9y^7)$ **92.** $(5a^2b^3)^2$

Graph.

93. $y = \frac{3}{2}x - 3$ **94.** $3x - 5y = 30$

SYNTHESIS

95. Write directions that would enable someone to construct a polynomial that contains a perfect-square trinomial, a difference of squares, and a common factor.

96. Leon concludes that since $x^2 - 9 = (x - 3)(x + 3)$, it must follow that $x^2 + 9 = (x + 3)(x - 3)$. What mistake(s) is he making?

Factor completely. If a polynomial is prime, state this.

97. $x^8 - 2^8$ **98.** $3x^2 - \frac{1}{3}$

99. $18x^3 - \frac{8}{25}x$ **100.** $0.49p - p^3$

101. $0.64x^2 - 1.21$ **102.** $(x + 3)^4 - 81$

103. $(y - 5)^4 - z^8$ **104.** $x^2 - \left(\frac{1}{x}\right)^2$

105. $a^{2n} - 49b^{2n}$ **106.** $81 - b^{4k}$

107. $x^4 - 8x^2 - 9$ **108.** $9b^{2n} + 12b^n + 4$

109. $16x^4 - 96x^2 + 144$

110. $(y + 3)^2 + 2(y + 3) + 1$

111. $49(x + 1)^2 - 42(x + 1) + 9$

112. $27x^3 - 63x^2 - 147x + 343$

113. Subtract $(x^2 + 1)^2$ from $x^2(x + 1)^2$.

Factor by grouping. Look for a grouping of three terms that is a perfect-square trinomial.

114. $a^2 + 2a + 1 - 9$

115. $y^2 + 6y + 9 - x^2 - 8x - 16$

Find c such that each polynomial is the square of a binomial.

116. $cy^2 + 6y + 1$ **117.** $cy^2 - 24y + 9$

118. Find the value of a if $x^2 + a^2x + a^2$ factors into $(x + a)^2$.

119. Show that the difference of the squares of two consecutive integers is the sum of the integers. (*Hint*: Use x for the smaller number.)

Factoring: A General Strategy	**5.5**
	Choosing the Right Method

CONNECTING THE CONCEPTS

Thus far, each section in this chapter has examined one or two different methods for factoring polynomials. In practice, when the need for factoring a polynomial arises, we must decide on our own which method to use. As preparation for such a situation, we now encounter polynomials of various types, in random order. Regardless of the polynomial we are faced with, the guidelines listed below can always be used.

To Factor a Polynomial

A. Always look for a common factor first. If there is one, factor out the largest common factor. Be sure to include it in your final answer.

B. Then look at the number of terms.

Two terms: If you have a difference of squares, factor accordingly. Do not try to factor a sum of squares: $A^2 + B^2$.

Three terms: Determine whether the trinomial is a perfect-square trinomial. If so, factor accordingly. If not, try trial and error, using the standard method or grouping.

Four terms: Try factoring by grouping.

C. Always *factor completely*. When a factor can itself be factored, be sure to factor it.

D. Check by multiplying.

Choosing the Right Method

Example 1

Factor: $5t^4 - 80$.

Solution

A. We look for a common factor:

$$5t^4 - 80 = 5(t^4 - 16). \qquad \text{5 is the largest common factor.}$$

B. The factor $t^4 - 16$ is a difference of squares: $(t^2)^2 - 4^2$. We factor it, being careful to rewrite the 5 from step (A):

$$5t^4 - 80 = 5(t^2 + 4)(t^2 - 4). \qquad t^4 - 16 = (t^2 + 4)(t^2 - 4)$$

C. Since $t^2 - 4$ is not prime, we continue factoring:

$$5t^4 - 80 = 5(t^2 + 4)(t^2 - 4) = 5(t^2 + 4)(t - 2)(t + 2)$$

This is a sum of squares with no common factor. It cannot be factored!

D. *Check:* $5(t^2 + 4)(t - 2)(t + 2) = 5(t^2 + 4)(t^2 - 4)$

$$= 5(t^4 - 16) = 5t^4 - 80.$$

The factorization is $5(t^2 + 4)(t - 2)(t + 2)$.

E x a m p l e   2

Factor: $2x^3 + 10x^2 + x + 5$.

Solution

A. We look for a common factor. There is none.

B. Because there are four terms, we try factoring by grouping:

$2x^3 + 10x^2 + x + 5$

$= (2x^3 + 10x^2) + (x + 5)$ Separating into two binomials

$= 2x^2(x + 5) + 1(x + 5)$ Factoring out the largest common factor from each binomial. The 1 serves as an aid.

$= (x + 5)(2x^2 + 1)$ Factoring out the common factor, $x + 5$

C. Nothing can be factored further, so we have factored completely.

D. *Check:* $(x + 5)(2x^2 + 1) = 2x^3 + x + 10x^2 + 5$

$$= 2x^3 + 10x^2 + x + 5.$$

The factorization is $(x + 5)(2x^2 + 1)$.

E x a m p l e   3

Factor: $x^5 - 2x^4 - 35x^3$.

Solution

A. We note that there is a common factor, x^3:

$$x^5 - 2x^4 - 35x^3 = x^3(x^2 - 2x - 35).$$

B. The factor $x^2 - 2x - 35$ is not a perfect-square trinomial. We factor it using trial and error:

$x^5 - 2x^4 - 35x^3 = x^3(x^2 - 2x - 35)$

$= x^3(x - 7)(x + 5).$

C. Nothing can be factored further, so we have factored completely.

D. *Check:* $x^3(x - 7)(x + 5) = x^3(x^2 - 2x - 35)$

$$= x^5 - 2x^4 - 35x^3.$$

The factorization is $x^3(x - 7)(x + 5)$.

Example 4

Factor: $x^2 - 20x + 100$.

Solution

A. We look first for a common factor. There is none.

B. This polynomial is a perfect-square trinomial. We factor it accordingly:

$$x^2 - 20x + 100 = x^2 - 2 \cdot x \cdot 10 + 10^2 \qquad \text{Try to do this step mentally.}$$
$$= (x - 10)^2.$$

C. Nothing can be factored further, so we have factored completely.

D. *Check:* $(x - 10)(x - 10) = x^2 - 20x + 100$.

The factorization is $(x - 10)(x - 10)$, or $(x - 10)^2$.

Example 5

Factor: $6x^2y^4 - 21x^3y^5 + 3x^2y^6$.

Solution

A. We first factor out the largest common factor, $3x^2y^4$:

$$6x^2y^4 - 21x^3y^5 + 3x^2y^6 = 3x^2y^4(2 - 7xy + y^2).$$

B. There are three terms in $2 - 7xy + y^2$. Since only y^2 is a square, we do not have a perfect-square trinomial. Can $2 - 7xy + y^2$ be factored by trial and error? A key to the answer is that x appears only in $-7xy$. If $2 - 7xy + y^2$ could be factored into a form like $(1 - y)(2 - y)$, there would be no x in the middle term. Thus, $2 - 7xy + y^2$ cannot be factored.

C. Nothing can be factored further, so we have factored completely.

D. *Check:* $3x^2y^4(2 - 7xy + y^2) = 6x^2y^4 - 21x^3y^5 + 3x^2y^6$.

The factorization is $3x^2y^4(2 - 7xy + y^2)$.

Example 6

Factor: $px + py + qx + qy$.

Solution

A. We look first for a common factor. There is none.

B. There are four terms. We try factoring by grouping:

$$px + py + qx + qy = p(x + y) + q(x + y)$$
$$= (x + y)(p + q).$$

C. Nothing can be factored further, so we have factored completely.

D. *Check:* $(x + y)(p + q) = xp + xq + yp + yq$
$$= px + py + qx + qy.$$

The factorization is $(x + y)(p + q)$.

E x a m p l e   7

Factor: $25x^2 + 20xy + 4y^2$.

Solution

A. We look first for a common factor. There is none.

B. There are three terms. Note that the first term and the last term are squares:

$$25x^2 = (5x)^2 \quad \text{and} \quad 4y^2 = (2y)^2.$$

We see that twice the product of $5x$ and $2y$ is the middle term,

$$2 \cdot 5x \cdot 2y = 20xy,$$

so the trinomial is a perfect square.

To factor, we write a binomial squared. The binomial is the sum of the terms being squared:

$$25x^2 + 20xy + 4y^2 = (5x + 2y)^2.$$

C. Nothing can be factored further, so we have factored completely.

D. *Check:* $(5x + 2y)(5x + 2y) = 25x^2 + 20xy + 4y^2$.

The factorization is $(5x + 2y)(5x + 2y)$, or $(5x + 2y)^2$.

E x a m p l e   8

Factor: $p^2q^2 + 7pq + 12$.

Solution

A. We look first for a common factor. There is none.

B. Since only one term is a square, we do not have a perfect-square trinomial. We use trial and error, thinking of the product pq as a single variable:

$$(pq + \quad)(pq + \quad).$$

We factor the last term, 12. All the signs are positive, so we consider only positive factors. Possibilities are 1, 12 and 2, 6 and 3, 4. The pair 3, 4 gives a sum of 7 for the coefficient of the middle term. Thus,

$$p^2q^2 + 7pq + 12 = (pq + 3)(pq + 4).$$

C. Nothing can be factored further, so we have factored completely.

D. *Check:* $(pq + 3)(pq + 4) = p^2q^2 + 7pq + 12$.

The factorization is $(pq + 3)(pq + 4)$.

Compare the variables appearing in Example 7 with those appearing in Example 8. Note that when one variable appears in the leading term and another variable appears in the last term, as in Example 7, each binomial contains two variable terms. When two variables appear in the leading term, as in Example 8, each binomial contains just one variable term.

Example 9

Factor: $a^4 - 16b^4$.

Solution

A. We look first for a common factor. There is none.

B. There are two terms. Since $a^4 = (a^2)^2$ and $16b^4 = (4b^2)^2$, we see that we do have a difference of squares. Thus,

$$a^4 - 16b^4 = (a^2 + 4b^2)(a^2 - 4b^2).$$

C. The factor $(a^2 - 4b^2)$ is itself a difference of squares. Thus,

$$a^4 - 16b^4 = (a^2 + 4b^2)(a + 2b)(a - 2b). \qquad \text{Factoring } a^2 - 4b^2$$

D. *Check:* $(a^2 + 4b^2)(a + 2b)(a - 2b) = (a^2 + 4b^2)(a^2 - 4b^2)$
$$= a^4 - 16b^4.$$

The factorization is $(a^2 + 4b^2)(a + 2b)(a - 2b)$.

FOR EXTRA HELP

Exercise Set 5.5

Digital Video Tutor CD 3
Videotape 10

InterAct Math

Math Tutor Center

MathXL

MyMathLab.com

Factor completely. If a polynomial is prime, state this.

1. $5x^2 - 45$

2. $10a^2 - 640$

3. $a^2 + 25 + 10a$

4. $y^2 + 49 - 14y$

5. $8t^2 - 18t - 5$

6. $2t^2 + 11t + 12$

7. $x^3 - 24x^2 + 144x$

8. $x^3 - 18x^2 + 81x$

9. $x^3 + 3x^2 - 4x - 12$

10. $x^3 - 5x^2 - 25x + 125$

11. $98t^2 - 18$

12. $27t^3 - 3t$

13. $20x^3 - 4x^2 - 72x$

14. $9x^3 + 12x^2 - 45x$

15. $x^2 + 4$

16. $t^2 + 25$

17. $a^4 + 8a^2 + 8a^3 + 64a$

18. $t^4 + 7t^2 - 3t^3 - 21t$

19. $x^5 - 14x^4 + 49x^3$

20. $2x^6 + 8x^5 + 8x^4$

21. $20 - 6x - 2x^2$

22. $45 - 3x - 6x^2$

23. $t^2 - 7t - 6$

24. $t^2 - 8t + 6$

25. $4x^4 - 64$

26. $5x^5 - 80x$

27. $9 + t^8$

28. $t^4 - 9$

29. $x^5 - 4x^4 + 3x^3$

30. $x^6 - 2x^5 + 7x^4$

31. $x^2 - y^2$

32. $p^2q^2 - r^2$

33. $12n^2 + 24n^3$

34. $ax^2 + ay^2$

35. $ab^2 - a^2b$

36. $36mn - 9m^2n^2$

37. $2\pi rh + 2\pi r^2$

38. $4\pi r^2 + 2\pi r$

Aha! **39.** $(a + b)(x - 3) + (a + b)(x + 4)$

40. $5c(a^3 + b) - (a^3 + b)$

41. $n^2 + 2n + np + 2p$

42. $x^2 + x + xy + y$

43. $2x^2 - 4x + xz - 2z$

44. $a^2 - 3a + ay - 3y$

45. $x^2 + y^2 + 2xy$

46. $3x^2 + 13xy - 10y^2$

47. $9c^2 - 6cd + d^2$

48. $4b^2 + a^2 - 4ab$

49. $7p^4 - 7q^4$

50. $16x^2 + 24xy + 9y^2$

51. $25z^2 + 10zy + y^2$

52. $4x^2y^2 + 12xyz + 9z^2$

53. $a^5 - 4a^4b - 5a^3b^2$

54. $a^4b^4 - 16$

55. $a^2 + ab + 2b^2$

56. $4p^2q - pq^2 + 4p^3$

57. $2mn - 360n^2 + m^2$

58. $3b^2 + 17ab - 6a^2$

59. $m^2n^2 - 4mn - 32$

60. $12 + x^2y^2 + 8xy$

61. $a^5b^2 + 3a^4b - 10a^3$

62. $p^2q^2 + 7pq + 6$

63. $8x^2 - 36x + 40$

64. $4ab^5 - 32b^4 + a^2b^6$

65. $2s^6t^2 + 10s^3t^3 + 12t^4$

66. $x^6 + x^5y - 2x^4y^2$

67. $a^2 + 2a^2bc + a^2b^2c^2$

68. $36a^2 - 15a + \frac{25}{16}$

69. $\frac{1}{81}x^2 - \frac{8}{27}x + \frac{16}{9}$

70. $\frac{1}{4}a^2 + \frac{1}{3}ab + \frac{1}{9}b^2$

71. $1 - 16x^{12}y^{12}$

72. $b^4a - 81a^5$

73. Kelly factored $16 - 8x + x^2$ as $(x - 4)^2$, while Tony factored it as $(4 - x)^2$. Are they both correct? Why or why not?

74. Describe in your own words a strategy for factoring polynomials.

SKILL MAINTENANCE

75. Show that the pairs $(-1, 11)$, $(0, 7)$, and $(3, -5)$ are solutions of $y = -4x + 7$.

Solve.

76. $5x - 4 = 0$

77. $3x + 7 = 0$

78. $2x + 9 = 0$

79. $4x - 9 = 0$

80. Graph: $y = -\frac{1}{2}x + 4$.

SYNTHESIS

81. There are third-degree polynomials in x that we are not yet able to factor, despite the fact that they are not prime. Explain how such a polynomial could be created.

82. Describe a method that could be used to find a binomial of degree 16 that can be expressed as the product of prime binomial factors.

Factor.

83. $-(x^5 + 7x^3 - 18x)$

84. $18 + a^3 - 9a - 2a^2$

85. $3a^4 - 15a^2 + 12$

86. $x^4 - 7x^2 - 18$

Aha! **87.** $y^2(y + 1) - 4y(y + 1) - 21(y + 1)$

88. $y^2(y - 1) - 2y(y - 1) + (y - 1)$

89. $6(x - 1)^2 + 7y(x - 1) - 3y^2$

90. $(y + 4)^2 + 2x(y + 4) + x^2$

91. $2(a + 3)^4 - (a + 3)^3(b - 2) - (a + 3)^2(b - 2)^2$

92. $5(t - 1)^5 - 6(t - 1)^4(s - 1) + (t - 1)^3(s - 1)^2$

COLLABORATIVE

CORNER

*Matching Factorizations**

Focus: Factoring

Time: 20 minutes

Group size: Begin with the entire class. The end result is pairs of students. If there is an odd number of students, the instructor should participate.

Materials: Prepared sheets of paper, pins or tape. On half of the sheets, the instructor writes a polynomial. On the remaining sheets, the instructor writes the factorization of those polynomials. The activity is more interesting if the polynomials and factorizations are similar; for example,

$$x^2 - 2x - 8, \quad (x - 2)(x - 4),$$
$$x^2 - 6x + 8, \quad (x - 1)(x - 8),$$
$$x^2 - 9x + 8, \quad (x + 2)(x - 4).$$

ACTIVITY

1. As class members enter the room, the instructor pins or tapes either a polynomial or a factorization to the back of each student. Class members are told only whether their sheet of paper contains a polynomial or a factorization. All students should remain quiet and not tell others what is on their sheet of paper.

2. After all students are wearing a sheet of paper, they should mingle with one another, attempting to match up their factorization with the appropriate polynomial or vice versa. They may ask "yes/no" questions of one another that relate to factoring and polynomials. Answers to the questions should be yes or no. For example, a legitimate question might be "Is my last term negative?", "Do my factors have opposite signs?", or "Does my factorization include the factors of 6?"

3. The game is over when all factorization/polynomial pairs have "found" one another.

*Thanks to Jann MacInnes for suggesting this activity.

Solving Quadratic Equations by Factoring	# 5.6
	The Principle of Zero Products • Factoring to Solve Equations • Graphing and Quadratic Equations

CONNECTING THE CONCEPTS

Chapter 4 and Sections 5.1–5.5 have been devoted to finding equivalent expressions. Whether we are adding, subtracting, multiplying, dividing, or factoring polynomials, the result of our work is an expression that is equivalent to the original expression.

Here in Section 5.6 we return to the task of solving equations. This time, however, the equations will contain a variable raised to a power greater than 1 and will often have more than one solution. Our ability to factor will play a pivotal role in solving such equations.

Second-degree equations like $4x^2 - 9 = 0$ and $x^2 + 6x + 5 = 0$ are said to be **quadratic**.

Quadratic Equation

A *quadratic equation* is an equation equivalent to one of the form

$$ax^2 + bx + c = 0, \quad \text{where } a \neq 0.$$

In order to solve quadratic equations, we need to develop a new principle.

The Principle of Zero Products

Suppose we are told that the product of two numbers is 6. On the basis of this information, it is impossible to know the value of either number—the product could be $2 \cdot 3$, $6 \cdot 1$, $12 \cdot \frac{1}{2}$, and so on. However, if we are told that the product of two numbers is 0, we know that at least one of the two numbers must itself be 0. For example, if $(x + 3)(x - 2) = 0$, we can conclude that either $x + 3$ is 0 or $x - 2$ is 0.

The Principle of Zero Products

An equation $AB = 0$ is true if and only if $A = 0$ or $B = 0$, or both. (A product is 0 if and only if at least one factor is 0.)

E x a m p l e 1

Solve: $(x + 3)(x - 2) = 0$.

Solution　We are told that the product of $x + 3$ and $x - 2$ is 0. In order for a product to be 0, at least one factor must be 0. We reason that either

$$x + 3 = 0 \quad or \quad x - 2 = 0. \qquad \text{Using the principle of zero products}$$

We solve each equation:

$$x + 3 = 0 \quad or \quad x - 2 = 0$$
$$x = -3 \quad or \quad x = 2.$$

Both -3 and 2 can be checked in the original equation.

Check:　For -3:

$$\frac{(x + 3)(x - 2) = 0}{\begin{array}{c} (-3 + 3)(-3 - 2) \overset{?}{\vphantom{|}} 0 \\ 0(-5) \\ 0 \end{array}} \Big| \; 0 \quad \text{TRUE}$$

For 2:

$$\frac{(x + 3)(x - 2) = 0}{\begin{array}{c} (2 + 3)(2 - 2) \overset{?}{\vphantom{|}} 0 \\ 5(0) \\ 0 \end{array}} \Big| \; 0 \quad \text{TRUE}$$

The solutions are -3 and 2.

When we are using the principle of zero products, the word "or" is meant to emphasize that any one of the factors could be the one that represents 0.

E x a m p l e 2

Solve: $(5x + 1)(x - 7) = 0$.

Solution　We have

$$(5x + 1)(x - 7) = 0$$
$$5x + 1 = 0 \quad or \quad x - 7 = 0 \qquad \text{Using the principle of zero products}$$
$$5x = -1 \quad or \quad x = 7 \qquad \text{Solving the two equations separately}$$
$$x = -\tfrac{1}{5} \quad or \quad x = 7.$$

Check:　For $-\tfrac{1}{5}$:

$$\frac{(5x + 1)(x - 7) = 0}{\begin{array}{c} \left(5\left(-\tfrac{1}{5}\right) + 1\right)\left(-\tfrac{1}{5} - 7\right) \overset{?}{\vphantom{|}} 0 \\ (-1 + 1)\left(-7\tfrac{1}{5}\right) \\ 0\left(-7\tfrac{1}{5}\right) \\ 0 \end{array}} \Big| \; 0 \quad \text{TRUE}$$

For 7:

$$\frac{(5x + 1)(x - 7) = 0}{\begin{array}{c} (5(7) + 1)(7 - 7) \overset{?}{\vphantom{|}} 0 \\ (35 + 1)0 \\ 0 \end{array}} \Big| \; 0 \quad \text{TRUE}$$

The solutions are $-\tfrac{1}{5}$ and 7.

The principle of zero products can be used whenever a product equals 0—even if a factor has only one term.

E x a m p l e   3

Solve: $3t(t - 5) = 0$.

Solution   We have

$$3t(t - 5) = 0 \qquad \text{The factors are } 3t \text{ and } t - 5.$$
$$3t = 0 \quad or \quad t - 5 = 0 \qquad \text{Using the principle of zero products}$$
$$t = 0 \quad or \qquad t = 5. \qquad \text{Solving the two equations separately}$$

The solutions are 0 and 5. The check is left to the student.

Factoring to Solve Equations

By factoring and using the principle of zero products, we can now solve a variety of quadratic equations.

E x a m p l e   4

Solve: $x^2 + 5x + 6 = 0$.

Solution   This equation differs from those solved in Chapter 2. There are no like terms to combine, and there is a squared term. We first factor the polynomial. Then we use the principle of zero products:

$$x^2 + 5x + 6 = 0$$
$$(x + 2)(x + 3) = 0 \qquad \text{Factoring}$$
$$x + 2 = 0 \quad or \quad x + 3 = 0 \qquad \text{Using the principle of zero products}$$
$$x = -2 \quad or \qquad x = -3.$$

Check: For -2:   For -3:

$$\begin{array}{c|c} x^2 + 5x + 6 = 0 \\ \hline (-2)^2 + 5(-2) + 6 \;?\; 0 \\ 4 - 10 + 6 \\ -6 + 6 \\ 0 \;\big|\; 0 \quad \text{TRUE} \end{array}$$

$$\begin{array}{c|c} x^2 + 5x + 6 = 0 \\ \hline (-3)^2 + 5(-3) + 6 \;?\; 0 \\ 9 - 15 + 6 \\ -6 + 6 \\ 0 \;\big|\; 0 \quad \text{TRUE} \end{array}$$

The solutions are -2 and -3.

The principle of zero products is used even if the factoring consists of only removing a common factor.

E x a m p l e   5

Solve: $x^2 + 7x = 0$.

Solution   Although there is no constant term, the equation is still quadratic. Thus the methods of Chapter 2 are not sufficient. We try factoring instead:

$$x^2 + 7x = 0$$
$$x(x + 7) = 0 \qquad \text{Removing the greatest common factor, } x$$
$$x = 0 \quad or \quad x + 7 = 0$$
$$x = 0 \quad or \qquad x = -7.$$

The solutions are 0 and -7. The check is left to the student.

> **Caution!**   We *must* have 0 on one side of the equation before the principle of zero products can be used. Get all nonzero terms on one side and 0 on the other.

Example 6

Solve: **(a)** $x^2 - 8x = -16$; **(b)** $4t^2 = 25$.

Solution

a) We first add 16 to get 0 on one side:

$$x^2 - 8x = -16$$

$$x^2 - 8x + 16 = 0 \qquad \text{Adding 16 to both sides to get 0 on one side}$$

$$(x - 4)(x - 4) = 0 \qquad \text{Factoring}$$

$$x - 4 = 0 \quad or \quad x - 4 = 0 \qquad \text{Using the principle of zero products}$$

$$x = 4 \quad or \qquad x = 4.$$

There is only one solution, 4. The check is left to the student.

b) We have

$$4t^2 = 25$$

$$4t^2 - 25 = 0 \qquad \text{Subtracting 25 from both sides to get 0 on one side}$$

$$(2t - 5)(2t + 5) = 0 \qquad \text{Factoring a difference of squares}$$

$$\left.\begin{array}{ll} 2t - 5 = 0 \quad or \quad 2t + 5 = 0 \\ 2t = 5 \quad or \qquad 2t = -5 \\ t = \frac{5}{2} \quad or \qquad t = -\frac{5}{2}. \end{array}\right\} \quad \begin{array}{l}\text{Solving the two equations}\\\text{separately}\end{array}$$

The solutions are $\frac{5}{2}$ and $-\frac{5}{2}$. The check is left to the student.

When solving quadratic equations by factoring, remember that a factorization is not useful unless 0 is on the other side of the equation.

Example 7

Solve: $(x + 3)(2x - 1) = 9$.

Solution   Be careful with an equation like this! Since we need 0 on one side, we multiply out the product on the left and then subtract 9 from both sides:

$$(x + 3)(2x - 1) = 9$$

$$2x^2 + 5x - 3 = 9 \qquad \text{Multiplying on the left}$$

$$2x^2 + 5x - 3 - 9 = 9 - 9 \qquad \begin{array}{l}\text{Subtracting 9 from both sides to}\\\text{get 0 on one side}\end{array}$$

$$2x^2 + 5x - 12 = 0$$

$$(2x - 3)(x + 4) = 0 \qquad \text{Factoring}$$

$$2x - 3 = 0 \quad or \quad x + 4 = 0 \qquad \text{Using the principle of zero products}$$

$$2x = 3 \quad or \qquad x = -4$$

$$x = \frac{3}{2} \quad or \qquad x = -4.$$

Check: For $\frac{3}{2}$:

$$(x + 3)(2x - 1) = 9$$

$$\frac{\left(\frac{3}{2} + 3\right)\left(2 \cdot \frac{3}{2} - 1\right) \; ? \; 9}{}$$

$$\left(\frac{9}{2}\right)(2)$$

$$9 \; \bigg| \; 9 \;\; \text{TRUE}$$

For -4:

$$(x + 3)(2x - 1) = 9$$

$$\frac{(-4 + 3)(2(-4) - 1) \; ? \; 9}{}$$

$$(-1)(-9)$$

$$9 \; \bigg| \; 9 \;\; \text{TRUE}$$

The solutions are $\frac{3}{2}$ and -4.

Graphing and Quadratic Equations

Recall from Chapter 3 that to find the x-intercept of a graph, we replace y with 0 and solve for x. This same procedure is used to find the x-intercepts of the graph of any equation of the form $y = ax^2 + bx + c$. Equations like this are graphed in Chapter 9. Their graphs are shaped like the following curves.

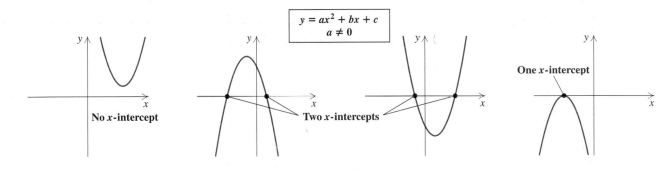

$$y = ax^2 + bx + c$$
$$a \neq 0$$

No x-intercept

Two x-intercepts

One x-intercept

E x a m p l e   8

Find the x-intercepts for the graph of the equation shown. (The grid is intentionally not included.)

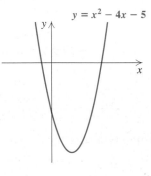

$$y = x^2 - 4x - 5$$

Solution   To find the x-intercepts, we let $y = 0$ and solve for x:

$$0 = x^2 - 4x - 5 \qquad \text{Substituting 0 for } y$$

$$0 = (x - 5)(x + 1) \qquad \text{Factoring}$$

$$x - 5 = 0 \;\; or \;\; x + 1 = 0 \qquad \text{Using the principle of zero products}$$

$$x = 5 \;\; or \;\;\;\;\; x = -1. \qquad \text{Solving for } x$$

The x-intercepts are $(5, 0)$ and $(-1, 0)$.

technology
connection

A grapher allows us to solve quadratic equations by locating any x-intercepts that might exist. This technique is especially useful when an equation cannot be solved by factoring. As an example, to determine the intercepts in the figure shown at right, we can utilize the ZERO or ROOT option of the CALC menu. This option requires you to enter an x-value slightly less than the x-intercept as a LEFT BOUND. An x-value slightly more than the x-intercept is then entered as a RIGHT BOUND. Finally, a GUESS value between the two bounds is entered and the x-intercept, or ZERO or ROOT, is displayed.

Use a grapher to find the solutions, if they exist, accurate to two decimal places.

1. $x^2 + 4x - 3 = 0$
2. $x^2 - 5x - 2 = 0$
3. $x^2 + 13.54x + 40.95 = 0$
4. $x^2 - 4.43x + 6.32 = 0$
5. $1.235x^2 - 3.409x = 0$

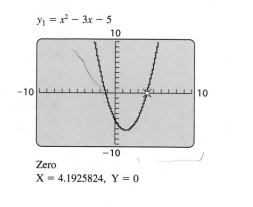

$y_1 = x^2 - 3x - 5$

Zero
X = 4.1925824, Y = 0

FOR EXTRA HELP

Exercise Set 5.6

Digital Video Tutor CD 3
Videotape 10

InterAct Math

Math Tutor Center

MathXL

MyMathLab.com

Solve using the principle of zero products.

1. $(x + 5)(x + 6) = 0$
2. $(x + 1)(x + 2) = 0$
3. $(x - 3)(x + 7) = 0$
4. $(x + 9)(x - 3) = 0$
5. $(2x - 9)(x + 4) = 0$
6. $(3x - 5)(x + 1) = 0$
7. $(10x - 9)(4x + 7) = 0$
8. $(2x - 7)(3x + 4) = 0$
9. $x(x + 6) = 0$
10. $t(t + 9) = 0$
11. $\left(\frac{2}{3}x - \frac{12}{11}\right)\left(\frac{7}{4}x - \frac{1}{12}\right) = 0$
12. $\left(\frac{1}{9} - 3x\right)\left(\frac{1}{5} + 2x\right) = 0$
13. $5x(2x + 9) = 0$
14. $12x(4x + 5) = 0$
15. $(20 - 0.4x)(7 - 0.1x) = 0$
16. $(1 - 0.05x)(1 - 0.3x) = 0$

Solve by factoring and using the principle of zero products.

17. $x^2 + 7x + 6 = 0$
18. $x^2 - 6x + 5 = 0$
19. $x^2 - 4x - 21 = 0$
20. $x^2 - 7x - 18 = 0$
21. $x^2 + 9x + 14 = 0$
22. $x^2 + 8x + 15 = 0$
23. $x^2 - 6x = 0$
24. $x^2 + 8x = 0$

25. $7t + t^2 = 0$

26. $4t + t^2 = 0$

27. $9x^2 = 4$

28. $4x^2 = 25$

29. $0 = 25 + x^2 + 10x$

30. $0 = 6x + x^2 + 9$

31. $1 + x^2 = 2x$

32. $x^2 + 16 = 8x$

33. $8x^2 = 5x$

34. $3x^2 = 7x$

35. $3x^2 - 7x = 20$

36. $6x^2 - 4x = 10$

37. $2y^2 + 12y = -10$

38. $12y^2 - 5y = 2$

39. $(x - 7)(x + 1) = -16$

40. $(x + 2)(x - 7) = -18$

41. $y(3y + 1) = 2$

42. $t(t - 5) = 14$

43. $81x^2 - 5 = 20$

44. $36m^2 - 9 = 40$

45. $(x - 1)(5x + 4) = 2$

46. $(x + 3)(3x + 5) = 7$

47. $x^2 - 2x = 18 + 5x$

48. $3x^2 - 2x = 9 - 8x$

49. $(6a + 1)(a + 1) = 21$

50. $(2t + 1)(4t - 1) = 14$

51. Use the following graph to solve $x^2 - 3x - 4 = 0$.

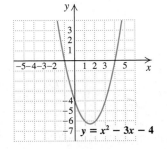

52. Use the following graph to solve $x^2 + x - 6 = 0$.

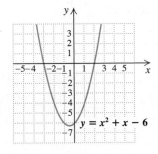

53. Use the following graph to solve $-x^2 + 2x + 3 = 0$.

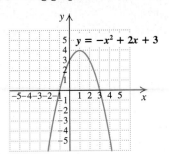

54. Use the following graph to solve $-x^2 - x + 6 = 0$.

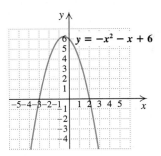

Find the x-intercepts for the graph of each equation.
Grids are intentionally not included.

55. $y = x^2 + 3x - 4$

56. $y = x^2 - x - 6$

57. $y = x^2 - 2x - 15$

58. $y = x^2 + 2x - 8$

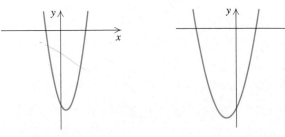

59. $y = 2x^2 + x - 10$ **60.** $y = 2x^2 + 3x - 9$

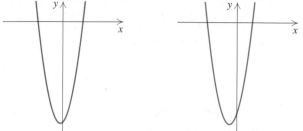

61. The equation $x^2 + 1$ has no real-number solutions. What implications does this have for the graph of $y = x^2 + 1$?

62. What is the difference between a quadratic polynomial and a quadratic equation?

SKILL MAINTENANCE

Translate to an algebraic expression.

63. The square of the sum of a and b

64. The sum of the squares of a and b

65. The sum of two consecutive integers

Translate to an inequality.

66. 5 more than twice a number is less than 19.

67. 7 less than half of a number exceeds 24.

68. 3 less than a number is at least 34.

SYNTHESIS

69. When the principle of zero products is used to solve a quadratic equation, will there always be two solutions? Why or why not?

70. What is wrong with solving $x^2 = 3x$ by dividing both sides of the equation by x?

Solve.

71. $(2x - 5)(x + 7)(3x + 8) = 0$

72. $(4x + 9)(3x - 2)(5x + 1) = 0$

73. Find an equation with integer coefficients that has the given numbers as solutions. For example, 3 and -2 are solutions to $x^2 - x - 6 = 0$.

a) $-4, 5$ **b)** $-1, 7$ **c)** $\frac{1}{4}, 3$

d) $\frac{1}{2}, \frac{1}{3}$ **e)** $\frac{2}{3}, \frac{3}{4}$ **f)** $-1, 2, 3$

Solve.

74. $16(x - 1) = x(x + 8)$

75. $a(9 + a) = 4(2a + 5)$

76. $(t - 5)^2 = 2(5 - t)$

77. $x^2 - \frac{9}{25} = 0$

78. $x^2 - \frac{25}{36} = 0$

Aha! **79.** $(t + 1)^2 = 9$

80. $\frac{27}{25}x^2 = \frac{1}{3}$

81. For each equation on the left, find an equivalent equation on the right.

a) $x^2 + 10x - 2 = 0$ $4x^2 + 8x + 36 = 0$

b) $(x - 6)(x + 3) = 0$ $(2x + 8)(2x - 5) = 0$

c) $5x^2 - 5 = 0$ $9x^2 - 12x + 24 = 0$

d) $(2x - 5)(x + 4) = 0$ $(x + 1)(5x - 5) = 0$

e) $x^2 + 2x + 9 = 0$ $x^2 - 3x - 18 = 0$

f) $3x^2 - 4x + 8 = 0$ $2x^2 + 20x - 4 = 0$

82. Explain how to construct an equation that has seven solutions.

83. Explain how the graph in Exercise 55 can be used to visualize the solutions of
$$x^2 + 3x - 4 = -6.$$

Use a grapher to find the solutions of each equation. Round solutions to the nearest hundredth.

84. $x^2 + 1.80x - 5.69 = 0$

85. $x^2 - 9.10x + 15.77 = 0$

86. $-x^2 + 0.63x + 0.22 = 0$

87. $x^2 + 13.74x + 42.00 = 0$

88. $6.4x^2 - 8.45x - 94.06 = 0$

89. $0.84x^2 - 2.30x = 0$

90. $1.23x^2 + 4.63x = 0$

5.7

Solving Applications

Applications • The Pythagorean Theorem

Applications

We can use the five-step problem-solving process and our new methods for solving quadratic equations to solve new types of problems.

Example 1

Study Tip

The best way to prepare for a final exam is to do so over a period of at least two weeks. First review each chapter, studying the formulas, problems, properties, and procedures in the chapter Summary and Review. Then retake your quizzes and tests. If you miss any questions, spend extra time reviewing the corresponding topics. Watch the videotapes that accompany the text or use the InterAct Math Tutorial Software. Also consider participating in a study group or attending a tutoring or review session.

Page numbers.   The product of the page numbers on two consecutive pages of a book is 156. Find the page numbers.

Solution

1. **Familiarize.**   Consecutive page numbers are one apart, like 49 and 50. Let $x =$ the first page number; then $x + 1 =$ the next page number.

2. **Translate.**   We reword the problem before translating:

    Rewording: The first page number times the next page number is 156.

    Translating: x $\cdot$ $(x + 1)$ $= 156$

3. **Carry out.**   We solve the equation as follows:

    $$x(x + 1) = 156$$
    $$x^2 + x = 156 \qquad \text{Multiplying}$$
    $$x^2 + x - 156 = 0 \qquad \text{Subtracting 156 to get 0 on one side}$$
    $$(x - 12)(x + 13) = 0 \qquad \text{Factoring}$$
    $$x - 12 = 0 \quad or \quad x + 13 = 0 \qquad \text{Using the principle of zero products}$$
    $$x = 12 \quad or \qquad x = -13. \qquad \text{Solving each equation}$$

4. **Check.**   The solutions of the equation are 12 and -13. Since page numbers cannot be negative, -13 must be rejected. On the other hand, if x is 12, then $x + 1$ is 13 and $12 \cdot 13 = 156$. Thus, 12 checks.

5. **State.**   The page numbers are 12 and 13.

Example 2

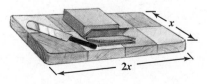

Manufacturing.   Wooden Work, Ltd., builds cutting boards that are twice as long as they are wide. The most popular board that Wooden Work makes has an area of 800 cm². What are the dimensions of the board?

Solution

1. **Familiarize.**   We first make a drawing. Recall that the area of any rectangle is Length · Width. We let $x =$ the width of the board, in centimeters. The length is then $2x$.

2. **Translate.** We reword and translate as follows:

Rewording: The area of the rectangle is 800 cm².

Translating: $2x \cdot x = 800$

3. **Carry out.** We solve the equation as follows:

$$2x \cdot x = 800$$
$$2x^2 = 800$$
$$2x^2 - 800 = 0 \qquad \text{Subtracting 800 to get 0 on one side of the equation}$$
$$2(x^2 - 400) = 0 \qquad \text{Removing a common factor of 2}$$
$$2(x - 20)(x + 20) = 0 \qquad \text{Factoring a difference of squares}$$
$$(x - 20)(x + 20) = 0 \qquad \text{Dividing both sides by 2}$$
$$x - 20 = 0 \quad or \quad x + 20 = 0 \qquad \text{Using the principle of zero products}$$
$$x = 20 \quad or \quad x = -20. \qquad \text{Solving each equation}$$

4. **Check.** The solutions of the equation are 20 and -20. Since the width must be positive, -20 cannot be a solution. To check 20 cm, we note that if the width is 20 cm, then the length is $2 \cdot 20\,\text{cm} = 40\,\text{cm}$ and the area is $20\,\text{cm} \cdot 40\,\text{cm} = 800\,\text{cm}^2$. Thus the solution 20 checks.

5. **State.** The cutting board is 20 cm wide and 40 cm long.

E x a m p l e 3

Dimensions of a sail. The mainsail of Stacey's Lightning-styled sailboat has an area of 125 ft². If the sail is 15 ft taller than it is wide, find the height and the width of the sail.

Solution

1. **Familiarize.** We first make a drawing. The formula for the area of a triangle is Area $= \frac{1}{2} \cdot$ (base) $\cdot$ (height). We let $b =$ the width, in feet, of the triangle's base and $b + 15 =$ the height, in feet.

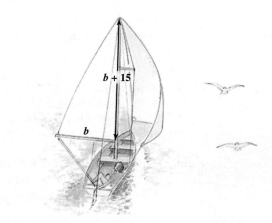

2. **Translate.** We reword and translate as follows:

Rewording: The area of the sail is 125 ft².

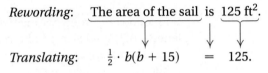

Translating: $\frac{1}{2} \cdot b(b + 15)$ = 125.

3. **Carry out.** We solve the equation as follows:

$$\frac{1}{2} \cdot b \cdot (b + 15) = 125$$

$$\frac{1}{2}(b^2 + 15b) = 125 \qquad \text{Multiplying}$$

$$b^2 + 15b = 250 \qquad \text{Multiplying by 2 to clear fractions}$$

$$b^2 + 15b - 250 = 0 \qquad \text{Subtracting 250 to get 0 on one side}$$

$$(b + 25)(b - 10) = 0 \qquad \text{Factoring}$$

$$b + 25 = 0 \quad or \quad b - 10 = 0 \quad \text{Using the principle of zero products}$$

$$b = -25 \quad or \qquad b = 10.$$

4. **Check.** The width must be positive, so -25 cannot be a solution. Suppose the base is 10 ft. The height would be $10 + 15$, or 25 ft, and the area $\frac{1}{2}(10)(25)$, or 125 ft². These numbers check in the original problem.

5. **State.** Stacey's mainsail is 25 ft tall and 10 ft wide.

E x a m p l e   4

Games in a league's schedule. In a sports league of x teams in which all teams play each other twice, the total number N of games played is given by

$$x^2 - x = N.$$

The Colchester Youth Soccer League plays a total of 240 games, with all teams playing each other twice. How many teams are in the league?

Solution

1. **Familiarize.** To familiarize yourself with this equation, reread Example 7 in Section 4.2, where we first considered it.

2. **Translate.** We are trying to find the number of teams x in a league in which 240 games are played and all teams play each other twice. We replace N with 240 in the formula above:

$$x^2 - x = 240. \qquad \text{Substituting 240 for } N. \text{ This is now an equation in one variable.}$$

3. **Carry out.** We solve the equation as follows:

$$x^2 - x = 240$$

$$x^2 - x - 240 = 0 \qquad \text{Subtracting 240 to get 0 on one side}$$

$$(x - 16)(x + 15) = 0 \qquad \text{Factoring}$$

$$x - 16 = 0 \quad or \quad x + 15 = 0 \quad \text{Using the principle of zero products}$$

$$x = 16 \quad or \qquad x = -15.$$

4. **Check.** Since the number of teams must be positive, -15 cannot be a solution. However, 16 checks, since $16^2 - 16 = 256 - 16 = 240$.

5. **State.** There are 16 teams in the league.

The Pythagorean Theorem

The following problems involve the Pythagorean theorem, which relates the lengths of the sides of a *right* triangle. A triangle is a **right triangle** if it has a 90°, or *right*, angle. The side opposite the 90° angle is called the **hypotenuse**. The other sides are called **legs**.

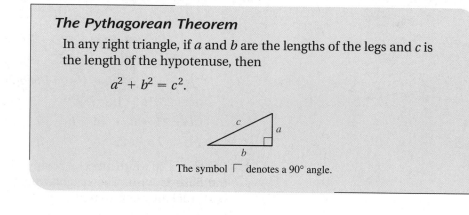

The Pythagorean Theorem

In any right triangle, if a and b are the lengths of the legs and c is the length of the hypotenuse, then

$$a^2 + b^2 = c^2.$$

The symbol ⌐ denotes a 90° angle.

E x a m p l e   5

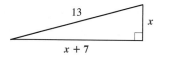

Right triangle geometry.   One leg of a right triangle is 7 m longer than the other. The length of the hypotenuse is 13 m. Find the lengths of the legs.

Solution

1. **Familiarize.**   We make a drawing and let $x =$ the length of one leg, in meters. Since the other leg is 7 m longer, we know that $x + 7 =$ the length of the other leg, in meters. The hypotenuse has length 13 m.

2. **Translate.**   Applying the Pythagorean theorem, we obtain the following translation:

$$a^2 + b^2 = c^2$$
$$x^2 + (x + 7)^2 = 13^2. \qquad \text{Substituting}$$

3. **Carry out.**   We solve the equation as follows:

$x^2 + (x^2 + 14x + 49) = 169$	Squaring the binomial and 13
$2x^2 + 14x + 49 = 169$	Combining like terms
$2x^2 + 14x - 120 = 0$	Subtracting 169 to get 0 on one side
$2(x^2 + 7x - 60) = 0$	Factoring out a common factor
$2(x + 12)(x - 5) = 0$	Factoring
$x + 12 = 0 \quad or \quad x - 5 = 0$	Using the principle of zero products
$x = -12 \quad or \qquad x = 5.$	

4. **Check.**   The integer -12 cannot be a length of a side because it is negative. When $x = 5$, $x + 7 = 12$, and $5^2 + 12^2 = 13^2$. So 5 checks.

5. **State.**   The lengths of the legs are 5 m and 12 m.

E x a m p l e   6

Roadway design. Elliott Street is 24 ft wide when it ends at Main Street in Brattleboro, Vermont. A 40-ft long diagonal crosswalk allows pedestrians to cross Main Street to or from either corner of Elliott Street (see the figure). Determine the width of Main Street.

Solution

1. **Familiarize.** A drawing has already been provided, but we can redraw and label the relevant part.

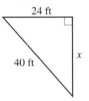

   Note that the two streets intersect at a right angle. We let x = the width of Main Street, in feet.

2. **Translate.** Since a right triangle is formed, we can use the Pythagorean theorem:

   $$a^2 + b^2 = c^2$$
   $$x^2 + 24^2 = 40^2. \quad \text{Substituting}$$

3. **Carry out.** We solve the equation as follows:

   $$x^2 + 576 = 1600 \qquad \text{Squaring 24 and 40}$$
   $$x^2 - 1024 = 0 \qquad \text{Subtracting 1600 from both sides}$$
   $$(x - 32)(x + 32) = 0 \qquad \text{Note that } 1024 = 32^2. \text{ A calculator might}$$
   $$\text{be helpful here.}$$

   $$x - 32 = 0 \quad or \quad x + 32 = 0 \quad \text{Using the principle of zero products}$$
   $$x = 32 \quad or \qquad x = -32.$$

4. **Check.** Since the width of a street must be positive, -32 is not a solution. If the width is 32 ft, we have $32^2 + 24^2 = 1024 + 576 = 1600$, which is 40^2. Thus, 32 checks.

5. **State.** The width of Main Street is 32 ft.

FOR EXTRA HELP

Exercise Set 5.7

Digital Video Tutor CD 3 Videotape 10   InterAct Math   Math Tutor Center   MathXL   MyMathLab.com

Solve. Use the five-step problem-solving approach.

1. A number is 6 less than its square. Find all such numbers.

2. A number is 2 less than its square. Find all such numbers.

3. One leg of a right triangle is 3 cm longer than the other leg. The length of the hypotenuse is 15 cm. Find the length of each side.

4. One leg of a right triangle is 2 cm shorter than the other leg. The length of the hypotenuse is 10 cm. Find the length of each side.

5. *Page numbers.* The product of the page numbers on two facing pages of a book is 110. Find the page numbers.

6. *Page numbers.* The product of the page numbers on two facing pages of a book is 210. Find the page numbers.

7. The product of two consecutive odd integers is 255. Find the integers.

8. The product of two consecutive even integers is 224. Find the integers.

9. *Framing.* A rectangular picture frame is twice as long as it is wide. If the area of the frame is 288 in², find its dimensions.

w

$2w$

10. *Furnishings.* A rectangular table in Arlo's House of Tunes is six times as long as it is wide. If the area of the table is 24 ft², find the length and the width of the table.

w $6w$

11. *Design.* The keypad and viewing window of the TI83 graphing calculator is rectangular. The length of the rectangle is 2 cm more than twice the width. If the area of the rectangle is 144 cm², find the length and the width.

w

$2w + 2$

12. *Area of a garden.* The length of a rectangular garden is 4 m greater than the width. The area of the garden is 96 m². Find the length and the width.

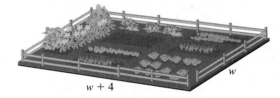

$w + 4$ w

13. *Dimensions of a triangle.* A triangle is 10 cm wider than it is tall. The area is 28 cm². Find the height and the base.

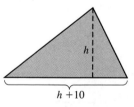

h

$h + 10$

14. *Dimensions of a triangle.* The height of a triangle is 3 cm less than the length of the base. If the area of the triangle is 35 cm², find the height and the length of the base.

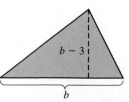

$b - 3$

b

15. *Dimensions of a sail.* The height of the jib sail on a Lightning sailboat is 5 ft greater than the length of its "foot." If the area of the sail is 42 ft², find the length of the foot and the height of the sail.

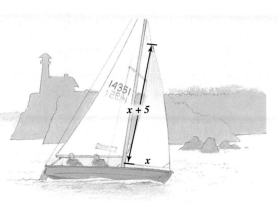

16. *Road design.* A triangular traffic island has a base half as long as its height. Find the base and the height if the island has an area of 64 m².

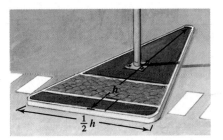

17. *Physical education.* An outdoor-education ropes course includes a cable that slopes downward from a height of 37 ft to a height of 30 ft. The trees that the cable connects are 24 ft apart. How long is the cable?

18. *Aviation.* Engine failure forced Geraldine to pilot her Cessna 150 to an emergency landing. To land, Geraldine's plane glided 17,000 ft over a 15,000-ft stretch of deserted highway. From what altitude did the descent begin?

Games in a league. Use the formula from Example 4, $x^2 - x = N$, for Exercises 19–22. Assume that in each league teams play each other twice.

19. A women's volleyball league has 20 teams. What is the total number of games to be played?

20. A chess league has 14 teams. What is the total number of games to be played?

21. A women's softball league plays a total of 132 games. How many teams are in the league?

22. A basketball league plays a total of 90 games. How many teams are in the league?

23. *Construction.* The diagonal braces in a lookout tower are 15 ft long and span a distance of 12 ft. How high does each brace reach vertically?

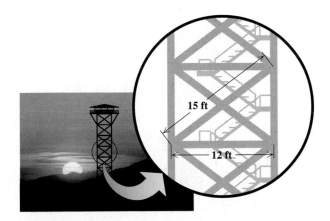

24. *Reach of a ladder.* Twyla has a 26-ft ladder leaning against her house. If the bottom of the ladder is 10 ft from the base of the house, how high does the ladder reach?

Number of handshakes. The number of possible handshakes H within a group of n people is given by $H = \frac{1}{2}(n^2 - n)$. Use this formula for Exercises 25–28.

25. At a meeting, there are 15 people. How many handshakes are possible?

26. At a party, there are 30 people. How many handshakes are possible?

27. *High-fives.* After winning the championship, all Los Angeles Laker teammates exchanged "high-fives." Altogether there were 66 high-fives. How many players were there?

28. *Toasting.* During a toast at a party, there were 190 "clicks" of glasses. How many people took part in the toast?

29. *Architecture.* An architect has allocated a rectangular space of 264 ft² for a square dining room and a 10-ft wide kitchen, as shown in the figure. Find the dimensions of each room.

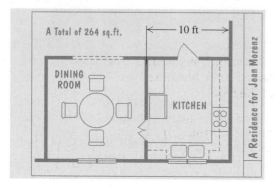

30. *Guy wire.* The guy wire on a TV antenna is 1 m longer than the height of the antenna. If the guy wire is anchored 3 m from the foot of the antenna, how tall is the antenna?

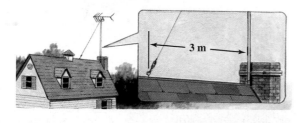

Height of a rocket. *For Exercises 31–34, assume that a water rocket is launched upward with an initial velocity of 48 ft/sec. Its height h, in feet, after t seconds, is given by* $h = 48t - 16t^2$.

31. Determine the height of the rocket $\frac{1}{2}$ sec after it has been launched.

32. Determine the height 1.5 sec after the rocket has been launched.

33. When will the rocket be exactly 32 ft above the ground?

34. When will the rocket crash into the ground?

35. Write a problem for a classmate to solve such that only one of two solutions of a quadratic equation can be used as an answer.

36. Can we solve any problem that translates to a quadratic equation? Why or why not?

SKILL MAINTENANCE

Simplify.

37. $-\dfrac{2}{3} \cdot \dfrac{4}{7}$

38. $-\dfrac{4}{5} \cdot \dfrac{2}{9}$

39. $\dfrac{5}{6}\left(\dfrac{-7}{9}\right)$

40. $\dfrac{3}{8}\left(-\dfrac{5}{6}\right)$

41. $-\dfrac{2}{3} + \dfrac{4}{7}$

42. $-\dfrac{4}{5} + \dfrac{2}{9}$

43. $\dfrac{5}{6} + \dfrac{-7}{9}$

44. $\dfrac{3}{8} + \left(-\dfrac{5}{6}\right)$

SYNTHESIS

The converse of the Pythagorean theorem is also true. That is, if $a^2 + b^2 = c^2$, *then the triangle is a right triangle (where a and b are the lengths of the legs and c is the length of the hypotenuse). Use this result to answer Exercises 45 and 46.*

45. An archaeologist has measuring sticks of 3 ft, 4 ft, and 5 ft. Explain how she could draw a 7-ft by 9-ft rectangle on a piece of land being excavated.

Aha!

46. Explain how measuring sticks of 5 cm, 12 cm, and 13 cm can be used to draw a right triangle that has two 45° angles.

47. *Telephone service.* Use the information in the figure below to determine the height of the telephone pole.

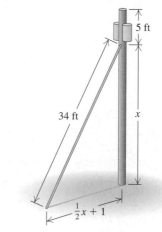

48. *Sailing.* The mainsail of a Lightning sailboat is a right triangle in which the hypotenuse is called the leech. If a 24-ft tall mainsail has a leech length of 26 ft and if Dacron® sailcloth costs $10 per square foot, find the cost of a new mainsail.

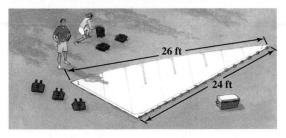

49. *Roofing.* A *square* of shingles covers 100 ft² of surface area. How many squares will be needed to reshingle the house shown?

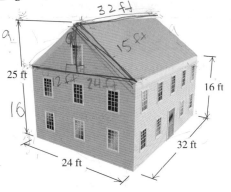

50. Solve for *x*.

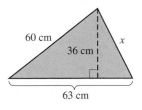

51. The ones digit of a number less than 100 is 4 greater than the tens digit. The sum of the number and the product of the digits is 58. Find the number.

52. *Pool sidewalk.* A cement walk of uniform width is built around a 20-ft by 40-ft rectangular pool. The total area of the pool and the walk is 1500 ft². Find the width of the walk.

53. *Dimensions of an open box.* A rectangular piece of cardboard is twice as long as it is wide. A 4-cm square is cut out of each corner, and the sides are turned up to make a box with an open top. The volume of the box is 616 cm³. Find the original dimensions of the cardboard.

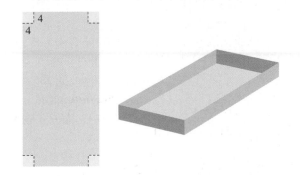

54. *Dimensions of a closed box.* The total surface area of a closed box is 350 m². The box is 9 m high and has a square base and lid. Find the length of a side of the base.

55. *Rain-gutter design.* An open rectangular gutter is made by turning up the sides of a piece of metal 20 in. wide. The area of the cross-section of the gutter is 48 in². Find the possible depths of the gutter.

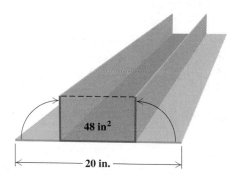

56. The length of each side of a square is increased by 5 cm to form a new square. The area of the new square is $2\frac{1}{4}$ times the area of the original square. Find the area of each square.

57. Find a polynomial for the shaded area in the figure below.

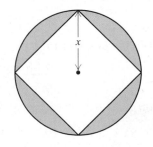

Summary and Review 5

Key Terms

Factor, p. 275

Common factor, p. 275

Factoring by grouping, p. 278

Prime polynomial, p. 285

Factor completely, p. 285

Perfect-square trinomial, p. 298

Difference of squares, p. 300

Quadratic equation, p. 311

Right triangle, p. 322

Hypotenuse, p. 322

Leg, p. 322

Important Properties and Formulas

Factoring Formulas

$A^2 + 2AB + B^2 = (A + B)^2;$
$A^2 - 2AB + B^2 = (A - B)^2;$
$A^2 - B^2 = (A + B)(A - B)$

To factor a polynomial:

A. Look first for a common factor. If there is one, factor out the largest common factor. Be sure to include it in your final answer.

B. Look at the number of terms.

   Two terms: If you have a difference of squares, factor accordingly.

   Three terms: If you have a perfect-square trinomial, factor accordingly. If not, try trial and error, using the standard method or grouping.

Four terms: Try factoring by grouping.

C. Always factor completely.

D. Check by multiplying.

The Principle of Zero Products: $AB = 0$ is true if and only if $A = 0$ or $B = 0$, or both.

The Pythagorean theorem: $a^2 + b^2 = c^2$

This indicates 90°.

Review Exercises

Find three factorizations of each monomial.

1. $36x^3$

2. $-20x^5$

Factor completely. If a polynomial is prime, state this.

3. $2x^4 + 6x^3$

4. $a^2 - 7a$

5. $4t^2 - 9$

6. $x^2 + 4x - 12$

7. $x^2 + 14x + 49$

8. $12x^3 + 12x^2 + 3x$

9. $6x^3 + 9x^2 + 2x + 3$

10. $6t^2 - 5t - 1$

11. $81a^4 - 1$

12. $9x^3 + 12x^2 - 45x$

13. $2x^2 - 50$

14. $x^4 + 4x^3 - 2x - 8$

15. $a^2b^4 - 36$

16. $8x^6 - 32x^5 + 4x^4$

17. $75 + 12x^2 + 60x$

18. $a^2 + 4$

19. $x^3 - x^2 - 30x$

20. $4x^2 - 25$

21. $9x^2 + 25 - 30x$

22. $6x^2 - 28x - 48$

23. $4t^2 - 13t + 10$

24. $2t^2 - 7t - 4$

25. $18x^2 - 12x + 2$

26. $3x^2 - 27$

27. $15 - 8x + x^2$

28. $25x^2 - 20x + 4$

29. $x^2y^2 + xy - 12$

30. $12a^2 + 84ab + 147b^2$

31. $m^2 + 5m + mt + 5t$

32. $32x^4 - 128y^4z^4$

Solve.

33. $(x - 1)(x + 3) = 0$

34. $x^2 + 2x - 35 = 0$

35. $9x^2 = 1$

36. $3x^2 + 2 = 5x$

37. $2x^2 + 5x = 12$

38. $(x + 1)(x - 2) = 4$

39. The square of a number is 12 more than the number. Find all such numbers.

40. Find the x-intercepts for the graph of
$$y = 2x^2 - 3x - 5.$$

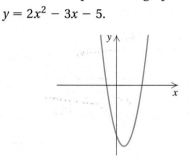

41. A triangular sign is as wide as it is tall. Its area is 800 cm². Find the height and the base.

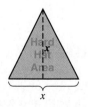

42. A guy wire from a ham radio antenna is 26 m long. It reaches from the top of the antenna to a point on the ground 10 m from the base of the antenna. How tall is the antenna?

SYNTHESIS

43. On a quiz, Edith writes the factorization of $4x^2 - 100$ as $(2x - 10)(2x + 10)$. If this were a 10-point question, how many points would you give Edith? Why?

44. How do the equations solved in this chapter differ from those solved in previous chapters?

Solve.

45. The pages of a book measure 15 cm by 20 cm. Margins of equal width surround the printing on each page and constitute one half of the area of the page. Find the width of the margins.

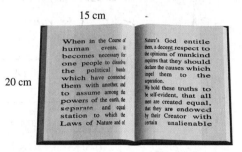

46. The cube of a number is the same as twice the square of the number. Find the number.

47. The length of a rectangle is two times its width. When the length is increased by 20 cm and the width is decreased by 1 cm, the area is 160 cm². Find the original length and width.

Solve.

48. $(x - 2)2x^2 + x(x - 2) - (x - 2)15 = 0$

Aha! **49.** $x^2 + 25 = 0$

Chapter Test 5

1. Find three factorizations of $8x^4$.

Factor completely.

2. $x^2 - 7x + 10$

3. $x^2 + 25 - 10x$

4. $6y^2 - 8y^3 + 4y^4$

5. $x^3 + x^2 + 2x + 2$

6. $x^2 - 5x$

7. $x^3 + 2x^2 - 3x$

8. $28x - 48 + 10x^2$

9. $4x^2 - 9$

10. $x^2 - x - 12$

11. $6m^3 + 9m^2 + 3m$

12. $3w^2 - 75$

13. $60x + 45x^2 + 20$

14. $3x^4 - 48$

15. $49x^2 - 84x + 36$

16. $5x^2 - 26x + 5$

17. $x^4 + 2x^3 - 3x - 6$

18. $80 - 5x^4$

19. $4x^2 - 4x - 15$

20. $6t^3 + 9t^2 - 15t$

21. $3m^2 - 9mn - 30n^2$

Solve.

22. $x^2 - x - 20 = 0$

23. $2x^2 + 7x = 15$

24. $x(x - 3) = 28$

25. Find the x-intercepts for the graph of $y = 3x^2 - 4x - 7$.

26. The length of a rectangle is 2 m more than the width. The area of the rectangle is 48 m^2. Find the length and the width.

27. A mason wants to be sure she has a right corner in a building's foundation. She marks a point 3 ft from the corner along one wall and another point 4 ft from the corner along the other wall. If the corner is a right angle, what should the distance be between the two marked points?

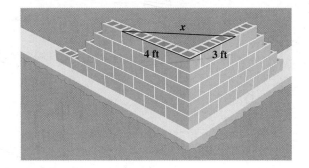

SYNTHESIS

28. The length of a rectangle is five times its width. When the length is decreased by 3 and the width is increased by 2, the area of the new rectangle is 60. Find the original length and width.

29. Factor: $(a + 3)^2 - 2(a + 3) - 35$.

30. Solve: $20x(x + 2)(x - 1) = 5x^3 - 24x - 14x^2$.

6
Rational Expressions and Equations

AN APPLICATION

In South Africa, the design of every woven handbag, or *gipatsi*, is created by repeating two or more patterns around the bag. If a weaver uses a four-strand, a six-strand, and an eight-strand pattern, what is the smallest number of strands needed in order for all three patterns to repeat a whole number of times?

This problem appears as Exercise 80 in Section 6.3.

*W*eavers use math when designing patterns, from calculating the size of the pattern and number of times it repeats to figuring the number of threads per inch and the weight of the thread. I also use math to convert between English and metric units when dyeing warps and woofs.

ESTELLE CARLSON
Handweaver and Designer
Los Angeles, California

*J*ust as fractions are needed to solve certain arithmetic problems, rational expressions *similar to those in the following pages are needed to solve certain algebra problems. We now learn how to simplify, as well as add, subtract, multiply, and divide, rational expressions. These skills will then be used to solve the equations that arise from real-life problems like the one on the preceding page.*

Rational Expressions

6.1

Simplifying Rational Expressions • Factors That Are Opposites

Just as a rational number is a quotient of two integers, a **rational expression** is a quotient of two polynomials. The following are examples of rational expressions:

$$\frac{7}{3}, \quad \frac{5}{x+6}, \quad \frac{t^2 - 5t + 6}{4t^2 - 7}.$$

Rational expressions are examples of *algebraic fractions*. They are also examples of *fractional expressions*.

Because rational expressions indicate division, we must be careful to avoid denominators that are 0. When a variable is replaced with a number that produces a denominator of 0, the rational expression is undefined. For example, in the expression

$$\frac{x+3}{x-7},$$

when x is replaced with 7, the denominator is 0, and the expression is undefined:

$$\frac{x+3}{x-7} = \frac{7+3}{7-7} = \frac{10}{0}. \longleftarrow \text{Division by 0 is undefined.}$$

When x is replaced with a number other than 7—say, 6—the expression *is* defined because the denominator is not zero:

$$\frac{x+3}{x-7} = \frac{6+3}{6-7} = \frac{9}{-1} = -9.$$

E x a m p l e 1 Find all numbers for which the rational expression

$$\frac{x+4}{x^2 - 3x - 10}$$

is undefined.

Solution The value of the numerator has no bearing on whether or not a rational expression is defined. To determine which numbers make the rational expression undefined, we set the *denominator* equal to 0 and solve:

$$x^2 - 3x - 10 = 0$$
$$(x - 5)(x + 2) = 0 \qquad \text{Factoring}$$
$$x - 5 = 0 \quad or \quad x + 2 = 0 \qquad \text{Using the principle of zero products}$$
$$x = 5 \quad or \qquad x = -2. \qquad \text{Solving each equation}$$

Check:

For $x = 5$:

$$\frac{x + 4}{x^2 - 3x - 10} = \frac{5 + 4}{5^2 - 3 \cdot 5 - 10} \qquad \begin{array}{l}\text{There are no restrictions on}\\ \text{the numerator.}\end{array}$$

$$= \frac{9}{25 - 15 - 10} = \frac{9}{0}. \qquad \text{This expression is undefined.}$$

For $x = -2$:

$$\frac{x + 4}{x^2 - 3x - 10} = \frac{-2 + 4}{(-2)^2 - 3(-2) - 10}$$

$$= \frac{2}{4 + 6 - 10} = \frac{2}{0}. \qquad \text{This expression is undefined.}$$

Thus, $\dfrac{x + 4}{x^2 - 3x - 10}$ is undefined for $x = 5$ and $x = -2$.

technology connection

To check Example 1 with a grapher, let $y_1 = x^2 - 3x - 10$ and $y_2 = (x + 4)/(x^2 - 3x - 10)$ or $(x + 4)/y_1$ and use the TABLE feature. Since $x^2 - 3x - 10$ is 0 for $x = -2$, we cannot evaluate y_2 for $x = -2$.

TBL MIN = −2 ΔTBL = 1

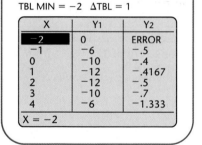

X	Y1	Y2
−2	0	ERROR
−1	−6	−.5
0	−10	−.4
1	−12	−.4167
2	−12	−.5
3	−10	−.7
4	−6	−1.333

X = −2

Simplifying Rational Expressions

Simplifying rational expressions is similar to simplifying the fractional expressions studied in Section 1.3. We saw, for example, that an expression like $\frac{15}{40}$ can be simplified as follows:

$$\frac{15}{40} = \frac{3 \cdot 5}{8 \cdot 5} \qquad \begin{array}{l}\text{Factoring the numerator and the denominator.}\\ \text{Note the common factor, 5.}\end{array}$$

$$= \frac{3}{8} \cdot \frac{5}{5} \qquad \text{Rewriting as a product of two fractions}$$

$$= \frac{3}{8} \cdot 1 \qquad \frac{5}{5} = 1$$

$$= \frac{3}{8}. \qquad \text{Using the identity property of 1 to remove the factor 1}$$

Similar steps are followed when simplifying rational expressions: We factor and remove a factor equal to 1, using the fact that

$$\frac{ab}{cb} = \frac{a}{c} \cdot \frac{b}{b}.$$

E x a m p l e 2 Simplify: $\dfrac{8x^2}{24x}$.

Solution

$$\frac{8x^2}{24x} = \frac{8 \cdot x \cdot x}{3 \cdot 8 \cdot x}$$ Factoring the numerator and the denominator.
Note the common factor of $8 \cdot x$.

$$= \frac{x}{3} \cdot \frac{8x}{8x}$$ Rewriting as a product of two rational expressions

$$= \frac{x}{3} \cdot 1 \qquad \frac{8x}{8x} = 1$$

$$= \frac{x}{3}$$ Removing the factor 1

We say that $8x^2/(24x)$ *simplifies* to $x/3$.* In the work that follows, we assume that all denominators are nonzero.

E x a m p l e 3 Simplify: $\dfrac{5a + 15}{10}$.

Solution

$$\frac{5a + 15}{10} = \frac{5(a + 3)}{5 \cdot 2}$$ Factoring the numerator and the denominator.
Note the common factor of 5.

$$= \frac{5}{5} \cdot \frac{a + 3}{2}$$ Rewriting as a product of two rational expressions

$$= 1 \cdot \frac{a + 3}{2} \qquad \frac{5}{5} = 1$$

$$= \frac{a + 3}{2}$$ Removing the factor 1

Sometimes the common factor has two or more terms.

E x a m p l e 4 Simplify.

a) $\dfrac{6x + 12}{7x + 14}$ **b)** $\dfrac{6a^2 + 4a}{2a^2 + 2a}$ **c)** $\dfrac{x^2 - 1}{x^2 + 3x + 2}$

Solution

a) $\dfrac{6x + 12}{7x + 14} = \dfrac{6(x + 2)}{7(x + 2)}$ Factoring the numerator and the denominator

$$= \frac{6}{7} \cdot \frac{x + 2}{x + 2}$$ Rewriting as a product of two rational expressions

---

*In more advanced courses, we would *not* say that $8x^2/(24x)$ simplifies to $x/3$, but would instead say that $8x^2/(24x)$ simplifies to $x/3$ *with the restriction that $x \neq 0$.*

$$\frac{6}{7} \cdot \frac{x+2}{x+2} = \frac{6}{7} \cdot 1 \qquad \frac{x+2}{x+2} = 1$$

$$= \frac{6}{7} \qquad\qquad \text{Removing the factor 1}$$

b) $\dfrac{6a^2 + 4a}{2a^2 + 2a} = \dfrac{2a(3a+2)}{2a(a+1)}$ Factoring the numerator and the denominator

$$= \frac{2a}{2a} \cdot \frac{3a+2}{a+1} \qquad \text{Rewriting as a product of two rational expressions}$$

$$= 1 \cdot \frac{3a+2}{a+1} \qquad \frac{2a}{2a} = 1$$

$$= \frac{3a+2}{a+1} \qquad \text{Removing the factor 1. Note in this step that you } cannot \text{ remove the remaining } a\text{'s because } a \text{ is not a factor.}$$

c) $\dfrac{x^2 - 1}{x^2 + 3x + 2} = \dfrac{(x+1)(x-1)}{(x+1)(x+2)}$ Factoring

$$= \frac{x+1}{x+1} \cdot \frac{x-1}{x+2} \qquad \text{Rewriting as a product of two rational expressions}$$

$$- 1 \cdot \frac{x-1}{x+2} \qquad \frac{x+1}{x+1} = 1$$

$$= \frac{x-1}{x+2} \qquad \text{Removing the factor 1}$$

Canceling is a shortcut that can be used—and easily *misused*—when working with rational expressions. As we stated in Section 1.3, canceling must be done with care and understanding. Essentially, canceling streamlines the steps in which we remove a factor equal to 1. Example 4(c) could have been streamlined as follows:

$$\frac{x^2 - 1}{x^2 + 3x + 2} = \frac{\cancel{(x+1)}(x-1)}{\cancel{(x+1)}(x+2)} \qquad \begin{array}{l} \text{When a factor equal to 1 is noted,} \\ \text{it is "canceled":} \ \dfrac{x+1}{x+1} = 1 \end{array}$$

$$= \frac{x-1}{x+2}. \qquad \text{Simplifying}$$

Caution!   Canceling is often used incorrectly. The following cancellations are *incorrect*:

$$\frac{\cancel{x}+2}{\cancel{x}+3}; \qquad \frac{a^2 - \cancel{3}}{\cancel{3}}; \qquad \frac{6\cancel{x}^2 + 5\cancel{x} + 1}{4\cancel{x}^2 - 3\cancel{x}}.$$

Wrong!   Wrong!   Wrong!

None of the above cancellations removes a factor equal to 1. Factors are parts of products. For example, in $x \cdot 2$, x and 2 are factors, but in $x + 2$, x and 2 are terms, *not* factors. If it is not a factor, then it cannot be canceled.

E x a m p l e   5

Simplify: $\dfrac{3x^2 - 2x - 1}{x^2 - 3x + 2}$.

Solution   We factor the numerator and the denominator and look for common factors:

$$\frac{3x^2 - 2x - 1}{x^2 - 3x + 2} = \frac{(3x + 1)\cancel{(x - 1)}}{(x - 2)\cancel{(x - 1)}}$$

Try to visualize this as
$$\frac{3x + 1}{x - 2} \cdot \frac{x - 1}{x - 1}.$$

$$= \frac{3x + 1}{x - 2}.$$

Removing a factor equal to 1:
$$\frac{x - 1}{x - 1} = 1$$

When a rational expression is simplified, the result is an equivalent expression. Example 3 says that

$$\frac{5a + 15}{10} \quad \text{is equivalent to} \quad \frac{a + 3}{2}.$$

This result can be partially checked using a value of a. For instance, if $a = 2$, then

$$\frac{5a + 15}{10} = \frac{5 \cdot 2 + 15}{10} = \frac{25}{10} = \frac{5}{2}$$

and

$$\frac{a + 3}{2} = \frac{2 + 3}{2} = \frac{5}{2}.$$

> To see why this check is not foolproof, see Exercise 61.

If evaluating both expressions yields differing results, we know that a mistake has been made. For example, if $(5a + 15)/10$ is incorrectly simplified as $(a + 15)/2$ and we evaluate using $a = 2$, we have

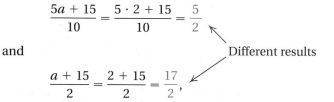

$$\frac{5a + 15}{10} = \frac{5 \cdot 2 + 15}{10} = \frac{5}{2}$$

and

Different results

$$\frac{a + 15}{2} = \frac{2 + 15}{2} = \frac{17}{2},$$

which demonstrates that a mistake has been made.

Factors That Are Opposites

Consider

$$\frac{x - 4}{8 - 2x}, \quad \text{or, equivalently,} \quad \frac{x - 4}{2(4 - x)}.$$

At first glance, the numerator and the denominator do not appear to have any common factors. But $x - 4$ and $4 - x$ are opposites, or additive inverses, of each other. Thus we can find a common factor by factoring out -1 in one expression.

E x a m p l e 6

Simplify $\dfrac{x-4}{8-2x}$ and check by evaluating.

Solution   We have

$$\dfrac{x-4}{8-2x} = \dfrac{x-4}{2(4-x)}$$ Factoring

$$= \dfrac{x-4}{2(-1)(x-4)}$$ Note that $4-x = -(x-4)$.

$$= \dfrac{x-4}{-2(x-4)}$$ Had we originally factored out -2, we could have gone directly to this step.

$$= \dfrac{1}{-2} \cdot \dfrac{x-4}{x-4}$$ Rewriting as a product. It is important to write the 1 in the numerator.

$$= -\dfrac{1}{2}.$$ Removing a factor equal to 1: $(x-4)/(x-4) = 1$

As a partial check, note that for any choice of x other than 4, the value of the rational expression is $-\frac{1}{2}$. For example, if $x = 6$, then

$$\dfrac{x-4}{8-2x} = \dfrac{6-4}{8-2 \cdot 6}$$

$$= \dfrac{2}{-4} = -\dfrac{1}{2}.$$

Exercise Set 6.1

FOR EXTRA HELP

Digital Video Tutor CD 4   InterAct Math   Math Tutor Center   MathXL   MyMathLab.com
Videotape 11

List all numbers for which each rational expression is undefined.

1. $\dfrac{25}{-7x}$

2. $\dfrac{14}{-5y}$

3. $\dfrac{t-3}{t+8}$

4. $\dfrac{a-8}{a+7}$

5. $\dfrac{a-4}{3a-12}$

6. $\dfrac{x^2-9}{4x-12}$

7. $\dfrac{x^2-16}{x^2-3x-28}$

8. $\dfrac{p^2-9}{p^2-7p+10}$

9. $\dfrac{m^3-2m}{m^2-25}$

10. $\dfrac{7-3x+x^2}{49-x^2}$

Simplify by removing a factor equal to 1. Show all steps.

11. $\dfrac{60a^2b}{40ab^3}$

12. $\dfrac{45x^3y^2}{9x^5y}$

13. $\dfrac{35x^2y}{14x^3y^5}$

14. $\dfrac{12a^5b^6}{18a^3b}$

15. $\dfrac{9x+15}{12x+20}$

16. $\dfrac{14x-7}{10x-5}$

17. $\dfrac{a^2-9}{a^2+4a+3}$

18. $\dfrac{a^2+5a+6}{a^2-9}$

Simplify, if possible. Then check by evaluating, as in Example 6.

19. $\dfrac{36x^6}{24x^9}$

20. $\dfrac{75a^5}{50a^3}$

21. $\dfrac{-2y + 6}{-8y}$

22. $\dfrac{4x - 12}{6x}$

23. $\dfrac{6a^2 - 3a}{7a^2 - 7a}$

24. $\dfrac{3m^2 + 3m}{6m^2 + 9m}$

25. $\dfrac{t^2 - 16}{t^2 + t - 20}$

26. $\dfrac{a^2 - 4}{a^2 + 5a + 6}$

27. $\dfrac{3a^2 + 9a - 12}{6a^2 - 30a + 24}$

28. $\dfrac{2t^2 - 6t + 4}{4t^2 + 12t - 16}$

29. $\dfrac{x^2 + 8x + 16}{x^2 - 16}$

30. $\dfrac{x^2 - 25}{x^2 - 10x + 25}$

31. $\dfrac{t^2 - 1}{t + 1}$

32. $\dfrac{a^2 - 1}{a - 1}$

33. $\dfrac{y^2 + 4}{y + 2}$

34. $\dfrac{x^2 + 1}{x + 1}$

35. $\dfrac{5x^2 - 20}{10x^2 - 40}$

36. $\dfrac{6x^2 - 54}{4x^2 - 36}$

37. $\dfrac{5y + 5}{y^2 + 7y + 6}$

38. $\dfrac{6t + 12}{t^2 - t - 6}$

39. $\dfrac{y^2 + 3y - 18}{y^2 + 2y - 15}$

40. $\dfrac{a^2 + 10a + 21}{a^2 + 11a + 28}$

41. $\dfrac{(a - 3)^2}{a^2 - 9}$

42. $\dfrac{t^2 - 4}{(t + 2)^2}$

43. $\dfrac{x - 8}{8 - x}$

44. $\dfrac{6 - x}{x - 6}$

45. $\dfrac{7t - 14}{2 - t}$

46. $\dfrac{4a - 12}{3 - a}$

47. $\dfrac{a - b}{3b - 3a}$

48. $\dfrac{q - p}{2p - 2q}$

49. $\dfrac{3x^2 - 3y^2}{2y^2 - 2x^2}$

50. $\dfrac{7a^2 - 7b^2}{3b^2 - 3a^2}$

Aha! **51.** $\dfrac{7s^2 - 28t^2}{28t^2 - 7s^2}$

52. $\dfrac{9m^2 - 4n^2}{4n^2 - 9m^2}$

53. Explain how simplifying is related to the identity property of 1.

54. If a rational expression is undefined for $x = 5$ and $x = -3$, what is the degree of the denominator? Why?

SKILL MAINTENANCE

Simplify.

55. $-\dfrac{2}{3} \cdot \dfrac{6}{7}$

56. $\dfrac{5}{9}\left(\dfrac{-6}{11}\right)$

57. $\dfrac{5}{8} \div \left(-\dfrac{1}{6}\right)$

58. $\dfrac{7}{10} \div \left(-\dfrac{8}{15}\right)$

59. $\dfrac{7}{9} - \dfrac{2}{3} \cdot \dfrac{6}{7}$

60. $\dfrac{2}{3} - \left(\dfrac{3}{4}\right)^2$

SYNTHESIS

61. Terry *incorrectly* simplifies

$$\dfrac{x^2 + x - 2}{x^2 + 3x + 2} \quad \text{as} \quad \dfrac{x - 1}{x + 2}.$$

He then checks his simplification by evaluating both expressions for $x = 1$. Use this situation to explain why evaluating is not a foolproof check.

62. How could you convince someone that $a - b$ and $b - a$ are opposites of each other?

Simplify.

63. $\dfrac{x^4 - y^4}{(y - x)^4}$

64. $\dfrac{16y^4 - x^4}{(x^2 + 4y^2)(x - 2y)}$

65. $\dfrac{(x - 1)(x^4 - 1)(x^2 - 1)}{(x^2 + 1)(x - 1)^2(x^4 - 2x^2 + 1)}$

66. $\dfrac{x^5 - 2x^3 + 4x^2 - 8}{x^7 + 2x^4 - 4x^3 - 8}$

67. $\dfrac{10t^4 - 8t^3 + 15t - 12}{8 - 10t + 12t^2 - 15t^3}$

68. $\dfrac{(t^4 - 1)(t^2 - 9)(t - 9)^2}{(t^4 - 81)(t^2 + 1)(t + 1)^2}$

69. $\dfrac{(t + 2)^3(t^2 + 2t + 1)(t + 1)}{(t + 1)^3(t^2 + 4t + 4)(t + 2)}$

70. $\dfrac{(x^2 - y^2)(x^2 - 2xy + y^2)}{(x + y)^2(x^2 - 4xy - 5y^2)}$

71. Select any number x, multiply by 2, add 5, multiply by 5, subtract 25, and divide by 10. What do you get? Explain how this procedure can be used for a number trick.

Multiplication and Division

6.2

Multiplication • Division

Multiplication and division of rational expressions is similar to multiplication and division with fractions. In this section, we again assume that all denominators are nonzero.

Multiplication

Recall that to multiply fractions, we simply multiply their numerators and multiply their denominators. Rational expressions are multiplied in a similar way.

> ### The Product of Two Rational Expressions
>
> To multiply rational expressions, multiply numerators and multiply denominators:
>
> $$\frac{A}{B} \cdot \frac{C}{D} = \frac{AC}{BD}.$$
>
> Then factor and simplify the result if possible.

For example,

$$\frac{3}{5} \cdot \frac{8}{11} = \frac{24}{55} \quad \text{and} \quad \frac{x}{3} \cdot \frac{x+2}{y} = \frac{x(x+2)}{3y}.$$

Fraction bars are grouping symbols, so parentheses are needed when writing some products. Because we generally simplify, we often leave parentheses in the product. There is no need to multiply further.

Example 1   Multiply and simplify.

a) $\dfrac{5a^3}{4} \cdot \dfrac{2}{5a}$

b) $\dfrac{x^2 + 6x + 9}{x^2 - 4} \cdot \dfrac{x - 2}{x + 3}$

c) $\dfrac{x^2 + x - 2}{15} \cdot \dfrac{5}{2x^2 - 3x + 1}$

Solution

a) $\dfrac{5a^3}{4} \cdot \dfrac{2}{5a} = \dfrac{5a^3(2)}{4(5a)}$ Forming the product of the numerators and the product of the denominators

$ = \dfrac{2 \cdot 5 \cdot a \cdot a \cdot a}{2 \cdot 2 \cdot 5 \cdot a}$ Factoring the numerator and the denominator

$ = \dfrac{2 \cdot \cancel{5} \cdot \cancel{a} \cdot a \cdot a}{2 \cdot 2 \cdot \cancel{5} \cdot \cancel{a}}$

$ = \dfrac{a^2}{2}$ Removing a factor equal to 1: $\dfrac{2 \cdot 5 \cdot a}{2 \cdot 5 \cdot a} = 1$

b) $\dfrac{x^2 + 6x + 9}{x^2 - 4} \cdot \dfrac{x - 2}{x + 3} = \dfrac{(x^2 + 6x + 9)(x - 2)}{(x^2 - 4)(x + 3)}$ Multiplying the numerators and the denominators

$ = \dfrac{(x + 3)(x + 3)(x - 2)}{(x + 2)(x - 2)(x + 3)}$ Factoring the numerator and the denominator

$ = \dfrac{\cancel{(x + 3)}(x + 3)\cancel{(x - 2)}}{(x + 2)\cancel{(x - 2)}\cancel{(x + 3)}}$ Removing a factor equal to 1: $\dfrac{(x + 3)(x - 2)}{(x + 3)(x - 2)} = 1$

$ = \dfrac{x + 3}{x + 2}$

c) $\dfrac{x^2 + x - 2}{15} \cdot \dfrac{5}{2x^2 - 3x + 1} = \dfrac{(x^2 + x - 2)5}{15(2x^2 - 3x + 1)}$ Multiplying the numerators and the denominators

$ = \dfrac{(x + 2)(x - 1)5}{5(3)(x - 1)(2x - 1)}$ Factoring the numerator and the denominator. Try to go directly to this step.

$ = \dfrac{(x + 2)\cancel{(x - 1)}\cancel{5}}{\cancel{5}(3)\cancel{(x - 1)}(2x - 1)}$ Removing a factor equal to 1: $\dfrac{5(x - 1)}{5(x - 1)} = 1$

$ = \dfrac{x + 2}{3(2x - 1)}$

There is no need to multiply out the numerator or the denominator in our results.

Division

As with fractions, reciprocals of rational expressions are found by interchanging the numerator and the denominator. For example,

the reciprocal of $\dfrac{2}{7}$ is $\dfrac{7}{2}$,   and   the reciprocal of $\dfrac{3x}{x + 5}$ is $\dfrac{x + 5}{3x}$.

> ### The Quotient of Two Rational Expressions
> To divide by a rational expression, multiply by its reciprocal:
> $$\frac{A}{B} \div \frac{C}{D} = \frac{A}{B} \cdot \frac{D}{C} = \frac{AD}{BC}.$$
> Then factor and simplify if possible.

Example 2

Divide: **(a)** $\dfrac{x}{5} \div \dfrac{7}{y}$; **(b)** $(x + 2) \div \dfrac{x - 1}{x + 3}$.

Solution

a) $\dfrac{x}{5} \div \dfrac{7}{y} = \dfrac{x}{5} \cdot \dfrac{y}{7}$ \qquad Multiplying by the reciprocal of the divisor

$\phantom{\dfrac{x}{5} \div \dfrac{7}{y}} = \dfrac{xy}{35}$ \qquad Multiplying rational expressions

b) $(x + 2) \div \dfrac{x - 1}{x + 3} = \dfrac{x + 2}{1} \cdot \dfrac{x + 3}{x - 1}$ \qquad Multiplying by the reciprocal of the divisor. Writing $x + 2$ as

$\phantom{(x + 2) \div \dfrac{x - 1}{x + 3}} = \dfrac{(x + 2)(x + 3)}{x - 1}$ \qquad $\dfrac{x + 2}{1}$ can be helpful.

As usual, we should simplify when possible. Often that will require that we factor one or more polynomials. Our hope is to discover a common factor that appears in both the numerator and the denominator.

Example 3

Divide and simplify: $\dfrac{x + 1}{x^2 - 1} \div \dfrac{x + 1}{x^2 - 2x + 1}$.

Solution

$\dfrac{x + 1}{x^2 - 1} \div \dfrac{x + 1}{x^2 - 2x + 1} = \dfrac{x + 1}{x^2 - 1} \cdot \dfrac{x^2 - 2x + 1}{x + 1}$ \qquad Multiplying by the reciprocal of the divisor

$ = \dfrac{(x + 1)(x - 1)(x - 1)}{(x + 1)(x - 1)(x + 1)}$ \qquad Multiplying rational expressions and factoring numerators and denominators

$ = \dfrac{\cancel{(x + 1)}\,\cancel{(x - 1)}(x - 1)}{\cancel{(x + 1)}\,\cancel{(x - 1)}(x + 1)}$ \qquad Removing a factor equal to 1: $\dfrac{(x + 1)(x - 1)}{(x + 1)(x - 1)} = 1$

$ = \dfrac{x - 1}{x + 1}$

Example 4

technology
connection

In performing a partial check of Example 4(b), care must be taken in placing parentheses. For example, we enter the original expression in Example 4(b) as $y_1 = ((x^2 - 2x - 3)/(x^2 - 4))/((x + 1)/(x + 5))$ and the simplified expression as $y_2 = ((x - 3)(x + 5))/((x - 2)(x + 2))$. Comparing values of y_1 and y_2, we see that the simplification is probably correct.

X	Y₁	Y₂
−5	ERROR	0
−4	−.5833	−.5833
−3	−2.4	−2.4
−2	ERROR	ERROR
−1	ERROR	5.3333
0	3.75	3.75
1	4	4
X = −5		

1. Check Example 4(a).
2. Why are there 3 ERROR messages shown for y_1 on the screen above, and only 1 for y_2?

Divide and, if possible, simplify.

a) $\dfrac{a^2 + 3a + 2}{a^2 + 4} \div (5a^2 + 10a)$

b) $\dfrac{x^2 - 2x - 3}{x^2 - 4} \div \dfrac{x + 1}{x + 5}$

Solution

a) $\dfrac{a^2 + 3a + 2}{a^2 + 4} \div (5a^2 + 10a)$

$= \dfrac{a^2 + 3a + 2}{a^2 + 4} \cdot \dfrac{1}{5a^2 + 10a}$ Multiplying by the reciprocal of the divisor

$= \dfrac{(a + 2)(a + 1)}{(a^2 + 4)5a(a + 2)}$ Multiplying rational expressions and factoring

$= \dfrac{\cancel{(a + 2)}(a + 1)}{(a^2 + 4)5a\cancel{(a + 2)}}$

$= \dfrac{a + 1}{(a^2 + 4)5a}$ Removing a factor equal to 1: $\dfrac{a + 2}{a + 2} = 1$

b) $\dfrac{x^2 - 2x - 3}{x^2 - 4} \div \dfrac{x + 1}{x + 5} = \dfrac{x^2 - 2x - 3}{x^2 - 4} \cdot \dfrac{x + 5}{x + 1}$ Multiplying by the reciprocal of the divisor

$= \dfrac{(x - 3)(x + 1)(x + 5)}{(x - 2)(x + 2)(x + 1)}$ Multiplying rational expressions and factoring

$= \dfrac{(x - 3)\cancel{(x + 1)}(x + 5)}{(x - 2)(x + 2)\cancel{(x + 1)}}$ Removing a factor equal to 1:

$= \dfrac{(x - 3)(x + 5)}{(x - 2)(x + 2)}$ $\dfrac{x + 1}{x + 1} = 1$

Exercise Set   6.2

FOR EXTRA HELP

Digital Video Tutor CD 4   InterAct Math   Math Tutor Center   MathXL   MyMathLab.com
Videotape 11

Multiply. Leave each answer in factored form.

1. $\dfrac{9x}{4} \cdot \dfrac{x - 5}{2x + 1}$

2. $\dfrac{3x}{4} \cdot \dfrac{5x + 2}{x - 1}$

3. $\dfrac{a - 4}{a + 6} \cdot \dfrac{a + 2}{a + 6}$

4. $\dfrac{a + 3}{a + 6} \cdot \dfrac{a + 3}{a - 1}$

5. $\dfrac{2x + 3}{4} \cdot \dfrac{x + 1}{x - 5}$

6. $\dfrac{x + 2}{3x - 4} \cdot \dfrac{4}{5x + 6}$

7. $\dfrac{a - 5}{a^2 + 1} \cdot \dfrac{a + 2}{a^2 - 1}$

8. $\dfrac{t + 3}{t^2 - 2} \cdot \dfrac{t + 3}{t^2 - 4}$

9. $\dfrac{x + 4}{2 + x} \cdot \dfrac{x - 1}{x + 1}$

10. $\dfrac{m + 4}{m + 8} \cdot \dfrac{2 + m}{m + 5}$

Multiply and, if possible, simplify.

11. $\dfrac{5a^4}{6a} \cdot \dfrac{2}{a}$

12. $\dfrac{10}{t^7} \cdot \dfrac{3t^2}{25t}$

13. $\dfrac{3c}{d^2} \cdot \dfrac{8d}{6c^3}$

14. $\dfrac{3x^2y}{2} \cdot \dfrac{4}{xy^3}$

15. $\dfrac{x^2 - 3x - 10}{(x - 2)^2} \cdot \dfrac{x - 2}{x - 5}$

16. $\dfrac{t + 2}{t - 2} \cdot \dfrac{t^2 - 5t + 6}{(t + 2)^2}$

17. $\dfrac{a^2 + 25}{a^2 - 4a + 3} \cdot \dfrac{a - 5}{a + 5}$

18. $\dfrac{x + 3}{x^2 + 9} \cdot \dfrac{x^2 + 5x + 4}{x + 9}$

19. $\dfrac{a^2 - 9}{a^2} \cdot \dfrac{5a}{a^2 + a - 12}$

20. $\dfrac{x^2 + 10x - 11}{5x} \cdot \dfrac{x^3}{x + 11}$

21. $\dfrac{4a^2}{3a^2 - 12a + 12} \cdot \dfrac{3a - 6}{2a}$

22. $\dfrac{5v + 5}{v - 2} \cdot \dfrac{2v^2 - 8v + 8}{v^2 - 1}$

23. $\dfrac{t^2 + 2t - 3}{t^2 + 4t - 5} \cdot \dfrac{t^2 - 3t - 10}{t^2 + 5t + 6}$

24. $\dfrac{x^2 + 5x + 4}{x^2 - 6x + 8} \cdot \dfrac{x^2 + 5x - 14}{x^2 + 8x + 7}$

25. $\dfrac{5a^2 - 180}{10a^2 - 10} \cdot \dfrac{20a + 20}{2a - 12}$

26. $\dfrac{2t^2 - 98}{4t^2 - 4} \cdot \dfrac{8t + 8}{16t - 112}$

Aha! **27.** $\dfrac{x^2 + 4x + 4}{(x - 1)^2} \cdot \dfrac{x^2 - 2x + 1}{(x + 2)^2}$

28. $\dfrac{x + 5}{(x + 2)^2} \cdot \dfrac{x^2 + 7x + 10}{(x + 5)^2}$

29. $\dfrac{t^2 + 8t + 16}{(t + 4)^3} \cdot \dfrac{(t + 2)^3}{t^2 + 4t + 4}$

30. $\dfrac{(y - 1)^3}{y^2 - 2y + 1} \cdot \dfrac{y^2 - 4y + 4}{(y - 2)^3}$

Find the reciprocal of each expression.

31. $\dfrac{3x}{7}$

32. $\dfrac{3 - x}{x^2 + 4}$

33. $a^3 - 8a$

34. $\dfrac{7}{a^2 - b^2}$

35. $\dfrac{x^2 + 2x - 5}{x^2 - 4x + 7}$

36. $\dfrac{x^2 - 3xy + y^2}{x^2 + 7xy - y^2}$

Divide and, if possible, simplify.

37. $\dfrac{3}{8} \div \dfrac{5}{2}$

38. $\dfrac{5}{9} \div \dfrac{2}{7}$

39. $\dfrac{x}{4} \div \dfrac{5}{x}$

40. $\dfrac{5}{x} \div \dfrac{x}{12}$

41. $\dfrac{a^5}{b^4} \div \dfrac{a^2}{b}$

42. $\dfrac{x^5}{y^2} \div \dfrac{x^2}{y}$

43. $\dfrac{y + 5}{4} \div \dfrac{y}{2}$

44. $\dfrac{a + 2}{a - 3} \div \dfrac{a - 1}{a + 3}$

45. $\dfrac{4y - 8}{y + 2} \div \dfrac{y - 2}{y^2 - 4}$

46. $\dfrac{x^2 - 1}{x} \div \dfrac{x + 1}{x - 1}$

47. $\dfrac{a}{a - b} \div \dfrac{b}{b - a}$

48. $\dfrac{x - y}{6} \div \dfrac{y - x}{3}$

49. $(y^2 - 9) \div \dfrac{y^2 - 2y - 3}{y^2 + 1}$

50. $(x^2 - 5x - 6) \div \dfrac{x^2 - 1}{x + 6}$

51. $\dfrac{7x - 7}{16} \div \dfrac{x - 1}{6}$

52. $\dfrac{-4 + 2x}{15} \div \dfrac{x - 2}{3}$

53. $\dfrac{-6 + 3x}{5} \div \dfrac{4x - 8}{25}$

54. $\dfrac{-12 + 4x}{12} \div \dfrac{-6 + 2x}{6}$

55. $\dfrac{a + 2}{a - 1} \div \dfrac{3a + 6}{a - 5}$

56. $\dfrac{t - 3}{t + 2} \div \dfrac{4t - 12}{t + 1}$

57. $(2x - 1) \div \dfrac{2x^2 - 11x + 5}{4x^2 - 1}$

58. $(a + 7) \div \dfrac{3a^2 + 14a - 49}{a^2 + 8a + 7}$

59. $\dfrac{x - 5}{x + 5} \div \dfrac{2x^2 - 50}{x^2 + 25}$

60. $\dfrac{3x^2 - 27}{x^2 + 1} \div \dfrac{x + 3}{x - 3}$

61. $\dfrac{a^2 - 10a + 25}{a^2 + 7a + 12} \div \dfrac{a^2 - a - 20}{a^2 + 6a + 9}$

62. $\dfrac{a^2 + 5a + 4}{a^2 - 2a + 1} \div \dfrac{a^2 + 8a + 16}{a^2 - 5a - 6}$

63. $\dfrac{c^2 + 10c + 21}{c^2 - 2c - 15} \div (c^2 + 2c - 35)$

64. $\dfrac{1 - z}{1 + 2z - z^2} \div (1 - z)$

65. $\dfrac{x - y}{x^2 + 2xy + y^2} \div \dfrac{x^2 - y^2}{x^2 - 5xy + 4y^2}$

66. $\dfrac{a^2 - b^2}{a^2 - 4ab + 4b^2} \div \dfrac{a^2 - 3ab + 2b^2}{a - 2b}$

67. A student claims to be able to divide, but not multiply, rational expressions. Why is this claim difficult to believe?

68. Why is it important to insert parentheses when multiplying rational expressions in which the numerators and the denominators contain more than one term?

SKILL MAINTENANCE

Simplify.

69. $\dfrac{3}{4} + \dfrac{5}{6}$

70. $\dfrac{7}{8} + \dfrac{5}{6}$

71. $\dfrac{2}{9} - \dfrac{1}{6}$

72. $\dfrac{3}{10} - \dfrac{7}{15}$

73. $\dfrac{2}{5} - \left(\dfrac{3}{2}\right)^2$

74. $\dfrac{5}{9} + \dfrac{2}{3} \cdot \dfrac{4}{5}$

SYNTHESIS

75. Is the reciprocal of a product the product of the two reciprocals? Why or why not?

76. Explain why the quotient

$$\dfrac{x + 3}{x - 5} \div \dfrac{x - 7}{x + 1}$$

is undefined for $x = 5$, $x = -1$, and $x = 7$.

Simplify.

Aha! 77. $\dfrac{3x - y}{2x + y} \div \dfrac{3x - y}{2x + y}$

78. $\dfrac{2a^2 - 5ab}{c - 3d} \div (4a^2 - 25b^2)$

79. $(x - 2a) \div \dfrac{a^2x^2 - 4a^4}{a^2x + 2a^3}$

80. $\dfrac{3a^2 - 5ab - 12b^2}{3ab + 4b^2} \div (3b^2 - ab)^2$

81. $\dfrac{3x^2 - 2xy - y^2}{x^2 - y^2} \div (3x^2 + 4xy + y^2)^2$

82. $\dfrac{y^2 - 4xy}{y - x} \div \dfrac{16x^2y^2 - y^4}{4x^2 - 3xy - y^2} \div \dfrac{4}{x^3y^3}$

Aha! 83. $\dfrac{a^2 - 3b}{a^2 + 2b} \cdot \dfrac{a^2 - 2b}{a^2 + 3b} \cdot \dfrac{a^2 + 2b}{a^2 - 3b}$

84. $\dfrac{z^2 - 8z + 16}{z^2 + 8z + 16} \div \dfrac{(z - 4)^5}{(z + 4)^5} \div \dfrac{3z + 12}{z^2 - 16}$

85. $\dfrac{x^2 - x + xy - y}{x^2 + 6x - 7} \div \dfrac{x^2 + 2xy + y^2}{4x + 4y}$

86. $\dfrac{3x + 3y + 3}{9x} \div \dfrac{x^2 + 2xy + y^2 - 1}{x^4 + x^2}$

87. $\dfrac{(t + 2)^3}{(t + 1)^3} \div \dfrac{t^2 + 4t + 4}{t^2 + 2t + 1} \cdot \dfrac{t + 1}{t + 2}$

88. $\dfrac{3y^3 + 6y^2}{y^2 - y - 12} \div \dfrac{y^2 - y}{y^2 - 2y - 8} \cdot \dfrac{y^2 + 5y + 6}{y^2}$

89. $\dfrac{6y - 4x}{(2x + 3y)^2} \cdot \dfrac{2x - 3y}{x^2 - 9y^2} \div \dfrac{4x^2 - 12xy + 9y^2}{9y^2 + 12xy + 4x^2}$

90. $\dfrac{a^4 - 81b^4}{a^2c - 6abc + 9b^2c} \cdot \dfrac{a + 3b}{a^2 + 9b^2} \div \dfrac{a^2 + 6ab + 9b^2}{(a - 3b)^2}$

CORNER

Currency Exchange

COLLABORATIVE

Focus: Least common multiples and
           proportions

Time: 20 minutes

Group size: 2

Travel between different countries usually ne-
cessitates an exchange of currencies. Recently
one Canadian dollar was worth 64 cents in U.S.
funds. Use this exchange rate for the activity that
follows.

ACTIVITY

1. Within each group of two students, one student should play the role of a U.S. citizen
   planning a visit to Canada. The other student
   should play the role of a Canadian planning a
   visit to the United States. Use the exchange
   rate of one Canadian dollar for 64 cents in
   U.S. funds.

2. Determine how much Canadian money the
   U.S. citizen would receive in exchange for $64
   of U.S. funds (this should be easy). Then determine how much U.S. money the Canadian
   would receive in exchange for $64 of Cana-
   dian funds. Finally, determine how much the
   Canadian would receive in exchange for $100
   of Canadian funds and how much the U.S.
   citizen would receive for $100 of U.S. funds.

3. The answers to part (2) should indicate that
   coins are needed to exchange $64 of Cana-
   dian money for U.S. funds, or to exchange
   $100 of U.S. funds for Canadian money. What
   is the smallest amount of Canadian dollars
   that can be exchanged for U.S. dollars without requiring coins? What is the smallest
   amount of U.S. dollars that can be exchanged
   for Canadian dollars without requiring coins?
   (*Hint*: See part 2.)

4. Use the results from part (3) to find two other
   amounts of U.S. currency that can be exchanged without requiring coins. Answers
   may vary.

5. Find the smallest number *a* for which neither
   conversion—from *a* Canadian dollars to U.S.
   funds or from *a* U.S. dollars to Canadian
   funds—will require coins. (*Hint*: Use LCMs
   and the results of part 2 above.)

6. At one time in 2000, one New Zealand dollar
   was worth about 40 cents in U.S. funds. Find
   the smallest number *a* for which neither conversion—from *a* New Zealand dollars to U.S.
   funds or from *a* U.S. dollars to New Zealand
   funds—will require coins. (*Hint*: See part 5.)

Addition, Subtraction, and Least Common Denominators	# 6.3

Addition When Denominators Are the Same • Subtraction When Denominators Are the Same • Least Common Multiples and Denominators

Addition When Denominators Are the Same

Recall that to add fractions having the same denominator, like $\frac{2}{7}$ and $\frac{3}{7}$, we add the numerators and keep the common denominator: $\frac{2}{7} + \frac{3}{7} = \frac{5}{7}$. The same procedure is used when rational expressions share a common denominator.

> ### The Sum of Two Rational Expressions
> To add when the denominators are the same, add the numerators and keep the common denominator:
> $$\frac{A}{B} + \frac{C}{B} = \frac{A+C}{B}.$$

Example 1

Add. Simplify the result, if possible.

a) $\dfrac{4}{a} + \dfrac{3+a}{a}$ **b)** $\dfrac{3x}{x-5} + \dfrac{2x+1}{x-5}$

c) $\dfrac{2x^2+3x-7}{2x+1} + \dfrac{x^2+x-8}{2x+1}$ **d)** $\dfrac{x-5}{x^2-9} + \dfrac{2}{x^2-9}$

Solution

a) $\dfrac{4}{a} + \dfrac{3+a}{a} = \dfrac{7+a}{a}$ When the denominators are alike, add the numerators and keep the common denominator.

b) $\dfrac{3x}{x-5} + \dfrac{2x+1}{x-5} = \dfrac{5x+1}{x-5}$ Adding the numerators and combining like terms

c) $\dfrac{2x^2+3x-7}{2x+1} + \dfrac{x^2+x-8}{2x+1} = \dfrac{(2x^2+3x-7)+(x^2+x-8)}{2x+1}$

$\qquad\qquad = \dfrac{3x^2+4x-15}{2x+1}$ Combining like terms in the numerator

d) $\dfrac{x-5}{x^2-9} + \dfrac{2}{x^2-9} = \dfrac{x-3}{x^2-9}$ Combining like terms in the numerator: $x-5+2 = x-3$

$\qquad\qquad = \dfrac{x-3}{(x-3)(x+3)}$ Factoring

$\qquad\qquad = \dfrac{1 \cdot (x-3)}{(x-3)(x+3)}$ Removing a factor equal to 1: $\dfrac{x-3}{x-3} = 1$

$\qquad\qquad = \dfrac{1}{x+3}$

Subtraction When Denominators Are the Same

When two fractions have the same denominator, we subtract one numerator from the other and keep the common denominator: $\frac{5}{7} - \frac{2}{7} = \frac{3}{7}$. The same procedure is used with rational expressions.

The Difference of Two Rational Expressions

To subtract when the denominators are the same, subtract the second numerator from the first and keep the common denominator:

$$\frac{A}{B} - \frac{C}{B} = \frac{A - C}{B}.$$

Caution!   A fraction bar is a grouping symbol, just like parentheses, under the numerator. When a numerator is subtracted, be sure to subtract *every* term in that numerator.

Example 2

Subtract: **(a)** $\dfrac{3x}{x + 2} - \dfrac{x - 5}{x + 2}$; **(b)** $\dfrac{x^2}{x - 4} - \dfrac{x + 12}{x - 4}$.

Solution

a) $\dfrac{3x}{x + 2} - \dfrac{x - 5}{x + 2} = \dfrac{3x - (x - 5)}{x + 2}$ The parentheses are needed to make sure that we subtract both terms.

$= \dfrac{3x - x + 5}{x + 2}$ Removing the parentheses and changing signs (using the distributive law)

$= \dfrac{2x + 5}{x + 2}$ Combining like terms

b) $\dfrac{x^2}{x - 4} - \dfrac{x + 12}{x - 4} = \dfrac{x^2 - (x + 12)}{x - 4}$ Remember the parentheses!

$= \dfrac{x^2 - x - 12}{x - 4}$ Removing parentheses (using the distributive law)

$= \dfrac{(x - 4)(x + 3)}{x - 4}$ Factoring, in hopes of simplifying

$= \dfrac{(x - 4)(x + 3)}{x - 4}$ Removing a factor equal to 1: $\dfrac{x - 4}{x - 4} = 1$

$= x + 3$

Least Common Multiples and Denominators

Thus far, every pair of rational expressions that we have added or subtracted shared a common denominator. To add or subtract rational expressions that lack a common denominator, we must first find equivalent rational expressions that *do* have a common denominator.

In algebra, we find a common denominator much as we do in arithmetic. Recall that to add $\frac{1}{12}$ and $\frac{7}{30}$, we first identify the smallest number that contains both 12 and 30 as factors. Such a number, the **least common multiple** (**LCM**) of the denominators, is then used as the **least common denominator** (**LCD**).

Let's find the LCM of 12 and 30 using a method that can also be used with polynomials. We begin by writing the prime factorization of 12:

$$12 = 2 \cdot 2 \cdot 3.$$

Next, we write the prime factorization of 30:

$$30 = 2 \cdot 3 \cdot 5.$$

The LCM must include the factors of each number, so it must include each prime factor the greatest number of times that it appears in either of the factorizations. To find the LCM for 12 and 30, we select one factorization, say

$$2 \cdot 2 \cdot 3,$$

and note that because it lacks a factor of 5, it does not contain the entire factorization of 30. If we multiply $2 \cdot 2 \cdot 3$ by 5, every prime factor occurs just often enough to contain both 12 and 30 as factors.

$$\text{LCM} = 2 \cdot 2 \cdot 3 \cdot 5$$

12 is a factor of the LCM.

30 is a factor of the LCM.

Note that each prime factor—2, 3, and 5—is used the greatest number of times that it appears in either of the individual factorizations. The factor 2 occurs twice and the factors 3 and 5 once each.

To Find the Least Common Denominator (LCD)

1. Write the prime factorization of each denominator.
2. Select one of the factorizations and inspect it to see if it contains the other.

   a) If it does, it represents the LCM of the denominators.

   b) If it does not, multiply that factorization by any factors of the other denominator that it lacks. The final product is the LCM of the denominators.

The LCD is the LCM of the denominators. It should contain each factor the greatest number of times that it occurs in any of the individual factorizations.

Let's finish adding $\dfrac{1}{12}$ and $\dfrac{7}{30}$:

$$\frac{1}{12} + \frac{7}{30} = \frac{1}{2 \cdot 2 \cdot 3} + \frac{7}{2 \cdot 3 \cdot 5}.$$ The least common denominator (LCD) is $2 \cdot 2 \cdot 3 \cdot 5$.

We found above that the LCD is $2 \cdot 2 \cdot 3 \cdot 5$. To get the LCD, we see that the first denominator needs a factor of 5, and the second denominator needs another factor of 2. This is accomplished by multiplying by different forms of 1. We can do this because $a \cdot 1 = a$, for any number a:

$$\frac{1}{12} + \frac{7}{30} = \frac{1}{2 \cdot 2 \cdot 3} \cdot \frac{5}{5} + \frac{7}{2 \cdot 3 \cdot 5} \cdot \frac{2}{2}$$ $\dfrac{5}{5} = 1$ and $\dfrac{2}{2} = 1$

$$= \frac{5}{2 \cdot 2 \cdot 3 \cdot 5} + \frac{14}{2 \cdot 3 \cdot 5 \cdot 2}$$ The denominators are now the LCD.

$$= \frac{19}{60}.$$ Adding the numerators and computing the LCD

Expressions like $\dfrac{5}{36x^2}$ and $\dfrac{7}{24x}$ are added in much the same manner.

Example 3

Find the LCD of $\dfrac{5}{36x^2}$ and $\dfrac{7}{24x}$.

Solution

1. We begin by writing the prime factorizations of $36x^2$ and $24x$:

$$36x^2 = 2 \cdot 2 \cdot 3 \cdot 3 \cdot x \cdot x;$$
$$24x = 2 \cdot 2 \cdot 2 \cdot 3 \cdot x.$$

2. Except for a third factor of 2, the factorization of $36x^2$ contains the entire factorization of $24x$. To find the smallest product that contains both $36x^2$ and $24x$ as factors, we multiply $36x^2$ by a third factor of 2:

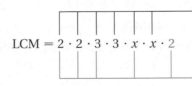

$36x^2$ is a factor of the LCM.

$$\text{LCM} = 2 \cdot 2 \cdot 3 \cdot 3 \cdot x \cdot x \cdot 2$$

Note that each factor appears the greatest number of times that it occurs in either of the above factorizations.

$24x$ is a factor of the LCM.

The LCM is thus $2^3 \cdot 3^2 \cdot x^2$, or $72x^2$, so the LCD is $72x^2$.

We can now add $\dfrac{5}{36x^2}$ and $\dfrac{7}{24x}$:

$$\frac{5}{36x^2} + \frac{7}{24x} = \frac{5}{2 \cdot 2 \cdot 3 \cdot 3 \cdot x \cdot x} + \frac{7}{2 \cdot 2 \cdot 2 \cdot 3 \cdot x}.$$

In Example 3, we found that the LCD is $2 \cdot 2 \cdot 2 \cdot 3 \cdot 3 \cdot x \cdot x$. To obtain equivalent expressions with this LCD, we multiply each expression by 1, using the missing factors of the LCD to write 1:

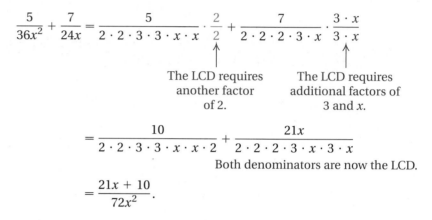

$$= \frac{10}{2 \cdot 2 \cdot 3 \cdot 3 \cdot x \cdot x \cdot 2} + \frac{21x}{2 \cdot 2 \cdot 2 \cdot 3 \cdot x \cdot 3 \cdot x}$$

Both denominators are now the LCD.

$$= \frac{21x + 10}{72x^2}.$$

You now have the "big" picture of why LCMs are needed when adding rational expressions. For the remainder of this section, we will practice finding LCMs and rewriting rational expressions so that they have the LCD as the denominator. In Section 6.4, we will return to the addition and subtraction of rational expressions.

Example 4

For each pair of polynomials, find the least common multiple.

a) $15a$ and $35b$
b) $21x^3y^6$ and $7x^5y^2$
c) $x^2 + 5x - 6$ and $x^2 - 1$

Solution

a) We write the prime factorizations and then construct the LCM:

$$15a = 3 \cdot 5 \cdot a$$
$$35b = 5 \cdot 7 \cdot b$$

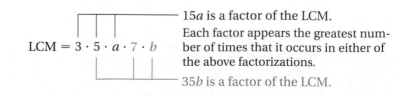

$15a$ is a factor of the LCM.

LCM $= 3 \cdot 5 \cdot a \cdot 7 \cdot b$ Each factor appears the greatest number of times that it occurs in either of the above factorizations.

$35b$ is a factor of the LCM.

The LCM is $3 \cdot 5 \cdot a \cdot 7 \cdot b$, or $105ab$.

b) $21x^3y^6 = 3 \cdot 7 \cdot x \cdot x \cdot x \cdot y \cdot y \cdot y \cdot y \cdot y \cdot y$ Try to visualize the factors
$7x^5y^2 = 7 \cdot x \cdot x \cdot x \cdot x \cdot x \cdot y \cdot y$ of x and y mentally.

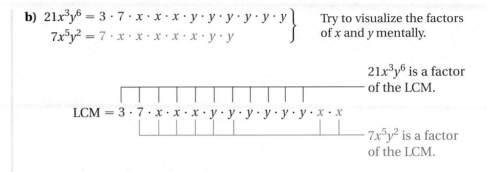

$21x^3y^6$ is a factor of the LCM.

$\text{LCM} = 3 \cdot 7 \cdot x \cdot x \cdot x \cdot y \cdot y \cdot y \cdot y \cdot y \cdot y \cdot x \cdot x$

$7x^5y^2$ is a factor of the LCM.

Note that we used the highest power of each factor in $21x^3y^6$ and $7x^5y^2$. The LCM is $21x^5y^6$.

c) $x^2 + 5x - 6 = (x - 1)(x + 6)$
$x^2 - 1 = (x - 1)(x + 1)$

$x^2 + 5x - 6$ is a factor of the LCM.

$\text{LCM} = (x - 1)(x + 6)(x + 1)$

$x^2 - 1$ is a factor of the LCM.

The LCM is $(x - 1)(x + 6)(x + 1)$. There is no need to multiply this out.

The above procedure can be used to find the LCM of three polynomials as well. We factor each polynomial and then construct the LCM using each factor the greatest number of times that it appears in any one factorization.

Example 5

For each group of polynomials, find the LCM.

a) $12x$, $16y$, and $8xyz$
b) $x^2 + 4$, $x + 1$, and 5

Solution

a) $12x = 2 \cdot 2 \cdot 3 \cdot x$
$16y = 2 \cdot 2 \cdot 2 \cdot 2 \cdot y$
$8xyz = 2 \cdot 2 \cdot 2 \cdot x \cdot y \cdot z$

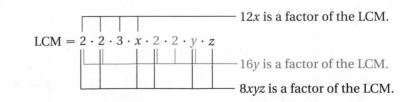

$12x$ is a factor of the LCM.

$\text{LCM} = 2 \cdot 2 \cdot 3 \cdot x \cdot 2 \cdot 2 \cdot y \cdot z$

$16y$ is a factor of the LCM.

$8xyz$ is a factor of the LCM.

The LCM is $2^4 \cdot 3 \cdot xyz$, or $48xyz$.

b) Since $x^2 + 4$, $x + 1$, and 5 are not factorable, the LCM is their product: $5(x^2 + 4)(x + 1)$.

To add or subtract rational expressions with different denominators, we must be able to write equivalent expressions that have the LCD.

E x a m p l e   6   Find equivalent expressions that have the LCD:

$$\frac{x + 3}{x^2 + 5x - 6}, \quad \frac{x + 7}{x^2 - 1}.$$

Solution   From Example 4(c), we know that the LCD is

$$(x + 6)(x - 1)(x + 1).$$

Since

$$x^2 + 5x - 6 = (x + 6)(x - 1),$$

the factor of the LCD that is missing from the first denominator is $x + 1$. We multiply by 1 using $(x + 1)/(x + 1)$:

$$\left.\begin{array}{l}\dfrac{x + 3}{x^2 + 5x - 6} = \dfrac{x + 3}{(x + 6)(x - 1)} \cdot \dfrac{x + 1}{x + 1} \\[2mm] \qquad = \dfrac{(x + 3)(x + 1)}{(x + 6)(x - 1)(x + 1)}.\end{array}\right\}$$ Finding an equivalent expression that has the least common denominator

For the second expression, we have $x^2 - 1 = (x + 1)(x - 1)$. The factor of the LCD that is missing is $x + 6$. We multiply by 1 using $(x + 6)/(x + 6)$:

$$\left.\begin{array}{l}\dfrac{x + 7}{x^2 - 1} = \dfrac{x + 7}{(x + 1)(x - 1)} \cdot \dfrac{x + 6}{x + 6} \\[2mm] \qquad = \dfrac{(x + 7)(x + 6)}{(x + 1)(x - 1)(x + 6)}.\end{array}\right\}$$ Finding an equivalent expression that has the least common denominator

We leave the results in factored form. In Section 6.4, we will carry out the actual addition and subtraction of such rational expressions.

Exercise Set 6.3

Perform the indicated operation. Simplify, if possible.

1. $\dfrac{3}{x} + \dfrac{9}{x}$

2. $\dfrac{4}{a^2} + \dfrac{9}{a^2}$

3. $\dfrac{x}{15} + \dfrac{2x+5}{15}$

4. $\dfrac{a}{7} + \dfrac{3a-4}{7}$

5. $\dfrac{4}{a+3} + \dfrac{5}{a+3}$

6. $\dfrac{5}{x+2} + \dfrac{8}{x+2}$

7. $\dfrac{9}{a+2} - \dfrac{3}{a+2}$

8. $\dfrac{8}{x+7} - \dfrac{2}{x+7}$

9. $\dfrac{3y+8}{2y} - \dfrac{y+1}{2y}$

10. $\dfrac{5+3t}{4t} - \dfrac{2t+1}{4t}$

11. $\dfrac{7x+8}{x+1} + \dfrac{4x+3}{x+1}$

12. $\dfrac{3a+13}{a+4} + \dfrac{2a+7}{a+4}$

13. $\dfrac{7x+8}{x+1} - \dfrac{4x+3}{x+1}$

14. $\dfrac{3a+13}{a+4} - \dfrac{2a+7}{a+4}$

15. $\dfrac{a^2}{a-4} + \dfrac{a-20}{a-4}$

16. $\dfrac{x^2}{x+5} + \dfrac{7x+10}{x+5}$

17. $\dfrac{x^2}{x-2} - \dfrac{6x-8}{x-2}$

18. $\dfrac{a^2}{a+3} - \dfrac{2a+15}{a+3}$

Aha! **19.** $\dfrac{t^2-5t}{t-1} + \dfrac{5t-t^2}{t-1}$

20. $\dfrac{y^2+6y}{y+2} + \dfrac{2y+12}{y+2}$

21. $\dfrac{x-4}{x^2+5x+6} + \dfrac{7}{x^2+5x+6}$

22. $\dfrac{x-5}{x^2-4x+3} + \dfrac{2}{x^2-4x+3}$

23. $\dfrac{3a^2+14}{a^2+5a-6} - \dfrac{13a}{a^2+5a-6}$

24. $\dfrac{2a^2+15}{a^2-7a+12} - \dfrac{11a}{a^2-7a+12}$

25. $\dfrac{t^2-3t}{t^2+6t+9} + \dfrac{2t-12}{t^2+6t+9}$

26. $\dfrac{y^2-7y}{y^2+8y+16} + \dfrac{6y-20}{y^2+8y+16}$

27. $\dfrac{2x^2+x}{x^2-8x+12} - \dfrac{x^2-2x+10}{x^2-8x+12}$

28. $\dfrac{2x^2+3}{x^2-6x+5} - \dfrac{3+2x^2}{x^2-6x+5}$

29. $\dfrac{3-2x}{x^2-6x+8} + \dfrac{7-3x}{x^2-6x+8}$

30. $\dfrac{1-2t}{t^2-5t+4} + \dfrac{4-3t}{t^2-5t+4}$

31. $\dfrac{x-7}{x^2+3x-4} - \dfrac{2x-3}{x^2+3x-4}$

32. $\dfrac{5-3x}{x^2-2x+1} - \dfrac{x+1}{x^2-2x+1}$

Find the LCM.

33. 15, 27   **34.** 10, 15   **35.** 8, 9

36. 12, 15   **37.** 6, 9, 21   **38.** 8, 36, 40

Find the LCM.

39. $12x^2,\ 6x^3$

40. $10t^3,\ 5t^4$

41. $15a^4b^7,\ 10a^2b^8$

42. $6a^2b^7,\ 9a^5b^2$

43. $2(y-3),\ 6(y-3)$

44. $4(x-1),\ 8(x-1)$

45. $x^2-4,\ x^2+5x+6$

46. $x^2+3x+2,\ x^2-4$

47. $t^3+4t^2+4t,\ t^2-4t$

48. $y^3-y^2,\ y^4-y^2$

49. $10x^2y,\ 6y^2z,\ 5xz^3$

50. $8x^3z,\ 12xy^2,\ 4y^5z^2$

51. $a+1,\ (a-1)^2,\ a^2-1$

52. $x^2-9,\ x+3,\ (x-3)^2$

53. $m^2-5m+6,\ m^2-4m+4$

54. $2x^2+5x+2,\ 2x^2-x-1$

Aha! **55.** $t-3,\ t+3,\ (t^2-9)^2$

56. $a-5,\ (a^2-10a+25)^2$

57. $6x^3-24x^2+18x,\ 4x^5-24x^4+20x^3$

58. $9x^3-9x^2-18x,\ 6x^5-24x^4+24x^3$

Find equivalent expressions that have the LCD.

59. $\dfrac{5}{6x^5},\ \dfrac{y}{12x^3}$

60. $\dfrac{3}{10a^3},\ \dfrac{b}{5a^6}$

61. $\dfrac{3}{2a^2b},\ \dfrac{7}{8ab^2}$

62. $\dfrac{7}{3x^4y^2},\ \dfrac{4}{9xy^3}$

63. $\dfrac{2x}{x^2 - 4}$, $\dfrac{4x}{x^2 + 5x + 6}$

64. $\dfrac{5x}{x^2 - 9}$, $\dfrac{2x}{x^2 + 11x + 24}$

65. If the LCM of two numbers is their product, what can you conclude about the two numbers?

66. Explain why the product of two numbers is not always their least common multiple.

SKILL MAINTENANCE

Write each number in two equivalent forms.

67. $\dfrac{7}{-9}$

68. $-\dfrac{3}{2}$

Simplify.

69. $\dfrac{5}{18} - \dfrac{7}{12}$

70. $\dfrac{8}{15} - \dfrac{13}{20}$

Find a polynomial that can represent the shaded area of each figure.

71. **72.**

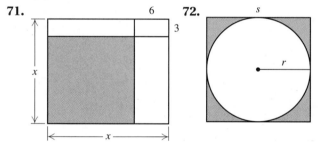

SYNTHESIS

73. If the LCM of a binomial and a trinomial is the trinomial, what relationship exists between the two expressions?

74. If the LCM of two third-degree polynomials is a sixth-degree polynomial, what can be concluded about the two polynomials?

Perform the indicated operations. Simplify, if possible.

75. $\dfrac{6x - 1}{x - 1} + \dfrac{3(2x + 5)}{x - 1} + \dfrac{3(2x - 3)}{x - 1}$

76. $\dfrac{2x + 11}{x - 3} \cdot \dfrac{3}{x + 4} + \dfrac{-1}{4 + x} \cdot \dfrac{6x + 3}{x - 3}$

77. $\dfrac{x^2}{3x^2 - 5x - 2} - \dfrac{2x}{3x + 1} \cdot \dfrac{1}{x - 2}$

78. $\dfrac{x + y}{x^2 - y^2} + \dfrac{x - y}{x^2 - y^2} - \dfrac{2x}{x^2 - y^2}$

South African artistry. **In South Africa, the design of every woven handbag, or gipatsi (plural, sipatsi) is created by repeating two or more geometric patterns. Each pattern encircles the bag, sharing the strands of fabric with any pattern above or below. The length, or period, of each pattern is the number of strands required to construct the pattern. For a gipatsi to be considered beautiful, each individual pattern must fit a whole number of times around the bag (Source: Gerdes, Paulus, Women, Art and Geometry in Southern Africa. Asmara, Eritrea: Africa World Press, Inc., p. 5).**

79. A weaver is using two patterns to create a gipatsi. Pattern A is 10 strands long, and pattern B is 3 strands long. What is the smallest number of strands that can be used to complete the gipatsi?

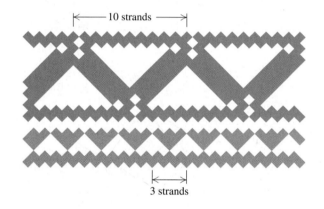

80. A weaver is using a four-strand pattern, a six-strand pattern, and an eight-strand pattern. What is the smallest number of strands that can be used to complete the gipatsi?

81. For technical reasons, the number of strands is generally a multiple of 4. Answer Exercise 79 with this additional requirement in mind.

Find the LCM.

82. 72, 90, 96

83. $8x^2 - 8$, $6x^2 - 12x + 6$, $10 - 10x$

84. $9x^2 - 16$, $6x^2 - x - 12$, $16 - 24x + 9x^2$

85. *Running.* Kim and Jed leave the starting point of a fitness loop at the same time. Kim jogs a lap in 6 min and Jed jogs one in 8 min. Assuming they continue to run at the same pace, when will they next meet at the starting place?

86. *Bus schedules.*   Beginning at 5:00 A.M., a hotel shuttle bus leaves Salton Airport every 25 min, and the downtown shuttle bus leaves the airport every 35 min. What time will it be when both shuttles again leave at the same time?

87. *Appliances.*   Dishwashers last an average of 10 yr, clothes washers an average of 14 yr, and refrigerators an average of 20 yr (*Source: Energy Savers: Tips on Saving Energy and Money at Home,* produced for the U.S. Department of Energy by the National Renewable Energy Laboratory, 1998). If an apartment house is equipped with new dishwashers, clothes washers, and refrigerators in 2002, in what year will all three appliances need to be replaced at once?

88. Explain how evaluating can be used to perform a partial check on the result of Example 1(d):
$$\frac{x-5}{x^2-9}+\frac{2}{x^2-9}=\frac{1}{x+3}.$$

89. On p. 348, the second step in finding an LCD is to select one of the factorizations of the denominators. Does it matter which one is selected? Why or why not?

Addition and Subtraction with Unlike Denominators

6.4

Adding and Subtracting with LCDs •
When Factors Are Opposites

Adding and Subtracting with LCDs

We now know how to rewrite two rational expressions in an equivalent form that uses the LCD. Once rational expressions share a common denominator, they can be added or subtracted just as in Section 6.3.

> ### To Add or Subtract Rational Expressions Having Different Denominators
> 1. Find the LCD.
> 2. Multiply each rational expression by a form of 1 made up of the factors of the LCD missing from that expression's denominator.
> 3. Add or subtract the numerators, as indicated. Write the sum or difference over the LCD.
> 4. Simplify, if possible.

Example 1   Add: $\dfrac{5x^2}{8} + \dfrac{7x}{12}$.

Solution

1. First, we find the LCD:

$$\left.\begin{array}{l} 8 = 2 \cdot 2 \cdot 2 \\ 12 = 2 \cdot 2 \cdot 3 \end{array}\right\} \quad \text{LCD} = 2 \cdot 2 \cdot 2 \cdot 3, \text{ or } 24.$$

2. The denominator 8 needs to be multiplied by 3 in order to obtain the LCD. The denominator 12 needs to be multiplied by 2 in order to obtain the LCD. Thus we multiply by $\frac{3}{3}$ and $\frac{2}{2}$ to get the LCD:

$$\begin{aligned} \frac{5x^2}{8} + \frac{7x}{12} &= \frac{5x^2}{2 \cdot 2 \cdot 2} + \frac{7x}{2 \cdot 2 \cdot 3} \\ &= \frac{5x^2}{2 \cdot 2 \cdot 2} \cdot \frac{3}{3} + \frac{7x}{2 \cdot 2 \cdot 3} \cdot \frac{2}{2} \qquad \text{Multiplying each expression} \\ & \hphantom{= \frac{5x^2}{2 \cdot 2 \cdot 2} \cdot \frac{3}{3} + \frac{7x}{2 \cdot 2 \cdot 3} \cdot \frac{2}{2} \quad} \text{by a form of 1 to get the LCD} \\ &= \frac{15x^2}{24} + \frac{14x}{24}. \end{aligned}$$

3. Next, we add the numerators:

$$\frac{15x^2}{24} + \frac{14x}{24} = \frac{15x^2 + 14x}{24}.$$

4. Since $15x^2 + 14x$ and 24 have no common factor,

$$\frac{15x^2 + 14x}{24}$$

cannot be simplified any further.

Subtraction is performed in much the same way.

Example 2   Subtract: $\dfrac{7}{8x} - \dfrac{5}{12x^2}$.

Solution   We follow, but do not list, the four steps shown above. First, we find the LCD:

$$\left.\begin{array}{l} 8x = 2 \cdot 2 \cdot 2 \cdot x \\ 12x^2 = 2 \cdot 2 \cdot 3 \cdot x \cdot x \end{array}\right\} \quad \text{LCD} = 2 \cdot 2 \cdot 3 \cdot x \cdot x \cdot 2, \text{ or } 24x^2.$$

The denominator $8x$ must be multiplied by $3x$ in order to obtain the LCD. The denominator $12x^2$ must be multiplied by 2 in order to obtain the LCD. Thus we multiply by $\frac{3x}{3x}$ and $\frac{2}{2}$ to get the LCD. Then we subtract and, if possible, simplify.

$$\frac{7}{8x} - \frac{5}{12x^2} = \frac{7}{8x} \cdot \frac{3x}{3x} - \frac{5}{12x^2} \cdot \frac{2}{2}$$

$$= \frac{21x}{24x^2} - \frac{10}{24x^2}$$

> *Caution!* Do not simplify *these* rational expressions or you will lose the LCD.

$$= \frac{21x - 10}{24x^2}$$ This cannot be simplified, so we are done.

When denominators contain polynomials with two or more terms, the same steps are used.

E x a m p l e   3

Add: $\dfrac{2a}{a^2 - 1} + \dfrac{1}{a^2 + a}$.

Solution   First, we find the LCD:

$$\left. \begin{array}{l} a^2 - 1 = (a - 1)(a + 1) \\ a^2 + a = a(a + 1) \end{array} \right\} \quad \text{LCD} = (a - 1)(a + 1)a.$$

We multiply by a form of 1 to get the LCD in each expression:

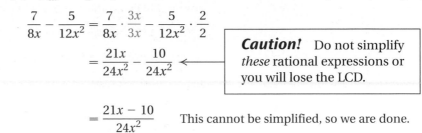

$$\frac{2a}{a^2 - 1} + \frac{1}{a^2 + a} = \frac{2a}{(a - 1)(a + 1)} \cdot \frac{a}{a} + \frac{1}{a(a + 1)} \cdot \frac{a - 1}{a - 1}$$

Multiplying by $\dfrac{a}{a}$ and $\dfrac{a - 1}{a - 1}$ to get the LCD

$$= \frac{2a^2}{(a - 1)(a + 1)a} + \frac{a - 1}{a(a + 1)(a - 1)}$$

$$= \frac{2a^2 + a - 1}{a(a - 1)(a + 1)}$$ Adding numerators

$$= \frac{(2a - 1)(a + 1)}{a(a - 1)(a + 1)}$$

Simplifying by factoring and removing a factor equal to 1: $\dfrac{a + 1}{a + 1} = 1$

$$= \frac{2a - 1}{a(a - 1)}.$$

E x a m p l e   4

Perform the indicated operations.

a) $\dfrac{x + 4}{x - 2} - \dfrac{x - 7}{x + 5}$

b) $\dfrac{x}{x^2 + 11x + 30} + \dfrac{-5}{x^2 + 9x + 20}$

c) $\dfrac{x}{x^2 + 5x + 6} - \dfrac{2}{x^2 + 3x + 2}$

Solution

a) First, we find the LCD. It is just the product of the denominators:

$$\text{LCD} = (x - 2)(x + 5).$$

We multiply by a form of 1 to get the LCD in each expression. Then we subtract and try to simplify.

$$\frac{x + 4}{x - 2} - \frac{x - 7}{x + 5} = \frac{x + 4}{x - 2} \cdot \frac{x + 5}{x + 5} - \frac{x - 7}{x + 5} \cdot \frac{x - 2}{x - 2}$$

$$= \frac{x^2 + 9x + 20}{(x - 2)(x + 5)} - \frac{x^2 - 9x + 14}{(x - 2)(x + 5)} \qquad \begin{array}{l}\text{Multiplying out}\\ \text{numerators (but}\\ \text{not denominators)}\end{array}$$

$$= \frac{x^2 + 9x + 20 - (x^2 - 9x + 14)}{(x - 2)(x + 5)} \qquad \begin{array}{l}\text{When subtracting}\\ \text{a numerator with}\\ \text{more than one}\\ \text{term, parentheses}\\ \text{are important.}\end{array}$$

$$= \frac{x^2 + 9x + 20 - x^2 + 9x - 14}{(x - 2)(x + 5)} \qquad \begin{array}{l}\text{Removing paren-}\\ \text{theses and sub-}\\ \text{tracting every}\\ \text{term}\end{array}$$

$$= \frac{18x + 6}{(x - 2)(x + 5)}$$

Although $18x + 6$ can be factored as $6(3x + 1)$, doing so will not enable us to simplify our result.

b) $\dfrac{x}{x^2 + 11x + 30} + \dfrac{-5}{x^2 + 9x + 20}$

$$= \frac{x}{(x + 5)(x + 6)} + \frac{-5}{(x + 5)(x + 4)} \qquad \begin{array}{l}\text{Factoring the denominators}\\ \text{in order to find the LCD.}\\ \text{The LCD is } (x + 5)(x + 6)(x + 4).\end{array}$$

$$= \frac{x}{(x + 5)(x + 6)} \cdot \frac{x + 4}{x + 4} + \frac{-5}{(x + 5)(x + 4)} \cdot \frac{x + 6}{x + 6} \qquad \begin{array}{l}\text{Multiplying to}\\ \text{get the LCD}\end{array}$$

$$= \frac{x^2 + 4x}{(x + 5)(x + 6)(x + 4)} + \frac{-5x - 30}{(x + 5)(x + 6)(x + 4)} \qquad \begin{array}{l}\text{Multiplying in}\\ \text{each numerator}\end{array}$$

$$= \frac{x^2 + 4x - 5x - 30}{(x + 5)(x + 6)(x + 4)} \qquad \text{Adding numerators}$$

$$= \frac{x^2 - x - 30}{(x + 5)(x + 6)(x + 4)} \qquad \text{Combining like terms in the numerator}$$

$$= \frac{\cancel{(x + 5)}(x - 6)}{\cancel{(x + 5)}(x + 6)(x + 4)}$$

$$= \frac{x - 6}{(x + 6)(x + 4)} \qquad \begin{array}{l}\text{Always simplify the result, if possible, by}\\ \text{removing a factor equal to 1; here } \dfrac{x + 5}{x + 5} = 1.\end{array}$$

c) $\dfrac{x}{x^2 + 5x + 6} - \dfrac{2}{x^2 + 3x + 2}$

$= \dfrac{x}{(x + 2)(x + 3)} - \dfrac{2}{(x + 2)(x + 1)}$ Factoring denominators.
The LCD is $(x + 2)(x + 3)(x + 1)$.

$= \dfrac{x}{(x + 2)(x + 3)} \cdot \dfrac{x + 1}{x + 1} - \dfrac{2}{(x + 2)(x + 1)} \cdot \dfrac{x + 3}{x + 3}$

$= \dfrac{x^2 + x}{(x + 2)(x + 3)(x + 1)} - \dfrac{2x + 6}{(x + 2)(x + 3)(x + 1)}$

$= \dfrac{x^2 + x - (2x + 6)}{(x + 2)(x + 3)(x + 1)}$ Don't forget the parentheses!

$= \dfrac{x^2 + x - 2x - 6}{(x + 2)(x + 3)(x + 1)}$ Remember to subtract each term in $2x + 6$.

$= \dfrac{x^2 - x - 6}{(x + 2)(x + 3)(x + 1)}$ Combining like terms in the numerator

$= \dfrac{(x + 2)(x - 3)}{(x + 2)(x + 3)(x + 1)}$

$= \dfrac{x - 3}{(x + 3)(x + 1)}$ Factoring and simplifying; $\dfrac{x + 2}{x + 2} = 1$

When Factors Are Opposites

Recall from Section 6.1 that expressions of the form $a - b$ and $b - a$ are opposites of each other. When either of these binomials is multiplied by -1, the result is the other binomial:

$$-1(a - b) = -a + b = b + (-a) = b - a;$$
$$-1(b - a) = -b + a = a + (-b) = a - b.$$

Multiplication by -1 reverses the order in which subtraction occurs.

E x a m p l e   5   Add: $\dfrac{x}{x - 5} + \dfrac{7}{5 - x}$.

Solution   Since the denominators are opposites of each other, we can find a common denominator by multiplying either rational expression by $-1/-1$. Because polynomials are most often written in descending order, we choose to reverse the subtraction in the second denominator:

$\dfrac{x}{x - 5} + \dfrac{7}{5 - x} = \dfrac{x}{x - 5} + \dfrac{7}{5 - x} \cdot \dfrac{-1}{-1}$ Writing 1 as $-1/-1$ and multiplying to obtain a common denominator

$= \dfrac{x}{x - 5} + \dfrac{-7}{-5 + x}$

$= \dfrac{x}{x - 5} + \dfrac{-7}{x - 5}$ Note that $-5 + x = x + (-5) = x - 5$.

$= \dfrac{x - 7}{x - 5}$.

Sometimes, after factoring to find the LCD, we find a factor in one denominator that is the opposite of a factor in the other denominator. When this happens, multiplication by $-1/-1$ can again be helpful.

E x a m p l e 6

Perform the indicated operations and simplify.

a) $\dfrac{x}{x^2 - 25} + \dfrac{3}{5 - x}$

b) $\dfrac{x + 9}{x^2 - 4} + \dfrac{6 - x}{4 - x^2} - \dfrac{1 + x}{x^2 - 4}$

Solution

a) $\dfrac{x}{x^2 - 25} + \dfrac{3}{5 - x} = \dfrac{x}{(x - 5)(x + 5)} + \dfrac{3}{5 - x}$ Factoring

$= \dfrac{x}{(x - 5)(x + 5)} + \dfrac{3}{5 - x} \cdot \dfrac{-1}{-1}$ Multiplying by $-1/-1$ changes $5 - x$ to $x - 5$.

$= \dfrac{x}{(x - 5)(x + 5)} + \dfrac{-3}{x - 5}$ $(5 - x)(-1) = x - 5$

$= \dfrac{x}{(x - 5)(x + 5)} + \dfrac{-3}{(x - 5)} \cdot \dfrac{x + 5}{x + 5}$ The LCD is $(x - 5)(x + 5)$.

$= \dfrac{x}{(x - 5)(x + 5)} + \dfrac{-3x - 15}{(x - 5)(x + 5)}$

$= \dfrac{-2x - 15}{(x - 5)(x + 5)}$

b) Since $4 - x^2$ is the opposite of $x^2 - 4$, multiplying the second rational expression by $-1/-1$ will lead to a common denominator:

$\dfrac{x + 9}{x^2 - 4} + \dfrac{6 - x}{4 - x^2} - \dfrac{1 + x}{x^2 - 4} = \dfrac{x + 9}{x^2 - 4} + \dfrac{6 - x}{4 - x^2} \cdot \dfrac{-1}{-1} - \dfrac{1 + x}{x^2 - 4}$

$= \dfrac{x + 9}{x^2 - 4} + \dfrac{x - 6}{x^2 - 4} - \dfrac{1 + x}{x^2 - 4}$

$= \dfrac{x + 9 + x - 6 - 1 - x}{x^2 - 4}$ Adding and subtracting numerators

$= \dfrac{x + 2}{x^2 - 4}$

$\left.\begin{array}{l} = \dfrac{(x + 2) \cdot 1}{(x + 2)(x - 2)} \\[2em] = \dfrac{1}{x - 2}. \end{array}\right\}$ Simplifying

Exercise Set   6.4

Perform the indicated operation. Simplify, if possible.

1. $\dfrac{3}{x} + \dfrac{7}{x^2}$

2. $\dfrac{5}{x} + \dfrac{6}{x^2}$

3. $\dfrac{1}{6r} - \dfrac{3}{8r}$

4. $\dfrac{4}{9t} - \dfrac{7}{6t}$

5. $\dfrac{4}{xy^2} + \dfrac{2}{x^2y}$

6. $\dfrac{2}{c^2d} + \dfrac{7}{cd^3}$

7. $\dfrac{8}{9t^3} - \dfrac{5}{6t^2}$

8. $\dfrac{-2}{3xy^2} - \dfrac{6}{x^2y^3}$

9. $\dfrac{x+5}{8} + \dfrac{x-3}{12}$

10. $\dfrac{x-4}{9} + \dfrac{x+5}{6}$

11. $\dfrac{a+2}{2} - \dfrac{a-4}{4}$

12. $\dfrac{x-2}{6} - \dfrac{x+1}{3}$

13. $\dfrac{2a-1}{3a^2} + \dfrac{5a+1}{9a}$

14. $\dfrac{a+4}{16a} + \dfrac{3a+4}{4a^2}$

15. $\dfrac{x-1}{4x} - \dfrac{2x+3}{x}$

16. $\dfrac{4z-9}{3z} - \dfrac{3z-8}{4z}$

17. $\dfrac{2c-d}{c^2d} + \dfrac{c+d}{cd^2}$

18. $\dfrac{x+y}{xy^2} + \dfrac{3x+y}{x^2y}$

19. $\dfrac{5x+3y}{2x^2y} - \dfrac{3x+4y}{xy^2}$

20. $\dfrac{4x+2t}{3xt^2} - \dfrac{5x-3t}{x^2t}$

21. $\dfrac{5}{x-1} + \dfrac{5}{x+1}$

22. $\dfrac{3}{x-2} + \dfrac{3}{x+2}$

23. $\dfrac{4}{z-1} - \dfrac{2}{z+1}$

24. $\dfrac{5}{x+5} - \dfrac{3}{x-5}$

25. $\dfrac{2}{x+5} + \dfrac{3}{4x}$

26. $\dfrac{3}{x+1} + \dfrac{2}{3x}$

27. $\dfrac{8}{3t^2-15t} - \dfrac{3}{2t-10}$

28. $\dfrac{3}{2t^2-2t} - \dfrac{5}{2t-2}$

29. $\dfrac{4x}{x^2-25} + \dfrac{x}{x+5}$

30. $\dfrac{2x}{x^2-16} + \dfrac{x}{x-4}$

31. $\dfrac{t}{t-3} - \dfrac{5}{4t-12}$

32. $\dfrac{6}{z+4} - \dfrac{2}{3z+12}$

33. $\dfrac{2}{x+3} + \dfrac{4}{(x+3)^2}$

34. $\dfrac{3}{x-1} + \dfrac{2}{(x-1)^2}$

35. $\dfrac{3}{x+2} - \dfrac{8}{x^2-4}$

36. $\dfrac{2t}{t^2-9} - \dfrac{3}{t-3}$

37. $\dfrac{3a}{4a-20} + \dfrac{9a}{6a-30}$

38. $\dfrac{4a}{5a-10} + \dfrac{3a}{10a-20}$

Aha! **39.** $\dfrac{x}{x-5} + \dfrac{x}{5-x}$

40. $\dfrac{x+4}{x} + \dfrac{x}{x+4}$

41. $\dfrac{7}{a^2+a-2} + \dfrac{5}{a^2-4a+3}$

42. $\dfrac{x}{x^2+2x+1} + \dfrac{1}{x^2+5x+4}$

43. $\dfrac{x}{x^2+9x+20} - \dfrac{4}{x^2+7x+12}$

44. $\dfrac{x}{x^2+5x+6} - \dfrac{2}{x^2+3x+2}$

45. $\dfrac{3z}{z^2-4z+4} + \dfrac{10}{z^2+z-6}$

46. $\dfrac{3}{x^2-9} + \dfrac{2}{x^2-x-6}$

Aha! **47.** $\dfrac{-5}{x^2+17x+16} - \dfrac{0}{x^2+9x+8}$

48. $\dfrac{x}{x^2+15x+56} - \dfrac{1}{x^2+13x+42}$

49. $\dfrac{2x}{5} - \dfrac{x-3}{-5}$

50. $\dfrac{x}{4} - \dfrac{3x-5}{-4}$

51. $\dfrac{y^2}{y-3} + \dfrac{9}{3-y}$

52. $\dfrac{t^2}{t-2} + \dfrac{4}{2-t}$

53. $\dfrac{b-7}{b^2-16} + \dfrac{7-b}{16-b^2}$

54. $\dfrac{a-3}{a^2-25} + \dfrac{a-3}{25-a^2}$

55. $\dfrac{y+2}{y-7} + \dfrac{3-y}{49-y^2}$

56. $\dfrac{4-p}{25-p^2} + \dfrac{p+1}{p-5}$

57. $\dfrac{5x}{x^2-9} - \dfrac{4}{3-x}$

58. $\dfrac{8x}{16-x^2} - \dfrac{5}{x-4}$

59. $\dfrac{3x+2}{3x+6} + \dfrac{x}{4-x^2}$

60. $\dfrac{a}{a^2-1} + \dfrac{2a}{a-a^2}$

61. $\dfrac{4-a^2}{a^2-9} - \dfrac{a-2}{3-a}$

62. $\dfrac{4x}{x^2-y^2} - \dfrac{6}{y-x}$

Perform the indicated operations. Simplify, if possible.

63. $\dfrac{x-3}{2-x} - \dfrac{x+3}{x+2} + \dfrac{x+6}{4-x^2}$

64. $\dfrac{t-5}{1-t} - \dfrac{t+4}{t+1} + \dfrac{t+2}{t^2-1}$

65. $\dfrac{x+5}{x+3} + \dfrac{x+7}{x+2} - \dfrac{7x+19}{(x+3)(x+2)}$

66. $\dfrac{2x+5}{x+1} + \dfrac{x+7}{x+5} - \dfrac{5x+17}{(x+1)(x+5)}$

67. $\dfrac{t}{s+t} - \dfrac{t}{s-t}$

68. $\dfrac{a}{b-a} - \dfrac{b}{b+a}$

69. $\dfrac{1}{x+y} + \dfrac{1}{x-y} - \dfrac{2x}{x^2-y^2}$

70. $\dfrac{2r}{r^2-s^2} + \dfrac{1}{r+s} - \dfrac{1}{r-s}$

71. What is the advantage of using the *least* common denominator—rather than just *any* common denominator—when adding or subtracting rational expressions?

72. Describe a procedure that can be used to add any two rational expressions.

SKILL MAINTENANCE

Simplify.

73. $-\dfrac{3}{7} \div \dfrac{6}{13}$

74. $\dfrac{5}{12} \div \left(-\dfrac{3}{4}\right)$

75. $\dfrac{\frac{2}{9}}{\frac{5}{3}}$

76. $\dfrac{\frac{7}{10}}{\frac{3}{5}}$

Graph.

77. $y = -\dfrac{1}{2}x - 5$

78. $y = \dfrac{1}{2}x - 5$

SYNTHESIS

79. How could you convince someone that

$$\dfrac{1}{3-x} \quad \text{and} \quad \dfrac{1}{x-3}$$

are opposites of each other?

80. Are parentheses as important for adding rational expressions as they are for subtracting rational expressions? Why or why not?

Write expressions for the perimeter and the area of each rectangle.

81.

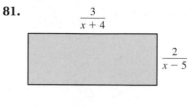

82.

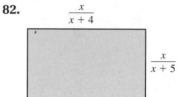

Perform the indicated operations.

83. $\dfrac{2x+11}{x-3} \cdot \dfrac{3}{x+4} + \dfrac{2x+1}{4+x} \cdot \dfrac{3}{3-x}$

84. $\dfrac{x^2}{3x^2-5x-2} - \dfrac{2x}{3x+1} \cdot \dfrac{1}{x-2}$

Aha! **85.** $\left(\dfrac{x}{x+7} - \dfrac{3}{x+2}\right)\left(\dfrac{x}{x+7} + \dfrac{3}{x+2}\right)$

86. $\dfrac{1}{ay-3a+2xy-6x} - \dfrac{xy+ay}{a^2-4x^2}\left(\dfrac{1}{y-3}\right)^2$

87. $\dfrac{2x^2+5x-3}{2x^2-9x+9} + \dfrac{x+1}{3-2x} + \dfrac{4x^2+8x+3}{x-3} \cdot \dfrac{x+3}{9-4x^2}$

88. $\left(\dfrac{a}{a-b} + \dfrac{b}{a+b}\right)\left(\dfrac{1}{3a+b} + \dfrac{2a+6b}{9a^2-b^2}\right)$

89. Express

$$\dfrac{a-3b}{a-b}$$

as a sum of two rational expressions with denominators that are opposites of each other. Answers may vary.

Complex Rational Expressions

6.5

Using Division to Simplify • Multiplying by the LCD

A **complex rational expression**, or **complex fractional expression**, is a rational expression that has one or more rational expressions within its numerator or denominator. Here are some examples:

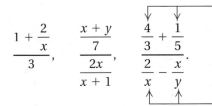

$$\frac{1 + \dfrac{2}{x}}{3}, \quad \frac{x + y}{\dfrac{2x}{x + 1}}, \quad \frac{\dfrac{4}{3} + \dfrac{1}{5}}{\dfrac{2}{x} - \dfrac{x}{y}}.$$

These are rational expressions within the complex rational expression.

We will consider two methods for simplifying complex rational expressions. Each method offers certain advantages.

Using Division to Simplify (Method 1)

Our first method for simplifying complex rational expressions involves rewriting the expression as a quotient of two rational expressions.

> **To Simplify a Complex Rational Expression by Dividing**
> 1. Add or subtract, as needed, to get a single rational expression in the numerator.
> 2. Add or subtract, as needed, to get a single rational expression in the denominator.
> 3. Divide the numerator by the denominator (invert and multiply).
> 4. If possible, simplify by removing a factor equal to 1.

The key here is to express a complex rational expression as one rational expression divided by another. We can then proceed as in Section 6.2.

Example 1

Simplify: $\dfrac{\dfrac{x}{x - 3}}{\dfrac{4}{5x - 15}}.$

Solution Here the numerator and denominator are already single rational expressions. This allows us to start by dividing (step 3), as in Section 6.2:

$$\frac{\dfrac{x}{x-3}}{\dfrac{4}{5x-15}} = \frac{x}{x-3} \div \frac{4}{5x-15}$$ Rewriting with a division symbol

$$= \frac{x}{x-3} \cdot \frac{5x-15}{4}$$ Multiplying by the reciprocal of the divisor (inverting and multiplying)

$$= \frac{x}{x-3} \cdot \frac{5(x-3)}{4}$$ Factoring and removing a factor equal to 1: $\dfrac{x-3}{x-3} = 1$

$$= \frac{5x}{4}.$$

Often we must add or subtract in the numerator and/or denominator before we can divide.

Example 2

Simplify.

a) $\dfrac{\dfrac{5}{2a} + \dfrac{1}{a}}{\dfrac{1}{4a} - \dfrac{5}{6}}$

b) $\dfrac{\dfrac{x^2}{y} - \dfrac{5}{x}}{xz}$

Solution

a) $\dfrac{\dfrac{5}{2a} + \dfrac{1}{a}}{\dfrac{1}{4a} - \dfrac{5}{6}} = \dfrac{\dfrac{5}{2a} + \dfrac{1}{a} \cdot \dfrac{2}{2}}{\dfrac{1}{4a} \cdot \dfrac{3}{3} - \dfrac{5}{6} \cdot \dfrac{2a}{2a}}$ ← Multiplying by 1 to get the LCD, 2a, for the numerator of the complex rational expression ← Multiplying by 1 to get the LCD, 12a, for the denominator of the complex rational expression

$$= \dfrac{\dfrac{5}{2a} + \dfrac{2}{2a}}{\dfrac{3}{12a} - \dfrac{10a}{12a}} = \dfrac{\dfrac{7}{2a}}{\dfrac{3-10a}{12a}}$$ ← Adding ← Subtracting

$$= \frac{7}{2a} \div \frac{3-10a}{12a}$$ Rewriting with a division symbol. This is often done mentally.

$$= \frac{7}{2a} \cdot \frac{12a}{3-10a}$$ Multiplying by the reciprocal of the divisor (inverting and multiplying)

$$= \frac{7}{2a} \cdot \frac{2a \cdot 6}{3-10a}$$ Removing a factor equal to 1: $\dfrac{2a}{2a} = 1$

$$= \frac{42}{3-10a}$$

b) $\dfrac{\dfrac{x^2}{y} - \dfrac{5}{x}}{xz} = \dfrac{\dfrac{x^2}{y} \cdot \dfrac{x}{x} - \dfrac{5}{x} \cdot \dfrac{y}{y}}{xz}$ ⟵ Multiplying by 1 to get the LCD, xy, for the numerator of the complex rational expression

$$= \dfrac{\dfrac{x^3}{xy} - \dfrac{5y}{xy}}{xz}$$

$$= \dfrac{\dfrac{x^3 - 5y}{xy}}{xz}$$ ⟵ Subtracting

⟵ If you prefer, write xz as $\dfrac{xz}{1}$.

$$= \dfrac{x^3 - 5y}{xy} \div (xz)$$ Rewriting with a division symbol

$$= \dfrac{x^3 - 5y}{xy} \cdot \dfrac{1}{xz}$$ Multiplying by the reciprocal of the divisor (inverting and multiplying)

$$= \dfrac{x^3 - 5y}{x^2 yz}$$

Multiplying by the LCD (Method 2)

A second method for simplifying complex rational expressions relies on multiplying by an expression equal to 1.

> ### To Simplify a Complex Rational Expression by Multiplying by the LCD
>
> 1. Find the LCD of *all* rational expressions within the complex rational expression.
> 2. Multiply the complex rational expression by a factor equal to 1. Write 1 as the LCD over itself (LCD/LCD).
> 3. Distribute and simplify. No fractional expressions should remain within the complex rational expression.
> 4. Factor and, if possible, simplify.

E x a m p l e   3   Simplify: $\dfrac{\dfrac{1}{2} + \dfrac{3}{4}}{\dfrac{5}{6} - \dfrac{3}{8}}$.

Solution

1. Unlike Method 1, in which $\frac{1}{2} + \frac{3}{4}$ would be treated separately from $\frac{5}{6} - \frac{3}{8}$, here we look for the LCD of *all* four fractions. That LCD is 24.

2. We multiply by a form of 1, using the LCD:

$$\frac{\dfrac{1}{2}+\dfrac{3}{4}}{\dfrac{5}{6}-\dfrac{3}{8}}=\frac{\dfrac{1}{2}+\dfrac{3}{4}}{\dfrac{5}{6}-\dfrac{3}{8}}\cdot\frac{24}{24} \qquad \text{Multiplying by a factor equal to 1,}$$
$$\text{using the LCD: } \frac{24}{24}=1$$

3. Using the distributive law, we perform the multiplication:

$$\frac{\dfrac{1}{2}+\dfrac{3}{4}}{\dfrac{5}{6}-\dfrac{3}{8}}\cdot\frac{24}{24}=\frac{\left(\dfrac{1}{2}+\dfrac{3}{4}\right)24}{\left(\dfrac{5}{6}-\dfrac{3}{8}\right)24}\;\longleftarrow\;\begin{array}{l}\text{Multiplying the numerator by 24}\\\text{Don't forget the parentheses!}\\\text{Multiplying the denominator by 24}\end{array}$$

$$=\frac{\dfrac{1}{2}(24)+\dfrac{3}{4}(24)}{\dfrac{5}{6}(24)-\dfrac{3}{8}(24)} \qquad \text{Using the distributive law}$$

$$=\frac{12+18}{20-9},\;\;\text{or}\;\;\frac{30}{11}. \qquad \text{Simplifying}$$

4. The result, $\frac{30}{11}$, cannot be factored or simplified, so we are done.

Multiplying like this effectively clears fractions in both the top and bottom of the complex rational expression. In Example 4 we follow, but do not list, the same four steps.

Example 4

Simplify.

a) $\dfrac{\dfrac{3}{x}+\dfrac{1}{2x}}{\dfrac{1}{3x}-\dfrac{3}{4x}}$

b) $\dfrac{1-\dfrac{1}{x}}{1-\dfrac{1}{x^2}}$

Solution

a) The denominators within the complex expression are x, $2x$, $3x$, and $4x$, so the LCD is $12x$. We multiply by 1 using $(12x)/(12x)$:

$$\frac{\dfrac{3}{x}+\dfrac{1}{2x}}{\dfrac{1}{3x}-\dfrac{3}{4x}}=\frac{\dfrac{3}{x}+\dfrac{1}{2x}}{\dfrac{1}{3x}-\dfrac{3}{4x}}\cdot\frac{12x}{12x}=\frac{\dfrac{3}{x}(12x)+\dfrac{1}{2x}(12x)}{\dfrac{1}{3x}(12x)-\dfrac{3}{4x}(12x)}. \qquad \begin{array}{l}\text{Using the}\\\text{distributive law}\end{array}$$

When we multiply by $12x$, all fractions in the numerator and the denominator of the complex rational expression are cleared:

$$\frac{\dfrac{3}{x}(12x)+\dfrac{1}{2x}(12x)}{\dfrac{1}{3x}(12x)-\dfrac{3}{4x}(12x)}=\frac{36+6}{4-9}=-\frac{42}{5}.$$

b) $\dfrac{1 - \dfrac{1}{x}}{1 - \dfrac{1}{x^2}} = \dfrac{1 - \dfrac{1}{x}}{1 - \dfrac{1}{x^2}} \cdot \dfrac{x^2}{x^2}$ The LCD is x^2 so we multiply by 1 using x^2/x^2.

$= \dfrac{1 \cdot x^2 - \dfrac{1}{x} \cdot x^2}{1 \cdot x^2 - \dfrac{1}{x^2} \cdot x^2}$ Using the distributive law

$= \dfrac{x^2 - x}{x^2 - 1}$ All fractions have been cleared within the complex rational expression.

$\left. \begin{array}{l} = \dfrac{x(x-1)}{(x+1)(x-1)} \\[2mm] = \dfrac{x}{x+1} \end{array} \right\}$ Factoring and simplifying: $\dfrac{x-1}{x-1} = 1$

It is important to understand both of the methods studied in this section. Sometimes, as in Example 1, the complex rational expression is either given as—or easily written as—a quotient of two rational expressions. In these cases, Method 1 (using division) is probably the easiest method to use. Other times, as in Example 4(a), it is not difficult to find the LCD of all denominators in the complex rational expression. When this occurs, it is usually easier to use Method 2 (multiplying by the LCD). The more practice you get using both methods, the better you will be at selecting the easier method for any given problem.

Exercise Set 6.5

Simplify. Use either method or the method specified by your instructor.

1. $\dfrac{1 + \dfrac{1}{2}}{1 + \dfrac{1}{4}}$

2. $\dfrac{1 + \dfrac{3}{4}}{1 + \dfrac{1}{2}}$

3. $\dfrac{1 + \dfrac{1}{3}}{5 - \dfrac{5}{27}}$

4. $\dfrac{3 + \dfrac{1}{5}}{1 - \dfrac{3}{5}}$

5. $\dfrac{\dfrac{s}{3} + s}{\dfrac{3}{s} + s}$

6. $\dfrac{\dfrac{1}{x} - 5}{\dfrac{1}{x} + 3}$

7. $\dfrac{\dfrac{2}{x}}{\dfrac{3}{x} + \dfrac{1}{x^2}}$

8. $\dfrac{\dfrac{4}{x} - \dfrac{1}{x^2}}{\dfrac{2}{x^2}}$

9. $\dfrac{\dfrac{2a - 5}{3a}}{\dfrac{a - 1}{6a}}$

10. $\dfrac{\dfrac{a + 4}{a^2}}{\dfrac{a - 2}{3a}}$

11. $\dfrac{\dfrac{x}{4} - \dfrac{4}{x}}{\dfrac{1}{4} + \dfrac{1}{x}}$

12. $\dfrac{\dfrac{3}{x} + \dfrac{3}{8}}{\dfrac{x}{8} - \dfrac{3}{x}}$

13. $\dfrac{\dfrac{1}{5} + \dfrac{1}{x}}{\dfrac{5 + x}{5}}$

14. $\dfrac{\dfrac{1}{3} - \dfrac{1}{a}}{\dfrac{3 - a}{3}}$

15. $\dfrac{\dfrac{1}{t^2} + 1}{\dfrac{1}{t} - 1}$

16. $\dfrac{2 + \dfrac{1}{x}}{2 - \dfrac{1}{x^2}}$

17. $\dfrac{\dfrac{x^2}{x^2 - y^2}}{\dfrac{x}{x + y}}$

18. $\dfrac{\dfrac{a^2}{a - 3}}{\dfrac{2a}{a^2 - 9}}$

19. $\dfrac{\dfrac{2}{a} + \dfrac{4}{a^2}}{\dfrac{5}{a^3} - \dfrac{3}{a}}$

20. $\dfrac{\dfrac{5}{x^3} - \dfrac{1}{x^2}}{\dfrac{2}{x} + \dfrac{3}{x^2}}$

21. $\dfrac{\dfrac{2}{7a^4} - \dfrac{1}{14a}}{\dfrac{3}{5a^2} + \dfrac{2}{15a}}$

22. $\dfrac{\dfrac{5}{4x^3} - \dfrac{3}{8x}}{\dfrac{3}{2x} + \dfrac{3}{4x^3}}$

Aha! **23.** $\dfrac{\dfrac{x}{5y^3} + \dfrac{3}{10y}}{\dfrac{3}{10y} + \dfrac{x}{5y^3}}$

24. $\dfrac{\dfrac{a}{6b^3} + \dfrac{4}{9b^2}}{\dfrac{5}{6b} - \dfrac{1}{9b^3}}$

25. $\dfrac{\dfrac{5}{ab^4} + \dfrac{2}{a^3b}}{\dfrac{5}{a^3b} - \dfrac{3}{ab}}$

26. $\dfrac{\dfrac{2}{x^2y} + \dfrac{3}{xy^2}}{\dfrac{3}{xy^2} + \dfrac{2}{x^2y}}$

27. $\dfrac{2 - \dfrac{3}{x^2}}{2 + \dfrac{3}{x^4}}$

28. $\dfrac{3 - \dfrac{2}{a^4}}{2 + \dfrac{3}{a^3}}$

29. $\dfrac{t - \dfrac{2}{t}}{t + \dfrac{5}{t}}$

30. $\dfrac{x + \dfrac{3}{x}}{x - \dfrac{2}{x}}$

31. $\dfrac{3 + \dfrac{4}{ab^3}}{\dfrac{3 + a}{a^2b}}$

32. $\dfrac{5 + \dfrac{3}{x^2y}}{\dfrac{3 + x}{x^3y}}$

33. $\dfrac{\dfrac{x + 5}{x^2}}{\dfrac{2}{x} - \dfrac{3}{x^2}}$

34. $\dfrac{\dfrac{a + 6}{a^3}}{\dfrac{2}{a^2} + \dfrac{3}{a}}$

35. $\dfrac{x - 3 + \dfrac{2}{x}}{x - 4 + \dfrac{3}{x}}$

36. $\dfrac{x - 2 + \dfrac{1}{x}}{x - 5 + \dfrac{4}{x}}$

37. Is it possible to simplify complex rational expressions without knowing how to divide rational expressions? Why or why not?

38. Why is the distributive law important when simplifying complex rational expressions?

SKILL MAINTENANCE

Solve.

39. $3x - 5 + 2(4x - 1) = 12x - 3$

40. $(x - 1)7 - (x + 1)9 = 4(x + 2)$

41. $\dfrac{3}{4}x - \dfrac{5}{8} = \dfrac{3}{8}x + \dfrac{7}{4}$

42. $\dfrac{5}{9} - \dfrac{2x}{3} = \dfrac{5x}{6} + \dfrac{4}{3}$

43. $x^2 - 7x - 30 = 0$

44. $x^2 + 8x - 20 = 0$

SYNTHESIS

45. Which of the two methods presented would you use to simplify Exercise 18? Why?

46. Which of the two methods presented would you use to simplify Exercise 26? Why?

In Exercises 47–50, find all x-values for which the given expression is undefined.

47. $\dfrac{\dfrac{x - 5}{x - 6}}{\dfrac{x - 7}{x - 8}}$

48. $\dfrac{\dfrac{x + 1}{x + 2}}{\dfrac{x + 3}{x + 4}}$

49. $\dfrac{\dfrac{2x + 3}{5x + 4}}{\dfrac{3}{7} - \dfrac{2x}{9}}$

50. $\dfrac{\dfrac{3x - 5}{2x - 7}}{\dfrac{4x}{5} - \dfrac{5}{6}}$

51. The formula

$$\dfrac{\dfrac{P\left(1 + \dfrac{i}{12}\right)^2}{\left(1 + \dfrac{i}{12}\right)^2 - 1}}{\dfrac{i}{12}},$$

where P is a loan amount and i is an interest rate, arises in certain business situations. Simplify this expression. (*Hint*: Expand the binomials.)

52. Find the simplified form for the reciprocal of

$$\dfrac{2}{x - 1} - \dfrac{1}{3x - 2}.$$

Simplify.

53. $\dfrac{\dfrac{5}{x + 2} - \dfrac{3}{x - 2}}{\dfrac{x}{x - 1} + \dfrac{x}{x + 1}}$

54. $\dfrac{\dfrac{x}{x + 5} + \dfrac{3}{x + 2}}{\dfrac{2}{x + 2} - \dfrac{x}{x + 5}}$

Aha! **55.** $\left[\dfrac{\dfrac{x - 1}{x - 1} - 1}{\dfrac{x + 1}{x - 1} + 1}\right]^5$

56. $1 + \dfrac{1}{1 + \dfrac{1}{1 + \dfrac{1}{x}}}$

57. $\dfrac{\dfrac{z}{1 - \dfrac{z}{2 + 2z}} - 2z}{\dfrac{2z}{5z - 2} - 3}$

58. Under what circumstance(s) will there be no restrictions on the variable appearing in a complex rational expression?

59. Use a grapher to check Example 2(a).

Solving Rational Equations

6.6

Solving a New Type of Equation • A Visual Interpretation

Our study of rational expressions allows us to solve a type of equation that we could not have solved prior to this chapter.

CONNECTING THE CONCEPTS

In Sections 6.1–6.5, we learned how to *simplify expressions* like

$$\frac{x^2-1}{x+2} \cdot \frac{3x}{x+1} \quad \text{and} \quad \frac{\dfrac{5}{x}}{\dfrac{2}{x^2}+\dfrac{3}{x}}.$$

In this section, we will return to *solving equations*. These equations will look like

the following:

$$x+\frac{6}{x}=-5 \quad \text{and} \quad \frac{3}{x-5}+\frac{1}{x+5}=\frac{2}{x^2-25}.$$

As always, be careful not to confuse simplifying an expression with solving an equation. When expressions are simplified, the result is an equivalent expression. When equations are solved, the result is a solution.

Solving a New Type of Equation

A **rational**, or **fractional**, **equation** is an equation containing one or more rational expressions, often with the variable in a denominator. Here are some examples:

$$\frac{2}{3}+\frac{5}{6}=\frac{x}{9}, \qquad t+\frac{7}{t}=-5, \qquad \frac{x^2}{x-1}=\frac{1}{x-1}.$$

> #### To Solve a Rational Equation
> 1. List any restrictions that exist. No solution can make a denominator equal 0.
> 2. Clear the equation of fractions by multiplying both sides by the LCD of all rational expressions in the equation.
> 3. Solve the resulting equation using the addition principle, the multiplication principle, and the principle of zero products, as needed.
> 4. Check the possible solution(s) in the original equation.

In the examples that follow, we *do not* use the LCD to add or subtract rational expressions. Instead, we use the LCD as a multiplier that will clear fractions.

Example 1

Solve: $\dfrac{x}{6} - \dfrac{x}{8} = \dfrac{1}{12}$.

Solution   Because no variable appears in a denominator, no restrictions exist. The LCD is 24, so we multiply both sides by 24:

$$24\left(\dfrac{x}{6} - \dfrac{x}{8}\right) = 24 \cdot \dfrac{1}{12}$$

Using the multiplication principle to multiply both sides by the LCD. Parentheses are important!

$$24 \cdot \dfrac{x}{6} - 24 \cdot \dfrac{x}{8} = 24 \cdot \dfrac{1}{12}$$

Using the distributive law

Be sure to multiply *each* term by the LCD.

$$\left.\begin{array}{c} \dfrac{24x}{6} - \dfrac{24x}{8} = \dfrac{24}{12} \\[2mm] 4x - 3x = 2 \end{array}\right\}$$

Simplifying. Note that all fractions have been cleared.

$$x = 2.$$

Check:

$$\begin{array}{c|c} \dfrac{x}{6} - \dfrac{x}{8} = \dfrac{1}{12} \\[2mm] \dfrac{2}{6} - \dfrac{2}{8} \ \Big|?\ \dfrac{1}{12} \\[2mm] \dfrac{1}{3} - \dfrac{1}{4} \\[2mm] \dfrac{4}{12} - \dfrac{3}{12} \\[2mm] \dfrac{1}{12} \ \Big|\ \dfrac{1}{12} \ \text{TRUE} \end{array}$$

This checks, so the solution is 2.

Up to now, the multiplication principle has been used only to multiply both sides of an equation by a nonzero constant. Because rational equations often contain variables in a denominator, clearing fractions may now require us to multiply both sides of an equation by a variable expression. Since a variable expression could represent 0, *multiplying both sides of an equation by a variable expression does not always produce an equivalent equation.* Thus checking in the original equation is very important.

Example 2

Solve.

a) $\dfrac{2}{3x} + \dfrac{1}{x} = 10$

b) $x + \dfrac{6}{x} = -5$

c) $1 + \dfrac{3x}{x+2} = \dfrac{-6}{x+2}$

d) $\dfrac{3}{x-5} + \dfrac{1}{x+5} = \dfrac{2}{x^2-25}$

Solution

a) Note that x cannot be 0. The LCD is $3x$, so we multiply both sides by $3x$:

$$\frac{2}{3x} + \frac{1}{x} = 10$$
The LCD is $3x$; $x \neq 0$.

$$3x\left(\frac{2}{3x} + \frac{1}{x}\right) = 3x \cdot 10$$
Using the multiplication principle to multiply both sides by the LCD. *Don't forget the parentheses!*

$$3\!\!\!/x \cdot \frac{2}{3\!\!\!/x} + 3\!\!\!/x \cdot \frac{1}{\!\!\!/x} = 3x \cdot 10$$
Using the distributive law

$$2 + 3 = 30x$$
Removing factors equal to 1: $3x/(3x) = 1$ and $x/x = 1$. This clears all fractions.

$$5 = 30x$$

$$\frac{5}{30} = x, \quad \text{so } x = \frac{1}{6}.$$
Since $\frac{1}{6} \neq 0$, which was the restriction in the first step, this *should* check.

Check:

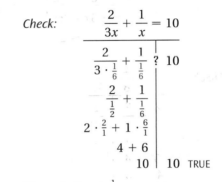

$$
\begin{array}{c|c}
\dfrac{2}{3x} + \dfrac{1}{x} = 10 & \\[2mm]
\hline
\dfrac{2}{3 \cdot \frac{1}{6}} + \dfrac{1}{\frac{1}{6}} \ ? \ 10 & \\[3mm]
\dfrac{2}{\frac{1}{2}} + \dfrac{1}{\frac{1}{6}} & \\[3mm]
2 \cdot \frac{2}{1} + 1 \cdot \frac{6}{1} & \\[2mm]
4 + 6 & \\[1mm]
10 & 10 \quad \text{TRUE}
\end{array}
$$

The solution is $\frac{1}{6}$.

b) Again, note that x cannot be 0. We multiply both sides of the equation by the LCD, x:

$$x + \frac{6}{x} = -5$$
We cannot have $x = 0$.

$$x\left(x + \frac{6}{x}\right) = x(-5)$$
Multiplying both sides by x. *Don't forget the parentheses!*

$$x \cdot x + \!\!\!/x \cdot \frac{6}{\!\!\!/x} = -5x$$
Using the distributive law

$$x^2 + 6 = -5x$$
Removing a factor equal to 1: $x/x = 1$. We are left with a quadratic equation.

$$x^2 + 5x + 6 = 0$$
Using the addition principle to add $5x$ to both sides

$$(x + 3)(x + 2) = 0$$
Factoring

$$x + 3 = 0 \quad or \quad x + 2 = 0$$
Using the principle of zero products

$$x = -3 \quad or \quad x = -2$$
Since neither solution is 0, the restriction in the first step, they should both check.

Check: For -3:

$$x + \frac{6}{x} = -5$$

$$-3 + \frac{6}{-3} \;?\; -5$$

$$-3 - 2$$

$$-5 \;\big|\; -5 \quad \text{TRUE}$$

For -2:

$$x + \frac{6}{x} = -5$$

$$-2 + \frac{6}{-2} \;?\; -5$$

$$-2 - 3$$

$$-5 \;\big|\; -5 \quad \text{TRUE}$$

Both of these check, so there are two solutions, -3 and -2.

c) To avoid division by 0, we must have $x + 2 \neq 0$, or $x \neq -2$. With this restriction in mind, we multiply both sides of the equation by the LCD, $x + 2$:

$$1 + \frac{3x}{x + 2} = \frac{-6}{x + 2} \qquad \text{We cannot have } x = -2.$$

$$(x + 2)\left(1 + \frac{3x}{x + 2}\right) = (x + 2)\frac{-6}{x + 2} \qquad \begin{array}{l}\text{Multiplying both sides} \\ \text{by } x + 2. \text{ Don't forget} \\ \text{the parentheses.}\end{array}$$

$$(x + 2) \cdot 1 + (x + 2)\frac{3x}{x + 2} = (x + 2)\frac{-6}{x + 2} \qquad \begin{array}{l}\text{Using the distributive} \\ \text{law}\end{array}$$

$$x + 2 + 3x = -6 \qquad \begin{array}{l}\text{Removing a factor equal to 1:} \\ (x + 2)/(x + 2) = 1\end{array}$$

$$4x + 2 = -6$$

$$4x = -8$$

$$x = -2 \qquad \text{Above, we stated that } x \neq -2.$$

Because of the above restriction, -2 must be rejected as a solution. The student can confirm that -2 results in division by 0. The equation has no solution.

d)

$$\frac{3}{x - 5} + \frac{1}{x + 5} = \frac{2}{x^2 - 25} \qquad \begin{array}{l}\text{Note that } x \neq 5 \text{ and} \\ x \neq -5. \text{ The LCD is} \\ (x - 5)(x + 5).\end{array}$$

$$(x - 5)(x + 5)\left(\frac{3}{x - 5} + \frac{1}{x + 5}\right) = (x - 5)(x + 5)\frac{2}{(x - 5)(x + 5)}$$

$$\frac{(x - 5)(x + 5)3}{x - 5} + \frac{(x - 5)(x + 5)}{x + 5} = \frac{2(x - 5)(x + 5)}{(x - 5)(x + 5)} \qquad \begin{array}{l}\text{Using the} \\ \text{distributive law}\end{array}$$

$$(x + 5)3 + (x - 5) = 2 \qquad \left\{\begin{array}{l}\text{Removing factors equal to 1:} \\ \dfrac{x - 5}{x - 5} = 1, \dfrac{x + 5}{x + 5} = 1, \text{ and} \\ \dfrac{(x - 5)(x + 5)}{(x - 5)(x + 5)} = 1\end{array}\right.$$

$$3x + 15 + x - 5 = 2 \qquad \text{Using the distributive law}$$

$$4x + 10 = 2$$

$$4x = -8$$

$$x = -2 \qquad \begin{array}{l}-2 \neq 5 \text{ and } -2 \neq -5, \\ \text{so } -2 \text{ *should* check.}\end{array}$$

We leave it to the student to check that -2 is the solution.

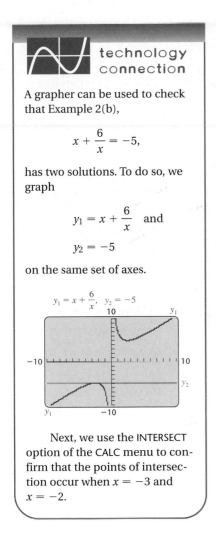

technology connection

A grapher can be used to check that Example 2(b),

$$x + \frac{6}{x} = -5,$$

has two solutions. To do so, we graph

$$y_1 = x + \frac{6}{x} \quad \text{and}$$

$$y_2 = -5$$

on the same set of axes.

Next, we use the INTERSECT option of the CALC menu to confirm that the points of intersection occur when $x = -3$ and $x = -2$.

A Visual Interpretation

It is possible to solve a rational equation by graphing. The procedure consists of graphing each side of the equation and then determining the first coordinate(s) of any point(s) of intersection. (Since the advent of the graphing calculator, producing such graphs requires little work.) For example, the equation

$$\frac{x}{4} + \frac{x}{2} = 6$$

can be solved by graphing the equations

$$y = \frac{x}{4} + \frac{x}{2} \quad \text{and} \quad y = 6$$

on the same set of axes.

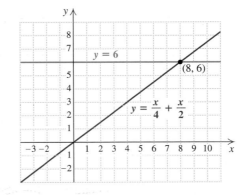

As we can see in the graph above, when $x = 8$, the value of $x/4 + x/2$ is 6. Thus, 8 is the solution of $x/4 + x/2 = 6$. We can check by substitution:

$$\frac{x}{4} + \frac{x}{2} = \frac{8}{4} + \frac{8}{2} = 2 + 4 = 6.$$

FOR EXTRA HELP

Exercise Set 6.6

Digital Video Tutor CD 4
Videotape 12

InterAct Math

Math Tutor Center

MathXL

MyMathLab.com

Solve. If no solution exists, state this.

1. $\dfrac{5}{8} - \dfrac{4}{5} = \dfrac{x}{20}$

2. $\dfrac{4}{5} - \dfrac{2}{3} = \dfrac{x}{9}$

3. $\dfrac{1}{3} + \dfrac{5}{6} = \dfrac{1}{x}$

4. $\dfrac{3}{5} + \dfrac{1}{8} = \dfrac{1}{x}$

5. $\dfrac{1}{6} + \dfrac{1}{8} = \dfrac{1}{t}$

6. $\dfrac{1}{8} + \dfrac{1}{10} = \dfrac{1}{t}$

7. $x + \dfrac{5}{x} = -6$

8. $x + \dfrac{6}{x} = -7$

9. $\dfrac{x}{6} - \dfrac{6}{x} = 0$

10. $\dfrac{x}{7} - \dfrac{7}{x} = 0$

11. $\dfrac{5}{x} = \dfrac{6}{x} - \dfrac{1}{3}$

12. $\dfrac{4}{t} = \dfrac{5}{t} - \dfrac{1}{2}$

13. $\dfrac{5}{3t} + \dfrac{3}{t} = 1$

14. $\dfrac{3}{4x} + \dfrac{5}{x} = 1$

15. $\dfrac{x-8}{x+3} = \dfrac{1}{4}$

16. $\dfrac{a-4}{a+5} = \dfrac{3}{8}$

17. $\dfrac{2}{x+1} = \dfrac{1}{x-2}$

18. $\dfrac{5}{x-1} = \dfrac{3}{x+2}$

19. $\dfrac{a}{6} - \dfrac{a}{10} = \dfrac{1}{6}$

20. $\dfrac{t}{8} - \dfrac{t}{12} = \dfrac{1}{8}$

21. $\dfrac{x+1}{3} - 1 = \dfrac{x-1}{2}$

22. $\dfrac{x+2}{5} - 1 = \dfrac{x-2}{4}$

23. $\dfrac{4}{t-5} = \dfrac{t-1}{t-5}$

24. $\dfrac{2}{t-9} = \dfrac{t-7}{t-9}$

25. $\dfrac{3}{x+4} = \dfrac{5}{x}$

26. $\dfrac{2}{x+3} = \dfrac{7}{x}$

27. $\dfrac{a-4}{a-1} = \dfrac{a+2}{a-2}$

28. $\dfrac{x-2}{x-3} = \dfrac{x-1}{x+1}$

29. $\dfrac{4}{x-3} + \dfrac{2x}{x^2-9} = \dfrac{1}{x+3}$

30. $\dfrac{x}{x+4} - \dfrac{4}{x-4} = \dfrac{x^2+16}{x^2-16}$

31. $\dfrac{5}{y-3} - \dfrac{30}{y^2-9} = 1$

32. $\dfrac{1}{x+3} + \dfrac{1}{x-3} = \dfrac{1}{x^2-9}$

33. $\dfrac{4}{8-a} = \dfrac{4-a}{a-8}$

34. $\dfrac{t+10}{7-t} = \dfrac{3}{t-7}$

Aha! **35.** $\dfrac{-2}{x+2} = \dfrac{x}{x+2}$

36. $\dfrac{3}{2x-6} = \dfrac{x}{2x-6}$

37. When solving rational equations, why do we multiply each side by the LCD?

38. Explain the difference between adding rational expressions and solving rational equations.

SKILL MAINTENANCE

39. The sum of two consecutive odd numbers is 276. Find the numbers.

40. The length of a rectangle is 3 yd greater than the width. The area of the rectangle is 10 yd². Find the perimeter.

41. The height of a triangle is 3 cm longer than its base. If the area of the triangle is 54 cm², find the measurements of the base and the height.

42. The product of two consecutive even integers is 48. Find the numbers.

43. *Human physiology.* Between June 9 and June 24, Seth's beard grew 0.9 cm. Find the rate at which Seth's beard grows.

44. *Gardening.* Between July 7 and July 12, Carla's string beans grew 1.4 in. Find the string beans' growth rate.

SYNTHESIS

45. Describe a method that can be used to create rational equations that have no solution.

46. How can a graph be used to determine how many solutions an equation has?

Solve.

47. $1 + \dfrac{x-1}{x-3} = \dfrac{2}{x-3} - x$

48. $\dfrac{4}{y-2} + \dfrac{3}{y^2-4} = \dfrac{5}{y+2} + \dfrac{2y}{y^2-4}$

49. $\dfrac{x}{x^2+3x-4} + \dfrac{x}{x^2+6x+8} =$

$\qquad\qquad\qquad \dfrac{2x}{x^2+x-2} - \dfrac{1}{x^2+6x+8}$

50. $\dfrac{12-6x}{x^2-4} = \dfrac{3x}{x+2} - \dfrac{3-2x}{2-x}$

51. $\dfrac{x^2}{x^2-4} = \dfrac{x}{x+2} - \dfrac{2x}{2-x}$

52. $7 - \dfrac{a-2}{a+3} = \dfrac{a^2-4}{a+3} + 5$

53. $\dfrac{1}{x-1} + x - 5 = \dfrac{5x-4}{x-1} - 6$

54. $\dfrac{5-3a}{a^2+4a+3} - \dfrac{2a+2}{a+3} = \dfrac{3-a}{a+1}$

55. Use a grapher to check the solutions to Examples 1 and 2(c).

56. Use a grapher to check your answers to Exercises 9, 25, and 47.

Applications Using Rational Equations and Proportions

6.7

Problem Solving • Problems Involving Work • Problems Involving Motion • Problems Involving Proportions

In many areas of study, applications involving rates, proportions, or reciprocals translate to rational equations. By using the five steps for problem solving and the lessons of Section 6.6, we can now solve such problems.

Problem Solving

Example 1

A number, plus three times its reciprocal, is -4. Find the number.

Solution

1. **Familiarize.** Let's try to guess the number. Try 2: $2 + 3 \cdot \frac{1}{2} = \frac{7}{2}$. Although $\frac{7}{2} \neq -4$, the guess helps us to better understand how the problem can be translated. We let $x =$ the number for which we are searching.

2. **Translate.** From the *Familiarize* step, we can translate directly:

   A number, plus three times its reciprocal, is -4.

   $$x + 3 \cdot \frac{1}{x} = -4$$

3. **Carry out.** We solve the equation:

   $$x + 3 \cdot \frac{1}{x} = -4$$
   We note the restriction that x cannot equal 0.

   $$x\left(x + \frac{3}{x}\right) = x(-4)$$
   Multiplying both sides of the equation by the LCD, x. Don't forget the parentheses.

   $$x \cdot x + x \cdot \frac{3}{x} = -4x$$
   Using the distributive law

   $$x^2 + 3 = -4x$$
   Simplifying

   $$x^2 + 4x + 3 = 0$$
   $$(x + 3)(x + 1) = 0$$
   $$x + 3 = 0 \quad or \quad x + 1 = 0$$
   Using the principle of zero products

   $$x = -3 \quad or \quad x = -1.$$

4. **Check.** Three times the reciprocal of -3 is $3 \cdot \frac{1}{-3}$, or -1. Since $-3 + (-1) = -4$, the number -3 is a solution.
   Three times the reciprocal of -1 is $3 \cdot \frac{1}{-1}$, or -3. Since $-1 + (-3) = -4$, the number -1 is also a solution.

5. **State.** The solutions are -3 and -1.

Problems Involving Work

E x a m p l e   2

Sorting recyclables.   Cecilia and Aaron work as volunteers at a town's recycling depot. Cecilia can sort a day's accumulation of recyclables in 4 hr, while Aaron requires 6 hr to do the same job. How long would it take them, working together, to sort the recyclables?

Solution

1. **Familiarize.**   We familiarize ourselves with the problem by exploring two common, but *incorrect*, approaches.

   a) One common incorrect approach is to simply add the two times:

   $$4\,\text{hr} + 6\,\text{hr} = 10\,\text{hr}.$$

   Let's think about this. If Cecilia can do the sorting *alone* in 4 hr, then Cecilia and Aaron *together* should take *less* than 4 hr. Thus we reject 10 hr as a solution and reason that the answer must be less than 4 hr.

   b) Another incorrect approach is to assume that Cecilia does half the sorting and Aaron does the other half. Then

   Cecilia sorts $\frac{1}{2}$ of the accumulation in $\frac{1}{2}(4\,\text{hr})$, or 2 hr, and
   Aaron sorts $\frac{1}{2}$ of the accumulation in $\frac{1}{2}(6\,\text{hr})$, or 3 hr.

   This would waste time since Cecilia would finish 1 hr earlier than Aaron. In reality, Cecilia would help Aaron after completing her half, so that Aaron would actually sort less than half of the accumulation. This tells us that the entire job will take them between 2 hr and 3 hr.

A correct approach is to consider how much of the sorting is finished in 1 hr, 2 hr, 3 hr, and so on. It takes Cecilia 4 hr to sort the recyclables alone, so her rate is $\frac{1}{4}$ of the job per hour. It takes Aaron 6 hr to do the sorting alone, so his rate is $\frac{1}{6}$ of the job per hour. Working together, they can complete

$$\frac{1}{4} + \frac{1}{6}, \quad \text{or } \frac{5}{12} \text{ of the sorting in 1 hr.}$$ Together, their rate is $\frac{5}{12}$ of the job per hour.

In 2 hr, Cecilia can do $\frac{1}{4} \cdot 2$ of the sorting and Aaron can do $\frac{1}{6} \cdot 2$ of the sorting. Working together, they can complete

$$\frac{1}{4} \cdot 2 + \frac{1}{6} \cdot 2, \quad \text{or } \frac{5}{6} \text{ of the sorting in 2 hr.}$$ Note that $\frac{5}{12} \cdot 2 = \frac{5}{6}$.

Continuing this reasoning, we can form a table.

	Fraction of the Sorting Completed		
Time	**Cecilia**	**Aaron**	**Together**
1 hr	$\dfrac{1}{4}$	$\dfrac{1}{6}$	$\dfrac{1}{4} + \dfrac{1}{6}$, or $\dfrac{5}{12}$
2 hr	$\dfrac{1}{4} \cdot 2$	$\dfrac{1}{6} \cdot 2$	$\left(\dfrac{1}{4} + \dfrac{1}{6}\right)2$, or $\dfrac{5}{12} \cdot 2$, or $\dfrac{5}{6}$ ← This is too little.
3 hr	$\dfrac{1}{4} \cdot 3$	$\dfrac{1}{6} \cdot 3$	$\left(\dfrac{1}{4} + \dfrac{1}{6}\right)3$, or $\dfrac{5}{12} \cdot 3$, or $1\dfrac{1}{4}$ ← This is too much.
t hr	$\dfrac{1}{4} \cdot t$	$\dfrac{1}{6} \cdot t$	$\left(\dfrac{1}{4} + \dfrac{1}{6}\right)t$, or $\dfrac{5}{12} \cdot t$

From the table, we see that if they work 3 hr, the fraction of the sorting that they complete is $1\frac{1}{4}$, which is more of the job than needs to be done. We need to find a number t for which the fraction of the sorting that is completed in t hours is exactly 1, no more and no less.

2. **Translate.** From the table, we see that the time we want is some number t for which

Portion of work done by Cecilia in t hr $\underbrace{\dfrac{1}{4} \cdot t}$ + $\underbrace{\dfrac{1}{6} \cdot t}$ = 1,   Portion of work done by Aaron in t hr

or

$$\underbrace{\left(\dfrac{1}{4} + \dfrac{1}{6}\right)t = 1, \quad \text{or} \quad \dfrac{5}{12} \cdot t = 1.}_{\text{Portion of work done together in } t \text{ hr}}$$

3. **Carry out.** We can choose any one of the above equations to solve:

$$\dfrac{5}{12} \cdot t = 1$$

$$\dfrac{12}{5} \cdot \dfrac{5}{12} \cdot t = \dfrac{12}{5} \cdot 1 \qquad \text{Multiplying both sides by } \tfrac{12}{5}$$

$$t = \dfrac{12}{5}, \quad \text{or} \quad 2\dfrac{2}{5} \text{ hr.}$$

4. **Check.** The check can be done following the pattern used in the table of the *Familiarize* step above:

$$\dfrac{1}{4} \cdot \dfrac{12}{5} + \dfrac{1}{6} \cdot \dfrac{12}{5} = \dfrac{3}{5} + \dfrac{2}{5} = \dfrac{5}{5} = 1.$$

A second, partial, check is that (as we predicted in step 1) the answer is between 2 hr and 3 hr.

5. **State.** Together, it takes Cecilia and Aaron $2\frac{2}{5}$ hr to complete the sorting.

> **The Work Principle**
>
> Suppose that A requires a units of time to complete a task and B requires b units of time to complete the same task. Then
>
> A works at a rate of $\dfrac{1}{a}$ tasks per unit of time,
>
> B works at a rate of $\dfrac{1}{b}$ tasks per unit of time, and
>
> A and B together work at a rate of $\dfrac{1}{a} + \dfrac{1}{b}$ tasks per unit of time.
>
> If A and B, working together, require t units of time to complete the task, then all three of the following equations hold:
>
> $$\frac{1}{a} \cdot t + \frac{1}{b} \cdot t = 1; \qquad \left(\frac{1}{a} + \frac{1}{b} \right) t = 1; \qquad \frac{1}{a} + \frac{1}{b} = \frac{1}{t}.$$

Problems Involving Motion

Problems that deal with distance, speed (or rate), and time are called **motion problems**. Translation of these problems involves the distance formula, $d = r \cdot t$, and/or the equivalent formulas $r = d/t$ and $t = d/r$.

E x a m p l e 3   ***Driving speed.***   Nancy drives 20 mph faster than her father, Greg. In the same time that Nancy travels 180 mi, her father travels 120 mi. Find their speeds.

Solution

1. **Familiarize.**   Suppose that Greg drives 30 mph. Nancy would then be driving at $30 + 20$, or 50 mph. Thus, if r is the speed of Greg's car, in miles per hour, then the speed of Nancy's car is $r + 20$.

   If Greg drove 30 mph, he would drive 120 mi in 120/30, or 4 hr. At 50 mph, Nancy would drive 180 mi in 180/50, or $3\frac{3}{5}$ hr. Because we know that both drivers spend the same amount of time traveling, and because $4 \text{ hr} \neq 3\frac{3}{5} \text{ hr}$, we see that our guess of 30 mph is incorrect. We let $t =$ the time, in hours, that is spent traveling and create a table.

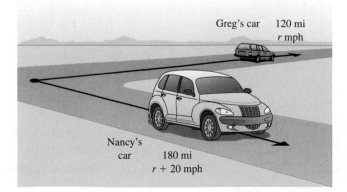

| d | $=$ | r | $\cdot$ | t |

	Distance (in miles)	Speed (in miles per hour)	Time (in hours)
Greg's Car	120	r	t
Nancy's Car	180	$r + 20$	t

2. **Translate.**   Examine how we checked our guess. We found, and then compared, the two driving times. The times were found by dividing the distances, 120 mi and 180 mi, by the rates, 30 mph and 50 mph, respectively. Thus the t's in the table above can be replaced, using the formula $t = d/r$. This yields a table that uses only one variable.

	Distance (in miles)	Speed (in miles per hour)	Time (in hours)
Greg's Car	120	r	$120/r$
Nancy's Car	180	$r + 20$	$180/(r + 20)$

← The times must ← be the same.

Since the times must be the same for both cars, we have the equation

$$\frac{120}{r} = \frac{180}{r + 20}.$$

Note that $\dfrac{\text{mi}}{\text{mph}} = \dfrac{\text{mi}}{\text{mi/hr}} = \cancel{\text{mi}} \cdot \dfrac{\text{hr}}{\cancel{\text{mi}}} = \text{hr}$, so we are indeed comparing two times.

3. **Carry out.**   To solve the equation, we first multiply both sides by the LCD, $r(r + 20)$:

$$r(r + 20) \cdot \frac{120}{r} = r(r + 20) \cdot \frac{180}{r + 20}$$

Multiplying both sides by the LCD, $r(r + 20)$. Note that we must have $r \neq 0$ and $r \neq -20$.

$$120(r + 20) = 180r \qquad \text{Simplifying}$$
$$120r + 2400 = 180r \qquad \text{Using the distributive law}$$
$$2400 = 60r \qquad \text{Subtracting } 120r \text{ from both sides}$$
$$40 = r. \qquad \text{Dividing both sides by 60}$$

We now have a possible solution. The speed of Greg's car is 40 mph, and the speed of Nancy's car is $40 + 20$, or 60 mph.

4. **Check.**   We first reread the problem to confirm that we were to find the speeds. Note that if Nancy drives 60 mph and Greg drives 40 mph, Nancy is indeed going 20 mph faster than her father. If Nancy travels 180 mi at 60 mph, she drives for 180/60, or 3 hr. If Greg travels 120 mi at 40 mph, he drives for 120/40, or 3 hr. Since the times are the same, the speeds check.

5. **State.**   Greg is driving at 40 mph, while Nancy is driving at 60 mph.

Problems Involving Proportions

A **ratio** of two quantities is their quotient. For example, 37% is the ratio of 37 to 100, or $\frac{37}{100}$. A **proportion** is an equation stating that two ratios are equal.

Proportion

An equality of ratios, $A/B = C/D$, is called a *proportion*. The numbers within a proportion are said to be *proportional* to each other.

Proportions can be used to solve a variety of applied problems.

E x a m p l e 4

Mileage.   In 2000, Honda introduced the Insight, a gasoline–electric car that travels 280 mi on 4 gal of gas. Find the amount of gas required for a 700-mi trip.

Solution   By assuming that the car always burns gas at the same rate, we can form a proportion in which the ratio of miles to gallons is expressed in two ways:

$$\text{Miles} \longrightarrow \frac{280}{4} = \frac{700}{x} \longleftarrow \text{Miles}$$
$$\text{Gallons} \longrightarrow \qquad\quad \longleftarrow \text{Gallons}$$

To solve for x, we multiply both sides of the equation by the LCD, $4x$:

$$4x \cdot \frac{280}{4} = 4x \cdot \frac{700}{x} \qquad \text{We could also simplify first and solve } 70 = \frac{700}{x}.$$

$$4 \cdot \frac{280x}{4} = x \cdot \frac{4 \cdot 700}{x} \qquad \text{Removing factors equal to 1: } \frac{4}{4} = 1 \text{ and } \frac{x}{x} = 1$$

$$280x = 4 \cdot 700$$

$$x = \frac{4 \cdot 700}{280} \qquad \text{Dividing both sides by 280}$$

$$x = 10. \qquad \text{Simplifying}$$

The trip will require 10 gal of gas.

Proportions arise in geometry when we are studying *similar triangles*. If two triangles are **similar**, then their corresponding angles have the same measure and their corresponding sides are proportional. To illustrate, if triangle *ABC* is similar to triangle *RST*, then angles *A* and *R* have the same measure, angles *B* and *S* have the same measure, angles *C* and *T* have the same measure, and

$$\frac{a}{r} = \frac{b}{s} = \frac{c}{t}.$$

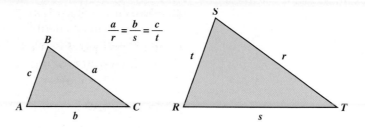

$$\frac{a}{r} = \frac{b}{s} = \frac{c}{t}$$

Example 5

Similar triangles.   Triangles *ABC* and *XYZ* are similar. Solve for *z* if *x* = 10, *a* = 8, and *c* = 5.

Solution   We make a drawing, write a proportion, and then solve. Note that side *a* is always opposite angle *A*, side *x* is always opposite angle *X*, and so on.

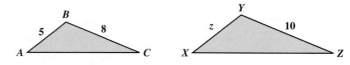

We have

$$\frac{z}{5} = \frac{10}{8}$$ The proportions $\frac{5}{z} = \frac{8}{10}, \frac{5}{8} = \frac{z}{10}$, or $\frac{8}{5} = \frac{10}{z}$ could also be used.

$$z = \frac{10}{8} \cdot 5$$ Multiplying both sides by 5

$$z = \frac{50}{8}, \text{ or } 6.25.$$

Example 6

Environmental science.   To determine the number of humpback whales in a pod, a marine biologist, using tail markings, identifies 27 members of the pod. Several weeks later, 40 whales from the pod are randomly sighted. Of the 40 sighted, 12 are from the 27 originally identified. Estimate the number of whales in the pod.

Solution

1. **Familiarize.**   If we knew that the 27 whales that were first identified constituted, say, 10% of the pod, we could easily calculate the pod's population from the proportion

$$\frac{27}{W} = \frac{10}{100},$$

where *W* is the size of the pod's population. Unfortunately, we are *not* told the percentage of the pod identified. We must reread the problem, looking for numbers that could be used to approximate this percentage.

2. **Translate.**   Since 12 of the 40 whales that were later sighted were among those originally identified, the ratio 12/40 estimates the percentage of the pod originally identified. We can then translate to a proportion:

Whales originally identified $\longrightarrow \dfrac{27}{W} = \dfrac{12}{40}$ $\longleftarrow$ Original whales sighted later
Entire pod $\longrightarrow$ $\phantom{\dfrac{27}{W}}$ $\longleftarrow$ Whales sighted later

3. **Carry out.**   To solve the proportion, we multiply by the LCD, $40W$:

$$40W \cdot \frac{27}{W} = 40W \cdot \frac{12}{40} \qquad \text{Multiplying both sides by } 40W$$

$$40 \cdot 27 = W \cdot 12 \qquad \begin{array}{l}\text{Removing factors equal to 1: } W/W = 1 \\ \text{and } 40/40 = 1\end{array}$$

$$\frac{40 \cdot 27}{12} = W \quad \text{or} \quad W = 90 \qquad \text{Dividing both sides by 12}$$

4. **Check.**   The check is left to the student.

5. **State.**   There are about 90 whales in the pod.

FOR EXTRA HELP

Exercise Set 6.7

Digital Video Tutor CD 4
Videotape 12   InterAct Math   Math Tutor Center   MathXL   MyMathLab.com

Solve.

1. A number, minus four times its reciprocal, is 3. Find the number.

2. A number, minus five times its reciprocal, is 4. Find the number.

3. The sum of a number and its reciprocal is 2. Find the number.

4. The sum of a number and five times its reciprocal is 6. Find the number.

5. *Construction.*   It takes Fontella 4 hr to put up paneling in a room. Omar takes 5 hr to do the same job. How long would it take them, working together, to panel the room?

6. *Carpentry.*   By checking work records, a carpenter finds that Juanita can build a small shed in 12 hr. Anton can do the same job in 16 hr. How long would it take if they worked together?

7. *Shoveling.*   Vern can shovel the snow from his driveway in 45 min. Nina can do the same job in 60 min. How long would it take Nina and Vern to shovel the driveway if they worked together?

8. *Raking.*   Zoë can rake her yard in 4 hr. Steffi does the same job in 3 hr. How long would it take the two of them, working together, to rake the yard?

9. *Masonry.*   By checking work records, a contractor finds that it takes Kenny Dewitt 8 hr to construct a wall of a certain size. It takes Betty Wohnt 6 hr to construct the same wall. How long would it take if they worked together?

10. *Plumbing.*   By checking work records, a plumber finds that Raul can plumb a house in 48 hr. Mira can do the same job in 36 hr. How long would it take if they worked together?

11. *Gardening.*   Nicole can weed her vegetable garden in 50 min, while Glen can weed the same garden in 40 min. How long would it take if they worked together?

12. *Harvesting.*   Bobbi can pick a quart of raspberries in 20 min. Blanche can pick a quart in 25 min. How long would it take if Bobbi and Blanche worked together?

13. *Computer printers.*   The HP OfficeJetG85 printer can copy Charlotte's dissertation in 12 min. The HP LaserJet 3200se can copy the same document in 20 min. If the two machines work together, how long would they take to copy the dissertation?

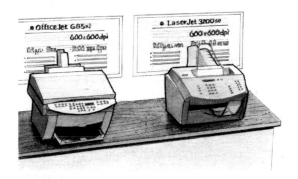

14. *Fax machines.*   The Brother MFC4500® can fax a year-end report in 10 min while the Xerox 850® can fax the same report in 8 min. How long would it take the two machines, working together, to fax the report? (Assume that the recipient has at least two machines for incoming faxes.)

15. *Speed of travel.*   A loaded Roadway truck is moving 40 mph faster than a New York Railways freight train. In the time that it takes the train to travel 150 mi, the truck travels 350 mi. Find their speeds. Complete the following table as part of the familiarization.

d	$=$	r	$\cdot$	t

	Distance (in miles)	Speed (in miles per hour)	Time (in hours)
Truck	350	r	$\dfrac{350}{r}$
Train	150		

16. *Train speeds.*   A B & M freight train is 14 km/h slower than an AMTRAK passenger train. The B & M train travels 330 km in the same time that it takes the AMTRAK train to travel 400 km. Find their speeds. Complete the following table as part of the familiarization.

d	$=$	r	$\cdot$	t

	Distance (in km)	Speed (in km/hr)	Time (in hours)
B & M	330		
AMTRAK	400	r	$\dfrac{400}{r}$

17. *Bicycle speed.*   Hank bicycles 5 km/h slower than Kelly. In the time that it takes Hank to bicycle 42 km, Kelly can bicycle 57 km. How fast does each bicyclist travel?

18. *Driving speed.*   Hillary's Lexus travels 30 mph faster than Bill's Harley. In the same time that Bill travels 75 mi, Hillary travels 120 mi. Find their speeds.

19. *Walking speed.*   Bonnie power walks 3 km/h faster than Ralph. In the time that it takes Ralph to walk 7.5 km, Bonnie walks 12 km. Find their speeds.

20. *Cross-country skiing.*   Gerard cross-country skis 4 km/h faster than Sally. In the time that it takes Sally to ski 18 km, Gerard skis 24 km. Find their speeds.

Aha! 21. *Tractor speed.*   Manley's tractor is just as fast as Caledonia's. It takes Manley 1 hr more than it takes Caledonia to drive to town. If Manley is 20 mi from town and Caledonia is 15 mi from town, how long does it take Caledonia to drive to town?

22. *Boat speed.* Tory and Emilio's motorboats both travel at the same speed. Tory pilots her boat 40 km before docking. Emilio continues for another 2 hr, traveling a total of 100 km before docking. How long did it take Tory to navigate the 40 km?

Geometry. *For each pair of similar triangles, find the value of the indicated letter.*

23. b

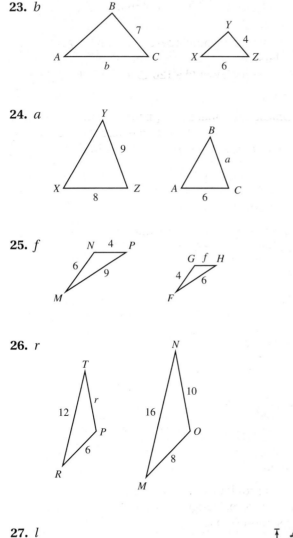

24. a

25. f

26. r

27. l

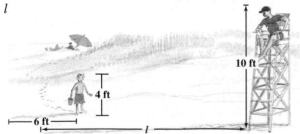

28. h

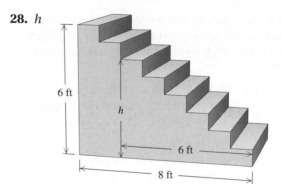

Geometry. *When three parallel lines are crossed by two or more lines (transversals), the lengths of corresponding segments are proportional (see the following figure).*

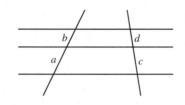

29. If a is 8 cm when b is 5 cm, find d when c is 6 cm.

30. If d is 7 cm when c is 10 cm, find a when b is 9 cm.

31. If c is 2 m longer than b and d is 2 m shorter than b, find all four lengths when a is 15 m.

32. If d is 2 m shorter than b and c is 3 m longer than b, find b, c, and d when a is 18 m.

33. *Coffee harvest.* The coffee beans from 14 trees are needed to produce 7.7 kg of coffee. (This is the average amount that each person in the United States consumes each year.) How many trees are needed to produce 308 kg of coffee?

34. *Walking speed.* Wanda walked 234 km in 14 days. At this rate, how far would she walk in 42 days?

35. *Hemoglobin.* A normal 10-cc specimen of human blood contains 1.2 g of hemoglobin. How much hemoglobin would 16 cc of the same blood contain?

36. *Baking.* In a potato bread recipe, the ratio of milk to flour is $\frac{3}{13}$. If 5 cups of milk are used, how many cups of flour are used?

37. *Wages.* For comparable work, U.S. women earn 77 cents for each dollar earned by a man (*Source*: *Burlington Free Press*, 12/7/98, p. 3, Business Monday). If a male sales manager earns $42,000, how much would a female earn for comparable work?

$^{Aha!}$ **38.** *Money.* The ratio of the weight of copper to the weight of zinc in a U.S. penny is $\frac{1}{39}$. If 50 kg of zinc is being turned into pennies, how much copper is needed?

39. *Deer population.* To determine the number of deer in the Great Gulf Wilderness, a game warden catches 318 deer, tags them, and lets them loose. Later, 168 deer are caught; 56 of them have tags. Estimate the number of deer in the preserve.

40. *Moose population.* To determine the size of Pine County's moose population, naturalists catch 69 moose, tag them, and then set them free. Months later, 40 moose are caught, of which 15 have tags. Estimate the size of the moose population.

41. *Light bulbs.* A sample of 184 light bulbs contained 6 defective bulbs. How many defective bulbs would you expect in a sample of 1288 bulbs?

42. *Fish population.* To determine the number of trout in a lake, a naturalist catches 112 trout, tags them, and throws them back into the lake. Later, 82 trout are caught; 32 of them have tags. Estimate the number of trout in the lake.

43. *Miles driven.* Emmanuel is allowed to drive his leased car for 45,000 mi in 4 yr without penalty. In the first $1\frac{1}{2}$ yr, Emmanuel has driven 16,000 mi. At this rate will he exceed the mileage allowed for 4 yr?

44. *Firecrackers.* A sample of 144 firecrackers contained 9 "duds." How many duds would you expect in a sample of 320 firecrackers?

45. *Fox population.* To determine the number of foxes in King County, a naturalist catches, tags, and then releases 25 foxes. Later, 36 foxes are caught; 4 of them have tags. Estimate the fox population of the county.

46. *Weight on the moon.* The ratio of the weight of an object on the moon to the weight of that object on Earth is 0.16 to 1.
a) How much would a 12-ton rocket weigh on the moon?
b) How much would a 180-lb astronaut weigh on the moon?

47. *Weight on Mars.* The ratio of the weight of an object on Mars to the weight of that object on Earth is 0.4 to 1.
a) How much would a 12-ton rocket weigh on Mars?
b) How much would a 120-lb astronaut weigh on Mars?

48. Simplest fractional notation for a rational number is $\frac{9}{17}$. Find an equivalent ratio where the sum of the numerator and the denominator is 104.

49. Is it correct to assume that two workers will complete a task twice as quickly as one person working alone? Why or why not?

50. If two triangles are exactly the same shape and size, are they similar? Why or why not?

SKILL MAINTENANCE

Graph.

51. $y = 2x - 6$

52. $y = -2x + 6$

53. $3x + 2y = 12$

54. $x - 3y = 6$

55. $y = -\dfrac{3}{4}x + 2$

56. $y = \dfrac{2}{5}x - 4$

SYNTHESIS

57. Write a problem similar to Example 2 for a classmate to solve. Design the problem so that the translation step is

$$\frac{t}{7} + \frac{t}{5} = 1.$$

58. Write a problem similar to Example 3 for a classmate to solve. Design the problem so that the translation step is

$$\frac{30}{r + 4} = \frac{18}{r}.$$

59. *Car cleaning.* Together, Michelle, Sal, and Kristen can wax a car in 1 hr 20 min. To complete the job alone, Michelle needs twice the time that Sal needs and 2 hr more than Kristen. How long would it take each to wax the car working alone?

60. *Programming.* Rosina, Ng, and Oscar can write a computer program in 3 days. Rosina can write the program in 8 days and Ng can do it in 10 days. How many days will it take Oscar to write the program?

61. *Wiring.* Janet can wire a house in 28 hr. Linus can wire a house in 34 hr. How long will it take Janet and Linus, working together, to wire *two* houses?

62. *Quilting.* Ann and Betty work together and sew a quilt in 4 hr. Working alone, Betty would need 6 hr more than Ann to sew a quilt. How long would it take each of them working alone?

63. *Grading.* Alma can grade a batch of placement exams in 3 hr. Kevin can grade a batch in 4 hr. If they work together to grade a batch of exams, what percentage of the exams will have been graded by Alma?

Aha! **64.** *Roofing.* Working alone, Russ can reshingle a roof in 12 hr. When Joan works with Russ, the job takes 6 hr. How long would it take Joan, working alone, to reshingle the roof?

65. *Boating.* The speed of a boat in still water is 10 mph. It travels 24 mi upstream and 24 mi downstream in a total time of 5 hr. What is the speed of the current?

66. *Elections.* Melanie beat her opponent for the presidency of the student senate by a 3-to-2 ratio. If 450 votes were cast, how many votes did Melanie receive?

67. *Commuting.* To reach an appointment 50 mi away, Dr. Wright allowed 1 hr. After driving 30 mi, she realized that her speed would have to be increased 15 mph for the remainder of the trip. What was her speed for the first 30 mi?

68. *Distances.* The shadow from a 40-ft cliff just reaches across a water-filled quarry at the same time that a 6-ft tall diver casts a 10-ft shadow. How wide is the quarry?

69. How soon after 5 o'clock will the hands on a clock first be together?

70. Given that

$$\frac{A}{B} = \frac{C}{D},$$

write three other proportions using A, B, C, and D.

71. If two triangles are similar, are their areas and perimeters proportional? Why or why not?

72. Are the equations

$$\frac{A + B}{B} = \frac{C + D}{D} \quad \text{and} \quad \frac{A}{B} = \frac{C}{D}$$

equivalent? Why or why not?

COLLABORATIVE

CORNER

Sharing the Workload

Focus: Modeling, estimation, and work problems

Time: 20–25 minutes

Group size: 3

Materials: Paper, pencils, textbooks, and a watch

Many tasks can be done by two people working together. If both people work at the same rate, each does half the task, and the project is completed in half the time. However, when the work rates differ, the faster worker performs more than half of the task.

ACTIVITY

1. The project is to write down (but not answer) Review Exercises 1–34 from Chapter 6 (p. 389) on a sheet of paper. The problems should be spaced apart and written clearly so that they can be used for studying in the future. Two of the members in each group should write down the exercises, each working at his or her own pace. One worker should begin with Exercise 1 and work forward, while the other worker should begin with Exercise 34 and work backward. The third group member should record how long it takes for them to reach a point where together the workers have copied all 34 exercises. The time required should be recorded in the appropriate box of the table below.

2. Next, one of the workers from part (1) should be timed copying all 34 exercises working alone. Record that time in the table. The group should then predict how long it will take the second worker, working alone, to copy the exercises.

3. The second worker should then be timed writing down the exercises. Record this time and compare it with the prediction from part (2). How far off was the group's prediction?

4. Let t_1, t_2, and t_3 represent the times required for the first worker, the second worker, and the two workers together, respectively, to complete a task. Then develop a model that can be used to find t_2 when t_1 and t_3 are known.

5. Compare the actual experimental time from part (3) with the time predicted by the model in part (4). List reasons that might account for any discrepancy.

Time Required Working Together	Time Required for One of the Workers, Working Alone	Estimated Time for the Other Worker, Working Alone	Actual Time Required for the Other Worker, Working Alone

Summary and Review 6

Key Terms

Rational expression, p. 332

Least Common Multiple, LCM, p. 348

Least Common Denominator, LCD, p. 348

Complex rational expression, p. 363

Rational equation, p. 369

Motion problem, p. 378

Ratio, p. 380

Proportion, p. 380

Similar triangles, p. 380

Important Properties and Formulas

To add, subtract, multiply, and divide rational expressions:

$$\frac{A}{B} \cdot \frac{C}{D} = \frac{AC}{BD}; \qquad \frac{A}{B} \div \frac{C}{D} = \frac{A}{B} \cdot \frac{D}{C} = \frac{AD}{BC};$$

$$\frac{A}{B} + \frac{C}{B} = \frac{A + C}{B}; \qquad \frac{A}{B} - \frac{C}{B} = \frac{A - C}{B}.$$

To find the least common denominator (LCD):

1. Write the prime factorization of each denominator.
2. Select one of the factorizations and inspect it to see if it contains the other.

    a) If it does, it represents the LCM of the denominators.
    b) If it does not, multiply that factorization by any factors of the other denominator that it lacks. The final product is the LCM of the denominators.

The LCD is the LCM of the denominators. It contains each factor the greatest number of times that it occurs in any of the individual factorizations.

To add or subtract rational expressions that have different denominators:

1. Find the LCD.
2. Multiply each rational expression by a form of 1 made up of the factors of the LCD missing from that expression's denominator.
3. Add or subtract the numerators, as indicated. Write the sum or difference over the LCD.
4. Simplify, if possible.

To simplify a complex rational expression by dividing:

1. Add or subtract, as needed, to get a single rational expression in the numerator.
2. Add or subtract, as needed, to get a single rational expression in the denominator.
3. Divide the numerator by the denominator (invert and multiply).
4. If possible, simplify by removing a factor equal to 1.

To simplify a complex rational expression by multiplying by the LCD:

1. Find the LCD of all rational expressions *within* the complex rational expression.
2. Multiply the complex rational expression by a factor equal to 1. Write 1 as the LCD over itself (LCD/LCD).
3. Distribute and simplify. No fractional expressions should remain within the complex rational expression.
4. Factor and, if possible, simplify.

To solve a rational equation:

1. List any restrictions that exist. No solution can make a denominator equal 0.
2. Clear the equation of fractions by multiplying both sides by the LCD of all rational expressions in the equation.
3. Solve the resulting equation using the addition principle, the multiplication principle, and the principle of zero products, as needed.
4. Check the possible solution(s) in the original equation.

The Work Principle

Suppose that a = the time it takes A to complete a task, b = the time it takes B to complete the task, and t = the time it takes them working together. Then all of the following hold:

$$\frac{1}{a} \cdot t + \frac{1}{b} \cdot t = 1; \quad \left(\frac{1}{a} + \frac{1}{b} \right) t = 1;$$

$$\frac{t}{a} + \frac{t}{b} = 1.$$

Review Exercises

List all numbers for which each expression is undefined.

1. $\dfrac{35}{-x^2}$

2. $\dfrac{9}{a-5}$

3. $\dfrac{x-7}{x^2-36}$

4. $\dfrac{x^2+3x+2}{x^2+x-30}$

5. $\dfrac{-6}{(t+2)^2}$

Simplify.

6. $\dfrac{4x^2-8x}{4x^2+4x}$

7. $\dfrac{14x^2-x-3}{2x^2-7x+3}$

8. $\dfrac{(y-5)^2}{y^2-25}$

9. $\dfrac{5x^2-20y^2}{2y-x}$

Multiply or divide and, if possible, simplify.

10. $\dfrac{a^2-36}{10a} \cdot \dfrac{2a}{a+6}$

11. $\dfrac{8t+8}{2t^2+t-1} \cdot \dfrac{t^2-1}{t^2-2t+1}$

12. $\dfrac{10-5t}{3} \div \dfrac{t-2}{12t}$

13. $\dfrac{4x^4}{x^2-1} \div \dfrac{2x^3}{x^2-2x+1}$

14. $\dfrac{x^2+1}{x-2} \cdot \dfrac{2x+1}{x+1}$

15. $(t^2+3t-4) \div \dfrac{t^2-1}{t+4}$

Find the LCM.

16. $8a^2b^7,\ 6a^5b^3$

17. $x^2-x,\ x^5-x^3,\ x^4$

18. $y^2-y-2,\ y^2-4$

Add or subtract and, if possible, simplify.

19. $\dfrac{x+8}{x+7} + \dfrac{10-4x}{x+7}$

20. $\dfrac{3}{3x-9} + \dfrac{x-2}{3-x}$

21. $\dfrac{6x-3}{x^2-x-12} - \dfrac{2x-15}{x^2-x-12}$

22. $\dfrac{3x-1}{2x} - \dfrac{x-3}{x}$

23. $\dfrac{x+3}{x-2} - \dfrac{x}{2-x}$

24. $\dfrac{2a}{a+1} - \dfrac{4a}{1-a^2}$

25. $\dfrac{d^2}{d-c} + \dfrac{c^2}{c-d}$

26. $\dfrac{1}{x^2-25} - \dfrac{x-5}{x^2-4x-5}$

27. $\dfrac{3x}{x+2} - \dfrac{x}{x-2} + \dfrac{8}{x^2-4}$

28. $\dfrac{2}{5x} + \dfrac{3}{2x+4}$

Simplify.

29. $\dfrac{\dfrac{1}{z}+1}{\dfrac{1}{z^2}-1}$

30. $\dfrac{2+\dfrac{1}{xy^2}}{\dfrac{1+x}{x^4y}}$

31. $\dfrac{\dfrac{c}{d}-\dfrac{d}{c}}{\dfrac{1}{c}+\dfrac{1}{d}}$

Solve.

32. $\dfrac{3}{y} - \dfrac{1}{4} = \dfrac{1}{y}$

33. $\dfrac{5}{x+3} = \dfrac{3}{x+2}$

34. $\dfrac{15}{x} - \dfrac{15}{x+2} = 2$

35. Rhetta can polish a helicopter rotor blade in 9 hr. Jason can do the same job in 12 hr. How long would it take if they worked together?

36. The distance by highway between Richmond and Waterbury is 70 km, and the distance by rail is 60 km. A car and a train leave Richmond at the same time and arrive in Waterbury at the same time, the car having traveled 15 km/h faster than the train. Find the speed of the car and the speed of the train.

37. The reciprocal of 1 more than a number is twice the reciprocal of the number itself. What is the number?

38. A sample of 25 doorknobs contained 2 defective doorknobs. How many defective doorknobs would you expect among 375 doorknobs?

39. Triangles ABC and XYZ are similar. Find the value of x.

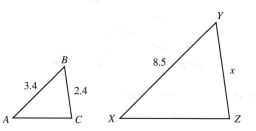

40. A game warden catches, tags, and then releases 15 zebras. A month later, a sample of 20 zebras is collected and 6 of them have tags. Use this information to estimate the size of the zebra population in that area.

SYNTHESIS

41. For what procedures in this chapter is the LCD used to clear fractions?

42. A student insists on finding a common denominator by always multiplying the denominators of the expressions being added. How could this approach be improved?

Simplify.

43. $\dfrac{2a^2 + 5a - 3}{a^2} \cdot \dfrac{5a^3 + 30a^2}{2a^2 + 7a - 4} \div \dfrac{a^2 + 6a}{a^2 + 7a + 12}$

44. $\dfrac{12a}{(a - b)(b - c)} - \dfrac{2a}{(b - a)(c - b)}$

Aha! **45.** $\dfrac{5(x - y)}{(x - y)(x + 2y)} - \dfrac{5(x - 3y)}{(x + 2y)(x - 3y)}$

Chapter Test 6

List all numbers for which each expression is undefined.

1. $\dfrac{8 - x}{3x}$

2. $\dfrac{5}{x + 8}$

3. $\dfrac{x - 7}{x^2 - 49}$

4. $\dfrac{x^2 + x - 30}{x^2 - 3x + 2}$

5. Simplify: $\dfrac{6x^2 + 17x + 7}{2x^2 + 7x + 3}$.

Multiply or divide and, if possible, simplify.

6. $\dfrac{a^2 - 25}{9a} \cdot \dfrac{6a}{5 - a}$

7. $\dfrac{25y^2 - 1}{9y^2 - 6y} \div \dfrac{5y^2 + 9y - 2}{3y^2 + y - 2}$

8. $\dfrac{4x^2 - 1}{x^2 - 2x + 1} \div \dfrac{x - 2}{x^2 + 1}$

9. $(x^2 + 6x + 9) \cdot \dfrac{(x - 3)^2}{x^2 - 9}$

10. Find the LCM:

$y^2 - 9,\ y^2 + 10y + 21,\ y^2 + 4y - 21.$

Add or subtract, and, if possible, simplify.

11. $\dfrac{16 + x}{x^3} + \dfrac{7 - 4x}{x^3}$

12. $\dfrac{5 - t}{t^2 + 1} - \dfrac{t - 3}{t^2 + 1}$

13. $\dfrac{x - 4}{x - 3} + \dfrac{x - 1}{3 - x}$

14. $\dfrac{x - 4}{x - 3} - \dfrac{x - 1}{3 - x}$

15. $\dfrac{5}{t - 1} + \dfrac{3}{t}$

16. $\dfrac{1}{x^2 - 16} - \dfrac{x + 4}{x^2 - 3x - 4}$

17. $\dfrac{1}{x - 1} + \dfrac{4}{x^2 - 1} - \dfrac{2}{x^2 - 2x + 1}$

Simplify.

18. $\dfrac{9 - \dfrac{1}{y^2}}{3 - \dfrac{1}{y}}$

19. $\dfrac{\dfrac{3}{a^2 b} - \dfrac{2}{ab^3}}{\dfrac{1}{ab} + \dfrac{2}{a^4 b}}$

Solve.

20. $\dfrac{7}{y} - \dfrac{1}{3} = \dfrac{1}{4}$

21. $\dfrac{15}{x} - \dfrac{15}{x-2} = -2$

22. Kopy Kwik has 2 copiers. One can copy a year-end report in 20 min. The other can copy the same document in 30 min. How long would it take both machines, working together, to copy the report?

23. A recipe for pizza crust calls for $3\frac{1}{2}$ cups of whole wheat flour and $1\frac{1}{4}$ cups of warm water. If 6 cups of whole wheat flour are used, how much water should be used?

24. Craig drives 20 km/h faster than Marilyn. In the same time that Marilyn drives 225 km, Craig drives 325 km. Find the speed of each car.

SYNTHESIS

25. Reggie and Rema work together to mulch the flower beds around an office complex in $2\frac{6}{7}$ hr. Working alone, it would take Reggie 6 hr more than it would take Rema. How long would it take each of them to complete the landscaping working alone?

26. Simplify: $1 - \dfrac{1}{1 - \dfrac{1}{1 - \dfrac{1}{a}}}$.

Cumulative Review 1–6

1. Use the commutative law of addition to write an expression equivalent to $a + 2b$.

2. Write a true sentence using either $<$ or $>$:
   $-3.1 \;\rule{1em}{0.8em}\; -3.15$.

3. Evaluate $(y - 1)^2$ for $y = -6$.

4. Simplify: $-4[2(x - 3) - 1]$.

Simplify.

5. $-\frac{1}{2} + \frac{3}{8} + (-6) + \frac{3}{4}$

6. $-\frac{72}{108} \div \left(-\frac{2}{3}\right)$

7. $-6.262 \div 1.01$

8. $4 \div (-2) \cdot 2 + 3 \cdot 4$

Solve.

9. $3(x - 2) = 24$

10. $49 = x^2$

11. $-6t = 20$

12. $5x + 7 = -3x - 9$

13. $4(y - 5) = -2(y + 2)$

14. $x^2 + 11x + 10 = 0$

15. $\dfrac{4}{9}t + \dfrac{2}{3} = \dfrac{1}{3}t - \dfrac{2}{9}$

16. $\dfrac{4}{x} + x = 5$

17. $3 - y \geq 2y + 5$

18. $\dfrac{2}{x - 3} = \dfrac{5}{3x + 1}$

19. $2x^2 + 7x = 4$

20. $4(x + 7) < 5(x - 3)$

21. $\dfrac{t^2}{t + 5} = \dfrac{25}{t + 5}$

22. $(2x + 7)(x - 5) = 0$

23. $\dfrac{2}{x^2 - 9} + \dfrac{5}{x - 3} = \dfrac{3}{x + 3}$

Solve each formula.

24. $3a - b + 9 = c$, for b

25. $\frac{3}{4}(x + 2y) = z$, for y

Combine like terms.

26. $x + 2y - 2z + \frac{1}{2}x - z$

27. $2x^3 - 7 + \frac{3}{7}x^2 - 6x^3 - \frac{4}{7}x^2 + 5$

Graph.

28. $y = \frac{3}{4}x + 5$

29. $x = -3$

30. $4x + 5y = 20$

31. $y = 6$

32. Find the slope of the line containing the points $(1, 5)$ and $(2, 3)$.

Simplify.

33. $\dfrac{x^{-5}}{x^{-3}}$

34. $y^2 \cdot y^{-10}$

35. $-(2a^2b^7)^2$

36. Subtract:
$$(-8y^2 - y + 2) - (y^3 - 6y^2 + y - 5).$$

Multiply.

37. $4(3x + 4y + z)$

38. $(2x^2 - 1)(x^3 + x - 3)$

39. $(6x - 5y)^2$

40. $(x + 3)(2x - 7)$

41. $(2x^3 + 1)(2x^3 - 1)$

Factor.

42. $6x - 2x^2 - 24x^4$ **43.** $16x^2 - 81$

44. $t^2 - 10t + 24$ **45.** $8x^2 + 10x + 3$

46. $6x^2 - 28x + 16$ **47.** $4t^2 - 36$

48. $25t^2 + 40t + 16$ **49.** $3x^2 + 10x - 8$

50. $x^4 + 2x^3 - 3x - 6$

Simplify.

51. $\dfrac{y^2 - 36}{2y + 8} \cdot \dfrac{y + 4}{y + 6}$ **52.** $\dfrac{x^2 - 1}{x^2 - x - 2} \div \dfrac{x - 1}{x - 2}$

53. $\dfrac{5ab}{a^2 - b^2} + \dfrac{a + b}{a - b}$ **54.** $\dfrac{x + 2}{4 - x} - \dfrac{x + 3}{x - 4}$

55. $\dfrac{1 + \dfrac{2}{x}}{1 - \dfrac{4}{x^2}}$ **56.** $\dfrac{\dfrac{1}{t} + 2t}{t - \dfrac{2}{t^2}}$

Divide.

57. $\dfrac{15x^4 - 12x^3 + 6x^2 + 2x + 18}{3x^2}$

58. $(15x^4 - 12x^3 + 6x^2 + 2x + 18) \div (x + 3)$

Solve.

59. Linnae has $36 budgeted for stationery. Engraved stationery costs $20 for the first 25 sheets and $0.08 for each additional sheet. How many sheets of stationery can Linnae order and still stay within her budget?

60. The price of a box of cereal increased 15% to $4.14. What was the price of the cereal before the increase?

61. If the sides of a square are increased by 2 ft, the area of the original square plus the area of the enlarged square is 452 ft^2. Find the length of a side of the original square.

62. The sum of two consecutive even integers is -554. Find the integers.

63. It takes Dina 50 min to shovel 9 in. of snow from her driveway. It takes Nell 75 min to do the same job. How long would it take if they worked together?

64. Phil's Ford Focus travels 10 km/h faster than Harley's. In the same time that Harley drives 120 km, Phil travels 150 km. Find the speed of each car.

65. A 78-in. board is to be cut into two pieces. One piece must be twice as long as the other. How long should the shorter piece be?

SYNTHESIS

66. Simplify: $(x + 7)(x - 4) - (x + 8)(x - 5)$.

67. Solve: $\frac{1}{3}|n| + 8 = 56$.

Aha! **68.** Multiply: $[4y^3 - (y^2 - 3)][4y^3 + (y^2 - 3)]$.

69. Factor: $2a^{32} - 13{,}122b^{40}$.

70. Solve: $x(x^2 + 3x - 28) - 12(x^2 + 3x - 28) = 0$.

71. Simplify: $-\left|0.875 - \left(-\frac{1}{8}\right) - 8\right|$.

72. Solve: $\dfrac{2}{x - 3} \cdot \dfrac{3}{x + 3} - \dfrac{4}{x^2 - 7x + 12} = 0$.

73. Jesse can peel a bushel of potatoes in 60 min. When Jesse and Priscilla work together, the job takes 20 min. When working together with Jesse, what percentage of the work is performed by Priscilla?

7

Systems and More Graphing

AN APPLICATION

New England Natural Bakers Muesli gets 20% of its calories from fat. Breadshop Supernatural granola gets 35% of its calories from fat. (*Source*: Onion River Cooperative, Burlington, VT) How much of each type should be used to create a 45-lb mixture that gets 30% of its calories from fat?

This problem appears as Exercise 32 in Section 7.4.

I use math not only to keep track of sales, but also to compare and calculate prices. When we purchase bulk herbs, we must calculate the amounts to package and price them to sell.

ANGELLEE KOERNER
Manager,
Natural Foods Store
Sandy, Utah

n fields ranging from business to zoology, problems arise that are most easily solved using a system of equations. Section 7.1 examines how graphing can be used to solve a system of equations, and Sections 7.2 and 7.3 discuss algebraic methods of solving systems. These three methods are then used in a variety of applications and as an aid in solving systems of inequalities.

Systems of Equations and Graphing

7.1

Solutions of Systems • Solving Systems of Equations by Graphing

Solutions of Systems

A **system of equations** is a set of two or more equations that are to be solved simultaneously. We will find that it is often easier to translate real-world situations to a system of two equations that use two variables, or unknowns, than it is to represent the situation with one equation using one variable. To see this, let's see how the following problem from Section 2.5 can be represented by two equations using two unknowns:

A rectangular community garden is to be enclosed with 92 m of fencing. In order to allow for compost storage, the garden must be 4 m longer than it is wide. Determine the dimensions of the garden.

If we let w = the width of the garden, in meters, and l = the length of the garden, in meters, the problem translates to the following system of equations:

$$2l + 2w = 92, \qquad \text{The perimeter is 92.}$$
$$l = w + 4. \qquad \text{The length is 4 more than the width.}$$

To solve a system of equations is to find all ordered pairs for which *both* equations are true.

l, or *w* + 4

Example 1

Consider the system from above:

$$2l + 2w = 92,$$
$$l = w + 4.$$

Determine if each pair is a solution of the system: **(a)** $(25, 21)$; **(b)** $(16, 12)$.

Solution

a) We check by substituting (alphabetically) 25 for l and 21 for w:

$2l + 2w = 92$
$2 \cdot 25 + 2 \cdot 21 \; ? \; 92$
$50 + 42$
$92 \; \vert \; 92$ TRUE

$l = w + 4$
$25 \; ? \; 21 + 4$
$25 \; \vert \; 25$ TRUE

Since $(25, 21)$ checks in *both* equations, it is a solution of the system.

b) We substitute 16 for l and 12 for w:

$$
\begin{array}{c}
2l + 2w = 92 \\ \hline
2 \cdot 16 + 2 \cdot 12 \;?\; 92 \\
32 + 24 \quad \big| \\
56 \quad \big| \quad 92 \quad \text{FALSE}
\end{array}
\qquad
\begin{array}{c}
l = w + 4 \\ \hline
16 \;?\; 12 + 4 \\
16 \quad \big| \quad 16 \qquad \text{TRUE}
\end{array}
$$

Since $(16, 12)$ is not a solution of $2l + 2w = 92$, it is *not* a solution of the system.

In Example 1, we demonstrated that $(25, 21)$ is a solution of the system, but we did not show how the pair $(25, 21)$ was found. One way to find such a solution uses graphs.

Solving Systems of Equations by Graphing

Recall that a graph of an equation is a set of points representing its solution set. Each point on the graph corresponds to an ordered pair that is a solution of the equation. By graphing two equations on the same set of axes, we can identify a solution of both equations by looking for a point of intersection.

Example 2

Solve this system of equations by graphing:

$$
\begin{aligned}
x + y &= 7, \\
y &= 3x - 1.
\end{aligned}
$$

Solution We graph the equations using any method studied earlier. The equation $x + y = 7$ can be graphed easily using the intercepts, $(0, 7)$ and $(7, 0)$. The equation $y = 3x - 1$ is in slope–intercept form, so we graph the line by plotting its y-intercept, $(0, -1)$, and "counting off" a slope of 3.

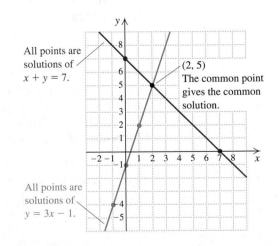

All points are solutions of $x + y = 7$.

$(2, 5)$
The common point gives the common solution.

All points are solutions of $y = 3x - 1$.

The "apparent" solution of the system, $(2, 5)$, should be checked in both equations.

Check:

$$\frac{x + y = 7}{2 + 5 \ ? \ 7}$$
$$7 \ | \ 7 \ \text{TRUE}$$

$$\frac{y = 3x - 1}{5 \ ? \ 3 \cdot 2 - 1}$$
$$5 \ | \ 5 \qquad \text{TRUE}$$

Since it checks in both equations, $(2, 5)$ is a solution of the system.

A system of equations that has at least one solution, like the systems in Examples 1 and 2, is said to be **consistent**. A system for which there is no solution is said to be **inconsistent**.

E x a m p l e   3

Solve this system of equations by graphing:

$$y = 3x + 4,$$
$$y = 3x - 3.$$

Solution   Both equations are in slope–intercept form so it is easy to see that both lines have the same slope, 3. The y-intercepts differ so the lines are parallel, as shown in the figure at right.

Because the lines are parallel, there is no point of intersection. Thus the system is inconsistent and has no solution.

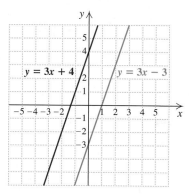

Sometimes both equations in a system have the same graph.

E x a m p l e   4

Solve this system of equations by graphing:

$$2x + 3y = 6,$$
$$-8x - 12y = -24.$$

Solution   Graphing the equations, we see that they both represent the same line. This can also be seen by solving each equation for y, obtaining the equivalent slope–intercept form, $y = -\frac{2}{3}x + 2$. Because the equations are equivalent, any solution of one equation is a solution of the other equation as well. We show four such solutions.

We check one solution, $(0, 2)$, in each of the original equations.

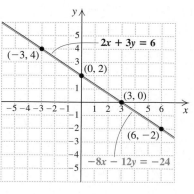

$$\frac{2x + 3y = 6}{2(0) + 3(2) \ ? \ 6}$$
$$0 + 6 \ |$$
$$6 \ | \ 6 \ \text{TRUE}$$

$$\frac{-8x - 12y = -24}{-8(0) - 12(2) \ ? \ -24}$$
$$0 - 24 \ |$$
$$-24 \ | \ -24 \ \text{TRUE}$$

On your own, check that $(3, 0)$ is also a solution of the system. If two points are solutions, then all points on the line containing them are solutions. The lines coincide, so there is an infinite number of solutions. Since a solution exists, the system is consistent.

When one equation can be obtained by multiplying both sides of another equation by a constant, the two equations are called **dependent**. Thus the equations in Example 4 are dependent, but those in Examples 2 and 3 are **independent**. For systems of two equations, when two equations are dependent, they are equivalent. When systems contain more than two equations, the definition of dependent must be slightly modified and it is possible for dependent equations to not be equivalent.

When a system of two linear equations in two variables is graphed, one of the following must occur:

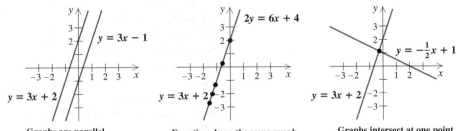

Graphs are parallel.
The system is *inconsistent* because there is no solution. Since neither equation is a multiple of the other, the equations are *independent*.

Equations have the same graph.
The system is *consistent* and has an infinite number of solutions. Since one equation is two times the other, the equations are *dependent*.

Graphs intersect at one point.
The system is *consistent* and has one solution. Since neither equation is a multiple of the other, the equations are *independent*.

Although graphing lets us "see" the solution of a system, it does not always allow us to find a precise solution. For example, the solution of the system

$$3x + 7y = 5,$$
$$6x - 7y = 1$$

is $\left(\frac{2}{3}, \frac{3}{7}\right)$, but finding that precise solution from a graph—*even with a computer or graphing calculator*—can be difficult. Fortunately, systems like this can be solved precisely using methods discussed in Sections 7.2 and 7.3.

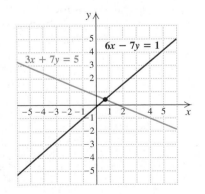

Exercise Set 7.1

Determine whether each ordered pair is a solution of the system of equations. Use alphabetical order of the variables.

1. $(3, 2)$; $2x + 3y = 12$,
$x - 4y = -5$

2. $(1, 5)$; $5x - 2y = -5$,
$3x - 7y = -32$

3. $(3, 2)$; $3b - 2a = 0$,
$b + 2a = 15$

4. $(2, -2)$; $b + 2a = 2$,
$b - a = -4$

5. $(15, 20)$; $3x - 2y = 5$,
$6x - 5y = -10$

6. $(-1, -3)$; $3r + s = -6$,
$2r = 1 + s$

Solve each system of equations by graphing. If there is no solution or an infinite number of solutions, state this.

7. $x + y = 7$,
$x - y = 1$

8. $x - y = 1$,
$x + y = 3$

9. $y = -2x + 5$,
$x + y = 4$

10. $y = 2x - 5$,
$x + y = 4$

11. $y = x - 2$,
$y = -3x + 2$

12. $y = -2x + 3$,
$y = x - 3$

13. $4x - 20 = 5y$,
$8x - 10y = 12$

14. $6x + 12 = 2y$,
$6 - y = -3x$

15. $x = 6$,
$y = -1$

16. $x = -2$,
$y = 5$

17. $2x + y = 8$,
$x - y = 7$

18. $3x + y = 4$,
$x - y = 4$

19. $x - y = 5$,
$2x + y = 4$

20. $x - y = 8$,
$3x + y = 12$

21. $x + 2y = 7$,
$3x + 6y = 21$

22. $x + 3y = 6$,
$4x + 12y = 24$

23. $2x = 3y - 6$,
$x = 3y$

24. $3y - 9 = 6x$,
$y = x$

Aha! 25. $y = \frac{1}{5}x + 4$,
$2y = \frac{2}{5}x + 8$

26. $y = \frac{1}{3}x + 2$,
$y = \frac{1}{3}x - 7$

27. $2x - y = 5$,
$y = \frac{1}{2}x + 1$

28. $2x + y = -1$,
$y = 3x - 6$

29. $3x + 4y = 8$,
$x + 2y = 10$

30. $2x - 3y = 5$,
$x - 2y = 6$

31. $4x + 2y = -2$,
$5x + 4y = 5$

32. $3x + 2y = 1$,
$2x + 5y = -14$

33. $x = \frac{1}{2}y$,
$x = 3$

34. $x = \frac{1}{3}y$,
$y = 6$

35. Suppose that the equations in a system of two linear equations are dependent. Does it follow that the system is consistent? Why or why not?

36. Why is slope–intercept form especially useful when solving systems of equations by graphing?

SKILL MAINTENANCE

Solve.

37. $4x - 5(9 - 2x) = 7$

38. $5x - (7 + 3x) = 8$

39. $3(4 - 2y) - 5y = 6$

40. $2(8 - 3y) - 4y = 7$

Simplify.

41. $5(2x + 3y) - 3(7x + 5y)$

42. $4(5x + 6y) - 5(4x + 7y)$

SYNTHESIS

43. Is it possible for a system of two linear equations to have exactly two solutions? Why or why not?

44. Explain how it is possible to determine whether or not the system

$Ax + By = C$,
$Dx + Ey = F$

is inconsistent by simply examining A, B, C, D, E, and F.

45. Which of the systems in Exercises 7–34 contain dependent equations?

46. Which of the systems in Exercises 7–34 are consistent?

47. Which of the systems in Exercises 7–34 are inconsistent?

48. Which of the systems in Exercises 7–34 contain independent equations?

49. Write an equation that can be paired with $5x + 2y = 3$ to form a system that has $(-1, 4)$ as the solution. Answers may vary.

50. Write an equation that can be paired with $4x + 3y = 6$ to form a system that has $(3, -2)$ as the solution. Answers may vary.

51. The solution of the following system is $(2, -3)$. Find A and B.

$$Ax - 3y = 13,$$
$$x - By = 8$$

52. Solve by graphing:

$$4x - 8y = -7,$$
$$2x + 3y = 7.$$

(*Hint*: Try different scales on your graph.)

53. *Copying costs.*   Shelby occasionally goes to Mailboxes Etc.® with small copying jobs. He can purchase a "copy card" for \$20 that will entitle him to 500 copies, or he can simply pay 6¢ per page.
   a) Create cost equations for each method of paying for a number (up to 500) of copies.
   b) Graph both cost equations on the same set of axes.
   c) Use the graph to determine how many copies Shelby must make if the card is to be more economical.

Aha! **54.** Solve:

$$3x - 4y = 2,$$
$$-6x + 8y = -6.$$

55. Use a grapher to solve the system

$$y = 1.2x - 32.7,$$
$$y = -0.7x + 46.15.$$

CORNER

Conserving Energy and Money

Focus: System of linear equations (two variables)

Time: 20 minutes

Group size: 2

Materials: Graph paper

Jean, a condo owner, has an old electric water heater that consumes \$100 of electricity per month. To replace it with a new gas water heater that consumes only \$25 of gas per month will cost \$250 plus \$150 for installation. Jean wants to know how long it will take for the new gas water heater to "pay for itself," in other words, to produce savings.

ACTIVITY

1. The "break-even" point occurs when the costs of the two water heaters are equal. Determine, by "guessing and checking," the number of months before Jean breaks even.

2. One of the group members should create a cost equation of the form $y = mx + b$ for the electric heater, where y is the cost, in dollars, and x is the time, in months. He or she should also graph the equation on the graph paper provided.

3. The second group member should create a cost equation of the form $y = mx + b$ for the gas heater, where y is the cost, in dollars, and x is the time, in months. This equation should be graphed on the same graph used in part (2).

4. Working together, the group should determine the coordinates of the point of intersection of the two lines. This is the break-even point. Which coordinate indicates the number of months before Jean breaks even? What does the other coordinate indicate? Compare the answer found graphically with the estimate made in part (1).

COLLABORATIVE

Systems of Equations and Substitution

7.2

The Substitution Method • Solving for the Variable First • Problem Solving

Near the end of Section 7.1, we mentioned that graphing can be an imprecise method for solving systems. In this section and the next, we develop methods of finding exact solutions using algebra.

The Substitution Method

One nongraphical method for solving systems is known as the **substitution method**. It uses algebra and is thus considered an *algebraic* method.

Example 1

Solve the system

$$x + y = 7, \qquad (1)$$
$$y = 3x - 1. \qquad (2)$$

We have numbered the equations (1) and (2) for easy reference.

Solution The second equation says that y and $3x - 1$ represent the same value. Thus, in the first equation, we can substitute $3x - 1$ for y:

$$x + y = 7, \qquad \text{Equation (1)}$$
$$x + 3x - 1 = 7. \qquad \text{Substituting } 3x - 1 \text{ for } y$$

The equation $x + 3x - 1 = 7$ has only one variable, for which we now solve:

$$4x - 1 = 7 \qquad \text{Combining like terms}$$
$$4x = 8 \qquad \text{Adding 1 to both sides}$$
$$x = 2. \qquad \text{Dividing both sides by 4}$$

We have found the x-value of the solution. To find the y-value, we return to the original pair of equations. Substituting into either equation will give us the y-value. We choose equation (1):

$$x + y = 7 \qquad \text{Equation (1)}$$
$$2 + y = 7 \qquad \text{Substituting 2 for } x$$
$$y = 5. \qquad \text{Subtracting 2 from both sides}$$

The ordered pair $(2, 5)$ appears to be a solution. We check:

$$\frac{x + y = 7}{2 + 5 \; ? \; 7} \qquad \qquad \frac{y = 3x - 1}{5 \; ? \; 3 \cdot 2 - 1}$$
$$7 \mid 7 \;\; \text{TRUE} \qquad \qquad 5 \mid 5 \qquad \text{TRUE}$$

Since $(2, 5)$ checks, it is the solution. For this particular system, we can also check by examining the graph from Example 2 in Section 7.1, as shown at left.

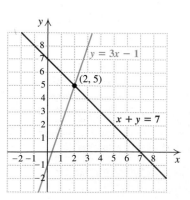

> ***Caution!***   A solution of a system of equations in two variables is an ordered *pair* of numbers. Once you have solved for one variable, don't forget the other. A common mistake is to solve for only one variable.

Example 2   Solve:

$$x = 13 - 3y, \qquad (1)$$
$$5 - x = y. \qquad (2)$$

Solution   We substitute $13 - 3y$ for x in the second equation:

$$5 - x = y, \qquad \text{Equation (2)}$$
$$5 - (13 - 3y) = y. \qquad \text{Substituting } 13 - 3y \text{ for } x. \text{ The parentheses are important.}$$

Now we solve for y:

$$5 - 13 + 3y = y \qquad \text{Removing parentheses and changing signs}$$
$$-8 + 3y = y$$
$$-8 = -2y \qquad \text{Solving for } y$$
$$4 = y.$$

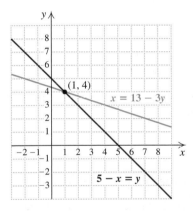

(1, 4)

$x = 13 - 3y$

$5 - x = y$

Next, we substitute 4 for y in equation (1) of the original system:

$$x = 13 - 3y \qquad \text{Equation (1)}$$
$$x = 13 - 3 \cdot 4 \qquad \text{Substituting 4 for } y$$
$$x = 1. \qquad \text{Simplifying}$$

The pair $(1, 4)$ is the solution. A graph is shown at left as a check.

Solving for the Variable First

Sometimes neither equation has a variable alone on one side. In that case, we solve one equation for one of the variables and then proceed as before.

Example 3   Solve:

$$x - 2y = 6, \qquad (1)$$
$$3x + 2y = 4. \qquad (2)$$

Solution   We can solve either equation for either variable. Since the coefficient of x is 1 in equation (1), it is easier to solve that equation for x:

$$x - 2y = 6 \qquad \text{Equation (1)}$$
$$x = 6 + 2y. \qquad \text{Adding } 2y \text{ to both sides} \qquad (3)$$

technology connection

To check Example 3 with a grapher, we must first solve each equation for y. When we do so, equation (1) becomes $y = (6 - x)/(-2)$ and equation (2) becomes $y = (4 - 3x)/2$.

1. Use the INTERSECT option of the CALC menu to determine the solution of the system.
2. What happens when parentheses are deleted from the two equations above?

We substitute $6 + 2y$ for x in equation (2) of the original pair and solve for y:

$$3x + 2y = 4 \qquad \text{Equation (2)}$$

$$3(6 + 2y) + 2y = 4 \qquad \text{Substituting } 6 + 2y \text{ for } x$$

> Remember to use parentheses when you substitute.

$$18 + 6y + 2y = 4 \qquad \text{Using the distributive law}$$

$$18 + 8y = 4 \qquad \text{Combining like terms}$$

$$8y = -14 \qquad \text{Subtracting 18 from both sides}$$

$$y = \frac{-14}{8} = -\frac{7}{4}. \qquad \text{Dividing both sides by 8}$$

To find x, we can substitute $-\frac{7}{4}$ for y in equation (1), (2), or (3). Because it is generally easier to use an equation that has already been solved for a specific variable, we decide to use equation (3):

$$x = 6 + 2y = 6 + 2\left(-\tfrac{7}{4}\right) = 6 - \tfrac{7}{2} = \tfrac{12}{2} - \tfrac{7}{2} = \tfrac{5}{2}.$$

We check the ordered pair $\left(\tfrac{5}{2}, -\tfrac{7}{4}\right)$.

Check:

$$\begin{array}{c|c}
x - 2y = 6 & \\
\hline
\tfrac{5}{2} - 2\left(-\tfrac{7}{4}\right) \; ? \; 6 & \\
\tfrac{5}{2} + \tfrac{7}{2} & \\
\tfrac{12}{2} & \\
6 & 6 \quad \text{TRUE}
\end{array}
\qquad
\begin{array}{c|c}
3x + 2y = 4 & \\
\hline
3 \cdot \tfrac{5}{2} + 2\left(-\tfrac{7}{4}\right) \; ? \; 4 & \\
\tfrac{15}{2} - \tfrac{7}{2} & \\
\tfrac{8}{2} & \\
4 & 4 \quad \text{TRUE}
\end{array}$$

Since $\left(\tfrac{5}{2}, -\tfrac{7}{4}\right)$ checks, it is the solution.

Some systems have no solution and some have an infinite number of solutions.

Example 4

Solve each system.

a) $y = 3x + 4,$ (1)
    $y = 3x - 3$ (2)

b) $2y = 6x + 4,$ (1)
    $y = 3x + 2$ (2)

Solution

a) We solved this system graphically in Example 3 of Section 7.1. The lines are parallel and the system has no solution. Let's see what happens if we try to solve this system by substituting $3x - 3$ for y in the first equation:

$$y = 3x + 4 \qquad \text{Equation (1)}$$

$$3x - 3 = 3x + 4 \qquad \text{Substituting } 3x - 3 \text{ for } y$$

$$-3 = 4. \qquad \text{Subtracting } 3x \text{ from both sides}$$

When we subtract $3x$ from both sides, we obtain a *false* equation. In such a case, when solving algebraically leads to a false equation, we state that the system has no solution and thus is inconsistent.

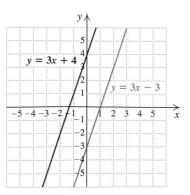

b) A graph of this system first appeared on p. 397. The lines coincide, so the system has an infinite number of solutions. If we use substitution to solve the system, we can substitute $3x + 2$ for y in equation (1):

$$2y = 6x + 4 \qquad \text{Equation (1)}$$
$$2(3x + 2) = 6x + 4 \qquad \text{Substituting } 3x + 2 \text{ for } y$$
$$6x + 4 = 6x + 4.$$

Since this last equation is true for *any* choice of x, the system has an infinite number of solutions. Whenever the algebraic solution of a system of two equations leads to an equation that is true for all real numbers, we state that the system has an infinite number of solutions.

Problem Solving

Now let's use the substitution method in problem solving.

Example 5

Supplementary angles.   Two angles are supplementary. One angle measures 30° more than twice the other. Find the measures of the two angles.

Solution

1. **Familiarize.**   Recall that two angles are supplementary if the sum of their measures is 180°. We could try to guess a solution, but instead we make a drawing and translate. Let x and y represent the measures of the two angles.

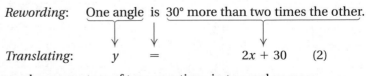

Supplementary angles

2. **Translate.**   Since we are told that the angles are supplementary, one equation is

$$x + y = 180. \qquad (1)$$

The second sentence can be rephrased and translated as follows:

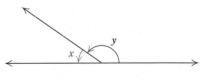

Rewording: One angle is 30° more than two times the other.

Translating: y $=$ $2x + 30$ (2)

We now have a system of two equations in two unknowns:

$$x + y = 180, \qquad (1)$$
$$y = 2x + 30. \qquad (2)$$

3. **Carry out.**   We substitute $2x + 30$ for y in equation (1):

$$x + y = 180 \qquad \text{Equation (1)}$$
$$x + (2x + 30) = 180 \qquad \text{Substituting}$$
$$3x + 30 = 180$$
$$3x = 150 \qquad \text{Subtracting 30 from both sides}$$
$$x = 50. \qquad \text{Dividing both sides by 3}$$

Substituting 50 for x in equation (1) then gives us

$$x + y = 180 \qquad \text{Equation (1)}$$
$$50 + y = 180 \qquad \text{Substituting 50 for } x$$
$$y = 130.$$

4. **Check.**   If one angle is 50° and the other is 130°, then the sum of the measures is 180°. Thus the angles are supplementary. If 30° is added to twice the measure of the smaller angle, we have $2 \cdot 50° + 30°$, or 130°, which is the measure of the other angle. The numbers check.

5. **State.**   One angle measures 50° and the other 130°.

FOR EXTRA HELP

Exercise Set **7.2**

Digital Video Tutor CD 4 Videotape 13   InterAct Math   Math Tutor Center   MathXL   MyMathLab.com

---

Solve each system using the substitution method. If a system has no solution or an infinite number of solutions, state this.

1. $x + y = 9$,
$\quad x = y + 1$

2. $x + y = 7$,
$\quad y = x + 3$

3. $\quad y = x - 3$,
$3x + y = 5$

4. $\quad x = y + 1$,
$x + 2y = 4$

5. $\quad y = 2x + 1$,
$x + y = 4$

6. $\quad y = 2x - 5$,
$3y - x = 5$

7. $\quad r = -3s$,
$r + 4s = 10$

8. $3x + y = 2$,
$\quad y = -2x$

9. $\quad x = y - 8$,
$3x + 2y = 1$

10. $2x + 3y = 8$,
$\quad x = y - 6$

11. $\quad y = 3x - 1$,
$6x - 2y = 2$

12. $\quad x = 2y + 1$,
$3x - 6y = 2$

13. $x - y = 6$,
$x + y = -2$

14. $s + t = -4$,
$s - t = 2$

15. $y - 2x = -6$,
$2y - x = 5$

16. $x - y = 5$,
$x + 2y = 7$

17. $x - 4y = 3$,
$2x - 6 = 8y$

18. $x - 2y = 7$,
$3x - 21 = 6y$

19. $y = -2x + 3$,
$3y = -6x + 9$

20. $\quad y = 2x + 5$,
$-2y = -4x - 10$

21. $x + 2y = 10$,
$3x + 4y = 8$

22. $2x + 3y = -2$,
$2x - y = 9$

23. $3a + 2b = 2$,
$-2a + b = 8$

24. $x - y = -3$,
$2x + 3y = -6$

25. $y - 2x = 0$,
$3x + 7y = 17$

26. $r - 2s = 0$,
$4r - 3s = 15$

27. $8x + 2y = 6$,
$y = 3 - 4x$

28. $x - 3y = 7$,
$-4x + 12y = 28$

29. $x - 3y = -1$,
$5y - 2x = 4$

30. $x - 2y = 5$,
$2y - 3x = 1$

Aha! **31.** $2x - y = 0$,
$2x - y = -2$

32. $5x = y - 3$,
$5x = y + 5$

Solve.

33. The sum of two numbers is 87. One number is 3 more than the other. Find the numbers.

34. The sum of two numbers is 76. One number is 2 more than the other. Find the numbers.

35. Find two numbers for which the sum is 58 and the difference is 14.

36. Find two numbers for which the sum is 76 and the difference is 12.

37. The difference between two numbers is 16. Three times the larger number is seven times the smaller. What are the numbers?

38. The difference between two numbers is 18. Twice the smaller number plus three times the larger is 74. What are the numbers?

39. *Supplementary angles.* Two angles are supplementary. One angle is 60° less than twice the other. Find the measure of each angle.

40. *Supplementary angles.* Two angles are supplementary. One angle is 8° less than three times the other. Find the measure of each angle.

41. *Complementary angles.* Two angles are complementary. Their difference is 42°. Find the measure of each angle. (*Complementary angles* are angles for which the sum is 90°.)

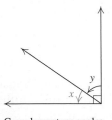

Complementary angles

42. *Complementary angles.* Two angles are complementary. One angle is 42° more than one-half the other. Find the measure of each angle.

43. *Perimeter of a rectangle.* The state of Colorado is a rectangle with a perimeter of 1300 mi (see the figure at the top of the next column). The length is 110 mi more than the width. Find the length and the width.

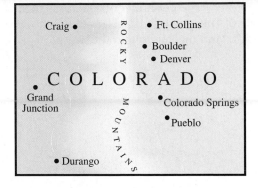

44. *Perimeter of a rectangle.* The state of Wyoming is a rectangle with a perimeter of 1280 mi. The width is 90 mi less than the length. Find the length and the width.

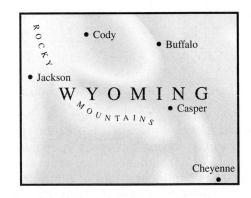

45. *Racquetball.* A regulation racquetball court should have a perimeter of 120 ft, with a length that is twice the width. Find the length and the width of a court.

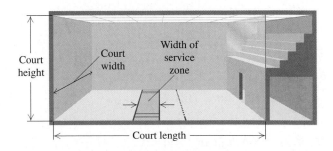

46. *Racquetball.* The height of the front wall of a standard racquetball court is four times the width of the service zone (see the figure). Together, these measurements total 25 ft. Find the height and the width.

47. *Lacrosse.* The perimeter of a lacrosse field is 340 yd. The length is 10 yd less than twice the width. Find the length and the width.

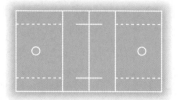

48. *Soccer.* The perimeter of a soccer field is 280 yd. The width is 5 more than half the length. Find the length and the width.

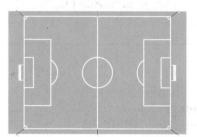

49. Joel solves every system of two equations (in x and y) by first solving for y in the first equation and then substituting into the second equation. Is he using the best approach? Why or why not?

50. Describe two advantages of the substitution method over the graphing method for solving systems of equations.

SKILL MAINTENANCE

Simplify.

51. $2(5x + 3y) - 3(5x + 3y)$

52. $4(7x + 5y) - 7(4x + 5y)$

53. $3(8x + 2y) - 2(7x + 3y)$

54. $2(7x + 5 - 3y) - 7(2x + 5)$

55. $2(5x - 3y) - 5(2x + y)$

56. $4(2x + 3y) + 3(5x - 4y)$

SYNTHESIS

57. Under what circumstances can a system of equations be solved more easily by graphing than by substitution?

58. Janine can tell by inspection that the system
$$x = 2y - 1,$$
$$x = 2y + 3$$
has no solution. How can she tell?

Solve by the substitution method.

Aha! **59.** $\dfrac{1}{6}(a + b) = 1,$

$\dfrac{1}{4}(a - b) = 2$

60. $\dfrac{x}{5} - \dfrac{y}{2} = 3,$

$\dfrac{x}{4} + \dfrac{3y}{4} = 1$

61. $y + 5.97 = 2.35x,$
$2.14y - x = 4.88$

62. $a + 4.2b = 25.1,$
$9a - 1.8b = 39.78$

Exercises 63 and 64 contain systems of three equations in three variables. A solution is an ordered triple of the form (x, y, z). Use the substitution method to solve.

63. $x + y + z = 4,$
$x - 2y - z = 1,$
$y = -1$

64. $x + y + z = 180,$
$x = z - 70,$
$2y - z = 0$

65. *Softball.* The perimeter of a softball diamond is two-thirds of the perimeter of a baseball diamond. Together, the two perimeters measure 200 yd. Find the distance between the bases in each sport.

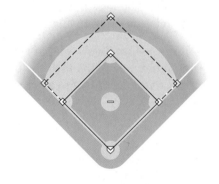

66. Solve Example 3 by first solving for $2y$ in equation (1) and then substituting for $2y$ in equation (2). Is this method easier than the procedure used in Example 3? Why or why not?

67. Write a system of two linear equations that can be solved more quickly—but still precisely—by a grapher than by substitution. Time yourself using both methods to solve the system.

Systems of Equations and Elimination

7.3

Solving by the Elimination Method • Problem Solving

We have seen that graphing is not always a precise method of solving a system of equations, especially when fractional solutions are involved. The substitution method, considered in Section 7.2, is precise but sometimes difficult to use. For example, to solve the system

$$2x + 3y = 13, \qquad (1)$$
$$4x - 3y = 17 \qquad (2)$$

by substitution, we would need to first solve for a variable in one of the equations. Were we to solve equation (1) for y, we would find (after several steps) that $y = \frac{13}{3} - \frac{2}{3}x$. We could then use the expression $\frac{13}{3} - \frac{2}{3}x$ in equation (2) as a replacement for y:

$$4x - 3\left(\frac{13}{3} - \frac{2}{3}x\right) = 17.$$

As you can see, although substitution *could* be used to solve this system, doing so is not easy. Fortunately, another method, *elimination*, can be used to solve systems and, on problems like this, is simpler to use.

Solving by the Elimination Method

The **elimination method** for solving systems of equations makes use of the addition principle. To see how it works, we use it to solve the system above.

Example 1   Solve the system

$$2x + 3y = 13, \qquad (1)$$
$$4x - 3y = 17. \qquad (2)$$

Solution   According to equation (2), $4x - 3y$ and 17 are the same number. Thus we can add $4x - 3y$ to the left side of equation (1) and 17 to the right side:

$$2x + 3y = 13 \qquad (1)$$
$$\underline{4x - 3y = 17} \qquad (2)$$
$$6x + 0y = 30. \qquad \text{Adding. Note that } y \text{ has been "eliminated."}$$

The resulting equation has just one variable:

$$6x = 30.$$

Dividing both sides of this equation by 6, we find that $x = 5$.

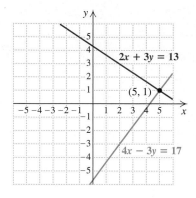

Next, we substitute 5 for x in either of the original equations:

$$2x + 3y = 13 \qquad \text{Equation (1)}$$
$$2 \cdot 5 + 3y = 13 \qquad \text{Substituting 5 for } x$$
$$10 + 3y = 13$$
$$3y = 3$$
$$y = 1. \qquad \text{Solving for } y$$

We check the ordered pair $(5, 1)$. The graph shown at left also serves as a check.

Check:

$$\begin{array}{c|c} 2x + 3y = 13 \\ \hline 2(5) + 3(1) \; ? \; 13 \\ 10 + 3 \; \Big| \\ 13 \; \Big| \; 13 \quad \text{TRUE} \end{array} \qquad \begin{array}{c|c} 4x - 3y = 17 \\ \hline 4(5) - 3(1) \; ? \; 17 \\ 20 - 3 \; \Big| \\ 17 \; \Big| \; 17 \quad \text{TRUE} \end{array}$$

Since $(5, 1)$ checks in both equations, it is the solution.

The system in Example 1 is easier to solve by elimination than by substitution because two terms, $-3y$ in equation (2) and $3y$ in equation (1), are opposites. Some systems have no pair of terms that are opposites. When this occurs, we need to multiply one or both of the equations by appropriate numbers to create a pair of terms that are opposites.

Example 2

Solve:

$$2x + 3y = 8, \qquad (1)$$
$$x + 3y = 7. \qquad (2)$$

Solution   For these equations, addition will not eliminate a variable. However, if the $3y$ were $-3y$ in one equation, we could eliminate y. We multiply both sides of equation (2) by -1 to find an equivalent equation that contains $-3y$, and then add:

$$\begin{array}{ll} 2x + 3y = 8 & \text{Equation (1)} \\ \underline{-x - 3y = -7} & \text{Multiplying both sides of equation (2) by } -1 \\ x = 1. & \text{Adding} \end{array}$$

Next, we substitute 1 for x in either of the original equations:

$$x + 3y = 7 \qquad \text{Equation (2)}$$
$$1 + 3y = 7 \qquad \text{Substituting 1 for } x$$
$$\left. \begin{array}{l} 3y = 6 \\ y = 2. \end{array} \right\} \quad \text{Solving for } y$$

We can check the ordered pair $(1, 2)$. The graph shown at left is also a check.

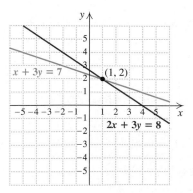

Check:

$$\begin{array}{c|c} 2x + 3y = 8 \\ \hline 2 \cdot 1 + 3 \cdot 2 \; ? \; 8 \\ 2 + 6 \; \Big| \\ 8 \; \Big| \; 8 \quad \text{TRUE} \end{array} \qquad \begin{array}{c|c} x + 3y = 7 \\ \hline 1 + 3 \cdot 2 \; ? \; 7 \\ 1 + 6 \; \Big| \\ 7 \; \Big| \; 7 \quad \text{TRUE} \end{array}$$

Since $(1, 2)$ checks in both equations, it is the solution.

In Example 2, we used the multiplication principle, multiplying by -1. We often need to multiply by a number other than -1.

Example 3

Solve:

$$3x + 6y = -6, \qquad (1)$$
$$5x - 2y = 14. \qquad (2)$$

Solution   No terms are opposites, but if both sides of equation (2) are multiplied by 3 $\left(\text{or if both sides of equation (1) are multiplied by } \frac{1}{3}\right)$, the coefficients of y will be opposites. Note that 6 is the LCM of 2 and 6:

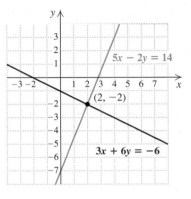

$$
\begin{aligned}
3x + 6y &= -6 &&\text{Equation (1)}\\
\underline{15x - 6y} &= \underline{42} &&\text{Multiplying both sides of equation (2) by 3}\\
18x &= 36 &&\text{Adding}\\
x &= 2. &&\text{Solving for } x
\end{aligned}
$$

We then go back to equation (1) and substitute 2 for x:

$$
\begin{aligned}
3 \cdot 2 + 6y &= -6 &&\text{Substituting 2 for } x \text{ in equation (1)}\\
6 + 6y &= -6 \\
6y &= -12 \\
y &= -2.
\end{aligned}\left.\vphantom{\begin{aligned}6+6y\\6y\\y\end{aligned}}\right\} \text{Solving for } y
$$

We leave it to the student to confirm that $(2, -2)$ checks and is the solution. The graph in the margin also serves as a check.

Example 4

Solve:

$$3y + 1 + 2x = 0, \qquad (1)$$
$$5x = 7 - 4y. \qquad (2)$$

Solution   It is often helpful to write both equations in the form $Ax + By = C$ before attempting to eliminate a variable:

$$
\begin{aligned}
2x + 3y &= -1, &&\text{Subtracting 1 from both sides and rearranging the terms of the first equation}\\
5x + 4y &= 7. &&\text{Adding } 4y \text{ to both sides of equation (2)}
\end{aligned}
$$

Since neither coefficient of x is a multiple of the other and neither coefficient of y is a multiple of the other, we use the multiplication principle with *both* equations:

$$
\begin{aligned}
2x + 3y &= -1, &&(3)\\
5x + 4y &= 7. &&(4)
\end{aligned}
\qquad \text{Note that the LCM of 2 and 5 is 10.}
$$

We can eliminate the x-term by multiplying both sides of equation (3) by 5 and both sides of equation (4) by -2:

$$
\begin{aligned}
10x + 15y &= -5 &&\text{Multiplying both sides of equation (3) by 5}\\
\underline{-10x - 8y} &= \underline{-14} &&\text{Multiplying both sides of equation (4) by } -2\\
7y &= -19 &&\text{Adding}\\
y &= \tfrac{-19}{7} = -\tfrac{19}{7}. &&\text{Dividing by 7}
\end{aligned}
$$

We substitute $-\frac{19}{7}$ for y in equation (3):

$$2x + 3y = -1 \qquad \text{Equation (3)}$$
$$2x + 3\left(-\tfrac{19}{7}\right) = -1 \qquad \text{Substituting } -\tfrac{19}{7} \text{ for } y$$
$$2x - \tfrac{57}{7} = -1$$
$$2x = -1 + \tfrac{57}{7} \qquad \text{Adding } \tfrac{57}{7} \text{ to both sides}$$
$$2x = -\tfrac{7}{7} + \tfrac{57}{7} = \tfrac{50}{7}$$
$$x = \tfrac{50}{7} \cdot \tfrac{1}{2} = \tfrac{25}{7}. \qquad \text{Solving for } x$$

We check the ordered pair $\left(\frac{25}{7}, -\frac{19}{7}\right)$.

Check:

$$\frac{3y + 1 + 2x = 0}{3\left(-\tfrac{19}{7}\right) + 1 + 2 \cdot \tfrac{25}{7} \; ? \; 0}$$
$$-\tfrac{57}{7} + \tfrac{7}{7} + \tfrac{50}{7} \quad \Big|$$
$$0 \; \Big| \; 0 \quad \text{TRUE}$$

$$\frac{5x = 7 - 4y}{5 \cdot \tfrac{25}{7} \; ? \; 7 - 4\left(-\tfrac{19}{7}\right)}$$
$$\tfrac{125}{7} \quad \Big| \quad \tfrac{49}{7} + \tfrac{76}{7}$$
$$\tfrac{125}{7} \quad \Big| \quad \tfrac{125}{7} \quad \text{TRUE}$$

The solution is $\left(\frac{25}{7}, -\frac{19}{7}\right)$.

Next, we consider a system with no solution and see what happens when the elimination method is used.

Example 5

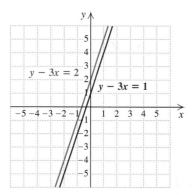

Solve:

$$y - 3x = 2, \qquad (1)$$
$$y - 3x = 1. \qquad (2)$$

Solution To eliminate y, we multiply both sides of equation (2) by -1. Then we add:

$$y - 3x = 2$$
$$\underline{-y + 3x = -1} \qquad \text{Multiplying both sides of equation (2) by } -1$$
$$0 = 1. \qquad \text{Adding}$$

Note that in eliminating y, we eliminated x as well. The resulting equation, $0 = 1$, is false for any pair (x, y), so there is *no solution*.

Sometimes there is an infinite number of solutions. Consider a system that we graphed in Example 4 of Section 7.1.

Example 6

Solve:

$$2x + 3y = 6, \qquad (1)$$
$$-8x - 12y = -24. \qquad (2)$$

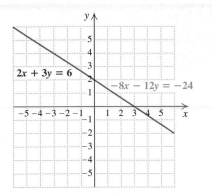

$2x + 3y = 6$

$-8x - 12y = -24$

Solution   To eliminate x, we multiply both sides of equation (1) by 4 and then add the two equations:

$$8x + 12y = 24 \qquad \text{Multiplying both sides of equation (1) by 4}$$
$$\underline{-8x - 12y = -24}$$
$$0 = 0. \qquad \text{Adding}$$

Again, we have eliminated *both* variables. The resulting equation, $0 = 0$, is always true, indicating that the equations are dependent. Such a system has an infinite number of solutions.

When decimals or fractions appear, we can first multiply to clear them. Then we proceed as before.

Example 7

Solve:

$$\tfrac{1}{2}x + \tfrac{3}{4}y = 2, \qquad (1)$$
$$x + 3y = 7. \qquad (2)$$

Solution   The number 4 is the LCD for equation (1). Thus we multiply both sides of equation (1) by 4 to clear fractions:

$$4\left(\tfrac{1}{2}x + \tfrac{3}{4}y\right) = 4 \cdot 2 \qquad \text{Multiplying both sides of equation (1) by 4}$$
$$4 \cdot \tfrac{1}{2}x + 4 \cdot \tfrac{3}{4}y = 8 \qquad \text{Using the distributive law}$$
$$2x + 3y = 8.$$

The resulting system is

$$2x + 3y = 8, \qquad \text{This equation is equivalent to equation (1).}$$
$$x + 3y = 7.$$

As we saw in Example 2, the solution of this system is $(1, 2)$.

Problem Solving

We now use the elimination method to solve a problem.

Example 8

Phone rates.   Recently, calls made from a pay phone using Five line cost 5 cents per minute plus a 50-cent connection fee. The same call, using an AT&T calling card, cost 5.9 cents per minute. For what length of phone call will the costs be the same?

Solution

1. **Familiarize.**   To become familiar with the problem, we make and check a guess of 30 min. A 30-min Five line call would cost $5(30) + 50$¢, or \$2. Using AT&T, a 30-min call would cost $5.9(30)$¢, or \$1.77. Because \$2 $\neq$ \$1.77, our guess is incorrect. However, from the check, we can see how algebra can be used to find a solution without more guessing. We let m = the length of a phone call, in minutes, and c = the cost of the call, in cents.

2. **Translate.**   We reword the problem and translate as follows:

Rewording: Five line's cost is 50¢ plus 5¢ times the length of the call, in minutes.

Translating: c $=$ 50 $+$ 5 $\cdot$ m

Note that cents are used as the monetary unit. We write a second equation representing AT&T.

Rewording: AT&T's cost is 5.9¢ times the length of the call, in minutes.

Translating: c $=$ 5.9 $\cdot$ m

We now have the system of equations

$c = 50 + 5m,$

$c = 5.9m.$

3. **Carry out.**   To solve the system, we multiply the second equation by -1 and add to eliminate c:

$$c = 50 + 5m$$
$$\underline{-c = - 5.9m}$$
$$0 = 50 - 0.9m.$$

We can now solve for m:

$$0.9m = 50$$
$$m = \frac{50}{0.9}$$
$$m \approx 55.6.$$

4. **Check.**   For 55.6 min, the cost of a Five line call would be

$$50 + 5 \cdot 55.6, \quad \text{or} \quad 50 + 278.0¢, \quad \text{or} \quad \$3.28$$

and the cost of an AT&T call would be

$$5.9 \cdot 55.6, \quad \text{or} \quad 328.04¢, \quad \text{or} \quad \$3.28. \qquad \text{Rounding to the nearest cent}$$

Thus the costs are the same for a 55.6-min call.

5. **State.**   For a 55.6-min call, the costs are the same.

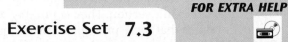

FOR EXTRA HELP

Digital Video Tutor CD 4
Videotape 13

InterAct Math

Math Tutor Center

MathXL

MyMathLab.com

Exercise Set 7.3

Solve using the elimination method. If a system has no solution or an infinite number of solutions, state this.

1. $x + y = 3,$
$x - y = 7$

2. $x - y = 6,$
$x + y = 12$

3. $x + y = 6,$
$-x + 2y = 15$

4. $x + y = 6,$
$-x + 3y = -2$

5. $3x - y = 9,$
$2x + y = 6$

6. $4x - y = 1,$
$3x + y = 13$

7. $5a + 4b = 7,$
$-5a + b = 8$

8. $7c + 4d = 16,$
$c - 4d = -4$

9. $8x - 5y = -9,$
$3x + 5y = -2$

10. $3a - 3b = -15,$
$-3a - 3b = -3$

11. $3a - 6b = 8,$
$-3a + 6b = -8$

12. $8x + 3y = 4,$
$-8x - 3y = -4$

13. $-x - y = 8,$
$2x - y = -1$

14. $x + y = -7,$
$3x + y = -9$

15. $x + 3y = 19,$
$x - y = -1$

16. $3x - y = 8,$
$x + 2y = 5$

17. $x + y = 5,$
$4x - 3y = 13$

18. $x - y = 7,$
$3x - 5y = 15$

19. $2w - 3z = -1,$
$3w + 4z = 24$

20. $7p + 5q = 2,$
$8p - 9q = 17$

21. $2a + 3b = -1,$
$3a + 5b = -2$

22. $3x - 4y = 16,$
$5x + 6y = 14$

23. $3y = x,$
$5x + 14 = y$

24. $5a = 2b,$
$2a + 11 = 3b$

25. $4x - 10y = 13,$
$-2x + 5y = 8$

26. $2p + 5q = 9,$
$3p - 2q = 4$

27. $8n + 6 - 3m = 0,$
$32 = m - n$

28. $6x - 8 + y = 0,$
$11 = 3y - 8x$

29. $3x + 5y = 4,$
$-2x + 3y = 10$

30. $2x + y = 13,$
$4x + 2y = 23$

31. $0.06x + 0.05y = 0.07,$
$0.4x - 0.3y = 1.1$

32. $x - \frac{3}{2}y = 13,$
$\frac{3}{2}x - y = 17$

33. $x + \frac{9}{2}y = \frac{15}{4},$
$\frac{9}{10}x - y = \frac{9}{20}$

34. $1.8x - 2y = 0.9,$
$0.04x + 0.18y = 0.15$

Solve.

35. *Local truck rentals.*   Budget rents a 16-ft truck for $49 plus 39¢ per mile. U-Haul rents a 17-ft truck for $29.95 plus 49¢ per mile. (*Source*: Budget Rent a Car and U-Haul Truck Rentals, July 2000) For what mileage is the cost the same?

36. *Local truck rentals.*   U-Haul rents a cargo van for $19.95 plus 39¢ per mile. Budget rents a cargo van for $39 plus 30¢ per mile. (*Source*: Budget Rent a Car and U-Haul Truck Rentals, July 2000) For what mileage is the cost the same?

37. *Complementary angles.*   Two angles are complementary. One angle is 12° more than twice the other. Find the measure of each angle.

38. *Complementary angles.*   Two angles are complementary. Their difference is 26°. Find the measure of each angle.

39. *Phone rates.*   Recently, MCI Worldcom offered two long-distance calling plans. One-Plus costs $3.95 per month plus 7¢ a minute. Another plan has no monthly fee, but costs 9¢ a minute. For what number of minutes will the two plans cost the same?

40. *Phone rates.*   Recently, Sprint offered one calling plan that charges 25¢ a minute for calling-card calls. Another plan charges 7¢ a minute for calling-card calls, but costs an additional $4 per month. For what number of minutes will the two plans cost the same?

41. *Supplementary angles.*   Two angles are supplementary. One angle measures 5° less than four times the measure of the other. Find the measure of each angle.

42. *Supplementary angles.*   Two angles are supplementary. One angle measures 45° more than twice the measure of the other. Find the measure of each angle.

43. *Planting grapes.*   South Wind Vineyards uses 820 acres to plant Chardonnay and Riesling grapes. The vintner knows the profits will be greatest by planting 140 more acres of Chardonnay than

Riesling. How many acres of each grape should be planted?

44. *Farming.* Sleek Meadows Horse Farm plants 31 acres of hay and oats. The owners know that their needs are best met if they plant 9 acres less of hay than oats. How many acres of each should they plant?

45. *Framing.* Angel has 18 ft of molding from which he needs to make a rectangular frame. Because of the dimensions of the mirror being framed, the frame must be twice as long as it is wide. What should the dimensions of the frame be?

46. *Gardening.* Patrice has 108 ft of fencing for a rectangular garden. If the garden's length is to be $1\frac{1}{2}$ times its width, what should the garden's dimensions be?

47. Describe a method that could be used for writing a system that contains dependent equations.

48. Describe a method that could be used for writing an inconsistent system of equations.

SKILL MAINTENANCE

Convert to decimal notation.

49. 8%   **50.** 7.3%   **51.** 0.4%

Solve.

52. What percent of 45 is 18?

53. What number is 9% of 350?

54. 24 is what percent of 150?

SYNTHESIS

55. If a system has an infinite number of solutions, does it follow that *any* ordered pair is a solution? Why or why not?

56. Explain how the multiplication and addition principles are used in this section. Then count the number of times that these principles are used in Example 4.

Solve using substitution, elimination, or graphing.

57. $x + y = 7,$
$3(y - x) = 9$

58. $y = 3x + 4,$
$3 + y = 2(y - x)$

59. $2(5a - 5b) = 10,$
$-5(2a + 6b) = 10$

60. $0.05x + y = 4,$
$\dfrac{x}{2} + \dfrac{y}{3} = 1\dfrac{1}{3}$

Aha! **61.** $y = -\dfrac{2}{7}x + 3,$
$y = \dfrac{4}{5}x + 3$

62. $y = \dfrac{2}{5}x - 7,$
$y = \dfrac{2}{5}x + 4$

Solve for x and y.

63. $y = ax + b,$
$y = x + c$

64. $ax + by + c = 0,$
$ax + cy + b = 0$

65. *Caged rabbits and pheasants.* Several ancient Chinese books included problems that can be solved by translating to systems of equations. *Arithmetical Rules in Nine Sections* is a book of 246 problems compiled by a Chinese mathematician, Chang Tsang, who died in 152 B.C. One of the problems is: Suppose there are a number of rabbits and pheasants confined in a cage. In all, there are 35 heads and 94 feet. How many rabbits and how many pheasants are there? Solve the problem.

66. *Age.* Patrick's age is 20% of his mother's age. Twenty years from now, Patrick's age will be 52% of his mother's age. How old are Patrick and his mother now?

67. *Age.* If 5 is added to a man's age and the total is divided by 5, the result will be his daughter's age. Five years ago, the man's age was eight times his daughter's age. Find their present ages.

68. *Dimensions of a triangle.* When the base of a triangle is increased by 1 ft and the height is increased by 2 ft, the height changes from being two thirds of the base to being four fifths of the base. Find the original dimensions of the triangle.

More Applications Using Systems

7.4

Total Value Problems • Mixture Problems

CONNECTING THE CONCEPTS

We now have three distinctly different ways to solve a system. Each method has certain strengths and weaknesses.

Method	Strengths	Weaknesses
Graphical	Solutions are displayed visually. Works with any system that can be graphed.	Inexact when solutions involve numbers that are not integers or are very large and off the graph.
Substitution	Always yields exact solutions. Easy to use when a variable is alone on one side of an equation.	Introduces extensive computations with fractions when solving more complicated systems. Solutions are not graphically displayed.
Elimination	Always yields exact solutions. Easy to use when fractions or decimals appcar in the system.	Solutions are not graphically displayed.

When selecting the best method to use for a particular system, consider the strengths and weaknesses listed above. As you gain experience with these methods, it will become easier to choose the best method for any given system.

The five steps for problem solving and our methods for solving systems of equations can be used in a variety of applications.

Total Value Problems

E x a m p l e 1

Basketball scores. In the final game of the 2000 basketball season, the Los Angeles Lakers scored 96 of their points on a combination of 43 two- and three-point baskets (*Source*: National Basketball Association). How many shots of each type were made?

Solution

1. **Familiarize.** Suppose that of the 43 baskets, 30 were two-pointers and 13 were three-pointers. These 43 baskets would then amount to a total of

$$30 \cdot 2 + 13 \cdot 3 = 60 + 39 = 99 \text{ points.}$$

Although our guess is incorrect, checking the guess has familiarized us with the problem. We let $w =$ the number of two-pointers made and $r =$ the number of three-pointers made.

2. **Translate.** Since a total of 43 baskets was made, we must have

$$w + r = 43.$$

To find a second equation, we reword some information and focus on the points scored, just as when we checked our guess above.

Rewording: The points scored from two-pointers plus the points scored from three-pointers totaled 96.

Translating: $w \cdot 2$ + $r \cdot 3$ = 96.

The problem has been translated to the following system of equations:

$$w + r = 43, \quad (1)$$
$$2w + 3r = 96. \quad (2)$$

3. **Carry out.** For purposes of review, we solve by substitution. First we solve equation (1) for w:

$$w + r = 43 \qquad \text{Equation (1)}$$
$$w = 43 - r. \qquad \text{Solving for } w \qquad (3)$$

Next, we replace w in equation (2) with $43 - r$:

$$2w + 3r = 96 \qquad \text{Equation (2)}$$
$$2(43 - r) + 3r = 96 \qquad \text{Substituting } 43 - r \text{ for } w$$
$$86 - 2r + 3r = 96$$
$$\left.\begin{array}{l} 86 + r = 96 \\ r = 10. \end{array}\right\} \text{Solving for } r$$

We find w by substituting 10 for r in equation (3):

$$w = 43 - r = 43 - 10 = 33.$$

4. **Check.** If the Lakers made 33 two-pointers and 10 three-pointers, they would have made 43 shots, for a total of

$$33 \cdot 2 + 10 \cdot 3 = 66 + 30 = 96 \text{ points.}$$

The numbers check.

5. **State.** The Lakers made 33 two-pointers and 10 three-pointers.

E x a m p l e   2

Film processing. Photoworks.com charges $7.00 for processing a 24-exposure roll and $10.00 for processing a 36-exposure roll. After their class trip, Ms. Barnes' fourth-grade class sent 19 rolls of film to Photoworks and paid $151 for processing. How many rolls of each type were processed?

Solution

1. **Familiarize.** When faced with a new problem, it is often useful to compare it to a similar problem that you have already solved. Here instead of counting two- and three-point baskets, as in Example 1, we are counting 24-exposure and 36-exposure rolls of film. We let $w =$ the number of 24-exposure rolls that were processed and $r =$ the number of 36-exposure rolls that were processed.

2. **Translate.** Since a total of 19 rolls was processed, we have

   $$w + r = 19.$$

   To find a second equation, we reword some information, focusing on the amount of money paid.

   Rewording: The money paid for the 24-exposure processing **plus** the money paid for the 36-exposure processing **totaled** $151

   Translating: $w \cdot 7.00$ $+$ $r \cdot 10.00$ $=$ $151.$

   Presenting the information in a table can be helpful.

	24-exposure	36-exposure	Total
Cost per Roll	$7.00	$10.00	
Number of Rolls	w	r	19
Money Paid	7.00w	10.00r	151

   $19 \longrightarrow w + r = 19$

   $151 \longrightarrow 7w + 10r = 151$

   We have translated to a system of equations:

   $$w + r = 19, \quad (1)$$
   $$7w + 10r = 151. \quad (2)$$

3. **Carry out.** The system can be solved using elimination:

   $$\begin{array}{rl} -7w - 7r = -133 & \text{Multiplying both sides of equation (1) by } -7 \\ \underline{7w + 10r = 151} & \text{Equation (2)} \\ 3r = 18 & \text{Adding} \\ r = 6. & \text{Dividing both sides by 3} \end{array}$$

   To solve for w, we substitute 6 for r:

   $$\begin{array}{ll} w + r = 19 & \text{Using equation (1)} \\ w + 6 = 19 & \\ w = 13. & \text{Subtracting 6 from both sides} \end{array}$$

4. **Check.** If $w = 13$ and $r = 6$, a total of 19 rolls was developed. The amount paid was 13($7.00), or $91, for the 24-exposure rolls, and 6($10.00), or $60, for the 36-exposure rolls. The total paid was therefore $91 + $60, or $151. The numbers check.

5. **State.** The class had 13 rolls of 24-exposure film and 6 rolls of 36-exposure film developed.

Mixture Problems

Example 3

Blending coffees. The Java Joint wants to mix Kenyan beans that sell for $8.25 per pound with Venezuelan beans that sell for $9.50 per pound to form a 50-lb batch of Morning Blend that sells for $9.00 per pound. How many pounds of Kenyan beans and how many pounds of Venezuelan beans should go into the blend?

Solution

1. **Familiarize.** This problem seems similar to Example 2. Instead of two types of film, we have Kenyan coffee and Venezuelan coffee. Instead of two different prices for processing, we have two different prices per pound. Finally, instead of having the total paid, we know the weight and price per pound of the batch of Morning Blend being made. Note that we can easily find the value of the batch of Morning Blend by multiplying 50 lb times $9.00 per pound. We let $k =$ the number of pounds of Kenyan coffee used and $v =$ the number of pounds of Venezuelan coffee used.

2. **Translate.** Since a 50-lb batch is being made, we must have

$$k + v = 50.$$

To find a second equation, we consider the total value of the 50-lb batch. That value must be the same as the value of the Kenyan beans and the value of the Venezuelan beans that go into the blend.

Rewording: The value of the Kenyan beans plus the value of the Venezuelan beans is the value of the Morning Blend.

Translating: $k \cdot 8.25$ $+$ $v \cdot 9.50$ $=$ $50 \cdot 9.00$

This information can be presented in a table.

	Kenyan	Venezuelan	Morning Blend	
Number of Pounds	k	v	50	$\rightarrow k + v = 50$
Price per Pound	$8.25	$9.50	$9.00	
Value of Beans	$8.25k$	$9.50v$	$50 \cdot 9$, or 450	$\rightarrow 8.25k + 9.50v = 450$

We have translated to a system of equations:

$$k + v = 50, \qquad (1)$$
$$8.25k + 9.50v = 450. \qquad (2)$$

3. **Carry out.** When equation (1) is solved for k, we have $k = 50 - v$. We then substitute $50 - v$ for k in equation (2):

$8.25(50 - v) + 9.50v = 450$	Solving by substitution
$412.50 - 8.25v + 9.50v = 450$	Using the distributive law
$1.25v = 37.50$	Combining like terms; subtracting 412.50 from both sides
$v = 30.$	Dividing both sides by 1.25

If $v = 30$, we see from equation (1) that $k = 20$.

4. **Check.** If 20 lb of Kenyan beans and 30 lb of Venezuelan beans are mixed, a 50-lb blend will result. The value of 20 lb of Kenyan beans is 20($8.25), or $165. The value of 30 lb of Venezuelan beans is 30($9.50), or $285, so the value of the blend is $165 + $285 = $450. A 50-lb blend priced at $9.00 a pound is also worth $450, so our answer checks.

5. **State.** The Morning Blend should be made by combining 20 lb of Kenyan beans with 30 lb of Venezuelan beans.

E x a m p l e   4

Paint colors.   At a local "paint swap," Gayle found large supplies of Skylite Pink (12.5% red pigment) and MacIntosh Red (20% red pigment). How many gallons of each color should be mixed in order to create a 10-gal batch of Summer Rose (17% red pigment)?

Solution

1. **Familiarize.** This problem is similar to Example 3. Instead of mixing two types of coffee and keeping an eye on the price of the mixture, we are mixing two types of paint and keeping an eye on the amount of pigment in the mixture.

   To visualize this problem, think of the pigment as a solid that, given time, would settle to the bottom of each can. Let's guess that 2 gal of Skylite Pink and 8 gal of MacIntosh Red are mixed. How much pigment would be in the mixture? The Skylite Pink would contribute 12.5% of 2 gal, or 0.25 gal of pigment, and the MacIntosh Red would contribute 20% of 8 gal, or 1.6 gal of pigment. Thus the 10-gal mixture would contain $0.25 + 1.6 = 1.85$ gal of pigment. Since Gayle wants the 10 gal of Summer Rose to be 17% pigment, and since 17% of 10 gal is 1.7 gal, our guess is incorrect. Rather than check another guess, we let $p = $ the number of gallons of Skylite Pink needed and $m = $ the number of gallons of MacIntosh Red needed.

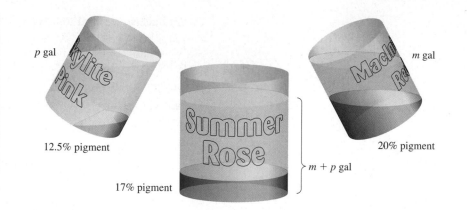

2. **Translate.** As in Example 3, the information given can be arranged in a table.

	Skylite Pink	MacIntosh Red	Summer Rose
Amount of Paint (in gallons)	p	m	10
Percent Pigment	12.5%	20%	17%
Amount of Pigment (in gallons)	$0.125p$	$0.2m$	0.17×10, or 1.7

A system of two equations can be formed by reading across the first and third rows of the table. Since Gayle needs 10 gal of mixture, we must have

$p + m = 10.$ ← Total amount of paint

Since the amount of pigment in the Summer Rose paint comes from the amount of pigment in the Skylite Pink and the MacIntosh Red paint, we have

$0.125p + 0.2m = 1.7.$ ← Total amount of pigment

We have translated to a system of equations:

$$p + m = 10, \quad (1)$$
$$0.125p + 0.2m = 1.7. \quad (2)$$

3. **Carry out.** We note that if we multiply both sides of equation (2) by -5, we can eliminate m (other approaches will also work):

$$p + m = 10$$
$$\underline{-0.625p - m = -8.5} \qquad \text{We observed that } (-5)(0.2m) = -m.$$
$$0.375p = 1.5$$
$$p = 4. \qquad \text{Dividing both sides by } 0.375$$

If $p = 4$, we see from equation (1) that $m = 6$.

4. **Check.** Clearly, 4 gal of Skylite Pink and 6 gal of MacIntosh Red do add up to a total of 10 gal. To see if the mixture is the right color, Summer Rose, we calculate the amount of pigment in the mixture: $0.125 \cdot 4 + 0.2 \cdot 6 = 0.5 + 1.2 = 1.7$. Since 1.7 is 17% of 10, the mixture is the correct color.

5. **State.** Gayle needs 4 gal of Skylite Pink and 6 gal of MacIntosh Red in order to make 10 gal of Summer Rose.

Re-examine Examples 1–4, looking for similarities. Examples 3 and 4 are often called *mixture problems*, but they have much in common with Examples 1 and 2.

> ### Problem-Solving Tip
>
> When solving a problem, see if it is patterned or modeled after a problem that you have already solved.

Solve. Use the five steps for problem solving.

1. *Basketball scoring.*   In a recent game, Jason Terry of the Atlanta Hawks scored 20 points on a combination of 9 two- and three-point baskets (*Source*: National Basketball Association). How many shots of each type were made?

2. *Basketball scoring.*   In a recent game, Bob Sura of the Cleveland Cavaliers scored 16 points on a combination of 7 two- and three-point baskets (*Source*: National Basketball Association). How many shots of each type were made?

3. *Basketball scoring.*   In their last game of the 2000 basketball season, the Indiana Pacers scored 84 of their points on a combination of 36 two- and three-point baskets (*Source*: National Basketball Association). How many shots of each type were made?

4. *Basketball scoring.*   In winning the 2000 conference finals, the Los Angeles Lakers scored 69 of their points on a combination of 31 two- and three-pointers (*Source*: National Basketball Association). How many shots of each type did they make?

5. *Film prices.*   Vitcom Photo charges $1.50 for a roll of 24-exposure film and $2.50 for a roll of 36-exposure film. Linda bought 17 rolls for $34.50. How many rolls of each type of film did she order?

6. *Film prices.*   Filmworks charges $1.75 for a 24-exposure roll of film and $2.25 for a 36-exposure roll of film. Stu bought 19 rolls of film for $39.25. How many rolls of each type did he buy?

7. *Returnable bottles.*   The Dixville Cub Scout troop collected 430 returnable bottles and cans, some worth 5 cents each and the rest worth 10 cents each. If the total value of the cans and bottles was $26.20, how many 5-cent bottles or cans and how many 10-cent bottles or cans were collected?

8. *Ice cream cones.* A busload of campers stopped at a dairy stand for ice cream. They ordered 40 cones, some soft-serve at $1.75 and the rest hard-pack at $2.00. If the total bill was $74, how many of each type of cone were ordered?

9. *Zoo admissions.* During the summer months, the Bronx Zoo charges $9 for adults and $5 for children and seniors (*Source*: Bronx Zoo). One July day, a total of $6320 was collected from 960 admissions. How many adult admissions were there?

10. *Zoo admissions.* From November 2 through January 3, the Bronx Zoo charges $6 for adults and $3 for children and seniors (*Source*: Bronx Zoo). One December day, a total of $1554 was collected from 394 admissions. How many adult admissions were there?

11. *Disneyland admissions.* A three-day "passport" to Disneyland costs $86 for adults and $65 for children. An outing club with 23 members paid $1684 for their three-day passports. How many children and how many adults are in the club?

12. *Yellowstone Park admissions.* Entering Yellowstone National Park costs $20 for a car and $15 for a motorcycle. On a typical day, 5950 cars or motorcycles enter and pay a total of $107,875 (*Source*: Yellowstone National Park). How many motorcycles enter on a typical day?

13. *Music lessons.* Alice charges $25 for a private guitar lesson and $18 for a group guitar lesson. One day in August, Alice earned $265 from 12 students. How many students of each type did Alice teach?

14. *Dance lessons.* Jean charges $20 for a private tap lesson and $12 for a group class. One Wednesday, Jean earned $216 from 14 students. How many students of each type did Jean teach?

15. *Coffee blends.* Cafe Europa mixes Brazilian coffee worth $19 per kilogram with Turkish coffee worth $22 per kilogram. The mixture should be worth $20 per kilogram. How much of each type of coffee should be used to make a 300-kg mixture?

16. *Seed mix.* Sunflower seed is worth $1.00 per pound and rolled oats are worth $1.35 per pound. How much of each would you use to make 50 lb of a mixture worth $1.14 per pound?

17. *Mixed nuts.* A grocer wishes to mix peanuts worth $2.52 per pound with Brazil nuts worth $3.80 per pound to make 480 lb of a mixture worth $3.44 per pound. How much of each should be used?

18. *Mixed nuts.* The Nuthouse has 10 kg of mixed cashews and pecans worth $8.40 per kilogram. Cashews alone sell for $8.00 per kilogram, and pecans sell for $9.00 per kilogram. How many kilograms of each are in the mixture?

19. *Acid mixtures.* Jerome's experiment requires him to mix a 50%-acid solution with an 80%-acid solution to create 200 mL of a 68%-acid solution. How much 50%-acid solution and how much 80%-acid solution should he use? Complete the following table as part of the *Translate* step.

Type of Solution	50%-Acid	80%-Acid	68%-Acid Mix
Amount of Solution	x	y	
Percent Acid	50%		68%
Amount of Acid in Solution		$0.8y$	

20. *Production.* Clear Shine window cleaner is 12% alcohol and Sunstream window cleaner is 30% alcohol. How much of each should be used to make 90 oz of a cleaner that is 20% alcohol?

21. *Horticulture.* A solution containing 28% fungicide is to be mixed with a solution containing 40% fungicide to make 300 L of a solution containing 36% fungicide. How much of each solution should be used?

22. *Chemistry.*   E-Chem Testing has a solution that is 80% base and another that is 30% base. A technician needs 200 L of a solution that is 62% base. The 200 L will be prepared by mixing the two solutions on hand. How much of each should be used?

23. *Octane ratings.*   The octane rating of a gasoline is a measure of the amount of isooctane in the gas (*Source*: Champlain Electric and Petroleum Equipment). How much 87-octane gas and 95-octane gas should be blended in order to mix a 10-gal batch of 93-octane gas?

24. *Octane ratings.*   The octane rating of a gasoline is a measure of the amount of isooctane in the gas (*Source*: Champlain Electric and Petroleum Equipment). How much 87-octane gas and 93-octane gas should be blended in order to make 12 gal of 91-octane gas?

25. *Printing.*   Using some pages that hold 1300 words per page and others that hold 1850 words per page, a typesetter is able to completely fill 12 pages with an 18,350-word document. How many pages of each kind were used?

26. *Coin value.*   A collection of quarters and nickels is worth $1.25. There are 13 coins in all. How many of each are there?

27. *Basketball scoring.*   Wilt Chamberlain once scored 100 points on a combination of 64 foul shots (each worth one point) and two-pointers. How many shots of each type did he make?

28. *Basketball scoring.*   Shaquille O'Neal recently scored 34 points on a combination of 21 foul shots and two-pointers. How many shots of each type did he make?

29. *Suntan lotion.*   Lisa has a tube of Kinney's suntan lotion that is rated 15 spf and a second tube of Coppertone that is 30 spf. How many fluid ounces of each type of lotion should be mixed in order to create 50 fluid ounces of sunblock that is rated 20 spf?

30. *Cough syrup.*   Dr. Zeke's cough syrup is 2% alcohol. Vitabrite cough syrup is 5% alcohol. How much of each type should be used in order to prepare an 80-oz batch of cough syrup that is 3% alcohol?

Aha! **31.** *Textile production.*   DRG Outdoor Products uses one insulation that is 20% goose down and another that is 34% goose down. How many pounds of each should be used to create 50 lb of insulation that is 27% goose down?

32. *Nutrition.*   New England Natural Bakers Muesli gets 20% of its calories from fat. Breadshop Supernatural granola gets 35% of its calories from fat. (*Source*: Onion River Cooperative, Burlington VT). How much of each type should be used to create a 45-lb mixture that gets 30% of its calories from fat?

33. Why might fractional answers be acceptable on problems like Examples 3 and 4, but not on problems like Examples 1 and 2?

34. Write a problem for a classmate to solve by translating to a system of two equations in two unknowns.

SKILL MAINTENANCE

Solve.

35. $7 - 3x < 22$

36. $4 - 5x \geq 39$

Graph on a number line.

37. $x + 2 \geq 6$

38. $3x + 3 < 9$

39. $6 < -\frac{1}{2}x + 1$

40. $4 > -\frac{1}{3}x + 2$

SYNTHESIS

41. In Exercise 28, suppose that some of O'Neal's 21 baskets were three-pointers. Could the problem still be solved? Why or why not?

42. In Exercise 26, suppose that some of the 13 coins may have been half-dollars. Could the problem still be solved? Why or why not?

43. *Coffee.*   Kona coffee, grown only in Hawaii, is highly desired and quite expensive. To create a blend of beans that is 30% Kona, the Brewtown Beanery is adding pure Kona beans to a 45-lb sack of Columbian beans. How many pounds of Kona should be added to the 45 lb of Columbian?

44. *Chemistry.*   A tank contains 8000 L of a solution that is 40% acid. How much water should be added in order to make a solution that is 30% acid?

45. *Automobile maintenance.*   The radiator in Candy's Honda Accord contains 6.3 L of antifreeze and water. This mixture is 30% antifreeze. How much should be drained and replaced with pure antifreeze so that the mixture will be 50% antifreeze?

6.3 liters

46. *Investing.*   One year Shannon made $288 from two investments: $1100 was invested at one yearly rate and $1800 at a rate that was 1.5% higher. Find the two rates of interest.

47. *Octane rating.*   Many cars need gasoline with an octane rating of at least 87. After mistakenly putting 5 gal of 85-octane gas in her empty gas tank, Kim plans to add 91-octane gas until the mixture's octane rating is 87. How much 91-octane gas should she add?

48. *Sporting-goods prices.*   Together, a bat, ball, and glove cost $99.00. The bat costs $9.95 more than the ball, and the glove costs $65.45 more than the bat. How much does each cost?

49. *Painting.*   Campus Painters has two kinds of paint. If 9 gal of the inexpensive paint is mixed with 7 gal of the expensive paint, the mixture will be worth $19.70 per gallon. If 3 gal of the inexpensive paint is mixed with 5 gal of the expensive paint, the mixture will be worth $19.825 per gallon. What is the price per gallon of each type of paint?

50. *Investing.*   Eduardo invested $54,000, part of it at 6% and the rest at 6.5%. The total yield after one year is $3385. How much was invested at each rate?

51. *Dairy farming.*   Farmer Benz has 100 L of milk that is 4.6% butterfat. How much skim milk (no butterfat) should be added to make milk that is 2% butterfat?

52. *Payroll.*   Ace Engineering pays a total of $325 an hour when employing some workers at $20 an hour and others at $25 an hour. When the number of $20 workers is increased by 50% and the number of $25 workers is decreased by 20%, the cost per hour is $400. How many workers were originally employed at each rate?

53. A two-digit number is six times the sum of its digits. The tens digit is 1 more than the ones digit. Find the number.

54. The sum of the digits of a two-digit number is 12. When the digits are reversed, the number is decreased by 18. Find the original number.

55. *Literature.*   In Lewis Carroll's *Through the Looking Glass,* Tweedledum says to Tweedledee, "The sum of your weight and twice mine is 361 pounds." Then Tweedledee says to Tweedledum, "Contrariwise, the sum of your weight and twice mine is 362 pounds." Find the weights of Tweedledum and Tweedledee.

CORNER

Sunoco's Custom Blending Pump

Focus: Mixture problems
Time: 30 minutes
Group size: 3
Materials: Calculators

Sunoco® gasoline stations pride themselves on offering customers the "custom blending pump." While most competitors offer just three octane levels—87, 93, and a blend that is 89—Sunoco customers can select an octane level of 86, 87, 89, 92, or 94. The Sunoco supplier brings 86- and 94-octane gasoline to each station and a computerized pump mixes the two to create the blend that a customer desires.

ACTIVITY

1. Assume that your group's gas station has a generous supply of 86-octane and 94-octane gasoline. Each group member should select a different one of the custom blends: 87, 89, or 92.
2. Each member will be asked to determine the number of gallons of 86-octane gas and

94-octane gas needed to form 100 gal of his or her selected blend. Before doing so, however, the group should outline a series of steps that each member will use to solve his or her problem. Group members should agree on the letters chosen as variables, the variable that will be solved for first, and the sequence of steps that will be followed. For the purposes of this activity, it is best to use substitution to solve the system of equations that will be created.

3. *Following the agreed-upon series of steps,* each group member should determine how many gallons of 86-octane gas and how many gallons of 94-octane gas should be mixed in order to form 100 gal of his or her selected blend. Check that all work is done correctly and consistently.
4. The pumps in use at most gas stations blend mixtures in $\frac{1}{10}$-gal "batches." How can the results of part (3) be adjusted so that $\frac{1}{10}$-gal, not 100-gal, blends are formulated?

Linear Inequalities in Two Variables

7.5

Graphing Linear Inequalities • Linear Inequalities in One Variable

Just as the solutions of linear equations like $5x + 4y = 13$ or $y = \frac{1}{2}x + 1$ can be graphed, so too can the solutions of *linear inequalities* like $5x + 4y < 13$ or $y > \frac{1}{2}x + 1$ be represented graphically.

Graphing Linear Inequalities

In Section 2.6, we found that solutions of inequalities like $5x + 9 \le 4x + 3$ can be represented by a shaded portion of a number line. When a solution included an endpoint, we drew a solid dot, and when the endpoint was excluded, we drew an open dot. To graph inequalities like $y > \frac{1}{2}x + 1$ or $2x + 3y \le 6$, we will shade a region of a plane. That region will be either above or below the graph of a "boundary line" (in this case, the graph of $y = \frac{1}{2}x + 1$ or $2x + 3y = 6$). If the symbol is $\le$ or $\ge$, we will draw the boundary line solid, since it is part of the solution. When the boundary is excluded—that is, if $<$ or $>$ is used—we will draw a dashed line.

E x a m p l e 1

Graph: $y > \frac{1}{2}x + 1$.

Solution We begin by graphing the boundary line $y = \frac{1}{2}x + 1$. The slope is $\frac{1}{2}$ and the y-intercept is $(0, 1)$. This line is drawn dashed since the symbol $>$ is used.

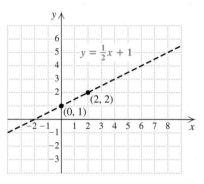

The plane is now split in two. If we consider the coordinates of a few points above the line, we will find that all are solutions of $y > \frac{1}{2}x + 1$.

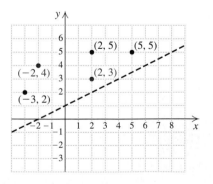

Here is a check for the points $(2, 3)$ and $(-2, 4)$:

$$\begin{array}{c|c}
\multicolumn{2}{l}{y > \frac{1}{2}x + 1} \\
\hline
3 \ ? & \frac{1}{2} \cdot 2 + 1 \\
& 1 + 1 \\
3 & 2 \qquad \text{TRUE}
\end{array}
\qquad
\begin{array}{c|c}
\multicolumn{2}{l}{y > \frac{1}{2}x + 1} \\
\hline
4 \ ? & \frac{1}{2}(-2) + 1 \\
& -1 + 1 \\
4 & 0 \qquad \text{TRUE}
\end{array}$$

The student can check that *any* point on the same side of the dashed line as $(2, 3)$ or $(-2, 4)$ is a solution. If one point in a region solves an inequality, then *all* points in that region are solutions. The graph of

$$y > \tfrac{1}{2}x + 1$$

is shown below. Note that the solution set consists of all points in the shaded region. Furthermore, note that for any inequality of the form $y > mx + b$ or $y \geq mx + b$, we shade the region *above* the boundary line.

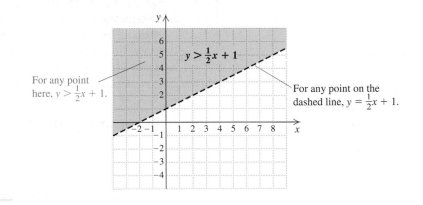

Example 2

Graph: $2x + 3y \leq 6$.

Solution   First, we establish the boundary line by graphing $2x + 3y = 6$. This can be done either by using the intercepts, $(0, 2)$ and $(3, 0)$, or by finding slope–intercept form, $y = -\tfrac{2}{3}x + 2$. Since the inequality contains the symbol $\leq$, we draw a solid boundary line to include all pairs on the line as part of the solution. The graph of $2x + 3y \leq 6$ also includes either the region above or below the line. By using a "test point" that is clearly above or below the line, we can determine which region to shade. The origin, $(0, 0)$, is often a convenient test point:

$$\begin{array}{c|c} \multicolumn{2}{c}{2x + 3y \leq 6} \\ \hline 2 \cdot 0 + 3 \cdot 0 \ ? \ 6 & \\ 0 \ | \ 6 & \text{TRUE} \end{array}$$

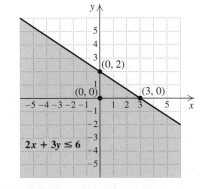

The point $(0, 0)$ is a solution and it appears in the region below the boundary line. Thus this region, along with the line itself, represents the solution.

The original inequality is equivalent to $y \leq -\tfrac{2}{3}x + 2$. Note that for any inequality of the form $y \leq mx + b$ or $y < mx + b$, we shade the region *below* the boundary line.

> **To Graph a Linear Inequality**
> 1. Draw the boundary line by replacing the inequality symbol with an equals sign and graphing the resulting equation. If the inequality symbol is $<$ or $>$, the line is dashed. If the symbol is $\leq$ or $\geq$, the line is solid.
> 2. Shade the region on one side of the boundary line. To determine which side, select a point not on the line as a test point. If that point's coordinates are a solution of the inequality, shade the region containing the point. If not, shade the other region.
> 3. Inequalities of the form $y < mx + b$ or $y \leq mx + b$ are shaded below the boundary line. Inequalities of the form $y > mx + b$ or $y \geq mx + b$ are shaded above the boundary line.

Linear Inequalities in One Variable

E x a m p l e   3

Graph $y \leq -2$ on a plane.

Solution   We graph $y = -2$ as a solid line to indicate that all points on the line are solutions. Again, we select $(0, 0)$ as a test point. It may help to write $y \leq -2$ as $y \leq 0 \cdot x - 2$:

$$\frac{y \leq 0 \cdot x - 2}{\begin{array}{c|c} 0 \; ? \; 0 \cdot 0 - 2 \\ 0 \; | \; -2 \quad \text{FALSE} \end{array}}$$

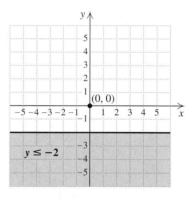

Since $(0, 0)$ is *not* a solution, we do not shade the region in which it appears. Instead, we shade below the boundary line as shown. The solution consists of all ordered pairs whose y-coordinates are less than or equal to -2.

E x a m p l e   4

Graph $x < 3$ on a plane.

Solution   We graph $x = 3$ using a dashed line. To determine which region to shade, we again use the test point, $(0, 0)$. It may help to write $x < 3$ as $x + 0 \cdot y < 3$:

$$\frac{x + 0 \cdot y < 3}{\begin{array}{c|c} 0 + 0 \cdot 0 \; ? \; 3 \\ 0 \; | \; 3 \quad \text{TRUE} \end{array}}$$

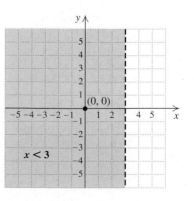

Since $(0, 0)$ is a solution, we shade to the left. The solution consists of all ordered pairs with first coordinates less than 3.

technology connection

To graph $2x + 3y \leq 6$ on a grapher, we must first solve for y. This is precisely what we did in Example 2, where we found that $y \leq -\frac{2}{3}x + 2$. On many graphers, this graph is drawn by entering $(-2/3)x + 2$ as y_1, moving the cursor to the GraphStyle icon just to the left of y_1, pressing ENTER until ◣ appears, and then pressing GRAPH. The symbol ◣ indicates that the area *below* the line is shaded. On some graphers, a SHADE option must be used.

Note that the boundary line always appears solid.

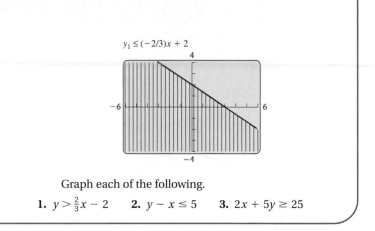

$y_1 \leq (-2/3)x + 2$

Graph each of the following.

1. $y > \frac{2}{3}x - 2$ **2.** $y - x \leq 5$ **3.** $2x + 5y \geq 25$

Exercise Set 7.5

1. Determine whether $(-3, -5)$ is a solution of
$x + 3y < -18$.

2. Determine whether $(5, -3)$ is a solution of
$-2x + 4y \leq -2$.

Aha! **3.** Determine whether $\left(\frac{7}{8}, \frac{1}{2}\right)$ is a solution of
$6y + 5x \geq -3$.

4. Determine whether $(-6, 5)$ is a solution of
$x + 0 \cdot y < 3$.

Graph on a plane.

5. $y \leq x + 4$ **6.** $y \leq x - 2$

7. $y < x - 1$ **8.** $y < x + 5$

9. $y \geq x - 3$ **10.** $y \geq x - 1$

11. $y \leq 2x - 1$ **12.** $y \leq 3x + 2$

13. $x + y \leq 4$ **14.** $x + y \leq 6$

15. $x - y > 7$ **16.** $x - y > 5$

17. $y \geq 1 - 2x$ **18.** $y > 2 - 3x$

19. $y - 3x > 0$ **20.** $y - 2x \geq 0$

21. $x \geq 3$ **22.** $x > -4$

23. $y \leq 3$ **24.** $y > -1$

25. $y \geq -5$ **26.** $y < 0$

27. $x < 4$ **28.** $x \leq 5$

29. $x - y < -10$ **30.** $y - 2x \leq -1$

31. $2x + 3y \leq 12$ **32.** $5x + 4y \geq 20$

33. Examine the solution of Example 2. Why is the point $(4.5, -1)$ *not* a good choice for a test point?

34. Why is $(0, 0)$ such a "convenient" test point to use?

SKILL MAINTENANCE

Evaluate.

35. $3x + 5y$, for $x = -2$ and $y = 4$

36. $2x + 4y$, for $x = 3$ and $y = -1$

The graph below shows the prices paid for a ton of newspaper and for a ton of corrugated cardboard by recyclers in the Boston market (Source: Chittenden Solid Waste District, Williston, VT).

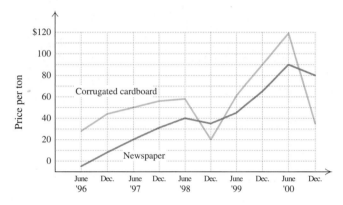

37. How much was being paid for corrugated cardboard in June 1999?

38. During what six-month period did the price paid for newspaper first match the price paid for corrugated cardboard?

39. When did the value of newspaper peak?

40. When was the price paid for corrugated cardboard $20 per ton?

41. During what six-month period did the price paid for corrugated cardboard drop the most?

42. During what six-month period did the price paid for corrugated cardboard increase the most?

SYNTHESIS

43. Under what circumstances is the graph of $Ax + By > 0$ shaded *above* the line $Ax + By = 0$?

44. Describe a procedure that could be used to graph any inequality of the form $Ax + By < C$.

45. *Elevators.* Many Otis elevators have a capacity of 1000 lb. Suppose that c children, each weighing 75 lb, and a adults, each weighing 150 lb, are on an elevator. Find and graph an inequality that asserts that the elevator is overloaded.

46. *Hockey wins and losses.* A hockey team needs at least 60 points for the season in order to make the playoffs. Suppose that a team finishes with w wins, each worth 2 points, and t ties, each worth 1 point. Find and graph an inequality that indicates whether a team made the playoffs.

47. *Photography.* Recently, Filmsense charged $10 to develop and ship a 24-exposure roll of film and $14 to develop and ship a 36-exposure roll. Find and graph an inequality indicating that w 24-exposure rolls and r 36-exposure rolls were developed at a cost exceeding $140.

48. *Architecture.* Most architects agree that the sum of a step's riser r and tread t, in inches, should not be less than 17 in. Find and graph an inequality that describes the situation.

Find an inequality for each graph shown.

49.

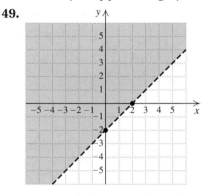

50.

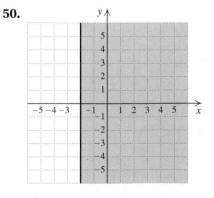

Graph on a plane. (Hint: Use several test points.)

51. $xy \le 0$

52. $xy \ge 0$

53. Graph: $y + 3x \le 4.9$.

54. Graph: $0.7x - y \le 2.3$.

Systems of Linear Inequalities

7.6

Graphing Systems of Inequalities •
Locating Solution Sets

Systems of linear equations were graphed in Section 7.1. We now consider **systems of linear inequalities** in two variables, such as

$$x + y \leq 3,$$
$$x - y < 3.$$

When systems of equations were solved graphically, we searched for any points common to both lines. To solve a system of inequalities graphically, we again look for points common to both graphs. This is accomplished by graphing each inequality and determining where the graphs overlap.

E x a m p l e   1   Graph the solutions of the system

$$x + y \leq 3,$$
$$x - y < 3.$$

Solution   To graph $x + y \leq 3$, we draw the graph of $x + y = 3$ using a solid line (see the graph on the left, below). Since $(0, 0)$ is a solution of $x + y \leq 3$, we shade (in red) all points on that side of the line. The arrows near the ends of the line also indicate the region that contains solutions.

Next, we superimpose the graph of $x - y < 3$, using a dashed line for $x - y = 3$ and again using $(0, 0)$ as a test point. Since $(0, 0)$ is a solution, we shade (in blue) the region on the same side of the dashed line as $(0, 0)$.

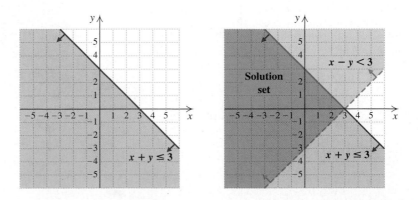

The solution set of the system is the region shaded purple along with the purple portion of the line $x + y = 3$.

Example 2

Graph the solutions of the system

$$x \geq 3,$$
$$x - 3y < 6.$$

Solution   We graph $x \geq 3$ using blue and $x - 3y < 6$ using red. The solution set is the purple region along with the purple portion of the solid line.

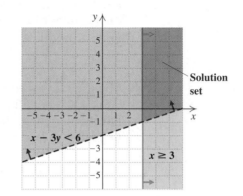

Example 3

Graph the solutions of the system

$$x - 2y < 0,$$
$$-2x + y > 2.$$

Solution   We graph $x - 2y < 0$ using red and $-2x + y > 2$ using blue. The region that is purple is the solution set of the system since those points solve both inequalities.

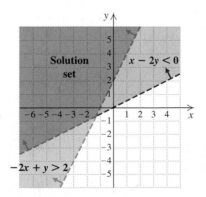

Many graphers can be used to display systems of linear inequalities. To do so, each inequality is entered along with the side that should be shaded. The grapher automatically uses a different shading for each common region.

▲ $y_1 = (-2/3)x + 5,$

▼ $y_2 = .5x - 3,$

▲ $y_3 = 3x + 4$

Exercise Set **7.6**

Graph the solutions of each system.

1. $x + y \leq 8,$
$x - y \leq 3$

2. $x + y \leq 3,$
$x - y \leq 4$

3. $y - 2x > 1,$
$y - 2x < 3$

4. $x + y < 6,$
$x + y > 0$

5. $y \geq -3,$
$x > 2 + y$

6. $x > 3,$
$x + y \leq 4$

7. $y > 3x - 2,$
$y < -x + 4$

8. $y \geq x,$
$y \leq 1 - x$

9. $x \leq 4,$
$y \leq 5$

10. $x \geq -5,$
$y \geq -2$

11. $x \leq 0,$
$y \leq 0$

12. $x \geq 0,$
$y \geq 0$

13. $2x - 3y \geq 9,$
$2y + x > 6$

14. $3x - 2y \leq 8,$
$2x + y > 6$

15. $y > 5x + 2,$
$y \leq 1 - x$

16. $\quad y > 4,$
$2y + x \leq 4$

17. $x + y \leq 5,$
$x \geq 0,$
$y \geq 0,$
$y \leq 3$

18. $x + 2y \leq 8,$
$x \leq 6,$
$x \geq 0,$
$y \geq 0$

19. $y - x \geq 1,$
$y - x \leq 3,$
$x \leq 5,$
$x \geq 2$

20. $x - 2y \leq 0,$
$y - 2x \leq 2,$
$x \leq 2,$
$y \leq 2$

21. $y \leq x,$
$x \geq -2,$
$x \leq -y$

22. $\quad y > 0,$
$2y + x \geq -6,$
$x + 2 \leq 2y$

23. Will a system of linear inequalities always have a solution? Why or why not?

24. If shadings are used to represent the inequalities in a system, will the most heavily shaded region always represent the solution set (assuming a solution set exists)?

SKILL MAINTENANCE

Solve.

25. $7 = \dfrac{k}{5}$

26. $4 = \dfrac{k}{6}$

27. $18 = k \cdot 3$

28. $10 = k \cdot 2$

29. $5 = k \cdot 45$

30. $7 = k \cdot 21$

SYNTHESIS

31. Explain how it would be possible for the solution of a system of linear inequalities to be a line.

32. Explain how it would be possible for the solution of a system of linear inequalities to be a single point.

Graph the solutions of each system. If no solution exists, state this.

33. $2x + 5y \geq 18,$
$4x + 3y \geq 22,$
$2x + y \geq 8,$
$x \geq 0,$
$y \geq 0$

34. $3r + 6t \geq 36,$
$2r + 3t \geq 21,$
$5r + 3t \geq 30,$
$t \geq 0,$
$r \geq 0$

35. $2x + 5y \geq 10,$
$x - 3y \leq 6,$
$4x + 10y \leq 20$

36. $x + 3y \leq 6,$
$2x + y \geq 4,$
$3x + 9y \geq 18$

Aha! **37.** $2x + 3y \leq 1,$
$4x + 6y > 9,$
$5x - 2y \leq 8,$
$x \leq 12,$
$y \geq -15$

38. $5x - 4y \geq 8,$
$2x + 3y < 9,$
$2y \geq -8,$
$x \leq -5,$
$y < -6$

39. Explain how it would be possible for the solution of a system of linear inequalities to be a line segment.

40. Use a grapher to solve Exercise 13.

41. Use a grapher to solve Exercise 14.

Direct and Inverse Variation

7.7

Equations of Direct Variation • Problem Solving with Direct Variation • Equations of Inverse Variation • Problem Solving with Inverse Variation

Many problems lead to equations of the form $y = kx$ or $y = k/x$, for some constant k. Such equations are called *equations of variation.*

Equations of Direct Variation

A bicycle tour is traveling at a rate of 15 km/h. In 1 hr, it goes 15 km. In 2 hr, it goes 30 km. In 3 hr, it goes 45 km, and so on. In the graph below, we use the number of hours as the first coordinate and the number of kilometers traveled as the second coordinate: $(1, 15)$, $(2, 30)$, $(3, 45)$, $(4, 60)$, and so on. Note that the second coordinate is always 15 times the first.

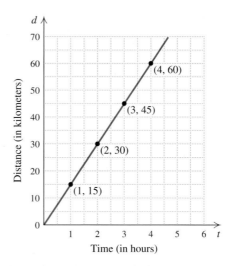

In this example, distance is a constant multiple of time, so we say that there is **direct variation** and that distance **varies directly** as time. The **equation of variation** is $d = 15t$.

> ### *Direct Variation*
>
> When a situation translates to an equation described by $y = kx$, with k a constant, we say that y *varies directly* as x. The equation $y = kx$ is called an *equation of direct variation.*

In direct variation, as one variable increases, the other variable increases as well.

The terminologies

"*y* varies as *x*,"

"*y* is directly proportional to *x*," and

"*y* is proportional to *x*"

also imply direct variation and are used in many situations. The constant *k* is called the **constant of proportionality** or the **variation constant**. It can be found if one pair of values of *x* and *y* is known. Once *k* is known, other pairs can be determined.

E x a m p l e   1

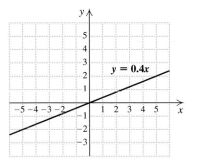

A visualization of Example 1

If *y* varies directly as *x* and $y = 2$ when $x = 5$, find the equation of variation.

Solution   We substitute to find *k*:

$$y = kx$$
$$2 = k \cdot 5 \qquad \text{Substituting to solve for } k$$
$$\tfrac{2}{5} = k, \quad \text{or} \quad k = 0.4. \qquad \text{Dividing both sides by 5}$$

Thus the equation of variation is $y = 0.4x$. A visualization of the situation is shown at left.

From these last two graphs, we see that when *y* varies directly as *x*, the constant of proportionality is also the slope of the associated graph—the rate at which *y* changes with respect to *x*.

E x a m p l e   2

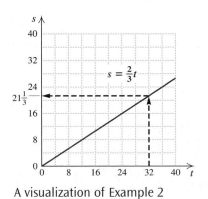

A visualization of Example 2

Find an equation in which *s* varies directly as *t* and $s = 10$ when $t = 15$. Then find the value of *s* when $t = 32$.

Solution   We have

$$s = kt \qquad \text{We know that } s \text{ varies directly as } t.$$
$$10 = k \cdot 15 \qquad \text{Substituting 10 for } s \text{ and 15 for } t$$
$$\tfrac{10}{15} = k, \quad \text{or} \quad k = \tfrac{2}{3}. \qquad \text{Solving for } k$$

Thus the equation of variation is $s = \tfrac{2}{3}t$. When $t = 32$, we have

$$s = \tfrac{2}{3}t$$
$$s = \tfrac{2}{3} \cdot 32 \qquad \text{Substituting 32 for } t \text{ in the equation of variation}$$
$$s = \tfrac{64}{3}, \text{ or } 21\tfrac{1}{3}.$$

The value of *s* is $21\tfrac{1}{3}$ when $t = 32$.

Problem Solving with Direct Variation

In applications, it is often necessary to find an equation of variation and then use it to find other values, much as we did in Example 2.

Example 3

The karat rating R of a gold object varies directly as the percentage P of gold in the object. A 14-karat gold ring is 58.25% gold. What is the percentage of gold in a 24-karat gold ring?

Solution

1., 2. Familiarize and **Translate.** The problem states that we have direct variation between R and P. Thus an equation $R = kP$ applies.

3. Carry out. We find an equation of variation:

$$R = kP$$
$$14 = k(0.5825) \qquad \text{Substituting 14 for } R \text{ and 58.25\%, or 0.5825, for } P$$
$$\frac{14}{0.5825} = k$$
$$24.03 \approx k. \qquad \text{Dividing and rounding to the nearest hundredth}$$

The equation of variation is $R = 24.03P$. When $R = 24$, we have

$$R = 24.03P$$
$$24 = 24.03P \qquad \text{Substituting 24 for } R$$
$$\frac{24}{24.03} = P \qquad \text{Solving for } P$$
$$0.999 \approx P$$
$$99.9\% \approx P.$$

4. Check. The check might be done by repeating the computations. You might also note that as the karat rating increased from 14 to 24, the percentage increased from 58.25% to 99.9%. The ratios 14/0.5825 and 24/0.999 are both about 24.03.

5. State. A 24-karat gold ring is 99.9% gold.

Equations of Inverse Variation

A car is traveling a distance of 20 mi. At a speed of 5 mph, the trip will take 4 hr. At 20 mph, it will take 1 hr. At 40 mph, it will take $\frac{1}{2}$ hr, and so on. This determines a set of pairs of numbers:

$$(5, 4), \qquad (20, 1), \qquad \left(40, \tfrac{1}{2}\right), \quad \text{and so on.}$$

Note that the product of speed and time for each of these pairs is 20. Note too that as the speed *increases*, the time *decreases*.

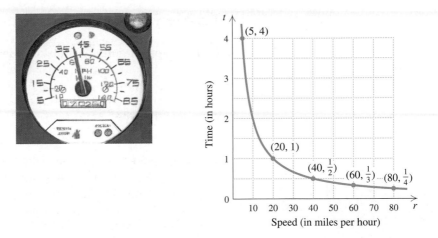

In this case, the product of speed and time is constant so we say that there is **inverse variation** and that time **varies inversely** as speed. The equation of variation is

$$rt = 20 \text{ (a constant)}, \quad \text{or} \quad t = \frac{20}{r}.$$

> ### Inverse Variation
>
> When a situation translates to an equation described by $y = k/x$, with k a constant, we say that y *varies inversely* as x. The equation $y = k/x$ is called an *equation of inverse variation*.

In inverse variation, as one variable increases, the other variable decreases. The terminology

"*y* is inversely proportional to *x*"

also implies inverse variation and is used in some situations. The constant k is again called the *constant of proportionality* or the *variation constant*.

E x a m p l e   4   If y varies inversely as x and $y = 145$ when $x = 0.8$, find the equation of variation.

Solution   We substitute to find k:

$$y = \frac{k}{x}$$

$$145 = \frac{k}{0.8} \qquad \text{Substituting to solve for } k$$

$$(0.8)145 = k \qquad \text{Multiplying both sides by 0.8}$$

$$116 = k.$$

The equation of variation is $y = \dfrac{116}{x}$.

Problem Solving with Inverse Variation

Often in applications, we must decide what kind of variation, if any, applies.

E x a m p l e   5

It takes 4 hr for 20 people to raise a barn. How long would it take 25 people to complete the job?

Solution

1. **Familiarize.**   Think about the situation. What kind of variation applies? It seems reasonable that the greater the number of people working on a job, the less time it will take. Thus we assume that inverse variation applies. We let $T =$ the time to complete the job, in hours, and $N =$ the number of people working.

2. **Translate.**   Since inverse variation applies, we have

$$T = \frac{k}{N}.$$

3. **Carry out.**   We find an equation of variation:

$$T = \frac{k}{N}$$

$$4 = \frac{k}{20} \qquad \text{Substituting 4 for } T \text{ and 20 for } N$$

$$20 \cdot 4 = k \qquad \text{Multiplying both sides by 20}$$

$$80 = k.$$

The equation of variation is $T = \dfrac{80}{N}$. When $N = 25$, we have

$$T = \frac{80}{25} \qquad \text{Substituting 25 for } N$$

$$T = 3.2.$$

4. **Check.**   A check might be done by repeating the computations or by noting that $(3.2)(25)$ and $(4)(20)$ are both 80. Also, as the number of people increases, the time needed to complete the job decreases, as expected.

5. **State.**   It should take 3.2 hr for 25 people to raise a barn.

For each of the following, find an equation of variation in which y varies directly as x and the following are true.

1. $y = 28$, when $x = 2$

2. $y = 30$, when $x = 8$ $y = \quad x$

3. $y = 0.7$, when $x = 0.4$

4. $y = 0.8$, when $x = 0.5$

5. $y = 400$, when $x = 75$

6. $y = 650$, when $x = 175$

7. $y = 200$, when $x = 300$

8. $y = 500$, when $x = 60$

For each of the following, find an equation of variation in which y varies inversely as x and the following are true.

9. $y = 45$, when $x = 2$

10. $y = 8$, when $x = 3$

11. $y = 7$, when $x = 10$

12. $y = 0.125$, when $x = 8$

13. $y = 6.25$, when $x = 25$

14. $y = 42$, when $x = 50$

15. $y = 42$, when $x = 5$

16. $y = 0.2$, when $x = 10$

Solve.

17. *Wages.*   Maureen's paycheck P varies directly as the number of hours worked H. For 15 hr of work, the pay is $135. Find the pay for 23 hr of work.

18. *Manufacturing.*   The number of bolts B that a machine can make varies directly as the time T that it operates. It can make 6578 bolts in 2 hr. How many can it make in 5 hr?

19. *Turkey servings.*   The number of servings S of meat that can be obtained from a turkey varies directly as its weight W. From a turkey weighing 15 kg, one can get 40 servings of meat. How many servings can be obtained from a 9-kg turkey?

20. *Gas volume.*   The volume V of a gas varies inversely as the pressure P on it. The volume of a gas is 200 cm^3 (cubic centimeters) under a pressure of 32 kg/cm^2. What will be its volume under a pressure of 20 kg/cm^2?

21. *Lunar weight.*   The weight M of an object on the moon varies directly as its weight E on Earth. One of the authors of this book, Marv Bittinger, weighs 192 lb, but would weigh only 32 lb on the moon. The other author, David Ellenbogen, weighs 185 lb. How much would he weigh on the moon?

22. *Electrical current.*   The current I in an electrical conductor varies inversely as the resistance R of the conductor. The current is 3 amperes when the resistance is 960 ohms. What is the current when the resistance is 540 ohms?

23. *Musical tones.*   The frequency, or pitch P, of a musical tone varies inversely as its wavelength W. A trumpet's concert A has a pitch of 440 vibrations per second and a wavelength of 2.4 ft. A trumpet's E above concert A has a frequency of 660 vibrations per second. What is its wavelength?

24. *Pumping time.*   The time t required to empty a tank varies inversely as the rate r of pumping. A pump can empty a tank in 90 min at the rate of 1200 L/min. How long will it take the pump to empty the tank at the rate of 2000 L/min?

25. *Cost of television.*   The cost c of operating a television varies directly as the number of hours n that it is in operation. It costs $14.00 to operate a standard-size color television continuously for 30 days. At this rate, how much would it cost to operate the television for 1 day? for 1 hr?

26. *Answering questions.*   The number of minutes m that a student should allow for each question on a quiz is inversely proportional to the number of questions n on the quiz. If a 16-question quiz means that students have 2.5 min per question, how many questions would appear on a quiz in which students have 4 min per question?

$^{Aha!}$ **27.** *Chartering a boat.* The cost per person c of a chartered fishing boat is inversely proportional to the number of people n who are chartering the boat. If it costs \$15.75 per person when 10 people charter a boat, how many people would be going fishing if the cost were \$31.50 per person?

28. *Weight on Mars.* The weight M of an object on Mars varies directly as its weight E on Earth. In 1999, Chen Yanqing, who weighs 128 lb, set a record for her weight class with a lift (snatch) of 231 lb (*Source*: 2001 *Guinness Book of World Records*). On Mars, this lift would be only 88 lb. How much would Yanqing weigh on Mars?

State whether each situation represents direct variation, inverse variation, or neither. Give reasons for your answers.

29. The cost of mailing a package in the United States and the distance that it travels

30. A runner's speed in a race and the time it takes to run the race

31. The weight of a turkey and the cooking time

32. The number of plays it takes to go 80 yd for a touchdown and the average gain per play

SKILL MAINTENANCE

Simplify.

33. $(-7)^2$ **34.** $(-5)^2$ **35.** 13^2 **36.** 15^2

37. $(3x)^2$ **38.** $(6a)^2$ **39.** $(a^2b)^2$ **40.** $(s^3t)^2$

SYNTHESIS

41. If x varies inversely as y and y varies inversely as z, how does x vary with regard to z? Why?

42. If a varies directly as b and b varies inversely as c, how does a vary with regard to c? Why?

Write an equation of variation for each situation. Leave k in each equation as the variation constant.

43. *Ecology.* In a stream, the amount of salt S carried varies directly as the sixth power of the speed of the stream v.

44. *Acoustics.* The square of the pitch P of a vibrating string varies directly as the tension t on the string.

45. *Lighting.* The intensity of illumination I from a light source varies inversely as the square of the distance d from the source.

46. *Peanut sales.* The number of bags of peanuts B sold at the circus varies directly as the number of people N in attendance.

47. *Wind energy.* The power P in a windmill varies directly as the cube of the wind speed v.

Write an equation of variation for each situation. Include a value for the variation constant in each equation.

48. *Geometry.* The perimeter P of an equilateral octagon varies directly as the length S of a side.

49. *Geometry.* The circumference C of a circle varies directly as the radius r.

50. *Geometry.* The area of a circle varies directly as the square of the length of the radius.

51. *Geometry.* The volume V of a sphere varies directly as the cube of the radius r.

Summary and Review 7

Key Terms	System of equations, p. 394	Substitution method, p. 400	Equation of variation, p. 434
	Consistent, p. 396	Elimination method, p. 407	Direct variation, p. 434
	Inconsistent, p. 396	Linear inequality, p. 425	Constant of proportionality, p. 435
	Dependent, p. 397	System of linear inequalities, p. 431	Inverse variation, p. 437
	Independent, p. 397		

Important Properties and Formulas

When graphing a system of two linear equations, one of the following must occur:

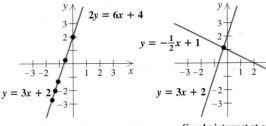

Graphs are parallel.

The system is *inconsistent* because there is no solution. Since the equations are not equivalent, they are *independent*.

Equations have the same graph.

The system is *consistent* and has an infinite number of solutions. The equations are *dependent* since they are equivalent.

Graphs intersect at one point.

The system is *consistent* and has one solution. Since the equations are not equivalent, they are *independent*.

A Comparison of Methods for Solving Systems of Linear Equations

Method	Strengths	Weaknesses
Graphical	Solutions are displayed visually. Works with any system that can be graphed.	Inexact when solutions involve numbers that are not integers or are very large and off the graph.
Substitution	Always yields exact solutions. Easy to use when a variable is alone on one side of an equation.	Introduces extensive computations with fractions when solving more complicated systems. Solutions are not graphically displayed.
Elimination	Always yields exact solutions. Easy to use when fractions or decimals appear in the system.	Solutions are not graphically displayed.

To graph a linear inequality:

1. Draw the boundary line by replacing the inequality symbol with an equals sign and graphing the resulting equation. If the inequality symbol is $<$ or $>$, the line is dashed. If the symbol is $\leq$ or $\geq$, the line is solid.
2. Shade the region on one side of the boundary line. To determine which side, select a point not on the line as a test point. If that point's coordinates are a solution of the inequality, shade the region containing the point. If not, shade the other region.
3. Inequalities of the form $y < mx + b$ or $y \leq mx + b$ are shaded below the boundary line. Inequalities of the form $y > mx + b$ or $y \geq mx + b$ are shaded above the boundary line.

Direct variation: $y = kx$, where k is a constant
Inverse variation: $y = k/x$, where k is a constant

Review Exercises

Determine whether each ordered pair is a solution of the system of equations.

1. $(-3, 2)$; $x + 2y = 1$,
$\qquad\quad x - 3y = -9$

2. $(4, -1)$; $3x - y = 13$,
$\qquad\quad 2x + y = -9$

Solve by graphing. If there is no solution or an infinite number of solutions, state this.

3. $y = 2x - 4$,
$\quad\; y = 3x - 5$

4. $x - y = 8$,
$\quad\; x + y = 4$

5. $3x - 4y = 8$,
$\quad\; 4y - 3x = 6$

6. $2x + y = 3$,
$\quad\; 4x + 2y = 6$

Solve using the substitution method. If there is no solution or an infinite number of solutions, state this.

7. $y = 4 - x$,
$\quad\; 3x + 4y = 21$

8. $x + 2y = 6$,
$\quad\; 2x + y = 8$

9. $x + y = 4$,
$\quad\; y = 2 - x$

10. $x + y = 6$,
$\quad\;\; y = 3 - 2x$

11. $3x - y = 7$,
$\quad\;\; 2x + 3y = 23$

12. $3x - y = 5$,
$\quad\;\; 6x = 2y + 10$

Solve using the elimination method. If there is no solution or an infinite number of solutions, state this.

13. $\;x + 2y = 9$,
$\quad\;\; 3x - 2y = -3$

14. $\;x - y = 8$,
$\quad\;\; 2x + y = 7$

15. $\;x - \frac{1}{3}y = -\frac{13}{3}$,
$\quad\;\; 3x - \;\;y = -13$

16. $2x + 3y = 8$,
$\quad\;\; 5x + 2y = -2$

17. $5x - 2y = 11$,
$\quad\;\; 3x - 7y = -5$

18. $-x - y = -5$,
$\quad\;\; 2x - y = 4$

19. $\;\;4x - 6y = 9$,
$\quad\; -2x + 3y = 6$

20. $2x + 6y = 4$,
$\quad\;\; 7x + 10y = -8$

Solve.

21. The sum of two numbers is 27. One half of the first number plus one third of the second number is 11. Find the numbers.

22. *Perimeter of a rectangle.* The perimeter of a rectangle is 96 cm. The length is 27 cm more than the width. Find the length and the width.

23. *Basketball scoring.* In a recent game, Allen Iverson of the Philadelphia 76ers scored 17 points on a combination of 8 two- and three-point baskets (*Source*: National Basketball Association). How many shots of each type were made?

24. *Meal prices.* The Silver Ranch Steak House charges $6.99 for an adult's buffet and $2.49 for a child's buffet. One evening the restaurant collected $2260.20 from 380 people. How many adults were served?

25. *Fat content.* Café Rich instant flavored coffee gets 40% of its calories from fat. Café Light coffee gets 25% of its calories from fat. How much of each brand of coffee should be mixed in order to make 200 g of instant coffee with 30% of its calories from fat?

Graph on a plane.

26. $x \le y$

27. $x - 2y \ge 4$

28. $y > -2$

29. $y \ge \frac{2}{3}x - 5$

30. $2x + y < 1$

31. $x < 4$

Graph the solutions of each system.

32. $x \ge 1$,
$\quad\; y \le -1$

33. $x - y > 2$,
$\quad\; x + y < 1$

34. If y varies inversely as x and $y = 81$ when $x = 3$, find the equation of variation.

35. *Catering.* The number of sandwiches S that can be made at a buffet varies directly as the number of pounds of cold cuts C in the buffet. From 6 lb of cold cuts, 25 sandwiches can be made. How many pounds of cold cuts are needed for 40 sandwiches?

SYNTHESIS

36. Explain why any solution of a system of equations is a point of intersection of the graphs of each equation in the system.

37. Monroe sketches the boundary lines of a system of two linear inequalities and notes that the lines are parallel. Since there is no point of intersection, he concludes that the solution set is empty. What is wrong with this conclusion?

38. The solution of the following system is $(6, 2)$. Find C and D.

$$2x - Dy = 6,$$
$$Cx + 4y = 14$$

39. Solve using the substitution method:

$$x - y + 2z = -3,$$
$$2x + y - 3z = 11,$$
$$z = -2.$$

40. Solve:

$$3(x - y) = 4 + x,$$
$$x = 5y + 2.$$

41. For a two-digit number, the sum of the ones digit and the tens digit is 6. When the digits are reversed, the new number is 18 more than the original number. Find the original number.

42. A stable boy agreed to work for one year. At the end of that time, he was to receive $240 and one horse. After 7 months, the boy quit the job, but still received the horse and $100. What was the value of the boy's yearly salary?

Chapter Test 7

1. Determine whether $(-1, -2)$ is a solution of the following system of equations:

$$3x - 4y = 5,$$
$$2x + 3y = -8.$$

Solve by graphing. If there is no solution or an infinite number of solutions, state this.

2. $y = -2x + 5,$
   $y = 4x - 1$

3. $2y - x = 7,$
   $2x - 4y = 4$

Solve using the substitution method. If there is no solution or an infinite number of solutions, state this.

4. $2x - 22 = 3y,$
   $y = 6 - x$

5. $x + y = 2,$
   $x + 2y = 5$

6. $x = 5y - 10,$
   $10y = 2x + 20$

Solve using the elimination method. If there is no solution or an infinite number of solutions, state this.

7. $x - y = 6,$
   $3x + y = -2$

8. $\frac{1}{2}x - \frac{1}{3}y = 8,$
   $\frac{2}{3}x + \frac{1}{2}y = 5$

9. $4x + 5y = 5,$
   $6x + 7y = 7$

10. $2x + 3y = 13,$
   $3x - 5y = 10$

Solve.

11. *Chemistry.* A chemist has one solution that is 25% acid and another solution that is 40% acid. How much of each is needed to make 60 L of a solution that is 30% acid?

12. *Complementary angles.* Two angles are complementary. One angle is 18° less than twice the other. Find the angles.

13. *Phone rates.* One calling plan offered by MCI WorldCom charges 12.9¢ per minute for daytime long-distance phone calls. A competing plan offered by AT&T charges a monthly fee of $4.95 plus 7¢ per minute for daytime long-distance phone calls. For how many minutes of long-distance calls per month are the costs of the two plans the same?

Graph on a plane.

14. $y > x - 1$

15. $2x - y \le 4$

16. $y < -2$

Graph the solutions of each system.

17. $y \ge x + 1,$
   $y > 2x$

18. $x + y \le 3,$
   $x \ge 0,$
   $y \ge 0$

19. If y varies directly as x and $y = 9$ when $x = 2$, find the equation of the variation.

20. *Work.* It takes 45 min for 2 people to shovel a driveway. How long would it take 5 people to shovel the same driveway?

SYNTHESIS

21. You are in line at a ticket window. There are two more people ahead of you in line than there are behind you. In the entire line, there are three times as many people as there are behind you. How many are in the line?

22. Graph on a plane: $|x| \leq 4$.

23. Find the numbers C and D such that $(-2, 3)$ is a solution of the system

$$Cx - 4y = 7,$$
$$3x + Dy = 8.$$

8
Radical Expressions and Equations

AN APPLICATION

Lamar and Nanci are building a rollerblade jump with a base that is 30 in. long and a ramp that is 33 in. long. How high will the back of the jump be?

This problem appears as Exercise 19 in Section 8.6.

I use geometry primarily when designing a skate park. Since all skate ramps are designed on an 8-ft radius, I use measurements that are multiples of 8. I make use of the concept of slope as well to design a bank ramp.

DAVID M. WOOD
Custom Skatepark
Developer
Burlington, Vermont

*M*any of us already have some familiarity with the notion of square roots. For example, 3 is a square root of 9 because $3^2 = 9$. In this chapter, we learn how to manipulate square roots of polynomials and rational expressions. Later in this chapter, these radical expressions *will appear in equations and in problem-solving situations.*

Introduction to Square Roots and Radical Expressions

8.1

Square Roots • Radicands and Radical Expressions • Irrational Numbers • Square Roots and Absolute Value • Problem Solving

We begin our study of square roots by examining square roots of numbers, square roots of variable expressions, and an application involving a formula.

Square Roots

Often in this text we have found the result of squaring a number. When the process is reversed, we say that we are looking for a number's *square root*.

> **Square Root**
> The number c is a *square root* of a if $c^2 = a$.

Every positive number has two square roots. For example, the square roots of 25 are 5 and -5 because $5^2 = 25$ and $(-5)^2 = 25$.

Example 1
Find the square roots of each number: **(a)** 81; **(b)** 100.

Solution

a) The square roots of 81 are 9 and -9. To check, note that $9^2 = 81$ and $(-9)^2 = (-9)(-9) = 81$.

b) The square roots of 100 are 10 and -10. To check, note that $10^2 = 100$ and $(-10)^2 = 100$.

The nonnegative square root of a number is called the **principal square root** of that number. A **radical sign**, $\sqrt{}$, is generally used when finding square roots and indicates the principal root. Thus, $\sqrt{25} = 5$ and $\sqrt{25} \neq -5$.

E x a m p l e   2   Find each of the following: **(a)** $\sqrt{225}$; **(b)** $-\sqrt{64}$.

Solution

a) The principal square root of 225 is its positive square root, so $\sqrt{225} = 15$.

b) The symbol $-\sqrt{64}$ represents the opposite of $\sqrt{64}$. Since $\sqrt{64} = 8$, we have $-\sqrt{64} = -8$.

Radicands and Radical Expressions

A **radical expression** is an algebraic expression that contains at least one radical sign. Here are some examples:

$$\sqrt{14}, \qquad 8 + \sqrt{2x}, \qquad \sqrt{t^2 + 4}, \qquad \sqrt{\frac{x^2 - 5}{2}}.$$

The expression under the radical is called the **radicand**.

E x a m p l e   3   Identify the radicand in each expression: **(a)** $\sqrt{x}$; **(b)** $\sqrt{y^2 - 5}$.

Solution

a) In $\sqrt{x}$, the radicand is x.

b) In $\sqrt{y^2 - 5}$, the radicand is $y^2 - 5$.

The square of any nonzero real number is always positive. For example, $8^2 = 64$ and $(-11)^2 = 121$. No real number, squared, is equal to a negative number. Thus the following expressions are not real numbers:

$$\sqrt{-100}, \qquad \sqrt{-49}, \qquad -\sqrt{-3}.$$

Numbers like $\sqrt{-100}$, $\sqrt{-49}$, and $-\sqrt{-3}$ are discussed in Chapter 9.

Irrational Numbers

In Section 1.4, we learned that numbers like $\sqrt{2}$ cannot be written as a ratio of two integers. These numbers are real but not rational. We call numbers like $\sqrt{2}$ *irrational*. A number that is the square of some rational number, like 9 or 64, is called a *perfect square*. The square root of any whole number that is not a perfect square is irrational.

E x a m p l e   4   Classify each of the following numbers as rational or irrational.

a) $\sqrt{3}$ **b)** $\sqrt{25}$

c) $\sqrt{35}$ **d)** $-\sqrt{9}$

Solution

a) $\sqrt{3}$ is irrational, since 3 is not a perfect square.

b) $\sqrt{25}$ is rational, since 25 is a perfect square: $\sqrt{25} = 5$.

c) $\sqrt{35}$ is irrational, since 35 is not a perfect square.

d) $-\sqrt{9}$ is rational, since 9 is a perfect square: $-\sqrt{9} = -3$.

For the following list, we have printed the irrational numbers in red: $\sqrt{1}$, $\sqrt{2}$, $\sqrt{3}$, $\sqrt{4}$, $\sqrt{5}$, $\sqrt{6}$, $\sqrt{7}$, $\sqrt{8}$, $\sqrt{9}$, $\sqrt{10}$, $\sqrt{11}$, $\sqrt{12}$, $\sqrt{13}$, $\sqrt{14}$, $\sqrt{15}$, $\sqrt{16}$, $\sqrt{17}$, $\sqrt{18}$, $\sqrt{19}$, $\sqrt{20}$, $\sqrt{21}$, $\sqrt{22}$, $\sqrt{23}$, $\sqrt{24}$, $\sqrt{25}$.

Often, when square roots are irrational, a calculator is used to find decimal approximations.

Example 5

Use a calculator to approximate $\sqrt{10}$ to three decimal places.

Solution   Calculators vary in their methods of operation. In most cases, however, we simply enter the number and then press $\boxed{\sqrt{}}$:

$$\sqrt{10} \approx 3.162277660. \qquad \text{Using a calculator with a 10-digit display}$$

Decimal representation of an irrational number would be nonrepeating and nonending. Rounding to three decimal places, we have $\sqrt{10} \approx 3.162$.

Square Roots and Absolute Value

Note that $\sqrt{(-5)^2} = \sqrt{25} = 5$ and $\sqrt{5^2} = \sqrt{25} = 5$, so it appears that squaring a number and then taking its square root is the same as taking the absolute value of the number: $|-5| = 5$ and $|5| = 5$. In short, the principal square root of the square of A is the absolute value of A:

For any real number A, $\sqrt{A^2} = |A|$.

Example 6

Simplify $\sqrt{(3x)^2}$ given that x can represent any real number.

Solution   If x represents a negative number, then $3x$ is negative. Since the principal square root is always positive, to write $\sqrt{(3x)^2} = 3x$ would be incorrect. Instead, we write

$$\sqrt{(3x)^2} = |3x|. \qquad \text{Note that } 3x \text{ could be negative.}$$

Fortunately, in many cases, it can be assumed that radicands that are variable expressions do not represent the square of a negative number. When this assumption is made, the need for absolute-value symbols disappears:

For $A \geq 0$, $\sqrt{A^2} = A$.

E x a m p l e   7

Simplify each expression. Assume that all variables represent nonnegative numbers.

a) $\sqrt{(3x)^2}$ **b)** $\sqrt{a^2b^2}$

Solution

a) $\sqrt{(3x)^2} = 3x$ Since $3x$ is assumed to be nonnegative, $|3x| = 3x$.

b) $\sqrt{a^2b^2} = \sqrt{(ab)^2} = ab$ Since ab is assumed to be nonnegative, $|ab| = ab$.

In Sections 8.2–8.6, we will often state assumptions that make absolute-value symbols unnecessary.

Problem Solving

Radical expressions often appear in applications.

E x a m p l e   8

Parking-lot arrival spaces. The attendants at a parking lot use spaces to leave cars before they are taken to long-term parking stalls. The required number N of such spaces is approximated by the formula

$$N = 2.5\sqrt{A},$$

where A is the average number of arrivals in peak hours. Find the number of spaces needed when an average of 43 cars arrive during peak hours.

Solution   We substitute 43 into the formula. We use a calculator to find an approximation:

$$N = 2.5\sqrt{43}$$
$$\approx 2.5(6.557)$$
$$\approx 16.393 \approx 17.$$

Note that we round *up* to 17 spaces because rounding down would create some overcrowding. Thus, for an average of 43 arrivals, 17 spaces are needed.

Calculator Note. Generally, when using a calculator, as in Example 8, we round at the *end* of our work. Doing so, we might find

$$N = 2.5\sqrt{43} \approx 2.5(6.557438524) = 16.39359631 \approx 16.394.$$

Note the discrepancy in the third decimal place. When using a calculator for approximation, be aware of possible variations in answers. You may get answers that differ from those given at the back of the book. Answers to the exercises have been found by rounding at the end of the calculations.

technology
connection

Graphing equations that contain radical expressions often involves approximating irrational numbers. Also, since the square root of a negative number is not real, such graphs may not exist for all choices of x. For example, the graph of $y = \sqrt{x - 1}$ does not exist for $x < 1$.

$y_1 = \sqrt{x - 1}$

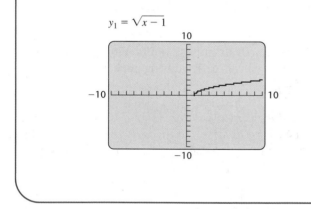

Similarly, the graph of $y = \sqrt{2 - x}$ does not exist for $x > 2$.

$y_1 = \sqrt{2 - x}$

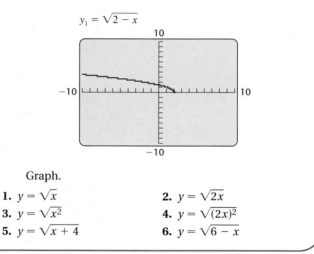

Graph.

1. $y = \sqrt{x}$ **2.** $y = \sqrt{2x}$

3. $y = \sqrt{x^2}$ **4.** $y = \sqrt{(2x)^2}$

5. $y = \sqrt{x + 4}$ **6.** $y = \sqrt{6 - x}$

FOR EXTRA HELP

Exercise Set 8.1

Digital Video Tutor CD 5
Videotape 15 | InterAct Math | Math Tutor Center | MathXL | MyMathLab.com

Find the square roots of each number.

1. 4 **2.** 9 **3.** 16

4. 1 **5.** 49 **6.** 121

7. 144 **8.** 169

Simplify.

9. $\sqrt{9}$ **10.** $\sqrt{4}$ **11.** $-\sqrt{1}$

12. $-\sqrt{25}$ **13.** $\sqrt{0}$ **14.** $-\sqrt{81}$

15. $-\sqrt{121}$ **16.** $\sqrt{361}$ **17.** $\sqrt{900}$

18. $\sqrt{441}$ **19.** $\sqrt{169}$ **20.** $\sqrt{144}$

21. $-\sqrt{625}$ **22.** $-\sqrt{400}$

Identify the radicand for each expression.

23. $\sqrt{a - 7}$ **24.** $\sqrt{t - 5}$

25. $5\sqrt{t^2 + 1}$

26. $8\sqrt{x^2 + 2}$

27. $x^2 y \sqrt{\dfrac{3}{x + 2}}$

28. $ab^2 \sqrt{\dfrac{a}{a - b}}$

Classify each number as rational or irrational.

29. $\sqrt{100}$

30. $\sqrt{6}$

31. $\sqrt{8}$

32. $\sqrt{10}$

33. $\sqrt{32}$

34. $\sqrt{49}$

35. $\sqrt{98}$

36. $\sqrt{75}$

37. $-\sqrt{4}$

38. $-\sqrt{1}$ Aha! **39.** $-\sqrt{19^2}$

40. $-\sqrt{14}$

Use a calculator or Table 2 to approximate each of the following numbers. Round to three decimal places.

41. $\sqrt{5}$

42. $\sqrt{6}$

43. $\sqrt{17}$

44. $\sqrt{19}$

45. $\sqrt{93}$

46. $\sqrt{43}$

Simplify. Assume that all variables represent nonnegative numbers.

47. $\sqrt{t^2}$

48. $\sqrt{x^2}$

49. $\sqrt{9x^2}$

50. $\sqrt{25a^2}$

51. $\sqrt{(7a)^2}$

52. $\sqrt{(4x)^2}$

53. $\sqrt{(17x)^2}$

54. $\sqrt{(8ab)^2}$

Parking spaces.   Solve. Use the formula $N = 2.5\sqrt{A}$ of Example 8.

55. Find the number of spaces needed when the average number of arrivals is **(a)** 36; **(b)** 29.

56. Find the number of spaces needed when the average number of arrivals is **(a)** 49; **(b)** 53.

Hang time.   An athlete's hang time (time airborne for a jump), T, in seconds, is given by $T = 0.144\sqrt{V}$, where V is the athlete's vertical leap, in inches.*

57. Kobe Bryant of the Los Angeles Lakers can jump 36 in. vertically. Find his hang time.

58. Brian Grant of the Miami Heat can jump 25 in. vertically. Find his hang time.

59. Which is the more exact way to write the square root of 12: 3.464101615 or $\sqrt{12}$? Why?

60. What is the difference between saying "*the* square root of 10" and saying "*a* square root of 10"?

SKILL MAINTENANCE

Simplify.

61. $(7x)^2$

62. $(-3a)^2$

63. $(4t^7)^2$

***

*Based on an article by Peter Brancazio, "The Mechanics of a Slam Dunk," Popular Mechanics, November 1991. Courtesy of Peter Brancazio, Brooklyn College.

64. $5x \cdot 9x^6$

65. $3a \cdot 16a^{10}$

66. $2t \cdot 4t^8 u^6$

SYNTHESIS

67. Explain in your own words why $\sqrt{A^2} \neq A$ when A is negative.

68. One number has only one square root. What is the number and why is it unique in this regard?

Simplify.

69. $\sqrt{\sqrt{81}}$

70. $\sqrt{3^2 + 4^2}$

Aha! **71.** Between what two consecutive integers is $-\sqrt{33}$?

72. Find a number that is the square of an integer and the cube of a different integer.

Solve. If no real-number solution exists, state this.

73. $\sqrt{t^2} = 7$

74. $\sqrt{y^2} = -5$

75. $-\sqrt{x^2} = -3$

76. $a^2 = 36$

Simplify. Assume that all variables represent positive numbers.

77. $\sqrt{(9a^3 b^4)^2}$

78. $\left(\sqrt{3a}\right)^2$

79. $\sqrt{\dfrac{144x^8}{36y^6}}$

80. $\sqrt{\dfrac{y^{12}}{8100}}$

81. $\sqrt{\dfrac{400}{m^{16}}}$

82. $\sqrt{\dfrac{p^2}{3600}}$

83. Use the graph of $y = \sqrt{x}$, shown below, to estimate each of the following to the nearest tenth: **(a)** $\sqrt{3}$; **(b)** $\sqrt{5}$; **(c)** $\sqrt{7}$. Be sure to check by multiplying.

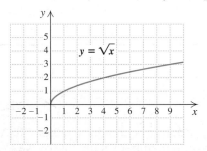

Speed of sound.   The speed V of sound traveling through air, in feet per second, is given by

$$V = \dfrac{1087\sqrt{273 + t}}{16.52},$$

where t is the temperature, in degrees Celsius. Using a calculator, find the speed of sound through air at each of the following temperatures. Round to the nearest tenth.

84. 28°C

85. 5°C

86. −10°C

87. 100°C

88. Use a grapher to draw the graphs of $y_1 = \sqrt{x-2}$, $y_2 = \sqrt{x+7}$, $y_3 = 5 + \sqrt{x}$, and $y_4 = -4 + \sqrt{x}$. If possible, graph all four equations using the SIMULTANEOUS mode and a $[-10, 10, -10, 10]$ window. Then determine which equation corresponds to each curve.

89. What restrictions on a are needed in order for

$$\sqrt{64a^{16}} = 8a^8$$

to be a true equation? Explain how you found your answer.

CORNER

Lengths and Cycles of a Pendulum

Focus: Square roots and modeling

Time: 25–35 minutes

Group size: 3

Materials: Rulers, clocks or watches (to measure seconds), pendulums (see below), calculators

L (in feet)	1	1.5	2	2.5	3	3.5
T (in seconds)						

A pendulum is simply a string, a rope, or a chain with a weight of some sort attached at one end. When the unweighted end is held, a pendulum can swing freely from side to side. A shoe hanging from a shoelace, a yo-yo, a pendant hanging from a chain, a fishing weight hanging from a fish line, or a hairbrush tied to a length of dental floss are all examples of a pendulum. In this activity, each group will develop a mathematical model (formula) that relates a pendulum's length L to the time T that it takes for one complete swing back and forth (one "cycle").

ACTIVITY

1. The group should design a pendulum (see above). One group member should hold the pendulum so that its length is 1 ft. A second group member should lift the weight to one side and then release (do not throw) the weight. The third group member should find the average time, in seconds, for one swing (cycle) back and forth, by timing *five* cycles and dividing by five. Repeat this for each pendulum length listed, so that the table above is completed.

2. Examine the table your group has created. Can you find one number, a, such that $T \approx aL$ for all pairs of values on the chart?

3. To see if a better model can be found, add a third row to the chart and fill in $\sqrt{L}$ for each value of L listed. Can you find one number, b, such that $T \approx b\sqrt{L}$? Does this appear to be a more accurate model than $T = aL$?

4. Use the model for part (3) to predict T when L is 4 ft. Then check your prediction by measuring T as you did in part (1) above. Was your prediction "acceptable"? Compare your results with those of other groups.

5. In Section 8.3, we use a formula equivalent to

$$T = \frac{2\pi}{\sqrt{32}} \cdot \sqrt{L}.$$

How does your value of b from part (3) compare with $2\pi/\sqrt{32}$?

Multiplying and Simplifying Radical Expressions

8.2

Multiplying • Simplifying and Factoring • Simplifying Square Roots of Powers • Multiplying and Simplifying

We now learn to multiply and simplify radical expressions.

Multiplying

To see how to multiply with radical notation, consider the following:

$$\sqrt{9} \cdot \sqrt{4} = 3 \cdot 2 = 6;\qquad \text{This is a product of square roots.}$$
$$\sqrt{9 \cdot 4} = \sqrt{36} = 6.\qquad \text{This is the square root of a product.}$$

Note that $\sqrt{9} \cdot \sqrt{4} = \sqrt{9 \cdot 4}$.

The Product Rule for Square Roots

For any real numbers $\sqrt{A}$ and $\sqrt{B}$,

$$\sqrt{A} \cdot \sqrt{B} = \sqrt{A \cdot B}.$$

(To multiply square roots, multiply the radicands and take the square root.)

E x a m p l e   1

Multiply: **(a)** $\sqrt{5}\,\sqrt{7}$; **(b)** $\sqrt{6}\,\sqrt{6}$; **(c)** $\sqrt{\frac{2}{3}}\,\sqrt{\frac{7}{5}}$; **(d)** $\sqrt{2x}\,\sqrt{3y}$.

Solution

a) $\sqrt{5}\,\sqrt{7} = \sqrt{5 \cdot 7} = \sqrt{35}$

b) $\sqrt{6}\,\sqrt{6} = \sqrt{6 \cdot 6} = \sqrt{36} = 6$ Try to do this one directly: $\sqrt{6}\,\sqrt{6} = 6$.

c) $\sqrt{\dfrac{2}{3}}\,\sqrt{\dfrac{7}{5}} = \sqrt{\dfrac{2}{3} \cdot \dfrac{7}{5}} = \sqrt{\dfrac{14}{15}}$

d) $\sqrt{2x}\,\sqrt{3y} = \sqrt{6xy}$ Note that in order for $\sqrt{2x}$ and $\sqrt{3y}$ to be real numbers, we must have $x, y \geq 0$.

Simplifying and Factoring

To factor a square root, we can use the product rule in reverse. That is,

$$\sqrt{AB} = \sqrt{A}\,\sqrt{B}.$$

This property is especially useful when a radicand that is not a perfect square contains a perfect-square factor. For example, the radicand 48 in $\sqrt{48}$ is not a perfect square, but one of its factors, 16, *is* a perfect square. Thus,

$$\sqrt{48} = \sqrt{16 \cdot 3} \qquad \text{16 is a perfect-square factor of 48.}$$
$$= \sqrt{16} \cdot \sqrt{3} \qquad \sqrt{16} \text{ is a rational factor of } \sqrt{48}.$$
$$= 4\sqrt{3}. \qquad \text{Simplifying } \sqrt{16}; \sqrt{3} \text{ cannot be simplified further.}$$

It is not always obvious that a radicand contains a factor that is a perfect square. In such a case, writing a prime factorization of the radicand may help. For example, if we did not immediately see that 50 contains a perfect square as a factor, we could write

$$\sqrt{50} = \sqrt{2 \cdot 5 \cdot 5} \qquad \text{Factoring into prime factors}$$
$$= \sqrt{2} \cdot \sqrt{5 \cdot 5} \qquad \text{Grouping pairs of like factors;} \\ \phantom{= \sqrt{2} \cdot \sqrt{5 \cdot 5} \qquad} 5 \cdot 5 \text{ is a perfect square.}$$
$$= \sqrt{2} \cdot 5.$$

To avoid any uncertainty as to what is under the radical sign, it is customary to write the radical factor last. Thus, $\sqrt{50} = 5\sqrt{2}$.

A radical expression, like $\sqrt{26}$, in which the radicand has no perfect-square factors, is considered to be in simplified form.

Simplified Form of a Square Root

A radical expression for a square root is simplified when its radicand has no factor other than 1 that is a perfect square.

E x a m p l e   2   Simplify by factoring (remember that all variables are assumed to represent nonnegative numbers).

a) $\sqrt{18}$ $\qquad\qquad\qquad\qquad\qquad\qquad$ **b)** $\sqrt{125x}$

c) $\sqrt{a^2 b}$ $\qquad\qquad\qquad\qquad\qquad\qquad$ **d)** $\sqrt{98t^2}$

Solution

a) $\sqrt{18} = \sqrt{9 \cdot 2} \qquad$ 9 is a perfect-square factor of 18.
$$\phantom{\sqrt{18}} = \sqrt{9}\,\sqrt{2} \qquad \sqrt{9} \text{ is a rational factor of } \sqrt{18}.$$
$$\phantom{\sqrt{18}} = 3\sqrt{2} \qquad \sqrt{2} \text{ cannot be simplified further.}$$

b) $\sqrt{125x} = \sqrt{25 \cdot 5x} \qquad$ 25 is a perfect-square factor of $125x$.
$$\phantom{\sqrt{125x}} = \sqrt{25}\,\sqrt{5x} \qquad \sqrt{25} \text{ is a rational factor of } \sqrt{125x}.$$
$$\phantom{\sqrt{125x}} = 5\sqrt{5x} \qquad \sqrt{5x} \text{ cannot be simplified further.}$$

c) $\sqrt{a^2 b} = \sqrt{a^2}\,\sqrt{b} \qquad$ Identifying a perfect-square factor and factoring into a product of radicals
$$\phantom{\sqrt{a^2 b}} = a\sqrt{b} \qquad \text{No absolute-value signs are necessary since } a \text{ is assumed to be nonnegative.}$$

d) $\sqrt{98t^2} = \sqrt{49 \cdot t^2 \cdot 2}$ 49 and t^2 are perfect-square factors of $98t^2$.

$= \sqrt{49} \cdot \sqrt{t^2} \cdot \sqrt{2}$

$= 7t\sqrt{2}$ Taking square roots. No absolute-value signs are necessary since t is assumed to be nonnegative.

Simplifying Square Roots of Powers

To take the square root of an even power such as x^{10}, note that $x^{10} = (x^5)^2$. Then

$$\sqrt{x^{10}} = \sqrt{(x^5)^2} = x^5. \quad \text{Remember that we assume } x \geq 0.$$

The exponent of the square root is half the exponent of the radicand. That is,

$$\sqrt{x^{10}} = x^5. \quad \tfrac{1}{2}(10) = 5$$

E x a m p l e 3

Simplify: **(a)** $\sqrt{x^6}$; **(b)** $\sqrt{x^8}$; **(c)** $\sqrt{t^{22}}$.

Solution

a) $\sqrt{x^6} = \sqrt{(x^3)^2} = x^3$ Half of 6 is 3.

b) $\sqrt{x^8} = \sqrt{(x^4)^2} = x^4$ Half of 8 is 4.

c) $\sqrt{t^{22}} = \sqrt{(t^{11})^2} = t^{11}$ Half of 22 is 11.

If a radicand is an odd power, we can simplify by factoring, as in the following example.

E x a m p l e 4

Simplify: **(a)** $\sqrt{x^9}$; **(b)** $\sqrt{32x^{15}}$.

Solution

a) $\sqrt{x^9} = \sqrt{x^8 \cdot x}$ x^8 is the largest perfect-square factor of x^9.

$= \sqrt{x^8}\sqrt{x}$

$= x^4\sqrt{x}$

> **Caution!** The square root of x^9 *is not* x^3.

b) $\sqrt{32x^{15}} = \sqrt{16x^{14} \cdot 2x}$ 16 is the largest perfect-square factor of 32; x^{14} is the largest perfect-square factor of x^{15}.

$= \sqrt{16}\sqrt{x^{14}}\sqrt{2x}$

$= 4x^7\sqrt{2x}$ Simplifying. Since $2x$ has no perfect-square factor, we are done.

Multiplying and Simplifying

Sometimes we can simplify after multiplying. To do so, we again try to identify any perfect-square factors of the radicand.

E x a m p l e   5

Multiply and, if possible, simplify. Remember that all variables are assumed to represent nonnegative numbers.

a) $\sqrt{2}\,\sqrt{14}$ **b)** $\sqrt{5x}\,\sqrt{3x}$ **c)** $\sqrt{2x^8}\,\sqrt{9x^3}$

Solution

a)
$$\begin{aligned}
\sqrt{2}\,\sqrt{14} &= \sqrt{2\cdot 14} && \text{Multiplying}\\
&= \sqrt{2\cdot 2\cdot 7} && \text{Writing the prime factorization}\\
&= \sqrt{2^2}\,\sqrt{7} && \text{Note that } 2\cdot 2,\text{ or }4,\text{ is a perfect-square factor.}\\
&= 2\sqrt{7} && \text{Simplifying}
\end{aligned}$$

b)
$$\left.\begin{aligned}
\sqrt{5x}\,\sqrt{3x} &= \sqrt{15x^2} && \text{Multiplying}\\
&= \sqrt{x^2}\,\sqrt{15}\\
&= x\sqrt{15}
\end{aligned}\right\} \text{ Simplifying}$$

c) To simplify $\sqrt{2x^8}\,\sqrt{9x^3}$, note that x^8 and 9 are perfect-square factors. This allows us to simplify *before* multiplying:

$$\begin{aligned}
\sqrt{2x^8}\,\sqrt{9x^3} &= \sqrt{2}\cdot\sqrt{(x^4)^2}\cdot\sqrt{9}\cdot\sqrt{x^3} && \begin{array}{l}(x^4)^2 \text{ and 9 are}\\ \text{perfect squares.}\end{array}\\
&= \sqrt{2}\cdot x^4\cdot 3\cdot\sqrt{x^3} && \text{Simplifying}\\
&= 3x^4\sqrt{2}\,\sqrt{x^3}. && \text{Using a commutative law}
\end{aligned}$$

The result, $3x^4\sqrt{2}\,\sqrt{x^3}$, or $3x^4\sqrt{2x^3}$, can be simplified further:

$$\begin{aligned}
3x^4\sqrt{2x^3} &= 3x^4\sqrt{x^2\cdot 2x}\\
&= 3x^4\sqrt{x^2}\,\sqrt{2x}\\
&= 3x^4 x\sqrt{2x}\\
&= 3x^5\sqrt{2x}.
\end{aligned}$$

x^3 has a perfect-square factor, x^2.

FOR EXTRA HELP

Exercise Set **8.2**

Digital Video Tutor CD 5   InterAct Math   Math Tutor Center   MathXL   MyMathLab.com
Videotape 15

Multiply.

1. $\sqrt{5}\,\sqrt{3}$ **2.** $\sqrt{7}\,\sqrt{5}$ **3.** $\sqrt{4}\,\sqrt{3}$

4. $\sqrt{2}\,\sqrt{9}$ **5.** $\sqrt{\frac{2}{5}}\,\sqrt{\frac{3}{4}}$ **6.** $\sqrt{\frac{3}{8}}\,\sqrt{\frac{1}{5}}$

7. $\sqrt{13}\cdot\sqrt{13}$ **8.** $\sqrt{17}\,\sqrt{17}$ **9.** $\sqrt{25}\,\sqrt{3}$

10. $\sqrt{36}\,\sqrt{2}$ **11.** $\sqrt{2}\,\sqrt{x}$ **12.** $\sqrt{3}\,\sqrt{a}$

13. $\sqrt{7}\,\sqrt{2a}$ **14.** $\sqrt{5}\,\sqrt{7t}$ **15.** $\sqrt{5x}\,\sqrt{7}$

16. $\sqrt{5m}\,\sqrt{2n}$ **17.** $\sqrt{3a}\,\sqrt{2c}$ **18.** $\sqrt{3x}\,\sqrt{yz}$

Simplify by factoring. Assume that all variables represent nonnegative numbers.

19. $\sqrt{20}$ **20.** $\sqrt{28}$ **21.** $\sqrt{50}$

22. $\sqrt{45}$ **23.** $\sqrt{700}$ **24.** $\sqrt{200}$

25. $\sqrt{9x}$ **26.** $\sqrt{4y}$ **27.** $\sqrt{75a}$

28. $\sqrt{40m}$ **29.** $\sqrt{16a}$ **30.** $\sqrt{49b}$

31. $\sqrt{64y^2}$ **32.** $\sqrt{9x^2}$ **33.** $\sqrt{13x^2}$

34. $\sqrt{29t^2}$ **35.** $\sqrt{28t^2}$ **36.** $\sqrt{125a^2}$

37. $\sqrt{80}$ **38.** $\sqrt{98}$ **39.** $\sqrt{288y}$

40. $\sqrt{363p}$ **41.** $\sqrt{a^{14}}$ **42.** $\sqrt{t^{20}}$

43. $\sqrt{x^{12}}$ **44.** $\sqrt{x^{16}}$ **45.** $\sqrt{r^7}$

46. $\sqrt{t^5}$ **47.** $\sqrt{t^{19}}$ **48.** $\sqrt{p^{17}}$

49. $\sqrt{90a^3}$

50. $\sqrt{250y^3}$

51. $\sqrt{8a^5}$

52. $\sqrt{12b^7}$

53. $\sqrt{104p^{17}}$

54. $\sqrt{90m^{23}}$

Multiply and, if possible, simplify.

55. $\sqrt{7} \cdot \sqrt{14}$

56. $\sqrt{3}\,\sqrt{6}$

57. $\sqrt{3} \cdot \sqrt{27}$

58. $\sqrt{2} \cdot \sqrt{8}$

59. $\sqrt{3x}\,\sqrt{12y}$

60. $\sqrt{5x}\,\sqrt{20y}$

61. $\sqrt{13}\,\sqrt{13x}$

62. $\sqrt{11}\,\sqrt{11x}$

63. $\sqrt{10b}\,\sqrt{50b}$

64. $\sqrt{6a}\,\sqrt{18a}$

Aha! **65.** $\sqrt{7x} \cdot \sqrt{7x}$

66. $\sqrt{3a}\,\sqrt{3a}$

67. $\sqrt{ab}\,\sqrt{ac}$

68. $\sqrt{xy}\,\sqrt{xz}$

69. $\sqrt{2x}\,\sqrt{14x^5}$

70. $\sqrt{15m^7}\,\sqrt{5m}$

71. $\sqrt{x^2y^3}\,\sqrt{xy^4}$

72. $\sqrt{x^3y^2}\,\sqrt{xy}$

73. $\sqrt{50ab}\,\sqrt{10a^2b^7}$

74. $\sqrt{10xy^2}\,\sqrt{5x^2y^3}$

Speed of a skidding car.   *The formula*

$$r = 2\sqrt{5L}$$

can be used to approximate the speed r, in miles per hour, of a car that has left a skid mark L feet long.

75. What was the speed of a car that left skid marks of 20 ft? of 150 ft?

76. What was the speed of a car that left skid marks of 30 ft? of 70 ft?

77. How would you convince someone that the following equation is not true?

$$\sqrt{x^2 - 49} = \sqrt{x^2} - \sqrt{49} = x - 7$$

This is not true.

78. Explain why the rules for manipulating exponents are important when simplifying radical expressions.

SKILL MAINTENANCE

Simplify.

79. $\dfrac{a^7b^3}{a^2b}$

80. $\dfrac{x^5y^9}{x^2y}$

81. $\dfrac{3x}{5y} \cdot \dfrac{2x}{7y}$

82. $\dfrac{7a}{3b} \cdot \dfrac{5a}{2b}$

83. $\dfrac{2r^3}{7t} \cdot \dfrac{rt}{rt}$

84. $\dfrac{5x^7}{11y} \cdot \dfrac{xy}{xy}$

SYNTHESIS

85. Explain why $\sqrt{16x^4} = 4x^2$, but $\sqrt{4x^{16}} \neq 2x^4$.

86. Simplify $\sqrt{49}$, $\sqrt{490}$, $\sqrt{4900}$, $\sqrt{49,000}$, and $\sqrt{490,000}$; then describe the pattern you see.

Simplify.

87. $\sqrt{0.01}$

88. $\sqrt{0.25}$

89. $\sqrt{0.0625}$

90. $\sqrt{0.000001}$

Use the proper symbol ($>$, $<$, or $=$) between each pair of values to make a true sentence. Do not use a calculator.

91. $15\sqrt{2}$ ▢ $\sqrt{450}$

92. 15 ▢ $4\sqrt{14}$

93. $3\sqrt{11}$ ▢ $7\sqrt{2}$

94. 16 ▢ $\sqrt{15}\,\sqrt{17}$

95. 8 ▢ $\sqrt{15} + \sqrt{17}$

96. $5\sqrt{7}$ ▢ $4\sqrt{11}$

Multiply and then simplify by factoring.

97. $\sqrt{27(x + 1)}\,\sqrt{12y(x + 1)^2}$

98. $\sqrt{18(x - 2)}\,\sqrt{20(x - 2)^3}$

99. $\sqrt{x^9}\,\sqrt{2x}\,\sqrt{10x^5}$

100. $\sqrt{2^{109}}\,\sqrt{x^{306}}\,\sqrt{x^{11}}$

Fill in the blank.

101. $\sqrt{21x^9} \cdot$ _____ $= 7x^{14}\sqrt{6x^7}$

102. $\sqrt{35x^7} \cdot$ _____ $= 5x^5\sqrt{14x^3}$

Simplify.

103. $\sqrt{x^{16n}}$

104. $\sqrt{0.04x^{4n}}$

105. Simplify $\sqrt{y^n}$, when n is an odd whole number greater than or equal to 3.

Quotients Involving Square Roots

8.3

Dividing Radical Expressions • Rationalizing Denominators

In this section, we divide radical expressions and simplify quotients containing radicals.

Dividing Radical Expressions

To see how to divide with radical notation, consider the following:

$$\frac{\sqrt{25}}{\sqrt{16}} = \frac{5}{4} \text{ since } \sqrt{25} = 5 \text{ and } \sqrt{16} = 4;$$

$$\sqrt{\frac{25}{16}} = \frac{5}{4} \text{ since } \frac{5}{4} \cdot \frac{5}{4} = \frac{25}{16}.$$

We see that $\dfrac{\sqrt{25}}{\sqrt{16}} = \sqrt{\dfrac{25}{16}}$.

> ### The Quotient Rule for Square Roots
> For any real numbers $\sqrt{A}$ and $\sqrt{B}$, $B \neq 0$,
>
> $$\frac{\sqrt{A}}{\sqrt{B}} = \sqrt{\frac{A}{B}}.$$
>
> (To divide two square roots, divide the radicands and take the square root.)

E x a m p l e 1

Divide and simplify: **(a)** $\dfrac{\sqrt{27}}{\sqrt{3}}$; **(b)** $\dfrac{\sqrt{8a^7}}{\sqrt{2a}}$.

Solution

a) $\dfrac{\sqrt{27}}{\sqrt{3}} = \sqrt{\dfrac{27}{3}}$ We can now simplify the radicand: $\dfrac{27}{3} = 9$.

$\qquad\qquad = \sqrt{9} = 3$

b) $\dfrac{\sqrt{8a^7}}{\sqrt{2a}} = \sqrt{\dfrac{8a^7}{2a}}$ We do this because $\dfrac{8a^7}{2a}$ can be simplified.

$\qquad\qquad = \sqrt{4a^6} = 2a^3$ We assume $a > 0$.

The quotient rule for square roots can also be read from right to left:

$$\sqrt{\frac{A}{B}} = \frac{\sqrt{A}}{\sqrt{B}}.$$

We generally select the form that makes simplification easier.

E x a m p l e   2   Simplify by taking square roots in the numerator and the denominator separately.

a) $\sqrt{\dfrac{25}{9}}$ **b)** $\sqrt{\dfrac{1}{16}}$ **c)** $\sqrt{\dfrac{49}{t^2}}$

Solution

a) $\sqrt{\dfrac{25}{9}} = \dfrac{\sqrt{25}}{\sqrt{9}} = \dfrac{5}{3}$ Taking the square root of the numerator and the square root of the denominator. This is sometimes done mentally, in one step.

b) $\sqrt{\dfrac{1}{16}} = \dfrac{\sqrt{1}}{\sqrt{16}} = \dfrac{1}{4}$ Taking the square root of the numerator and the square root of the denominator

c) $\sqrt{\dfrac{49}{t^2}} = \dfrac{\sqrt{49}}{\sqrt{t^2}} = \dfrac{7}{t}$ We assume $t > 0$.

Sometimes a rational expression can be simplified to one that has a perfect-square numerator and/or a perfect-square denominator.

E x a m p l e   3   Simplify: **(a)** $\sqrt{\dfrac{14}{50}}$; **(b)** $\sqrt{\dfrac{48x^3}{3x^7}}$.

Solution

a) $\sqrt{\dfrac{14}{50}} = \sqrt{\dfrac{7 \cdot 2}{25 \cdot 2}}$

$= \sqrt{\dfrac{7 \cdot \cancel{2}}{25 \cdot \cancel{2}}}$ Removing a factor equal to 1: $\dfrac{2}{2} = 1$

$= \dfrac{\sqrt{7}}{\sqrt{25}} = \dfrac{\sqrt{7}}{5}$

b) $\sqrt{\dfrac{48x^3}{3x^7}} = \sqrt{\dfrac{16 \cdot \cancel{3x^3}}{x^4 \cdot \cancel{3x^3}}}$ Removing a factor equal to 1: $\dfrac{3x^3}{3x^3} = 1$

$= \dfrac{\sqrt{16}}{\sqrt{x^4}} = \dfrac{4}{x^2}$ We assume $x \neq 0$.

Rationalizing Denominators

A procedure for finding an equivalent expression without a radical in the denominator is sometimes useful. This makes long division involving decimal approximations easier to perform. The procedure, called **rationalizing the denominator**, relies on multiplying by a carefully selected form of 1 and the fact that $\sqrt{x} \cdot \sqrt{x} = x$.

Example 4

Rationalize each denominator: **(a)** $\dfrac{8}{\sqrt{7}}$; **(b)** $\sqrt{\dfrac{5}{a}}$.

Solution

a) By writing 1 as $\sqrt{7}/\sqrt{7}$, we can find an expression that is equivalent to $8/\sqrt{7}$ without a radical in the denominator:

$$\frac{8}{\sqrt{7}} = \frac{8}{\sqrt{7}} \cdot \frac{\sqrt{7}}{\sqrt{7}} \qquad \text{Multiplying by 1, using the denominator,}$$
$$\sqrt{7}, \text{ to write 1}$$

$$= \frac{8\sqrt{7}}{7}. \qquad \sqrt{7} \cdot \sqrt{7} = 7$$

b)
$$\sqrt{\frac{5}{a}} = \frac{\sqrt{5}}{\sqrt{a}} \qquad \begin{array}{l}\text{The square root of a quotient is the}\\ \text{quotient of the square roots.}\end{array}$$

$$= \frac{\sqrt{5}}{\sqrt{a}} \cdot \frac{\sqrt{a}}{\sqrt{a}} \qquad \text{Multiplying by 1, using the denominator, } \sqrt{a}, \text{ to write 1}$$

$$= \frac{\sqrt{5a}}{a} \qquad \sqrt{a} \cdot \sqrt{a} = a. \text{ We assume } a > 0.$$

It is usually easiest to rationalize a denominator after the expression has been simplified.

Example 5

Rationalize each denominator.

a) $\sqrt{\dfrac{5}{18}}$ **b)** $\dfrac{\sqrt{7a}}{\sqrt{20}}$

Solution

a)
$$\sqrt{\frac{5}{18}} = \frac{\sqrt{5}}{\sqrt{18}} \qquad \begin{array}{l}\text{The square root of a quotient is the}\\ \text{quotient of the square roots.}\end{array}$$

$$= \frac{\sqrt{5}}{\sqrt{9}\sqrt{2}} \qquad \begin{array}{l}\\ \text{Simplifying the denominator.}\\ \text{Note that 9 is a perfect square.}\end{array}$$

$$= \frac{\sqrt{5}}{3\sqrt{2}}$$

$$= \frac{\sqrt{5}}{3\sqrt{2}} \cdot \frac{\sqrt{2}}{\sqrt{2}} \qquad \text{Multiplying by 1, using } \sqrt{2} \text{ to write 1}$$

$$= \frac{\sqrt{10}}{3 \cdot 2} = \frac{\sqrt{10}}{6}$$

b) $\dfrac{\sqrt{7a}}{\sqrt{20}} = \dfrac{\sqrt{7a}}{\sqrt{4}\,\sqrt{5}}$ ⎫   Simplifying the denominator.
Note that 4 is a perfect square.

$= \dfrac{\sqrt{7a}}{2\sqrt{5}}$

$= \dfrac{\sqrt{7a}}{2\sqrt{5}} \cdot \dfrac{\sqrt{5}}{\sqrt{5}}$ Multiplying by 1

$= \dfrac{\sqrt{35a}}{2 \cdot 5}$

$= \dfrac{\sqrt{35a}}{10}$

> ***Caution!***   Our solutions in Example 5 cannot be simplified any further. A common mistake is to remove a factor of 1 that does not exist. For example, $\dfrac{\sqrt{10}}{6}$ *cannot* be simplified to $\dfrac{\sqrt{5}}{3}$ because $\sqrt{10}$ and 6 do not share a common factor.

Exercise Set 8.3

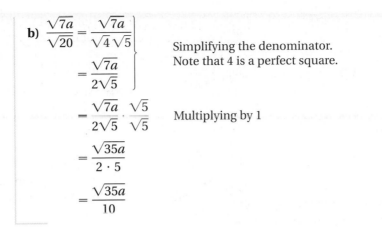

FOR EXTRA HELP

Digital Video Tutor CD 5   InterAct Math   Math Tutor Center   MathXL   MyMathLab.com
Videotape 15

Simplify. Assume that all variables represent positive numbers.

1. $\dfrac{\sqrt{12}}{\sqrt{3}}$

2. $\dfrac{\sqrt{20}}{\sqrt{5}}$

3. $\dfrac{\sqrt{75}}{\sqrt{3}}$

4. $\dfrac{\sqrt{72}}{\sqrt{2}}$

5. $\dfrac{\sqrt{35}}{\sqrt{5}}$

6. $\dfrac{\sqrt{18}}{\sqrt{3}}$

7. $\dfrac{\sqrt{7}}{\sqrt{63}}$

8. $\dfrac{\sqrt{3}}{\sqrt{48}}$

9. $\dfrac{\sqrt{18}}{\sqrt{32}}$

10. $\dfrac{\sqrt{12}}{\sqrt{75}}$

11. $\dfrac{\sqrt{8x}}{\sqrt{2x}}$

12. $\dfrac{\sqrt{18b}}{\sqrt{2b}}$

13. $\dfrac{\sqrt{48y^3}}{\sqrt{3y}}$

14. $\dfrac{\sqrt{63x^3}}{\sqrt{7x}}$

15. $\dfrac{\sqrt{27x^5}}{\sqrt{3x}}$

16. $\dfrac{\sqrt{20a^8}}{\sqrt{5a^2}}$

17. $\dfrac{\sqrt{21a^9}}{\sqrt{7a^3}}$

18. $\dfrac{\sqrt{35t^{11}}}{\sqrt{5t^5}}$

19. $\sqrt{\dfrac{36}{25}}$

20. $\sqrt{\dfrac{9}{49}}$

21. $\sqrt{\dfrac{49}{16}}$

22. $\sqrt{\dfrac{100}{49}}$

23. $-\sqrt{\dfrac{25}{81}}$

24. $-\sqrt{\dfrac{25}{64}}$

25. $\sqrt{\dfrac{2a^3}{50a}}$

26. $\sqrt{\dfrac{7a^5}{28a}}$

27. $\sqrt{\dfrac{6x^7}{32x}}$

28. $\sqrt{\dfrac{4x^3}{50x}}$

29. $\sqrt{\dfrac{21t^9}{28t^3}}$

30. $\sqrt{\dfrac{10t^9}{18t^5}}$

Rationalize each denominator.

31. $\dfrac{5}{\sqrt{3}}$

32. $\dfrac{7}{\sqrt{2}}$

33. $\dfrac{\sqrt{3}}{\sqrt{7}}$

34. $\dfrac{\sqrt{7}}{\sqrt{11}}$

35. $\dfrac{\sqrt{4}}{\sqrt{27}}$

36. $\dfrac{\sqrt{9}}{\sqrt{8}}$

37. $\dfrac{\sqrt{2}}{\sqrt{3}}$

38. $\dfrac{\sqrt{5}}{\sqrt{7}}$

39. $\dfrac{\sqrt{3}}{\sqrt{50}}$

40. $\dfrac{\sqrt{5}}{\sqrt{18}}$

41. $\dfrac{\sqrt{2a}}{\sqrt{45}}$

42. $\dfrac{\sqrt{3a}}{\sqrt{32}}$

43. $\sqrt{\dfrac{12}{5}}$ **44.** $\sqrt{\dfrac{8}{3}}$ **45.** $\sqrt{\dfrac{2}{x}}$

46. $\sqrt{\dfrac{3}{x}}$ **47.** $\sqrt{\dfrac{t}{32}}$ **48.** $\sqrt{\dfrac{a}{12}}$

49. $\sqrt{\dfrac{x}{40}}$ **50.** $\sqrt{\dfrac{y}{90}}$ *Aha!* **51.** $\sqrt{\dfrac{3a}{25}}$

52. $\sqrt{\dfrac{7t}{16}}$ **53.** $\sqrt{\dfrac{5x^3}{12x}}$ **54.** $\sqrt{\dfrac{7t^3}{32t}}$

Period of a swinging pendulum. *The period T of a pendulum is the time it takes to move from one side to the other and back. A formula for the period is*

$$T = 2\pi\sqrt{\dfrac{L}{32}},$$

where T is in seconds and L is in feet. Use 3.14 *for* π.

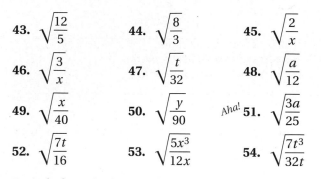

55. Find the periods of pendulums of lengths 32 ft and 50 ft.

56. Find the periods of pendulums of lengths 8 ft and 2 ft.

57. The pendulum of a mantle clock is $2/\pi^2$ ft long. How long does it take to swing from one side to the other and back?

58. The pendulum of a grandfather clock is $98/\pi^2$ ft long. How long does it take to swing from one side to the other and back?

59. Ingrid is swinging at the bottom of a 72-ft–long bungee cord. How long does it take her to complete one swing back and forth? (*Hint*: $\sqrt{2.5} \approx 1.6$.)

60. Don is swinging back and forth on an 18-ft–long rope swing near the Green River Reservoir. How long does it take him to complete one swing back and forth?

61. Why is it important to know how to multiply radical expressions before learning how to divide them?

62. Describe a method that could be used to rationalize the *numerator* of a radical expression.

SKILL MAINTENANCE

Simplify.

63. $5x + 9 + 7x + 4$ **64.** $7a + 2 + 3a + 4$

65. $2a^3 - a^2 - 3a^3 - 7a^2$ **66.** $4t - 3t^2 - t - 8t^2$

Multiply.

67. $9x(2x - 7)$ **68.** $2x(3x - 5)$

69. $(3 + 4x)(2 + 5x)$ **70.** $(7 - 3x)(2 - x)$

SYNTHESIS

71. Is it always best to rewrite an expression of the form $\sqrt{a}/\sqrt{b}$ as $\sqrt{a/b}$ before simplifying? Why or why not?

72. When using long division, why is it easier to approximate $\sqrt{2}/2$ than it is to approximate $1/\sqrt{2}$?

Rationalize each denominator and, if possible, simplify.

73. $\sqrt{\dfrac{7}{1000}}$ **74.** $\sqrt{\dfrac{3}{800}}$

75. $\sqrt{\dfrac{5x^2}{8x^7y^3}}$ **76.** $\sqrt{\dfrac{3x^2y}{a^2x^5}}$

77. $\sqrt{\dfrac{2a}{5b^3c^9}}$ **78.** $\sqrt{\dfrac{1}{5zw^2}}$

Aha! **79.** $\dfrac{3}{\sqrt{\sqrt{7}}}$ **80.** $\dfrac{2}{\sqrt{\sqrt{5}}}$

Simplify. Assume that $0 \le x \le y$ *and* $0 < z \le 1$.

81. $\sqrt{\dfrac{1}{x^2} - \dfrac{2}{xy} + \dfrac{1}{y^2}}$ **82.** $\sqrt{2 - \dfrac{4}{z^2} + \dfrac{2}{z^4}}$

More Operations with Radicals

8.4

Adding and Subtracting Radical Expressions • More with Multiplication • More with Rationalizing Denominators

We now consider addition and subtraction of radical expressions as well as some new types of multiplication and simplification.

Adding and Subtracting Radical Expressions

The sum of a rational number and an irrational number, like $5 + \sqrt{2}$, *cannot* be simplified. However, the sum of **like radicals**—that is, radical expressions that have a common radical factor—*can* be simplified.

E x a m p l e   1

Add: $3\sqrt{5} + 4\sqrt{5}$.

Solution   Recall that to simplify an expression like $3x + 4x$, we use the distributive law, as follows:

$$3x + 4x = (3 + 4)x = 7x.$$ The middle step is usually performed mentally.

In this example, x is replaced with $\sqrt{5}$:

$$3\sqrt{5} + 4\sqrt{5} = (3 + 4)\sqrt{5}$$ Using the distributive law to factor out $\sqrt{5}$
$$= 7\sqrt{5}.$$ $3\sqrt{5}$ and $4\sqrt{5}$ are like radicals.

To simplify in this manner, the radical factors must be the same.

E x a m p l e   2

Simplify.

a) $9\sqrt{17} - 3\sqrt{17}$ **b)** $7\sqrt{x} + \sqrt{x}$

c) $5\sqrt{2} - \sqrt{18}$ **d)** $\sqrt{5} + \sqrt{20} + \sqrt{7}$

Solution

a) $9\sqrt{17} - 3\sqrt{17} = (9 - 3)\sqrt{17}$ Using the distributive law.
$$= 6\sqrt{17}$$ Try to do this mentally.

b) $7\sqrt{x} + \sqrt{x} = (7 + 1)\sqrt{x}$ Using the distributive law.
$$= 8\sqrt{x}$$ Try to do this mentally.

c) $5\sqrt{2} - \sqrt{18} = 5\sqrt{2} - \sqrt{9 \cdot 2}$
$$= 5\sqrt{2} - \sqrt{9}\sqrt{2}$$ Simplifying $\sqrt{18}$

$$= 5\sqrt{2} - 3\sqrt{2}$$ We now have like radicals.

$$= 2\sqrt{2}$$ Using the distributive law mentally:
$$5\sqrt{2} - 3\sqrt{2} = (5 - 3)\sqrt{2}$$

d) $\sqrt{5} + \sqrt{20} + \sqrt{7} = \sqrt{5} + \sqrt{4}\sqrt{5} + \sqrt{7}$ Simplifying $\sqrt{20}$

$\qquad\qquad\qquad\qquad = \sqrt{5} + 2\sqrt{5} + \sqrt{7}$ We now have like radicals.

$\qquad\qquad\qquad\qquad = 3\sqrt{5} + \sqrt{7}$ Adding like radicals; $3\sqrt{5} + \sqrt{7}$ cannot be simplified.

> **Caution!** It is *not true* that the sum of two square roots is the square root of the sum: $\sqrt{A} + \sqrt{B} \neq \sqrt{A + B}$. For example, $\sqrt{9} + \sqrt{16} \neq \sqrt{9 + 16}$ since $3 + 4 \neq 5$.

More with Multiplication

Radical expressions with more than one term are multiplied in much the same way that polynomials with more than one term are multiplied.

E x a m p l e **3**

Multiply.

a) $\sqrt{2}(\sqrt{3} + \sqrt{5})$ **b)** $(4 + \sqrt{7})(2 + \sqrt{7})$

c) $(2 - \sqrt{5})(2 + \sqrt{5})$ **d)** $(2 + \sqrt{3})(5 - 4\sqrt{3})$

Solution

a) $\sqrt{2}(\sqrt{3} + \sqrt{5}) = \sqrt{2}\sqrt{3} + \sqrt{2}\sqrt{5}$ Using the distributive law

$\qquad\qquad\qquad\quad = \sqrt{6} + \sqrt{10}$ Using the product rule for square roots

b) $(4 + \sqrt{7})(2 + \sqrt{7}) = 4 \cdot 2 + 4 \cdot \sqrt{7} + \sqrt{7} \cdot 2 + \sqrt{7} \cdot \sqrt{7}$ Using FOIL

$\qquad\qquad\qquad\qquad = 8 + 4\sqrt{7} + 2\sqrt{7} + 7$

$\qquad\qquad\qquad\qquad = 15 + 6\sqrt{7}$ Combining like terms

c) Note that $(2 - \sqrt{5})(2 + \sqrt{5})$ is of the form $(A - B)(A + B)$:

$(2 - \sqrt{5})(2 + \sqrt{5}) = 2 \cdot 2 + 2 \cdot \sqrt{5} - \sqrt{5} \cdot 2 - \sqrt{5} \cdot \sqrt{5}$

$\qquad\qquad\qquad\qquad = 4 + 2\sqrt{5} - 2\sqrt{5} - 5$ As expected, the

$\qquad\qquad\qquad\qquad = 4 - 5$ middle terms are

$\qquad\qquad\qquad\qquad = -1$ opposites.

d) $(2 + \sqrt{3})(5 - 4\sqrt{3}) = 2 \cdot 5 - 2 \cdot 4\sqrt{3} + \sqrt{3} \cdot 5 - \sqrt{3} \cdot 4\sqrt{3}$

$\qquad\qquad\qquad\qquad = 10 - 8\sqrt{3} + 5\sqrt{3} - 4 \cdot 3$ $2 \cdot 5 = 10; 2 \cdot 4 = 8;$ and $\sqrt{3} \cdot \sqrt{3} = 3$

$\qquad\qquad\qquad\qquad = 10 - 3\sqrt{3} - 12$ Adding like radicals

$\qquad\qquad\qquad\qquad = -2 - 3\sqrt{3}$

More with Rationalizing Denominators

Note in Example 3(c) that the result has no radicals. This will happen whenever expressions like $\sqrt{a} + \sqrt{b}$ and $\sqrt{a} - \sqrt{b}$ are multiplied:

$$(\sqrt{a} + \sqrt{b})(\sqrt{a} - \sqrt{b}) = (\sqrt{a})^2 - (\sqrt{b})^2 = a - b.$$

Expressions such as $\sqrt{3} - \sqrt{5}$ and $\sqrt{3} + \sqrt{5}$ are said to be **conjugates** of each other. So too are expressions like $2 + \sqrt{7}$ and $2 - \sqrt{7}$. Once the conjugate of a denominator has been found, it can be used to rationalize the denominator.

Example 4

Rationalize each denominator and, if possible, simplify.

a) $\dfrac{3}{2 + \sqrt{5}}$

b) $\dfrac{2}{\sqrt{7} - \sqrt{3}}$

Solution

a) We multiply by a form of 1, using the conjugate of $2 + \sqrt{5}$, which is $2 - \sqrt{5}$, as the numerator and the denominator:

$$\frac{3}{2 + \sqrt{5}} = \frac{3}{2 + \sqrt{5}} \cdot \frac{2 - \sqrt{5}}{2 - \sqrt{5}} \qquad \text{Multiplying by 1}$$

$$= \frac{3(2 - \sqrt{5})}{(2 + \sqrt{5})(2 - \sqrt{5})}$$

$$= \frac{3(2 - \sqrt{5})}{2^2 - (\sqrt{5})^2} \qquad \text{Using } (A + B)(A - B) = A^2 - B^2$$

$$= \frac{3(2 - \sqrt{5})}{-1} \qquad \begin{array}{l}\text{Simplifying the denominator.}\\ \text{See Example 3(c).}\end{array}$$

$$= \frac{6 - 3\sqrt{5}}{-1}$$

$$= -6 + 3\sqrt{5}. \qquad \begin{array}{l}\text{Dividing } both \text{ terms in the}\\ \text{numerator by } -1\end{array}$$

b) $\dfrac{2}{\sqrt{7} - \sqrt{3}} = \dfrac{2}{\sqrt{7} - \sqrt{3}} \cdot \dfrac{\sqrt{7} + \sqrt{3}}{\sqrt{7} + \sqrt{3}}$ $\quad \begin{array}{l}\text{Multiplying by 1, using } \sqrt{7} + \sqrt{3},\\ \text{the conjugate of } \sqrt{7} - \sqrt{3}\end{array}$

$$= \frac{2(\sqrt{7} + \sqrt{3})}{(\sqrt{7} - \sqrt{3})(\sqrt{7} + \sqrt{3})}$$

$$= \frac{2(\sqrt{7} + \sqrt{3})}{(\sqrt{7})^2 - (\sqrt{3})^2} \qquad \text{Using } (A - B)(A + B) = A^2 - B^2$$

$$= \frac{2(\sqrt{7} + \sqrt{3})}{7 - 3} \qquad \text{The denominator is free of radicals.}$$

$$= \frac{2(\sqrt{7} + \sqrt{3})}{4} \qquad \begin{array}{l}\text{Since 2 is a common factor of both}\\ \text{the numerator and denominator,}\\ \text{we simplify.}\end{array}$$

$$\left.\begin{array}{l} = \dfrac{2(\sqrt{7} + \sqrt{3})}{2 \cdot 2} \\[2mm] = \dfrac{\sqrt{7} + \sqrt{3}}{2} \end{array}\right\} \qquad \begin{array}{l}\text{Factoring and removing a factor}\\ \text{equal to 1: } \dfrac{2}{2} = 1\end{array}$$

Exercise Set 8.4

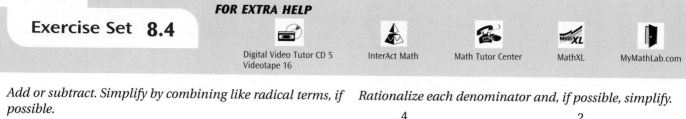
Add or subtract. Simplify by combining like radical terms, if possible.

1. $4\sqrt{3} + 8\sqrt{3}$

2. $7\sqrt{2} + 4\sqrt{2}$

3. $7\sqrt{2} - 5\sqrt{2}$

4. $9\sqrt{5} - 6\sqrt{5}$

5. $9\sqrt{y} + 3\sqrt{y}$

6. $6\sqrt{x} + 7\sqrt{x}$

7. $6\sqrt{a} - 12\sqrt{a}$

8. $9\sqrt{x} - 11\sqrt{x}$

9. $5\sqrt{6x} + 2\sqrt{6x}$

10. $5\sqrt{2a} + 3\sqrt{2a}$

11. $12\sqrt{14y} - \sqrt{14y}$

12. $9\sqrt{10y} - \sqrt{10y}$

13. $2\sqrt{5} + 7\sqrt{5} + 5\sqrt{5}$

14. $5\sqrt{7} + 2\sqrt{7} + 4\sqrt{7}$

15. $3\sqrt{6} - 7\sqrt{6} + 2\sqrt{6}$

16. $7\sqrt{2} - 9\sqrt{2} + 4\sqrt{2}$

17. $2\sqrt{5} + \sqrt{45}$

18. $5\sqrt{3} + \sqrt{8}$

19. $\sqrt{25a} - \sqrt{a}$

20. $\sqrt{x} - \sqrt{9x}$

21. $7\sqrt{50} - 3\sqrt{2}$

22. $2\sqrt{3} - 4\sqrt{75}$

23. $3\sqrt{12} + 2\sqrt{300}$

24. $6\sqrt{18} + 5\sqrt{8}$

25. $\sqrt{45} + \sqrt{80}$

26. $\sqrt{72} + \sqrt{98}$

Aha! **27.** $9\sqrt{8} + \sqrt{72} - 9\sqrt{8}$

28. $4\sqrt{12} + \sqrt{27} - \sqrt{12}$

29. $7\sqrt{12} - 2\sqrt{27} + \sqrt{75}$

30. $5\sqrt{18} - 2\sqrt{32} - \sqrt{50}$

31. $\sqrt{9x} + \sqrt{49x} - 9\sqrt{x}$

32. $\sqrt{16a} - 4\sqrt{a} + \sqrt{25a}$

Multiply.

33. $\sqrt{2}(\sqrt{5} + \sqrt{7})$

34. $\sqrt{5}(\sqrt{2} + \sqrt{11})$

35. $\sqrt{5}(\sqrt{6} - \sqrt{10})$

36. $\sqrt{6}(\sqrt{15} - \sqrt{7})$

37. $(3 + \sqrt{2})(4 + \sqrt{2})$

38. $(5 + \sqrt{11})(3 + \sqrt{11})$

39. $(\sqrt{7} - 2)(\sqrt{7} - 5)$

40. $(\sqrt{10} + 4)(\sqrt{10} - 7)$

41. $(\sqrt{5} + 4)(\sqrt{5} - 4)$

42. $(1 + \sqrt{5})(1 - \sqrt{5})$

43. $(\sqrt{6} - \sqrt{3})(\sqrt{6} + \sqrt{3})$

44. $(\sqrt{2} + \sqrt{6})(\sqrt{2} - \sqrt{6})$

45. $(4 + 3\sqrt{2})(1 - \sqrt{2})$

46. $(8 - \sqrt{7})(3 + 2\sqrt{7})$

47. $(7 + \sqrt{3})^2$

48. $(2 + \sqrt{5})^2$

49. $(1 - 2\sqrt{3})^2$

50. $(6 - 3\sqrt{5})^2$

51. $(\sqrt{x} - \sqrt{10})^2$

52. $(\sqrt{a} - \sqrt{6})^2$

Rationalize each denominator and, if possible, simplify.

53. $\dfrac{4}{7 + \sqrt{2}}$

54. $\dfrac{2}{3 + \sqrt{5}}$

55. $\dfrac{6}{2 - \sqrt{7}}$

56. $\dfrac{3}{4 - \sqrt{2}}$

57. $\dfrac{2}{\sqrt{7} + 5}$

58. $\dfrac{6}{\sqrt{10} + 3}$

59. $\dfrac{\sqrt{6}}{\sqrt{6} - 5}$

60. $\dfrac{\sqrt{10}}{\sqrt{10} - 7}$

61. $\dfrac{\sqrt{5}}{\sqrt{5} + \sqrt{3}}$

62. $\dfrac{\sqrt{7}}{\sqrt{7} - \sqrt{5}}$

63. $\dfrac{\sqrt{3}}{\sqrt{5} - \sqrt{3}}$

64. $\dfrac{\sqrt{6}}{\sqrt{7} + \sqrt{6}}$

65. $\dfrac{2}{\sqrt{7} + \sqrt{2}}$

66. $\dfrac{6}{\sqrt{5} - \sqrt{3}}$

67. $\dfrac{\sqrt{7} + \sqrt{5}}{\sqrt{7} - \sqrt{5}}$

68. $\dfrac{\sqrt{10} - \sqrt{6}}{\sqrt{10} + \sqrt{6}}$

69. What is the purpose of having the signs differ within each pair of conjugates?

70. Describe a method that could be used to rationalize a numerator that is the sum of two radical expressions.

SKILL MAINTENANCE

Solve.

71. $3x + 5 + 2(x - 3) = 4 - 6x$

72. $3(x - 4) - 2 = 8(2x + 3)$

73. $x^2 - 5x = 6$

74. $x^2 + 10 = 7x$

75. *Mixing juice.* Jolly Juice is 3% real fruit juice, and Real Squeeze is 6% real fruit juice. How many liters of each should be used in order to make an 8-L mixture that is 5.4% real fruit juice?

76. *Commuting costs.* Thelma says it costs at least $7.50 every time she drives to work: $3 in tolls and $1.50 per hour for parking. For how long does Thelma park?

SYNTHESIS

77. Is it possible to square the sum of two radical expressions without adding like radicals? Why or why not?

78. Why must you know how to add and subtract radical expressions before you can rationalize denominators with two terms?

Add or subtract and, if possible, simplify.

79. $5\sqrt{\dfrac{1}{2}} + \dfrac{7}{2}\sqrt{18} - 4\sqrt{98}$

80. $\sqrt{\dfrac{25}{x}} + \dfrac{\sqrt{x}}{2x} - \dfrac{5}{\sqrt{2}}$

81. $a\sqrt{a^{17}b^9} - b\sqrt{a^{13}b^{11}} + a\sqrt{a^9b^{15}}$

82. $\sqrt{8x^6y^3} - x\sqrt{2y^7} - \dfrac{x}{3}\sqrt{18x^2y^9}$

83. $7x\sqrt{12xy^2} - 9y\sqrt{27x^3} + 5\sqrt{300x^3y^2}$

84. Can you find any pairs of nonnegative numbers a and b for which $\sqrt{a} + \sqrt{b} = \sqrt{a + b}$? If so, name them.

85. Three students were asked to simplify $\sqrt{10} + \sqrt{50}$. Their answers were $\sqrt{10}\left(1 + \sqrt{5}\right)$, $\sqrt{10} + 5\sqrt{2}$, and $\sqrt{2}\left(5 + \sqrt{5}\right)$. Which answer(s), if any, is correct?

8.5

Radical Equations

Solving Radical Equations • Problem Solving and Applications

CONNECTING THE CONCEPTS

In Sections 8.1–8.4, we learned how to simplify and manipulate radical expressions. We performed this work to find *equivalent expressions*.

Now that we know how to manipulate radical notation, it is time to learn how to solve a new type of equation. As in our earlier work

with equations, finding *equivalent equations* will be part of our strategy. What is different, however, is that now we will use a step that does not always produce equivalent equations. Checking solutions will therefore be more important than ever.

An equation in which a variable appears in a radicand is called a **radical equation**. The following are examples:

$$\sqrt{2x} - 4 = 7, \quad 2\sqrt{x + 2} = \sqrt{x + 10}, \quad \text{and} \quad 3 + \sqrt{27 - 3x} = x.$$

We now learn to solve such equations and use them in problem solving.

Solving Radical Equations

An equation with a square root can be rewritten without the radical by using *the principle of squaring*.

> ### The Principle of Squaring
> If $a = b$, then $a^2 = b^2$.

The principle of squaring does *not* say that if $a^2 = b^2$, then $a = b$. Indeed, if a is replaced with -5 and b with 5, then $a^2 = b^2$ is true (since $(-5)^2 = 5^2$), but the equation $a = b$ is false (since $-5 \neq 5$). Thus, although the principle of squaring can lead us to any solutions that a radical equation might have, it can also lead us to numbers that are not solutions.

E x a m p l e   1   Solve: (a) $\sqrt{x} + 3 = 7$; (b) $\sqrt{2x} = -5$.

Solution

a) Our plan is to isolate the radical term on one side of the equation and then use the principle of squaring:

$$\sqrt{x} + 3 = 7$$
$$\sqrt{x} = 4 \qquad \text{Subtracting 3 to get the radical alone on one side}$$
$$(\sqrt{x})^2 = 4^2 \qquad \text{Squaring both sides (using the principle of squaring)}$$
$$x = 16$$

Check:
$$\begin{array}{c} \sqrt{x} + 3 = 7 \\ \hline \sqrt{16} + 3 \ ? \ 7 \\ 4 + 3 \ \Big| \\ 7 \ \Big| \ 7 \quad \text{TRUE} \end{array}$$

The solution is 16.

b)
$$\sqrt{2x} = -5$$
$$(\sqrt{2x})^2 = (-5)^2 \qquad \text{Squaring both sides (using the principle of squaring)}$$
$$\left. \begin{array}{c} 2x = 25 \\ x = \frac{25}{2}. \end{array} \right\} \quad \text{Solving for } x$$

Check:
$$\begin{array}{c} \sqrt{2x} = -5 \\ \hline \sqrt{2 \cdot \frac{25}{2}} \ ? \ -5 \\ \sqrt{25} \ \Big| \\ 5 \ \Big| \ -5 \quad \text{FALSE} \end{array}$$

There are no solutions. You might have suspected this from the start since no number has a principal square root that is negative.

> ***Caution!***   When the principle of squaring is used to solve an equation, all possible solutions *must* be checked in the original equation!

E x a m p l e   2

Solve: $3\sqrt{x} = \sqrt{x + 32}$.

Solution   We have

$$3\sqrt{x} = \sqrt{x + 32}$$

$$\left(3\sqrt{x}\right)^2 = \left(\sqrt{x + 32}\right)^2 \quad \text{Squaring both sides (using the principle}$$
$$\text{of squaring)}$$

$$3^2\left(\sqrt{x}\right)^2 = x + 32 \quad \text{Squaring the product on the left;}$$
$$\text{simplifying on the right}$$

$$9x = x + 32 \quad \text{Simplifying on the left}$$

$$\left.\begin{array}{r} 8x = 32 \\ x = 4. \end{array}\right\} \quad \text{Solving for } x$$

Check:
$$\begin{array}{c|c} \multicolumn{2}{c}{3\sqrt{x} = \sqrt{x + 32}} \\ \hline 3\sqrt{4} \ ? \ \sqrt{4 + 32} & \\ 3 \cdot 2 & \sqrt{36} \\ 6 & 6 \quad \text{TRUE} \end{array}$$

The number 4 checks. The solution is 4.

We have been using the following strategy.

> ### To Solve a Radical Equation
> 1. Isolate a radical term.
> 2. Use the principle of squaring (square both sides).
> 3. Solve the new equation.
> 4. Check all possible solutions in the original equation.

In some cases, we may need to apply the principle of zero products (see Section 5.6) after squaring.

E x a m p l e   3

Solve: **(a)** $x - 5 = \sqrt{x + 7}$; **(b)** $3 + \sqrt{27 - 3x} = x$.

Solution

a)
$$x - 5 = \sqrt{x + 7}$$

$$(x - 5)^2 = \left(\sqrt{x + 7}\right)^2 \quad \text{Using the principle of squaring}$$

$$x^2 - 10x + 25 = x + 7 \quad \text{Squaring a binomial on the left side}$$

$$x^2 - 11x + 18 = 0 \quad \text{Adding } -x - 7 \text{ to both sides}$$

$$(x - 9)(x - 2) = 0 \quad \text{Factoring}$$

$$x - 9 = 0 \quad or \quad x - 2 = 0 \quad \text{Using the principle of zero products}$$

$$x = 9 \quad or \quad x = 2$$

technology connection

Solutions of radical equations can be visualized on a grapher. To "see" that 9 is the solution of Example 3(a), we let $y_1 = x - 5$ (the left side of $x - 5 = \sqrt{x + 7}$) and $y_2 = \sqrt{x + 7}$ (the right side of $x - 5 = \sqrt{x + 7}$). Using the INTERSECT option of CALC, we see that $(9, 4)$ is on both curves. This indicates that when $x = 9$, y_1 and y_2 are both 4. No intersection occurs at $x = 2$, which confirms our check.

$y_1 = x - 5, \quad y_2 = \sqrt{x + 7}$

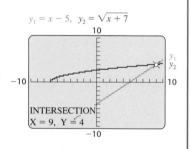

1. Use a grapher to visualize the solution of Example 2.
2. Use a grapher to visualize the solution of Example 3(b).

Check: For 9:

$$x - 5 = \sqrt{x + 7}$$
$$\overline{9 - 5 \,\,?\,\, \sqrt{9 + 7}}$$
$$4 \,\mid\, 4 \qquad \text{TRUE}$$

For 2:

$$x - 5 = \sqrt{x + 7}$$
$$\overline{2 - 5 \,\,?\,\, \sqrt{2 + 7}}$$
$$-3 \,\mid\, 3 \qquad \text{FALSE}$$

The number 9 checks, but 2 does not. Thus the solution is 9.

b) $3 + \sqrt{27 - 3x} = x$

$\sqrt{27 - 3x} = x - 3$ Subtracting 3 to isolate the radical

$\left(\sqrt{27 - 3x}\right)^2 = (x - 3)^2$ Using the principle of squaring

$27 - 3x = x^2 - 6x + 9$

$0 = x^2 - 3x - 18$ Adding $3x - 27$ to both sides

$0 = (x - 6)(x + 3)$ Factoring

$x - 6 = 0 \quad or \quad x + 3 = 0$ Using the principle of zero products

$x = 6 \quad or \qquad x = -3$

Check: For 6:

$$3 + \sqrt{27 - 3x} = x$$
$$\overline{3 + \sqrt{27 - 3 \cdot 6} \,\,?\,\, 6}$$
$$3 + \sqrt{9}$$
$$3 + 3$$
$$6 \,\mid\, 6 \quad \text{TRUE}$$

For -3:

$$3 + \sqrt{27 - 3x} = x$$
$$\overline{3 + \sqrt{27 - 3 \cdot (-3)} \,\,?\,\, -3}$$
$$3 + \sqrt{27 + 9}$$
$$3 + \sqrt{36}$$
$$3 + 6$$
$$9 \,\mid\, -3 \quad \text{FALSE}$$

The number 6 checks, but -3 does not. The solution is 6.

Problem Solving and Applications

Many applications translate to radical equations. For example, at a temperature of t degrees Fahrenheit, sound travels s feet per second, where

$$s = 21.9\sqrt{5t + 2457}.$$

E x a m p l e   4

Musical performances.   The band U2 regularly performs outdoors for crowds in excess of 20,000 people. An officer stationed at a U2 concert used a radar gun to determine that the sound of the band was traveling at a rate of 1170 ft/sec. What was the air temperature at the concert?

Solution

1. **Familiarize.** If we did not already know the formula relating speed of sound to air temperature, we could consult a reference book, a person in the field, or perhaps the Internet. There is no need to memorize this formula.

2. **Translate.**   We substitute 1170 for s in the formula $s = 21.9\sqrt{5t + 2457}$:

$$1170 = 21.9\sqrt{5t + 2457}.$$

3. **Carry out.** We solve the equation for t:

$$1170 = 21.9\sqrt{5t + 2457}$$

$$\frac{1170}{21.9} = \sqrt{5t + 2457} \qquad \text{Dividing both sides by 21.9}$$

$$\left(\frac{1170}{21.9}\right)^2 = \left(\sqrt{5t + 2457}\right)^2 \qquad \text{Using the principle of squaring}$$

$$2854.2 \approx 5t + 2457 \qquad \text{Simplifying}$$

$$397.2 \approx 5t \qquad \text{Subtracting 2457 from both sides}$$

$$79.4 \approx t. \qquad \text{Dividing both sides by 5}$$

4. **Check.** A temperature of 79.4°F sounds reasonable. A complete check is left to the student.

5. **State.** The temperature at the concert was about 79.4°F.

Exercise Set 8.5

Solve.

1. $\sqrt{x} = 5$

2. $\sqrt{x} = 8$

3. $\sqrt{x} + 4 = 12$

4. $\sqrt{x} + 3 = 15$

5. $\sqrt{2x + 1} = 13$

6. $\sqrt{2x + 4} = 9$

7. $2 + \sqrt{3 - y} = 9$

8. $3 + \sqrt{1 - a} = 5$

9. $8 - 4\sqrt{5n} = 0$

10. $6 - 2\sqrt{3n} = 0$

11. $\sqrt{4x - 6} = \sqrt{x + 9}$

12. $\sqrt{4x + 7} = \sqrt{2x + 13}$

Aha! 13. $\sqrt{x} = -7$

14. $\sqrt{x} = -5$

15. $\sqrt{2t + 5} = \sqrt{3t + 7}$

16. $\sqrt{4t + 7} = \sqrt{5t + 8}$

17. $\sqrt{3x - 2} = x - 4$

18. $x - 7 = \sqrt{x - 5}$

19. $a - 9 = \sqrt{a - 3}$

20. $\sqrt{t + 18} = t - 2$

21. $x - 1 = 6\sqrt{x - 9}$

22. $x - 5 = \sqrt{15 - 3x}$

23. $\sqrt{5x + 21} = x + 3$

24. $\sqrt{22 - x} = x - 2$

25. $t + 4 = 4\sqrt{t + 1}$

26. $1 + 2\sqrt{y - 1} = y$

27. $\sqrt{x^2 + 5} - x + 2 = 0$

28. $\sqrt{x^2 + 6} - x + 3 = 0$

29. $\sqrt{(p + 6)(p + 1)} - 2 = p + 1$

30. $\sqrt{(4x + 5)(x + 4)} = 2x + 5$

31. $\sqrt{2 - 7x} = \sqrt{5 - 2x}$

32. $\sqrt{7 - 3x} = \sqrt{12 + x}$

33. $x - 1 = \sqrt{(x + 1)(x - 2)}$

34. $x = 1 + \sqrt{1 - x}$

Speed of a skidding car. *The formula* $r = 2\sqrt{5L}$ *can be used to approximate the speed r, in miles per hour, of a car that has left a skid mark of length L, in feet.*

35. How far will a car skid at 48 mph? at 80 mph?

36. How far will a car skid at 40 mph? at 60 mph?

Temperature and the speed of sound. *Solve, using the formula* $s = 21.9\sqrt{5t + 2457}$ *from Example 4.*

37. During blasting for avalanche control in Utah's Wasatch Mountains, sound traveled at a rate of 1113 ft/sec. What was the temperature at the time?

38. At a recent concert by the Dave Matthews Band, sound traveled at a rate of 1176 ft/sec. What was the temperature at the time?

Sighting to the horizon. *At a height of h meters, one can see V kilometers to the horizon, where* $V = 3.5\sqrt{h}$.

39. Ahab can see 99.4 km to the horizon from the top of Cobble Hill. What is the height of the hill?

40. A steeplejack can see 21 km to the horizon from the top of a building. What is the altitude of the steeplejack's eyes?

41. A scout can see 84 km to the horizon from atop a firetower. What is the altitude of the scout's eyes?

42. Amelia can see 378 km to the horizon from an airplane window. How high is the airplane?

Period of a swinging pendulum. *The formula* $T = 2\pi\sqrt{L/32}$ *can be used to find the period T, in seconds, of a pendulum of length L, in feet.*

43. What is the length of a pendulum that has a period of 1.0 sec? Use 3.14 for π.

44. What is the length of a pendulum that has a period of 2.0 sec? Use 3.14 for π.

45. Do you believe that the principle of squaring can be extended to powers other than 2? That is, if $a = b$, does it follow that $a^n = b^n$ for any integer n? Why or why not?

46. Explain in your own words why possible solutions of radical equations must be checked.

SKILL MAINTENANCE

47. *Test scores.* Amy is taking a test in which items of type A are worth 10 points and items of type B are worth 15 points. She answers 16 questions correctly and scores 180 points. How many questions of each type did Amy answer correctly?

48. *Metallurgy.* Francine stocks two alloys that are different purities of gold. The first is three-fourths pure gold and the second is five-twelfths pure gold. How many ounces of each should she melt and mix in order to obtain a 60-oz mixture that is two-thirds pure gold?

49. *Driving distances.* A car leaves Hereford traveling north at a speed of 56 km/h. Another car leaves Hereford one hour later, traveling north at 84 km/h. How far from Hereford will the second car overtake the first?

50. *Coin value.* A collection of dimes and quarters is worth $15.25. There are 103 coins in all. How many of each are there?

SYNTHESIS

51. Review Example 3 and explain how someone could create a radical equation for which one "solution" checks and one does not.

52. Explain what would have happened in Example 1(a) if we had not isolated the radical before squaring. Could we still have solved the equation? Why or why not?

53. Find a number such that 1 less than the square root of twice the number is 7.

54. Find a number such that the opposite of three times its square root is −33.

Sometimes the principle of squaring must be used more than once in order to solve an equation. Solve Exercises 55–62 by using the principle of squaring as often as necessary.

55. $5 - \sqrt{x} = \sqrt{x - 5}$

56. $1 + \sqrt{x} = \sqrt{x + 9}$

57. $\sqrt{t + 4} = 1 - \sqrt{3t + 1}$

58. $\sqrt{y + 8} - \sqrt{y} = 2$

59. $3 + \sqrt{19 - x} = 5 + \sqrt{4 - x}$

60. $\sqrt{y + 1} - \sqrt{y - 2} = \sqrt{2y - 5}$

61. $2\sqrt{x - 1} - \sqrt{x - 9} = \sqrt{3x - 5}$

62. $x + (2 - x)\sqrt{x} = 0$

63. *Changing elevations.*   A mountain climber pauses to rest and view the horizon. Using the formula $V = 3.5\sqrt{h}$, the climber computes the distance to the horizon and then climbs another 100 m. At this higher elevation, the horizon is 20 km farther than before. At what height was the climber when the first computation was made? (*Hint*: Use a system of equations.)

64. Solve $A = \sqrt{1 + \sqrt{a/b}}$ for b.

Graph. Use at least three ordered pairs.

65. $y = \sqrt{x}$

66. $y = \sqrt{x - 2}$

67. $y = \sqrt{x - 3}$

68. $y - \sqrt{x + 1}$

69. Graph $y = x - 7$ and $y = \sqrt{x - 5}$ using the same set of axes. Determine where the graphs intersect in order to estimate a solution of $x - 7 = \sqrt{x - 5}$.

70. Graph $y = 1 + \sqrt{x}$ and $y = \sqrt{x + 9}$ using the same set of axes. Determine where the graphs intersect in order to estimate a solution of $1 + \sqrt{x} = \sqrt{x + 9}$.

Use a grapher to solve Exercises 71 and 72. Round answers to the nearest hundredth when appropriate.

71. $\sqrt{x + 3} = 2x - 1$

72. $-\sqrt{x + 3} = 2x - 1$

73. In Example 4, it was stated that the sound was traveling at a rate of 1170 ft/sec. How could this speed be determined by someone in attendance at the concert?

COLLABORATIVE

CORNER

How Close Is Too Close?

Focus: Radical equations and problem solving

Time: 15 minutes

Group size: 2

Materials: Calculators

The faster a car is traveling, the more distance it needs to stop. Thus it is important for drivers to allow sufficient space between their vehicle and the vehicle in front of them. Police recommend that for each 10 miles per hour of speed, a driver allow 1 car length. Thus a driver traveling 30 mph should have at least 3 car lengths between his or her vehicle and the one in front.

The formula $r = 2\sqrt{5L}$ can be used to find the speed, in miles per hour, that a car was traveling when it left skid marks L feet long.

ACTIVITY

1. The group should estimate the average length of a car in which one of them frequently travels.
2. Using a calculator as needed, each group member should complete three rows in the table below. Column 1 lists a car's speed s, and column 2 lists the minimum amount of space between cars traveling s miles per hour, as recommended by police. Column 3 lists the speed that a vehicle *could* travel were it forced to stop in the distance listed in column 2, using the formula $r = 2\sqrt{5L}$.

Column 1 s (in miles per hour)	Column 2 L (in feet)	Column 3 r (in miles per hour)
20		
30		
40		
50		
60		
70		

3. The group should determine whether there are any speeds at which the "1 car length per 10 mph" guideline might not suffice. On what reasoning do you base your answer? Compare tables to determine how car length affects the results. What recommendations would your group make to a new driver?

Applications Using Right Triangles

8.6

Right Triangles • Problem Solving

Radicals frequently occur in problem-solving situations in which the Pythagorean theorem is used. In Section 5.7, when we first used the Pythagorean theorem, we had not yet studied square roots. We now know that if $x^2 = n$, then x is a square root of n.

Right Triangles

For convenience, we restate the Pythagorean theorem.

> ### The Pythagorean Theorem*
>
> In any right triangle, if a and b are the lengths of the legs and c is the length of the hypotenuse, then
>
> $$a^2 + b^2 = c^2.$$
>
> Hypotenuse
> c a Leg
> $90°$
> b
> Leg

When the Pythagorean theorem is used to find a length, we need not concern ourselves with negative square roots, since length cannot be negative.

E x a m p l e   1

Find the length of the hypotenuse of the triangle shown. Give an exact answer and an approximation to three decimal places.

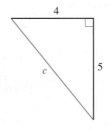

4

c 5

---

*Recall that the converse of the Pythagorean theorem also holds. That is, if a, b, and c are the lengths of the sides of a triangle and $a^2 + b^2 = c^2$, then the triangle is a right triangle.

Solution   We have

$$a^2 + b^2 = c^2$$
$$4^2 + 5^2 = c^2$$ Substituting the lengths of the legs
$$16 + 25 = c^2$$
$$41 = c^2.$$

We now use the fact that if $x^2 = n$, then $x = \sqrt{n}$ or $x = -\sqrt{n}$. In this case, since c is a length, it follows that c is the positive square root of 41:

$$c = \sqrt{41}$$ This is an exact answer.
$$c \approx 6.403.$$ Using a calculator or Table 2 for an approximation

Example 2

Find the length of the indicated leg in each triangle. In each case, give an exact answer and an approximation to three decimal places.

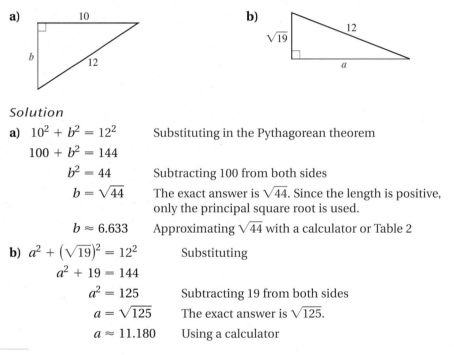

a) 10

b

12

b) $\sqrt{19}$

12

a

Solution

a) $10^2 + b^2 = 12^2$ Substituting in the Pythagorean theorem
$$100 + b^2 = 144$$
$$b^2 = 44$$ Subtracting 100 from both sides
$$b = \sqrt{44}$$ The exact answer is $\sqrt{44}$. Since the length is positive, only the principal square root is used.
$$b \approx 6.633$$ Approximating $\sqrt{44}$ with a calculator or Table 2

b) $a^2 + \left(\sqrt{19}\right)^2 = 12^2$ Substituting
$$a^2 + 19 = 144$$
$$a^2 = 125$$ Subtracting 19 from both sides
$$a = \sqrt{125}$$ The exact answer is $\sqrt{125}$.
$$a \approx 11.180$$ Using a calculator

Problem Solving

The five-step process and the Pythagorean theorem can be used for problem solving.

Example 3

Reach of a ladder.   A 32-ft ladder is leaning against a house. The bottom of the ladder is 7 ft from the house. How high is the top of the ladder? Give an exact answer and an approximation to three decimal places.

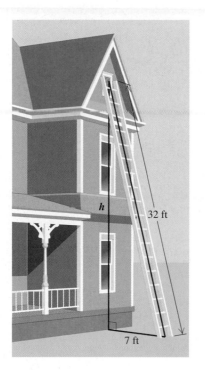

Solution

1. **Familiarize.** First we make a drawing. In it, there is a right triangle. We label the unknown height h.

2. **Translate.** We use the Pythagorean theorem, substituting 7 for a, h for b, and 32 for c:

$$7^2 + h^2 = 32^2.$$

3. **Carry out.** We solve the equation:

$$7^2 + h^2 = 32^2$$
$$49 + h^2 = 1024$$
$$h^2 - 975$$
$$h = \sqrt{975} \qquad \text{This answer is exact.}$$
$$h \approx 31.225. \qquad \text{Approximating with a calculator}$$

4. **Check.** We check by substituting 7, $\sqrt{975}$, and 32:

$$\frac{a^2 + b^2 = c^2}{7^2 + \left(\sqrt{975}\right)^2 \ ? \ 32^2}$$

$$\begin{array}{c|c} 49 + 975 & 1024 \\ 1024 & 1024 \quad \text{TRUE} \end{array}$$

5. **State.** The top of the ladder is $\sqrt{975}$, or about 31.225 ft from the ground.

E x a m p l e 4

Softball dimensions.   A softball diamond is a square 65 ft on a side. How far is it from home plate to second base? Give an exact answer and an approximation to three decimal places.

Solution

1. **Familiarize.**   We first make a drawing. Note that the first- and second-base lines, together with a line from home to second, form a right triangle. We label the unknown distance d.

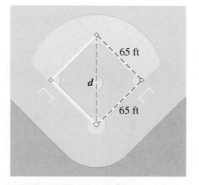

2. **Translate.**   We substitute 65 for a, 65 for b, and d for c in the Pythagorean theorem:

$$65^2 + 65^2 = d^2.$$

3. Carry out. We solve the equation:

$$65^2 + 65^2 = d^2$$
$$4225 + 4225 = d^2$$
$$8450 = d^2$$
$$\sqrt{8450} = d \qquad \text{This is exact.}$$
$$91.924 \approx d. \qquad \text{This is approximate.}$$

4. Check. We check by substituting 65, 65, and $\sqrt{8450}$:

$$\frac{a^2 + b^2 = c^2}{}$$

$$65^2 + 65^2 \overset{?}{} \left(\sqrt{8450}\right)^2$$

$4225 + 4225$	8450
8450	$8450 \qquad$ TRUE

5. State. From home plate to second base is $\sqrt{8450}$, or about 91.924 ft.

FOR EXTRA HELP

Exercise Set 8.6

Digital Video Tutor CD 5
Videotape 16

InterAct Math

Math Tutor Center

MathXL

MyMathLab.com

Find the length of the third side of each triangle. Give an exact answer and, where appropriate, an approximation to three decimal places.

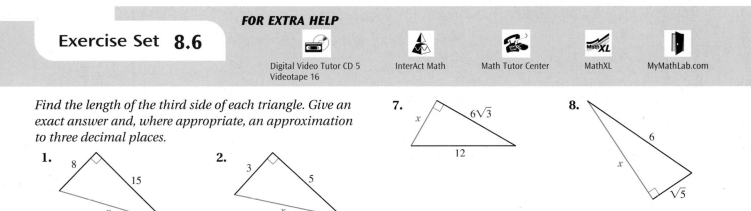

1. 8, 15, x

2. 3, 5, x

3. x, 6, 6

4. 7, 7, x

5. x, 13, 5

6. 13, x, 12

7. x, $6\sqrt{3}$, 12

8. 6, x, $\sqrt{5}$

In a right triangle, find the length of the side not given. Give an exact answer and, where appropriate, an approximation to three decimal places. Keep in mind that a and b are the lengths of the legs and c is the length of the hypotenuse.

9. $a = 12, b = 5$ **10.** $a = 24, b = 10$

11. $a = 18, c = 30$ **12.** $a = 9, c = 15$

13. $b = 1, c = \sqrt{5}$ **14.** $b = 1, c = \sqrt{2}$

15. $a = 1, c = \sqrt{3}$ **16.** $a = \sqrt{3}, b = \sqrt{5}$

17. $c = 10, b = 5\sqrt{3}$ **18.** $a = 5, b = 5$

Solve. Don't forget to make drawings. Give an exact answer and an approximation to three decimal places.

19. *Roller blading.* Lamar and Nanci are building a roller blade jump with a base that is 30 in. long and a ramp that is 33 in. long. How high will the back of the jump be?

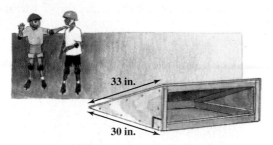

20. *Masonry.* Find the length of a diagonal of a square tile that has sides 4 cm long.

21. *Plumbing.* A new water pipe is being prepared so that it will run diagonally under a kitchen floor. If the kitchen is 8 ft wide and 12 ft long, how long should the pipe be?

22. *Guy wires.* How long must a guy wire be to reach from the top of a 13-m telephone pole to a point on the ground 9 m from the foot of the pole?

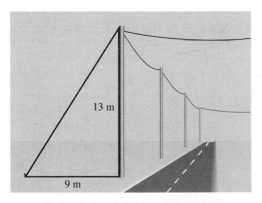

23. *Wiring.* JR Electric is installing a security system with a wire that will run diagonally over the "drop" ceiling of a 10-ft by 16-ft room. How long will that section of wire need to be?

24. *Soccer fields.* The smallest regulation soccer field is 50 yd wide and 100 yd long. Find the length of a diagonal of such a field.

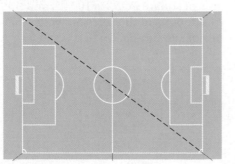

25. *Soccer fields.* The largest regulation soccer field is 100 yd wide and 130 yd long. Find the length of a diagonal of such a field.

26. *Baseball.* A baseball diamond is a square 90 ft on a side. How far is it from first base to third base?

27. *Lacrosse.* A regulation lacrosse field is 60 yd wide and 110 yd long. Find the length of a diagonal of such a field.

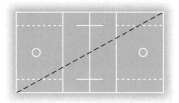

28. *Surveying.* A surveyor had poles located at points *P*, *Q*, and *R*. The distances that she was able to measure are marked in the figure. What is the approximate length of the lake?

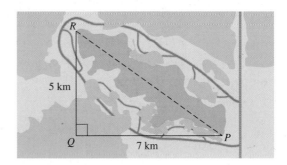

29. In an *equilateral triangle,* all sides have the same length. Can a right triangle be equilateral? Why or why not?

30. In an *isosceles triangle,* two sides have the same length. Can a right triangle be isosceles? Why or why not?

SKILL MAINTENANCE

For Exercises 31–34, consider the following list:

$$-45, -9.7, -\sqrt{5}, 0, \tfrac{2}{7}, \sqrt{7}, \pi, 5.09, 19.$$

31. List all rational numbers.

32. List all irrational numbers.

33. List all real numbers.

34. List all integers.

Simplify.

35. $(-2)^5$

36. $\left(\dfrac{5}{3}\right)^2$

37. $(2x)^3$

38. $(-2x)^3$

SYNTHESIS

39. Should a homeowner use a 28-ft ladder to repair clapboard that is 27 ft above ground level? Why or why not?

40. Can the length of a triangle's hypotenuse ever equal the combined lengths of the two legs? Why or why not?

41. *Cordless telephones.* The Panasonic KXTG-2500B cordless phone has a range of one quarter mile. Vance has a corner office in the Empire State Building, 900 ft above street level. Can Vance locate the Panasonic's phone base in his office and use the handset at a restaurant at street level on the opposite corner? Use the figure below and show your work.

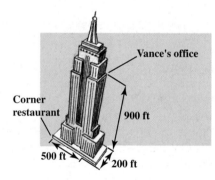

42. *Cordless telephones.* Virginia's AT&T 9002 cordless phone has a range of 1000 ft. Her apartment is a corner unit, located as shown in the figure below. Will Virginia be able to use the phone at the community pool?

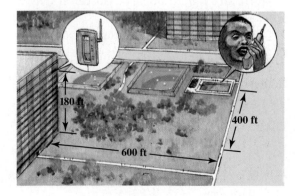

43. *Aviation.* A pilot is instructed to descend from 32,000 ft to 21,000 ft over a horizontal distance of 5 mi. What distance will the plane travel during this descent?

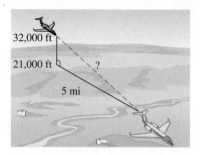

44. The diagonal of a square has a length of $8\sqrt{2}$ ft. Find the length of a side of the square.

45. Find the length of a side of a square that has an area of 7 m^2.

46. A right triangle has sides with lengths that are consecutive even integers. Find the lengths of the sides.

47. Find the length of the diagonal of a cube with sides of length s.

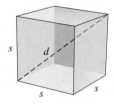

48. Figure $ABCD$ is a square. Find the length of a diagonal, $\overline{AC}$.

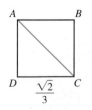

49. The area of square $PQRS$ is 100 ft^2, and A, B, C, and D are midpoints of the sides on which they lie. Find the area of square $ABCD$.

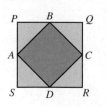

50. Express the height h of an equilateral triangle in terms of the length of a side a.

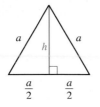

51. *Racquetball.*   A racquetball court is 20 ft by 20 ft by 40 ft. What is the longest straight-line distance that can be measured in this racquetball court?

52. *Distance driven.*   Two cars leave a service station at the same time. One car travels east at a speed of 50 mph, and the other travels south at a speed of 60 mph. After one half hour, how far apart are they?

53. Solve for x.

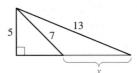

54. *Ranching.*   If 2 mi of fencing encloses a square plot of land with an area of 160 acres, how large a square, in acres, will 4 mi of fencing enclose?

CORNER

Pythagorean Triples

Focus: Pythagorean theorem

Time: 15 minutes

Group size: 2–4

Materials: Tape measure and chalk; string and scissors

We mentioned in the footnote on p. 475 that the converse of the Pythagorean theorem is also true: If a, b, and c are the lengths of the sides of a triangle and $a^2 + b^2 = c^2$, then the triangle is a right triangle. Each such set of three numbers is called a *Pythagorean triple.* Since $3^2 + 4^2 = 5^2$ and $5^2 + 12^2 = 13^2$, then (3, 4, 5) and (5, 12, 13) are both Pythagorean triples. Such numbers provide carpenters, masons, archaeologists, and others with a handy way of locating a line that forms a 90° angle with another line.

ACTIVITY

1. Suppose that the group is in the process of building a deck. They determine that they need to position an 8-ft piece of lumber so that it forms a 90° angle with a wall. Use a tape measure, chalk, Pythagorean triples, and—if desired—string and scissors to construct a right angle at a specified point at the base of some wall in or near your classroom.

2. To check that the angle formed in part (1) is truly 90°, repeat the procedure using a different Pythagorean triple.

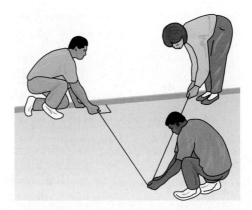

Higher Roots and Rational Exponents

8.7

Higher Roots • Products and Quotients Involving Higher Roots • Rational Exponents • Calculators

In this section, we study *higher* roots, such as cube roots or fourth roots, and exponents that are not integers.

Higher Roots

Recall that c is a square root of a if $c^2 = a$. A similar definition exists for *cube roots*.

> ### Cube Root
> The number c is the *cube root* of a if $c^3 = a$.

The symbolism $\sqrt[3]{a}$ is used to represent the cube root of a. In the radical $\sqrt[3]{a}$, the number 3 is called the **index**, and a is called the **radicand**.

E x a m p l e   1   Find the cube root of each number: **(a)** 8; **(b)** -125.

Solution

a) The cube root of 8 is the number whose cube is 8. Since $2^3 = 2 \cdot 2 \cdot 2 = 8$, the cube root of 8 is 2: $\sqrt[3]{8} = 2$.

b) The cube root of -125 is the number whose cube is -125. Since $(-5)^3 = (-5)(-5)(-5) = -125$, the cube root of -125 is -5: $\sqrt[3]{-125} = -5$.

> ### nth Root
> The number c is an *n*th *root* of a if $c^n = a$.
>
> If n is odd, there is only one *n*th root and $\sqrt[n]{a}$ represents that root.
>
> If n is even, there are two *n*th roots, and $\sqrt[n]{a}$ represents the non-negative *n*th root.
>
> Even roots of negative numbers are not real numbers.

Positive numbers always have *two* *n*th roots when n is even but when we refer to *the* *n*th root of a positive number a, denoted $\sqrt[n]{a}$, we mean the *positive* *n*th root. Thus, although -3 and 3 are both fourth roots of 81 (since $(-3)^4 = 3^4 = 81$), 3 is considered *the* fourth root of 81. In symbols:

$$\sqrt[4]{81} = 3.$$

E x a m p l e   2   Find each root: **(a)** $\sqrt[4]{16}$; **(b)** $\sqrt[5]{-32}$; **(c)** $\sqrt[4]{-16}$; **(d)** $-\sqrt[3]{64}$.

Solution

a) $\sqrt[4]{16} = 2$ Since $2^4 = 2 \cdot 2 \cdot 2 \cdot 2 = 16$

b) $\sqrt[5]{-32} = -2$ Since $(-2)^5 = (-2)(-2)(-2)(-2)(-2) = -32$

c) $\sqrt[4]{-16}$ is not a real number, because it is an even root of a negative number.

d) $-\sqrt[3]{64} = -\left(\sqrt[3]{64}\right)$ This is the opposite of $\sqrt[3]{64}$.

$\qquad\quad = -4$ $\qquad 4^3 = 4 \cdot 4 \cdot 4 = 64$

Some roots occur so frequently that you may want to memorize them.

Square Roots		Cube Roots	Fourth Roots	Fifth Roots
$\sqrt{1} = 1$	$\sqrt{4} = 2$	$\sqrt[3]{1} = 1$	$\sqrt[4]{1} = 1$	$\sqrt[5]{1} = 1$
$\sqrt{9} = 3$	$\sqrt{16} = 4$	$\sqrt[3]{8} = 2$	$\sqrt[4]{16} = 2$	$\sqrt[5]{32} = 2$
$\sqrt{25} = 5$	$\sqrt{36} = 6$	$\sqrt[3]{27} = 3$	$\sqrt[4]{81} = 3$	$\sqrt[5]{243} = 3$
$\sqrt{49} = 7$	$\sqrt{64} = 8$	$\sqrt[3]{64} = 4$	$\sqrt[4]{256} = 4$	
$\sqrt{81} = 9$	$\sqrt{100} = 10$	$\sqrt[3]{125} = 5$	$\sqrt[4]{625} = 5$	
$\sqrt{121} = 11$	$\sqrt{144} = 12$	$\sqrt[3]{216} = 6$		

Products and Quotients Involving Higher Roots

The rules for working with products and quotients of square roots can be extended to products and quotients of *n*th roots. Prime factorizations can again be useful when no simplification is readily apparent.

> **The Product and Quotient Rules**
>
> $$\sqrt[n]{AB} = \sqrt[n]{A}\,\sqrt[n]{B}, \qquad \sqrt[n]{\dfrac{A}{B}} = \dfrac{\sqrt[n]{A}}{\sqrt[n]{B}}$$

E x a m p l e   3   Simplify: **(a)** $\sqrt[3]{40}$; **(b)** $\sqrt[3]{\dfrac{125}{27}}$; **(c)** $\sqrt[4]{1250}$; **(d)** $\sqrt[5]{\dfrac{2}{243}}$.

Solution

a) $\sqrt[3]{40} = \sqrt[3]{8 \cdot 5}$ Note that $40 = 2 \cdot 2 \cdot 2 \cdot 5 = 8 \cdot 5$ and 8 is a perfect cube.

$\qquad\quad = \sqrt[3]{8} \cdot \sqrt[3]{5}$

$\qquad\quad = 2\sqrt[3]{5}$

b) $\sqrt[3]{\dfrac{125}{27}} = \dfrac{\sqrt[3]{125}}{\sqrt[3]{27}} = \dfrac{5}{3}$ $125 = 5 \cdot 5 \cdot 5$ and $27 = 3 \cdot 3 \cdot 3$, so 125 and 27 are perfect cubes. See the chart on p. 483.

c) $\sqrt[4]{1250} = \sqrt[4]{625 \cdot 2}$ Note that $1250 = 2 \cdot 5 \cdot 5 \cdot 5 \cdot 5 = 2 \cdot 625$ and 625 is a perfect fourth power.

$ \quad = \sqrt[4]{625} \cdot \sqrt[4]{2}$

$ \quad = 5\sqrt[4]{2}$

d) $\sqrt[5]{\dfrac{2}{243}} = \dfrac{\sqrt[5]{2}}{\sqrt[5]{243}}$ $243 = 3 \cdot 3 \cdot 3 \cdot 3 \cdot 3$, so 243 is a perfect fifth power. See the chart on p. 483.

$ \quad = \dfrac{\sqrt[5]{2}}{3}$

Rational Exponents

Expressions containing rational exponents, like $8^{1/3}$, $4^{5/2}$, and $81^{-3/4}$, are defined in a manner that ensures that the laws of exponents still hold. For example, if the product rule, $a^m \cdot a^n = a^{m+n}$, is to hold, then

$$a^{1/2} \cdot a^{1/2} = a^{1/2+1/2}$$
$$= a^1 = a.$$

This says that the square of $a^{1/2}$ is a, which suggests that $a^{1/2}$ is a square root of a:

$$a^{1/2} \quad \text{means} \quad \sqrt{a}.$$

This idea is generalized as follows.

> **The Exponent $1/n$**
> $a^{1/n}$ means $\sqrt[n]{a}$. If a is negative, then the index n must be odd.
> (Note that $a^{1/2}$ is written $\sqrt{a}$, in which the index 2 is understood.)

Example 4   Simplify: **(a)** $8^{1/3}$; **(b)** $100^{1/2}$; **(c)** $81^{1/4}$; **(d)** $(-243)^{1/5}$.

Solution

a) $8^{1/3} = \sqrt[3]{8} = 2$

b) $100^{1/2} = \sqrt{100} = 10$

c) $81^{1/4} = \sqrt[4]{81} = 3$

d) $(-243)^{1/5} = \sqrt[5]{-243} = -3$

If we still wish to multiply exponents when raising a power to a power, we must have $a^{2/3} = (a^{1/3})^2$ and $a^{2/3} = (a^2)^{1/3}$. This suggests both that $a^{2/3} = \left(\sqrt[3]{a}\right)^2$ and that $a^{2/3} = \sqrt[3]{a^2}$.

> **Positive Rational Exponents**
>
> For any natural numbers m and n ($n \neq 1$) and any real number a for which $\sqrt[n]{a}$ exists,
>
> $a^{m/n}$ means $\left(\sqrt[n]{a}\right)^m$, or equivalently, $a^{m/n}$ means $\sqrt[n]{a^m}$.

In most cases, it is easiest to simplify using $\left(\sqrt[n]{a}\right)^m$.

E x a m p l e 5   Simplify: (a) $27^{2/3}$; (b) $8^{5/3}$; (c) $81^{3/4}$.

Solution

a) $27^{2/3} = (27^{1/3})^2 = \left(\sqrt[3]{27}\right)^2 = 3^2 = 9$

b) $8^{5/3} = (8^{1/3})^5 = \left(\sqrt[3]{8}\right)^5 = 2^5 = 32$

c) $81^{3/4} = (81^{1/4})^3 = \left(\sqrt[4]{81}\right)^3 = 3^3 = 27$

Negative rational exponents are defined in much the same way that negative integer exponents are.

> **Negative Rational Exponents**
>
> For any rational number m/n and any nonzero real number a for which $a^{m/n}$ exists,
>
> $$a^{-m/n} = \frac{1}{a^{m/n}}.$$

E x a m p l e 6   Simplify: (a) $16^{-1/2}$; (b) $27^{-1/3}$; (c) $32^{-2/5}$; (d) $64^{-3/2}$.

Solution

a) $16^{-1/2} = \dfrac{1}{16^{1/2}} = \dfrac{1}{\sqrt{16}} = \dfrac{1}{4}$

b) $27^{-1/3} = \dfrac{1}{27^{1/3}} = \dfrac{1}{\sqrt[3]{27}} = \dfrac{1}{3}$

c) $32^{-2/5} = \dfrac{1}{32^{2/5}} = \dfrac{1}{(32^{1/5})^2} = \dfrac{1}{\left(\sqrt[5]{32}\right)^2} = \dfrac{1}{2^2} = \dfrac{1}{4}$

d) $64^{-3/2} = \dfrac{1}{64^{3/2}} = \dfrac{1}{\left(\sqrt{64}\right)^3} = \dfrac{1}{8^3} = \dfrac{1}{512}$

> ***Caution!***   A negative exponent does not indicate that the expression in which it appears is negative.

Calculators

A calculator with a key for finding powers can be used to approximate numbers like $\sqrt[5]{8}$. Generally such keys are labeled $\boxed{x^y}$, $\boxed{a^x}$, or $\boxed{\wedge}$. We find approximations by entering the radicand, pressing the power key, entering the exponent, and then pressing $\boxed{=}$ or $\boxed{\text{ENTER}}$. Thus, $\sqrt[5]{8}$ can be approximated by entering 8, pressing the power key, entering 0.2 or $(1 \div 5)$, and pressing $\boxed{=}$ to get $\sqrt[5]{8} \approx 1.515716567$. Consult an owner's manual or your instructor if your calculator works differently.

Exercise Set 8.7

FOR EXTRA HELP

Digital Video Tutor CD 5
Videotape 16

InterAct Math

Math Tutor Center

MathXL

MyMathLab.com

Simplify. If an expression does not represent a real number, state this.

1. $\sqrt[3]{-64}$ **2.** $\sqrt[3]{-8}$ **3.** $\sqrt[3]{-125}$

4. $\sqrt[3]{-27}$ **5.** $\sqrt[3]{1000}$ **6.** $\sqrt[3]{8}$

7. $-\sqrt[3]{216}$ **8.** $-\sqrt[3]{-343}$ **9.** $\sqrt[4]{81}$

10. $\sqrt[4]{625}$ **11.** $\sqrt[5]{0}$ **12.** $\sqrt[5]{1}$

13. $\sqrt[5]{-1}$ **14.** $\sqrt[5]{-243}$ **15.** $\sqrt[4]{-81}$

16. $\sqrt[4]{-1}$ *Aha!* **17.** $\sqrt[4]{10,000}$ **18.** $\sqrt[5]{100,000}$

19. $\sqrt[3]{7^3}$ **20.** $\sqrt[4]{5^4}$ **21.** $\sqrt[6]{64}$

22. $\sqrt[6]{1}$ *Aha!* **23.** $\sqrt[9]{a^9}$ **24.** $\sqrt[3]{n^3}$

25. $\sqrt[3]{54}$ **26.** $\sqrt[3]{32}$ **27.** $\sqrt[4]{48}$

28. $\sqrt[5]{160}$ **29.** $\sqrt[3]{\dfrac{64}{125}}$ **30.** $\sqrt[3]{\dfrac{125}{27}}$

31. $\sqrt[5]{\dfrac{32}{243}}$ **32.** $\sqrt[4]{\dfrac{625}{256}}$ **33.** $\sqrt[3]{\dfrac{7}{8}}$

34. $\sqrt[5]{\dfrac{17}{32}}$ **35.** $\sqrt[4]{\dfrac{14}{81}}$ **36.** $\sqrt[3]{\dfrac{10}{27}}$

Simplify.

37. $25^{1/2}$ **38.** $9^{1/2}$ **39.** $125^{1/3}$

40. $1000^{1/3}$ **41.** $32^{1/5}$ **42.** $16^{1/4}$

43. $16^{3/4}$ **44.** $8^{4/3}$ **45.** $9^{5/2}$

46. $4^{3/2}$ **47.** $64^{2/3}$ **48.** $32^{2/5}$

49. $8^{5/3}$ **50.** $16^{5/4}$ **51.** $4^{5/2}$

52. $36^{-1/2}$ **53.** $25^{-1/2}$ **54.** $32^{-1/5}$

55. $256^{-1/4}$ **56.** $100^{-3/2}$ **57.** $16^{-3/4}$

58. $81^{-3/4}$ **59.** $81^{-5/4}$ **60.** $32^{-2/5}$

61. $8^{-2/3}$ **62.** $625^{-3/4}$

63. Expressions of the form $a^{m/n}$ can be rewritten as $(\sqrt[n]{a})^m$ or $\sqrt[n]{a^m}$. Which radical expression would you use when simplifying $25^{3/2}$ and why?

64. Explain in your own words why $\sqrt[n]{a}$ is negative when n is odd and a is negative.

SKILL MAINTENANCE

Solve.

65. $x^2 - 5x - 6 = 0$ **66.** $x^2 + 4x - 5 = 0$

67. $4t^2 - 9 = 0$ **68.** $25t^2 - 4 = 0$

69. $3x^2 + 8x + 4 = 0$ **70.** $5x^2 + 13x - 6 = 0$

SYNTHESIS

71. If $a > b$, does it follow that $a^{1/n} > b^{1/n}$? Why or why not?

72. Under what condition(s) will $a^{-3/5}$ be negative?

Using a calculator, approximate each of the following to three decimal places.

73. $10^{4/5}$ **74.** $24^{1/4}$

75. $36^{3/8}$ **76.** $10^{3/2}$

Simplify.

77. $(x^{2/3})^{7/3}$

78. $a^{1/4}a^{3/2}$

79. $\dfrac{p^{5/6}}{p^{2/3}}$

80. $m^{-2/3}m^{1/4}m^{3/2}$

Graph.

81. $y = \sqrt[3]{x}$

82. $y = \sqrt[4]{x}$

83. Use a grapher to draw the graphs of $y_1 = x^{2/3}$, $y_2 = x^1$, $y_3 = x^{5/4}$, and $y_4 = x^{3/2}$. Use the window $[-1, 17, -1, 32]$ and the SIMULTANEOUS mode. Then determine which curve corresponds to each equation.

Summary and Review 8

Key Terms

Square root, p. 446
Principal square root, p. 446
Radical sign, p. 446
Radical expression, p. 447
Radicand, p. 447
Irrational number, p. 447
Perfect square, p. 447

Rationalize the denominator, p. 460
Like radicals, p. 463
Conjugates, p. 465
Radical equation, p. 467
Principle of squaring, p. 467
Pythagorean theorem, p. 475

Higher root, p. 482
Cube root, p. 482
Index, p. 482
nth root, p. 482
Rational exponent, p. 484

Important Properties and Formulas

For any real number A, $\sqrt{A^2} = |A|$.   For $A \ge 0$, $\sqrt{A^2} = A$.

The Product Rule: $\sqrt{A}\sqrt{B} = \sqrt{AB}$; $\sqrt[n]{AB} = \sqrt[n]{A}\sqrt[n]{B}$

The Quotient Rule: $\dfrac{\sqrt{A}}{\sqrt{B}} = \sqrt{\dfrac{A}{B}}$; $\sqrt[n]{\dfrac{A}{B}} = \dfrac{\sqrt[n]{A}}{\sqrt[n]{B}}$

Simplified Form of a Square Root

A radical expression for a square root is simplified when its radicand has no factor other than 1 that is a perfect square.

The principle of squaring: If $a = b$, then $a^2 = b^2$.

To solve a radical equation:

1. Isolate a radical term.
2. Use the principle of squaring (square both sides).
3. Solve the new equation.
4. Check all possible solutions in the original equation.

The Pythagorean theorem: $a^2 + b^2 = c^2$, where a and b are the lengths of the legs of a right triangle and c is the length of the hypotenuse.

The exponent $1/n$: $a^{1/n} = \sqrt[n]{a}$

Rational exponents: $a^{m/n} = \left(\sqrt[n]{a}\right)^m = \sqrt[n]{a^m}$;

$$a^{-m/n} = \frac{1}{a^{m/n}}$$

Review Exercises

Find the square roots of each number.

1. 25

2. 196

3. 900

4. 225

Simplify.

5. $-\sqrt{144}$

6. $\sqrt{81}$

7. $\sqrt{49}$

8. $-\sqrt{169}$

Identify each radicand.

9. $5x\sqrt{2x^2y}$

10. $a\sqrt{\dfrac{a}{b}}$

Determine whether each square root is rational or irrational.

11. $-\sqrt{36}$

12. $-\sqrt{12}$

13. $\sqrt{18}$

14. $\sqrt{25}$

Use a calculator or Table 2 to approximate each square root. Round to three decimal places.

15. $\sqrt{3}$

16. $\sqrt{99}$

Simplify. Assume for Exercises 17–37 that all variables represent nonnegative numbers.

17. $\sqrt{p^2}$

18. $\sqrt{(7x)^2}$

19. $\sqrt{16m^2}$

20. $\sqrt{(ac)^2}$

Simplify by factoring.

21. $\sqrt{48}$

22. $\sqrt{98t^2}$

23. $\sqrt{32p}$

24. $\sqrt{x^6}$

25. $\sqrt{12a^{13}}$

26. $\sqrt{36m^{15}}$

Multiply and, if possible, simplify.

27. $\sqrt{3}\,\sqrt{11}$

28. $\sqrt{6}\,\sqrt{10}$

29. $\sqrt{5x}\,\sqrt{7t}$

30. $\sqrt{3a}\,\sqrt{8a}$

31. $\sqrt{5x}\,\sqrt{10xy^2}$

32. $\sqrt{20a^3b}\,\sqrt{5a^2b^2}$

Simplify.

33. $\dfrac{\sqrt{35}}{\sqrt{45}}$

34. $\dfrac{\sqrt{30y^9}}{\sqrt{54y}}$

35. $\sqrt{\dfrac{25}{64}}$

36. $\sqrt{\dfrac{20}{45}}$

37. $\sqrt{\dfrac{49}{t^2}}$

38. $10\sqrt{5} + 3\sqrt{5}$

39. $\sqrt{80} - \sqrt{45}$

40. $2\sqrt{x} - \sqrt{25x}$

41. $\left(2 + \sqrt{3}\right)^2$

42. $\left(2 + \sqrt{3}\right)\left(2 - \sqrt{3}\right)$

43. $\left(1 + 2\sqrt{7}\right)\left(3 - \sqrt{7}\right)$

Rationalize the denominator.

44. $\sqrt{\dfrac{1}{2}}$

45. $\dfrac{\sqrt{5}}{\sqrt{8}}$

46. $\sqrt{\dfrac{5}{y}}$

47. $\dfrac{2}{\sqrt{3}}$

48. $\dfrac{4}{2 + \sqrt{3}}$

49. $\dfrac{1 + \sqrt{5}}{2 - \sqrt{5}}$

Solve.

50. $\sqrt{x - 3} = 7$

51. $\sqrt{5x + 3} = \sqrt{2x - 1}$

52. $\sqrt{x + 5} = x - 1$

53. $1 + x = \sqrt{1 + 5x}$

54. *Speed of a skidding car.*   The formula $r = 2\sqrt{5L}$ can be used to approximate the speed r, in miles per hour, of a car that has left a skid mark of length L, in feet. How far will a car skid at a speed of 90 mph?

In a right triangle, find the length of the side not given. Give an exact answer and, where appropriate, an approximation to three decimal places. Keep in mind that a and b are the lengths of the legs and c is the length of the hypotenuse.

55. $a = 15, c = 25$

56. $a = 1, b = \sqrt{2}$

57. *Guy wires.*   One wire steadying a radio tower stretches from a point 100 ft high on the tower to a point on the ground 25 ft from the base of the tower. How long is the wire?

100 ft

25 ft

Simplify. If an expression does not represent a real number, state this.

58. $\sqrt[5]{32}$

59. $\sqrt[4]{-16}$

60. $\sqrt[3]{-27}$

61. $\sqrt[4]{32}$

Simplify.

62. $100^{1/2}$

63. $9^{1/2}$

64. $16^{3/2}$

65. $81^{-3/4}$

SYNTHESIS

66. Jesse automatically eliminates any possible solutions of a radical equation that are negative. What mistake is he making?

67. Why should you simplify each term in a radical expression before attempting to combine like radical terms?

68. Simplify: $\sqrt{\sqrt{\sqrt{256}}}$.

69. Solve: $\sqrt{x^2} = -10$.

70. Use square roots to factor $x^2 - 5$.

71. Solve $A = \sqrt{a^2 + b^2}$ for b.

Chapter Test 8

1. Find the square roots of 81.

Simplify.

2. $\sqrt{64}$

3. $-\sqrt{25}$

4. Identify the radicand in $3\sqrt{x + 4}$.

Determine whether each square root is rational or irrational.

5. $\sqrt{10}$

6. $\sqrt{16}$

Approximate using a calculator or Table 2. Round to three decimal places.

7. $\sqrt{87}$

8. $\sqrt{7}$

Simplify. Assume that a, y ≥ 0.

9. $\sqrt{a^2}$

10. $\sqrt{36y^2}$

Simplify by factoring. Assume that all variables represent nonnegative numbers.

11. $\sqrt{24}$

12. $\sqrt{27x^6}$

13. $\sqrt{4t^5}$

Perform the indicated operation and, if possible, simplify.

14. $\sqrt{5}\,\sqrt{6}$ **15.** $\sqrt{5}\,\sqrt{10}$

16. $\sqrt{7x}\,\sqrt{2y}$ **17.** $\sqrt{2t}\,\sqrt{8t}$

18. $\sqrt{3ab}\,\sqrt{6ab^3}$ **19.** $\dfrac{\sqrt{18}}{\sqrt{32}}$

20. $\dfrac{\sqrt{35x}}{\sqrt{80xy^2}}$ **21.** $\sqrt{\dfrac{27}{12}}$

22. $\sqrt{\dfrac{144}{a^2}}$ **23.** $3\sqrt{18} - 5\sqrt{18}$

24. $\sqrt{27} + 2\sqrt{12}$ **25.** $\left(4 - \sqrt{5}\right)^2$

26. $\left(4 - \sqrt{5}\right)\left(4 + \sqrt{5}\right)$

Rationalize each denominator.

27. $\sqrt{\dfrac{2}{5}}$ **28.** $\dfrac{2x}{\sqrt{y}}$

29. $\dfrac{10}{4 - \sqrt{5}}$

30. The legs of a right triangle are 8 cm and 4 cm long. Find the length of the hypotenuse. Give an exact answer and an approximation to three decimal places.

Solve.

31. $\sqrt{3x} + 2 = 14$ **32.** $\sqrt{6x + 13} = x + 3$

33. Valerie calculates that she can see 247.49 km to the horizon from an airplane window. How high is the airplane? Use the formula $V = 3.5\sqrt{h}$, where h is the altitude, in meters, and V is the distance to the horizon, in kilometers.

Simplify. If an expression does not represent a real number, state this.

34. $\sqrt[4]{16}$ **35.** $-\sqrt[6]{1}$ **36.** $\sqrt[3]{-64}$

37. $\sqrt[4]{-81}$ **38.** $9^{1/2}$ **39.** $27^{-1/3}$

40. $100^{3/2}$ **41.** $16^{-5/4}$

SYNTHESIS

42. Solve: $\sqrt{1 - x} + 1 = \sqrt{6 - x}$.

43. Simplify: $\sqrt{y^{16n}}$.

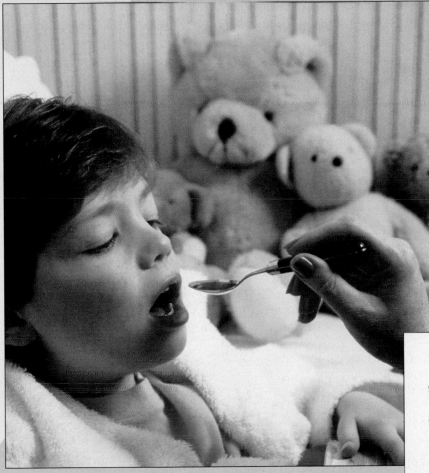

9
Quadratic Equations

AN APPLICATION

Young's rule for determining the size of a particular child's medicine dosage c is

$$c = \frac{a}{a + 12} \cdot d,$$

where a is the child's age and d is the typical adult dosage (*Source*: Olsen, June Looby, Leon J. Ablon, and Anthony Patrick Giangrasso, *Medical Dosage Calculations*, 6th ed.) If a child receives 8 mg of antihistamine when the typical adult dosage is 24 mg, how old is the child?

This problem appears as Exercise 65 in Section 9.4.

*N*urses dispense medicine on a regular basis. Since medicine often comes in a dosage different from what a doctor prescribes, I must calculate the correct dosage. Math also helps in critical thinking, such as determining dosage on the basis of body weight.

RENA J. WEBSTER
Registered Nurse
Indianapolis, Indiana

*Q*uadratic equations first appeared in Section 5.6. At that time, we used the principle of zero products because all of the equations could be solved by factoring. In this chapter, we will learn methods for solving any quadratic equation. These methods are then used in applications and in graphing.

Solving Quadratic Equations: The Principle of Square Roots

9.1

The Principle of Square Roots • Solving Quadratic Equations of the Type $(x + k)^2 = p$

The following are examples of quadratic equations:

$$x^2 - 7x + 9 = 0, \qquad 5t^2 - 4t = 8, \qquad 6y^2 = -9y, \qquad m^2 = 49.$$

We saw in Chapter 5 that one way to solve an equation like $m^2 = 49$ is to subtract 49 from both sides, factor, and then use the principle of zero products:

$$m^2 - 49 = 0$$

$$(m + 7)(m - 7) = 0$$

$$m + 7 = 0 \quad or \quad m - 7 = 0$$

$$m = -7 \quad or \quad m = 7.$$

This approach relies on our ability to factor. By using the *principle of square roots*, we can develop a method for solving equations like $m^2 = 49$ that allows us to solve equations when factoring is impractical.

The Principle of Square Roots

It is possible to solve $m^2 = 49$ by using the definition of square root: If $c^2 = a$, then c is a square root of a. Thus if $m^2 = 49$, then m is a square root of 49, namely, -7 or 7. This approach was used to solve right triangles in Section 8.6, but there only positive square roots appeared, since length is never negative.

> ### *The Principle of Square Roots*
>
> For any nonnegative real number k, if $x^2 = k$, then $x = \sqrt{k}$ or $x = -\sqrt{k}$.

E x a m p l e   1

Solve: $x^2 = 16$.

Solution   We use the principle of square roots:

$$x^2 = 16$$
$$x = \sqrt{16} \quad or \quad x = -\sqrt{16} \qquad \text{Using the principle of square roots}$$
$$x = 4 \qquad or \quad x = -4. \qquad \text{Simplifying}$$

We check mentally that $4^2 = 16$ and $(-4)^2 = 16$. The solutions are 4 and -4.

Unlike the principle of zero products, the principle of square roots can be used to solve quadratic equations that have irrational solutions.

E x a m p l e   2

Solve: **(a)** $x^2 = 17$; **(b)** $5x^2 = 15$; **(c)** $-3x^2 + 7 = 0$.

Solution

a)
$$x^2 = 17$$
$$x = \sqrt{17} \quad or \quad x = -\sqrt{17} \qquad \text{Using the principle of square roots}$$

Check: For $\sqrt{17}$:

$$\frac{x^2 = 17}{(\sqrt{17})^2 \overset{?}{\,} 17}$$
$$17 \mid 17 \;\; \text{TRUE}$$

For $-\sqrt{17}$:

$$\frac{x^2 = 17}{(-\sqrt{17})^2 \overset{?}{\,} 17}$$
$$17 \mid 17 \;\; \text{TRUE}$$

The solutions are $\sqrt{17}$ and $-\sqrt{17}$.

b)
$$5x^2 = 15$$
$$x^2 = 3 \qquad \text{Dividing both sides by 5 to isolate } x^2$$
$$x = \sqrt{3} \quad or \quad x = -\sqrt{3} \qquad \text{Using the principle of square roots}$$

We leave the check to the student. The solutions are $\sqrt{3}$ and $-\sqrt{3}$.

c)
$$-3x^2 + 7 = 0$$
$$7 = 3x^2 \qquad \text{Adding } 3x^2 \text{ to both sides}$$
$$\frac{7}{3} = x^2 \qquad \text{Dividing both sides by 3}$$
$$x = \sqrt{\frac{7}{3}} \quad or \quad x = -\sqrt{\frac{7}{3}} \qquad \text{Using the principle of square roots}$$

$$\left(\text{If we rationalize denominators, these answers can also be written } \frac{\sqrt{21}}{3} \right.$$
$$\left. \text{and } -\frac{\sqrt{21}}{3}. \right)$$

technology connection

We can visualize Example 2(a) on a grapher by letting $y_1 = x^2$ and $y_2 = 17$ and using the INTERSECT option of the CALC menu to find the x-coordinate at each point of intersection. Below, we show the intersection when $x \approx -\sqrt{17}$.

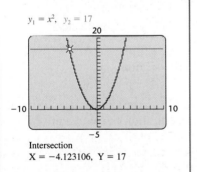

$y_1 = x^2, \; y_2 = 17$

Intersection
X = -4.123106, Y = 17

You can confirm that $(-4.123106)^2 \approx 17$. This visualization serves as a check of the algebraic approach.

1. Use a grapher to visualize and check the solutions of Examples 2(b) and 2(c).

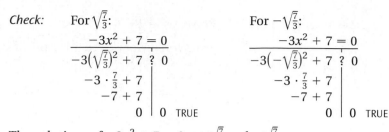

Check: For $\sqrt{\frac{7}{3}}$:

$$-3x^2 + 7 = 0$$

$$\overline{-3\left(\sqrt{\frac{7}{3}}\right)^2 + 7 \ ? \ 0}$$

$$-3 \cdot \frac{7}{3} + 7$$

$$-7 + 7$$

$$0 \ \big| \ 0 \quad \text{TRUE}$$

For $-\sqrt{\frac{7}{3}}$:

$$-3x^2 + 7 = 0$$

$$\overline{-3\left(-\sqrt{\frac{7}{3}}\right)^2 + 7 \ ? \ 0}$$

$$-3 \cdot \frac{7}{3} + 7$$

$$-7 + 7$$

$$0 \ \big| \ 0 \quad \text{TRUE}$$

The solutions of $-3x^2 + 7 = 0$ are $\sqrt{\frac{7}{3}}$ and $-\sqrt{\frac{7}{3}}$.

Solving Quadratic Equations of the Type $(x + k)^2 = p$

Equations like $(x - 5)^2 = 9$ or $(x + 2)^2 = 7$ are of the form $(x + k)^2 = p$. The principle of square roots can be used to solve such equations.

E x a m p l e   3   Solve: **(a)** $(x - 5)^2 = 9$; **(b)** $(x + 2)^2 = 7$.

Solution

a) $(x - 5)^2 = 9$

   $x - 5 = 3 \quad or \quad x - 5 = -3$ Using the principle of square roots; $\sqrt{9} = 3$ and $-\sqrt{9} = -3$

   $x = 8 \quad or \qquad x = 2$ Adding 5 to both sides

The solutions are 8 and 2. We leave the check to the student.

b) $(x + 2)^2 = 7$

   $x + 2 = \sqrt{7} \quad or \quad x + 2 = -\sqrt{7}$ Using the principle of square roots

   $x = -2 + \sqrt{7} \quad or \qquad x = -2 - \sqrt{7}$ Adding -2 to both sides

The solutions are $-2 + \sqrt{7}$ and $-2 - \sqrt{7}$, or simply $-2 \pm \sqrt{7}$ (read "-2 plus or minus $\sqrt{7}$"). We leave the check to the student.

In Example 3, the left sides of the equations are squares of binomials. Sometimes factoring can be used to express an equation in that form.

E x a m p l e   4   Solve by factoring and using the principle of square roots.

a) $x^2 + 8x + 16 = 49$ **b)** $x^2 + 6x + 9 = 10$

Solution

a) $x^2 + 8x + 16 = 49$ The left side is a perfect-square trinomial.

   $(x + 4)^2 = 49$ Factoring

   $x + 4 = 7 \quad or \quad x + 4 = -7$ Using the principle of square roots

   $x = 3 \quad or \qquad x = -11$

The solutions are 3 and -11.

b) $x^2 + 6x + 9 = 10$ The left side is a perfect-square trinomial.

$(x + 3)^2 = 10$ Factoring

$x + 3 = \sqrt{10}$ *or* $x + 3 = -\sqrt{10}$ Using the principle of square roots

$x = -3 + \sqrt{10}$ *or* $x = -3 - \sqrt{10}$

The solutions are $-3 + \sqrt{10}$ and $-3 - \sqrt{10}$, or simply $-3 \pm \sqrt{10}$.

Exercise Set 9.1

FOR EXTRA HELP

Digital Video Tutor CD 5 Videotape 17   InterAct Math   Math Tutor Center   MathXL   MyMathLab.com

Solve. Use the principle of square roots.

1. $x^2 = 49$

2. $x^2 = 100$

3. $a^2 = 25$

4. $a^2 = 36$

5. $t^2 = 17$

6. $n^2 = 13$

7. $3x^2 = 27$

8. $5x^2 = 20$

Aha! **9.** $9t^2 = 0$

10. $7t^2 = 21$

11. $4 - 9x^2 = 0$

12. $25 - 4a^2 = 0$

13. $49y^2 - 5 = 15$

14. $4y^2 - 3 = 9$

15. $8x^2 - 28 = 0$

16. $25x^2 - 35 = 0$

17. $(x - 1)^2 = 25$

18. $(x - 2)^2 = 49$

19. $(x + 4)^2 = 81$

20. $(x + 3)^2 = 36$

21. $(m + 3)^2 = 6$

22. $(m - 4)^2 = 21$

Aha! **23.** $(a - 7)^2 = 0$

24. $(a + 12)^2 = 81$

25. $(x - 5)^2 = 14$

26. $(x - 7)^2 = 12$

27. $(t + 3)^2 = 25$

28. $(x + 5)^2 = 49$

29. $\left(y - \frac{3}{4}\right)^2 = \frac{17}{16}$

30. $\left(x + \frac{3}{2}\right)^2 = \frac{13}{4}$

31. $x^2 - 10x + 25 = 100$

32. $x^2 - 6x + 9 = 64$

33. $p^2 + 8p + 16 = 1$

34. $y^2 + 14y + 49 = 4$

35. $t^2 - 6t + 9 = 13$

36. $m^2 - 2m + 1 = 5$

37. $x^2 + 12x + 36 = 18$

38. $x^2 + 4x + 4 = 12$

39. If a quadratic equation can be solved by factoring, what type of number(s) will be solutions? Why?

40. Under what conditions is it easier to use the principle of square roots rather than the principle of zero products to solve a quadratic equation?

SKILL MAINTENANCE

Factor.

41. $3x^2 + 12x + 3$

42. $5t^2 + 20t + 25$

43. $t^2 + 16t + 64$

44. $x^2 + 14x + 49$

45. $x^2 - 10x + 25$

46. $t^2 - 8t + 16$

SYNTHESIS

47. Is it possible for a quadratic equation to have $5 + \sqrt{2}$ as a solution, but not $5 - \sqrt{2}$? Why or why not?

48. If a quadratic equation can be solved using the principle of square roots, but not by factoring, what type of number(s) will be solutions? Why?

Factor the left side of each equation. Then solve.

49. $x^2 + \frac{7}{3}x + \frac{49}{36} = \frac{7}{36}$

50. $x^2 - 5x + \frac{25}{4} = \frac{13}{4}$

51. $m^2 - \frac{3}{2}m + \frac{9}{16} = \frac{17}{16}$

52. $t^2 + 3t + \frac{9}{4} = \frac{49}{4}$

53. $x^2 + 2.5x + 1.5625 = 9.61$

54. $a^2 - 3.8a + 3.61 = 27.04$

Use the graph of
$$y = (x + 3)^2$$
to solve each equation.

55. $(x + 3)^2 = 1$

56. $(x + 3)^2 = 4$

57. $(x + 3)^2 = 9$

58. $(x + 3)^2 = 0$

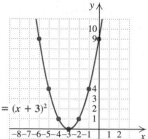

59. *Gravitational force.* Newton's law of gravitation states that the gravitational force f between objects of mass M and m, at a distance d from each other, is given by

$$f = \frac{kMm}{d^2},$$

where k is a constant. Solve for d.

CORNER

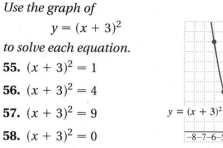

How Big is Half a Pizza?

COLLABORATIVE

Focus: Principle of square roots; formulas

Time: 15–25 minutes

Group size: 3

Materials: Calculators are optional.

Frankie, Johnnie, & Luigi, Too! has the best pizza in the Palo Alto area, according to a poll of Stanford University students. Pizzas there have 12-in., 14-in., and 16-in. diameters.

ACTIVITY

1. Let each member of the group play the role of either Frankie, Johnnie, or Luigi. As part of a promotion, Frankie is offering a "mini" pie that has half the area of a 12-in. pie, Johnnie is offering a "personal" pie that has half the area of a 14-in. pie, and Luigi is offering a "junior" pie that has half the area of a 16-in. pie. Each group member, according to the selected role, should calculate the area of both the original pie and the new offering.

2. Each group member, according to the chosen role, should calculate the radius and then the diameter of the "new" pie. Summarize the group's findings in a table similar to the one below.

	Diameter of Original Pie	Area of Original Pie	Area of New Pie	Radius of New Pie	Diameter of New Pie
Frankie					
Johnnie					
Luigi					

3. After checking each other's work, group members should develop a formula that can be used to determine the diameter of a circle that has half the area of a given circle. That is, if a circle has diameter d, what is the diameter of a circle with half the area?

4. If the diameter of one pizza is twice the diameter of another pizza, should the pizza cost twice as much? Why or why not?

Solving Quadratic Equations: Completing the Square

9.2

Completing the Square • Solving by Completing the Square

In Section 9.1, we solved equations like $(x - 5)^2 = 7$ using the principle of square roots. Equations like $x^2 + 8x + 16 = 12$ were similarly solved because the left side of the equation is a perfect-square trinomial. We now learn to solve equations like $x^2 - 8x = 2$, in which the left side is not (yet) a perfect-square trinomial. The new procedure involves *completing the square* and enables us to solve *any* quadratic equation.

Completing the Square

Recall that

$$(x + 3)^2 = (x + 3)(x + 3)$$
$$= x^2 + 3x + 3x + 9$$
$$= x^2 + 6x + 9 \qquad \text{This is a perfect-square trinomial.}$$

and, in general,

$$(x + a)^2 = x^2 + 2ax + a^2. \qquad \text{This is also a perfect-square trinomial.}$$

In $x^2 + 6x + 9$, note that 9 is the square of half of the coefficient of x: $\frac{1}{2} \cdot 6 = 3$, and $3^2 = 9$. Similarly, in $x^2 + 2ax + a^2$, note that a^2 is also the square of half of the coefficient of x: $\frac{1}{2} \cdot 2a = a$ and a^2 completes the pattern.

Consider the following quadratic equation:

$$x^2 + 10x = 4.$$

We need a number that can be added to both sides and that will make the left side a perfect-square trinomial. Such a number is described above as the square of half of the coefficient of x: $\frac{1}{2} \cdot 10 = 5$, and $5^2 = 25$. Thus we add 25 to both sides:

$$x^2 + 10x = 4 \qquad \text{\textit{Think:} Half of 10 is 5; } 5^2 = 25.$$
$$x^2 + 10x + 25 = 4 + 25 \qquad \text{Adding 25 to both sides}$$
$$(x + 5)^2 = 29. \qquad \text{Factoring the perfect-square trinomial}$$

By adding 25 to $x^2 + 10x$, we have *completed the square*. The resulting equation contains the square of a binomial on one side. Solutions can then be found using the principle of square roots, as in Section 9.1:

$$(x + 5)^2 = 29$$
$$x + 5 = \sqrt{29} \qquad or \quad x + 5 = -\sqrt{29} \qquad \text{Using the principle of square roots}$$
$$x = -5 + \sqrt{29} \quad or \qquad x = -5 - \sqrt{29}.$$

The solutions are $-5 \pm \sqrt{29}$.

technology connection

One way to check that $-5 - \sqrt{29}$ and $-5 + \sqrt{29}$ are both solutions of $x^2 + 10x = 4$ is to store $-5 - \sqrt{29}$ (or $-5 + \sqrt{29}$) as X in the calculator's memory. We then press $X^2 + 10X$ ENTER . The result should be 4.

> ### Completing the Square
>
> To *complete the square* for an expression like $x^2 + bx$, add half of the coefficient of x, squared. That is, add $(b/2)^2$.

A visual interpretation of completing the square is sometimes helpful.

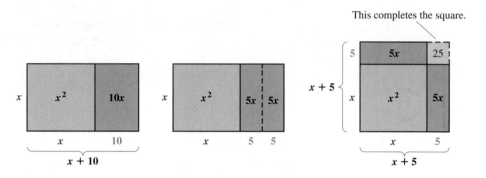

In each figure above, the sum of the pink and purple areas is $x^2 + 10x$. However, by splitting the purple area in half, we can "complete" a square by adding the blue area. The blue area is $5 \cdot 5$, or 25 square units.

Example 1

Complete the square and check by multiplying: **(a)** $x^2 - 12x$; **(b)** $x^2 + 5x$.

Solution

a) To complete the square for $x^2 - 12x$, note that the coefficient of x is -12. Half of -12 is -6 and $(-6)^2$ is 36. Thus, $x^2 - 12x$ becomes a perfect-square trinomial when 36 is added:

$$x^2 - 12x + 36 \quad \text{is the square of} \quad x - 6.$$

The number 36 completes the square.

Check: $\quad (x - 6)^2 = (x - 6)(x - 6) = x^2 - 6x - 6x + 36 = x^2 - 12x + 36$

b) To complete the square for $x^2 + 5x$, we take half of the coefficient of x and square it:

$$\left(\tfrac{5}{2}\right)^2 = \tfrac{25}{4}. \qquad \text{Half of 5 is } \tfrac{5}{2}; \left(\tfrac{5}{2}\right)^2 = \tfrac{5}{2} \cdot \tfrac{5}{2} = \tfrac{25}{4}.$$

Thus, $x^2 + 5x + \tfrac{25}{4}$ is the square of $x + \tfrac{5}{2}$. The number $\tfrac{25}{4}$ completes the square.

Check: $\quad \left(x + \tfrac{5}{2}\right)^2 = \left(x + \tfrac{5}{2}\right)\left(x + \tfrac{5}{2}\right) = x^2 + \tfrac{5}{2}x + \tfrac{5}{2}x + \tfrac{25}{4} = x^2 + 5x + \tfrac{25}{4}$

> *Caution!*   In Example 1, we are neither solving an equation nor writing an equivalent expression. Instead, we are learning how to find the number that completes the square. In Examples 2 and 3, we use the number that completes the square, along with the addition principle, to solve equations.

Solving by Completing the Square

The concept of completing the square can now be used to solve equations like $x^2 + 10x = 4$ much as we did on p. 497, prior to Example 1.

E x a m p l e   2   Solve by completing the square.

a) $x^2 + 6x = -8$
b) $x^2 - 10x + 14 = 0$

Solution

a) To solve $x^2 + 6x = -8$, we take half of 6 and square it, to get 9. Then we add 9 to both sides of the equation. This makes the left side the square of a binomial:

$$x^2 + 6x + 9 = -8 + 9 \qquad \text{Adding 9 to both sides to complete the square}$$

$$(x + 3)^2 = 1 \qquad \text{Factoring}$$

$$x + 3 = 1 \quad or \quad x + 3 = -1 \qquad \text{Using the principle of square roots}$$

$$x = -2 \quad or \qquad x = -4.$$

The solutions are -2 and -4. The check is left to the student.

b) We have

$$x^2 - 10x + 14 = 0$$

$$x^2 - 10x \qquad = -14 \qquad \text{Subtracting 14 from both sides to prepare the left side for completing the square}$$

$$x^2 - 10x + 25 = -14 + 25 \qquad \text{Adding 25 to both sides to complete the square: } (-10/2)^2 = 25$$

$$(x - 5)^2 = 11 \qquad \text{Factoring}$$

$$x - 5 = \sqrt{11} \qquad or \quad x - 5 = -\sqrt{11} \qquad \text{Using the principle of square roots}$$

$$x = 5 + \sqrt{11} \quad or \qquad x = 5 - \sqrt{11}.$$

The solutions are $5 + \sqrt{11}$ and $5 - \sqrt{11}$, or simply $5 \pm \sqrt{11}$. The check is left to the student.

To complete the square, the coefficient of x^2 must be 1. When the x^2-coefficient is not 1, we can multiply or divide on both sides to find an equivalent equation with an x^2-coefficient of 1.

E x a m p l e   3

Solve by completing the square.

a) $3x^2 + 24x = 3$
b) $2x^2 - 3x - 1 = 0$

Solution

a)
$$3x^2 + 24x = 3$$

$$\left.\begin{array}{c} \frac{1}{3}(3x^2 + 24x) = \frac{1}{3} \cdot 3 \\ x^2 + 8x = 1 \end{array}\right\}$$

We multiply by $\frac{1}{3}$ (or divide by 3) on both sides of the equation to ensure an x^2-coefficient of 1.

$$x^2 + 8x + 16 = 1 + 16$$

Adding 16 to both sides to complete the square: $\left(\frac{8}{2}\right)^2 = 16$

$$(x + 4)^2 = 17$$

Factoring

$$x + 4 = \sqrt{17} \qquad or \quad x + 4 = -\sqrt{17}$$
$$x = -4 + \sqrt{17} \quad or \qquad x = -4 - \sqrt{17}$$

The solutions are $-4 \pm \sqrt{17}$. The check is left to the student.

b)
$$2x^2 - 3x - 1 = 0$$

$$\frac{1}{2}(2x^2 - 3x - 1) = \frac{1}{2} \cdot 0$$

Multiplying by $\frac{1}{2}$ to make the x^2-coefficient 1

$$x^2 - \frac{3}{2}x - \frac{1}{2} = 0$$

$$x^2 - \frac{3}{2}x \qquad = \frac{1}{2}$$

Adding $\frac{1}{2}$ to both sides

$$x^2 - \frac{3}{2}x + \frac{9}{16} = \frac{1}{2} + \frac{9}{16}$$

Adding $\frac{9}{16}$ to both sides: $\left[\frac{1}{2}\left(-\frac{3}{2}\right)\right]^2 = \left[-\frac{3}{4}\right]^2 = \frac{9}{16}$. This completes the square on the left side.

$$\left(x - \frac{3}{4}\right)^2 = \frac{8}{16} + \frac{9}{16}$$

Factoring and finding a common denominator

$$\left(x - \frac{3}{4}\right)^2 = \frac{17}{16}$$

$$x - \frac{3}{4} = \frac{\sqrt{17}}{4} \qquad or \quad x - \frac{3}{4} = -\frac{\sqrt{17}}{4}$$

Using the principle of square roots

$$x = \frac{3}{4} + \frac{\sqrt{17}}{4} \quad or \qquad x = \frac{3}{4} - \frac{\sqrt{17}}{4}$$

The solutions of $2x^2 - 3x - 1 = 0$ are $\dfrac{3 \pm \sqrt{17}}{4}$. The checks are left to the student.

The steps in Example 3 can be used to solve *any* quadratic equation.

> **To Solve $ax^2 + bx + c = 0$ by Completing the Square**
>
> 1. If $a \neq 1$, multiply both sides of the equation by $1/a$ or divide both sides by a so that the x^2-coefficient is 1.
> 2. When the x^2-coefficient is 1, rewrite the equation in the form
>
> $$x^2 + bx = -c, \quad \text{or, if step (1) was needed,}$$
>
> $$x^2 + \frac{b}{a}x = -\frac{c}{a}.$$
>
> 3. Take half of the x-coefficient and square it. Add the result to both sides of the equation.
> 4. Express the left side as the square of a binomial. (Factor.)
> 5. Use the principle of square roots and complete the solution.

FOR EXTRA HELP

Digital Video Tutor CD 5
Videotape 17

InterAct Math

Math Tutor Center

MathXL

MyMathLab.com

Exercise Set 9.2

Complete the square. Check by multiplying.

1. $x^2 + 8x$

2. $x^2 + 4x$

3. $x^2 - 12x$

4. $x^2 - 14x$

5. $x^2 - 3x$

6. $x^2 - 9x$

7. $t^2 + t$

8. $y^2 - y$

9. $x^2 + \frac{5}{4}x$

10. $x^2 + \frac{4}{3}x$

11. $m^2 - \frac{9}{2}m$

12. $r^2 - \frac{2}{5}r$

Solve by completing the square.

13. $x^2 + 8x + 12 = 0$

14. $x^2 + 6x - 7 = 0$

15. $x^2 - 24x + 21 = 0$

16. $x^2 - 12x + 40 = 0$

17. $3x^2 - 6x - 15 = 0$

18. $3x^2 - 12x - 33 = 0$

19. $x^2 - 22x + 102 = 0$

20. $t^2 - 18t + 74 = 0$

21. $t^2 + 8t - 5 = 0$

22. $x^2 - 7x - 2 = 0$

23. $2x^2 + 10x - 12 = 0$

24. $2x^2 + 6x - 56 = 0$

25. $x^2 + \frac{3}{2}x - 2 = 0$

26. $x^2 - \frac{3}{2}x - 2 = 0$

27. $2x^2 + 3x - 16 = 0$

28. $2t^2 - 3t - 8 = 0$

29. $3t^2 + 6t - 1 = 0$

30. $3x^2 - 4x - 3 = 0$

31. $2x^2 = 9 + 5x$

32. $2x^2 = 5 + 9x$

33. $4x^2 + 12x = 7$

34. $6x^2 + 11x = 10$

35. How does completing the square allow us to solve equations that we could not have otherwise solved?

36. Explain how the addition principle, the multiplication principle, and the square-root principle were used in this section.

SKILL MAINTENANCE

Simplify.

37. $\dfrac{3 + 6x}{3}$

38. $\dfrac{4 + 2x}{2}$

39. $\dfrac{15 - 10x}{5}$

40. $\dfrac{18 - 15x}{3}$

41. $\dfrac{24 - 3\sqrt{5}}{9}$

42. $\dfrac{35 - 7\sqrt{6}}{14}$

SYNTHESIS

43. Sal states that "since solving a quadratic equation by completing the square relies on the principle of square roots, the solutions are always opposites of each other." Is Sal correct? Why or why not?

44. When completing the square, what determines if the number being added is a whole number or a fraction?

Find b such that each trinomial is a square.

45. $x^2 + bx + 49$

46. $x^2 + bx + 36$

47. $x^2 + bx + 50$

48. $x^2 + bx + 45$

49. $x^2 - bx + 48$

50. $4x^2 + bx + 16$

Solve each of the following by letting y_1 represent the left side of each equation, letting y_2 represent the right side, and graphing y_1 and y_2 on the same set of axes. INTERSECT *can then be used to determine the x-coordinate at any point of intersection. Find solutions accurate to two decimal places.*

51. $(x + 4)^2 = 13$

52. $(x + 6)^2 = 2$

53. $x^2 - 24x + 21 = 0$
(see Exercise 15)

54. $x^2 - 7x - 2 = 0$
(see Exercise 22)

55. $2x^2 = 9 + 5x$
(see Exercise 31)

56. $2x^2 = 5 + 9x$
(see Exercise 32)

Aha! **57.** What is the best way to solve $x^2 + 8x = 0$? Why?

CORNER

Using Areas to Complete the Square

COLLABORATIVE

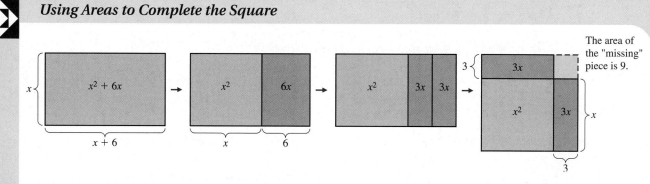

The area of the "missing" piece is 9.

Focus: Visualizing completion of the square

Time: 15–25 minutes

Group size: 2

Materials: Rulers and graph paper may be helpful.

It is not difficult to draw a representation of completing the square. To do so, we use areas and the fact that the area of any rectangle is given by multiplying the length and the width. For example, the above sequence of figures can be drawn to explain why 9 completes the square for $x^2 + 6x$.

ACTIVITY

1. Draw a sequence of four figures, similar to those shown above, to complete the square for $x^2 + 8x$. Group members should take turns, so that each person draws and labels two of the figures.

2. Repeat part (1) to complete the square for $x^2 + 14x$. The person who drew the first drawing in part (1) should take the second turn this time.

3. Keep in mind that when we complete the square, we are not forming an equivalent expression. For this reason, completing the square is generally performed by using the addition principle to form an equivalent *equation*. Use the work in parts (1) and (2) to solve the equations $x^2 + 8x = 9$ and $x^2 + 14x = 15$.

4. Each equation in part (3) has two solutions. Can both be represented geometrically? Why or why not?

9.3

The Quadratic Formula • Problem Solving

We now derive the *quadratic formula*. In many cases, this formula enables us to solve quadratic equations more quickly than the method of completing the square.

The Quadratic Formula

When mathematicians use a procedure repeatedly, they often try to find a formula for the procedure. The quadratic formula condenses into one calculation the many steps used to solve a quadratic equation by completing the square.

Consider a quadratic equation in *standard form*, $ax^2 + bx + c = 0$, with $a > 0$. Our plan is to solve this equation for x by completing the square. As the steps are performed, compare them with those in Example 3(b) on p. 500:

$$ax^2 + bx + c = 0$$

$$\frac{1}{a}(ax^2 + bx + c) = \frac{1}{a} \cdot 0 \qquad \text{Multiplying by } \frac{1}{a} \text{ to make the } x^2\text{-coefficient 1}$$

$$x^2 + \frac{b}{a}x + \frac{c}{a} = 0$$

$$x^2 + \frac{b}{a}x = -\frac{c}{a} \qquad \text{Adding } -\frac{c}{a} \text{ to both sides}$$

$$x^2 + \frac{b}{a}x + \frac{b^2}{4a^2} = -\frac{c}{a} + \frac{b^2}{4a^2} \qquad \text{Adding } \frac{b^2}{4a^2} \text{ to both sides:} \quad \left[\frac{1}{2} \cdot \frac{b}{a}\right]^2 = \left[\frac{b}{2a}\right]^2 = \frac{b^2}{4a^2}$$

$$\left(x + \frac{b}{2a}\right)^2 = -\frac{4ac}{4a^2} + \frac{b^2}{4a^2} \qquad \text{Factoring and finding a common denominator}$$

$$\left(x + \frac{b}{2a}\right)^2 = \frac{b^2 - 4ac}{4a^2}$$

$$x + \frac{b}{2a} = \sqrt{\frac{b^2 - 4ac}{4a^2}} \quad \text{or} \quad x + \frac{b}{2a} = -\sqrt{\frac{b^2 - 4ac}{4a^2}}. \qquad \text{Using the principle of square roots}$$

We assumed $a > 0$, so $\sqrt{4a^2} = 2a$. Thus we can simplify as follows:

$$x + \frac{b}{2a} = \frac{\sqrt{b^2 - 4ac}}{2a} \quad \text{or} \quad x + \frac{b}{2a} = -\frac{\sqrt{b^2 - 4ac}}{2a}$$

and

$$x = -\frac{b}{2a} + \frac{\sqrt{b^2 - 4ac}}{2a} \quad \text{or} \quad x = -\frac{b}{2a} - \frac{\sqrt{b^2 - 4ac}}{2a}. \qquad \text{Adding } -b/(2a) \text{ to both sides}$$

Thus,

$$x = -\frac{b}{2a} \pm \frac{\sqrt{b^2 - 4ac}}{2a},$$

or

$$x = \frac{-b \pm \sqrt{b^2 - 4ac}}{2a}.$$ Unless $b^2 - 4ac$ is 0, this represents two solutions.

This last equation is the result we sought. It is so useful that it is worth memorizing.

The Quadratic Formula

The solutions of $ax^2 + bx + c = 0$ are given by

$$x = \frac{-b \pm \sqrt{b^2 - 4ac}}{2a}.$$

The quadratic formula also holds when $a < 0$. A similar proof would show this, but we will not consider it here.

Example 1

Solve using the quadratic formula.

a) $4x^2 + 5x - 6 = 0$
b) $x^2 = 4x + 7$
c) $x^2 + x = -1$

Solution

a) We identify a, b, and c and substitute into the quadratic formula:

$$4x^2 + 5x - 6 = 0;$$

$\qquad a \quad\ b \quad c$

$$x = \frac{-b \pm \sqrt{b^2 - 4ac}}{2a}$$

$$x = \frac{-5 \pm \sqrt{5^2 - 4 \cdot 4(-6)}}{2 \cdot 4}$$ Substituting for a, b, and c

Be sure to write the fraction bar all the way across.

$$x = \frac{-5 \pm \sqrt{25 - (-96)}}{8}$$

$$x = \frac{-5 \pm \sqrt{121}}{8}$$

$$x = \frac{-5 \pm 11}{8}$$

$$x = \frac{-5 + 11}{8} \quad or \quad x = \frac{-5 - 11}{8}$$

$$x = \frac{6}{8} \quad\quad or \quad x = \frac{-16}{8}$$

$$x = \frac{3}{4} \quad\quad or \quad x = -2.$$

The solutions are $\frac{3}{4}$ and -2.

b) We rewrite $x^2 = 4x + 7$ in standard form, identify a, b, and c, and solve using the quadratic formula:

$$1x^2 - 4x - 7 = 0; \quad\quad \text{Subtracting } 4x + 7 \text{ from both sides}$$

$$\begin{matrix} \uparrow & \uparrow & \uparrow \\ a & b & c \end{matrix}$$

$$x = \frac{-(-4) \pm \sqrt{(-4)^2 - 4(1)(-7)}}{2 \cdot 1} \quad\quad \begin{array}{l}\text{Substituting into the}\\ \text{quadratic formula}\end{array}$$

$$x = \frac{4 \pm \sqrt{16 + 28}}{2} = \frac{4 \pm \sqrt{44}}{2}.$$

Since $\sqrt{44}$ can be simplified, we have

$$x = \frac{4 \pm \sqrt{4}\,\sqrt{11}}{2} - \frac{4 \pm 2\sqrt{11}}{2}.$$

Finally, since 2 is a common factor of 4 and $2\sqrt{11}$, we can simplify the fraction by removing a factor equal to 1:

$$x = \frac{2\left(2 \pm \sqrt{11}\right)}{2 \cdot 1} \quad\quad \text{Factoring}$$

$$x = \frac{2}{2} \cdot \frac{2 \pm \sqrt{11}}{1}. \quad\quad \text{Removing a factor equal to 1: } \frac{2}{2} = 1$$

The solutions are $2 + \sqrt{11}$ and $2 - \sqrt{11}$, or $2 \pm \sqrt{11}$.

c) We rewrite $x^2 + x = -1$ in standard form and use the quadratic formula:

$$1x^2 + 1x + 1 = 0; \quad\quad \text{Adding 1 to both sides}$$

$$\begin{matrix} \uparrow & \uparrow & \uparrow \\ a & b & c \end{matrix}$$

$$x = \frac{-1 \pm \sqrt{1^2 - 4 \cdot 1 \cdot 1}}{2 \cdot 1} \quad\quad \text{Substituting into the quadratic formula}$$

$$x = \frac{-1 \pm \sqrt{1 - 4}}{2}$$

$$x = \frac{-1 \pm \sqrt{-3}}{2}.$$

Since the radicand, -3, is negative, there are no real-number solutions. In Section 9.5, we will study a number system in which solutions of this equation can be found. For now we simply state, "No real-number solution exists."

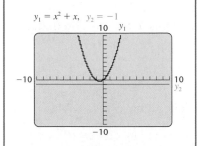

technology connection

To see that no real solutions exist for Example 1(c), we let $y_1 = x^2 + x$ and $y_2 = -1$.

$y_1 = x^2 + x, \quad y_2 = -1$

    The absence of any point of intersection supports the conclusion that no real-number solution exists.

1. What happens when the INTERSECT feature is used with the graph above?

2. How can the graph of $y = x^2 + x + 1$ be used to provide still another check of Example 1(c)?

CONNECTING THE CONCEPTS

We have now studied three different ways of solving quadratic equations. Each of these methods has certain advantages and disadvantages, as outlined in the chart at right. Note that although the quadratic formula can be used to solve *any* quadratic equation, the other methods are sometimes faster and easier to use.

Method	Advantages	Disadvantages
The quadratic formula	Can be used to solve *any* quadratic equation.	Can be slower than factoring or the principle of square roots.
The principle of square roots	Fastest way to solve equations of the form $ax^2 = p$, or $(x + k)^2 = p$. Can be used to solve *any* quadratic equation.	Can be slow when completing the square is required.
Factoring	Can be very fast.	Can be used only on certain equations. Many equations are difficult or impossible to solve by factoring.

Problem Solving

E x a m p l e   2

Diagonals in a polygon.   The number of diagonals d in a polygon that has n sides is given by the formula

$$d = \frac{n^2 - 3n}{2}.$$

If a polygon has 27 diagonals, how many sides does it have?

Solution

1. **Familiarize.**   A sketch can help us to become familiar with the problem. We draw a hexagon (6 sides) and count the diagonals. As the formula predicts, for $n = 6$, there are

$$\frac{6^2 - 3 \cdot 6}{2} = \frac{36 - 18}{2}$$

$$= \frac{18}{2} = 9 \text{ diagonals.}$$

Clearly, the polygon in question must have more than 6 sides. We might suspect that tripling the number of diagonals requires tripling the number of sides. Evaluating the above formula for $n = 18$, you can confirm that this is *not* the case. Rather than continue guessing, we proceed to a translation.

2. **Translate.**   Since the number of diagonals is 27, we substitute 27 for d:

$$27 = \frac{n^2 - 3n}{2}.$$

This gives us a translation.

3. **Carry out.**   We solve the equation for n, first reversing the equation for convenience:

$$\frac{n^2 - 3n}{2} = 27$$

$$n^2 - 3n = 54 \qquad \text{Multiplying both sides by 2 to clear fractions}$$

$$n^2 - 3n - 54 = 0 \qquad \text{Subtracting 54 from both sides}$$

$$(n - 9)(n + 6) = 0 \qquad \text{Factoring. There is no need for the quadratic formula here.}$$

$$n - 9 = 0 \quad or \quad n + 6 = 0$$

$$n = 9 \quad or \qquad n = -6.$$

4. **Check.**   Since the number of sides cannot be negative, -6 cannot be a solution. We leave it to the student to show by substitution that 9 checks.

5. **State.**   The polygon has 9 sides (it is a nonagon).

Example 3

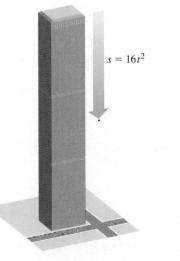

$s = 16t^2$

Free-falling objects.   A skyscraper is 1368 ft tall. How many seconds will it take a golf ball to fall from the top? Round to the nearest hundredth.

Solution

1. **Familiarize.**   If we did not know anything about this problem, we might consider looking up a formula in a mathematics or physics book. A formula that fits this situation is

$$s = 16t^2, \qquad \text{It is useful to remember this formula.}$$

where s is the distance, in feet, traveled by a body falling freely from rest in t seconds. This formula is actually an approximation because it does not account for air resistance. In this problem, we know the distance s to be 1368. We want to determine the time t that it takes the golf ball to reach the ground.

2. **Translate.**   The distance is 1368 ft and we need to solve for t. We substitute 1368 for s in the formula above to get the following translation:

$$1368 = 16t^2.$$

3. **Carry out.** Because there is no t-term, we can use the principle of square roots to solve:

$$1368 = 16t^2$$

$$\frac{1368}{16} = t^2 \qquad \text{Solving for } t^2$$

$$\sqrt{\frac{1368}{16}} = t \quad or \quad -\sqrt{\frac{1368}{16}} = t \qquad \text{Using the principle of square roots}$$

$$\frac{\sqrt{1368}}{4} = t \quad or \quad \frac{-\sqrt{1368}}{4} = t$$

$$9.25 \approx t \quad or \quad -9.25 \approx t. \qquad \text{Using a calculator and rounding to the nearest hundredth}$$

4. **Check.** The number -9.25 cannot be a solution because time cannot be negative in this situation. We substitute 9.25 in the original equation:

$$s = 16(9.25)^2 = 16(85.5625) = 1369.$$

   This is close. Remember that we approximated a solution.

5. **State.** It takes about 9.25 sec for the golf ball to fall to the ground from the top of the skyscraper.

E x a m p l e   4   ***Right triangles.*** The hypotenuse of a right triangle is 6 m long. One leg is 1 m longer than the other. Find the lengths of the legs. Round to the nearest hundredth.

Solution

1. **Familiarize.** We first make a drawing and label it. We let $s =$ the length, in meters, of one leg. Then $s + 1 =$ the length, in meters, of the other leg.

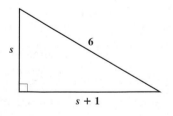

   Note that if $s = 3$, then $s + 1 = 4$ and $3^2 + 4^2 = 25 \neq 6^2$. Thus, because of the Pythagorean theorem, we see that $s \neq 3$. Another guess, $s = 4$, is too big since $4^2 + (4 + 1)^2 = 41 \neq 6^2$. Although we have not guessed the solution, we expect s to be between 3 and 4.

2. **Translate.** To translate, we use the Pythagorean theorem:

$$s^2 + (s + 1)^2 = 6^2.$$

3.   **Carry out.**   We solve the equation:

$$s^2 + (s + 1)^2 = 6^2$$

$$s^2 + s^2 + 2s + 1 = 36$$

$$2s^2 + 2s - 35 = 0 \qquad \text{This cannot be factored so we use the}$$
$$\qquad\qquad\qquad\qquad \text{quadratic formula.}$$

$$\uparrow \quad \uparrow \quad \uparrow$$
$$\boldsymbol{a} \quad \boldsymbol{b} \quad \boldsymbol{c}$$

$$s = \frac{-2 \pm \sqrt{2^2 - 4 \cdot 2(-35)}}{2 \cdot 2} \qquad \text{Remember:}$$
$$\qquad\qquad\qquad\qquad\qquad\qquad s = \frac{-b \pm \sqrt{b^2 - 4ac}}{2a}.$$

$$s = \frac{-2 \pm \sqrt{4 + 280}}{4} = \frac{-2 \pm \sqrt{284}}{4}$$

$$s \approx 3.71 \quad or \quad s \approx -4.71. \qquad \text{Using a calculator and rounding}$$
$$\qquad\qquad\qquad\qquad\qquad\qquad\qquad \text{to the nearest hundredth}$$

4.   **Check.**   Length cannot be negative, so -4.71 does not check. Note that if the smaller leg is 3.71 m, the other leg is 4.71 m. Then

$$(3.71)^2 + (4.71)^2 = 13.7641 + 22.1841 = 35.9482$$

and since $35.9482 \approx 6^2$, our approximation checks. Also, note that the value of s, 3.71, is between 3 and 4, as predicted in step (1).

5.   **State.**   One leg is about 3.71 m long; the other is about 4.71 m long.

FOR EXTRA HELP

Exercise Set 9.3

Digital Video Tutor CD 5   InterAct Math   Math Tutor Center   MathXL   MyMathLab.com
Videotape 17

Solve. Try to use factoring or the principle of square roots first. If both of these methods prove difficult, use the quadratic formula. If no real-number solutions exist, state this.

15. $x^2 + 2x + 1 = 7$ **16.** $x^2 - 4x + 4 = 5$

17. $3t^2 + 8t + 2 = 0$ **18.** $3t^2 - 4t - 2 = 0$

19. $2x^2 - 5x = 1$ **20.** $2x^2 + 2x = 3$

1. $x^2 - 7x = 18$ **2.** $x^2 + 4x = 21$ **21.** $4y^2 + 2y - 3 = 0$ **22.** $4y^2 - 4y - 3 = 0$

3. $x^2 = 8x - 16$ **4.** $x^2 = 6x - 9$ **23.** $2t^2 - 3t + 2 = 0$ **24.** $4y^2 + 2y + 3 = 0$

5. $3y^2 + 7y + 4 = 0$ **6.** $3y^2 + 2y - 8 = 0$ **25.** $3x^2 - 5x = 4$ **26.** $2x^2 + 3x = 1$

7. $4x^2 - 12x = 7$ **8.** $4x^2 + 4x = 15$ **27.** $2y^2 - 6y = 10$ **28.** $5m^2 = 3 + 11m$

9. $t^2 = 64$ **10.** $t^2 = 81$ **29.** $10x^2 - 15x = 0$ **30.** $7x^2 + 2 = 6x$

11. $x^2 + 4x - 7 = 0$ **12.** $x^2 + 2x - 2 = 0$ **31.** $5t^2 - 7t = -4$ **32.** $15t^2 + 10t = 0$

13. $y^2 - 10y + 19 = 0$ **14.** $y^2 + 6y - 2 = 0$ **33.** $9y^2 = 162$ **34.** $5t^2 = 100$

Solve using the quadratic formula. Use a calculator or Table 2 to approximate the solutions to the nearest thousandth.

35. $x^2 - 4x - 7 = 0$ **36.** $x^2 + 2x - 2 = 0$

37. $y^2 - 5y - 1 = 0$ **38.** $y^2 + 7y + 3 = 0$

39. $4x^2 + 4x = 1$ **40.** $4x^2 = 4x + 1$

Solve. If an irrational answer occurs, round to the nearest hundredth.

41. A polygon has 35 diagonals. How many sides does it have? See Example 2.

42. A polygon has 20 diagonals. How many sides does it have? See Example 2.

43. *Free-fall time.*   At 1490 ft, the Universal Financial Center in China is the world's tallest office building. How long would it take a marble to fall from the top? See Example 3.

44. *Free-fall time.*   The height of the Sears Tower in Chicago is 1454 ft. How long would it take a golf ball to fall from the top? See Example 3.

45. *Free-fall record.*   Stuntman Dar Robinson once fell 700 ft from the top of the CN Tower in Toronto before opening a parachute (*Source: Guinness World Records 2000 Millennium Edition*). How long did the free-fall portion of his jump last? See Example 3.

46. *Free-fall record.*   The world record for free-fall to the ground without a parachute by a woman is 175 ft and is held by Kitty O'Neill. Approximately how long did the fall take? See Example 3.

47. *Right triangles.*   The hypotenuse of a right triangle is 25 ft long. One leg is 17 ft longer than the other. Find the lengths of the legs.

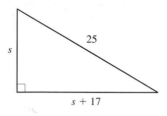

48. *Right triangles.*   The hypotenuse of a right triangle is 26 yd long. One leg is 14 yd longer than the other. Find the lengths of the legs.

49. *Area of a rectangle.*   The length of a rectangle is 4 cm greater than the width. The area is 60 cm^2. Find the length and the width.

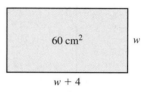

50. *Area of a rectangle.*   The length of a rectangle is 3 m greater than the width. The area is 70 m^2. Find the length and the width.

51. *Plumbing.*   A water pipe runs diagonally under a rectangular yard that is 6 m longer than it is wide. If the pipe is 30 m long, determine the dimensions of the yard.

52. *Guy wires.*   A 26-ft long guy wire is anchored 10 ft from the base of a telephone pole. How far up the pole does the wire reach?

53. *Right triangles.*   The area of a right triangle is 31 m^2. One leg is 2.4 m longer than the other. Find the lengths of the legs.

54. *Right triangles.*   The area of a right triangle is 26 cm^2. One leg is 5 cm longer than the other. Find the lengths of the legs.

55. *Area of a rectangle.*   The length of a rectangle is 3 in. greater than the width. The area is 30 in^2. Find the length and the width.

56. *Area of a rectangle.*   The length of a rectangle is 5 ft greater than the width. The area is 25 ft^2. Find the length and the width.

57. *Area of a rectangle.* The length of a rectangle is twice the width. The area is 16 m². Find the length and the width.

58. *Area of a rectangle.* The length of a rectangle is twice the width. The area is 20 cm². Find the length and the width.

Investments. *The formula* $A = P(1 + r)^t$ *is used to find the value A to which P dollars grows when invested for t years at an annual interest rate r. In Exercises 59–62, find the interest rate for the given information.*

59. $2000 grows to $2880 in 2 years

60. $2560 grows to $3610 in 2 years

61. $6000 grows to $6615 in 2 years

62. $3125 grows to $3645 in 2 years

63. *Environmental science.* In 2000, a circular oil slick about 30 km from Cape Town, South Africa, was 20 km² in area (*Source*: Environment News Service). How wide was the oil slick?

64. *Gardening.* Laura has enough mulch to cover 250 ft² of garden space. How wide is the largest circular flower garden that Laura can cover with mulch?

65. Under what condition(s) is the quadratic formula *not* the easiest way to solve a quadratic equation?

66. Roy claims to be able to solve any quadratic equation by completing the square. He also claims to be incapable of understanding why the quadratic formula works. Does this strike you as odd? Why or why not?

SKILL MAINTENANCE

Solve.

67. $2x - 7 = 43$

68. $9 - 2x = 1$

69. $\frac{3}{5}t + 6 = 15$

70. $\frac{2}{3}t - 4 = 2$

71. $\sqrt{4x} - 3 = 5$

72. $8 = \sqrt{2x} - 2$

SYNTHESIS

73. Is it possible for a quadratic equation to have one solution that is rational and one that is irrational? Why or why not?

74. Where does the ± symbol in the quadratic formula come from?

Solve.

75. $5x = -x(x - 7)$

76. $x(3x + 7) = 3x$

77. $3 - x(x - 3) = 4$

78. $x(5x - 7) = 1$

79. $(y + 4)(y + 3) = 15$

80. $x^2 + (x + 2)^2 = 7$

81. $\dfrac{x^2}{x + 3} - \dfrac{5}{x + 3} = 0$

82. $\dfrac{x^2}{x + 5} - \dfrac{7}{x + 5} = 0$

83. $\dfrac{1}{x} + \dfrac{1}{x + 1} = \dfrac{1}{3}$

84. $\dfrac{1}{x} + \dfrac{1}{x + 6} = \dfrac{1}{5}$

85. Find *r* in this figure. Round to the nearest hundredth.

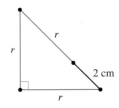

86. *Area of a square.* Find the area of a square for which the diagonal is 3 units longer than the length of the sides.

87. *Golden rectangle.* The so-called *golden rectangle* is said to be extremely pleasing visually and was used often by ancient Greek and Roman architects. The length of a golden rectangle is approximately 1.6 times the width. Find the dimensions of a golden rectangle if its area is 9000 m².

88. *Flagpoles.* A 20-ft flagpole is struck by lightning and, while not completely broken, falls over and touches the ground 10 ft from the bottom of the pole. How high up did the pole break?

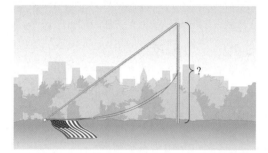

89. *Investments.* $4000 is invested at interest rate r for 2 yr. After 1 yr, an additional $2000 is invested, again at interest rate r. What is the interest rate if $6510 is in the account at the end of 2 yr?

90. *Investments.* In 2 yr, you want to have $3000. How much do you need to invest now if you can get an interest rate of 5.75% compounded annually?

91. *Enlarged strike zone.* In baseball, a batter's strike zone is a rectangular area about 15 in. wide and 40 in. high. Many batters subconsciously enlarge this area by 40% when fearful that if they don't swing, the umpire will call the pitch a strike. Assuming that the strike zone is enlarged by an invisible band of uniform width around the actual zone, find the dimensions of the enlarged strike zone.

92. Use a graph to approximate to the nearest thousandth the solutions of Exercises 35–40. Compare your answers with those found using a grapher or Table 2.

Formulas and Equations

9.4

Solving Formulas

Formulas arise frequently in the natural and social sciences, business, engineering, and health care. In Section 2.3, we saw that the same steps that are used to solve linear equations can be used to solve a formula that appears in this form. Similarly, the steps that are used to solve a rational, radical, or quadratic equation can also be used to solve a formula that appears in one of these forms. Before turning our attention to these formulas, let's briefly review the steps used to solve each type of equation.

CONNECTING THE CONCEPTS

We have now completed our study of the various types of equations that appear in elementary algebra. Below are examples of each type, along with their solutions. We leave all checks to the student.

LINEAR EQUATIONS

$$5x + 3 = 2x + 9$$

$3x = 6$	Adding $-2x - 3$ to both sides
$x = 2$	Dividing both sides by 3

The solution is 2.

RADICAL EQUATIONS

$$\sqrt{2x + 1} - 5 = 2$$

$\sqrt{2x + 1} = 7$	Adding 5 to both sides
$2x + 1 = 49$	Squaring both sides
$2x = 48$	Subtracting 1 from both sides
$x = 24$	Dividing both sides by 2

The solution is 24.

RATIONAL EQUATIONS

$$\frac{5}{2x} + \frac{4}{3x} = 2 \qquad \text{Note that } x \neq 0.$$

$6x\left(\dfrac{5}{2x} + \dfrac{4}{3x}\right) = 6x \cdot 2$	Multiplying both sides by the LCD, $6x$
$\dfrac{6x \cdot 5}{2x} + \dfrac{6x \cdot 4}{3x} = 12x$	Using the distributive law
$15 + 8 = 12x$	Simplifying
$23 = 12x$	
$\dfrac{23}{12} = x$	

The solution is $\frac{23}{12}$.

QUADRATIC EQUATIONS

$$2x^2 + 3x = 1$$

$2x^2 + 3x - 1 = 0$	Subtracting 1 from both sides
$x = \dfrac{-3 \pm \sqrt{3^2 - 4(2)(-1)}}{2 \cdot 2}$	Using the quadratic formula
$x = \dfrac{-3 \pm \sqrt{17}}{4}$	

The solutions are $\dfrac{-3}{4} + \dfrac{\sqrt{17}}{4}$ and $\dfrac{-3}{4} - \dfrac{\sqrt{17}}{4}$.

It is always wise to check solutions in the original equation, but this is especially important with radical or rational equations. The checks for the equations above are left to the student.

Example 1   Solve: **(a)** $3x + 17 = 8x$; **(b)** $\dfrac{3}{4x} + \dfrac{x}{6} = 1$; **(c)** $5 + \sqrt{2x} = 8$.

Solution

a)
$3x + 17 = 8x$	Recognizing this as a linear equation
$17 = 5x$	Subtracting $3x$ from both sides
$\dfrac{17}{5} = x$	Dividing both sides by 5

The solution is $\frac{17}{5}$. The check is left to the student.

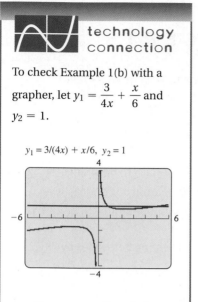

technology connection

To check Example 1(b) with a grapher, let $y_1 = \dfrac{3}{4x} + \dfrac{x}{6}$ and $y_2 = 1$.

$y_1 = 3/(4x) + x/6, \ y_2 = 1$

It appears as though the graph of $y_2 = 1$ intersects the curve in two points.

1. Use the INTERSECT feature to determine the solutions of Example 1(b) to the nearest hundredth.
2. How do your solutions compare with those found using the quadratic formula?

b)

$$\frac{3}{4x} + \frac{x}{6} = 1 \qquad \text{Recognizing this as a rational equation}$$

$$12x\left(\frac{3}{4x} + \frac{x}{6}\right) = 12x \cdot 1 \qquad \text{Multiplying both sides by the LCD, } 12x$$

$$\frac{12x \cdot 3}{4x} + \frac{12x \cdot x}{6} = 12x \qquad \text{Using the distributive law}$$

$$9 + 2x^2 = 12x \qquad \text{Simplifying: } \frac{12x \cdot 3}{4x} = 3 \cdot 3; \ \frac{12x \cdot x}{6} = 2x \cdot x$$

$$2x^2 - 12x + 9 = 0 \qquad \text{Subtracting } 12x \text{ from both sides}$$

$$x = \frac{-(-12) \pm \sqrt{(-12)^2 - 4 \cdot 2 \cdot 9}}{2 \cdot 2} \qquad \begin{array}{l}\text{Using the}\\ \text{quadratic formula}\end{array}$$

$$x = \frac{12 \pm \sqrt{72}}{4} = \frac{12 \pm \sqrt{36}\sqrt{2}}{4} \qquad \text{Simplifying}$$

$$x = \frac{12 \pm 6\sqrt{2}}{4} = \frac{2(6 \pm 3\sqrt{2})}{2 \cdot 2} \qquad \begin{array}{l}\text{Removing a factor}\\ \text{equal to 1: } 2/2 = 1\end{array}$$

$$x = \frac{6}{2} \pm \frac{3}{2}\sqrt{2} = 3 \pm \frac{3}{2}\sqrt{2} \qquad \text{Simplifying}$$

The solutions are $3 - \frac{3}{2}\sqrt{2}$ and $3 + \frac{3}{2}\sqrt{2}$. Note that neither solution creates a denominator equal to 0. A complete check is left to the student.

c) $\ 5 + \sqrt{2x} = 8 \qquad \text{Recognizing this as a radical equation}$

$\qquad \sqrt{2x} = 3 \qquad \text{Subtracting 5 from both sides}$

$\qquad \ \ 2x = 9 \qquad \text{Using the principle of powers: squaring both sides}$

$\qquad \ \ \ x = \frac{9}{2} \qquad \text{Dividing both sides by 2}$

Check:
$$\begin{array}{r} 5 + \sqrt{2x} = 8 \\ \hline 5 + \sqrt{2 \cdot \frac{9}{2}} \ ? \ 8 \\ 5 + \sqrt{9} \ \Big| \\ 5 + 3 \ \Big| \ 8 \quad \text{TRUE} \end{array}$$

The solution is $\frac{9}{2}$.

Solving Formulas

Formulas can be linear, rational, radical, or quadratic equations, as well as other types of equations that are beyond the scope of this text. To solve formulas, we use the same steps that we use to solve equations. Probably the greatest difference is that while the solution of an equation is a number, the solution of a formula is generally a variable expression.

E x a m p l e 2

Intelligence quotient. The formula $Q = \dfrac{100m}{c}$ is used to determine the intelligence quotient, Q, of a person of mental age m and chronological age c. Solve for c.

Solution We have

$$Q = \frac{100m}{c}$$

$$c \cdot Q = c \cdot \frac{100m}{c} \qquad \text{Multiplying both sides by } c$$

$$cQ = 100m \qquad \text{Removing a factor equal to 1: } \frac{c}{c} = 1$$

$$c = \frac{100m}{Q}. \qquad \text{Dividing both sides by } Q$$

This formula can be used to determine a person's chronological, or actual, age from his or her mental age and intelligence quotient.

E x a m p l e   3

A work formula. The formula $t/a + t/b = 1$ was used in Section 6.7. Solve this formula for t.

Solution We have

$$\frac{t}{a} + \frac{t}{b} = 1$$

$$ab\left(\frac{t}{a} + \frac{t}{b}\right) = ab \cdot 1 \qquad \text{Multiplying by the LCD, } ab, \text{ to clear fractions}$$

$$\left.\begin{array}{l} \dfrac{abt}{a} + \dfrac{abt}{b} = ab \\[2mm] bt + at = ab. \end{array}\right\rbrace \qquad \begin{array}{l} \text{Multiplying to remove parentheses and} \\ \text{removing factors equal to 1: } \dfrac{a}{a} = 1 \text{ and } \dfrac{b}{b} = 1 \end{array}$$

If the last equation were $3t + 2t = 6$, we would simply combine like terms ($3t + 2t$ is $5t$) and then divide by the coefficient of t (which would be 5). Since $bt + at$ *cannot* be combined, we factor instead:

$$(b + a)t = ab \qquad \text{Factoring out } t, \text{ the letter for which we are solving}$$

$$t = \frac{ab}{b + a}. \qquad \text{Dividing both sides by } b + a$$

The answer to Example 3 can be used when the times required to do a job independently (a and b) are known and t represents the time required to complete the task working together.

E x a m p l e   4

Temperature of a fluid. The temperature T of a fluid being heated by a steady flame or burner can be determined by the formula $T = a + b\sqrt{ct}$, where a is the initial temperature of the fluid, b and c are constants determined by the intensity of the heat and the type of fluid used, and t is time. Solve for t.

Solution   We have

$$T = a + b\sqrt{ct}$$

$$T - a = b\sqrt{ct} \qquad \text{Subtracting } a \text{ from both sides}$$

$$\frac{T - a}{b} = \sqrt{ct} \qquad \text{Dividing both sides by } b$$

$$\left(\frac{T - a}{b}\right)^2 = ct \qquad \begin{array}{l}\text{Using the principle of powers:}\\ \text{squaring both sides}\end{array}$$

$$\frac{T^2 - 2Ta + a^2}{b^2} = ct \qquad \text{Squaring } T - a \text{ and } b$$

$$\frac{T^2 - 2Ta + a^2}{b^2 c} = t. \qquad \text{Dividing both sides by } c$$

This equation can be used to determine the heating time required for an experiment.

In some cases, the quadratic formula is needed.

E x a m p l e   5

A *physics formula.* The distance s of an object from some established point can be determined using the formula $s = s_0 + v_0 t + \frac{1}{2}at^2$, where s_0 is the object's original distance from the point, v_0 is the object's original velocity, a is the object's acceleration, and t is time. Solve for t.

Solution   Note first that t appears in two terms, raised to the first and then the second powers. Thus the equation is "quadratic in t," and the quadratic formula is needed:

$$\tfrac{1}{2}at^2 + v_0 t + s_0 = s \qquad \text{Rewriting the equation from left to right}$$

$$\underbrace{\tfrac{1}{2}at^2}_{a} + \underbrace{v_0 t}_{b} + \underbrace{s_0 - s}_{c} = 0 \qquad \begin{array}{l}\text{Subtracting } s \text{ from both sides and}\\ \text{identifying the coefficients needed}\\ \text{for the quadratic formula}\end{array}$$

$$t = \frac{-v_0 \pm \sqrt{v_0^2 - 4\left(\frac{1}{2}a\right)(s_0 - s)}}{2 \cdot \frac{1}{2}a} \qquad \begin{array}{l}\text{Substituting into}\\ \text{the quadratic}\\ \text{formula}\end{array}$$

$$t = \frac{-v_0 \pm \sqrt{v_0^2 - 2a(s_0 - s)}}{a}$$

$$t = \frac{-v_0 \pm \sqrt{v_0^2 - 2as_0 + 2as}}{a}. \qquad \text{Simplifying}$$

Exercise Set 9.4

Solve.

1. $3x + 5 = 7x + 1$

2. $x - 9 = 5x + 4$

3. $x^2 = 5x - 6$

4. $3x + 4 = x^2$

5. $4 + \sqrt{x} = 9$

6. $7 = \sqrt{x} - 5$

7. $\dfrac{5}{x} + \dfrac{3}{4} = 2$

8. $\dfrac{2}{5} + \dfrac{3}{x} = 4$

9. $4(2x - 1) = 3x - 5$

10. $4x - 7 = 2(5x - 3)$

11. $\dfrac{3}{10x} - \dfrac{4}{5x} = 6$

12. $\dfrac{2}{9x} - \dfrac{5}{6} = 1$

13. $2t^2 - 7t + 3 = 0$

14. $3t^2 - 7t + 2 = 0$

15. $5\sqrt{2t} - 7 = 3$

16. $11 = 4\sqrt{3t} - 5$

17. $\dfrac{1}{4t} + \dfrac{t}{6} = 2$

18. $\dfrac{3}{10t} + \dfrac{t}{5} = 1$

19. $7t - 1 = 2(3 - t) + 5$

20. $5t - 3 = 4(1 - t) + 2$

21. $3\sqrt{2t - 1} = 2\sqrt{5t + 3}$

22. $5\sqrt{3t + 2} = 4\sqrt{2t - 3}$

23. $2n - 1 = 3n^2$

24. $3 - 5n = 2n^2$

25. $1 - \dfrac{3}{7n} = \dfrac{5}{14}$

26. $4 - \dfrac{5}{6n} = \dfrac{1}{12}$

Solve each formula for the specified variable.

27. $s = \dfrac{1}{2}gt^2$, for g

(A physics formula for distance)

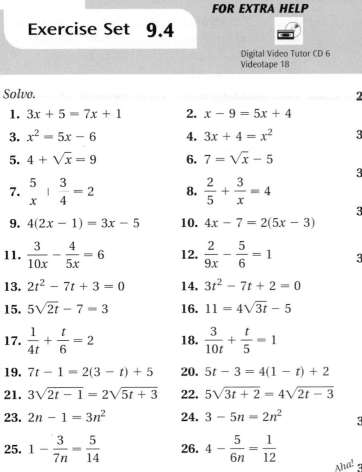

28. $A = \frac{1}{2}bh$, for h

(The area of a triangle)

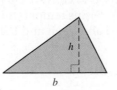

29. $S = 2\pi rh$, for h

(A formula for surface area)

30. $A = P(1 + rt)$, for t

(An interest formula)

31. $d = c\sqrt{h}$, for h

(A formula for distance to the horizon)

32. $n = c + \sqrt{t}$, for t

(A formula for population)

33. $\dfrac{1}{R} = \dfrac{1}{r_1} + \dfrac{1}{r_2}$, for R

(An electricity formula)

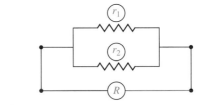

34. $\dfrac{1}{p} + \dfrac{1}{q} = \dfrac{1}{f}$, for f

(An optics formula)

Aha! **35.** $ax^2 + bx + c = 0$, for x

36. $(x - d)^2 = k$, for x

37. $\dfrac{m}{n} = p - q$, for n

38. $\dfrac{M - g}{t} = r + s$, for t

39. $rl + rS = L$, for r

40. $T = mg + mf$, for m

41. $S = 2\pi r(r + h)$, for h

(The surface area of a right circular cylinder)

42. $ab = ac + d$, for a

43. $\dfrac{s}{h} = \dfrac{h}{t}$, for h
(A geometry formula)

44. $\dfrac{x}{y} = \dfrac{z}{x}$, for x

45. $mt^2 + nt - p = 0$, for t
(*Hint*: Use the quadratic formula.)

46. $rs^2 - ts + p = 0$, for s

47. $\dfrac{m}{n} = r$, for n

48. $s + t = \dfrac{r}{v}$, for v

49. $m + t = \dfrac{n}{m}$, for m

50. $x - y = \dfrac{z}{y}$, for y

51. $n = p - 3\sqrt{t + c}$, for t

52. $M - m\sqrt{ct} = r$, for t

53. $\sqrt{m - n} = \sqrt{3t}$, for n

54. $\sqrt{2t + s} = \sqrt{s - t}$, for t

55. Is it easier to solve
$$\frac{1}{25} + \frac{1}{23} = \frac{1}{x} \text{ for } x,$$
or to solve
$$\frac{1}{p} + \frac{1}{q} = \frac{1}{f} \text{ for } f?$$
Explain why.

56. Explain why someone might want to solve
$A = \frac{1}{2}bh$ for h.
(See Exercise 28.)

SKILL MAINTENANCE

Simplify.

57. $\sqrt{6} \cdot \sqrt{10}$

58. $\sqrt{21} \cdot \sqrt{15}$

59. $\sqrt{150}$

60. $\sqrt{63}$

61. $\sqrt{4a^7b^4}$

62. $\sqrt{9x^8y^5}$

SYNTHESIS

63. As a step in solving a formula for a certain variable, a student takes the reciprocal on both sides of the equation. Is a mistake being made? Why or why not?

64. Describe a situation in which the result of Example 3,
$$t = \frac{ab}{b + a},$$
would be especially useful.

65. *Health care.* Young's rule for determining the size of a particular child's medicine dosage c is
$$c = \frac{a}{a + 12} \cdot d,$$
where a is the child's age and d is the typical adult dosage (*Source*: Olsen, June Looby, Leon J. Ablon, and Anthony Patrick Giangrasso, *Medical Dosage Calculations,* 6th ed., p. A-31). If a child receives 8 mg of antihistamine when the typical adult receives 24 mg, how old is the child?

Solve.

66. $Sr = \dfrac{rl - a}{r - l}$, for r

67. $fm = \dfrac{gm - t}{m}$, for m

68. $\dfrac{n_1}{p_1} + \dfrac{n_2}{p_2} = \dfrac{n_2 - n_1}{R}$, for n_2

69. $u = -F\left(E - \dfrac{P}{T}\right)$, for T

70. *Marine biology.* The formula
$$N = \frac{(b + d)f_1 - v}{(b - v)f_2}$$
is used when monitoring the water in fisheries. Solve for v.

71. *Meteorology.* The formula
$$C = \tfrac{5}{9}(F - 32)$$
is used to convert the Fahrenheit temperature F to the Celsius temperature C. At what temperature are the Fahrenheit and Celsius readings the same?

Complex Numbers as Solutions of Quadratic Equations

9.5

The Complex-Number System • Solutions of Equations

The Complex-Number System

Because negative numbers do not have square roots that are real numbers, mathematicians have devised a larger set of numbers known as *complex numbers*. In the complex-number system, the number i is used to represent the square root of -1.

> ### The Number i
> We define the number $i = \sqrt{-1}$. That is, $i = \sqrt{-1}$ and $i^2 = -1$.

E x a m p l e   1

Express in terms of i.

a) $\sqrt{-3}$ **b)** $\sqrt{-25}$

c) $-\sqrt{-10}$ **d)** $\sqrt{-24}$

Solution

a) $\sqrt{-3} = \sqrt{-1 \cdot 3} = \sqrt{-1} \cdot \sqrt{3} = i\sqrt{3}$, or $\sqrt{3}i$ ⟵——— i is *not* under the radical.

b) $\sqrt{-25} = \sqrt{-1 \cdot 25} = \sqrt{-1} \cdot \sqrt{25} = i \cdot 5 = 5i$

c) $-\sqrt{-10} = -\sqrt{-1 \cdot 10} = -\sqrt{-1} \cdot \sqrt{10} = -i\sqrt{10}$, or $-\sqrt{10}i$

d) $\sqrt{-24} = \sqrt{4(-1)6} = \sqrt{4}\,\sqrt{-1}\sqrt{6} = 2i\sqrt{6}$, or $2\sqrt{6}i$

> ### Imaginary Numbers
> An *imaginary* number* is a number that can be written in the form $a + bi$, where a and b are real numbers and $b \neq 0$.

The following are examples of imaginary numbers:

$$3 + 8i, \quad \sqrt{7} - 2i, \quad 4 + \sqrt{6}i, \quad \text{and} \quad 4i \text{ (here } a = 0\text{).}^{\dagger}$$

The imaginary numbers together with the real numbers form the set of **complex numbers**.

---

*The name "imaginary" should not lead you to believe that these numbers are not useful. Imaginary numbers have important applications in engineering and the physical sciences.
†Numbers like $4i$ are often called *pure imaginary* because they are of the form $0 + bi$.

> ### Complex Numbers
>
> A *complex number* is any number that can be written as $a + bi$, where a and b are real numbers. (Note that a and b both can be 0.)

It may help to remember that every real number is a complex number ($a + bi$ with $b = 0$), but not every complex number is real ($a + bi$ with $b \neq 0$). For example, numbers like $2 + 3i$ and $-7i$ are complex but not real.

Solutions of Equations

As we saw in Example 1(c) of Section 9.3, not all quadratic equations have real-number solutions. All quadratic equations *do* have complex-number solutions. These solutions are usually written in the form $a + bi$ unless a or b is zero.

E x a m p l e   2

Solve: **(a)** $x^2 + 3x + 4 = 0$; **(b)** $x^2 + 2 = 2x$.

Solution

a) We use the quadratic formula:

$$1x^2 + 3x + 4 = 0$$

$$x = \frac{-3 \pm \sqrt{3^2 - 4 \cdot 1 \cdot 4}}{2 \cdot 1} \qquad \text{Remember:} \quad x = \frac{-b \pm \sqrt{b^2 - 4ac}}{2a}.$$

$$x = \frac{-3 \pm \sqrt{-7}}{2} \qquad \text{Simplifying}$$

$$x = \frac{-3 \pm \sqrt{-1}\,\sqrt{7}}{2}$$

$$\left. x = \frac{-3 \pm i\sqrt{7}}{2} \right\} \qquad \text{Rewriting } \sqrt{-7} \text{ as } i\sqrt{7}$$

$$x = -\frac{3}{2} \pm \frac{\sqrt{7}}{2}i. \qquad \begin{array}{l} \text{Writing in the} \\ \text{form } a + bi \end{array}$$

The solutions are $-\dfrac{3}{2} + \dfrac{\sqrt{7}}{2}i$ and $-\dfrac{3}{2} - \dfrac{\sqrt{7}}{2}i$.

b) We have

$$x^2 + 2 = 2x$$

$$1x^2 - 2x + 2 = 0 \qquad \begin{array}{l} \text{Rewriting in standard form and} \\ \text{identifying } a, b, \text{ and } c \end{array}$$

$$x = \frac{-(-2) \pm \sqrt{(-2)^2 - 4 \cdot 1 \cdot 2}}{2 \cdot 1} \qquad x = \frac{-b \pm \sqrt{b^2 - 4ac}}{2a}$$

$$x = \frac{2 \pm \sqrt{-4}}{2} \qquad \text{Simplifying}$$

$$x = \frac{2 \pm \sqrt{-1}\,\sqrt{4}}{2}$$

$$x = \frac{2 \pm i2}{2} \qquad \sqrt{-1} = i \text{ and } \sqrt{4} = 2$$

$$x = \frac{2}{2} \pm \frac{2i}{2} \qquad \text{Rewriting in the form } a + bi$$

$$x = 1 \pm i. \qquad \text{Simplifying}$$

The solutions are $1 + i$ and $1 - i$.

Exercise Set **9.5**

FOR EXTRA HELP

Digital Video Tutor CD 6   InterAct Math   Math Tutor Center   MathXL   MyMathLab.com
Videotape 18

Express in terms of i.

1. $\sqrt{-1}$

2. $\sqrt{-36}$

3. $\sqrt{-16}$

4. $\sqrt{-81}$

5. $\sqrt{-50}$

6. $\sqrt{-44}$

7. $-\sqrt{-45}$

8. $-\sqrt{-20}$

9. $-\sqrt{-18}$

10. $-\sqrt{-28}$

11. $3 + \sqrt{-49}$

12. $7 + \sqrt{-4}$

13. $5 + \sqrt{-9}$

14. $-8 - \sqrt{-36}$

15. $2 - \sqrt{-98}$

16. $-2 + \sqrt{-125}$

Solve.

17. $x^2 + 4 = 0$

18. $x^2 + 9 = 0$

19. $x^2 = -28$

20. $x^2 = -48$

21. $t^2 + 4t + 5 = 0$

22. $t^2 - 4t + 6 = 0$

23. $(t + 3)^2 = -16$

24. $(t - 2)^2 = -25$

25. $x^2 + 5 = 2x$

26. $x^2 + 3 = -2x$

27. $t^2 + 7 - 4t = 0$

28. $t^2 + 8 + 4t = 0$

29. $2t^2 + 6t + 5 = 0$

30. $4y^2 + 3y + 2 = 0$

31. $1 + 2m + 3m^2 = 0$

32. $4p^2 + 3 = 6p$

33. Is it possible for a quadratic equation to have one imaginary-number solution and one real-number solution? Why or why not?

34. Under what condition(s) will an equation of the form $x^2 = c$ have imaginary-number solutions?

SKILL MAINTENANCE

Graph.

35. $y - \frac{3}{5}x$

36. $2x - 3y = 10$

37. $y = -4x$

38. $x = 2$

Simplify.

Aha! **39.** $(-17)^2 - (8 + 9)^2$

40. $-1(-4)^2(43)0$

SYNTHESIS

41. When using the quadratic formula to solve an equation, if $b^2 < 4ac$, are the solutions imaginary? Why or why not?

42. Can imaginary-number solutions of a quadratic equation be found using the method of completing the square? Why or why not?

Solve.

43. $(x + 1)^2 + (x + 3)^2 = 0$

44. $(p + 5)^2 + (p + 1)^2 = 0$

45. $\dfrac{2x - 1}{5} - \dfrac{2}{x} = \dfrac{x}{2}$

46. $\dfrac{1}{a - 1} - \dfrac{2}{a - 1} = 3a$

47. Use a grapher to confirm that there are no real-number solutions of Examples 2(a) and 2(b).

Graphs of Quadratic Equations

9.6

Graphing Equations of the Form $y = ax^2$ •
Graphing Equations of the Form $y = ax^2 + bx + c$

In this section, we will graph quadratic equations like

$$y = \tfrac{1}{2}x^2, \qquad y = x^2 + 2x - 3, \quad \text{and} \quad y = -5x^2 + 4.$$

Such equations, of the form $y = ax^2 + bx + c$ with $a \neq 0$, have graphs that are either cupped upward or downward. These graphs are symmetric with respect to an **axis of symmetry**, as shown below. When folded along its axis, the graph has two halves that match exactly.

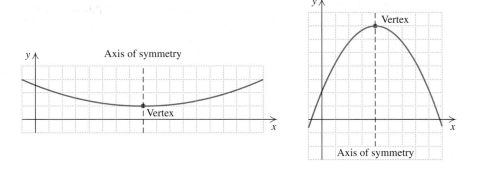

The point at which the graph of a quadratic equation crosses its axis of symmetry is called the **vertex** (plural, vertices). The y-coordinate of the vertex is the graph's largest value of y (if the curve opens downward) or smallest value of y (if the curve opens upward). Graphs of quadratic equations are called **parabolas**.

Graphing Equations of the Form $y = ax^2$

The simplest parabolas to sketch are given by equations of the form $y = ax^2$.

Example 1

Graph: $y = x^2$.

Solution   We choose numbers for x and find the corresponding values for y.

$\quad$ If $x = -2$, then $y = (-2)^2 = 4$. $\qquad$ We get the pair $(-2, 4)$.
$\quad$ If $x = -1$, then $y = (-1)^2 = 1$. $\qquad$ We get the pair $(-1, 1)$.
$\quad$ If $x = 0$, then $y = 0^2 = 0$. $\qquad$ We get the pair $(0, 0)$.
$\quad$ If $x = 1$, then $y = 1^2 = 1$. $\qquad$ We get the pair $(1, 1)$.
$\quad$ If $x = 2$, then $y = 2^2 = 4$. $\qquad$ We get the pair $(2, 4)$.

The following table lists these solutions of $y = x^2$. After several ordered pairs are found, we plot them and connect them with a smooth curve.

x	y $y = x^2$	(x, y)
-2	4	$(-2, 4)$
-1	1	$(-1, 1)$
0	0	$(0, 0)$
1	1	$(1, 1)$
2	4	$(2, 4)$

In Example 1, the vertex is $(0, 0)$ and the axis of symmetry is the y-axis. This will be the case for any parabola having an equation of the form $y = ax^2$.

E x a m p l e 2   Graph: $y = -\frac{1}{2}x^2$.

Solution   We select numbers for x, find the corresponding y-values, plot the resulting ordered pairs, and connect them with a smooth curve.

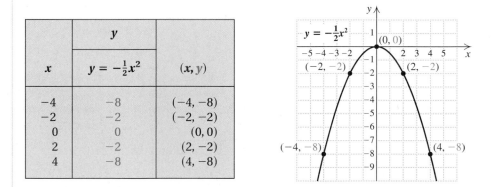

x	y $y = -\frac{1}{2}x^2$	(x, y)
-4	-8	$(-4, -8)$
-2	-2	$(-2, -2)$
0	0	$(0, 0)$
2	-2	$(2, -2)$
4	-8	$(4, -8)$

Graphing Equations of the Form $y = ax^2 + bx + c$

Recall from our work with lines that it is often very useful to know the x- and y-intercepts of a line. Intercepts are also useful when graphing parabolas. To find the intercepts of a parabola, we use the same approach we used with lines.

> ### The Intercepts of a Parabola
>
> To find the y-intercept of the graph of $y = ax^2 + bx + c$, replace x with 0 and solve for y.
>
> To find any x-intercept(s) of the graph of $y = ax^2 + bx + c$, replace y with 0 and solve for x.

E x a m p l e   3

Find all y- and x-intercepts of the graph of $y = 2x^2 - x - 28$.

Solution   To find the y-intercept, we replace x with 0 and solve for y:

$$y = 2 \cdot 0^2 - 0 - 28$$
$$y = 0 - 0 - 28$$
$$y = -28.$$

When x is 0, we have $y = -28$. Thus the y-intercept is $(0, -28)$.

To find the x-intercept(s), we replace y with 0 and solve for x:

$$0 = 2x^2 - x - 28.$$

The quadratic formula could be used, but factoring is faster:

$$0 = (2x + 7)(x - 4) \qquad \text{Factoring}$$

$$2x + 7 = 0 \quad or \quad x - 4 = 0$$
$$2x = -7 \quad or \qquad x = 4$$
$$x = -\tfrac{7}{2} \quad or \qquad x = 4.$$

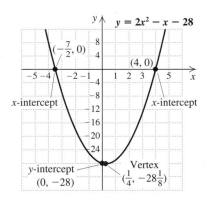

The x-intercepts are $(4, 0)$ and $\left(-\tfrac{7}{2}, 0\right)$. The y-intercept is $(0, -28)$.

Although we were not asked to graph the equation in Example 3, we did so to show that the x-coordinate of the vertex, $\tfrac{1}{4}$, is exactly midway between the x-intercepts. The quadratic formula, $x = \dfrac{-b \pm \sqrt{b^2 - 4ac}}{2a}$, can be used to locate x-intercepts when an equation of the form $y = ax^2 + bx + c$ is graphed. If one x-intercept is determined by $\dfrac{-b - \sqrt{b^2 - 4ac}}{2a}$ and the other by $\dfrac{-b + \sqrt{b^2 - 4ac}}{2a}$, then the average of these two values can be used to find the x-coordinate of the vertex. In Exercise 41, you are asked to show that this x-value is $\dfrac{-b}{2a}$. A more complicated approach can be used to show that the x-value of the vertex is $\dfrac{-b}{2a}$ even when no x-intercepts exist.

> ### The Vertex of a Parabola
>
> For a parabola given by the quadratic equation $y = ax^2 + bx + c$:
>
> **1.** The x-coordinate of the vertex is $-\dfrac{b}{2a}$.
>
> **2.** The y-coordinate of the vertex is found by substituting $-\dfrac{b}{2a}$ for x and solving for y.

E x a m p l e   4

Graph: $y = x^2 + 2x - 3$.

Solution   Our plan is to plot the vertex and some points on either side of the vertex. We will then draw a parabola passing through these points.

To locate the vertex, we use $-b/(2a)$ to find its x-coordinate:

$$x\text{-coordinate of the vertex} = -\frac{b}{2a} = -\frac{2}{2 \cdot (1)}$$
$$= -1.$$

We substitute -1 for x to find the y-coordinate of the vertex:

$$y\text{-coordinate of the vertex} = (-1)^2 + 2(-1) - 3 = 1 - 2 - 3$$
$$= -4.$$

The vertex is $(-1, -4)$. The axis of symmetry is $x = -1$.

We choose some x-values on both sides of the vertex and graph the parabola.

x	y $y = x^2 + 2x - 3$	(x, y)	
-4	5	$(-4, 5)$	
-3	0	$(-3, 0)$	← x-intercept
-2	-3	$(-2, -3)$	
-1	-4	$(-1, -4)$	← Vertex
0	-3	$(0, -3)$	← y-intercept
1	0	$(1, 0)$	← x-intercept
2	5	$(2, 5)$	

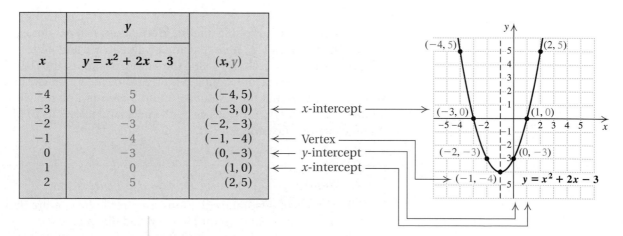

One tip for graphing quadratic equations involves the coefficient of x^2. Note that a in $y = ax^2 + bx + c$ tells us whether the graph opens upward or downward. When a is positive, as in Examples 1, 3, and 4, the graph opens upward; when a is negative, as in Example 2, the graph opens downward.

E x a m p l e   5   Graph: $y = -2x^2 + 4x + 1$.

Solution   Since the coefficient of x^2 is negative, we know that the graph opens downward. To locate the vertex, we first find its x-coordinate:

$$x\text{-coordinate of the vertex} = -\frac{b}{2a} = -\frac{4}{2 \cdot (-2)}$$

$$= 1.$$

We substitute 1 for x to find the y-coordinate of the vertex:

$$y\text{-coordinate of the vertex} = -2 \cdot 1^2 + 4 \cdot 1 + 1 = -2 + 4 + 1$$

$$= 3.$$

The vertex is $(1, 3)$. The axis of symmetry is $x = 1$.

We choose some x-values on both sides of the vertex, calculate their corresponding y-values, and graph the parabola.

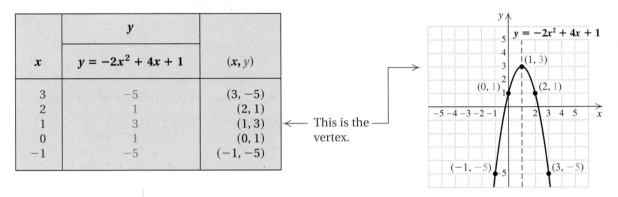

	y	
x	$y = -2x^2 + 4x + 1$	(x, y)
3	-5	$(3, -5)$
2	1	$(2, 1)$
1	3	$(1, 3)$ ← This is the vertex.
0	1	$(0, 1)$
-1	-5	$(-1, -5)$

A second tip for graphing quadratic equations can cut our calculation time in half. In Examples 1–5, note that any x-value to the left of the vertex is paired with the same y-value as an x-value the same distance to the right of the vertex. Thus, since the vertex for Example 5 is $(1, 3)$ and since -1 and 3 are both 2 units from 1, we know that the x-values -1 and 3 are both paired with the same y-value. This symmetry provides a useful check and allows us to plot *two* points after calculating just *one*.

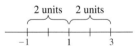

2 units   2 units

> ### Guidelines for Graphing Quadratic Equations
> 1. Graphs of quadratic equations, $y = ax^2 + bx + c$, are parabolas. They are cupped upward if $a > 0$ and downward if $a < 0$.
> 2. Use the formula $x = -b/(2a)$ to find the x-coordinate of the vertex. After calculating the y-coordinate, plot the vertex and some points on either side of it.
> 3. After a point has been graphed, a second point with the same y-coordinate can be found on the opposite side of the axis of symmetry.
> 4. Graph the y-intercept and, if requested, any x-intercepts.

Graph each quadratic equation, labeling the vertex and the y-intercept.

1. $y = x^2 + 1$

2. $y = x^2 + 2$

3. $y = -2x^2$

4. $y = -1 \cdot x^2$

5. $y = -x^2 + 2x$

6. $y = x^2 + x - 6$

7. $y = x^2 + 2x - 3$

8. $y = x^2 + 2x + 1$

9. $y = 3x^2 - 12x + 11$

10. $y = 2x^2 - 12x + 13$

11. $y = -2x^2 - 4x + 1$

12. $y = -3x^2 - 2x + 8$

13. $y = \frac{1}{4}x^2$

14. $y = -\frac{1}{3}x^2$

15. $y = -\frac{1}{2}x^2 + 5$

16. $y = \frac{1}{2}x^2 - 7$

17. $y = x^2 - 3x$

18. $y = x^2 + 4x$

Graph each equation, labeling the vertex, the y-intercept, and any x-intercepts. If an x-intercept is irrational, round to three decimal places.

19. $y = x^2 + 2x - 8$

20. $y = x^2 + x - 6$

21. $y = 2x^2 - 5x$

22. $y = 2x^2 - 7x$

23. $y = -x^2 - x + 12$

24. $y = -x^2 - 3x + 10$

25. $y = 3x^2 - 6x + 1$

26. $y = 3x^2 + 12x + 11$

27. $y = x^2 + 2x + 3$

28. $y = -x^2 - 2x - 3$

29. $y = 3 - 4x - 2x^2$

30. $y = 1 - 4x - 2x^2$

31. Why is it helpful to know the coordinates of the vertex when graphing a parabola?

32. Suppose that both x-intercepts of a parabola are known. What is the easiest way to find the coordinates of the vertex?

SKILL MAINTENANCE

Evaluate.

33. $3a^2 - 5a$, for $a = -1$

34. $4t^2 + 3t$, for $t = 2$

35. $5x^3 - 2x$, for $x = -1$

36. $3t^4 + 5t^2$, for $t = 2$

37. $-9.8x^5 + 3.2x$, for $x = 1$

38. $-32t^2 - 3t$, for $t = 0$

SYNTHESIS

39. Describe a method that could be used to find an equation for a parabola that has x-intercepts at r_1 and r_2.

40. What effect does the size of $|a|$ have on the graph of $y = ax^2 + bx + c$?

41. Show that the average of

$$\frac{-b - \sqrt{b^2 - 4ac}}{2a} \quad \text{and} \quad \frac{-b + \sqrt{b^2 - 4ac}}{2a}$$

is $-\dfrac{b}{2a}$.

42. *Height of a projectile.*   The height H, in feet, of a projectile with an initial velocity of 96 ft/sec is given by the equation

$$H = -16t^2 + 96t,$$

where t is the number of seconds from launch. Use the graph of this equation, shown below, or any equation-solving technique to answer the following.

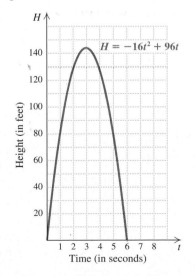

a) How many seconds after launch is the projectile 128 ft above ground?

b) When does the projectile reach its maximum height?

c) How many seconds after launch does the projectile return to the ground?

43. *Stopping distance.* In how many feet can a car stop if it is traveling at a speed of r mph? One estimate, developed in Britain, is as follows. The distance d, in feet, is given by

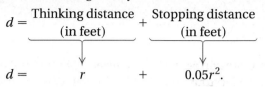

$$d = \underbrace{\text{Thinking distance}}_{\text{(in feet)}} + \underbrace{\text{Stopping distance}}_{\text{(in feet)}}$$

$$d = r + 0.05r^2.$$

a) How many feet would it take to stop a car traveling 25 mph? 40 mph? 55 mph? 65 mph? 75 mph? 100 mph?

b) Graph the equation, assuming $r \geq 0$.

44. On one set of axes, graph $y = x^2$, $y = (x - 3)^2$, and $y = (x + 1)^2$. Describe the effect that h has on the graph of $y = (x - h)^2$.

45. On one set of axes, graph $y = x^2$, $y = x^2 - 5$, and $y = x^2 + 2$. Describe the effect that k has on the graph of $y = x^2 + k$.

46. *Seller's supply.* As the price of a product increases, the seller is willing to sell, or *supply,* more of the product. Suppose that the supply for a certain product is given by

$$S = p^2 + p + 10,$$

where p is the price in dollars and S is the number sold, in thousands, at that price. Graph the equation for values of p such that $0 \leq p \leq 6$.

47. *Consumer's demand.* As the price of a product increases, consumers purchase, or *demand,* less of the product. Suppose that the demand for a certain product is given by

$$D = (p - 6)^2,$$

where p is the price in dollars and D is the number sold, in thousands, at that price. Graph the equation for values of p such that $0 \leq p \leq 6$.

48. *Equilibrium point.* The price p at which the consumer and the seller agree determines the *equilibrium point.* Find p such that

$$D = S$$

for the demand and supply curves in Exercises 46 and 47. How many units of the product will be sold at that price?

49. Use a grapher to draw the graph of $y = x^2 - 5$ and then, using the graph, estimate $\sqrt{5}$ to four decimal places.

9.7

Functions

Identifying Functions • **Functions Written as Formulas** • **Function Notation** • **Graphs of Functions** • **Recognizing Graphs of Functions**

Functions are enormously important in modern mathematics. The more mathematics you study, the more you will use functions.

Identifying Functions

Functions appear regularly in magazines and newspapers although they are usually not referred to as such. Consider the following table.

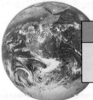

YEAR	1990	1991	1992	1993	1994	1995	1996	1997	1998	1999
Global temperature (in degrees Fahrenheit)	59.85°	59.74°	59.23°	59.36°	59.56°	59.72°	59.58°	59.74°	60.26°	59.81°

Source. National Oceanic and Atmospheric Administration

Note that to each year there corresponds *exactly one* temperature. A correspondence of this sort is called a **function**.

> ### Function
>
> A *function* is a correspondence (or rule) that assigns to each member of some set (called the *domain*) exactly one member of a set (called the *range*).

Sometimes the members of the domain are called **inputs**, and the members of the range **outputs**.

Domain (Set of Inputs)	Range (Set of Outputs)
1990	59.85
1991	59.74
1992	59.23
1993	59.36
1994	59.56
1995	59.72
1996	59.58
1997	
1998	60.26
1999	59.81

Note that each input has exactly one output, even though one of the outputs, 59.74, is used twice.

Example 1   Determine whether or not each of the following correspondences is a function.

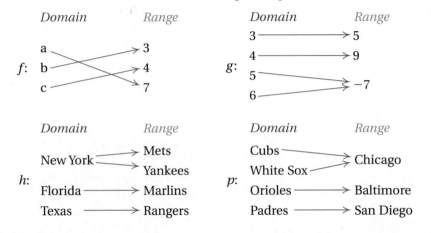

Solution   Correspondence f is a function because each member of the domain is matched to just one member of the range.

Correspondence g is also a function because each member of the domain is matched to just one member of the range.

Correspondence h is *not* a function because one member of the domain, New York, is matched to more than one member of the range.

Correspondence p is a function because each member of the domain is paired with just one member of the range.

Functions Written as Formulas

Many functions are described by formulas. Equations like $y = x + 3$ and $y = 4x^2$ are examples of such formulas. Outputs are found by substituting members of the domain for x.

E x a m p l e   2   **Thunderstorm distance.**   During a thunderstorm, it is possible to calculate how far away lightning is by using the formula

$$M = \tfrac{1}{5}t.$$

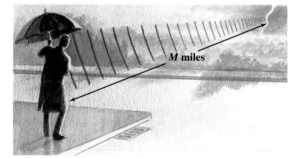

M miles

Here M is the distance, in miles, that a storm is from an observer when the sound of thunder arrives t seconds after the lightning has been sighted.

Complete the following table for this function.

t **(in seconds)**	0	1	2	3	4	5	6	10
M **(in miles)**	0	$\frac{1}{5}$						

Solution   To complete the table, we substitute values of t and compute M.

For $t = 2$, $M = \frac{1}{5} \cdot 2 = \frac{2}{5}$;   For $t = 3$, $M = \frac{1}{5} \cdot 3 = \frac{3}{5}$;

For $t = 4$, $M = \frac{1}{5} \cdot 4 = \frac{4}{5}$;   For $t = 5$, $M = \frac{1}{5} \cdot 5 = 1$;

For $t = 6$, $M = \frac{1}{5} \cdot 6 = \frac{6}{5}$, or $1\frac{1}{5}$;   For $t = 10$, $M = \frac{1}{5} \cdot 10 = 2$.

Function Notation

In Example 2, it was somewhat time-consuming to repeatedly write "For $t = \blacksquare$, $M = \frac{1}{5} \cdot \blacksquare$." Function notation clearly and concisely presents inputs and outputs together. The notation $M(t)$, read "M of t," denotes the output that is paired with the input t by the function M. Thus, for Example 2,

$$M(2) = \tfrac{1}{5} \cdot 2 = \tfrac{2}{5}, \qquad M(3) = \tfrac{1}{5} \cdot 3 = \tfrac{3}{5}, \quad \text{and, in general,} \quad M(t) = \tfrac{1}{5} \cdot t.$$

The notation $M(4) = \frac{4}{5}$ makes clear that when 4 is the input, $\frac{4}{5}$ is the output.

> **Caution!** $M(4)$ *does not* mean M times 4 and should not be read that way.

Equations for nonvertical lines can be written in function notation. For example, $f(x) = x + 2$, read "f of x equals x plus 2," can be used instead of $y = x + 2$ when we are discussing functions, although both equations describe the same correspondence.

E x a m p l e   3   For the function given by $f(x) = x + 2$, find each of the following.

a) $f(8)$ **b)** $f(-3)$ **c)** $f(0)$

Solution

a) $f(8) = 8 + 2$, or 10 $f(8)$ is read "f of 8"; $f(8)$ does not mean "f times 8"!

b) $f(-3) = -3 + 2$, or -1 $f(-3)$ is the output corresponding to the input -3.

c) $f(0) = 0 + 2$, or 2

It is sometimes helpful to think of a function as a machine that gives an output for each input that enters the machine. The following diagram is one way in which the function given by $g(t) = 2t^2 + t$ can be illustrated.

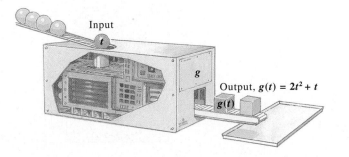

E x a m p l e 4

technology connection

There are several ways in which to check Example 4 with a grapher. Once $y_1 = 2x^2 + x$ has been entered, we can use TRACE, and then (on many calculators) enter the desired input. We can also use CALC and VALUE, or the TABLE feature. Yet another way to check is to use the VARS or Y-VARS key to access the function y_1. Parentheses can then be used to complete the writing of $y_1(3)$. When ENTER is pressed, we see the output 21. Use this approach to find $y_1(0)$ and $y_1(-2)$.

For the function $g(t) = 2t^2 + t$, find each of the following.

a) $g(3)$
b) $g(0)$
c) $g(-2)$

Solution

a) $g(3) = 2 \cdot 3^2 + 3$ Using 3 for each occurrence of t
$\qquad = 2 \cdot 9 + 3$
$\qquad = 21$

b) $g(0) = 2 \cdot 0^2 + 0$ Using 0 for each occurrence of t
$\qquad = 0$

c) $g(-2) = 2(-2)^2 + (-2)$ Using -2 for each occurrence of t
$\qquad = 2 \cdot 4 - 2$
$\qquad = 6$

Outputs are also called **function values**. In Example 4, $g(-2) = 6$. We can say that the "function value is 6 at -2," or "when x is -2, the value of the function is 6." Most often we say "g of -2 is 6."

Graphs of Functions

To graph a function, we find ordered pairs (x, y) or $(x, f(x))$, plot them, and connect the points. Note that y and $f(x)$ are often used interchangeably when working with functions and their graphs.

E x a m p l e 5

Graph: $f(x) = x + 2$.

Solution   A list of some function values is shown in this table. We plot the points and connect them. The graph is a straight line.

x	$f(x)$
-4	-2
-3	-1
-2	0
-1	1
0	2
1	3
2	4
3	5
4	6

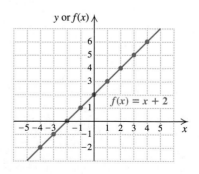

Example 6

Graph: $g(x) = 4 - x^2$.

Solution   Recall from Section 9.6 that the graph is a parabola. We calculate some function values and draw the curve.

$$g(0) = 4 - 0^2 = 4 - 0 = 4,$$
$$g(-1) = 4 - (-1)^2 = 4 - 1 = 3,$$
$$g(2) = 4 - (2)^2 = 4 - 4 = 0,$$
$$g(-3) = 4 - (-3)^2 = 4 - 9 = -5$$

x	$g(x)$
-3	-5
-2	0
-1	3
0	4
1	3
2	0
3	-5

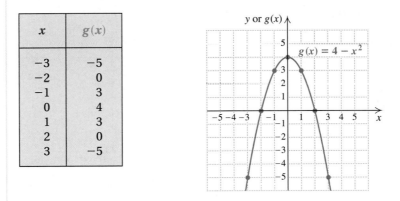

Example 7

Graph: $h(x) = |x|$.

Solution   A list of some function values is shown in the following table. We plot the points and connect them. The graph is V-shaped and symmetric, rising on either side of the vertical axis.

x	$h(x)$
-3	3
-2	2
-1	1
0	0
1	1
2	2
3	3

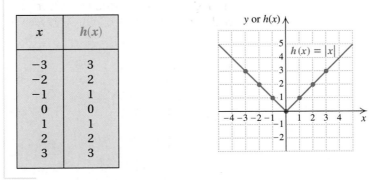

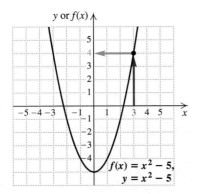

Recognizing Graphs of Functions

Consider the function f described by $f(x) = x^2 - 5$. Its graph is shown at left. It is also the graph of the equation $y = x^2 - 5$.

To find a function value, like $f(3)$, from a graph, we locate the input on the horizontal axis, move vertically to the graph of the function, and then horizontally to find the output on the vertical axis, where members of the range are found.

Recall that when one member of the domain is paired with two or more different members of the range, the correspondence is not a function. Thus, when a graph contains two or more points with the same first coordinate, it cannot represent a function. Points sharing a common first coordinate are vertically above and below each other, as shown in the following figure.

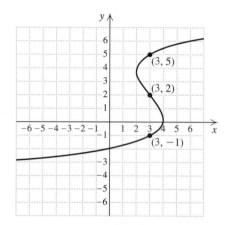

Since 3 is paired with more than one member of the range, the graph does not represent a function.

This observation leads to the *vertical-line test*.

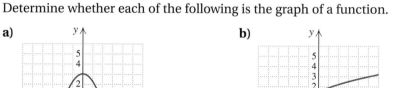

The Vertical-Line Test

A graph represents a function if it is impossible to draw a vertical line that intersects the graph more than once.

E x a m p l e   8   Determine whether each of the following is the graph of a function.

a)

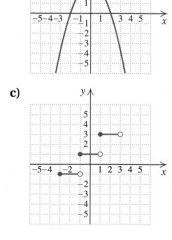

b)

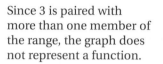

c)

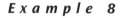

d)

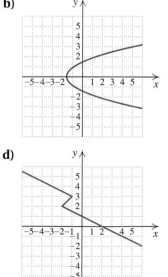

Solution

a) The graph *is* that of a function because no vertical line can cross the graph at more than one point. This can be confirmed with a ruler or straightedge.

b) The graph is *not* that of a function because a vertical line—say, $x = 1$—crosses the graph at more than one point.

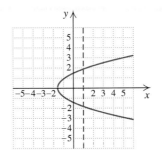

c) The graph *is* that of a function because no vertical line can cross the graph at more than one point. Note that the open dots indicate the absence of a point.

d) The graph is *not* that of a function because a vertical line—say, $x = -2$—crosses the graph at more than one point.

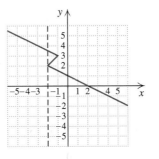

FOR EXTRA HELP

Exercise Set 9.7

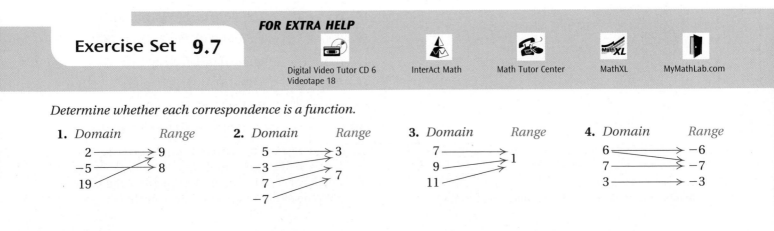

Digital Video Tutor CD 6   InterAct Math   Math Tutor Center   MathXL   MyMathLab.com
Videotape 18

Determine whether each correspondence is a function.

1. Domain   Range

2 ———→ 9
−5 ——→ 8
19 ——

2. Domain   Range

5 ———→ 3
−3 ———
7 ———→ 7
−7 ———

3. Domain   Range

7 ———→
9 ———→ 1
11 ———

4. Domain   Range

6 ———→ −6
7 ———→ −7
3 ———→ −3

5. *Domain*   *Range*

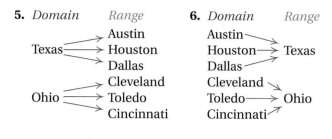

6. *Domain*   *Range*

Austin
Houston → Texas
Dallas
Cleveland
Toledo → Ohio
Cincinnati

7. *Domain*
(Where college spending money goes, nationally*)

Range
(Percentage of spending money)

Food ——————————→ 78%
Transportation ——————→ 7%
Books
Clothes ——————————→ 3%
Cigarettes ————————→ 1%
Social activities ——————→ 10%
Personal items

8. *Domain*
(Brand of single-serving pizza)

Range
(Number of calories)

Old Chicago Pizza-lite ——————→ 324
Weight Watchers Cheese
Banquet Zap Cheese ——————→ 310
Lean Cuisine Cheese
Pizza Hut Supreme Personal Pan ——→ 647
Celeste Suprema Pizza-For-One ——→ 678

Find the indicated outputs.

9. $f(4)$, $f(7)$, and $f(-2)$

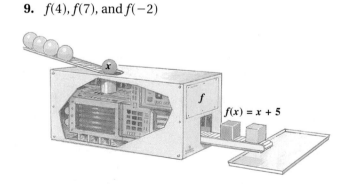

$f(x) = x + 5$

*Due to rounding, total exceeds 100%. *Source*: USA Today.

10. $g(1)$, $g(6)$, and $g(13)$

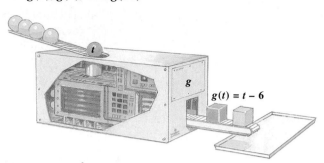

$g(t) = t - 6$

11. $h(-7)$, $h(5)$, and $h(10)$

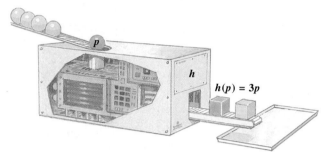

$h(p) = 3p$

12. $f(6)$, $f\left(-\frac{1}{2}\right)$, and $f(20)$

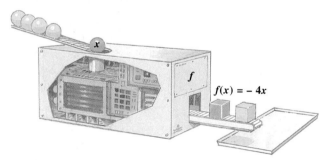

$f(x) = -4x$

13. $g(s) = 3s + 4$; find $g(1)$, $g(-5)$, and $g(6.7)$.

14. $h(x) = 19$; find $h(4)$, $h(-16)$, and $h(12.5)$.

15. $F(x) = 2x^2 - 3x$; find $F(0)$, $F(-1)$, and $F(2)$.

16. $P(x) = 3x^2 - 2x$; find $P(0)$, $P(-2)$, and $P(3)$.

17. $f(t) = (t + 1)^2$; find $f(-5)$, $f(0)$, and $f\left(-\frac{9}{4}\right)$.

18. $f(x) = (x - 1)^2$; find $f(-3)$, $f(93)$, and $f(-100)$.

19. $g(t) = t^3 + 3$; find $g(1)$, $g(-5)$, and $g(0)$.

20. $h(x) = x^4 - 3$; find $h(0)$, $h(-1)$, and $h(3)$.

21. *Predicting heights.* An anthropologist can estimate the height of a male or a female, given the lengths of certain bones. A *humerus* is the bone from the elbow to the shoulder. The height, in centimeters, of a female with a humerus of x centimeters is given by

$$F(x) = 2.75x + 71.48.$$

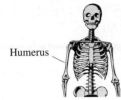

Humerus

If a humerus is known to be from a female, how tall was the female if the bone is **(a)** 32 cm long? **(b)** 30 cm long?

22. When a humerus (see Exercise 21) is from a male, the function given by $M(x) = 2.89x + 70.64$ is used to find the male's height, in centimeters. If a humerus is known to be from a male, how tall was the male if the bone is **(a)** 30 cm long? **(b)** 35 cm long?

23. *Temperature as a function of depth.*   The function given by $T(d) = 10d + 20$ gives the temperature, in degrees Celsius, inside the earth as a function of the depth d, in kilometers. Find the temperature at 5 km, 20 km, and 1000 km.

24. *Pressure at sea depth.*   The function given by $P(d) = 1 + (d/33)$ gives the pressure, in *atmospheres* (atm), at a depth of d feet, in the sea. Note that $P(0) = 1$ atm, $P(33) = 2$ atm, and so on. Find the pressure at 20 ft, 30 ft, and 100 ft.

25. *Melting snow.*   The function given by $W(d) = 0.112d$ approximates the amount, in centimeters, of water that results from d centimeters of snow melting. Find the amount of water that results from snow melting from depths of 16 cm, 25 cm, and 100 cm.

26. *Temperature conversions.*   The function given by $C(F) = \frac{5}{9}(F - 32)$ determines the Celsius temperature that corresponds to F degrees Fahrenheit. Find the Celsius temperature that corresponds to 62°F, 77°F, and 23°F.

Graph each function.

27. $f(x) = 3x - 2$

28. $g(x) = 2x + 5$

29. $g(x) = -2x + 1$

30. $f(x) = -\frac{1}{2}x + 2$

31. $f(x) = \frac{1}{2}x + 1$

32. $f(x) = -\frac{3}{4}x - 2$

33. $g(x) = 2|x|$

34. $h(x) = -|x|$

35. $g(x) = x^2$

36. $f(x) = x^2 - 1$

37. $f(x) = x^2 - x - 2$

38. $g(x) = x^2 + 6x + 5$

Determine whether each graph is that of a function.

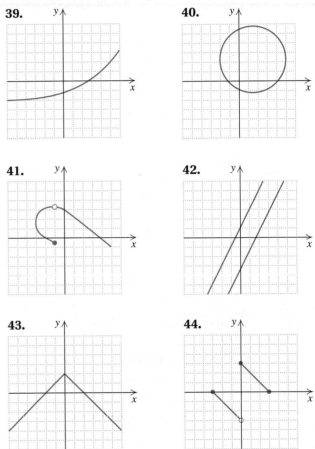

39.

40.

41.

42.

43.

44.

📓 45. Is it possible for a function to have more numbers in the range than in the domain? Why or why not?

📓 46. Is it possible for a function to have more numbers in the domain than in the range? Why or why not?

SKILL MAINTENANCE

Determine whether each pair of equations represents parallel lines.

47. $y = \frac{3}{4}x - 7,$
    $3x + 4y = 7$

48. $y = \frac{3}{5},$
    $y = -\frac{5}{3}$

49. $2x = 3y,$
    $4x = 6y - 1$

Solve each system using the substitution method.

50. $2x - y = 6,$
$4x - 2y = 5$

51. $x - 3y = 2,$
$3x - 9y = 6$

52. $y = \frac{2}{3}x - 1,$
$y = -3x + 2$

SYNTHESIS

53. Explain in your own words how the vertical-line test works.

54. If $f(x) = g(x) + 2$, how do the graphs of f and g compare?

Graph.

55. $g(x) = x^3$

56. $f(x) = 2 + \sqrt{x}$

57. $f(x) = |x| + x$

58. $g(x) = |x| - x$

59. Sketch a graph that is not that of a function.

60. If $f(-1) = -7$ and $f(3) = 8$, find a linear equation for $f(x)$.

61. If $g(0) = -4$, $g(-2) = 0$, and $g(2) = 0$, find a quadratic equation for $g(x)$.

Find the range of each function for the given domain.

62. $f(x) = 3x + 5$, when the domain is the set of whole numbers less than 4

63. $g(t) = t^2 - 5$, when the domain is the set of integers between -4 and 2

64. $h(x) = |x| - x$, when the domain is the set of integers between -2 and 20

65. $f(m) = m^3 + 1$, when the domain is the set of integers between -3 and 3

66. Use a grapher to check your answers to Exercises 55–58 and 60–65.

Summary and Review 9

Key Terms	Quadratic equation, p. 492 Complex number, p. 519 Range, p. 529 Principle of square roots, p. 492 Axis of symmetry, p. 522 Input, p. 529 Completing the square, p. 497 Vertex (plural, vertices), p. 522 Output, p. 529 Quadratic formula, p. 503 Parabola, p. 522 Function value, p. 532 Standard form, p. 503 Function, p. 528 Vertical-line test, p. 534 Imaginary number, p. 519 Domain, p. 529

Important Properties and Formulas

The principle of square roots:

If $x^2 = k$, then $x = \sqrt{k}$ or $x = -\sqrt{k}$.

The quadratic formula:

$$x = \frac{-b \pm \sqrt{b^2 - 4ac}}{2a}$$

To Solve a Quadratic Equation

1. If it is in the form $ax^2 = p$ or $(x + k)^2 = p$, use the principle of square roots.
2. When it is not in the form of (1), write it in standard form, $ax^2 + bx + c = 0$.
3. Try to factor and use the principle of zero products.
4. If it is not possible to factor or if factoring seems difficult, use the quadratic formula.

The solutions of a quadratic equation can always be found using the quadratic formula, but not always by factoring. When the radicand, $b^2 - 4ac$, is nonnegative, the equation has real-number solutions. When $b^2 - 4ac$ is negative, the equation has no real-number solutions.

The Intercepts of a Parabola

To find the y-intercept of the graph of $y = ax^2 + bx + c$, replace x with 0 and solve for y.

To find any x-intercept(s) of the graph of $y = ax^2 + bx + c$, replace y with 0 and solve for x.

The Vertex of a Parabola

For a parabola given by the quadratic equation $y = ax^2 + bx + c$:

1. The x-coordinate of the vertex is $-b(2a)$.
2. The y-coordinate of the vertex is found by substituting $-(b/2a)$ for x and solving for y.

Guidelines for Graphing Quadratic Equations

1. For $a \neq 0$, the graph of $y = ax^2 + bx + c$ is a parabola. It is cupped upward for $a > 0$ and downward for $a < 0$.
2. The formula $x = -b/(2a)$ can be used to find the x-coordinate of the vertex. Calculate the y-coordinate and plot the vertex as well as some points on either side of it.
3. Once a point has been graphed, a second point with the same y-coordinate can be found on the opposite side of the axis of symmetry.
4. Graph the y-intercept and, if requested, any x-intercepts.

The Vertical-Line Test

A graph represents a function if it is impossible to draw a vertical line that intersects the graph more than once.

Review Exercises

Solve by completing the square.

1. $3x^2 + 2x - 5 = 0$
2. $x^2 = 3x - 1$

Solve.

3. $8x^2 = 24$
4. $5x^2 - 8x + 3 = 0$
5. $x^2 - 2x - 10 = 0$
6. $x^2 + 64 = 0$
7. $3y^2 + 5y = 2$
8. $(x + 8)^2 = 13$
9. $(x - 3)^2 = -4$
10. $3x^2 - 6x - 9 = 0$
11. $9x^2 = 0$
12. $x^2 + 6x = 9$
13. $x^2 + x + 1 = 0$
14. $1 + 4x^2 = 8x$
15. $x^2 + 1 = 2x$
16. $40 = 5y^2$
17. $3m = 4 + 5m^2$
18. $6x^2 + 11x = 35$

Solve.

19. $2t^2 - s = r$, for t
20. $\sqrt{3m - n} - 1 = 9$, for m
21. $2a - b = c$, for a
22. $\dfrac{1}{r} + \dfrac{1}{s} = \dfrac{1}{t}$, for t

Approximate the solutions to the nearest thousandth.

23. $x^2 = 3x - 1$
24. $4y^2 + 8y + 1 = 0$

25. *Right triangles.* The hypotenuse of a right triangle is 7 m long. One leg is 3 m longer than the other. Find the lengths of the legs. Round to the nearest tenth.

26. *Investments.* $1000 is invested at interest rate r, compounded annually. In 2 years, it grows to $1102.50. What is the interest rate?

27. *Area of a rectangle.* The length of a rectangle is 3 m greater than the width. The area is 108 m². Find the length and the width.

28. *Free-fall time.* The height of the Lake Point Towers in Chicago is 645 ft. How long would it take an object to fall from the top?

Express in terms of i.

29. $2 - \sqrt{-64}$ **30.** $\sqrt{-24}$

Graph. Label the vertex, the y-intercept, and any x-intercepts.

31. $y = 2 - x^2$ **32.** $y = x^2 + 4x - 1$

33. $y = 3x^2 - 10x$ **34.** $y = x^2 - 2x + 1$

35. If $g(x) = 2x - 5$, find $g(2)$, $g(-1)$, and $g(3.5)$.

36. If $h(x) = |x| - 1$, find $h(1)$, $h(-1)$, and $h(-20)$.

37. *Calories.* Each day a moderately active person needs about 15 calories per pound of body weight. The function given by $C(p) = 15p$ approximates the number of calories that are needed to maintain body weight p, in pounds. How many calories are needed to maintain a body weight of 130 lb?

Graph.

38. $g(x) = x - 6$ **39.** $f(x) = x^2 + 2$

40. $h(x) = 3|x|$

Determine whether each graph is that of a function.

41.   **42.**

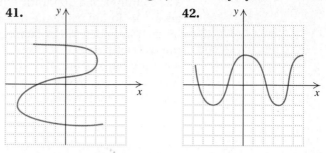

43. How can $b^2 - 4ac$ be used to determine how many x-intercepts the graph of $y = ax^2 + bx + c$ has?

SYNTHESIS

44. Explain why it is helpful to know the general shape of a graph before connecting plotted points.

45. Is it possible for a rational equation to have more solutions than a radical equation? Why or why not?

46. Two consecutive integers have squares that differ by 63. Find the integers.

47. Find b such that the trinomial $x^2 + bx + 49$ is a square.

48. Solve: $x - 4\sqrt{x} - 5 = 0$.

49. A square with sides of length s has the same area as a circle with radius 5 in. Find s.

Chapter Test 9

Solve.

1. $9x^2 = 45$ **2.** $3x^2 - 7x = 0$

3. $x^2 = x + 3$ **4.** $3y^2 + 4y = 15$

5. $48 = t^2 + 2t$ **6.** $x^2 = -49$

7. $(x - 2)^2 = 5$ **8.** $x^2 - 4x = -4$

9. $x^2 - 4x = -8$ **10.** $m^2 - 3m = 7$

11. $10 = 4x + x^2$ **12.** $3x^2 - 7x + 1 = 0$

13. Solve by completing the square:
$$x^2 - 4x - 10 = 0.$$

Solve.

14. $3 = n + 2\sqrt{p + 5}$, for p

15. $1 + x = \dfrac{a}{b}$, for b

16. Approximate the solutions to the nearest thousandth:
$$x^2 - 3x - 8 = 0.$$

17. *Area of a rectangle.* The width of a rectangle is 4 m less than the length. The area is 16.25 m². Find the length and the width.

18. *Diagonals of a polygon.* A polygon has 44 diagonals. How many sides does it have? (*Hint*: $d = (n^2 - 3n)/2$.)

Express in terms of i.

19. $\sqrt{-49}$

20. $3 - \sqrt{-32}$

Graph. Label the vertex, the y-intercept, and any x-intercepts.

21. $y = -x^2 + x - 5$

22. $y = x^2 + 2x - 15$

23. If $f(x) = \frac{1}{2}x + 1$, find $f(0)$, $f(1)$, and $f(2)$.

24. If $g(t) = -2|t| + 3$, find $g(-1)$, $g(0)$, and $g(3)$.

25. *World records.* The world record for the 10,000-m run has been decreasing steadily since 1940. The function given by $R(t) = 30.18 - 0.06t$ estimates the record, in minutes, t years after 1940. Predict what the record will be in 2010.

Graph.

26. $h(x) = x - 4$

27. $g(x) = x^2 - 4$

Determine whether each graph is that of a function.

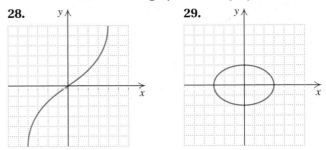

28.

29.

SYNTHESIS

30. Find the side of a square whose diagonal is 5 ft longer than a side.

31. Solve this system for x. Use the substitution method.

$$x - y = 2,$$
$$xy = 4$$

Cumulative Review 1–9

1. Write exponential notation for $x \cdot x \cdot x$.

2. Evaluate $(x - 3)^2 + 5$ for $x = 10$.

3. Use the commutative and associative laws to write an expression equivalent to $6(xy)$.

4. Find the LCM of 15 and 48.

5. Find the absolute value: $|-7|$.

Compute and simplify.

6. $-6 + 12 + (-4) + 7$

7. $2.8 - (-12.2)$

8. $-\frac{3}{8} \div \frac{5}{2}$

9. $13 \cdot 6 \div 3 \cdot 2 \div 13$

10. Remove parentheses and simplify:

$$4m + 9 - (6m + 13).$$

Solve.

11. $-5x = 45$

12. $-5x > 45$

13. $3(y - 1) - 2(y + 2) = 0$

14. $x^2 - 8x + 15 = 0$

15. $y - x - 1,$
    $y = 3 - x$

16. $\dfrac{x}{x + 1} = \dfrac{3}{2x} + 1$

17. $4x - 3y = 3,$
    $3x - 2y = 4$

18. $x^2 - x - 6 = 0$

19. $x^2 + 3x = 5$

20. $3 - x = \sqrt{x^2 - 3}$

21. $5 - 9x \le 19 + 5x$

22. $-\frac{7}{8}x + 7 = \frac{3}{8}x - 3$

23. $0.6x - 1.8 = 1.2x$

24. $x + y = 15,$
    $x - y = 15$

25. $x^2 + 2x + 5 = 0$

26. $3y^2 = 30$

27. $(x - 3)^2 = 6$

28. $\dfrac{6x - 2}{2x - 1} = \dfrac{9x}{3x + 1}$

29. $12x^2 + x = 20$

30. $\dfrac{2x}{x + 3} + \dfrac{6}{x} + 7 = \dfrac{18}{x^2 + 3x}$

31. $\sqrt{x + 9} = \sqrt{2x - 3}$

Solve the formula for the given letter.

32. $A = \dfrac{4s + 3}{t}$, for t

33. $\dfrac{1}{t} = \dfrac{1}{m} - \dfrac{1}{n}$, for m

Simplify. Write scientific notation for the result.

34. $(2.1 \times 10^7)(1.3 \times 10^{-12})$

35. $\dfrac{5.2 \times 10^{-1}}{2.6 \times 10^{-15}}$

Simplify.

36. $x^{-6} \cdot x^2$ **37.** $\dfrac{y^3}{y^{-4}}$ **38.** $(2y^6)^2$

39. Combine like terms and arrange in descending order:
$$2x - 3 + 5x^3 - 2x^3 + 7x^3 + x.$$

Perform the indicated operation and simplify.

40. $(4x^3 + 3x^2 - 5) + (3x^3 - 5x^2 + 4x - 12)$

41. $(6x^2 - 4x + 1) - (-2x^2 + 7)$

42. $-2y^2(4y^2 - 3y + 1)$

43. $(2t - 3)(3t^2 - 4t + 2)$

44. $\left(t - \tfrac{1}{4}\right)\left(t + \tfrac{1}{4}\right)$

45. $(3m - 2)^2$

46. $(15x^2y^3 + 10xy^2 + 5) - (5xy^2 - x^2y^2 - 2)$

47. $(x^2 - 0.2y)(x^2 + 0.2y)$

48. $(3p + 4q^2)^2$

49. $\dfrac{4}{2x - 6} \cdot \dfrac{x - 3}{x + 3}$

50. $\dfrac{3a^4}{a^2 - 1} \div \dfrac{2a^3}{a^2 - 2a + 1}$

51. $\dfrac{3}{3x - 1} + \dfrac{4}{5x}$

52. $\dfrac{2}{x^2 - 16} - \dfrac{x - 3}{x^2 - 9x + 20}$

53. $(x^3 + 7x^2 - 2x + 3) \div (x - 2)$

Factor.

54. $49x^2 - 4$ **55.** $x^3 - 8x^2 - 5x + 40$

56. $16x^3 - x$ **57.** $m^2 - 8m + 16$

58. $15x^2 + 14x - 8$ **59.** $18x^5 + 4x^4 - 10x$

60. $6p^2 - 5p - 6$

61. $2ac - 6ab - 3db + dc$

62. $9x^2 + 30xy + 25y^2$

63. $3t^4 + 6t^2 - 72$

64. $49a^2b^2 - 9$

Simplify.

65. $\dfrac{\dfrac{3}{x} + \dfrac{1}{2x}}{\dfrac{1}{3x} - \dfrac{3}{4x}}$ **66.** $\sqrt{49}$

67. $-\sqrt[4]{625}$ **68.** $\sqrt{64x^2}$
(Assume $x \geq 0$.)

69. Multiply: $\sqrt{a + b}\,\sqrt{a - b}$.

70. Multiply and simplify: $\sqrt{32ab}\,\sqrt{6a^4b^2}$ $(a, b \geq 0)$.

Simplify.

71. $8^{-2/3}$

72. $\sqrt{243x^3y^2}$ $(x, y \geq 0)$

73. $\sqrt{\dfrac{100}{81}}$

74. $\left(2\sqrt{3} + \sqrt{5}\right)\left(2\sqrt{3} - \sqrt{5}\right)$

75. $4\sqrt{12} + 2\sqrt{48}$

76. Divide and simplify: $\dfrac{\sqrt{72}}{\sqrt{45}}$.

77. The hypotenuse of a right triangle is 41 cm long and the length of a leg is 9 cm. Find the length of the other leg.

Graph on a plane.

78. $y = \tfrac{1}{3}x - 2$ **79.** $2x + 3y = -6$

80. $y = -3$ **81.** $4x - 3y > 12$

82. $y = x^2 + 2x + 1$ **83.** $x \geq -3$

84. Solve $9x^2 - 12x - 2 = 0$ by completing the square.

85. Approximate the solutions of $4x^2 = 4x + 1$ to the nearest thousandth.

86. Graph $y = x^2 + 2x - 5$. Label the vertex, the y-intercept, and the x-intercepts.

Solve.

87. *Work force.*   In 1998, 59% of the 3.6 million women who gave birth in the previous year returned to the work force (*Sources*: U.S. Bureau of the Census and the *Daily Reporter*, Hancock County, Indiana, 24 October 2000). How many women returned to the work force within a year of giving birth?

88. *Two-year college tuition.*   The average tuition at a public two-year college was $1239 in 1996 and $1705 in 2000 (*Sources*: *Statistical Abstract of the United States*, 1999 and *Daily Reporter*, Hancock County, Indiana, 24 October 2000). Find the rate at which tuition was increasing.

89. The product of two consecutive even integers is 224. Find the integers.

90. The length of a rectangle is 7 m more than the width. The length of a diagonal is 13 m. Find the length.

91. *Parking.*   Three-fifths of the automobiles entering Dalton each morning will fill the city parking lots. There are 3654 such parking spaces. How many cars enter Dalton each morning?

92. *Candy blends.*   Good's Candies, in Kennard, Indiana, mixes chocolate creams worth $11.50 per pound with nut candies worth $12.50 per pound in order to make 25 lb of a mixture worth $11.74 per pound. How many pounds of each kind of candy do they use?

93. *Warehouse.*   Kathy and Ken work in a textbook warehouse. Kathy can fill an order of 100 books in 40 min. Ken can fill the same order in 60 min. How long would it take them, working together, to fill the order?

94. *Cable.*   The cumulative cost c of cable television, in dollars, is given by $c = 20t + 25$, where t is the number of months that a household has had cable service. Graph the equation and use the graph to estimate the cost of cable service for 10 months.

95. For the function f described by
$$f(x) = 2x^2 + 7x - 4,$$
find $f(0)$, $f(-4)$, and $f(\frac{1}{2})$.

96. Determine whether the graphs of the following equations are parallel:
$$y - x = 4,$$
$$3y = 5 + 3x.$$

97. Graph the following system of inequalities:
$$x + y \leq 6,$$
$$x + y \geq 2,$$
$$x \leq 3,$$
$$x \geq 1.$$

98. Find the slope and the y-intercept:
$$-6x + 3y = -24.$$

99. Find the slope of the line containing the points $(-5, -6)$ and $(-4, 9)$.

100. Find a point–slope equation for the line containing $(1, -3)$ and having slope $m = -\frac{1}{2}$.

101. Simplify: $-\sqrt{-25}$.

SYNTHESIS

102. Find b such that the trinomial $x^2 - bx + 225$ is a square.

103. Find x.

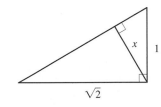

Determine whether each pair of expressions is equivalent.

104. $x^2 - 9$, $(x - 3)(x + 3)$

105. $\dfrac{x + 3}{3}$, x

106. $(x + 5)^2$, $x^2 + 25$

107. $\sqrt{x^2 + 16}$, $x + 4$

108. $\sqrt{x^2}$, $|x|$

Appendixes

Naming Sets • Membership • Subsets • Intersections • Unions

The notion of a "set" is used frequently in mathematics. We provide a basic introduction to sets in this appendix.

Naming Sets

To name the set of whole numbers less than 6, we can use *roster notation*, as follows:

$$\{0, 1, 2, 3, 4, 5\}.$$

The set of real numbers x for which x is less than 6 cannot be named by listing all its members because there is an infinite number of them. We name such a set using *set-builder notation*, as follows:

$$\{x \mid x < 6\}.$$

This is read

"The set of all x such that x is less than 6."

See Section 2.6 for more on this notation.

Membership

The symbol $\in$ means *is a member of* or *belongs to*, or *is an element of*. Thus,

$$x \in A$$

means

x is a member of A, or x belongs to A, or x is an element of A.

E x a m p l e   1

Classify each of the following as true or false.

a) $1 \in \{1, 2, 3\}$
b) $1 \in \{2, 3\}$
c) $4 \in \{x \mid x \text{ is an even whole number}\}$
d) $5 \in \{x \mid x \text{ is an even whole number}\}$

Solution

a) Since 1 is listed as a member of the set, $1 \in \{1, 2, 3\}$ is true.
b) Since 1 is *not* a member of $\{2, 3\}$, the statement $1 \in \{2, 3\}$ is false.
c) Since 4 is an even whole number, $4 \in \{x \mid x \text{ is an even whole number}\}$ is true.
d) Since 5 is *not* even, $5 \in \{x \mid x \text{ is an even whole number}\}$ is false.

Set membership can be illustrated with a diagram, as shown below.

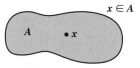

Subsets

If every element of A is also an element of B, then A is a *subset* of B. This is denoted $A \subseteq B$.

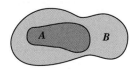

The set of whole numbers is a subset of the set of integers. The set of rational numbers is a subset of the set of real numbers.

E x a m p l e   2

Classify each of the following as true or false.

a) $\{1, 2\} \subseteq \{1, 2, 3, 4\}$
b) $\{p, q, r, w\} \subseteq \{a, p, r, z\}$
c) $\{x \mid x < 6\} \subseteq \{x \mid x \leq 11\}$

Solution

a) Since every element of $\{1, 2\}$ is in the set $\{1, 2, 3, 4\}$, it follows that $\{1, 2\} \subseteq \{1, 2, 3, 4\}$ is true.
b) Since $q \in \{p, q, r, w\}$, but $q \notin \{a, p, r, z\}$, it follows that $\{p, q, r, w\} \subseteq \{a, p, r, z\}$ is false.
c) Since every number that is less than 6 is also less than 11, the statement $\{x \mid x < 6\} \subseteq \{x \mid x \leq 11\}$ is true.

Intersections

The *intersection* of sets A and B, denoted $A \cap B$, is the set of members common to both sets.

Example 3

Find each intersection.

a) $\{0, 1, 3, 5, 25\} \cap \{2, 3, 4, 5, 6, 7, 9\}$ **b)** $\{a, p, q, w\} \cap \{p, q, t\}$

Solution

a) $\{0, 1, 3, 5, 25\} \cap \{2, 3, 4, 5, 6, 7, 9\} = \{3, 5\}$
b) $\{a, p, q, w\} \cap \{p, q, t\} = \{p, q\}$

Set intersection can be illustrated with a diagram, as shown below.

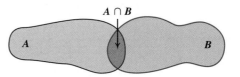

$A \cap B$

A B

The set without members is known as the *empty set*, and is written $\varnothing$, and sometimes { }. Each of the following is a description of the empty set:

The set of all 12-ft–tall people;

$\{2, 3\} \cap \{5, 6, 7\}$;

$\{x \mid x$ is an even natural number$\} \cap \{x \mid x$ is an odd natural number$\}$.

Unions

Two sets A and B can be combined to form a set that contains the members of both A and B. The new set is called the *union* of A and B, denoted $A \cup B$.

Example 4

Find each union.

a) $\{0, 5, 7, 13, 27\} \cup \{0, 2, 3, 4, 5\}$ **b)** $\{a, c, e, g\} \cup \{b, d, f\}$

Solution

a) $\{0, 5, 7, 13, 27\} \cup \{0, 2, 3, 4, 5\} = \{0, 2, 3, 4, 5, 7, 13, 27\}$

Note that the 0 and the 5 are *not* listed twice in the solution.

b) $\{a, c, e, g\} \cup \{b, d, f\} = \{a, b, c, d, e, f, g\}$

Set union can be illustrated with a diagram, as shown below.

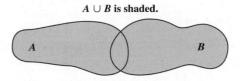

$A \cup B$ is shaded.

A B

Exercise Set A

Name each set using the roster method.

1. The set of whole numbers 3 through 7

2. The set of whole numbers 83 through 89

3. The set of odd numbers between 40 and 50

4. The set of multiples of 5 between 10 and 40

5. $\{x \mid \text{the square of } x \text{ is } 9\}$

6. $\{x \mid x \text{ is the cube of } 0.2\}$

Classify each statement as true or false.

7. $2 \in \{x \mid x \text{ is an odd number}\}$

8. $7 \in \{x \mid x \text{ is an odd number}\}$

9. Bruce Springsteen $\in$ The set of all rock stars

10. Apple $\in$ The set of all fruit

11. $-3 \in \{-4, -3, 0, 1\}$

12. $0 \in \{-4, -3, 0, 1\}$

13. $\frac{2}{3} \in \{x \mid x \text{ is a rational number}\}$

14. Heads $\in$ The set of outcomes of flipping a penny

15. $\{4, 5, 8\} \subseteq \{1, 3, 4, 5, 6, 7, 8, 9\}$

16. The set of vowels $\subseteq$ The set of consonants

17. $\{-1, -2, -3, -4, -5\} \subseteq \{-1, 2, 3, 4, 5\}$

18. The set of integers $\subseteq$ The set of rational numbers

Find each intersection.

19. $\{a, b, c, d, e\} \cap \{c, d, e, f, g\}$

20. $\{a, e, i, o, u\} \cap \{q, u, i, c, k\}$

21. $\{1, 2, 5, 10\} \cap \{0, 1, 7, 10\}$

22. $\{0, 1, 7, 10\} \cap \{0, 1, 2, 5\}$

23. $\{1, 2, 5, 10\} \cap \{3, 4, 7, 8\}$

24. $\{a, e, i, o, u\} \cap \{m, n, f, g, h\}$

Find each union.

25. $\{a, e, i, o, u\} \cup \{q, u, i, c, k\}$

26. $\{a, b, c, d, e\} \cup \{c, d, e, f, g\}$

27. $\{0, 1, 7, 10\} \cup \{0, 1, 2, 5\}$

28. $\{1, 2, 5, 10\} \cup \{0, 1, 7, 10\}$

29. $\{a, e, i, o, u\} \cup \{m, n, f, g, h\}$

30. $\{1, 2, 5, 10\} \cup \{a, b\}$

 31. What advantage(s) does set-builder notation have over roster notation?

 32. What advantage(s) does roster notation have over set-builder notation?

SYNTHESIS

33. Find the union of the set of integers and the set of whole numbers.

34. Find the intersection of the set of odd integers and the set of even integers.

35. Find the union of the set of rational numbers and the set of irrational numbers.

36. Find the intersection of the set of even integers and the set of positive rational numbers.

37. Find the intersection of the set of rational numbers and the set of irrational numbers.

38. Find the union of the set of negative integers, the set of positive integers, and the set containing 0.

39. For a set A, find each of the following.
    a) $A \cup \varnothing$ b) $A \cup A$
    c) $A \cap A$ d) $A \cap \varnothing$

40. A set is *closed* under an operation if, when the operation is performed on its members, the result is in the set. For example, the set of real numbers is closed under the operation of addition since the sum of any two real numbers is a real number.
    a) Is the set of even numbers closed under addition?
    b) Is the set of odd numbers closed under addition?
    c) Is the set $\{0, 1\}$ closed under addition?
    d) Is the set $\{0, 1\}$ closed under multiplication?
    e) Is the set of real numbers closed under multiplication?
    f) Is the set of integers closed under division?

41. Experiment with sets of various types and determine whether the following distributive law for sets is true:
$$A \cap (B \cup C) = (A \cap B) \cup (A \cap C).$$

Factoring Sums or Differences of Cubes

B

Factoring a Sum of Two Cubes • Factoring a Difference of Two Cubes

Although a sum of two squares cannot be factored unless a common factor exists, a sum of two *cubes* can always be factored. A difference of two cubes can also be factored.

Consider the following products:

$$(A + B)(A^2 - AB + B^2) = A(A^2 - AB + B^2) + B(A^2 - AB + B^2)$$
$$= A^3 - A^2B + AB^2 + A^2B - AB^2 + B^3$$
$$= A^3 + B^3$$

and

$$(A - B)(A^2 + AB + B^2) = A(A^2 + AB + B^2) - B(A^2 + AB + B^2)$$
$$= A^3 + A^2B + AB^2 - A^2B - AB^2 - B^3$$
$$= A^3 - B^3.$$

These equations show how we can factor a sum or a difference of two cubes.

To Factor a Sum or Difference of Cubes

$$A^3 + B^3 = (A + B)(A^2 - AB + B^2),$$
$$A^3 - B^3 = (A - B)(A^2 + AB + B^2)$$

This table of cubes will help in the examples that follow.

N	0.2	0.1	0	1	2	3	4	5	6	7	8
N^3	0.008	0.001	0	1	8	27	64	125	216	343	512

Example 1

Factor: $x^3 - 8$.

Solution   We have

$$x^3 - 8 = x^3 - 2^3 = (x - 2)(x^2 + x \cdot 2 + 2^2).$$
$$A^3 - B^3 = (A - B)(A^2 + A \ B + B^2)$$

This tells us that $x^3 - 8 = (x - 2)(x^2 + 2x + 4)$. Note that we cannot factor $x^2 + 2x + 4$. (It is not a perfect-square trinomial nor can it be factored by trial and error or grouping.) The check is left to the student.

Example 2

Factor: $x^3 + 125$.

Solution   We have

$$x^3 + 125 = x^3 + 5^3 = (x + 5)(x^2 - x \cdot 5 + 5^2).$$

$$A^3 + B^3 = (A + B)(A^2 - A\ B + B^2)$$

Thus, $x^3 + 125 = (x + 5)(x^2 - 5x + 25)$. We leave the check to the student.

Example 3

Factor: $16a^7b + 54ab^7$.

Solution   We first look for a common factor:

$$16a^7b + 54ab^7 = 2ab[8a^6 + 27b^6]$$
$$= 2ab[(2a^2)^3 + (3b^2)^3] \quad \text{This is of the form } A^3 + B^3,$$
$$\text{where } A = 2a^2 \text{ and } B = 3b^2.$$
$$= 2ab[(2a^2 + 3b^2)(4a^4 - 6a^2b^2 + 9b^4)].$$

We check using the distributive law:

$$2ab(2a^2 + 3b^2)(4a^4 - 6a^2b^2 + 9b^4)$$
$$= 2ab[2a^2(4a^4 - 6a^2b^2 + 9b^4) + 3b^2(4a^4 - 6a^2b^2 + 9b^4)]$$
$$= 2ab[8a^6 - 12a^4b^2 + 18a^2b^4 + 12b^2a^4 - 18a^2b^4 + 27b^6]$$
$$= 2ab[8a^6 + 27b^6]$$
$$= 16a^7b + 54ab^7.$$

We have a check. The factorization is

$$2ab(2a^2 + 3b^2)(4a^4 - 6a^2b^2 + 9b^4).$$

Example 4

Factor: $y^3 - 0.001$.

Solution   Since $0.001 = (0.1)^3$, we have a difference of cubes:

$$y^3 - 0.001 = (y - 0.1)(y^2 + 0.1y + 0.01).$$

The check is left to the student.

Remember the following about factoring sums or differences of squares and cubes:

Difference of cubes: $A^3 - B^3 = (A - B)(A^2 + AB + B^2),$

Sum of cubes: $A^3 + B^3 = (A + B)(A^2 - AB + B^2),$

Difference of squares: $A^2 - B^2 = (A + B)(A - B),$

Sum of squares: $A^2 + B^2$ **cannot be factored apart from factoring out a common factor if one exists.**

Exercise Set B

FOR EXTRA HELP

Digital Video Tutor CD: none   InterAct Math   Math Tutor Center   MathXL   MyMathLab.com
Videotape: none

Factor completely.

1. $t^3 + 8$

2. $p^3 + 27$

3. $a^3 - 64$

4. $w^3 - 1$

5. $z^3 + 125$

6. $x^3 + 1$

7. $8a^3 - 1$

8. $27x^3 - 1$

9. $y^3 - 27$

10. $p^3 - 8$

11. $64 + 125x^3$

12. $8 + 27b^3$

13. $125p^3 - 1$

14. $64w^3 - 1$

15. $27m^3 + 64$

16. $8t^3 + 27$

17. $p^3 - q^3$

18. $a^3 + b^3$

19. $x^3 + \frac{1}{8}$

20. $y^3 + \frac{1}{27}$

21. $2y^3 - 128$

22. $3z^3 - 3$

23. $24a^3 + 3$

24. $54x^3 + 2$

25. $rs^3 - 64r$

26. $ab^3 + 125a$

27. $5x^3 + 40z^3$

28. $2y^3 - 54z^3$

29. $x^3 + 0.001$

30. $y^3 + 0.125$

SYNTHESIS

31. Dino incorrectly believes that
$$a^3 - b^3 = (a - b)(a^2 + b^2).$$
How could you convince him that he is wrong?

32. If $x^3 + c$ is prime, what can you conclude about c? Why?

Factor. Assume that variables in exponents represent natural numbers.

33. $125c^6 + 8d^6$

34. $64x^6 + 8t^6$

35. $3x^{3a} - 24y^{3b}$

36. $\frac{8}{27}x^3 - \frac{1}{64}y^3$

37. $\frac{1}{24}x^3y^3 + \frac{1}{3}z^3$

38. $\frac{1}{16}x^{3a} + \frac{1}{2}y^{6a}z^{9b}$

Tables

TABLE 1 Fractional and Decimal Equivalents

Fractional Notation	$\frac{1}{10}$	$\frac{1}{8}$	$\frac{1}{6}$	$\frac{1}{5}$	$\frac{1}{4}$	$\frac{3}{10}$	$\frac{1}{3}$	$\frac{3}{8}$	$\frac{2}{5}$	$\frac{1}{2}$
Decimal Notation	0.1	0.125	$0.16\overline{6}$	0.2	0.25	0.3	$0.333\overline{3}$	0.375	0.4	0.5
Percent Notation	10%	12.5% or $12\frac{1}{2}\%$	$16.6\overline{6}\%$ or $16\frac{2}{3}\%$	20%	25%	30%	$33.3\overline{3}\%$ or $33\frac{1}{3}\%$	37.5% or $37\frac{1}{2}\%$	40%	50%
Fractional Notation	$\frac{3}{5}$	$\frac{5}{8}$	$\frac{2}{3}$	$\frac{7}{10}$	$\frac{3}{4}$	$\frac{4}{5}$	$\frac{5}{6}$	$\frac{7}{8}$	$\frac{9}{10}$	$\frac{1}{1}$
Decimal Notation	0.6	0.625	$0.666\overline{6}$	0.7	0.75	0.8	$0.83\overline{3}$	0.875	0.9	1
Percent Notation	60%	62.5% or $62\frac{1}{2}\%$	$66.6\overline{6}\%$ or $66\frac{2}{3}\%$	70%	75%	80%	$83.3\overline{3}\%$ or $83\frac{1}{3}\%$	87.5% or $87\frac{1}{2}\%$	90%	100%

TABLE 2 Squares and Square Roots with Approximations to Three Decimal Places

N	$\sqrt{N}$	N^2	N	$\sqrt{N}$	N^2	N	$\sqrt{N}$	N^2	N	$\sqrt{N}$	N^2
1	1	1	26	5.099	676	51	7.141	2601	76	8.718	5776
2	1.414	4	27	5.196	729	52	7.211	2704	77	8.775	5929
3	1.732	9	28	5.292	784	53	7.280	2809	78	8.832	6084
4	2	16	29	5.385	841	54	7.348	2916	79	8.888	6241
5	2.236	25	30	5.477	900	55	7.416	3025	80	8.944	6400
6	2.449	36	31	5.568	961	56	7.483	3136	81	9	6561
7	2.646	49	32	5.657	1024	57	7.550	3249	82	9.055	6724
8	2.828	64	33	5.745	1089	58	7.616	3364	83	9.110	6889
9	3	81	34	5.831	1156	59	7.681	3481	84	9.165	7056
10	3.162	100	35	5.916	1225	60	7.746	3600	85	9.220	7225
11	3.317	121	36	6	1296	61	7.810	3721	86	9.274	7396
12	3.464	144	37	6.083	1369	62	7.874	3844	87	9.327	7569
13	3.606	169	38	6.164	1444	63	7.937	3969	88	9.381	7744
14	3.742	196	39	6.245	1521	64	8	4096	89	9.434	7921
15	3.873	225	40	6.325	1600	65	8.062	4225	90	9.487	8100
16	4	256	41	6.403	1681	66	8.124	4356	91	9.539	8281
17	4.123	289	42	6.481	1764	67	8.185	4489	92	9.592	8464
18	4.243	324	43	6.557	1849	68	8.246	4624	93	9.644	8649
19	4.359	361	44	6.633	1936	69	8.307	4761	94	9.695	8836
20	4.472	400	45	6.708	2025	70	8.367	4900	95	9.747	9025
21	4.583	441	46	6.782	2116	71	8.426	5041	96	9.798	9216
22	4.690	484	47	6.856	2209	72	8.485	5184	97	9.849	9409
23	4.796	529	48	6.928	2304	73	8.544	5329	98	9.899	9604
24	4.899	576	49	7	2401	74	8.602	5476	99	9.950	9801
25	5	625	50	7.071	2500	75	8.660	5625	100	10	10,000

TABLE 3   Geometric Formulas

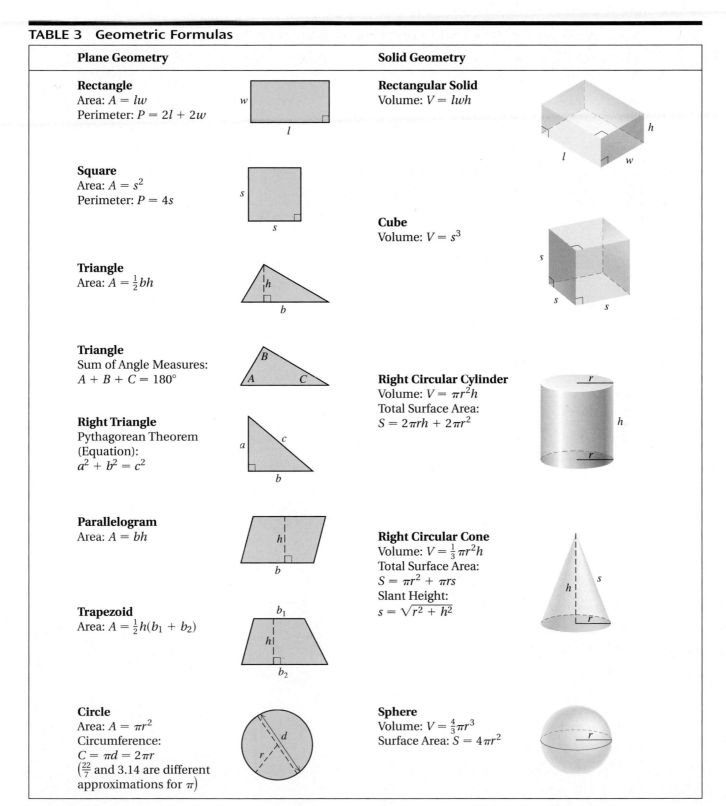

Plane Geometry	Solid Geometry
Rectangle Area: $A = lw$ Perimeter: $P = 2l + 2w$	**Rectangular Solid** Volume: $V = lwh$
Square Area: $A = s^2$ Perimeter: $P = 4s$	
	Cube Volume: $V = s^3$
Triangle Area: $A = \frac{1}{2}bh$	
Triangle Sum of Angle Measures: $A + B + C = 180°$	**Right Circular Cylinder** Volume: $V = \pi r^2 h$ Total Surface Area: $S = 2\pi rh + 2\pi r^2$
Right Triangle Pythagorean Theorem (Equation): $a^2 + b^2 = c^2$	
Parallelogram Area: $A = bh$	**Right Circular Cone** Volume: $V = \frac{1}{3}\pi r^2 h$ Total Surface Area: $S = \pi r^2 + \pi rs$ Slant Height: $s = \sqrt{r^2 + h^2}$
Trapezoid Area: $A = \frac{1}{2}h(b_1 + b_2)$	
Circle Area: $A = \pi r^2$ Circumference: $C = \pi d = 2\pi r$ $\left(\frac{22}{7}\text{ and } 3.14 \text{ are different}\right.$ approximations for π)	**Sphere** Volume: $V = \frac{4}{3}\pi r^3$ Surface Area: $S = 4\pi r^2$

Answers

CHAPTER 1

Technology Connection, p. 7

1. 3438   **2.** 47,531

Exercise Set 1.1, pp. 7–10

1. 27   **3.** 8   **5.** 4   **7.** 7   **9.** 5   **11.** 3   **13.** 24 ft^2
15. 15 cm^2 **17.** 0.270   **19.** Let j represent Jan's age;
$8 + j$, or $j + 8$ **21.** $6 + b$, or $b + 6$ **23.** $c - 9$
25. $6 + q$, or $q + 6$ **27.** Let p represent Phil's speed; $9p$,
or $p9$ **29.** $y - x$ **31.** $x \div w$, or $\dfrac{x}{w}$ **33.** $n - m$
35. Let l represent the length of the box and h represent
the height; $l + h$, or $h + l$ **37.** $9 \cdot 2m$, or $2m \cdot 9$
39. Let y represent "some number"; $\dfrac{1}{4}y$, or $\dfrac{y}{4}$
41. Let w represent the number of women attending;
64% of w, or $0.64w$ **43.** $50 - x$ **45.** Yes
47. No   **49.** Yes   **51.** Yes   **53.** Let x represent the
unknown number; $73 + x = 201$ **55.** Let x represent
the unknown number; $42x = 2352$ **57.** Let s represent
the number of squares your opponent controls;
$s + 35 = 64$ **59.** Let w represent the amount of
solid waste generated; 27% $\cdot w = 56$, or $0.27w = 56$
61. ▨   **63.** ▨   **65.** $337.50   **67.** 2   **69.** 6
71. $w + 4$ **73.** Let x and y represent the two
numbers; $\frac{1}{3} \cdot \frac{1}{2} \cdot xy$ **75.** $s + s + s + s$, or $4s$
77. ▨

Exercise Set 1.2, pp. 16–17

1. $x + 7$ **3.** $c + ab$ **5.** $3y + 9x$ **7.** $5(1 + a)$
9. $a \cdot 2$ **11.** ts **13.** $5 + ba$ **15.** $(a + 1)5$
17. $a + (5 + b)$ **19.** $(r + t) + 7$ **21.** $ab + (c + d)$
23. $8(xy)$ **25.** $(2a)b$ **27.** $(3 \cdot 2)(a + b)$
29. $(r + t) + 6$; $(t + 6) + r$ **31.** $17(ab)$; $b(17a)$
33. $(5 + x) + 2 = (x + 5) + 2$ Commutative law
$= x + (5 + 2)$ Associative law
$= x + 7$ Simplifying
35. $(m3)7 = m(3 \cdot 7)$ Associative law
$= m21$ Simplifying
$= 21m$ Commutative law
37. $4a + 12$ **39.** $6 + 6x$ **41.** $3x + 3$ **43.** $24 + 8y$
45. $18x + 54$ **47.** $5r + 10 + 15t$ **49.** $2a + 2b$
51. $5x + 5y + 10$ **53.** $x, xyz, 19$ **55.** $2a, \dfrac{a}{b}, 5b$
57. $2(a + b)$ **59.** $7(1 + y)$ **61.** $3(6x + 1)$
63. $5(x + 2 + 3y)$ **65.** $3(4x + 3)$ **67.** $3(a + 3b)$
69. $11(4x + y + 2z)$ **71.** ▨   **73.** [1.1] Let k represent
Kara's salary; $2k$ **74.** [1.1] $\dfrac{1}{2} \cdot m$, or $\dfrac{m}{2}$ **75.** ▨
77. Yes; distributive law   **79.** Yes; distributive law and
commutative law of multiplication   **81.** No; for example,
let $x = 1$ and $y = 2$. Then $30 \cdot 2 + 1 \cdot 15 = 60 + 15 = 75$
and $5[2(1 + 3 \cdot 2)] = 5[2(7)] = 5 \cdot 14 = 70$.   **83.** ▨

Exercise Set 1.3, pp. 24–26

1. $2 \cdot 25$; $5 \cdot 10$; 1, 2, 5, 10, 25, 50   **3.** $3 \cdot 14$; $6 \cdot 7$; 1, 2, 3,
6, 7, 14, 21, 42   **5.** $2 \cdot 13$ **7.** $2 \cdot 3 \cdot 5$ **9.** $2 \cdot 2 \cdot 5$

11. $3 \cdot 3 \cdot 3$ **13.** $2 \cdot 3 \cdot 3$ **15.** $2 \cdot 2 \cdot 2 \cdot 5$
17. Prime   **19.** $2 \cdot 3 \cdot 5 \cdot 7$ **21.** $5 \cdot 23$ **23.** $\frac{5}{7}$
25. $\frac{2}{7}$ **27.** $\frac{1}{8}$ **29.** 7   **31.** $\frac{1}{4}$ **33.** 6   **35.** $\frac{15}{16}$ **37.** $\frac{60}{41}$
39. $\frac{15}{7}$ **41.** $\frac{3}{14}$ **43.** $\frac{27}{8}$ **45.** $\frac{1}{2}$ **47.** $\frac{7}{6}$ **49.** $\frac{3b}{7a}$
51. $\frac{7}{a}$ **53.** $\frac{5}{6}$ **55.** 1   **57.** $\frac{5}{18}$ **59.** 0   **61.** $\frac{35}{18}$
63. $\frac{10}{3}$ **65.** 28   **67.** 1   **69.** $\frac{6}{35}$ **71.** 18   **73.** 🗒
75. [1.2] $5(3 + x)$; answers may vary
76. [1.2] $7 + (b + a)$, or $(a + b) + 7$ **77.** 🗒
79. 24 in.   **81.** $\frac{2}{5}$ **83.** $\frac{3q}{t}$ **85.** $\frac{6}{25}$ **87.** $\frac{5ap}{2cm}$
89. $\frac{28}{45}\,\text{m}^2$ **91.** $14\frac{2}{9}$ m   **93.** 🗒

Technology Connection, p. 31

1. 10.1   **2.** 1.5

Exercise Set 1.4, pp. 33–34

1. $-19, 59$ **3.** $-150, 65$ **5.** $-1286, 29{,}029$
7. $750, -125$ **9.** $20, -150, 300$
11.
13.
15.
17. 0.875   **19.** -0.75
21. $1.1\overline{6}$ **23.** $0.\overline{6}$ **25.** -0.5 **27.** 0.13   **29.** $<$
31. $>$ **33.** $<$ **35.** $<$ **37.** $>$ **39.** $<$ **41.** $<$
43. $x < -7$ **45.** $y \geq -10$ **47.** True   **49.** False
51. True   **53.** 23   **55.** 17   **57.** 5.6   **59.** 329   **61.** $\frac{9}{7}$
63. 0   **65.** 8   **67.** $-83, -4.7, 0, \frac{5}{9}, 8.31, 62$
69. $-83, 0, 62$ **71.** $-83, -4.7, 0, \frac{5}{9}, \pi, \sqrt{17}, 8.31, 62$
73. 🗒   **75.** [1.1] 42   **76.** [1.2] $ba + 5$, or $5 + ab$
77. 🗒   **79.** 🗒   **81.** $-23, -17, 0, 4$
83. $-\frac{5}{6}, -\frac{3}{4}, -\frac{2}{3}, \frac{1}{6}, \frac{3}{8}, \frac{1}{2}$ **85.** $<$ **87.** $=$ **89.** $<$
91. $-2, -1, 0, 1, 2$ **93.** $\frac{1}{9}$ **95.** $\frac{50}{9}$ **97.** 🗒

Exercise Set 1.5, pp. 39–40

1. -3 **3.** 4   **5.** 0   **7.** -8 **9.** -15 **11.** -8
13. 0   **15.** -41 **17.** 0   **19.** 7   **21.** -2 **23.** 11
25. -33 **27.** 0   **29.** 18   **31.** -32 **33.** 0   **35.** 20
37. -1.7 **39.** -9.1 **41.** $\frac{1}{5}$ **43.** $\frac{-6}{7}$ **45.** $-\frac{1}{15}$
47. $\frac{2}{9}$ **49.** -3 **51.** 0   **53.** Lost 3 students, or
-3 students   **55.** \$4300 loss, or $-\$4300$ **57.** $-\$195$
59. Fell $\$\frac{1}{16}$, or $-\$\frac{1}{16}$ **61.** $12a$ **63.** $9x$ **65.** $13t$
67. $-2m$ **69.** $-7a$ **71.** $1 - 2x$ **73.** $12x + 17$
75. $18n + 16$ **77.** 🗒   **79.** [1.2] $21z + 7y + 14$
80. [1.3] $\frac{28}{3}$ **81.** 🗒   **83.** $\$65\frac{1}{4}$ **85.** $-5y$ **87.** $-7m$
89. $-7t, -23$ **91.** 1 under par

Exercise Set 1.6, p. 46–47

1. -39 **3.** 9   **5.** 3.14   **7.** -23 **9.** $\frac{14}{3}$ **11.** -0.101
13. 72   **15.** $-\frac{2}{5}$ **17.** 1   **19.** -7 **21.** Negative three
minus five; -8 **23.** Two minus negative nine; 11
25. Four minus six; -2 **27.** Negative five minus
negative seven; 2   **29.** -2 **31.** -5 **33.** -6
35. -10 **37.** -6 **39.** 0   **41.** -5 **43.** -10 **45.** 2
47. 0   **49.** 0   **51.** 8   **53.** -11 **55.** 16   **57.** -16
59. -1 **61.** 11   **63.** -6 **65.** -2 **67.** -25 **69.** 1
71. -9 **73.** 17   **75.** -45 **77.** -81 **79.** -49
81. -7.9 **83.** -0.175 **85.** $-\frac{2}{7}$ **87.** $-\frac{7}{9}$ **89.** $\frac{3}{13}$
91. $3.8 - (-5.2)$; 9   **93.** $114 - (-79)$; 193   **95.** -58
97. 34   **99.** 41   **101.** -62 **103.** -139 **105.** 0
107. $-7x, -4y$ **109.** $9, -5t, -3st$ **111.** $-3x$
113. $-5a + 4$ **115.** $-7n - 9$ **117.** $-6x + 5$
119. $-8t - 7$ **121.** $-12x + 3y + 9$ **123.** $8x + 66$
125. $100°F$ **127.** 30,340 ft   **129.** 116 m   **131.** 🗒
133. [1.1] 432 ft^2 **134.** [1.2] $2 \cdot 2 \cdot 2 \cdot 2 \cdot 2 \cdot 3 \cdot 3 \cdot 3$
135. 🗒   **137.** True. For example, for $m = 5$ and $n = 3$,
$5 > 3$ and $5 - 3 > 0$, or $2 > 0$. For $m = -4$ and $n = -9$,
$-4 > -9$ and $-4 - (-9) > 0$, or $5 > 0$.
139. False. For example, let $m = 2$ and $n = -2$. Then 2
and -2 are opposites, but $2 - (-2) = 4 \neq 0$.   **141.** 🗒

Exercise Set 1.7, pp. 53–55

1. -36 **3.** -56 **5.** -24 **7.** -72 **9.** 42   **11.** 45
13. -170 **15.** -144 **17.** 1200   **19.** 98   **21.** -78
23. 21.7   **25.** $-\frac{2}{5}$ **27.** $\frac{1}{12}$ **29.** -11.13 **31.** $-\frac{5}{12}$
33. 252   **35.** 0   **37.** $\frac{1}{28}$ **39.** 150   **41.** 0   **43.** -720
45. $-30{,}240$ **47.** -4 **49.** -4 **51.** -2 **53.** 4
55. -8 **57.** 2   **59.** -12 **61.** -8 **63.** Undefined
65. -4 **67.** 0   **69.** 0   **71.** $-\frac{8}{3}; \frac{8}{-3}$ **73.** $-\frac{29}{35}; \frac{-29}{35}$
75. $\frac{-7}{3}; \frac{7}{-3}$ **77.** $-\frac{x}{2}; \frac{x}{-2}$ **79.** $-\frac{5}{4}$ **81.** $-\frac{13}{47}$
83. $-\frac{1}{10}$ **85.** $\frac{1}{4.3}$, or $\frac{10}{43}$ **87.** $-\frac{4}{9}$ **89.** -1 **91.** $\frac{21}{20}$
93. $\frac{12}{55}$ **95.** -1 **97.** 1   **99.** $-\frac{9}{11}$ **101.** $-\frac{7}{4}$
103. -12 **105.** -3 **107.** 1   **109.** 7   **111.** $-\frac{1}{9}$
113. $\frac{1}{10}$ **115.** $-\frac{7}{6}$ **117.** $\frac{6}{7}$ **119.** $-\frac{14}{15}$ **121.** 🗒
123. [1.3] $\frac{22}{39}$ **124.** [1.5] $12x - 2y - 9$ **125.** 🗒
127. For 2 and 3, the reciprocal of the sum is $1/(2 + 3)$, or
$1/5$. But $1/5 \neq 1/2 + 1/3$.   **129.** Negative
131. Negative   **133.** Negative   **135.** **(a)** m and n have
different signs; **(b)** either m or n is zero; **(c)** m and n have
the same sign   **137.** 🗒

Exercise Set 1.8, pp. 62–63

1. 4^3 **3.** x^7 **5.** $(3t)^5$ **7.** 16   **9.** 9   **11.** -9
13. 64   **15.** 625   **17.** 7   **19.** $81t^4$ **21.** $-343x^3$

23. 26   **25.** 86   **27.** 7   **29.** 5   **31.** 0   **33.** 9
35. 10   **37.** -7 **39.** -4 **41.** 1291   **43.** 14
45. 152   **47.** 36   **49.** 1   **51.** -26 **53.** -2 **55.** $-\frac{9}{2}$
57. -8 **59.** -3 **61.** -15 **63.** 9   **65.** 1   **67.** 6
69. -17 **71.** $-9x - 1$ **73.** $-7 + 2x$
75. $-4a + 3b - 7c$ **77.** $-3x^2 - 5x + 1$ **79.** $3x - 7$
81. $-3a + 9$ **83.** $5x - 6$ **85.** $-3t - 11r$
87. $9y - 25z$ **89.** $x^2 + 2$ **91.** $-t^3 - 2t$
93. $37a^2 - 23ab + 35b^2$ **95.** $-22t^3 - t^2 + 9t$
97. $2x - 25$ **99.** 🗍   **101.** [1.1] Let n represent the
number; $9 + 2n$ **102.** [1.1] Let m and n represent the
two numbers; $\frac{1}{2}(m + n)$ **103.** 🗍   **105.** $-6r - 5t + 21$
107. $-2x - f$ **109.** 🗍   **111.** False   **113.** False
115. True   **117.** 0   **119.** 39,000

Review Exercises: Chapter 1, p. 66–67

1. [1.1] 15   **2.** [1.1] 4   **3.** [1.8] -6 **4.** [1.8] -5
5. [1.1] $z - 7$ **6.** [1.1] xz **7.** [1.1] Let m and n represent the numbers; $mn + 1$, or $1 + mn$ **8.** [1.1] No
9. [1.1] Let t represent the value of wholesale tea sold in the U.S. in 1990, in billions of dollars; $4.6 = 2.8 + t$
10. [1.2] $x2 + y$ **11.** [1.2] $2x + (y + z)$
12. [1.2] $(4x)y$, $4(yx)$, $(4y)x$; answers may vary
13. [1.2] $18x + 30y$ **14.** [1.2] $40x + 24y + 16$
15. [1.2] $3(7x + 5y)$ **16.** [1.2] $7(5x + 2 + y)$
17. [1.3] $2 \cdot 2 \cdot 13$ **18.** [1.3] $\frac{5}{12}$ **19.** [1.3] $\frac{9}{4}$
20. [1.3] $\frac{31}{36}$ **21.** [1.3] $\frac{3}{16}$ **22.** [1.3] $\frac{3}{5}$ **23.** [1.3] $\frac{72}{25}$
24. [1.4] $-45, 72$ **25.** [1.4]

$$\xleftarrow{\hspace{0.5cm}}\underset{-5\,-4\,-3\,-2\,-1\ \ 0\ \ 1\ \ 2\ \ 3\ \ 4\ \ 5}{\overset{\frac{-1}{3}}{\bullet}}\xrightarrow{\hspace{0.5cm}}$$

26. [1.4] $x > -3$ **27.** [1.4] True   **28.** [1.4] False
29. [1.4] -0.875 **30.** [1.4] 1   **31.** [1.6] -7
32. [1.5] -3 **33.** [1.5] $-\frac{7}{12}$ **34.** [1.5] 0   **35.** [1.5] -5
36. [1.6] 5   **37.** [1.6] $-\frac{7}{5}$ **38.** [1.6] -7.9 **39.** [1.7] 54
40. [1.7] -9.18 **41.** [1.7] $-\frac{2}{7}$ **42.** [1.7] -140
43. [1.7] -7 **44.** [1.7] -3 **45.** [1.7] $\frac{3}{4}$ **46.** [1.8] 92
47. [1.8] 62   **48.** [1.8] 48   **49.** [1.8] 168   **50.** [1.8] $\frac{21}{8}$
51. [1.8] $\frac{103}{17}$ **52.** [1.5] $7a - 3b$ **53.** [1.6] $-2x + 5y$
54. [1.6] 7   **55.** [1.7] $-\frac{1}{7}$ **56.** [1.8] $(2x)^4$
57. [1.8] $-125x^3$ **58.** [1.8] $-3a + 9$
59. [1.8] $-2b + 21$ **60.** [1.8] $-3x + 9$
61. [1.8] $12y - 34$ **62.** [1.8] $5x + 24$
63. [1.1] 🗍 The value of a constant never varies. A variable can represent a variety of numbers.   **64.** [1.2] 🗍 A term is one of the parts of an expression that is separated from the other parts by plus signs. A factor is part of a product.
65. [1.2], [1.5], [1.8] 🗍 The distributive law is used in factoring algebraic expressions, multiplying algebraic expressions, combining like terms, finding the opposite of a sum, and subtracting algebraic expressions.
66. [1.8] 🗍 A negative quantity raised to an even power is

positive; a negative quantity raised to an odd power is negative.
67. [1.8] 25,281   **68.** [1.4] **(a)** $\frac{3}{11}$; **(b)** $\frac{10}{11}$ **69.** [1.8] $-\frac{5}{8}$
70. [1.8] -2.1 **71.** [1.4] I   **72.** [1.7] J   **73.** [1.8] A
74. [1.8] H   **75.** [1.6] K   **76.** [1.2] B   **77.** [1.4] C
78. [1.4] E   **79.** [1.7] D   **80.** [1.8] F   **81.** [1.4] G

Test: Chapter 1, pp. 67–68

1. [1.1] 4   **2.** [1.1] Let x represent the number; $x - 9$.
3. [1.1] 240 ft^2 **4.** [1.2] $q + 3p$ **5.** [1.2] $(x \cdot 4) \cdot y$
6. [1.1] Yes   **7.** [1.1] Let p represent the maximum production capability; $p - 282 = 2518$ **8.** [1.2] $30 - 5x$
9. [1.7] $-5y + 5$ **10.** [1.2] $11(1 - 4x)$
11. [1.2] $7(x + 3 + 2y)$ **12.** [1.3] $2 \cdot 2 \cdot 3 \cdot 5 \cdot 5$
13. [1.3] $\frac{2}{7}$ **14.** [1.4] $<$ **15.** [1.4] $>$ **16.** [1.4] $\frac{9}{4}$
17. [1.4] 2.7   **18.** [1.6] $-\frac{2}{3}$ **19.** [1.7] $-\frac{7}{4}$ **20.** [1.6] 8
21. [1.4] $-2 \geq x$ **22.** [1.6] 7.8   **23.** [1.5] -8
24. [1.5] $\frac{31}{40}$ **25.** [1.6] 10   **26.** [1.6] -2.5 **27.** [1.6] $\frac{7}{8}$
28. [1.7] -48 **29.** [1.7] $\frac{3}{16}$ **30.** [1.7] -9 **31.** [1.7] $\frac{3}{4}$
32. [1.7] -9.728 **33.** [1.8] -173 **34.** [1.6] 12
35. [1.8] -4 **36.** [1.8] 448   **37.** [1.6] $21a + 22y$
38. [1.8] $16x^4$ **39.** [1.8] $2x + 7$ **40.** [1.8] $9a - 12b - 7$
41. [1.8] $68y - 8$ **42.** [1.1] 15   **43.** [1.3] $\frac{23}{70}$
44. [1.8] 15   **45.** [1.8] $4a$

CHAPTER 2

Exercise Set 2.1, pp. 75–77

1. 15   **3.** -13 **5.** -10 **7.** -13 **9.** 15   **11.** -8
13. -6 **15.** 19   **17.** -4 **19.** $\frac{7}{3}$ **21.** $-\frac{13}{10}$ **23.** $\frac{41}{24}$
25. $-\frac{1}{20}$ **27.** 1.5   **29.** -5 **31.** 16   **33.** 4   **35.** 12
37. -23 **39.** 8   **41.** -7 **43.** -6 **45.** 6   **47.** 52
49. 36   **51.** -45 **53.** $\frac{6}{7}$ **55.** 1   **57.** $\frac{9}{2}$ **59.** -7.6
61. -2.5 **63.** -17 **65.** $-\frac{1}{2}$ **67.** -15 **69.** 12
71. 310.756   **73.** 🗍   **75.** [1.8] -34 **76.** [1.8] 41
77. [1.8] 1   **78.** [1.8] -16 **79.** 🗍   **81.** Identity
83. 0   **85.** Contradiction   **87.** Contradiction   **89.** 9.4
91. 6   **93.** 8   **95.** 11,074   **97.** 🗍

Technology Connection, p. 80

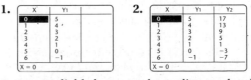

3. 4; not reliable because, depending on the choice of ΔTbl, it is easy to scroll past a solution without realizing it.

Exercise Set 2.2, pp. 83–84

1. 7 **3.** 8 **5.** 5 **7.** 14 **9.** -7 **11.** -5 **13.** -7
15. 15 **17.** 4 **19.** -12 **21.** 6 **23.** 8 **25.** 1
27. -20 **29.** 6 **31.** 7 **33.** 3 **35.** 0 **37.** 10
39. 4 **41.** 0 **43.** $-\frac{2}{5}$ **45.** $\frac{64}{3}$ **47.** $\frac{2}{5}$ **49.** 3
51. -4 **53.** $1.\overline{6}$ **55.** $-\frac{40}{37}$ **57.** 2 **59.** 2 **61.** 6
63. 8 **65.** 2 **67.** -4 **69.** -8 **71.** 2 **73.** 8
75. 1 **77.** -5 **79.** $\frac{11}{18}$ **81.** $-\frac{51}{31}$ **83.** 2 **85.** 🖉
87. [1.8] -7 **88.** [1.8] 15 **89.** [1.8] -15
90. [1.8] -28 **91.** 🖉 **93.** $\frac{1136}{909}$, or $1.\overline{2497}$
95. Contradiction **97.** Identity **99.** $\frac{2}{3}$ **101.** 0
103. 0 **105.** -2

Technology Connection, p. 85

1. 72,930

Exercise Set 2.3, pp. 88–91

1. 2 mi **3.** 1423 students **5.** 10.5 cal/oz **7.** 255 mg
9. $b = \dfrac{A}{h}$ **11.** $r = \dfrac{d}{t}$ **13.** $P = \dfrac{I}{rt}$ **15.** $m = 65 - H$
17. $l = \dfrac{P - 2w}{2}$, or $l = \dfrac{P}{2} - w$ **19.** $\pi = \dfrac{A}{r^2}$
21. $h = \dfrac{2A}{b}$ **23.** $m = \dfrac{E}{c^2}$ **25.** $d = 2Q - c$
27. $b = 3A - a - c$ **29.** $A = Ms$ **31.** $t = \dfrac{A}{a + b}$
33. $h = \dfrac{2A}{a + b}$ **35.** $L = W - \dfrac{N(R - r)}{400}$, or
$L = \dfrac{400W - NR + Nr}{400}$ **37.** 🖉 **39.** [1.7] 0
40. [1.7] 9.18 **41.** [1.8] -13 **42.** [1.8] 65 **43.** 🖉
45. 35 yr **47.** $a = \dfrac{w}{c} \cdot d$ **49.** $y = \dfrac{z^2}{t}$ **51.** $t = \dfrac{rs}{q - r}$
53. $S = 20a$, where S is the number of Btu's saved
55. $K = 8.70w + 17.78h - 9.52a + 92.4$

Exercise Set 2.4, pp. 95–97

1. 0.82 **3.** 0.09 **5.** 0.437 **7.** 0.0046 **9.** 29%
11. 99.8% **13.** 192% **15.** 210% **17.** 0.68%
19. 37.5% **21.** 28% **23.** $66\frac{2}{3}$% **25.** 25% **27.** 24%
29. $46\frac{2}{3}$, or $\frac{140}{3}$ **31.** 2.5 **33.** 84 **35.** 125% **37.** 0.8
39. 50% **41.** $280 **43.** 100 million voters
45. 24 women **47.** 0.5625 lb **49.** 27 bowlers
51. $86\frac{4}{11}$% **53.** $36 **55.** $148.50 **57.** $18/hr
59. 165 calories **61.** 🖉 **63.** [1.1] Let n represent the
number; $5 + n$ **64.** [1.1] Let t represent Tino's weight;
$t - 4$ **65.** [1.1] $8 \cdot 2a$ **66.** [1.1] Let x and y represent
the two numbers; $1 + xy$ **67.** 🖉 **69.** 18,500 people
71. 20% **73.** About 1.55%

Exercise Set 2.5, pp. 105–108

1. 11 **3.** 11 **5.** $75 **7.** $85
9. Approximately $62\frac{2}{3}$ mi **11.** 19, 20, 21 **13.** 29, 31
15. 62, 64 **17.** Bride: 84 yr; groom: 103 yr
19. 30°, 90°, 60° **21.** 95° **23.** 192, 193
25. 63 mm, 65 mm, 67 mm **27.** Width: 275 mi;
length: 365 mi **29.** $350 **31.** $852.94 **33.** 12 mi
35. $128\frac{1}{3}$ mi **37.** 65°, 25° **39.** 160 **41.** 🖉
43. [1.4] $<$ **44.** [1.4] $<$ **45.** [1.4] $<$ **46.** [1.4] $>$
47. 🖉 **49.** $37 **51.** 20 **53.** Length: 12 cm;
width: 9 cm **55.** 104°, 106°, 108°, 110°, 112°
57. $95.99 **59.** 10 **61.** 6 mi **63.** 🖉
65. 5.34 cm, 8.59 cm, 12.94 cm

Exercise Set 2.6, pp. 115–117

1. **(a)** Yes; **(b)** yes; **(c)** yes; **(d)** no; **(e)** yes
3. **(a)** No; **(b)** no; **(c)** yes; **(d)** yes; **(e)** no
5.
7.
9.
11.
13. **15.** $\{x \mid x > -4\}$
17. $\{x \mid x \le 2\}$ **19.** $\{x \mid x < -1\}$ **21.** $\{x \mid x \ge 0\}$
23. $\{y \mid y > 7\}$,
25. $\{x \mid x \le -18\}$,
27. $\{x \mid x < 10\}$,
29. $\{t \mid t \ge -3\}$,
31. $\{y \mid y > -5\}$,
33. $\{x \mid x \le 5\}$, **35.** $\{x \mid x \ge 5\}$
37. $\{y \mid y \le \frac{1}{2}\}$ **39.** $\{t \mid t > \frac{5}{8}\}$ **41.** $\{x \mid x < 0\}$
43. $\{t \mid t < 23\}$ **45.** $\{x \mid x < 7\}$,
47. $\{y \mid y \le 9\}$,
49. $\{x \mid x > -\frac{13}{7}\}$,
51. $\{t \mid t < -3\}$,
53. $\{y \mid y \ge -\frac{2}{7}\}$ **55.** $\{y \mid y \ge -\frac{1}{10}\}$ **57.** $\{x \mid x > \frac{4}{5}\}$
59. $\{x \mid x < 9\}$ **61.** $\{y \mid y \ge 4\}$ **63.** $\{t \mid t \le 7\}$
65. $\{x \mid x < -3\}$ **67.** $\{y \mid y < -4\}$ **69.** $\{x \mid x > -4\}$
71. $\{y \mid y < -\frac{10}{3}\}$ **73.** $\{x \mid x > -10\}$ **75.** $\{y \mid y < 2\}$

77. $\{y \mid y \geq 3\}$ **79.** $\{x \mid x > -4\}$ **81.** $\{x \mid x > -4\}$
83. $\{n \mid n \geq 70\}$ **85.** $\{x \mid x \leq 15\}$ **87.** $\{t \mid t < 14\}$
89. $\{y \mid y < 6\}$ **91.** $\{t \mid t \leq -4\}$ **93.** $\{r \mid r > -3\}$
95. $\{x \mid x \geq 8\}$ **97.** $\{x \mid x < \frac{11}{18}\}$ **99.** 🖃
101. [1.1] Let n represent "some number"; $3 + n$
102. [1.1] Let x and y represent the two numbers; $2(x + y)$
103. [1.1] Let x represent the number; $2x - 3$
104. [1.1] Let y represent the number; $5 + 2y$ **105.** 🖃
107. $\{t \mid t > -\frac{27}{19}\}$ **109.** $\{x \mid x \leq -4a\}$
111. $\left\{x \mid x > \dfrac{y - b}{a}\right\}$ **113.** (a) No; (b) yes; (c) no;
(d) yes; (e) no; (f) yes   **115.** $\{x \mid x \text{ is a real number}\}$,
or $(-\infty, \infty)$

Exercise Set 2.7, pp. 120–123

1. Let n represent the number; $n \geq 7$
3. Let b represent the baby's weight; $b > 2$
5. Let s represent the train's speed; $90 < s < 110$
7. Let m represent the number of people who attended the Million Man March; $m \leq 1,200,000$
9. Let c represent the cost of gasoline; $c \geq 1.50$
11. 15 or fewer copies   **13.** Mileages less than or equal to 341.4 mi   **15.** 3.5 hr or more   **17.** 2
19. Scores greater than or equal to 84
21. 4 servings or more   **23.** 27 min or more
25. Lengths greater than or equal to 92 ft; lengths less than or equal to 92 ft   **27.** Widths less than or equal to $11\frac{2}{3}$ ft   **29.** Times less than 4 units, or 1 hr
31. Lengths less than 21.5 cm
33. Blue-book value is greater than or equal to $10,625.
35. Temperatures greater than 37°C
37. It contains at least 16 g of fat per serving.
39. Depths less than 437.5 ft   **41.** Years after 2003
43. Dates at least 6 weeks after July 1
45. Heights greater than or equal to 4 ft
47. Mileages less than or equal to 215.2 mi   **49.** 🖃
51. [1.8] 2   **52.** [1.8] $\frac{1}{2}$ **53.** [1.8] $-\frac{10}{3}$ **54.** [1.8] $-\frac{1}{5}$
55. 🖃   **57.** More than 6 hr   **59.** Lengths less than or equal to 8 cm   **61.** They contain at least 7.5 g of fat per serving.   **63.** At least $42   **65.** 🖃

Review Exercises: Chapter 2, pp. 126–127

1. [2.1] -25 **2.** [2.1] 7   **3.** [2.1] -68 **4.** [2.1] 1
5. [2.1] -4 **6.** [2.1] 1.11   **7.** [2.1] $\frac{1}{2}$ **8.** [2.1] $-\frac{15}{64}$
9. [2.2] $\frac{38}{5}$ **10.** [2.2] -8 **11.** [2.2] -5 **12.** [2.2] $-\frac{1}{3}$
13. [2.2] 4   **14.** [2.2] 3   **15.** [2.2] 4   **16.** [2.2] 16
17. [2.2] 7   **18.** [2.2] $-\frac{7}{5}$ **19.** [2.2] 12   **20.** [2.2] 4
21 [2.3] $d = \dfrac{C}{\pi}$ **22.** [2.3] $B = \dfrac{3V}{h}$ **23.** [2.3] $a = 2A - b$

24. [2.4] 0.009   **25.** [2.4] 44%   **26.** [2.4] 20%
27. [2.4] 140   **28.** [2.6] Yes   **29.** [2.6] No
30. [2.6] Yes   **31.** [2.6]

$$5x - 6 < 2x + 3$$
$$-5\ -4\ -3\ -2\ -1\ \ 0\ \ 1\ \ 2\ \ 3\ \ 4\ \ 5$$

32. [2.6]

$$-2 < x \leq 5$$
$$-5\ -4\ -3\ -2\ -1\ \ 0\ \ 1\ \ 2\ \ 3\ \ 4\ \ 5$$

33. [2.6]

$$y > 0$$
$$-5\ -4\ -3\ -2\ -1\ \ 0\ \ 1\ \ 2\ \ 3\ \ 4\ \ 5$$

34. [2.6] $\{t \mid t \geq -\frac{1}{2}\}$

35. [2.6] $\{x \mid x \geq 7\}$ **36.** [2.6] $\{y \mid y > 3\}$
37. [2.6] $\{y \mid y \leq -4\}$ **38.** [2.6] $\{x \mid x < -11\}$
39. [2.6] $\{y \mid y > -7\}$ **40.** [2.6] $\{x \mid x > -6\}$
41. [2.6] $\{x \mid x > -\frac{9}{11}\}$ **42.** [2.6] $\{y \mid y \leq 12\}$
43. [2.6] $\{x \mid x \geq -\frac{1}{12}\}$ **44.** [2.5] $167   **45.** [2.5] 20
46. [2.5] 5 ft, 7 ft   **47.** [2.4] $144.2 billion
48. [2.5] 57, 59   **49.** [2.5] Width: 11 cm; length: 17 cm
50. [2.4] $160   **51.** [2.4] $42,038.22
52. [2.5] 35°, 85°, 60°   **53.** [2.7] $105 or less
54. [2.7] Widths greater than 17 cm
55. [2.1], [2.6] 🖃 Multiplying on both sides of an equation by *any* nonzero number results in an equivalent equation. When multiplying on both sides of an inequality, the sign of the number being multiplied by must be considered. If the number is positive, the direction of the inequality symbol remains unchanged; if the number is negative, the direction of the inequality symbol must be reversed to produce an equivalent inequality.
56. [2.1], [2.6] 🖃 The solutions of an equation can usually each be checked. The solutions of an inequality are normally too numerous to check. Checking a few numbers from the solution set found cannot guarantee that the answer is correct, although if any number does not check, the answer found is incorrect.
57. [2.5] Nile: 6671 km; Amazon: 6437 km
58. [2.4], [2.5] $18,600   **59.** [1.4], [2.2] -23, 23
60. [1.4], [2.1] -20, 20   **61.** [2.3] $a = \dfrac{y - 3}{2 - b}$

Test: Chapter 2, p. 127

1. [2.1] 9   **2.** [2.1] 15   **3.** [2.1] -6 **4.** [2.1] 49
5. [2.1] -12 **6.** [2.2] 2   **7.** [2.1] -8 **8.** [2.1] $-\frac{7}{20}$
9. [2.2] 7   **10.** [2.2] -5 **11.** [2.2] $\frac{23}{3}$
12. [2.6] $\{x \mid x > -5\}$ **13.** [2.6] $\{x \mid x > -13\}$
14. [2.6] $\{x \mid x < \frac{21}{8}\}$ **15.** [2.6] $\{y \mid y \leq -13\}$
16. [2.6] $\{y \mid y \leq -8\}$ **17.** [2.6] $\{x \mid x \leq -\frac{1}{20}\}$
18. [2.6] $\{x \mid x < -6\}$ **19.** [2.6] $\{x \mid x \leq -1\}$
20. [2.3] $r = \dfrac{A}{2\pi h}$ **21.** [2.3] $l = 2w - P$ **22.** [2.4] 2.3
23. [2.4] 5.4%   **24.** [2.4] 16   **25.** [2.4] 44%

26. [2.6]

$y < 4$

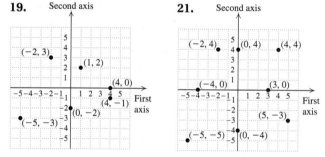

27. [2.6]

$-2 \le x \le 2$

28. [2.5] Width: 7 cm; length: 11 cm   **29.** [2.5] 60 mi
30. [2.5] 81 mm, 83 mm, 85 mm   **31.** [2.4] $4.59 billion
32. [2.7] All numbers greater than 6, or $\{n \mid n > 6\}$
33. [2.7] Lengths greater than or equal to 174 yd

34. [2.3] $d = \dfrac{a}{3}$ **35.** [1.4], [2.2] $-15, 15$

36. [2.5] 60 tickets

CHAPTER 3

Exercise Set 3.1, pp. 137–141

1. 2 drinks   **3.** The person weighs more than 200 lb.
5. About 2,720,000   **7.** About 1,360,000
9. 20.79 million tons   **11.** About 3 million tons
13. 70%   **15.** 1995   **17.** 1994 to 1995
19.

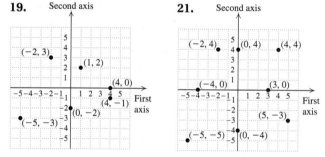

21.

23. $A(-4, 5)$; $B(-3, -3)$; $C(0, 4)$; $D(3, 4)$; $E(3, -4)$
25. $A(4, 1)$; $B(0, -5)$; $C(-4, 0)$; $D(-3, -2)$; $E(3, 0)$
27. IV   **29.** III   **31.** I   **33.** II   **35.** I and IV
37. I and III   **39.** (a) Approximately 56 births per 1000
females; (b) approximately 43 births per 1000 females
41.

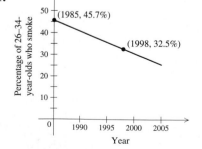

(a) About 40%; (b) about 27.5%

43.

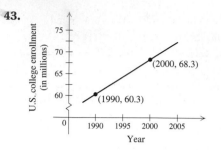

(a) About 65 million students; (b) about 72 million students
45.

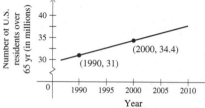

(a) About 32.7 million; (b) about 38 million   **47.** 📝
49. [1.8] -18 **50.** [1.8] -31 **51.** [1.8] 6   **52.** [1.8] 1
53. [2.3] $y = \frac{3}{2}x - 3$ **54.** [2.3] $y = \frac{7}{4}x - \frac{7}{2}$ **55.** 📝
57. I or III   **59.** $(-1, -5)$
61.

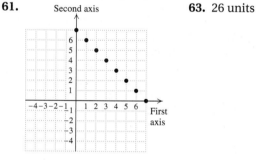

63. 26 units

65. Latitude 32.5° North; longitude 64.5° West   **67.** 📝

Technology Connection, p. 149

1. $y = -5x + 6.5$ **2.** $y = 3x + 4.5$

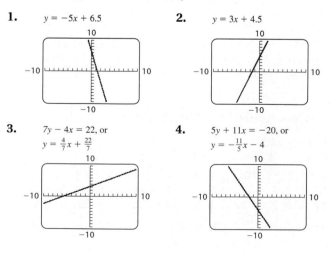

3. $7y - 4x = 22$, or
$y = \frac{4}{7}x + \frac{22}{7}$

4. $5y + 11x = -20$, or
$y = -\frac{11}{5}x - 4$

5. $2y - x^2 = 0$, or
$y = 0.5x^2$

6. $y + x^2 = 8$, or
$y = -x^2 + 8$

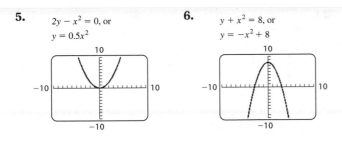

Exercise Set 3.2, pp. 149–151

1. No   **3.** No   **5.** Yes

7.

$$\frac{y = x - 2}{1\ ?\ 3 - 2}$$
$$1 \mid 1 \qquad \text{TRUE}$$

$$\frac{y = x - 2}{-4\ ?\ -2 - 2}$$
$$-4 \mid -4 \qquad \text{TRUE}$$

$(5, 3)$; answers may vary

9.

$$\frac{y = \frac{1}{2}x + 3}{5\ ?\ \frac{1}{2} \cdot 4 + 3}$$
$$\qquad 2 + 3$$
$$5 \mid 5 \qquad \text{TRUE}$$

$$\frac{y = \frac{1}{2}x + 3}{2\ ?\ \frac{1}{2}(-2) + 3}$$
$$\qquad -1 + 3$$
$$2 \mid 2 \qquad \text{TRUE}$$

$(0, 3)$; answers may vary

11.

$$\frac{y + 3x = 7}{1 + 3 \cdot 2\ ?\ 7}$$
$$1 + 6 \mid$$
$$7 \mid 7 \ \text{TRUE}$$

$$\frac{y + 3x = 7}{-5 + 3 \cdot 4\ ?\ 7}$$
$$-5 + 12 \mid$$
$$7 \mid 7 \ \text{TRUE}$$

$(1, 4)$; answers may vary

13.

$$\frac{4x - 2y = 10}{4 \cdot 0 - 2(-5)\ ?\ 10}$$
$$0 + 10 \mid$$
$$10 \mid 10 \ \text{TRUE}$$

$$\frac{4x - 2y = 10}{4 \cdot 4 - 2 \cdot 3\ ?\ 10}$$
$$16 - 6 \mid$$
$$10 \mid 10 \ \text{TRUE}$$

$(2, -1)$; answers may vary

15. **17.** **19.** **21.**

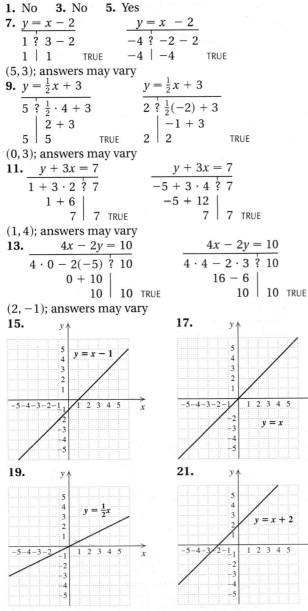

23. **25.** **27.** **29.** **31.** **33.** **35.** **37.**

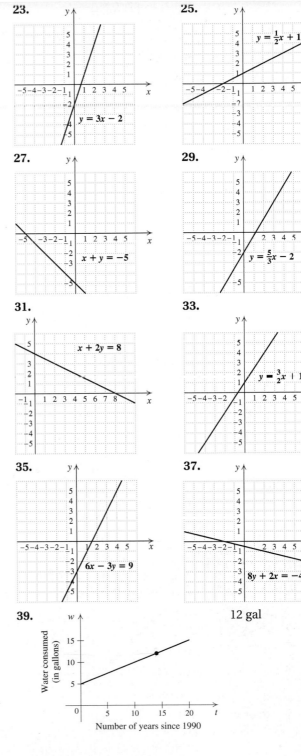

39.

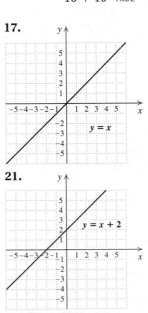

12 gal

41.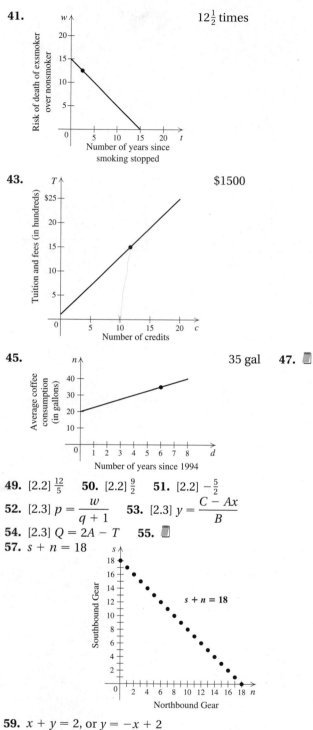

12½ times

43. $1500

45. 35 gal **47.**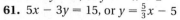

49. [2.2] $\frac{12}{5}$ **50.** [2.2] $\frac{9}{2}$ **51.** [2.2] $-\frac{5}{2}$

52. [2.3] $p = \dfrac{w}{q + 1}$ **53.** [2.3] $y = \dfrac{C - Ax}{B}$

54. [2.3] $Q = 2A - T$ **55.**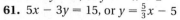

57. $s + n = 18$

59. $x + y = 2$, or $y = -x + 2$

61. $5x - 3y = 15$, or $y = \frac{5}{3}x - 5$

63.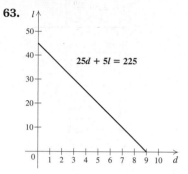

$$25d + 5l = 225$$

Answers may vary. 1 dinner, 40 lunches; 5 dinners, 20 lunches; 8 dinners, 5 lunches

65. $y = -|x|$ **67.** $y = -|x| + 2$

69. $y = -2.8x + 3.5$ **71.** $y = 2.8x - 3.5$

73. $y = x^2 + 4x + 1$

75. No; only whole-number values of the variables have meaning in these applications, so only those points for which both coordinates are whole numbers are solutions of the given problems.

Technology Connection, p. 156

1. $y = -0.72x - 15$

Xscl = 5, Yscl = 5

2. $y - 2.13x = 27$, or $y = 2.13x + 27$

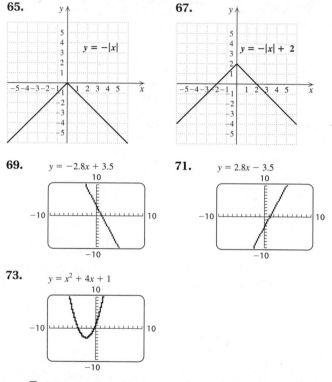

Xscl = 5, Yscl = 5

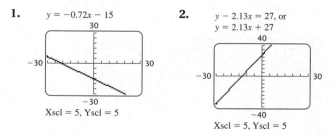

3. $5x + 6y = 84$, or
$y = -\frac{5}{6}x + 14$

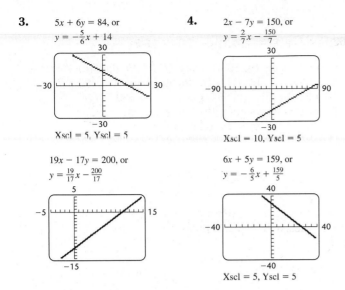

Xscl = 5, Yscl = 5

4. $2x - 7y = 150$, or
$y = \frac{2}{7}x - \frac{150}{7}$

Xscl = 10, Yscl = 5

$19x - 17y = 200$, or
$y = \frac{19}{17}x - \frac{200}{17}$

$6x + 5y = 159$, or
$y = -\frac{6}{5}x + \frac{159}{5}$

Xscl = 5, Yscl = 5

Exercise Set 3.3, pp. 159–160

1. (a) $(0, 5)$; **(b)** $(2, 0)$ **3. (a)** $(0, -4)$; **(b)** $(3, 0)$
5. (a) $(0, -2)$; **(b)** $(-3, 0)$, $(3, 0)$
7. (a) $(0, 4)$; **(b)** $(-3, 0)$, $(3, 0)$, $(5, 0)$ **9. (a)** $(0, 5)$; **(b)** $(3, 0)$
11. (a) $(0, -14)$; **(b)** $(4, 0)$ **13. (a)** $\left(0, \frac{10}{3}\right)$; **(b)** $\left(-\frac{5}{2}, 0\right)$
15. (a) $(0, 9)$; **(b)** none

17.

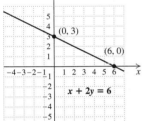

$x + 2y = 6$

19.

$6x + 9y = 36$

21.

$-x + 3y = 9$

23.

$2x - y = 8$

25.

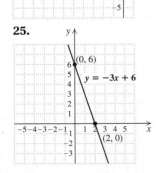

$y = -3x + 6$

27.

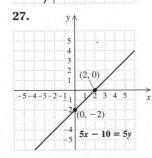

$5x - 10 = 5y$

29.

$2x - 5y = 10$

31.

$6x + 2y = 12$

33.

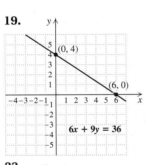

$x - 1 = y$

35.

$2x - 6y = 18$

37.

$4x - 3y = 12$

39.

$-3x = 6y - 2$

41.

$3 = 2x - 5y$

43.

$x + 2y = 0$

45.

$y = 5$

47.

$x = 4$

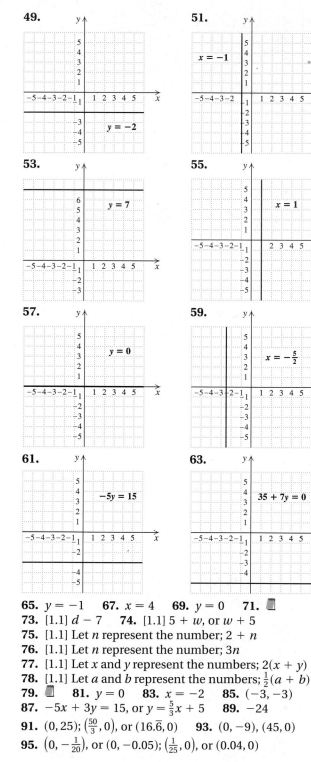

65. $y = -1$ **67.** $x = 4$ **69.** $y = 0$ **71.**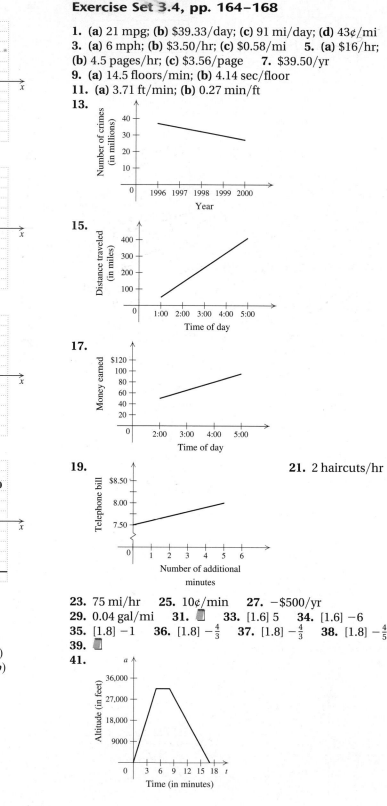
73. [1.1] $d - 7$ **74.** [1.1] $5 + w$, or $w + 5$
75. [1.1] Let n represent the number; $2 + n$
76. [1.1] Let n represent the number; $3n$
77. [1.1] Let x and y represent the numbers; $2(x + y)$
78. [1.1] Let a and b represent the numbers; $\frac{1}{2}(a + b)$
79. 🗒   **81.** $y = 0$ **83.** $x = -2$ **85.** $(-3, -3)$
87. $-5x + 3y = 15$, or $y = \frac{5}{3}x + 5$ **89.** -24
91. $(0, 25)$; $\left(\frac{50}{3}, 0\right)$, or $(16.\overline{6}, 0)$ **93.** $(0, -9)$, $(45, 0)$
95. $\left(0, -\frac{1}{20}\right)$, or $(0, -0.05)$; $\left(\frac{1}{25}, 0\right)$, or $(0.04, 0)$

Exercise Set 3.4, pp. 164–168

1. **(a)** 21 mpg; **(b)** \$39.33/day; **(c)** 91 mi/day; **(d)** 43¢/mi
3. **(a)** 6 mph; **(b)** \$3.50/hr; **(c)** \$0.58/mi   **5.** **(a)** \$16/hr;
(b) 4.5 pages/hr; **(c)** \$3.56/page   **7.** \$39.50/yr
9. **(a)** 14.5 floors/min; **(b)** 4.14 sec/floor
11. **(a)** 3.71 ft/min; **(b)** 0.27 min/ft

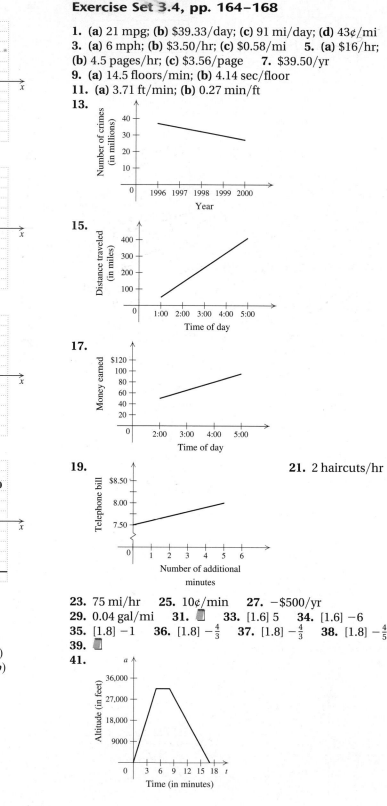

21. 2 haircuts/hr

23. 75 mi/hr   **25.** 10¢/min   **27.** $-$500/yr
29. 0.04 gal/mi   **31.** 🗒   **33.** [1.6] 5   **34.** [1.6] -6
35. [1.8] -1 **36.** [1.8] $-\frac{4}{3}$ **37.** [1.8] $-\frac{4}{3}$ **38.** [1.8] $-\frac{4}{5}$
39. 🗒
41.

43.

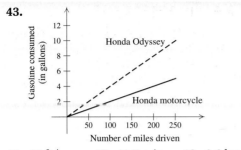

45. 13 ft/sec   **47.** 41.7 min   **49.** 3.6 bu/hr

Exercise Set 3.5, pp. 177–182

1. 15 calories/min   **3.** 1 point/$1000 income
5. -0.75%/yr, or $-3/4\%$/yr   **7.** $\frac{3}{4}$ **9.** $\frac{3}{2}$ **11.** $\frac{1}{3}$
13. -1 **15.** 0   **17.** $-\frac{1}{3}$ **19.** Undefined   **21.** $-\frac{1}{4}$
23. $\frac{3}{2}$ **25.** 0   **27.** -3 **29.** $\frac{3}{2}$ **31.** $-\frac{4}{5}$ **33.** $\frac{7}{9}$
35. $-\frac{2}{3}$ **37.** $-\frac{1}{2}$ **39.** 0   **41.** $-\frac{11}{6}$ **43.** Undefined
45. Undefined   **47.** 0   **49.** Undefined   **51.** 0
53. 8%   **55.** 8.$\overline{3}$%   **57.** $\frac{12}{41}$, or about 29%

59. About 29%   **61.** 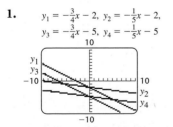   **63.** [2.3] $y = \dfrac{c - ax}{b}$

64. [2.3] $r - \dfrac{p + mn}{x}$ **65.** [2.3] $y = \dfrac{ax - c}{b}$

66. [2.3] $t = \dfrac{q - rs}{n}$ **67.** [1.8] 3   **68.** [1.8] 2   **69.**

71. $\left\{m \mid -\frac{7}{4} \le m \le 0\right\}$ **73.** $\dfrac{18 - x}{x}$ **75.** $\frac{1}{4}$

Technology Connection, p. 188

1. $y_1 = -\frac{3}{4}x - 2,\ y_2 = -\frac{1}{5}x - 2,$
$y_3 = -\frac{3}{4}x - 5,\ y_4 = -\frac{1}{5}x - 5$

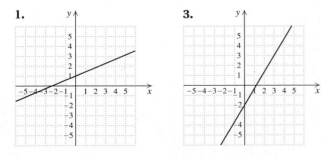

Exercise Set 3.6, pp. 188–190

1.   **3.**

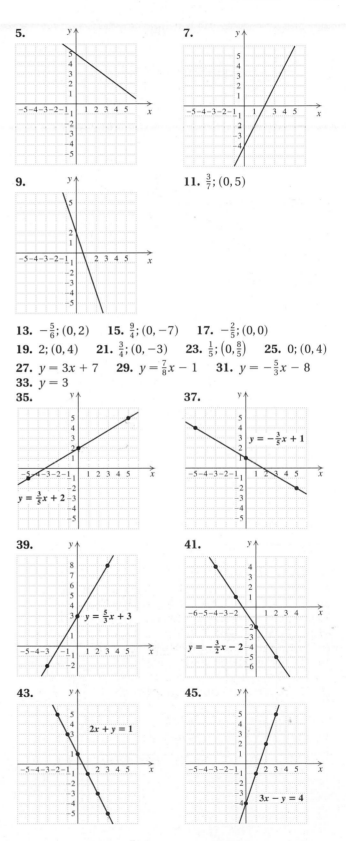

5.   **7.**

9.   **11.** $\frac{3}{7}$; $(0, 5)$

13. $-\frac{5}{6}$; $(0, 2)$ **15.** $\frac{9}{4}$; $(0, -7)$ **17.** $-\frac{2}{5}$; $(0, 0)$
19. 2; $(0, 4)$ **21.** $\frac{3}{4}$; $(0, -3)$ **23.** $\frac{1}{5}$; $\left(0, \frac{8}{5}\right)$ **25.** 0; $(0, 4)$
27. $y = 3x + 7$ **29.** $y = \frac{7}{8}x - 1$ **31.** $y = -\frac{5}{3}x - 8$
33. $y = 3$
35.   **37.**
39.   **41.**
43.   **45.**

47.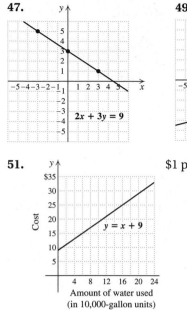

49.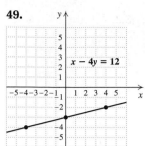

51. $1 per 10,000 gallons

53. $y = 1.5x + 16$ **55.** Yes   **57.** No   **59.** Yes
61. 🖵   **63.** [2.3] $y = m(x - h) + k$
64. [2.3] $y = -2(x + 4) + 9$ **65.** [1.6] 2   **66.** [1.6] 16
67. [1.6] -9 **68.** [1.6] -10 **69.** 🖵   **71.** Yes
73. No   **75.** Yes   **77.** When $x = 0$, $y = b$, so $(0, b)$ is on
the line. When $x = 1$, $y = m + b$, so $(1, m + b)$ is on the
line. Then

$$\text{slope} = \frac{(m + b) - b}{1 - 0} = m.$$

79. $y = \frac{1}{3}x + 3$ **81.** $y = \frac{5}{2}x + 1$

Exercise Set 3.7, pp. 194–195

1. $y - 7 = 6(x - 2)$ **3.** $y - 2 = \frac{3}{5}(x - 9)$
5. $y - 1 = -4(x - 3)$ **7.** $y - (-4) = \frac{3}{2}(x - 5)$
9. $y - 6 = \frac{5}{4}(x - (-2))$ **11.** $y - (-1) = -2(x - (-4))$
13. $y - 8 = 1(x - (-2))$ **15.** $y = 2x - 3$
17. $y = \frac{7}{4}x - 9$ **19.** $y = -3x - 2$ **21.** $y = -4x - 9$
23. $y = \frac{2}{3}x + 1$ **25.** $y = -\frac{5}{6}x + \frac{9}{2}$
27. 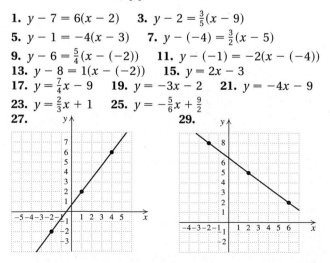   **29.**

31.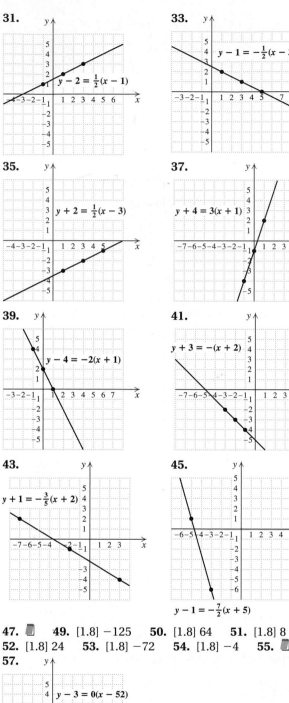

33.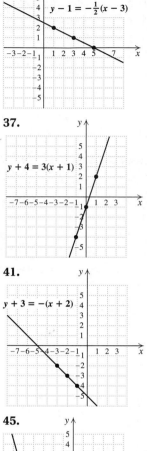

35.

37.

39.

41.

43.

45.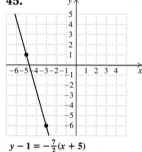

47. 🖵   **49.** [1.8] -125 **50.** [1.8] 64   **51.** [1.8] 8
52. [1.8] 24   **53.** [1.8] -72 **54.** [1.8] -4 **55.** 🖵
57.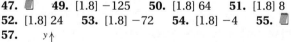

59. $y - 7 = \frac{5}{2}(x - 3)$; $y - 2 = \frac{5}{2}(x - 1)$

61. $y - 8 = \frac{3}{2}(x - 3)$; $y - 2 = \frac{3}{2}(x - (-1))$

63. $y - 8 = -\frac{5}{2}(x - (-3))$; $y - (-2) = -\frac{5}{2}(x - 1)$

65. $y = 2x - 9$ **67.** $y = -\frac{4}{3}x + \frac{23}{3}$ **69.** $y = -x + 6$

71. $y = \frac{2}{3}x + 3$ **73.** $y = \frac{2}{5}x - 2$ **75.** $y = \frac{3}{4}x - \frac{5}{2}$

77. $y - 7 = -\frac{2}{3}(x - (-4))$ **79.** $y = -4x + 7$

81. $y = \frac{10}{3}x + \frac{25}{3}$ **83.**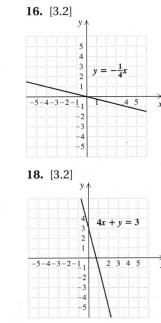

Review Exercises: Chapter 3, pp. 197–198

1. [3.1] $306,000   **2.** [3.1] $318.50

3.–5. [3.1]   **6.** [3.1] IV

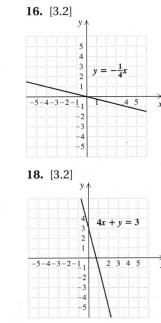

7. [3.1] III   **8.** [3.1] II   **9.** [3.1] $(-5, -1)$

10. [3.1] $(\;2, 5)$ **11.** [3.1] $(3, 0)$ **12.** [3.2] No

13. [3.2] Yes

14. [3.2]

$2x - y = 3$	$2x - y = 3$
$2 \cdot 0 - (-3) \;?\; 3$	$2 \cdot 2 - 1 \;?\; 3$
$0 + 3$	$4 - 1$
$3 \mid 3$ TRUE	$3 \mid 3$ TRUE

$(-1, -5)$; answers may vary

15. [3.2]

16. [3.2]

$y = x - 5$

$y = -\frac{1}{4}x$

17. [3.2]

18. [3.2]

$y = -x + 4$

$4x + y = 3$

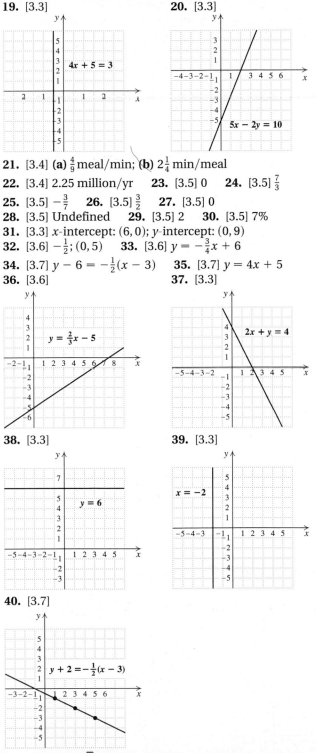

19. [3.3]

$4x + 5 = 3$

20. [3.3]

$5x - 2y = 10$

21. [3.4] **(a)** $\frac{4}{9}$ meal/min; **(b)** $2\frac{1}{4}$ min/meal

22. [3.4] 2.25 million/yr   **23.** [3.5] 0   **24.** [3.5] $\frac{7}{3}$

25. [3.5] $-\frac{3}{7}$ **26.** [3.5] $\frac{3}{2}$ **27.** [3.5] 0

28. [3.5] Undefined   **29.** [3.5] 2   **30.** [3.5] 7%

31. [3.3] x-intercept: $(6, 0)$; y-intercept: $(0, 9)$

32. [3.6] $-\frac{1}{2}$; $(0, 5)$ **33.** [3.6] $y = -\frac{3}{4}x + 6$

34. [3.7] $y - 6 = -\frac{1}{2}(x - 3)$ **35.** [3.7] $y = 4x + 5$

36. [3.6]   **37.** [3.3]

$y = \frac{2}{3}x - 5$ $2x + y = 4$

38. [3.3]   **39.** [3.3]

$y = 6$ $x = -2$

40. [3.7]

$y + 2 = -\frac{1}{2}(x - 3)$

41. [3.1], [3.4] A business might use a graph to plot how total sales changes from year to year or to visualize rate of growth or production.   **42.** [3.2] The y-intercept is the

point at which the graph crosses the *y*-axis. Since a point on the *y*-axis is neither left nor right of the origin, the first or *x*-coordinate of the point is 0.   **43.** [3.2] -1
44. [3.2] 19   **45.** [3.1] Area: 45 sq units; perimeter: 28 units   **46.** [3.2] $(0, 4), (1, 3), (-1, 3)$; answers may vary

Test: Chapter 3, pp. 198–199

1. [3.1] $112.32   **2.** [3.1] $777.60   **3.** [3.1] II   **4.** [3.1] III
5. [3.1] $(3, 4)$ **6.** [3.1] $(0, -4)$ **7.** [3.1] $(-5, 2)$
8. [3.2]   **9.** [3.3]

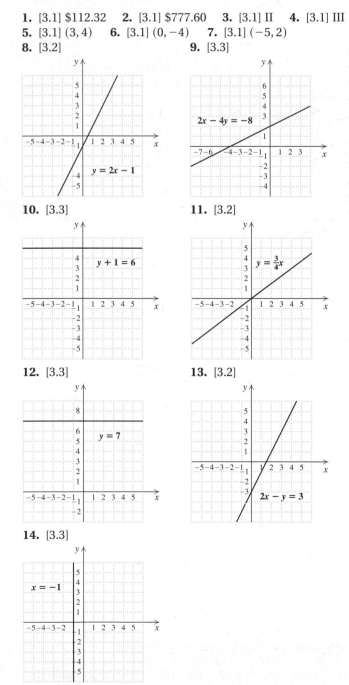

10. [3.3]   **11.** [3.2]

12. [3.3]   **13.** [3.2]

14. [3.3]

15. [3.3] *x*-intercept: $(6, 0)$; *y*-intercept: $(0, -10)$

16. [3.3] *x*-intercept: $(10, 0)$; *y*-intercept: $\left(0, \frac{5}{2}\right)$
17. [3.5] $\frac{9}{2}$ **18.** [3.5] $\frac{7}{12}$ **19.** [3.4] $\frac{1}{3}$ km/min
20. [3.6] 3; $(0, 7)$
21. [3.6]   **22.** [3.7]

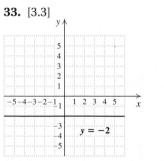
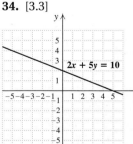

23. [3.7] $y - 8 = -3(x - 6)$ **24.** [3.6] $y = \frac{2}{5}x + 9$
25. [3.1] Area: 25 sq units; perimeter: 20 units

Cumulative Review: Chapters 1–3, pp. 199–200

1. [1.1] 15   **2.** [1.2] $12x - 15y + 21$
3. [1.2] $3(5x - 3y + 1)$ **4.** [1.3] $2 \cdot 3 \cdot 7$ **5.** [1.4] 0.45
6. [1.4] 4   **7.** [1.6] $\frac{1}{4}$ **8.** [1.7] -4 **9.** [1.6] $-x - y$
10. [2.4] 0.785   **11.** [1.3] $\frac{11}{60}$ **12.** [1.5] 2.6
13. [1.7] 7.28   **14.** [1.7] $-\frac{5}{12}$ **15.** [1.8] -2
16. [1.8] 27   **17.** [1.8] $-2y - 7$ **18.** [1.8] $5x + 11$
19. [2.1] -1.2 **20.** [2.1] -21 **21.** [2.2] 9
22. [2.1] $-\frac{20}{3}$ **23.** [2.2] 2   **24.** [2.1] $\frac{13}{8}$ **25.** [2.2] $-\frac{17}{21}$
26. [2.2] -17 **27.** [2.2] 2   **28.** [2.6] $\{x \mid x < 16\}$
29. [2.6] $\left\{x \mid x \le -\frac{11}{8}\right\}$ **30.** [2.3] $h = \dfrac{A - \pi r^2}{2\pi r}$
31. [3.1] IV   **32.** [2.6] $-1 < x \le 2$

33. [3.3]   **34.** [3.3]

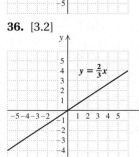

35. [3.2]   **36.** [3.2]

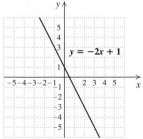

37. [3.3] $(10.5, 0); (0, -3)$ **38.** [3.3] $(-1.25, 0); (0, 5)$
39. [2.4] 160 million   **40.** [2.5] 15.6 million
41. [2.4] $120   **42.** [2.5] 50 m, 53 m, 40 m   **43.** [2.5] 8
44. [3.4] $40/person   **45.** [3.5] $-\frac{1}{3}$
46. [3.6] $y = \frac{2}{7}x - 4$ **47.** [3.6] $-\frac{1}{3}; (0, 3)$
48. [3.6]   **49.** [3.3]

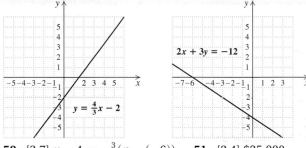

50. [3.7] $y - 4 = -\frac{3}{8}(x - (-6))$ **51.** [2.4] $25,000
52. [1.4], [2.2] $-4, 4$ **53.** [2.2] 2   **54.** [2.2] -5
55. [2.2] 3   **56.** [2.2] No solution
57. [2.3] $Q = \dfrac{2 - pm}{p}$
58. [3.6] $y = -\frac{7}{3}x + 7; y = -\frac{7}{3}x - 7; y = \frac{7}{3}x - 7;$
$y = \frac{7}{3}x + 7$

CHAPTER 4

Exercise Set 4.1, pp. 208–209

1. r^{10} **3.** 9^8 **5.** a^7 **7.** 5^{15} **9.** $(3y)^{12}$ **11.** $(5t)^7$
13. $a^5 b^9$ **15.** $(x + 1)^{12}$ **17.** r^{10} **19.** $x^4 y^7$ **21.** 7^3
23. x^{12} **25.** t^4 **27.** $5a$ **29.** 1   **31.** $3m^3$ **33.** $a^7 b^6$
35. $m^9 n^4$ **37.** 1   **39.** 5   **41.** 2   **43.** -4 **45.** x^{28}
47. 5^{16} **49.** m^{35} **51.** t^{80} **53.** $49x^2$ **55.** $-8a^3$
57. $16m^6$ **59.** $a^{14} b^7$ **61.** $x^8 y^7$ **63.** $24x^{19}$ **65.** $\dfrac{a^3}{64}$
67. $\dfrac{49}{25a^2}$ **69.** $\dfrac{a^{20}}{b^{15}}$ **71.** $\dfrac{y^6}{4}$ **73.** $\dfrac{x^8 y^4}{z^{12}}$ **75.** $\dfrac{a^{12}}{16b^{20}}$
77. $\dfrac{125 x^{21} y^3}{8 z^{12}}$ **79.** 1   **81.** 📝   **83.** [1.2] $3(s - r + t)$
84. [1.2] $-7(x - y + z)$ **85.** [1.6] $8x$
86. [1.6] $-3a - 6b$ **87.** [1.2] $2y + 3x$
88. [1.2] $5z + 2xy$ **89.** 📝   **91.** 📝
93. Let $a = 1$; then $(a + 5)^2 = 36$, but $a^2 + 5^2 = 26$.
95. Let $a = 0$; then $\dfrac{a + 7}{7} = 1$, but $a = 0$.   **97.** a^{8k}
99. $\frac{16}{375}$ **101.** 13   **103.** $<$ **105.** $<$ **107.** $>$
109. 4,000,000; 4,194,304; 194,304
111. 2,000,000,000; 2,147,483,648; 147,483,648
113. 65,536

Technology Connection, p. 215

1. 20.75

Exercise Set 4.2, pp. 215–218

1. $7x^4, x^3, -5x, 8$ **3.** $-t^4, 7t^3, -3t^2, 6$
5. Coefficients: 4, 7; degrees: 5, 1
7. Coefficients: 9, -3, 4; degrees: 2, 1, 0
9. Coefficients: 7, 9, 1; degrees: 4, 1, 3
11. Coefficients: 1, -1, 4, -3; degrees: 4, 3, 1, 0
13. (a) 3, 5, 2; (b) $7a^5, 7$; (c) 5   **15.** (a) 1, 0, 2; (b) $4t^2, 4$;
(c) 2   **17.** (a) 4, 2, 7, 0; (b) $x^7, 1$; (c) 7   **19.** (a) 1, 4, 0, 3;
(b) $-a^4, -1$; (c) 4
21.

Term	Coefficient	Degree of the Term	Degree of the Polynomial
$8x^5$	8	5	
$-\frac{1}{2}x^4$	$-\frac{1}{2}$	4	
$-4x^3$	-4	3	5
$7x^2$	7	2	
6	6	0	

23. Trinomial   **25.** None of these   **27.** Binomial
29. Monomial   **31.** $11x^2 + 3x$ **33.** $4a^4$
35. $6x^2 - 3x$ **37.** $5x^3 - x + 5$ **39.** $-x^4 - x^3$
41. $\frac{1}{15}x^4 + 10$ **43.** $3.4x^2 + 1.3x + 5.5$
45. $10t^4 - 12t^3 + 17t$ **47.** -16 **49.** 16   **51.** -3
53. 11   **55.** 55   **57.** About 9   **59.** About 6
61. About 15   **63.** 1112 ft   **65.** Approximately 449
67. $9200   **69.** $28,000   **71.** 62.8 cm   **73.** 153.86 m²
75. 📝   **77.** [1.5] 5   **78.** [1.6] -9 **79.** [1.2] $5(x + 3)$
80. [1.2] $7(a - 3)$ **81.** [3.4] 6.25¢/mi
82. [2.5] 274 and 275   **83.** 📝
85. $-6x^5 + 14x^4 - x^2 + 11$; answers may vary   **87.** 10
89. $5x^9 + 4x^8 + x^2 + 5x$ **91.** $x^3 - 2x^2 - 6x + 3$
93. 85.0   **95.** 50 yr
97.

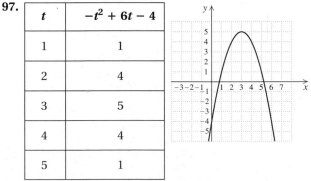

t	$-t^2 + 6t - 4$
1	1
2	4
3	5
4	4
5	1

Technology Connection, p. 223

1. In each case, let $y_1 = $ the expression before the addition
or subtraction has been performed, $y_2 = $ the simplified

sum or difference, and $y_3 = y_2 - y_1$; and note that the graph of y_3 coincides with the x-axis. That is, $y_3 = 0$.

Exercise Set 4.3, pp. 224–226

1. $-5x + 9$ **3.** $x^2 - 5x - 1$ **5.** $9t^2 + 5t - 3$
7. $6m^3 + 3m^2 - 3m - 9$ **9.** $7 + 13a + 6a^2 + 14a^3$
11. $9x^8 + 8x^7 - 3x^4 + 2x^2 - 2x + 5$
13. $-\frac{1}{2}x^4 + \frac{2}{3}x^3 + x^2$ **15.** $4.2t^3 + 3.5t^2 - 6.4t - 1.8$
17. $-3x^4 + 3x^2 + 4x$
19. $1.05x^4 + 0.36x^3 + 14.22x^2 + x + 0.97$
21. $-(-t^3 + 4t^2 - 9);\ t^3 - 4t^2 + 9$
23. $-(12x^4 - 3x^3 + 3);\ -12x^4 + 3x^3 - 3$ **25.** $-8x + 9$
27. $-3a^4 + 5a^2 - 9$ **29.** $4x^4 - 6x^2 - \frac{3}{4}x + 8$
31. $9x + 3$ **33.** $-t^2 - 7t + 5$
35. $6x^4 + 3x^3 - 4x^2 + 3x - 4$
37. $4.6x^3 + 9.2x^2 - 3.8x - 23$ **39.** 0
41. $4 + 2a + 7a^2 - 3a^3$ **43.** $\frac{3}{4}x^3 - \frac{1}{2}x$
45. $0.05t^3 - 0.07t^2 + 0.01t + 1$ **47.** $3x + 5$
49. $11x^4 + 12x^3 - 10x^2$ **51.** (a) $5x^2 + 4x$; (b) 145; 273
53. $14y + 25$ **55.** $(r + 11)(r + 9);\ 9r + 99 + 11r + r^2$
57. $(x + 3)^2;\ x^2 + 3x + 9 + 3x$ **59.** $\pi r^2 - 25\pi$
61. $18z - 64$ **63.** $y^2 - 4y + 4$ **65.** 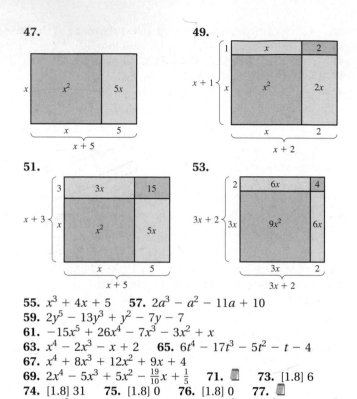 **67.** [1.8] 0
68. [1.8] 0 **69.** [1.8] $13t + 14$ **70.** [1.8] $14t - 15$
71. [2.6] $\{x \mid x < \frac{14}{3}\}$ **72.** [2.6] $\{x \mid x \le 12\}$ **73.**
75. $9t^2 - 20t + 11$ **77.** $-10y^2 - 2y - 10$
79. $-3y^4 - y^3 + 5y - 2$ **81.** $250.591x^3 + 2.812x$
83. $20w + 42$ **85.** $2x^2 + 20x$ **87.**

Technology Connection, p. 232

1. Let $y_1 = (-2x^2 - 3)(5x^3 - 3x + 4)$ and $y_2 = -10x^5 - 9x^3 - 8x^2 + 9x - 12$. With the table set in AUTO mode, note that the values in the Y1- and Y2-columns match, regardless of how far we scroll up or down.
2. Use TRACE, a Table, or a boldly drawn graph to confirm that Y3 is always 0.

Exercise Set 4.4, pp. 232–234

1. $30x^4$ **3.** x^3 **5.** $-x^8$ **7.** $28t^8$ **9.** $-0.02x^{10}$
11. $\frac{1}{15}x^4$ **13.** 0 **15.** $-28x^{11}$ **17.** $-3x^2 + 15x$
19. $4x^2 + 4x$ **21.** $3a^2 + 27a$ **23.** $x^5 + x^2$
25. $6x^3 - 18x^2 + 3x$ **27.** $15t^3 + 30t^2$ **29.** $-6x^4 - 6x^3$
31. $4a^9 - 8a^7 - \frac{5}{12}a^4$ **33.** $x^2 + 9x + 18$
35. $x^2 + 3x - 10$ **37.** $a^2 - 13a + 42$ **39.** $x^2 - 9$
41. $25 - 15x + 2x^2$ **43.** $t^2 + \frac{17}{6}t + 2$
45. $\frac{3}{16}a^2 + \frac{5}{4}a - 2$

47.

49.

51.

53.

55. $x^3 + 4x + 5$ **57.** $2a^3 - a^2 - 11a + 10$
59. $2y^5 - 13y^3 + y^2 - 7y - 7$
61. $-15x^5 + 26x^4 - 7x^3 - 3x^2 + x$
63. $x^4 - 2x^3 - x + 2$ **65.** $6t^4 - 17t^3 - 5t^2 - t - 4$
67. $x^4 + 8x^3 + 12x^2 + 9x + 4$
69. $2x^4 - 5x^3 + 5x^2 - \frac{19}{10}x + \frac{1}{5}$ **71.** **73.** [1.8] 6
74. [1.8] 31 **75.** [1.8] 0 **76.** [1.8] 0 **77.**
79. $75y^2 - 45y$ **81.** 5 **83.** $V = 4x^3 - 48x^2 + 144x$ in³;
$S = -4x^2 + 144$ in² **85.** $x^3 + 2x^2 - 210$ m³
87. 16 ft by 8 ft **89.** 0 **91.** 0

Exercise Set 4.5, pp. 241–243

1. $x^3 + 4x^2 + 3x + 12$ **3.** $x^4 + 2x^3 + 6x + 12$
5. $y^2 - y - 6$ **7.** $9x^2 + 21x + 10$ **9.** $5x^2 + 4x - 12$
11. $2 + 3t - 9t^2$ **13.** $2x^2 - 9x + 7$ **15.** $p^2 - \frac{1}{16}$
17. $x^2 - 0.01$ **19.** $2x^3 + 2x^2 + 6x + 6$
21. $-2x^2 - 11x + 6$ **23.** $a^2 + 18a + 81$
25. $1 - 2t - 15t^2$ **27.** $x^5 + 3x^3 - x^2 - 3$
29. $3x^6 - 2x^4 - 6x^2 + 4$ **31.** $4t^6 + 16t^3 + 15$
33. $8x^5 + 16x^3 + 5x^2 + 10$ **35.** $4x^3 - 12x^2 + 3x - 9$
37. $x^2 - 64$ **39.** $4x^2 - 1$ **41.** $25m^2 - 4$ **43.** $4x^4 - 9$
45. $9x^8 - 1$ **47.** $x^8 - 49$ **49.** $t^2 - \frac{9}{16}$
51. $x^2 + 4x + 4$ **53.** $9x^{10} + 6x^5 + 1$ **55.** $a^2 - \frac{4}{5}a + \frac{4}{25}$
57. $t^6 + 6t^3 + 9$ **59.** $4 - 12x^4 + 9x^8$
61. $25 + 60t^2 + 36t^4$ **63.** $49x^2 - 4.2x + 0.09$
65. $10a^5 - 5a^3$ **67.** $a^3 - a^2 - 10a + 12$
69. $9 - 12x^3 + 4x^6$ **71.** $4x^3 + 24x^2 - 12x$
73. $t^6 - 2t^3 + 1$ **75.** $15t^5 - 3t^4 + 3t^3$
77. $36x^8 - 36x^4 + 9$ **79.** $12x^3 + 8x^2 + 15x + 10$
81. $25 - 60x^4 + 36x^8$ **83.** $a^3 + 1$ **85.** $a^2 + 2a + 1$
87. $x^2 + 7x + 10$ **89.** $x^2 + 14x + 49$
91. $t^2 + 10t + 24$ **93.** $t^2 + 13t + 36$

95. $9x^2 + 24x + 16$ **97.**

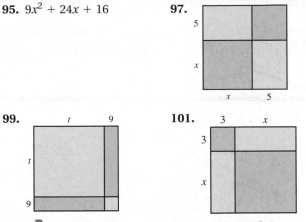

99.   **101.**

103. 🔲   **105.** [2.5] Lamps: 500 watts; air conditioner: 2000 watts; television: 50 watts   **106.** [3.1] II

107. [2.3] $y = \dfrac{8}{5x}$ **108.** [2.3] $a = \dfrac{c}{3b}$

109. [2.3] $x = \dfrac{b+c}{a}$ **110.** [2.3] $t = \dfrac{u-r}{s}$ **111.** 🔲

113. $16x^4 - 81$ **115.** $81t^4 - 72t^2 + 16$
117. $t^{24} - 4t^{18} + 6t^{12} - 4t^6 + 1$ **119.** 396   **121.** -7
123. $l^3 - l$ **125.** $Q(Q - 14) - 5(Q - 14)$,
$(Q - 5)(Q - 14)$; other equivalent expressions are possible.
127. $(y + 1)(y - 1)$, $y(y + 1) - y - 1$; other equivalent
expressions are possible.   **129.** 〰️

Technology Connection, p. 248

1. 36.22   **2.** 22,312

Exercise Set 4.6, pp. 248–251

1. -7 **3.** -92 **5.** 2.97 L   **7.** 360.4 m
9. 63.78125 in^2 **11.** Coefficients: 1, -2, 3, -5; degrees:
4, 2, 2, 0; 4   **13.** Coefficients: 17, -3, -7; degrees: 5, 5, 0; 5
15. $3a - 2b$ **17.** $3x^2y - 2xy^2 + x^2 + 5x$
19. $8u^2v - 5uv^2 + 7u^2$ **21.** $6a^2c - 7ab^2 + a^2b$
23. $3x^2 - 4xy + 3y^2$ **25.** $-6a^4 - 8ab + 7ab^2$
27. $-6r^2 - 5rt - t^2$ **29.** $3x^3 - x^2y + xy^2 - 3y^3$
31. $10y^4x^2 - 8y^3x$ **33.** $-8x + 8y$ **35.** $6z^2 + 7uz - 3u^2$
37. $x^2y^2 + 3xy - 28$ **39.** $4a^2 - b^2$ **41.** $15r^2t^2 - rt - 2$
43. $m^6n^2 + 2m^3n - 48$ **45.** $30x^2 - 28xy + 6y^2$
47. $0.4p^2q^2 - 0.02pq - 0.02$ **49.** $x^2 + 2xh + h^2$
51. $16a^2 + 40ab + 25b^2$ **53.** $c^4 - d^2$ **55.** $a^2b^2 - c^2d^4$
57. $a^2 + 2ab + b^2 - c^2$ **59.** $a^2 - b^2 - 2bc - c^2$
61. $a^2 + ab + ac + bc$ **63.** $x^2 - z^2$
65. $x^2 + y^2 + z^2 + 2xy + 2xz + 2yz$ **67.** $\frac{1}{2}x^2 + \frac{1}{2}xy - y^2$

69. We draw a rectangle with dimensions $r + s$ by $u + v$.

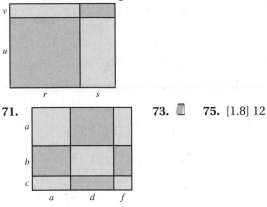

71.   **73.** 🔲   **75.** [1.8] 12

76. [1.8] 5   **77.** [1.8] 27   **78.** [1.8] 36   **79.** [1.8] 7
80. [1.8] 5   **81.** 🔲   **83.** $2\pi ab - \pi b^2$ **85.** $a^2 - 4b^2$
87. (a) $A^2 - B^2$; (b) $(A - B)^2 + (A - B)B + (A - B)B =$
$A^2 - 2AB + B^2 + AB - B^2 + AB - B^2 = A^2 - B^2$
89. $2x^2 - 2\pi r^2 + 4xh + 2\pi rh$
91. $x^4 - b^2x^2 - a^2x^2 + a^2b^2$ **93.** $P - 2Pr + Pr^2$
95. \$52,756.35

Exercise Set 4.7, pp. 257–258

1. $5x^5 - 2x$ **3.** $1 - 2u + u^6$ **5.** $5t^2 - 8t + 2$
7. $-5x^4 + 4x^2 + 1$ **9.** $6t^2 - 10t + \frac{3}{2}$ **11.** $3x^2 - 5x + \frac{1}{2}$
13. $2x + 3 + \dfrac{2}{x}$ **15.** $-3rs - r + 2s$ **17.** $x + 6$
19. $t - 5 + \dfrac{-45}{t - 5}$ **21.** $2x - 1 + \dfrac{1}{x + 6}$
23. $a^2 - 2a + 4$ **25.** $t + 4 + \dfrac{1}{t - 4}$ **27.** $x + 4$
29. $3a + 1 + \dfrac{3}{2a + 5}$ **31.** $t^2 - 3t + 1$
33. $t^2 - 2t + 3 + \dfrac{-4}{t + 1}$ **35.** $t^2 - 1 + \dfrac{3t - 1}{t^2 + 5}$
37. $2x^2 + 1 - \dfrac{x}{2x^2 - 3}$ **39.** 🔲   **41.** [1.5] -17
42. [1.5] -23 **43.** [1.6] -2 **44.** [1.6] 5
45. [2.5] 167.5 ft   **46.** [2.2] 2
47. [3.2]   **48.** [3.1]

49. 🔲   **51.** $5x^{6k} - 16x^{3k} + 14$ **53.** $3t^{2h} + 2t^h - 5$

55. $a + 3 + \dfrac{5}{5a^2 - 7a - 2}$ **57.** $2x^2 + x - 3$ **59.** 3
61. -1

Technology Connection, p. 265

1. 1.71×10^{17} **2.** $5.\overline{370} \times 10^{-15}$ **3.** 3.68×10^{16}

Exercise Set 4.8, pp. 265–267

1. $\dfrac{1}{5^2} = \dfrac{1}{25}$ **3.** $\dfrac{1}{10^4} = \dfrac{1}{10,000}$ **5.** $\dfrac{1}{(-2)^6} = \dfrac{1}{64}$ **7.** $\dfrac{1}{x^8}$
9. $\dfrac{x}{y^2}$ **11.** $\dfrac{t}{r^5}$ **13.** t^7 **15.** h^8 **17.** $\dfrac{1}{7}$
19. $\left(\dfrac{5}{2}\right)^2 = \dfrac{25}{4}$ **21.** $\left(\dfrac{2}{a}\right)^3 = \dfrac{8}{a^3}$ **23.** $\left(\dfrac{t}{s}\right)^7 = \dfrac{t^7}{s^7}$
25. 7^{-2} **27.** t^{-6} **29.** a^{-4} **31.** p^{-8} **33.** 5^{-1}
35. t^{-1} **37.** 2^3, or 8   **39.** $\dfrac{1}{x^9}$ **41.** $\dfrac{1}{t^2}$ **43.** $\dfrac{1}{a^{18}}$
45. t^{18} **47.** $\dfrac{1}{t^{12}}$ **49.** x^8 **51.** $\dfrac{1}{a^3 b^3}$ **53.** $\dfrac{1}{m^7 n^7}$
55. $\dfrac{9}{x^8}$ **57.** $\dfrac{25 t^6}{r^8}$ **59.** t^{10} **61.** $\dfrac{1}{y^4}$ **63.** y^5 **65.** x^5
67. $\dfrac{b^9}{a^7}$ **69.** 1 **71.** $\dfrac{3y^6 z^2}{x^5}$ **73.** $3s^2 t^4 u^4$ **75.** $\dfrac{1}{x^{12} y^{15}}$
77. $x^{24} y^8$ **79.** $\dfrac{b^5 c^4}{a^8}$ **81.** $\dfrac{9}{a^8}$ **83.** $49x^6$ **85.** $\dfrac{n^{12}}{m^3}$
87. $\dfrac{27 b^{12}}{8 a^6}$ **89.** 1 **91.** $71,200$ **93.** 0.00892
95. $904,000,000$ **97.** 0.0000000002764 **99.** $42,090,000$
101. 4.9×10^5 **103.** 5.83×10^{-3} **105.** 7.8×10^{10}
107. 9.07×10^{17} **109.** 5.27×10^{-7} **111.** 1.8×10^{-8}
113. 1.094×10^{15} **115.** 8×10^{12} **117.** 2.47×10^8
119. 3.915×10^{-16} **121.** 2.5×10^{13} **123.** 5×10^{-4}
125. 3×10^{-21} **127.** 🖩   **129.** [1.8] 15   **130.** [1.8] 49
131. [1.8] 78   **132.** [1.8] 6
133. [3.1]

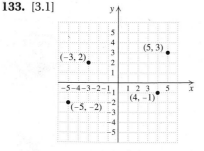

134. [2.3] $t = \dfrac{r - cx}{b}$ **135.** 🖩   **137.** 7×10^{23}
139. 4×10^{-10} **141.** 3^{11} **143.** 5 **145. (a)** False;
(b) false; **(c)** false   **147.** $6.304347826 \times 10^{25}$
149. $1.19140625 \times 10^{-15}$ **151.** 2.5×10^{12}
153. 1.15385×10^{12} times

Review Exercises: Chapter 4, pp. 269–270

1. [4.1] y^{11} **2.** [4.1] $(3x)^{14}$ **3.** [4.1] t^8 **4.** [4.1] 4^3
5. [4.1] 1 **6.** [4.1] $\dfrac{9t^8}{4s^6}$ **7.** [4.1] $-8x^3 y^6$ **8.** [4.1] $18x^5$
9. [4.1] $a^7 b^6$ **10.** [4.2] $3x^2, 6x, \frac{1}{2}$
11. [4.2] $-4y^5, 7y^2, -3y, -2$ **12.** [4.2] $7, -1, 7$
13. [4.2] $4, 6, -5, \frac{5}{3}$ **14.** [4.2] **(a)** 2, 0, 5; **(b)** $15t^5$, 15; **(c)** 5
15. [4.2] **(a)** 5, 4, 2, 1; **(b)** $-2x^5$, -2; **(c)** 5
16. [4.2] Binomial   **17.** [4.2] None of these
18. [4.2] Monomial   **19.** [4.2] $-x^2 + 9x$
20. [4.2] $-\frac{1}{4}x^3 + 4x^2 + 7$ **21.** [4.2] $-3x^5 + 25$
22. [4.2] $-2x^2 - 3x + 2$ **23.** [4.2] $10x^4 - 7x^2 - x - \frac{1}{2}$
24. [4.2] -17 **25.** [4.2] 10
26. [4.3] $x^5 + 3x^4 + 6x^3 - 2x - 9$
27. [4.3] $-x^5 + 3x^4 - x^3 - 2x^2$ **28.** [4.3] $2x^2 - 4x - 6$
29. [4.3] $x^5 - 3x^3 - 2x^2 + 8$
30. [4.3] $\frac{3}{4}x^4 + \frac{1}{4}x^3 - \frac{1}{3}x^2 - \frac{7}{4}x + \frac{3}{8}$
31. [4.3] $-x^5 + x^4 - 5x^3 - 2x^2 + 2x$
32. (a) [4.3] $4w + 6$; **(b)** [4.4] $w^2 + 3w$ **33.** [4.4] $-12x^3$
34. [4.5] $49x^2 + 14x + 1$ **35.** [4.5] $a^2 - 3a - 28$
36. [4.5] $m^2 - 25$ **37.** [4.4] $12x^3 - 23x^2 + 13x - 2$
38. [4.5] $x^2 - 18x + 81$ **39.** [4.4] $15t^5 - 6t^4 + 12t^3$
40. [4.4] $a^2 - 49$ **41.** [4.5] $x^2 - 1.05x + 0.225$
42. [4.4] $x^7 + x^5 - 3x^4 + 3x^3 - 2x^2 + 5x - 3$
43. [4.5] $9x^2 - 30x + 25$ **44.** [4.5] $2t^4 - 11t^2 - 21$
45. [4.5] $a^2 + \frac{1}{6}a - \frac{1}{3}$ **46.** [4.5] $9x^4 - 16$
47. [4.5] $4 - x^2$ **48.** [4.6] $2x^2 - 7xy - 15y^2$
49. [4.6] 49   **50.** [4.6] Coefficients: $1, -7, 9, -8$;
degrees: 6, 2, 2, 0; 6   **51.** [4.6] Coefficients: $1, -1, 1$;
degrees: 16, 40, 23; 40   **52.** [4.6] $-y + 9w - 5$
53. [4.6] $6m^3 + 4m^2 n - mn^2$ **54.** [4.6] $-x^2 - 10xy$
55. [4.6] $11x^3 y^2 - 8x^2 y - 6x^2 - 6x + 6$ **56.** [4.6] $p^3 - q^3$
57. [4.6] $9a^8 - 2a^4 b^3 + \frac{1}{9}b^6$ **58.** [4.6] $\frac{1}{2}x^2 - \frac{1}{2}y^2$

59. [4.7] $5x^2 - \frac{1}{2}x + 3$ **60.** [4.7] $3x^2 - 7x + 4 + \dfrac{1}{2x + 3}$

61. [4.7] $t^3 + 2t - 3$ **62.** [4.8] $\dfrac{1}{m^7}$ **63.** [4.8] t^{-8}

64. [4.8] $\dfrac{1}{7^2}$, or $\dfrac{1}{49}$ **65.** [4.8] $\dfrac{1}{a^{13} b^7}$ **66.** [4.8] $\dfrac{1}{x^{12}}$

67. [4.8] $\dfrac{x^6}{4y^2}$ **68.** [4.8] $\dfrac{y^3}{8x^3}$ **69.** [4.8] $8,300,000$
70. [4.8] 3.28×10^{-5} **71.** [4.8] 2.09×10^4
72. [4.8] 5.12×10^{-5} **73.** [4.8] 2.28×10^{11} platelets
74. [4.1] 🖩 In the expression $5x^3$, the exponent refers only to the x. In the expression $(5x)^3$, the entire expression within the parentheses is cubed.   **75.** [4.3] 🖩 The sum of two polynomials of degree n will also have degree n, since only the coefficients are added and the variables remain unchanged. An exception to this occurs when the leading terms of the two polynomials are opposites. The sum of those terms is then zero and the sum of the polynomials will have a degree less than n.   **76.** [4.2], [4.5] **(a)** 3; **(b)** 2

77. [4.1], [4.2] $-28x^8$ **78.** [4.2] $8x^4 + 4x^3 + 5x - 2$
79. [4.5] $-16x^6 + x^2 - 10x + 25$ **80.** [2.2], [4.5] $\frac{94}{13}$

Test: Chapter 4, p. 271

1. [4.1] t^8 **2.** [4.1] $(x + 3)^{11}$ **3.** [4.1] 3^3, or 27
4. [4.1] 1   **5.** [4.1] x^6 **6.** [4.1] $-27y^6$ **7.** [4.1] $-24x^{17}$
8. [4.1] a^6b^5 **9.** [4.2] Binomial   **10.** [4.2] $\frac{1}{3}$, -1, 7
11. [4.2] Degrees of terms: 3, 1, 5; 0; leading term: $7t^5$;
leading coefficient: 7; degree of polynomial: 5
12. [4.2] -7 **13.** [4.2] $5a^2 - 6$ **14.** [4.2] $\frac{7}{4}y^2 - 4y$
15. [4.2] $x^5 + 2x^3 + 4x^2 - 8x + 3$
16. [4.3] $4x^5 + x^4 + 5x^3 - 8x^2 + 2x - 7$
17. [4.3] $5x^4 + 5x^2 + x + 5$
18. [4.3] $-4x^4 + x^3 - 8x - 3$
19. [4.3] $x^5 + 1.3x^3 - 0.8x^2 - 3$
20. [4.4] $-12x^4 + 9x^3 + 15x^2$ **21.** [4.5] $x^2 - \frac{2}{3}x + \frac{1}{9}$
22. [4.5] $25t^2 - 49$ **23.** [4.5] $3b^2 - 4b - 15$
24. [4.5] $x^{14} - 4x^8 + 4x^6 - 16$ **25.** [4.5] $48 + 34y - 5y^2$
26. [4.4] $6x^3 - 7x^2 - 11x - 3$ **27.** [4.5] $64a^2 + 48a + 9$
28. [4.6] $-5x^3y - x^2y^2 + xy^3 - y^3 + 19$
29. [4.6] $8a^2b^2 + 6ab + 6ab^2 + ab^3 - 4b^3$
30. [4.6] $9x^{10} - 16y^{10}$ **31.** [4.7] $4x^2 + 3x - 5$
32. [4.7] $2x^2 - 4x - 2 + \dfrac{17}{3x + 2}$ **33.** [4.8] $\dfrac{1}{5^3}$
34. [4.8] y^{-8} **35.** [4.8] $\dfrac{1}{t^6}$ **36.** [4.8] $\dfrac{y^5}{x^5}$ **37.** [4.8] $\dfrac{b^4}{16a^{12}}$
38. [4.8] $\dfrac{c^3}{a^3b^3}$ **39.** [4.8] 3.9×10^9 **40.** [4.8] 0.00000005
41. [4.8] 1.75×10^{17} **42.** [4.8] 1.296×10^{22}
43. [4.8] 1.5×10^4
44. [4.4], [4.5] $V = l(l - 2)(l - 1) = l^3 - 3l^2 + 2l$
45. [2.2], [4.5] $\frac{100}{21}$

CHAPTER 5

Technology Connection, p. 279

1. Let $y_1 = 8x^4 + 6x - 28x^3 - 21$ and $y_2 = (4x^3 + 3)(2x - 7)$. Note that the Y1- and Y2-columns of the table match regardless of how far we scroll up or down.

Exercise Set 5.1, pp. 279–280

1. Answers may vary. $(10x)(x^2)$, $(5x^2)(2x)$, $(-2)(-5x^3)$
3. Answers may vary. $(-15)(a^4)$, $(-5a)(3a^3)$, $(-3a^2)(5a^2)$
5. Answers may vary. $(2x)(13x^4)$, $(13x^5)(2)$, $(-x^3)(-26x^2)$
7. $x(x + 8)$ **9.** $5t(2t - 1)$ **11.** $x^2(x + 6)$
13. $8x^2(x^2 - 3)$ **15.** $2(x^2 + x - 4)$
17. $a^2(7a^4 - 10a^2 - 14)$ **19.** $2x^2(x^6 + 2x^4 - 4x^2 + 5)$
21. $x^2y^2(x^3y^3 + x^2y + xy - 1)$ **23.** $5a^2b^2(ab^2 + 2b - 3a)$

25. $(y - 2)(y + 7)$ **27.** $(x + 3)(x^2 - 7)$
29. $(y + 8)(y^2 + 1)$ **31.** $(x + 3)(x^2 + 4)$
33. $(a + 3)(3a^2 + 2)$ **35.** $(3x - 4)(3x^2 + 1)$
37. $(t - 5)(4t^2 + 3)$ **39.** $(7x + 2)(x^2 - 2)$
41. $(6a - 7)(a^2 + 1)$ **43.** $(x + 8)(x^2 - 3)$
45. $(x + 6)(2x^2 - 5)$ **47.** $(w - 7)(w^2 + 4)$
49. Not factorable by grouping   **51.** $(x - 4)(2x^2 - 9)$
53. 🔲   **55.** [4.4] $x^2 + 8x + 15$ **56.** [4.4] $x^2 + 9x + 14$
57. [4.4] $a^2 - 4a - 21$ **58.** [4.4] $a^2 - 3a - 40$
59. [4.4] $6x^2 + 7x - 20$ **60.** [4.4] $12t^2 - 13t - 14$
61. [4.4] $9t^2 - 30t + 25$ **62.** [4.4] $4t^2 - 36t + 81$
63. 🔲   **65.** $(2x^2 + 3)(2x^3 + 3)$ **67.** $(x^5 + 1)(x^7 + 1)$
69. $(x - 1)(5x^4 + x^2 + 3)$
71. Answers may vary. $3x^4y^3 - 9x^3y^3 + 27x^2y^4$

Exercise Set 5.2, pp. 286–287

1. $(x + 5)(x + 1)$ **3.** $(x + 5)(x + 2)$
5. $(y + 4)(y + 7)$ **7.** $(a + 5)(a + 6)$
9. $(x - 1)(x - 4)$ **11.** $(z - 1)(z - 7)$
13. $(x - 5)(x - 3)$ **15.** $(y - 1)(y - 10)$
17. $(x + 7)(x - 6)$ **19.** $2(x + 2)(x - 9)$
21. $x(x + 2)(x - 8)$ **23.** $(y - 5)(y + 9)$
25. $(x - 11)(x + 9)$ **27.** $c^2(c + 8)(c - 7)$
29. $2(a + 5)(a - 7)$ **31.** Prime   **33.** Prime
35. $(x + 10)^2$ **37.** $3x(x - 25)(x + 4)$
39. $(x - 24)(x + 3)$ **41.** $(x - 9)(x - 16)$
43. $a^2(a + 12)(a - 11)$ **45.** $\left(x - \frac{1}{5}\right)^2$
47. $(9 + y)(3 + y)$ **49.** $(t + 0.2)(t - 0.5)$
51. $(p + 5q)(p - 2q)$ **53.** Prime
55. $(s - 5t)(s + 3t)$ **57.** $6a^8(a + 2)(a - 7)$ **59.** 🔲
61. [2.1] $\frac{8}{3}$ **62.** [2.1] $-\frac{7}{2}$ **63.** [4.4] $3x^2 + 22x + 24$
64. [4.4] $49w^2 + 84w + 36$ **65.** [2.4] 29,443
66. [2.5] 100°, 25°, 55°   **67.** 🔲
69. $-5, 5, -23, 23, -49, 49$
71. $-1(3 + x)(-10 + x)$, or $-1(x + 3)(x - 10)$
73. $-1(-2 + a)(12 + a)$, or $-1(a - 2)(a + 12)$
75. $-1(-6 + t)(14 + t)$, or $-1(t - 6)(t + 14)$
77. $\left(x - \frac{1}{4}\right)\left(x + \frac{1}{2}\right)$ **79.** $\frac{1}{3}a(a - 3)(a + 2)$
81. $(x^m + 4)(x^m + 7)$ **83.** $(a + 1)(x + 2)(x + 1)$
85. $2x^2(4 - \pi)$ **87.** $(x + 3)^3$, or $x^3 + 9x^2 + 27x + 27$

Exercise Set 5.3, pp. 296–297

1. $(2x - 1)(x + 4)$ **3.** $(3t - 5)(t + 3)$
5. $(3x - 1)(2x - 7)$ **7.** $(7x + 1)(x + 2)$
9. $(3a + 2)(3a - 4)$ **11.** $(3x + 1)(x - 2)$
13. $6(2t + 1)(t - 1)$ **15.** $(6t + 5)(3t - 2)$
17. $(5x + 3)(3x + 2)$ **19.** $(7x + 4)(5x + 2)$
21. Prime   **23.** $(5x + 4)^2$ **25.** $(8a + 3)(2a + 9)$
27. $2(3t - 1)(3t + 5)$ **29.** $(x - 3)(2x + 5)$
31. $3(2x + 1)(x + 5)$ **33.** $5(4x - 1)(x - 1)$
35. $4(3x - 1)(x + 6)$ **37.** $(3x + 1)(x + 1)$

39. $(y + 4)(y - 2)$ **41.** $(4t - 3)(2t - 7)$
43. $(3x + 2)(2x + 3)$ **45.** $(t + 3)(2t - 1)$
47. $(a - 4)(3a - 1)$ **49.** $(9t + 5)(t + 1)$
51. $(4x + 1)(4x + 7)$ **53.** $5(2a - 1)(a + 3)$
55. $2(x^2 + 3x - 7)$ **57.** $3x(3x - 1)(2x + 3)$
59. $(x + 1)(25x + 64)$ **61.** $3x(7x + 1)(8x + 1)$
63. $t^2(2t - 3)(7t + 1)$ **65.** $3(5x - 3)(3x + 2)$
67. $(3a + 2b)(3a + 4b)$ **69.** $(7p + 4q)(5p + 2q)$
71. $6(3x - 4y)(x + y)$ **73.** $2(3a - 2b)(4a - 3b)$
75. $x^2(4x + 7)(2x + 5)$ **77.** $a^6(3a + 4)(3a + 2)$
79. 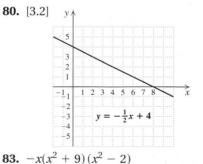 **81.** [2.5] 6369 km, 3949 mi **82.** [2.5] 40°
83. [4.4] $9x^2 + 6x + 1$ **84.** [4.4] $25x^2 - 20x + 4$
85. [4.4] $16t^2 - 40t + 25$ **86.** [4.4] $49a^2 + 14a + 1$
87. [4.4] $25x^2 - 4$ **88.** [4.4] $4x^2 - 9$ **89.** [4.4] $4t^2 - 49$
90. [4.4] $16a^2 - 49$ **91.** **93.** Prime
95. $2y(4xy + 1)(xy + 1)$ **97.** $(3t^5 + 2)^2$
99. $-(5x^m - 2)(3x^m - 4)$ **101.** $a(a^n - 1)^2$
103. $-2(a + 1)^n(a + 3)^2(a + 6)$

Exercise Set 5.4, pp. 303–304

1. Yes **3.** No **5.** No **7.** No **9.** $(x - 8)^2$
11. $(x + 7)^2$ **13.** $3(x - 1)^2$ **15.** $(2 + x)^2$, or $(x + 2)^2$
17. $2(3x - 1)^2$ **19.** $(7 + 4y)^2$, or $(4y + 7)^2$
21. $x^3(x - 9)^2$ **23.** $2x(x - 1)^2$ **25.** $5(2x + 5)^2$
27. $(7 - 3x)^2$, or $(3x - 7)^2$ **29.** $(4x + 3)^2$
31. $2(1 + 5x)^2$, or $2(5x + 1)^2$ **33.** $(2p + 3q)^2$
35. Prime **37.** $(8m + n)^2$ **39.** $2(4s - 5t)^2$ **41.** Yes
43. No **45.** No **47.** Yes **49.** $(y + 2)(y - 2)$
51. $(p + 3)(p - 3)$ **53.** $(7 + t)(-7 + t)$, or
$(t + 7)(t - 7)$ **55.** $6(a + 3)(a - 3)$ **57.** $(7x - 1)^2$
59. $2(10 - t)(10 + t)$ **61.** $5(4a - 3)(4a + 3)$
63. $5(t + 4)(t - 4)$ **65.** $2(2x + 7)(2x - 7)$
67. $x(6 + 7x)(6 - 7x)$ **69.** Prime
71. $(t - 1)(t + 1)(t^2 + 1)$ **73.** $3x(x - 4)^2$
75. $3(4t + 3)(4t - 3)$ **77.** $a^6(a - 1)^2$
79. $7(a + b)(a - b)$ **81.** $(5x + 2y)(5x - 2y)$
83. $(1 + a^2b^2)(1 + ab)(1 - ab)$ **85.** $2(3t + 2s)(3t - 2s)$
87. **89.** [2.5] 3.125 L **90.** [2.7] Scores ≥ 77
91. [4.1] $x^{12}y^{12}$ **92.** [4.1] $25a^4b^6$
93. [3.2] **94.** [3.2]

95. **97.** $(x^4 + 2^4)(x^2 + 2^2)(x + 2)(x - 2)$, or
$(x^4 + 16)(x^2 + 4)(x + 2)(x - 2)$ **99.** $2x\left(3x - \frac{2}{5}\right)\left(3x + \frac{2}{5}\right)$
101. $(0.8x - 1.1)(0.8x + 1.1)$

103. $(y^2 - 10y + 25 + z^4)(y - 5 + z^2)(y - 5 - z^2)$
105. $(a^n + 7b^n)(a^n - 7b^n)$ **107.** $(x^2 + 1)(x + 3)(x - 3)$
109. $16(x^2 - 3)^2$ **111.** $(7x + 4)^2$ **113.** $2x^3 - x^2 - 1$
115. $(y + x + 7)(y - x - 1)$ **117.** 16
119. $(x + 1)^2 - x^2 = [(x + 1) + x][(x + 1) - x] = 2x + 1$

Exercise Set 5.5, pp. 309–310

1. $5(x + 3)(x - 3)$ **3.** $(a + 5)^2$ **5.** $(4t + 1)(2t - 5)$
7. $x(x - 12)^2$ **9.** $(x + 3)(x + 2)(x - 2)$
11. $2(7t + 3)(7t - 3)$ **13.** $4x(5x + 9)(x - 2)$
15. Prime **17.** $a(a^2 + 8)(a + 8)$ **19.** $x^3(x - 7)^2$
21. $2(5 + x)(2 - x)$ **23.** Prime
25. $4(x^2 + 4)(x + 2)(x - 2)$ **27.** Prime
29. $x^3(x - 3)(x - 1)$ **31.** $(x + y)(x - y)$
33. $12n^2(1 + 2n)$ **35.** $ab(b - a)$ **37.** $2\pi r(h + r)$
39. $(a + b)(2x + 1)$ **41.** $(n + 2)(n + p)$
43. $(x - 2)(2x + z)$ **45.** $(x + y)^2$ **47.** $(3c - d)^2$
49. $7(p^2 + q^2)(p + q)(p - q)$ **51.** $(5z + y)^2$
53. $a^3(a - 5b)(a + b)$ **55.** Prime
57. $(m + 20n)(m - 18n)$ **59.** $(mn - 8)(mn + 4)$
61. $a^3(ab - 2)(ab + 5)$ **63.** $4(2x - 5)(x - 2)$
65. $2t^2(s^3 + 2t)(s^3 + 3t)$ **67.** $a^2(1 + bc)^2$
69. $\left(\frac{1}{9}x - \frac{4}{3}\right)^2$ **71.** $(1 + 4x^6y^6)(1 + 2x^3y^3)(1 - 2x^3y^3)$
73.
75. [3.2] For $(-1, 11)$:   For $(0, 7)$:

$y = -4x + 7$		$y = -4x + 7$	
11 ? $-4(-1) + 7$		7 ? $-4 \cdot 0 + 7$	
$\mid$ $4 + 7$		$\mid$ $0 + 7$	
11 $\mid$ 11	TRUE	7 $\mid$ 7	TRUE

For $(3, -5)$:

$y = -4x + 7$	
-5 ? $-4 \cdot 3 + 7$	
$\mid$ $-12 + 7$	
-5 $\mid$ -5	TRUE

76. [2.1] $\frac{4}{5}$ **77.** [2.1] $-\frac{7}{3}$ **78.** [2.1] $-\frac{9}{2}$ **79.** [2.1] $\frac{9}{4}$

80. [3.2]   **81.**

$y = -\frac{1}{2}x + 4$

83. $-x(x^2 + 9)(x^2 - 2)$
85. $3(a + 1)(a - 1)(a + 2)(a - 2)$
87. $(y + 1)(y - 7)(y + 3)$
89. $(2x - 2 + 3y)(3x - 3 - y)$
91. $(a + 3)^2(2a + b + 4)(a - b + 5)$

Technology Connection, p. 316

1. $-4.65, 0.65$ **2.** $-0.37, 5.37$ **3.** $-8.98, -4.56$
4. No solution   **5.** $0, 2.76$

Exercise Set 5.6, pp. 316–318

1. $-6, -5$ **3.** $-7, 3$ **5.** $-4, \frac{9}{2}$ **7.** $-\frac{7}{4}, \frac{9}{10}$ **9.** $-6, 0$
11. $\frac{18}{11}, \frac{1}{21}$ **13.** $-\frac{9}{2}, 0$ **15.** $50, 70$ **17.** $-1, -6$
19. $-3, 7$ **21.** $-7, -2$ **23.** $0, 6$ **25.** $-7, 0$
27. $-\frac{2}{3}, \frac{2}{3}$ **29.** -5 **31.** 1 **33.** $0, \frac{5}{8}$ **35.** $-\frac{5}{3}, 4$
37. $-1, -5$ **39.** 3 **41.** $-1, \frac{2}{3}$ **43.** $-\frac{5}{9}, \frac{5}{9}$ **45.** $-1, \frac{6}{5}$
47. $-2, 9$ **49.** $-\frac{5}{2}, \frac{4}{3}$ **51.** $-1, 4$ **53.** $-1, 3$
55. $(-4, 0), (1, 0)$ **57.** $(-3, 0), (5, 0)$ **59.** $\left(-\frac{5}{2}, 0\right), (2, 0)$
61. ▧   **63.** [1.1] $(a + b)^2$ **64.** [1.1] $a^2 + b^2$
65. [1.1] Let x represent the first integer; $x + (x + 1)$
66. [2.7] Let x represent the number; $5 + 2x < 19$
67. [2.7] Let x represent the number; $\frac{1}{2}x - 7 > 24$
68. [2.7] Let x represent the number; $x - 3 \geq 34$
69. ▧   **71.** $-7, -\frac{8}{3}, \frac{5}{2}$ **73.** (a) $x^2 - x - 20 = 0$;
(b) $x^2 - 6x - 7 = 0$; **(c)** $4x^2 - 13x + 3 = 0$;
(d) $6x^2 - 5x + 1 = 0$; **(e)** $12x^2 - 17x + 6 = 0$;
(f) $x^3 - 4x^2 + x + 6 = 0$ **75.** $-5, 4$ **77.** $-\frac{3}{5}, \frac{3}{5}$
79. $-4, 2$ **81.** (a) $2x^2 + 20x - 4 = 0$;
(b) $x^2 - 3x - 18 = 0$; **(c)** $(x + 1)(5x - 5) = 0$;
(d) $(2x + 8)(2x - 5) = 0$; **(e)** $4x^2 + 8x + 36 = 0$;
(f) $9x^2 - 12x + 24 = 0$ **83.** ▧   **85.** $2.33, 6.77$
87. $-9.15, -4.59$ **89.** $0, 2.74$

Exercise Set 5.7, pp. 324–327

1. $-2, 3$ **3.** 9 cm, 12 cm, 15 cm   **5.** 10, 11
7. -17 and -15, 15 and 17   **9.** Length: 24 in.;
width: 12 in.   **11.** Length: 18 cm; width: 8 cm
13. Base: 14 cm; height: 4 cm   **15.** Foot: 7 ft; height: 12 ft
17. 25 ft   **19.** 380   **21.** 12   **23.** 9 ft   **25.** 105
27. 12   **29.** Dining room: 12 ft by 12 ft;
kitchen: 12 ft by 10 ft   **31.** 20 ft   **33.** 1 sec, 2 sec
35. ▧   **37.** [1.7] $-\frac{8}{21}$ **38.** [1.7] $-\frac{8}{45}$ **39.** [1.7] $-\frac{35}{54}$
40. [1.7] $-\frac{5}{16}$ **41.** [1.5] $-\frac{2}{21}$ **42.** [1.5] $-\frac{26}{45}$
43. [1.5] $\frac{1}{18}$ **44.** [1.5] $-\frac{11}{24}$ **45.** ▧   **47.** 35 ft   **49.** 10
51. 37   **53.** 15 cm by 30 cm   **55.** 4 in., 6 in.
57. $(\pi - 2)x^2$

Review Exercises: Chapter 5, pp. 328–329

1. [5.1] Answers may vary. $(12x)(3x^2), (-9x^2)(-4x),$
$(6x)(6x^2)$ **2.** [5.1] Answers may vary. $(-4x^3)(5x^2),$
$(2x^4)(-10x), (-5x)(4x^4)$ **3.** [5.1] $2x^3(x + 3)$
4. [5.1] $a(a - 7)$ **5.** [5.4] $(2t - 3)(2t + 3)$
6. [5.2] $(x - 2)(x + 6)$ **7.** [5.4] $(x + 7)^2$
8. [5.4] $3x(2x + 1)^2$ **9.** [5.1] $(2x + 3)(3x^2 + 1)$
10. [5.3] $(6t + 1)(t - 1)$
11. [5.4] $(9a^2 + 1)(3a + 1)(3a - 1)$
12. [5.3] $3x(3x - 5)(x + 3)$ **13.** [5.4] $2(x - 5)(x + 5)$
14. [5.1] $(x + 4)(x^3 - 2)$ **15.** [5.4] $(ab^2 - 6)(ab^2 + 6)$
16. [5.1] $4x^4(2x^2 - 8x + 1)$ **17.** [5.4] $3(2x + 5)^2$
18. [5.4] Prime   **19.** [5.2] $x(x - 6)(x + 5)$
20. [5.4] $(2x + 5)(2x - 5)$ **21.** [5.4] $(3x - 5)^2$
22. [5.3] $2(3x + 4)(x - 6)$ **23.** [5.3] $(4t - 5)(t - 2)$
24. [5.3] $(2t + 1)(t - 4)$ **25.** [5.4] $2(3x - 1)^2$
26. [5.4] $3(x + 3)(x - 3)$ **27.** [5.2] $(5 - x)(3 - x)$
28. [5.4] $(5x - 2)(5x - 2)$ **29.** [5.2] $(xy - 3)(xy + 4)$
30. [5.4] $3(2a + 7b)^2$ **31.** [5.1] $(m + 5)(m + t)$
32. [5.4] $32(x^2 + 2y^2z^2)(x^2 - 2y^2z^2)$ **33.** [5.6] $-3, 1$
34. [5.6] $-7, 5$ **35.** [5.6] $-\frac{1}{3}, \frac{1}{3}$ **36.** [5.6] $\frac{2}{3}, 1$
37. [5.6] $-4, \frac{3}{2}$ **38.** [5.6] $-2, 3$ **39.** [5.6] $-3, 4$
40. [5.6] $-1, \frac{5}{2}$ **41.** [5.7] Height: 40 cm; base: 40 cm
42. [5.7] 24 m   **43.** [5.4] ▧ Answers may vary. Because
Edith did not first factor out the largest common factor, 4,
her factorization will not be "complete" until she removes
a common factor of 2 from each binomial. Awarding 3 to
7 points would seem reasonable.   **44.** [5.6] ▧ The equations solved in this chapter have an x^2-term (are quadratic), whereas those solved previously have no x^2-term
(are linear). The principle of zero products is used to
solve quadratic equations and is not used to solve linear
equations.   **45.** [5.7] 2.5 cm   **46.** [5.7] 0, 2
47. [5.7] Length: 12 cm; width: 6 cm   **48.** [5.6] $-3, 2, \frac{5}{2}$
49. [5.6] No real solution

Test: Chapter 5, p. 330

1. [5.1] $(2x)(4x^3), (-4x^2)(-2x^2), (8x)(x^3)$
2. [5.2] $(x - 2)(x - 5)$ **3.** [5.2] $(x - 5)^2$
4. [5.1] $2y^2(2y^2 - 4y + 3)$ **5.** [5.1] $(x + 1)(x^2 + 2)$
6. [5.1] $x(x - 5)$ **7.** [5.2] $x(x + 3)(x - 1)$
8. [5.3] $2(5x - 6)(x + 4)$ **9.** [5.4] $(2x - 3)(2x + 3)$
10. [5.2] $(x - 4)(x + 3)$ **11.** [5.3] $3m(2m + 1)(m + 1)$
12. [5.4] $3(w + 5)(w - 5)$ **13.** [5.4] $5(3x + 2)^2$
14. [5.4] $3(x^2 + 4)(x + 2)(x - 2)$ **15.** [5.4] $(7x - 6)^2$
16. [5.3] $(5x - 1)(x - 5)$ **17.** [5.1] $(x + 2)(x^3 - 3)$
18. [5.4] $5(4 + x^2)(2 + x)(2 - x)$
19. [5.3] $(2x + 3)(2x - 5)$ **20.** [5.3] $3t(2t + 5)(t - 1)$
21. [5.3] $3(m - 5n)(m + 2n)$ **22.** [5.6] $-4, 5$
23. [5.6] $-5, \frac{3}{2}$ **24.** [5.6] $-4, 7$ **25.** [5.6] $-1, \frac{7}{3}$
26. [5.7] Length: 8 m; width: 6 m   **27.** [5.7] 5 ft
28. [5.7] Width: 3; length: 15   **29.** [5.2] $(a - 4)(a + 8)$
30. [5.6] $-\frac{8}{3}, 0, \frac{2}{5}$

CHAPTER 6

Exercise Set 6.1, pp. 337–338

1. 0   **3.** -8 **5.** 4   **7.** $-4, 7$ **9.** $-5, 5$ **11.** $\dfrac{3a}{2b^2}$

13. $\dfrac{5}{2xy^4}$ **15.** $\dfrac{3}{4}$ **17.** $\dfrac{a-3}{a+1}$ **19.** $\dfrac{3}{2x^3}$ **21.** $\dfrac{y-3}{4y}$

23. $\dfrac{3(2a-1)}{7(a-1)}$ **25.** $\dfrac{t+4}{t+5}$ **27.** $\dfrac{a+4}{2(a-4)}$ **29.** $\dfrac{x+4}{x-4}$

31. $t-1$ **33.** $\dfrac{y^2+4}{y+2}$ **35.** $\dfrac{1}{2}$ **37.** $\dfrac{5}{y+6}$

39. $\dfrac{y+6}{y+5}$ **41.** $\dfrac{a-3}{a+3}$ **43.** -1 **45.** -7 **47.** $-\dfrac{1}{3}$

49. $-\dfrac{3}{2}$ **51.** -1 **53.** 🔲   **55.** [1.7] $-\dfrac{4}{7}$

56. [1.7] $-\dfrac{10}{33}$ **57.** [1.7] $-\dfrac{15}{4}$ **58.** [1.7] $-\dfrac{21}{16}$

59. [1.8] $\dfrac{13}{63}$ **60.** [1.8] $\dfrac{5}{48}$ **61.** 🔲

63. $\dfrac{(x^2+y^2)(x+y)}{(x-y)^3}$ **65.** $\dfrac{1}{x-1}$ **67.** $-\dfrac{2t^3+3}{3t^2+2}$ **69.** 1

71. 🔲

Technology Connection, p. 342

1. Let $y_1 = ((x^2 + 3x + 2)/(x^2 + 4))/(5x^2 + 10x)$ and $y_2 = (x + 1)/((x^2 + 4)(5x))$. With the table set in AUTO mode, note that the values in the Y1- and Y2-columns match except for $x = -2$.   **2.** ERROR messages occur when division by 0 is attempted. Since the simplified expression has no factor of $x + 5$ or $x + 1$ in a denominator, no ERROR message occurs in Y2 for $x = -5$ or -1.

Exercise Set 6.2, pp. 342–344

1. $\dfrac{9x(x-5)}{4(2x+1)}$ **3.** $\dfrac{(a-4)(a+2)}{(a+6)^2}$ **5.** $\dfrac{(2x+3)(x+1)}{4(x-5)}$

7. $\dfrac{(a-5)(a+2)}{(a^2+1)(a^2-1)}$ **9.** $\dfrac{(x+4)(x-1)}{(2+x)(x+1)}$ **11.** $\dfrac{5a^2}{3}$

13. $\dfrac{4}{c^2 d}$ **15.** $\dfrac{x+2}{x-2}$ **17.** $\dfrac{(a^2+25)(a-5)}{(a-3)(a-1)(a+5)}$

19. $\dfrac{5(a+3)}{a(a+4)}$ **21.** $\dfrac{2a}{a-2}$ **23.** $\dfrac{t-5}{t+5}$ **25.** $\dfrac{5(a+6)}{a-1}$

27. 1   **29.** $\dfrac{t+2}{t+4}$ **31.** $\dfrac{7}{3x}$ **33.** $\dfrac{1}{a^3-8a}$

35. $\dfrac{x^2-4x+7}{x^2+2x-5}$ **37.** $\dfrac{3}{20}$ **39.** $\dfrac{x^2}{20}$ **41.** $\dfrac{a^3}{b^3}$

43. $\dfrac{y+5}{2y}$ **45.** $4(y-2)$ **47.** $-\dfrac{a}{b}$

49. $\dfrac{(y+3)(y^2+1)}{y+1}$ **51.** $\dfrac{21}{8}$ **53.** $\dfrac{15}{4}$ **55.** $\dfrac{a-5}{3(a-1)}$

57. $\dfrac{(2x-1)(2x+1)}{x-5}$ **59.** $\dfrac{x^2+25}{2(x+5)^2}$

61. $\dfrac{(a-5)(a+3)}{(a+4)^2}$ **63.** $\dfrac{1}{(c-5)^2}$ **65.** $\dfrac{(x-4y)(x-y)}{(x+y)^3}$

67. 🔲   **69.** [1.3] $\dfrac{19}{12}$ **70.** [1.3] $\dfrac{41}{24}$ **71.** [1.3] $\dfrac{1}{18}$

72. [1.3] $-\dfrac{1}{6}$ **73.** [1.8] $-\dfrac{37}{20}$ **74.** [1.8] $\dfrac{49}{45}$ **75.** 🔲

77. 1   **79.** 1   **81.** $\dfrac{1}{(x+y)^3(3x+y)}$ **83.** $\dfrac{a^2-2b}{a^2+3b}$

85. $\dfrac{4}{x+7}$ **87.** 1   **89.** $-\dfrac{2}{(x-3y)(x+3y)}$

Exercise Set 6.3, pp. 353–355

1. $\dfrac{12}{x}$ **3.** $\dfrac{3x+5}{15}$ **5.** $\dfrac{9}{a+3}$ **7.** $\dfrac{6}{a+2}$ **9.** $\dfrac{2y+7}{2y}$

11. 11   **13.** $\dfrac{3x+5}{x+1}$ **15.** $a+5$ **17.** $x-4$ **19.** 0

21. $\dfrac{1}{x+2}$ **23.** $\dfrac{(3a-7)(a-2)}{(a+6)(a-1)}$ **25.** $\dfrac{t-4}{t+3}$

27. $\dfrac{x+5}{x-6}$ **29.** $-\dfrac{5}{x-4}$, or $\dfrac{5}{4-x}$

31. $-\dfrac{1}{x-1}$, or $\dfrac{1}{1-x}$ **33.** 135   **35.** 72   **37.** 126

39. $12x^3$ **41.** $30a^4 b^8$ **43.** $6(y-3)$

45. $(x-2)(x+2)(x+3)$ **47.** $t(t-4)(t+2)^2$

49. $30x^2 y^2 z^3$ **51.** $(a+1)(a-1)^2$

53. $(m-3)(m-2)^2$ **55.** $(t^2-9)^2$

57. $12x^3(x-5)(x-3)(x-1)$ **59.** $\dfrac{10}{12x^5}, \dfrac{x^2 y}{12x^5}$

61. $\dfrac{12b}{8a^2 b^2}, \dfrac{7a}{8a^2 b^2}$ **63.** $\dfrac{2x(x+3)}{(x-2)(x+2)(x+3)}$,

$\dfrac{4x(x-2)}{(x-2)(x+2)(x+3)}$ **65.** 🔲   **67.** [1.7] $-\dfrac{7}{9}, \dfrac{-7}{9}$

68. [1.7] $\dfrac{-3}{2}, \dfrac{3}{-2}$ **69.** [1.3] $-\dfrac{11}{36}$ **70.** [1.3] $-\dfrac{7}{60}$

71. [4.6] $x^2 - 9x + 18$ **72.** [4.6] $s^2 - \pi r^2$ **73.** 🔲

75. $\dfrac{18x+5}{x-1}$ **77.** $\dfrac{x}{3x+1}$ **79.** 30   **81.** 60

83. $120(x-1)^2(x+1)$ **85.** 24 min   **87.** 2142

89. 🔲

Exercise Set 6.4, pp. 361–362

1. $\dfrac{3x+7}{x^2}$ **3.** $-\dfrac{5}{24r}$ **5.** $\dfrac{4x+2y}{x^2 y^2}$ **7.** $\dfrac{16-15t}{18t^3}$

9. $\dfrac{5x+9}{24}$ **11.** $\dfrac{a+8}{4}$ **13.** $\dfrac{5a^2+7a-3}{9a^2}$

15. $\dfrac{-7x - 13}{4x}$ **17.** $\dfrac{c^2 + 3cd - d^2}{c^2 d^2}$ **19.** $\dfrac{3y^2 - 3xy - 6x^2}{2x^2 y^2}$

21. $\dfrac{10x}{(x - 1)(x + 1)}$ **23.** $\dfrac{2z + 6}{(z - 1)(z + 1)}$ **25.** $\dfrac{11x + 15}{4x(x + 5)}$

27. $\dfrac{16 - 9t}{6t(t - 5)}$ **29.** $\dfrac{x^2 - x}{(x - 5)(x + 5)}$ **31.** $\dfrac{4t - 5}{4(t - 3)}$

33. $\dfrac{2x + 10}{(x + 3)^2}$ **35.** $\dfrac{3x - 14}{(x - 2)(x + 2)}$ **37.** $\dfrac{9a}{4(a - 5)}$

39. 0   **41.** $\dfrac{12a - 11}{(a - 3)(a - 1)(a + 2)}$ **43.** $\dfrac{x - 5}{(x + 5)(x + 3)}$

45. $\dfrac{3z^2 + 19z - 20}{(z - 2)^2 (z + 3)}$ **47.** $\dfrac{-5}{x^2 + 17x + 16}$ **49.** $\dfrac{3x - 3}{5}$

51. $y + 3$ **53.** $\dfrac{2b - 14}{b^2 - 16}$ **55.** $\dfrac{y^2 + 10y + 11}{(y - 7)(y + 7)}$

57. $\dfrac{9x + 12}{(x - 3)(x + 3)}$ **59.** $\dfrac{3x^2 - 7x - 4}{3(x - 2)(x + 2)}$

61. $\dfrac{a - 2}{(a - 3)(a + 3)}$ **63.** $\dfrac{2x - 3}{2 - x}$ **65.** 2

67. $\dfrac{-2t^2}{(s + t)(s - t)}$ **69.** 0   **71.** 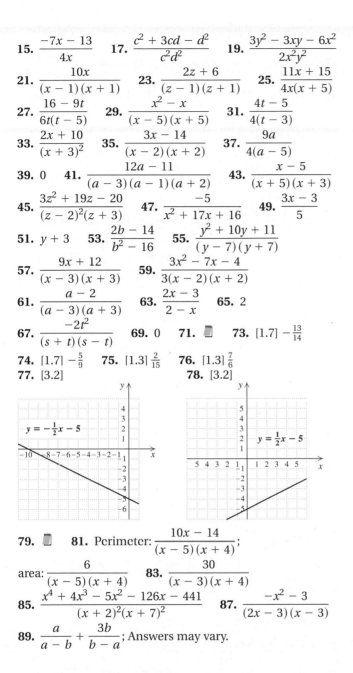   **73.** $[1.7] -\dfrac{13}{14}$

74. $[1.7] -\dfrac{5}{9}$ **75.** $[1.3] \dfrac{2}{15}$ **76.** $[1.3] \dfrac{7}{6}$

77. $[3.2]$ **78.** $[3.2]$

$y = -\frac{1}{2}x - 5$

$y = \frac{1}{2}x - 5$

79. 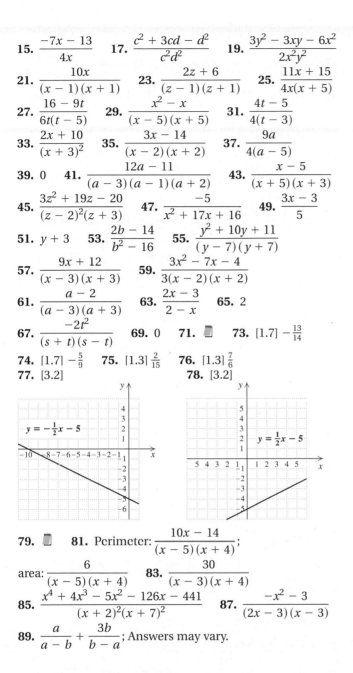   **81.** Perimeter: $\dfrac{10x - 14}{(x - 5)(x + 4)}$;

area: $\dfrac{6}{(x - 5)(x + 4)}$ **83.** $\dfrac{30}{(x - 3)(x + 4)}$

85. $\dfrac{x^4 + 4x^3 - 5x^2 - 126x - 441}{(x + 2)^2 (x + 7)^2}$ **87.** $\dfrac{-x^2 - 3}{(2x - 3)(x - 3)}$

89. $\dfrac{a}{a - b} + \dfrac{3b}{b - a}$; Answers may vary.

Technology Connection, p. 367

1. $(1 - 1 \div x) \div (1 - 1 \div x^2)$ **2.** Parentheses are needed to group separate terms into factors. When a fraction bar is replaced with a division sign, we need parentheses to preserve the groupings that had been created by the fraction bar. This holds for denominators and numerators alike.

Exercise Set 6.5, pp. 367–368

1. $\dfrac{6}{5}$ **3.** $\dfrac{18}{65}$ **5.** $\dfrac{4s^2}{9 + 3s^2}$ **7.** $\dfrac{2x}{3x + 1}$ **9.** $\dfrac{4a - 10}{a - 1}$

11. $x - 4$ **13.** $\dfrac{1}{x}$ **15.** $\dfrac{1 + t^2}{t - t^2}$ **17.** $\dfrac{x}{x - y}$

19. $\dfrac{2a^2 + 4a}{5 - 3a^2}$ **21.** $\dfrac{60 - 15a^3}{126a^2 + 28a^3}$ **23.** 1

25. $\dfrac{5a^2 + 2b^3}{5b^3 - 3a^2 b^3}$ **27.** $\dfrac{2x^4 - 3x^2}{2x^4 + 3}$ **29.** $\dfrac{t^2 - 2}{t^2 + 5}$

31. $\dfrac{3a^2 b^3 + 4a}{3b^2 + ab^2}$ **33.** $\dfrac{x + 5}{2x - 3}$ **35.** $\dfrac{x - 2}{x - 3}$ **37.**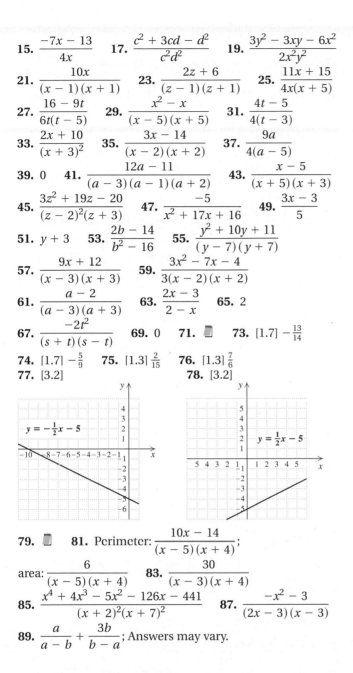

39. $[2.2] -4$ **40.** $[2.2] -4$ **41.** $[2.2] \dfrac{19}{3}$ **42.** $[2.2] -\dfrac{14}{27}$

43. $[5.6] -3, 10$ **44.** $[5.6] -10, 2$ **45.** 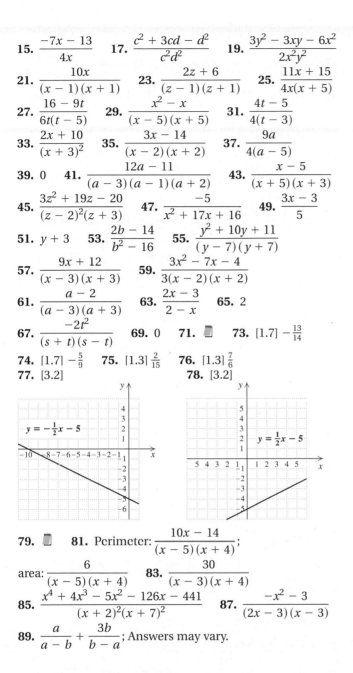   **47.** 6, 7, 8

49. $-\dfrac{4}{5}, \dfrac{27}{14}$ **51.** $\dfrac{P(i + 12)^2}{12(i + 24)}$

53. $\dfrac{(x - 8)(x - 1)(x + 1)}{x^2(x - 2)(x + 2)}$ **55.** 0

57. $\dfrac{2z(5z - 2)}{(z + 2)(13z - 6)}$ **59.**

Exercise Set 6.6, pp. 373–374

1. $-\dfrac{7}{2}$ **3.** $\dfrac{6}{7}$ **5.** $\dfrac{24}{7}$ **7.** $-5, -1$ **9.** $-6, 6$ **11.** 3

13. $\dfrac{14}{3}$ **15.** $\dfrac{35}{3}$ **17.** 5   **19.** $\dfrac{5}{2}$ **21.** -1 **23.** No solution   **25.** -10 **27.** $\dfrac{10}{7}$ **29.** No solution   **31.** 2

33. No solution   **35.** No solution   **37.**

39. $[2.5]$ 137, 139   **40.** $[1.2]$ 14 yd   **41.** $[1.2]$ Base: 9 cm; height: 12 cm   **42.** $[1.2] -8, -6; 6, 8$ **43.** $[3.4]$ 0.06 cm per day   **44.** $[3.4]$ 0.28 in. per day   **45.**   **47.** -2

49. $-\dfrac{1}{6}$ **51.** $-1, 0$ **53.** 4   **55.**

Exercise Set 6.7, pp. 382–386

1. $-1, 4$ **3.** 1   **5.** $\dfrac{20}{9}$ hr, or 2 hr $13\frac{1}{3}$ min

7. $\dfrac{180}{7}$ min, or $25\frac{5}{7}$ min   **9.** $\dfrac{24}{7}$ hr, or 3 hr $25\frac{5}{7}$ min

11. $\dfrac{200}{9}$ min, or $22\frac{2}{9}$ min   **13.** 7.5 min

15.

	Distance (in miles)	Speed (in miles per hour)	Time (in hours)
Truck	350	r	$\dfrac{350}{r}$
Train	150	$r - 40$	$\dfrac{150}{r - 40}$

Truck: 70 mph; train: 30 mph
17. Hank: 14 km/h; Kelly: 19 km/h   **19.** Ralph: 5 km/h; Bonnie: 8 km/h   **21.** 3 hr   **23.** 10.5   **25.** $\dfrac{8}{3}$ **27.** 15 ft

29. 3.75 cm   **31.** $b = 3$ m, $c = 5$ m, $d = 1$ m; $b = 10$ m, $c = 12$ m, $d = 8$ m   **33.** 560   **35.** 1.92 g   **37.** $32,340
39. 954   **41.** 42   **43.** No   **45.** 225
47. (a) 4.8 T; **(b)** 48 lb   **49.**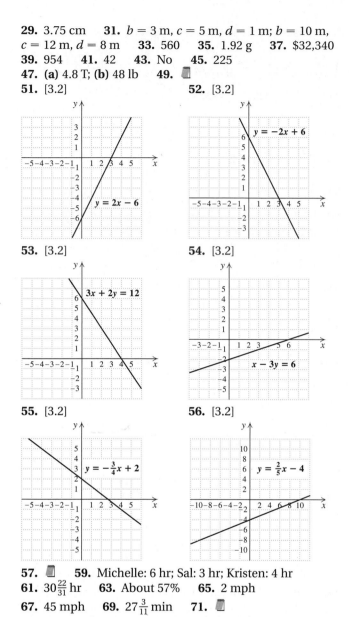
51. [3.2]   **52.** [3.2]

53. [3.2]   **54.** [3.2]

55. [3.2]   **56.** [3.2]

57. 📝   **59.** Michelle: 6 hr; Sal: 3 hr; Kristen: 4 hr
61. $30\frac{22}{31}$ hr   **63.** About 57%   **65.** 2 mph
67. 45 mph   **69.** $27\frac{3}{11}$ min   **71.** 📝

Review Exercises: Chapter 6, pp. 389–390

1. [6.1] 0   **2.** [6.1] 5   **3.** [6.1] $-6, 6$ **4.** [6.1] $-6, 5$
5. [6.1] -2 **6.** [6.1] $\dfrac{x - 2}{x + 1}$ **7.** [6.1] $\dfrac{7x + 3}{x - 3}$
8. [6.1] $\dfrac{y - 5}{y + 5}$ **9.** [6.1] $-5(x + 2y)$ **10.** [6.2] $\dfrac{a - 6}{5}$
11. [6.2] $\dfrac{8(t + 1)}{(2t - 1)(t - 1)}$ **12.** [6.2] $-20t$
13. [6.2] $\dfrac{2x(x - 1)}{x + 1}$ **14.** [6.2] $\dfrac{(x^2 + 1)(2x + 1)}{(x - 2)(x + 1)}$
15. [6.2] $\dfrac{(t + 4)^2}{t + 1}$ **16.** [6.3] $24a^5b^7$

17. [6.3] $x^4(x - 1)(x + 1)$
18. [6.3] $(y - 2)(y + 2)(y + 1)$ **19.** [6.3] $\dfrac{18 - 3x}{x + 7}$
20. [6.4] -1 **21.** [6.3] $\dfrac{4}{x - 4}$ **22.** [6.4] $\dfrac{x + 5}{2x}$
23. [6.4] $\dfrac{2x + 3}{x - 2}$ **24.** [6.4] $\dfrac{2a}{a - 1}$ **25.** [6.4] $d + c$
26. [6.4] $\dfrac{-x^2 + x + 26}{(x + 1)(x - 5)(x + 5)}$ **27.** [6.4] $\dfrac{2(x - 2)}{x + 2}$
28. [6.4] $\dfrac{19x + 8}{10x(x + 2)}$ **29.** [6.5] $\dfrac{z}{1 - z}$
30. [6.5] $\dfrac{x^3(2xy^2 + 1)}{y(1 + x)}$ **31.** [6.5] $c - d$ **32.** [6.6] 8
33. [6.6] $-\frac{1}{2}$ **34.** [6.6] $-5, 3$ **35.** [6.7] $\frac{36}{7}$ hr, or $5\frac{1}{7}$ hr
36. [6.7] Car: 105 km/h; train: 90 km/h   **37.** [6.7] -2
38. [6.7] 30   **39.** [6.7] 6   **40.** [6.7] 50
41. [6.3], [6.6] 📝 A student should master factoring before beginning a study of rational equations because it is necessary to factor when finding the LCD of the rational expressions. It may also be necessary to factor to use the principle of zero products after fractions have been cleared.   **42.** [6.3] 📝 Although multiplying the denominators of the expressions being added results in a common denominator, it is often not the *least* common denominator. Using a common denominator other than the LCD makes the expressions more complicated, requires additional simplifying after the addition has been performed, and leaves more room for error.
43. [6.2] $\dfrac{5(a + 3)^2}{a}$ **44.** [6.4] $\dfrac{10a}{(a - b)(b - c)}$
45. [6.3] 0

Test: Chapter 6, pp. 390–391

1. [6.1] 0   **2.** [6.1] -8 **3.** [6.1] $-7, 7$ **4.** [6.1] 1, 2
5. [6.1] $\dfrac{3x + 7}{x + 3}$ **6.** [6.2] $\dfrac{-2(a + 5)}{3}$
7. [6.2] $\dfrac{(5y + 1)(y + 1)}{3y(y + 2)}$
8. [6.2] $\dfrac{(2x + 1)(2x - 1)(x^2 + 1)}{(x - 1)^2(x - 2)}$
9. [6.2] $(x + 3)(x - 3)$ **10.** [6.3] $(y - 3)(y + 3)(y + 7)$
11. [6.3] $\dfrac{-3x + 23}{x^3}$ **12.** [6.3] $\dfrac{-2t + 8}{t^2 + 1}$ **13.** [6.4] $\dfrac{3}{3 - x}$
14. [6.4] $\dfrac{2x - 5}{x - 3}$ **15.** [6.4] $\dfrac{8t - 3}{t(t - 1)}$
16. [6.4] $\dfrac{-x^2 - 7x - 15}{(x + 4)(x - 4)(x + 1)}$
17. [6.4] $\dfrac{x^2 + 2x - 7}{(x + 1)(x - 1)^2}$ **18.** [6.5] $\dfrac{3y + 1}{y}$
19. [6.5] $\dfrac{a^2(3b^2 - 2a)}{b^2(a^3 + 2)}$ **20.** [6.6] 12   **21.** [6.6] $-3, 5$

22. [6.7] 12 min   **23.** [6.7] $2\frac{1}{7}$ cups
24. [6.7] Craig: 65 km/h; Marilyn: 45 km/h
25. [6.7] Rema: 4 hr; Reggie: 10 hr   **26.** [6.5] a

Cumulative Review: Chapters 1–6, pp. 391–392

1. [1.2] $2b + a$ **2.** [1.4] $>$ **3.** [1.8] 49
4. [1.8] $-8x + 28$ **5.** [1.5] $-\frac{43}{8}$ **6.** [1.7] 1
7. [1.7] -6.2 **8.** [1.8] 8   **9.** [2.2] 10   **10.** [5.6] $-7, 7$
11. [2.1] $-\frac{10}{3}$ **12.** [2.2] -2 **13.** [2.2] $\frac{8}{3}$
14. [5.6] $-10, -1$ **15.** [2.2] -8 **16.** [6.6] 1, 4
17. [2.6] $\left\{ y \mid y \le -\frac{2}{3} \right\}$ **18.** [6.6] -17 **19.** [5.6] $-4, \frac{1}{2}$
20. [2.6] $\{x \mid x > 43\}$ **21.** [6.6] 5   **22.** [5.6] $-\frac{7}{2}, 5$
23. [6.6] -13 **24.** [2.3] $3a - c + 9$ **25.** [2.3] $\dfrac{4z - 3x}{6}$
26. [1.6] $\frac{3}{2}x + 2y - 3z$ **27.** [1.6] $-4x^3 - \frac{1}{7}x^2 - 2$
28. [3.2]   **29.** [3.3]

30. [3.3]   **31.** [3.3]

32. [3.5] -2 **33.** [4.8] $\dfrac{1}{x^2}$ **34.** [4.8] y^{-8}, or $\dfrac{1}{y^8}$
35. [4.1] $-4a^4b^{14}$ **36.** [4.3] $-y^3 - 2y^2 - 2y + 7$
37. [1.2] $12x + 16y + 4z$
38. [4.4] $2x^5 + x^3 - 6x^2 - x + 3$
39. [4.5] $36x^2 - 60xy + 25y^2$ **40.** [4.5] $2x^2 - x - 21$
41. [4.5] $4x^6 - 1$ **42.** [5.1] $2x(3 - x - 12x^3)$
43. [5.4] $(4x + 9)(4x - 9)$ **44.** [5.2] $(t - 4)(t - 6)$
45. [5.3] $(4x + 3)(2x + 1)$ **46.** [5.3] $2(3x - 2)(x - 4)$
47. [5.4] $4(t + 3)(t - 3)$ **48.** [5.4] $(5t + 4)^2$
49. [5.3] $(3x - 2)(x + 4)$ **50.** [5.1] $(x + 2)(x^3 - 3)$

51. [6.2] $\dfrac{y - 6}{2}$ **52.** [6.2] 1   **53.** [6.4] $\dfrac{a^2 + 7ab + b^2}{(a + b)(a - b)}$
54. [6.4] $\dfrac{2x + 5}{4 - x}$ **55.** [6.5] $\dfrac{x}{x - 2}$ **56.** [6.5] $\dfrac{t(2t^2 + 1)}{t^3 - 2}$
57. [4.7] $5x^2 - 4x + 2 + \dfrac{2}{3x} + \dfrac{6}{x^2}$
58. [4.7] $15x^3 - 57x^2 + 177x - 529 + \dfrac{1605}{x + 3}$
59. [2.7] At most 225   **60.** [2.4] \$3.60   **61.** [5.7] 14 ft
62. [2.5] $-278, -276$ **63.** [6.7] 30 min
64. [6.7] Phil's: 50 km/h; Harley's: 40 km/h
65. [2.5] 26 in.   **66.** [4.3], [4.5] 12
67. [1.4], [2.2] $-144, 144$ **68.** [4.5] $16y^6 - y^4 + 6y^2 - 9$
69. [5.4] $2(a^{16} + 81b^{20})(a^8 + 9b^{10})(a^4 + 3b^5)(a^4 - 3b^5)$
70. [5.6] $-7, 4, 12$ **71.** [1.4], [1.6] -7 **72.** [6.6] 18
73. [6.7] $66\frac{2}{3}\%$

CHAPTER 7

Technology Connection, p. 397

1.   **2.** $y_1 - y_2 = y_3 = -2$

Exercise Set 7.1, pp. 398–399

1. Yes   **3.** No   **5.** Yes   **7.** $(4, 3)$ **9.** $(1, 3)$
11. $(1, -1)$ **13.** No solution   **15.** $(6, -1)$
17. $(5, -2)$ **19.** $(3, -2)$
21. Infinite number of solutions   **23.** $(-6, -2)$
25. Infinite number of solutions   **27.** $(4, 3)$
29. $(-12, 11)$ **31.** $(-3, 5)$ **33.** $(3, 6)$ **35.**
37. [2.2] $\frac{26}{7}$ **38.** [2.2] $\frac{15}{2}$ **39.** [2.2] $\frac{6}{11}$ **40.** [2.2] $\frac{9}{10}$
41. [1.8] $-11x$ **42.** [1.8] $-11y$ **43.**
45. Exercises 14, 21, 22, 25   **47.** Exercises 13 and 26
49. $2x + y = 2$. Answers may vary.   **51.** $A = 2; B = 2$
53. **(a)** Copy card: $y = 20$; per page: $y = 0.06x$
(b)   **(c)** more than 333 copies

55. $(41.5, 17.1)$

Technology Connection, p. 402

1. 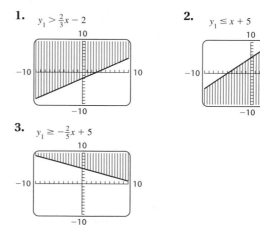 **2.** Both equations change: The first becomes $y = 6 - \dfrac{x}{-2}$ and the second becomes $y = 4 - \dfrac{3x}{2}$.

Exercise Set 7.2, pp. 404–406

1. $(5, 4)$ **3.** $(2, -1)$ **5.** $(1, 3)$ **7.** $(-30, 10)$ **9.** $(-3, 5)$ **11.** Infinite number of solutions **13.** $(2, -4)$ **15.** $\left(\frac{17}{3}, \frac{16}{3}\right)$ **17.** Infinite number of solutions **19.** Infinite number of solutions **21.** $(-12, 11)$ **23.** $(-2, 4)$ **25.** $(1, 2)$ **27.** Infinite number of solutions **29.** $(-7, -2)$ **31.** No solution **33.** 42, 45 **35.** 22, 36 **37.** 12, 28 **39.** 80°, 100° **41.** 24°, 66° **43.** Length: 380 mi; width: 270 mi **45.** Length: 40 ft; width: 20 ft **47.** Length: 110 yd; width: 60 yd **49.** **51.** [1.8] $-5x - 3y$ **52.** [1.8] $-15y$ **53.** [1.8] $10x$ **54.** [1.8] $-6y - 25$ **55.** [1.8] $-11y$ **56.** [1.8] $23x$ **57.** **59.** $(7, -1)$ **61.** $(4.38, 4.33)$ **63.** $(2, -1, 3)$ **65.** Baseball: 30 yd; softball: 20 yd **67.** Answers may vary. $2x + 3y = 5,$ $5x + 4y = 2$

Exercise Set 7.3, pp. 413–415

1. $(5, -2)$ **3.** $(-1, 7)$ **5.** $(3, 0)$ **7.** $(-1, 3)$ **9.** $\left(-1, \frac{1}{5}\right)$ **11.** Infinite number of solutions **13.** $(-3, -5)$ **15.** $(4, 5)$ **17.** $(4, 1)$ **19.** $(4, 3)$ **21.** $(1, -1)$ **23.** $(-3, -1)$ **25.** No solution **27.** $(50, 18)$ **29.** $(-2, 2)$ **31.** $(2, -1)$ **33.** $\left(\frac{231}{202}, \frac{117}{202}\right)$ **35.** 190.5 mi **37.** 26°, 64° **39.** 197.5 min **41.** 37°, 143° **43.** Riesling: 340 acres; Chardonnay: 480 acres **45.** Length: 6 ft; width: 3 ft **47.** **49.** [2.4] 0.08 **50.** [2.4] 0.073 **51.** [2.4] 0.004 **52.** [2.4] 40% **53.** [2.4] 31.5 **54.** [2.4] 16% **55.** **57.** $(2, 5)$ **59.** $\left(\frac{1}{2}, -\frac{1}{2}\right)$ **61.** $(0, 3)$ **63.** $x = \dfrac{c - b}{a - 1}$; $y = \dfrac{ac - b}{a - 1}$ **65.** Rabbits: 12; pheasants: 23 **67.** Man: 45 yr; daughter: 10 yr

Exercise Set 7.4, pp. 421–424

1. Two-pointers: 7; three-pointers: 2 **3.** Two-pointers: 24; three-pointers: 12 **5.** 24-exposure: 8; 36-exposure: 9 **7.** 5-cent bottles or cans: 336; 10-cent bottles or cans: 94 **9.** 380 **11.** Adults: 9; children: 14 **13.** Students receiving private lessons: 7; students receiving group lessons: 5 **15.** Brazilian: 200 kg; Turkish: 100 kg **17.** Peanuts: 135 lb; Brazil nuts: 345 lb

19.

Type of Solution	50%-acid	80%-acid	68%-acid mix
Amount of Solution	x	y	200
Percent Acid	50%	80%	68%
Amount of Acid in Solution	$0.5x$	$0.8y$	136

80 mL of 50%; 120 mL of 80%
21. 100 L of 28%; 200 L of 40% **23.** 87-octane: 2.5 gal; 95-octane: 7.5 gal **25.** 1300-word pages: 7; 1850-word pages: 5 **27.** Foul shots: 28; two-pointers: 36 **29.** Kinney's: $33\frac{1}{3}$ fl. oz; Coppertone: $16\frac{2}{3}$ fl. oz **31.** 25 lb of each **33.** **35.** [2.6] $\{x \mid x > -5\}$ **36.** [2.6] $\{x \mid x \le -7\}$ **37.** [2.6] **38.** [2.6] **39.** [2.6] **40.** [2.6] **41.** **43.** $19\frac{2}{7}$ lb **45.** 1.8 L **47.** 2.5 gal **49.** Inexpensive paint: \$19.41/gal; expensive paint: \$20.08/gal **51.** 130 L **53.** 54 **55.** Tweedledum: 120 lb; Tweedledee: 121 lb

Technology Connection, p. 429

1. $y_1 > \frac{2}{3}x - 2$

2. $y_1 \le x + 5$

3. $y_1 \ge -\frac{2}{5}x + 5$

Exercise Set 7.5, pp. 429–430

1. No **3.** Yes

5.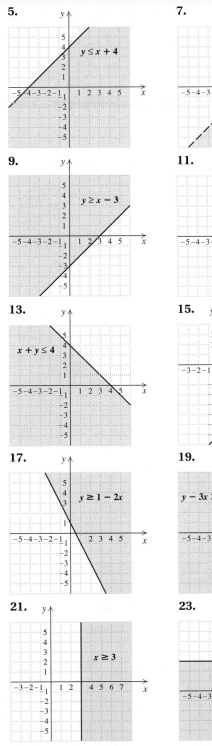
$y \leq x + 4$

7.
$y < x - 1$

9.
$y \geq x - 3$

11.
$y \leq 2x - 1$

13.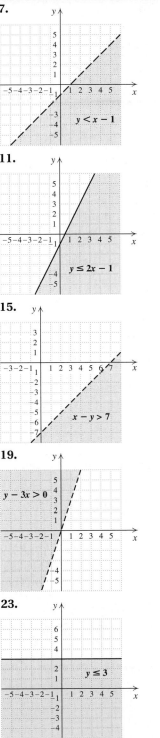
$x + y \leq 4$

15.
$x - y > 7$

17.
$y \geq 1 - 2x$

19.
$y - 3x > 0$

21.
$x \geq 3$

23.
$y \leq 3$

25.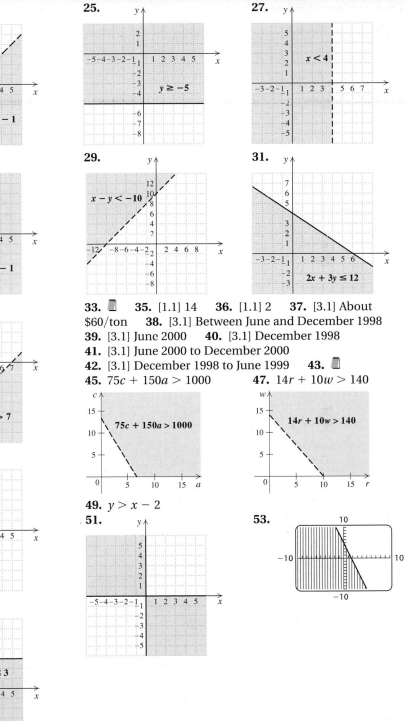
$y \geq -5$

27.
$x < 4$

29.
$x - y < -10$

31.
$2x + 3y \leq 12$

33. 🗒   **35.** [1.1] 14   **36.** [1.1] 2   **37.** [3.1] About $60/ton   **38.** [3.1] Between June and December 1998
39. [3.1] June 2000   **40.** [3.1] December 1998
41. [3.1] June 2000 to December 2000
42. [3.1] December 1998 to June 1999   **43.** 🗒
45. $75c + 150a > 1000$ **47.** $14r + 10w > 140$

$75c + 150a > 1000$

$14r + 10w > 140$

49. $y > x - 2$
51.

53.

Exercise Set 7.6, p. 433

1.
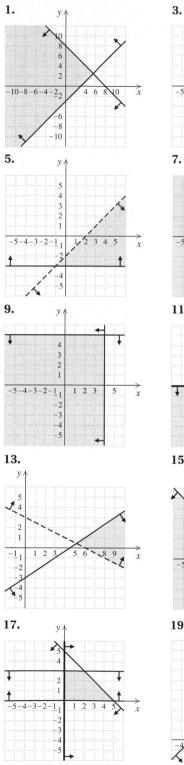

3.

5.

7.

9.

11.

13.
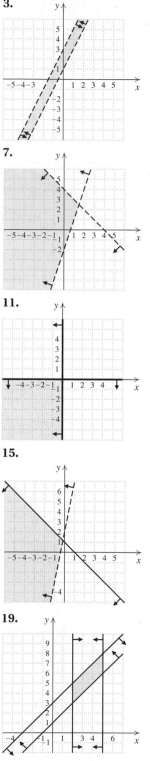

15.

17.

19.

21.
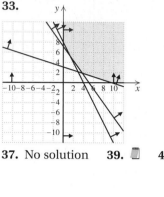

23. 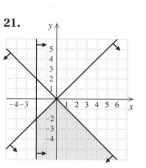 **25.** [2.1] 35

26. [2.1] 24 **27.** [2.1] 6 **28.** [2.1] 5 **29.** [2.1] $\frac{1}{9}$

30. [2.1] $\frac{1}{3}$ **31.**

33.

35.

$(5\frac{5}{11}, -\frac{2}{11})$

37. No solution **39.** **41.**
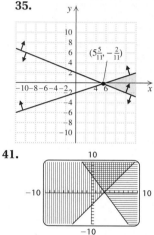

Exercise Set 7.7, pp. 439–440

1. $y = 14x$ **3.** $y = \frac{7}{4}x$ **5.** $y = \frac{16}{3}x$ **7.** $y = \frac{2}{3}x$

9. $y = \dfrac{90}{x}$ **11.** $y = \dfrac{70}{x}$ **13.** $y = \dfrac{156.25}{x}$ **15.** $y = \dfrac{210}{x}$

17. \$207 **19.** 24 **21.** About 30.8 lb **23.** 1.6 ft
25. \$0.47; \$0.02 **27.** 5 **29.** **31.** **33.** [1.8] 49
34. [1.8] 25 **35.** [1.8] 169 **36.** [1.8] 225
37. [4.1] $9x^2$ **38.** [4.1] $36a^2$ **39.** [4.1] a^4b^2

40. [4.1] s^6t^2 **41.** **43.** $S = kv^6$ **45.** $I = \dfrac{k}{d^2}$

47. $P = kv^3$ **49.** $C = 2\pi r$ **51.** $V = \frac{4}{3}\pi r^3$

Review Exercises, Chapter 7, pp. 442–443

1. [7.1] Yes **2.** [7.1] No **3.** [7.1] $(1, -2)$
4. [7.1] $(6, -2)$ **5.** [7.1] No solution
6. [7.1] Infinite number of solutions **7.** [7.2] $(-5, 9)$
8. [7.2] $\left(\frac{10}{3}, \frac{4}{3}\right)$ **9.** [7.2] No solution **10.** [7.2] $(-3, 9)$
11. [7.2] $(4, 5)$ **12.** [7.2] Infinite number of solutions
13. [7.3] $\left(\frac{3}{2}, \frac{15}{4}\right)$ **14.** [7.3] $(5, -3)$
15. [7.3] Infinite number of solutions **16.** [7.3] $(-2, 4)$

17. [7.3] $(3, 2)$ **18.** [7.3] $(3, 2)$ **19.** [7.3] No solution
20. [7.3] $(-4, 2)$ **21.** [7.2] 12 and 15
22. [7.2] $37\frac{1}{2}$ cm; $10\frac{1}{2}$ cm   **23.** [7.4] Two-pointers: 7; three-pointers: 1   **24.** [7.4] 292
25. [7.4] Café Rich: $66\frac{2}{3}$ g; Café Light: $133\frac{1}{3}$ g
26. [7.5]

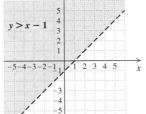

27. [7.5]

28. [7.5]

29. [7.5]

30. [7.5]

31. [7.5]

32. [7.6]

33. [7.6]

34. [7.7] $y = \dfrac{243}{x}$ **35.** [7.7] 9.6 lb

36. [7.1] 🖉 A solution of a system of two equations is an ordered pair that makes both equations true. The graph of an equation represents all ordered pairs that make that equation true. So in order for an ordered pair to make *both* equations true, it must be on both graphs.
37. [7.6] 🖉 The solution sets of linear inequalities are

regions, not lines. Thus the solution sets can intersect even if the boundary lines do not.   **38.** [7.1] $C = 1, D = 3$
39. [7.2] $(2, 1, -2)$ **40.** [7.2] $(2, 0)$ **41.** [7.4] 24
42. [7.4] $336

Test: Chapter 7, pp. 443–444

1. [7.1] Yes   **2.** [7.1] $(1, 3)$ **3.** [7.1] No solution
4. [7.2] $(8, -2)$ **5.** [7.2] $(-1, 3)$
6. [7.2] Infinite number of solutions   **7.** [7.3] $(1, -5)$
8. [7.3] $(12, -6)$ **9.** [7.3] $(0, 1)$ **10.** [7.3] $(5, 1)$
11. [7.4] 25%: 40 L; 40%: 20 L   **12.** [7.2] 36°, 54°
13. [7.3] About 84 min
14. [7.5]

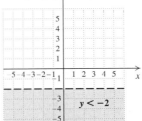

15. [7.5]

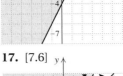

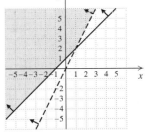

16. [7.5]

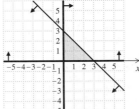

17. [7.6]

18. [7.6]

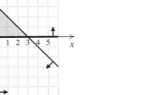

19. [7.7] $y = 4.5x$

20. [7.7] 18 min   **21.** [7.4] 9
22. [7.6]

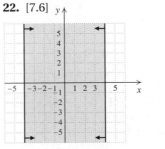

23. [7.1] $C = -\dfrac{19}{2}, D = \dfrac{14}{3}$

CHAPTER 8

Technology Connection, p. 450

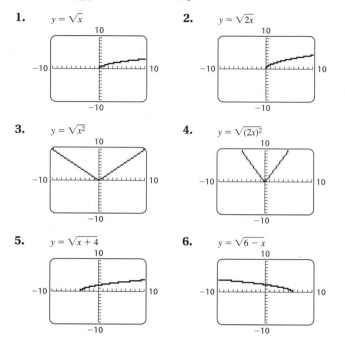

1. $y = \sqrt{x}$

2. $y = \sqrt{2x}$

3. $y = \sqrt{x^2}$

4. $y = \sqrt{(2x)^2}$

5. $y = \sqrt{x+4}$

6. $y = \sqrt{6-x}$

Exercise Set 8.1, pp. 450–452

1. $-2, 2$ **3.** $-4, 4$ **5.** $-7, 7$ **7.** $-12, 12$ **9.** 3
11. -1 **13.** 0 **15.** -11 **17.** 30 **19.** 13

21. -25 **23.** $a - 7$ **25.** $t^2 + 1$ **27.** $\dfrac{3}{x+2}$

29. Rational **31.** Irrational **33.** Irrational
35. Irrational **37.** Rational **39.** Rational **41.** 2.236
43. 4.123 **45.** 9.644 **47.** t **49.** $3x$ **51.** $7a$
53. $17x$ **55.** (a) 15; (b) 14 **57.** 0.864 sec **59.** ▨
61. [4.1] $49x^2$ **62.** [4.1] $9a^2$ **63.** [4.1] $16t^{14}$
64. [4.1] $45x^7$ **65.** [4.1] $48a^{11}$ **66.** [4.1] $8t^9u^6$ **67.** ▨
69. 3 **71.** $-6, -5$ **73.** $-7, 7$ **75.** $-3, 3$ **77.** $9a^3b^4$

79. $\dfrac{2x^4}{y^3}$ **81.** $\dfrac{20}{m^8}$ **83.** (a) 1.7; (b) 2.2; (c) 2.6

85. 1097.1 ft/sec **87.** 1270.8 ft/sec **89.** ▨

Exercise Set 8.2, pp. 456–457

1. $\sqrt{15}$ **3.** $\sqrt{12}$, or $2\sqrt{3}$ **5.** $\sqrt{\tfrac{3}{10}}$ **7.** 13
9. $\sqrt{75}$, or $5\sqrt{3}$ **11.** $\sqrt{2x}$ **13.** $\sqrt{14a}$ **15.** $\sqrt{35x}$
17. $\sqrt{6ac}$ **19.** $2\sqrt{5}$ **21.** $5\sqrt{2}$ **23.** $10\sqrt{7}$
25. $3\sqrt{x}$ **27.** $5\sqrt{3a}$ **29.** $4\sqrt{a}$ **31.** $8y$ **33.** $x\sqrt{13}$
35. $2t\sqrt{7}$ **37.** $4\sqrt{5}$ **39.** $12\sqrt{2y}$ **41.** a^7 **43.** x^6
45. $r^3\sqrt{r}$ **47.** $t^9\sqrt{t}$ **49.** $3a\sqrt{10a}$ **51.** $2a^2\sqrt{2a}$

53. $2p^8\sqrt{26p}$ **55.** $7\sqrt{2}$ **57.** 9 **59.** $6\sqrt{xy}$
61. $13\sqrt{x}$ **63.** $10b\sqrt{5}$ **65.** $7x$ **67.** $a\sqrt{bc}$
69. $2x^3\sqrt{7}$ **71.** $xy^3\sqrt{xy}$ **73.** $10ab^4\sqrt{5a}$
75. 20 mph; 54.8 mph **77.** ▨ **79.** [4.1] a^5b^2

80. [4.1] x^3y^8 **81.** [4.1] $\dfrac{6x^2}{35y^2}$ **82.** [4.1] $\dfrac{35a^2}{6b^2}$

83. [4.1] $\dfrac{2r^3}{7t}$ **84.** [4.1] $\dfrac{5x^7}{11y}$ **85.** ▨ **87.** 0.1

89. 0.25 **91.** $=$ **93.** $>$ **95.** $>$
97. $18(x+1)\sqrt{y(x+1)}$ **99.** $2x^7\sqrt{5x}$ **101.** $\sqrt{14x^{26}}$
103. x^{8n} **105.** $y^{(1/2)(n-1)}\sqrt{y}$

Exercise Set 8.3, pp. 461–462

1. 2 **3.** 5 **5.** $\sqrt{7}$ **7.** $\tfrac{1}{3}$ **9.** $\tfrac{3}{4}$ **11.** 2 **13.** $4y$

15. $3x^2$ **17.** $a^3\sqrt{3}$ **19.** $\tfrac{6}{5}$ **21.** $\tfrac{7}{4}$ **23.** $-\tfrac{5}{9}$ **25.** $\dfrac{a}{5}$

27. $\dfrac{x^3\sqrt{3}}{4}$ **29.** $\dfrac{t^3\sqrt{3}}{2}$ **31.** $\dfrac{5\sqrt{3}}{3}$ **33.** $\dfrac{\sqrt{21}}{7}$ **35.** $\dfrac{2\sqrt{3}}{9}$

37. $\dfrac{\sqrt{6}}{3}$ **39.** $\dfrac{\sqrt{6}}{10}$ **41.** $\dfrac{\sqrt{10a}}{15}$ **43.** $\dfrac{2\sqrt{15}}{5}$ **45.** $\dfrac{\sqrt{2x}}{x}$

47. $\dfrac{\sqrt{2t}}{8}$ **49.** $\dfrac{\sqrt{10x}}{20}$ **51.** $\dfrac{\sqrt{3a}}{5}$ **53.** $\dfrac{x\sqrt{15}}{6}$

55. 6.28 sec; 7.85 sec **57.** 0.5 sec **59.** 9.42 sec
61. ▨ **63.** [4.2] $12x + 13$ **64.** [4.2] $10a + 6$
65. [4.2] $-a^3 - 8a^2$ **66.** [4.2] $-11t^2 + 3t$
67. [4.4] $18x^2 - 63x$ **68.** [4.4] $6x^2 - 10x$
69. [4.4] $6 + 23x + 20x^2$ **70.** [4.4] $14 - 13x + 3x^2$

71. ▨ **73.** $\dfrac{\sqrt{70}}{100}$ **75.** $\dfrac{\sqrt{10xy}}{4x^3y^2}$ **77.** $\dfrac{\sqrt{10abc}}{5b^2c^5}$

79. $\dfrac{3\sqrt{7\sqrt{7}}}{7}$ **81.** $\dfrac{1}{x} - \dfrac{1}{y}$

Technology Connection, p. 465

1. -0.086 **2.** 1.217 **3.** 2.765

Exercise Set 8.4, pp. 466–467

1. $12\sqrt{3}$ **3.** $2\sqrt{2}$ **5.** $12\sqrt{y}$ **7.** $-6\sqrt{a}$ **9.** $7\sqrt{6x}$
11. $11\sqrt{14y}$ **13.** $14\sqrt{5}$ **15.** $-2\sqrt{6}$ **17.** $5\sqrt{5}$
19. $4\sqrt{a}$ **21.** $32\sqrt{2}$ **23.** $26\sqrt{3}$ **25.** $7\sqrt{5}$
27. $6\sqrt{2}$ **29.** $13\sqrt{3}$ **31.** $\sqrt{x}$ **33.** $\sqrt{10} + \sqrt{14}$
35. $\sqrt{30} - 5\sqrt{2}$ **37.** $14 + 7\sqrt{2}$ **39.** $17 - 7\sqrt{7}$
41. -11 **43.** 3 **45.** $-2 - \sqrt{2}$ **47.** $52 + 14\sqrt{3}$

49. $13 - 4\sqrt{3}$ **51.** $x - 2\sqrt{10x} + 10$ **53.** $\dfrac{28 - 4\sqrt{2}}{47}$

55. $-4 - 2\sqrt{7}$ **57.** $\dfrac{5 - \sqrt{7}}{9}$ **59.** $\dfrac{-6 - 5\sqrt{6}}{19}$

61. $\dfrac{5 - \sqrt{15}}{2}$ **63.** $\dfrac{\sqrt{15} + 3}{2}$ **65.** $\dfrac{2\sqrt{7} - 2\sqrt{2}}{5}$

67. $6 + \sqrt{35}$ **69.** 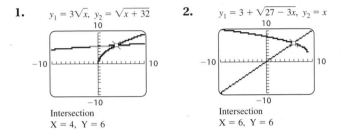   **71.** [2.2] $\frac{5}{11}$ **72.** [2.2] $-\frac{38}{13}$
73. [5.6] $-1, 6$ **74.** [5.6] $2, 5$
75. [7.4] Jolly Juice: 1.6 L; Real Squeeze: 6.4 L
76. [2.5] At least 3 hr   **77.** 🔲   **79.** $-15\sqrt{2}$
81. $a^5 b^4 \sqrt{ab}\,(a^4 - ab^2 + b^3)$ **83.** $37xy\sqrt{3x}$
85. All three are correct.

Technology Connection, p. 470

1. $y_1 = 3\sqrt{x},\ y_2 = \sqrt{x+32}$

2. $y_1 = 3 + \sqrt{27 - 3x},\ y_2 = x$

Intersection
X = 4, Y = 6

Intersection
X = 6, Y = 6

Exercise Set 8.5, pp. 471–473

1. 25   **3.** 64   **5.** 84   **7.** -46 **9.** $\frac{4}{5}$ **11.** 5
13. No solution   **15.** -2 **17.** 9   **19.** 12   **21.** 13, 25
23. 3   **25.** 0, 8   **27.** No solution   **29.** 3   **31.** $-\frac{3}{5}$
33. 3   **35.** 115.2 ft; 320 ft   **37.** 25.2°F   **39.** 806.56 m
41. 576 m   **43.** 0.81 ft   **45.** 🔲   **47.** [7.4] Type A: 12;
type B: 4   **48.** [7.4] First type: 45 oz; second type: 15 oz
49. [6.3] 168 km   **50.** [7.4] Dimes: 70; quarters: 33
51. 🔲   **53.** 32   **55.** 9   **57.** No solution   **59.** $-\frac{57}{16}$
61. 10   **63.** $34\frac{569}{784}$ m
65.

$y = \sqrt{x}$

67.

$y = \sqrt{x-3}$

69.

$y = \sqrt{x-5}$ $(9, 2)$

$y = x - 7$

The solution is 9.

71. 1.57   **73.** 🔲

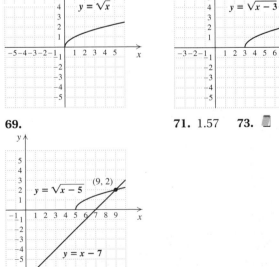

Exercise Set 8.6, pp. 478–481

1. 17   **3.** $6\sqrt{2} \approx 8.485$ **5.** 12   **7.** 6   **9.** 13
11. 24   **13.** 2   **15.** $\sqrt{2} \approx 1.414$ **17.** 5
19. $\sqrt{189}$ in. ≈ 13.748 in.   **21.** $\sqrt{208}$ ft ≈ 14.422 ft
23. $\sqrt{356}$ ft ≈ 18.868 ft   **25.** $\sqrt{26,900}$ yd ≈ 164.012 yd
27. $\sqrt{15,700}$ yd ≈ 125.300 yd   **29.** 🔲
31. [1.4] $-45, -9.7, 0, \frac{2}{7}, 5.09, 19$ **32.** [1.4] $-\sqrt{5}, \sqrt{7}, \pi$
33. [1.4] $-45, -9.7, -\sqrt{5}, 0, \frac{2}{7}, \sqrt{7}, \pi, 5.09, 19$
34. [1.4] $-45, 0, 19$ **35.** [4.1] -32 **36.** [4.1] $\frac{25}{9}$
37. [4.1] $8x^3$ **38.** [4.1] $-8x^3$ **39.** 🔲   **41.** Yes
43. 28,600 ft   **45.** $\sqrt{7}$ m   **47.** $s\sqrt{3}$ **49.** 50 ft^2
51. $20\sqrt{6}$ ft ≈ 48.990 ft   **53.** $12 - 2\sqrt{6} \approx 7.101$

Exercise Set 8.7, pp. 486–487

1. -4 **3.** -5 **5.** 10   **7.** -6 **9.** 3   **11.** 0
13. -1 **15.** Not a real number   **17.** 10   **19.** 7
21. 2   **23.** a **25.** $3\sqrt[3]{2}$ **27.** $2\sqrt[4]{3}$ **29.** $\frac{4}{5}$ **31.** $\frac{2}{3}$
33. $\frac{\sqrt[3]{7}}{2}$ **35.** $\frac{\sqrt[4]{14}}{3}$ **37.** 5   **39.** 5   **41.** 2   **43.** 8
45. 243   **47.** 16   **49.** 32   **51.** 32   **53.** $\frac{1}{5}$ **55.** $\frac{1}{4}$
57. $\frac{1}{8}$ **59.** $\frac{1}{243}$ **61.** $\frac{1}{4}$ **63.** 🔲   **65.** [5.6] $-1, 6$
66. [5.6] $-5, 1$ **67.** [5.6] $-\frac{3}{2}, \frac{3}{2}$ **68.** [5.6] $-\frac{2}{5}, \frac{2}{5}$
69. [5.6] $-2, -\frac{2}{3}$ **70.** [5.6] $-3, \frac{2}{5}$ **71.** 🔲   **73.** 6.310
75. 3.834   **77.** $x^{14/9}$ **79.** $p^{1/6}$
81.

$y = \sqrt[3]{x}$

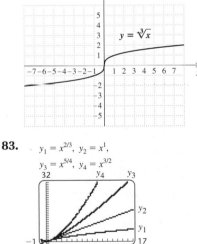

83. $y_1 = x^{2/3},\ y_2 = x^1,$
$y_3 = x^{5/4},\ y_4 = x^{3/2}$

Review Exercises: Chapter 8, pp. 488–489

1. [8.1] $5, -5$ **2.** [8.1] $14, -14$ **3.** [8.1] $30, -30$
4. [8.1] $15, -15$ **5.** [8.1] -12 **6.** [8.1] 9   **7.** [8.1] 7
8. [8.1] -13 **9.** [8.1] $2x^2 y$ **10.** [8.1] $\dfrac{a}{b}$

11. [8.1] Rational **12.** [8.1] Irrational
13. [8.1] Irrational **14.** [8.1] Rational **15.** [8.1] 1.732
16. [8.1] 9.950 **17.** [8.1] p **18.** [8.1] $7x$ **19.** [8.1] $4m$
20. [8.1] ac **21.** [8.2] $4\sqrt{3}$ **22.** [8.2] $7t\sqrt{2}$
23. [8.2] $4\sqrt{2p}$ **24.** [8.2] x^3 **25.** [8.2] $2a^6\sqrt{3a}$
26. [8.2] $6m^7\sqrt{m}$ **27.** [8.2] $\sqrt{33}$ **28.** [8.2] $2\sqrt{15}$
29. [8.2] $\sqrt{35xt}$ **30.** [8.2] $2a\sqrt{6}$ **31.** [8.2] $5xy\sqrt{2}$
32. [8.2] $10a^2b\sqrt{ab}$ **33.** [8.3] $\dfrac{\sqrt{7}}{3}$ **34.** [8.3] $\dfrac{y^4\sqrt{5}}{3}$

35. [8.3] $\frac{5}{8}$ **36.** [8.3] $\frac{2}{3}$ **37.** [8.3] $\dfrac{7}{t}$ **38.** [8.4] $13\sqrt{5}$

39. [8.4] $\sqrt{5}$ **40.** [8.4] $-3\sqrt{x}$ **41.** [8.4] $7 + 4\sqrt{3}$

42. [8.4] 1 **43.** [8.4] $-11 + 5\sqrt{7}$ **44.** [8.3] $\dfrac{\sqrt{2}}{2}$

45. [8.3] $\dfrac{\sqrt{10}}{4}$ **46.** [8.3] $\dfrac{\sqrt{5y}}{y}$ **47.** [8.3] $\dfrac{2\sqrt{3}}{3}$

48. [8.4] $8 - 4\sqrt{3}$ **49.** [8.4] $-7 - 3\sqrt{5}$ **50.** [8.5] 52
51. [8.5] No solution **52.** [8.5] 4 **53.** [8.5] 0, 3
54. [8.5] 405 ft **55.** [8.6] 20 **56.** [8.6] $\sqrt{3} \approx 1.732$
57. [8.6] $\sqrt{10{,}625} \approx 103.078$ ft **58.** [8.7] 2
59. [8.7] Not a real number **60.** [8.7] -3
61. [8.7] $2\sqrt[4]{2}$ **62.** [8.7] 10 **63.** [8.7] $\frac{1}{3}$ **64.** [8.7] 64
65. [8.7] $\frac{1}{27}$ **66.** [8.5] 🔲 He could be mistakenly
assuming that negative solutions are not possible because
principal square roots are never negative. He could also be
assuming that substituting a negative value for the variable
in a radicand always produces a negative radicand.
67. [8.4] 🔲 Some radical terms that are like terms may not
appear to be so until they are in simplified form.
68. [8.1] 2 **69.** [8.1] No solution
70. [8.4] $\left(x + \sqrt{5}\right)\left(x - \sqrt{5}\right)$
71. [8.5] $b = \sqrt{A^2 - a^2}$ or $b = -\sqrt{A^2 - a^2}$

Test: Chapter 8, pp. 489–490

1. [8.1] 9, -9 **2.** [8.1] 8 **3.** [8.1] -5 **4.** [8.1] $x + 4$
5. [8.1] Irrational **6.** [8.1] Rational **7.** [8.1] 9.327
8. [8.1] 2.646 **9.** [8.1] a **10.** [8.1] $6y$ **11.** [8.2] $2\sqrt{6}$
12. [8.2] $3x^3\sqrt{3}$ **13.** [8.2] $2t^2\sqrt{t}$ **14.** [8.2] $\sqrt{30}$
15. [8.2] $5\sqrt{2}$ **16.** [8.2] $\sqrt{14xy}$ **17.** [8.2] $4t$
18. [8.2] $3ab^2\sqrt{2}$ **19.** [8.3] $\frac{3}{4}$ **20.** [8.3] $\dfrac{\sqrt{7}}{4y}$

21. [8.3] $\frac{3}{2}$ **22.** [8.3] $\dfrac{12}{a}$ **23.** [8.4] $-6\sqrt{2}$

24. [8.4] $7\sqrt{3}$ **25.** [8.4] $21 - 8\sqrt{5}$ **26.** [8.4] 11
27. [8.3] $\dfrac{\sqrt{10}}{5}$ **28.** [8.3] $\dfrac{2x\sqrt{y}}{y}$ **29.** [8.4] $\dfrac{40 + 10\sqrt{5}}{11}$

30. [8.6] $\sqrt{80} \approx 8.944$ cm **31.** [8.5] 48 **32.** [8.5] $-2, 2$
33. [8.5] About 5000 m **34.** [8.7] 2 **35.** [8.7] -1
36. [8.7] -4 **37.** [8.7] Not a real number **38.** [8.7] 3
39. [8.7] $\frac{1}{3}$ **40.** [8.7] 1000 **41.** [8.7] $\frac{1}{32}$ **42.** [8.5] -3
43. [8.2] y^{8n}

CHAPTER 9

Technology Connection, p. 493

1.

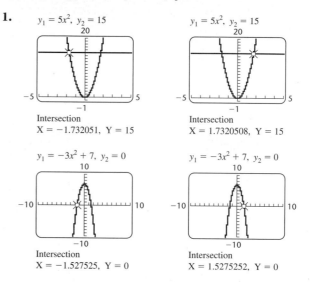

$y_1 = 5x^2,\ y_2 = 15$

Intersection
X = -1.732051, Y = 15

$y_1 = 5x^2,\ y_2 = 15$

Intersection
X = 1.7320508, Y = 15

$y_1 = -3x^2 + 7,\ y_2 = 0$

Intersection
X = -1.527525, Y = 0

$y_1 = -3x^2 + 7,\ y_2 = 0$

Intersection
X = 1.5275252, Y = 0

Exercise Set 9.1, pp. 495–496

1. $-7, 7$ **3.** $-5, 5$ **5.** $-\sqrt{17}, \sqrt{17}$ **7.** $-3, 3$ **9.** 0
11. $-\frac{2}{3}, \frac{2}{3}$ **13.** $-\dfrac{2\sqrt{5}}{7}, \dfrac{2\sqrt{5}}{7}$ **15.** $-\dfrac{\sqrt{14}}{2}, \dfrac{\sqrt{14}}{2}$
17. $-4, 6$ **19.** $-13, 5$ **21.** $-3 - \sqrt{6}, -3 + \sqrt{6}$, or $-3 \pm \sqrt{6}$ **23.** 7 **25.** $5 - \sqrt{14}, 5 + \sqrt{14}$, or $5 \pm \sqrt{14}$
27. $-8, 2$ **29.** $\dfrac{3 - \sqrt{17}}{4}, \dfrac{3 + \sqrt{17}}{4}$, or $\dfrac{3 \pm \sqrt{17}}{4}$
31. $-5, 15$ **33.** $-5, -3$ **35.** $3 - \sqrt{13}, 3 + \sqrt{13}$, or $3 \pm \sqrt{13}$ **37.** $-6 - 3\sqrt{2}, -6 + 3\sqrt{2}$, or $-6 \pm 3\sqrt{2}$
39. 🔲 **41.** [5.2] $3(x^2 + 4x + 1)$
42. [5.2] $5(t^2 + 4t + 5)$ **43.** [5.2] $(t + 8)^2$
44. [5.2] $(x + 7)^2$ **45.** [5.2] $(x - 5)^2$
46. [5.2] $(t - 4)^2$ **47.** 🔲 **49.** $\dfrac{-7 - \sqrt{7}}{6}, \dfrac{-7 + \sqrt{7}}{6}$,
or $\dfrac{-7 \pm \sqrt{7}}{6}$ **51.** $\dfrac{3 - \sqrt{17}}{4}, \dfrac{3 + \sqrt{17}}{4}$, or $\dfrac{3 \pm \sqrt{17}}{4}$
53. $-4.35, 1.85$ **55.** $-4, -2$ **57.** $-6, 0$
59. $d = \dfrac{\sqrt{kMmf}}{f}$

Exercise Set 9.2, pp. 501–502

1. $x^2 + 8x + 16$ **3.** $x^2 - 12x + 36$ **5.** $x^2 - 3x + \frac{9}{4}$
7. $t^2 + t + \frac{1}{4}$ **9.** $x^2 + \frac{5}{4}x + \frac{25}{64}$
11. $m^2 - \frac{9}{2}m + \frac{81}{16}$ **13.** $-6, -2$ **15.** $12 \pm \sqrt{123}$
17. $1 \pm \sqrt{6}$ **19.** $11 \pm \sqrt{19}$ **21.** $-4 \pm \sqrt{21}$
23. $-6, 1$ **25.** $\dfrac{-3 \pm \sqrt{41}}{4}$ **27.** $\dfrac{-3 \pm \sqrt{137}}{4}$

29. $\dfrac{-3 \pm 2\sqrt{3}}{3}$ **31.** $\dfrac{5 \pm \sqrt{97}}{4}$ **33.** $-\dfrac{7}{2}, \dfrac{1}{2}$ **35.**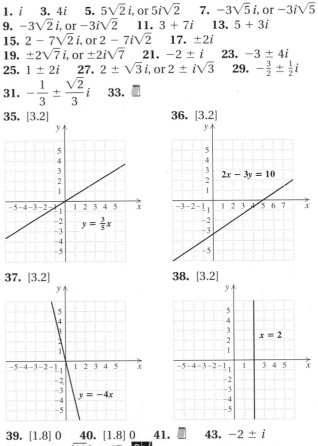
37. [6.1] $1 + 2x$ **38.** [6.1] $2 + x$ **39.** [6.1] $3 - 2x$
40. [6.1] $6 - 5x$ **41.** [8.4] $\dfrac{8 - \sqrt{5}}{3}$ **42.** [8.4] $\dfrac{5 - \sqrt{6}}{2}$
43.   **45.** $-14, 14$ **47.** $-10\sqrt{2}, 10\sqrt{2}$ **49.** $\pm 8\sqrt{3}$
51. $-7.61, -0.39$ **53.** $0.91, 23.09$ **55.** $-1.21, 3.71$
57.

Technology Connection, p. 505

1. An error message appears.
2. The graph has no x-intercepts, so there is no value of x
for which $x^2 + x + 1 = 0$ or, equivalently, for which
$x^2 + x = -1$.

Exercise Set 9.3, pp. 509–512

1. $-2, 9$ **3.** 4 **5.** $-\dfrac{4}{3}, -1$ **7.** $-\dfrac{1}{2}, \dfrac{7}{2}$ **9.** $-8, 8$
11. $-2 \pm \sqrt{11}$ **13.** $5 \pm \sqrt{6}$ **15.** $-1 \pm \sqrt{7}$
17. $\dfrac{-4 \pm \sqrt{10}}{3}$ **19.** $\dfrac{5 \pm \sqrt{33}}{4}$ **21.** $\dfrac{-1 \pm \sqrt{13}}{4}$
23. No real-number solution   **25.** $\dfrac{5 \pm \sqrt{73}}{6}$
27. $\dfrac{3 \pm \sqrt{29}}{2}$ **29.** $0, \dfrac{3}{2}$ **31.** No real-number solution
33. $\pm 3\sqrt{2}$ **35.** $-1.317, 5.317$ **37.** $-0.193, 5.193$
39. $-1.207, 0.207$ **41.** 10 **43.** 9.65 scc   **45.** 6.61 sec
47. 7 ft, 24 ft   **49.** Length: 10 cm; width: 6 cm
51. Length: 24 m; width: 18 m   **53.** 6.76 m, 9.16 m
55. 4.18 in., 7.18 in.   **57.** Length: 5.66 m; width: 2.83 m
59. 20% **61.** 5% **63.** 5.05 km   **65.**   **67.** [2.2] 25
68. [2.2] 4 **69.** [2.2] 15 **70.** [2.2] 9 **71.** [8.5] 16
72. [8.5] 50 **73.**   **75.** $0, 2$ **77.** $\dfrac{3 \pm \sqrt{5}}{2}$
79. $\dfrac{-7 \pm \sqrt{61}}{2}$ **81.** $\pm \sqrt{5}$ **83.** $\dfrac{5 \pm \sqrt{37}}{2}$
85. 4.83 cm   **87.** Length: 120 m; width: 75 m   **89.** 5%
91. Length: 44.06 in.; width: 19.06 in.

Technology Connection, p. 514

1. $.88; 5.12$ **2.** They are rational approximations of the
answers yielded by the quadratic formula.

Exercise Set 9.4, pp. 517–518

1. 1 **3.** $2, 3$ **5.** 25 **7.** 4 **9.** $-\dfrac{1}{5}$ **11.** $-\dfrac{1}{12}$
13. $\dfrac{1}{2}, 3$ **15.** 2 **17.** $\dfrac{12 \pm \sqrt{138}}{2}$ **19.** $\dfrac{4}{3}$
21. No real-number solution   **23.** No real-number

solution   **25.** $\dfrac{2}{3}$ **27.** $g = \dfrac{2s}{t^2}$ **29.** $h = \dfrac{S}{2\pi r}$
31. $h = \dfrac{d^2}{c^2}$ **33.** $R = \dfrac{r_1 r_2}{r_1 + r_2}$ **35.** $x = \dfrac{-b \pm \sqrt{b^2 - 4ac}}{2a}$
37. $n = \dfrac{m}{p - q}$ **39.** $r = \dfrac{L}{l + S}$ **41.** $h = \dfrac{S - 2\pi r^2}{2\pi r}$
43. $h = \pm \sqrt{st}$ **45.** $t = \dfrac{-n \pm \sqrt{n^2 + 4mp}}{2m}$
47. $n = \dfrac{m}{r}$ **49.** $m = \dfrac{-t \pm \sqrt{t^2 + 4n}}{2}$
51. $t = \dfrac{p^2 - 2pn + n^2 - 9c}{9}$ **53.** $n = m - 3t$ **55.**
57. [8.2] $2\sqrt{15}$ **58.** [8.2] $3\sqrt{35}$ **59.** [8.2] $5\sqrt{6}$
60. [8.2] $3\sqrt{7}$ **61.** [8.2] $2a^3 b^2 \sqrt{a}$ **62.** [8.2] $3x^4 y^2 \sqrt{y}$
63.   **65.** 6 yr   **67.** $m = \dfrac{g \pm \sqrt{g^2 - 4ft}}{2f}$
69. $T = \dfrac{FP}{EF + u}$ **71.** $-40°$

Exercise Set 9.5, p. 521

1. i **3.** $4i$ **5.** $5\sqrt{2}\,i$, or $5i\sqrt{2}$ **7.** $-3\sqrt{5}\,i$, or $-3i\sqrt{5}$
9. $-3\sqrt{2}\,i$, or $-3i\sqrt{2}$ **11.** $3 + 7i$ **13.** $5 + 3i$
15. $2 - 7\sqrt{2}\,i$, or $2 - 7i\sqrt{2}$ **17.** $\pm 2i$
19. $\pm 2\sqrt{7}\,i$, or $\pm 2i\sqrt{7}$ **21.** $-2 \pm i$ **23.** $-3 \pm 4i$
25. $1 \pm 2i$ **27.** $2 \pm \sqrt{3}\,i$, or $2 \pm i\sqrt{3}$ **29.** $-\dfrac{3}{2} \pm \dfrac{1}{2}i$
31. $-\dfrac{1}{3} \pm \dfrac{\sqrt{2}}{3}i$ **33.**
35. [3.2]   **36.** [3.2]

$y = \dfrac{3}{5}x$ $2x - 3y = 10$

37. [3.2]   **38.** [3.2]

$y = -4x$ $x = 2$

39. [1.8] 0 **40.** [1.8] 0 **41.**   **43.** $-2 \pm i$
45. $-1 \pm \sqrt{19}i$ **47.**

Exercise Set 9.6, pp. 527–528

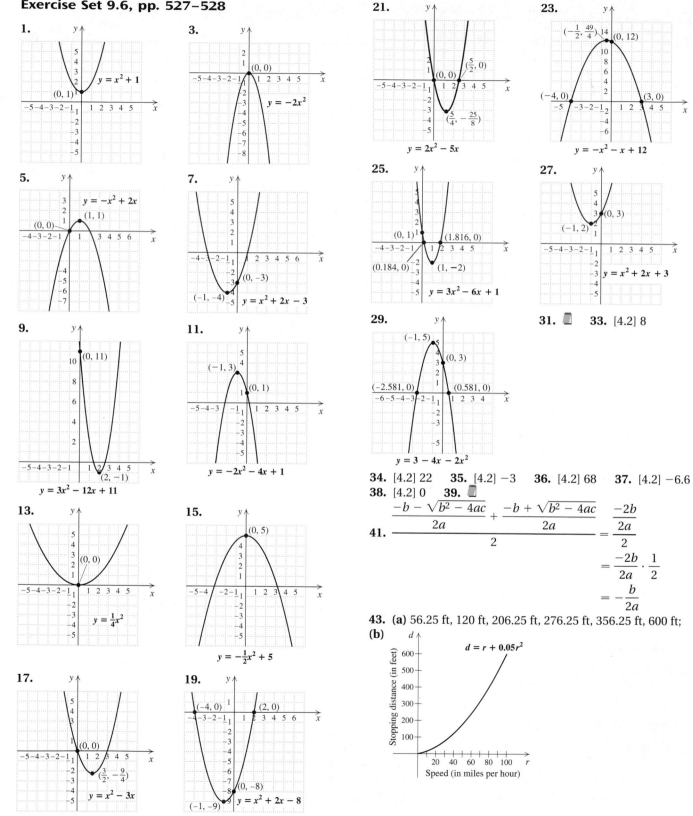

1. $y = x^2 + 1$

3. $y = -2x^2$, $(0, 0)$

5. $y = -x^2 + 2x$, $(0, 0)$, $(1, 1)$

7. $y = x^2 + 2x - 3$, $(0, -3)$, $(-1, -4)$

9. $y = 3x^2 - 12x + 11$, $(0, 11)$, $(2, -1)$

11. $y = -2x^2 - 4x + 1$, $(-1, 3)$, $(0, 1)$

13. $y = \frac{1}{4}x^2$, $(0, 0)$

15. $y = -\frac{1}{2}x^2 + 5$, $(0, 5)$

17. $y = x^2 - 3x$, $(0, 0)$, $\left(\frac{3}{2}, -\frac{9}{4}\right)$

19. $y = x^2 + 2x - 8$, $(-4, 0)$, $(2, 0)$, $(0, -8)$, $(-1, -9)$

21. $y = 2x^2 - 5x$, $(0, 0)$, $\left(\frac{5}{2}, 0\right)$, $\left(\frac{5}{4}, -\frac{25}{8}\right)$

23. $y = -x^2 - x + 12$, $\left(-\frac{1}{2}, \frac{49}{4}\right)$, $(0, 12)$, $(-4, 0)$, $(3, 0)$

25. $y = 3x^2 - 6x + 1$, $(0, 1)$, $(1.816, 0)$, $(0.184, 0)$, $(1, -2)$

27. $y = x^2 + 2x + 3$, $(0, 3)$, $(-1, 2)$

29. $y = 3 - 4x - 2x^2$, $(-1, 5)$, $(0, 3)$, $(-2.581, 0)$, $(0.581, 0)$

31. 🗒   **33.** [4.2] 8

34. [4.2] 22   **35.** [4.2] −3   **36.** [4.2] 68   **37.** [4.2] −6.6

38. [4.2] 0   **39.** 🗒

41.
$$\frac{\dfrac{-b - \sqrt{b^2 - 4ac}}{2a} + \dfrac{-b + \sqrt{b^2 - 4ac}}{2a}}{2} = \frac{\dfrac{-2b}{2a}}{2}$$
$$= \frac{-2b}{2a} \cdot \frac{1}{2}$$
$$= -\frac{b}{2a}$$

43. (a) 56.25 ft, 120 ft, 206.25 ft, 276.25 ft, 356.25 ft, 600 ft;

(b) $d = r + 0.05r^2$

Stopping distance (in feet)

Speed (in miles per hour)

45.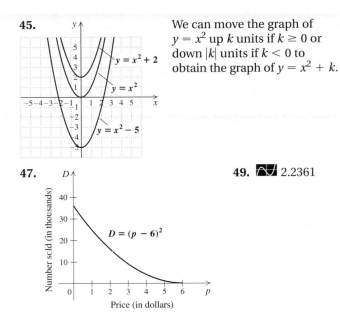

We can move the graph of $y = x^2$ up k units if $k \geq 0$ or down $|k|$ units if $k < 0$ to obtain the graph of $y = x^2 + k$.

47.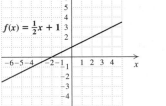

49. 2.2361

35.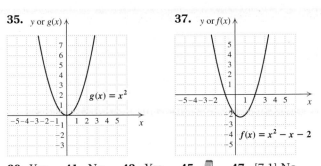

37.

39. Yes   **41.** No   **43.** Yes   **45.**   **47.** [7.1] No
48. [7.1] Yes   **49.** [7.1] Yes   **50.** [7.2] No solution
51. [7.2] Infinite number of solutions   **52.** [7.2] $\left(\frac{9}{11}, -\frac{5}{11}\right)$
53.

55.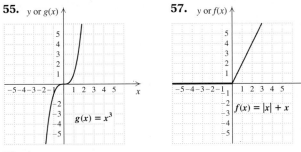

57.

59. Answers may vary.
61. $g(x) = x^2 - 4$ **63.** $\{-5, -4, -1, 4\}$
65. $\{-7, 0, 1, 2, 9\}$

Exercise Set 9.7, pp. 535–538

1. Yes   **3.** Yes   **5.** No   **7.** Yes   **9.** 9; 12; 3
11. -21; 15; 30   **13.** 7; -11; 24.1   **15.** 0; 5; 2
17. 16; 1; $\frac{25}{16}$ **19.** 4; -122; 3   **21.** (a) 159.48 cm;
(b) 153.98 cm   **23.** 70°C; 220°C; 10,020°C
25. 1.792 cm; 2.8 cm; 11.2 cm

27.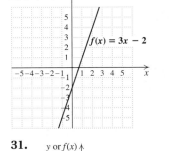

29.

31.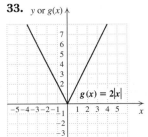

33.

Review Exercises: Chapter 9, pp. 539–540

1. $[9.2] -\frac{5}{3}, 1$ **2.** $[9.2] \dfrac{3 \pm \sqrt{5}}{2}$ **3.** $[9.1] \pm\sqrt{3}$

4. $[9.2] \frac{3}{5}, 1$ **5.** $[9.3] 1 \pm \sqrt{11}$ **6.** $[9.5] \pm 8i$

7. $[9.2] -2, \frac{1}{3}$ **8.** $[9.2] -8 \pm \sqrt{13}$ **9.** $[9.5] 3 \pm 2i$
10. $[9.2] -1, 3$ **11.** $[9.1] 0$ **12.** $[9.3] -3 \pm 3\sqrt{2}$

13. $[9.5] -\dfrac{1}{2} \pm \dfrac{\sqrt{3}}{2}i$ **14.** $[9.3] \dfrac{2 \pm \sqrt{3}}{2}$ **15.** $[9.2] 1$

16. $[9.1] \pm 2\sqrt{2}$ **17.** $[9.5] \dfrac{3}{10} \pm \dfrac{\sqrt{71}}{10}i$ **18.** $[9.2] -\frac{7}{2}, \frac{5}{3}$

19. $[9.4] t = \sqrt{\dfrac{r + s}{2}}$, or $t = \dfrac{\sqrt{2(r + s)}}{2}$

20. [9.4] $m = \dfrac{100 + n}{3}$

21. [9.4] $a = \dfrac{b + c}{2}$ **22.** [9.4] $t = \dfrac{rs}{r + s}$

23. [9.3] 0.382, 2.618   **24.** [9.3] -1.866, -0.134

25. [9.3] 3.217 m, 6.217 m   **26.** [9.3] 5%

27. [9.3] Length: 12 m; width: 9 m   **28.** [9.3] 6.349 sec

29. [9.5] $2 - 8i$ **30.** [9.5] $2\sqrt{6}\,i$, or $2i\sqrt{6}$

31. [9.6]   **32.** [9.6]

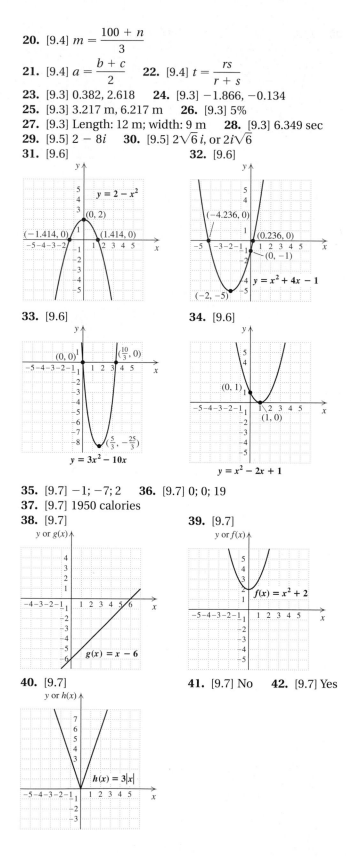

33. [9.6]   **34.** [9.6]

35. [9.7] -1; -7; 2   **36.** [9.7] 0; 0; 19

37. [9.7] 1950 calories

38. [9.7]   **39.** [9.7]

40. [9.7]

41. [9.7] No   **42.** [9.7] Yes

43. [9.6] 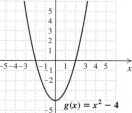 If $b^2 - 4ac$ is 0, then the quadratic formula becomes $x = -b/(2a)$; thus there is only one x-intercept. If $b^2 - 4ac$ is negative, then there are no real-number solutions and thus no x-intercepts. If $b^2 - 4ac$ is positive, then $x = \dfrac{-b + \sqrt{b^2 - 4ac}}{2a}$ or $x = \dfrac{-b - \sqrt{b^2 - 4ac}}{2a}$ so there must be two x-intercepts.

44. [9.6] 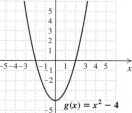 The graph can be drawn more accurately and with fewer points plotted when the general shape is known.   **45.** [9.4] Yes; a rational equation like $x + \dfrac{6}{x} = -5$ has two solutions whereas a quadratic equation like $x^2 + 6x + 9 = 0$ has just one solution.

46. [9.3] -32 and -31; 31 and 32   **47.** [9.1] ± 14

48. [9.3] 25   **49.** [9.3] $5\sqrt{\pi}$

Test: Chapter 9, pp. 540–541

1. [9.1] $\pm\sqrt{5}$ **2.** [9.3] $0, \dfrac{7}{3}$ **3.** [9.3] $\dfrac{1 \pm \sqrt{13}}{2}$

4. [9.3] $\dfrac{5}{3}, -3$ **5.** [9.3] $-8, 6$ **6.** [9.5] $\pm 7i$

7. [9.1] $2 \pm \sqrt{5}$ **8.** [9.1] 2   **9.** [9.5] $2 \pm 2i$

10. [9.3] $\dfrac{3 \pm \sqrt{37}}{2}$ **11.** [9.3] $-2 \pm \sqrt{14}$

12. [9.3] $\dfrac{7 \pm \sqrt{37}}{6}$ **13.** [9.2] $2 \pm \sqrt{14}$

14. [9.4] $p = \dfrac{n^2 - 6n - 11}{4}$ **15.** [9.4] $b = \dfrac{a}{1 + x}$

16. [9.3] 4.702, -1.702 **17.** [9.3] 6.5 m, 2.5 m

18. [9.3] 11   **19.** [9.5] $7i$ **20.** [9.5] $3 - 4\sqrt{2}\,i$

21. [9.6]   **22.** [9.6]

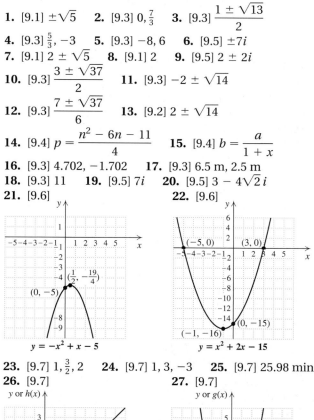

23. [9.7] $1, \dfrac{3}{2}, 2$ **24.** [9.7] 1, 3, -3 **25.** [9.7] 25.98 min

26. [9.7]   **27.** [9.7]

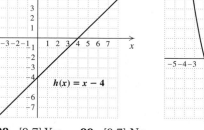

28. [9.7] Yes   **29.** [9.7] No

30. [9.3] $5 + 5\sqrt{2} \approx 12.071$ ft   **31.** [9.3] $1 \pm \sqrt{5}$

Cumulative Review: Chapters 1–9, pp. 541–543

1. [1.8] x^3 **2.** [1.8] 54   **3.** [1.2] $(6x)y$, $x(6y)$; there are other answers.   **4.** [6.3] 240   **5.** [1.4] 7   **6.** [1.5] 9
7. [1.6] 15   **8.** [1.7] $-\frac{3}{20}$ **9.** [1.8] 4
10. [1.8] $-2m - 4$ **11.** [2.1] -9 **12.** [2.6] $\{x \mid x < -9\}$
13. [2.2] 7   **14.** [5.6] 3, 5   **15.** [7.2] $(1, 2)$
16. [6.6] $-\frac{3}{5}$ **17.** [7.3] $(6, 7)$ **18.** [5.6] 3, -2
19. [9.3] $\dfrac{-3 \pm \sqrt{29}}{2}$ **20.** [8.5] 2   **21.** [2.6] $\{x \mid x \geq -1\}$
22. [2.2] 8   **23.** [2.2] -3 **24.** [7.3] $(15, 0)$
25. [9.5] $-1 \pm 2i$ **26.** [9.1] $\pm\sqrt{10}$ **27.** [9.1] $3 \pm \sqrt{6}$
28. [6.6] $\frac{2}{9}$ **29.** [5.6] $\frac{5}{4}, -\frac{4}{3}$ **30.** [6.6] No solution
31. [8.5] 12   **32.** [9.4] $t = \dfrac{4s + 3}{A}$ **33.** [9.4] $m = \dfrac{tn}{t + n}$
34. [4.8] 2.73×10^{-5} **35.** [4.8] 2.0×10^{14}
36. [4.8] x^{-4} **37.** [4.8] y^7 **38.** [4.1] $4y^{12}$
39. [4.2] $10x^3 + 3x - 3$ **40.** [4.3] $7x^3 - 2x^2 + 4x - 17$
41. [4.3] $8x^2 - 4x - 6$ **42.** [4.4] $-8y^4 + 6y^3 - 2y^2$
43. [4.4] $6t^3 - 17t^2 + 16t - 6$ **44.** [4.5] $t^2 - \frac{1}{16}$
45. [4.5] $9m^2 - 12m + 4$
46. [4.6] $15x^2y^3 + x^2y^2 + 5xy^2 + 7$ **47.** [4.6] $x^4 - 0.04y^2$
48. [4.6] $9p^2 + 24pq^2 + 16q^4$ **49.** [6.2] $\dfrac{2}{x + 3}$
50. [6.2] $\dfrac{3a(a - 1)}{2(a + 1)}$ **51.** [6.4] $\dfrac{27x - 4}{5x(3x - 1)}$
52. [6.4] $\dfrac{-x^2 + x + 2}{(x + 4)(x - 4)(x - 5)}$
53. [4.7] $x^2 + 9x + 16 + \dfrac{35}{x - 2}$
54. [5.4] $(7x + 2)(7x - 2)$ **55.** [5.1] $(x - 8)(x^2 - 5)$
56. [5.4] $x(4x + 1)(4x - 1)$ **57.** [5.4] $(m - 4)^2$
58. [5.3] $(5x - 2)(3x + 4)$ **59.** [5.1] $2x(9x^4 + 2x^3 - 5)$
60. [5.3] $(3p + 2)(2p - 3)$ **61.** [5.1] $(c - 3b)(2a + d)$
62. [5.4] $(3x + 5y)^2$ **63.** [5.5] $3(t^2 + 6)(t + 2)(t - 2)$
64. [5.4] $(7ab + 3)(7ab - 3)$ **65.** [6.5] $-\frac{42}{5}$ **66.** [8.1] 7
67. [8.7] -5 **68.** [8.2] $8x$ **69.** [8.2] $\sqrt{a^2 - b^2}$
70. [8.2] $8a^2b\sqrt{3ab}$ **71.** [8.7] $\frac{1}{4}$ **72.** [8.2] $9xy\sqrt{3x}$
73. [8.3] $\frac{10}{9}$ **74.** [8.4] 7   **75.** [8.4] $16\sqrt{3}$
76. [8.3] $\dfrac{2\sqrt{10}}{5}$ **77.** [8.6] 40 cm

78. [3.6]

79. [3.3]
$2x + 3y = -6$
$y = \frac{1}{3}x - 2$

80. [3.3]
$y = -3$

81. [7.5]
$4x - 3y > 12$

82. [9.6]
$y = x^2 + 2x + 1$

83. [7.6]
$x \geq -3$

84. [9.2] $\dfrac{2 \pm \sqrt{6}}{3}$ **85.** [9.3] 1.207, -0.207

86. [9.6]

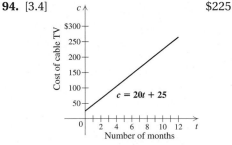

$(-3.449, 0)$ $(1.449, 0)$ $(0, -5)$ $(-1, -6)$ $y = x^2 + 2x - 5$

87. [2.4] 2.124 million   **88.** [3.4] $116.50/year
89. [5.7] 14, 16; $-16, -14$ **90.** [9.3] 12 m
91. [2.5] 6090   **92.** [7.4] Chocolate creams: 19 lb; nut candies: 6 lb   **93.** [6.7] 24 min
94. [3.4] $225

$c = 20t + 25$

Cost of cable TV / Number of months

95. [9.7] $-4, 0, 0$ **96.** [3.6] Parallel

97. [7.6]

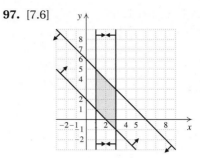

98. [3.6] $2; (0, -8)$ **99.** [3.5] 15

100. [3.7] $y + 3 = -\frac{1}{2}(x - 1)$ **101.** [9.5] $-5i$

102. [9.2] $30, -30$ **103.** [8.6] $\dfrac{\sqrt{6}}{3}$ **104.** [4.5] Yes

105. [6.1] No   **106.** [4.5] No   **107.** [8.1] No
108. [8.1] Yes

APPENDIXES

Exercise Set A, p. 548

1. $\{3, 4, 5, 6, 7\}$ **3.** $\{41, 43, 45, 47, 49\}$ **5.** $\{-3, 3\}$
7. False   **9.** True   **11.** True   **13.** True   **15.** True

17. False   **19.** $\{c, d, e\}$ **21.** $\{1, 10\}$ **23.** $\varnothing$
25. $\{a, e, i, o, u, q, c, k\}$ **27.** $\{0, 1, 2, 5, 7, 10\}$
29. $\{a, e, i, o, u, m, n, f, g, h\}$ **31.** 🗒
33. The set of integers   **35.** The set of real numbers
37. $\varnothing$ **39.** **(a)** A; **(b)** A; **(c)** A; **(d)** $\varnothing$ **41.** True

Exercise Set B, p. 551

1. $(t + 2)(t^2 - 2t + 4)$ **3.** $(a - 4)(a^2 + 4a + 16)$
5. $(z + 5)(z^2 - 5z + 25)$ **7.** $(2a - 1)(4a^2 + 2a + 1)$
9. $(y - 3)(y^2 + 3y + 9)$ **11.** $(4 + 5x)(16 - 20x + 25x^2)$
13. $(5p - 1)(25p^2 + 5p + 1)$
15. $(3m + 4)(9m^2 - 12m + 16)$
17. $(p - q)(p^2 + pq + q^2)$ **19.** $\left(x + \frac{1}{2}\right)\left(x^2 - \frac{1}{2}x + \frac{1}{4}\right)$
21. $2(y - 4)(y^2 + 4y + 16)$
23. $3(2a + 1)(4a^2 - 2a + 1)$ **25.** $r(s - 4)(s^2 + 4s + 16)$
27. $5(x + 2z)(x^2 - 2xz + 4z^2)$
29. $(x + 0.1)(x^2 - 0.1x + 0.01)$ **31.** 🗒
33. $(5c^2 + 2d^2)(25c^4 - 10c^2d^2 + 4d^4)$
35. $3(x^a - 2y^b)(x^{2a} + 2x^a y^b + 4y^{2b})$
37. $\frac{1}{3}\left(\frac{1}{2}xy + z\right)\left(\frac{1}{4}x^2y^2 - \frac{1}{2}xyz + z^2\right)$

Glossary

A

absolute value [1.4] The absolute value of a number is its distance from 0 on the number line.

additive inverse [1.6] Additive inverses are two numbers whose sum is 0.

algebraic expression [1.1] An algebraic expression is an expression that contains one or more variables and may contain any number of constants.

ascending order [4.3] When a polynomial is written with the terms arranged according to degree, from least to greatest, it is said to be in ascending order.

associative law of addition [1.2] The associative law of addition states that when three numbers are being added, regrouping the addends gives the same sum.

associative law of multiplication [1.2] The associative law of multiplication states that when three numbers are being multiplied, regrouping the factors gives the same product.

axes [3.1] Two perpendicular number lines used to identify points in a plane are axes.

axis of symmetry [9.6] The axis of symmetry is a line that can be drawn through a graph such that the part of the graph on one side of the line is an exact reflection of the part on the opposite side.

B

bar graph [3.1] A bar graph is a graphic means of displaying data using bars proportional in length to the numbers represented.

base [1.8] The base in exponential notation is the number being raised to a power.

binomial [4.2] A polynomial that is composed of two terms is called a binomial.

C

circle graph [3.1] A circle graph is a graphic means of displaying data using sectors of a circle to represent percents.

circumference [2.3] The circumference is the distance around a circle.

coefficient [2.1] The coefficient is the number preceding a variable in an algebraic expression.

commutative law of addition [1.2] The commutative law of addition states that changing the order in which two numbers are added does not affect the sum.

commutative law of multiplication [1.2] The commutative law of multiplication states that changing the order in which two numbers are multiplied does not affect the product.

completing the square [9.2] To complete the square for an expression like $x^2 + bx$, add half of the coefficient of x, squared.

complex number [9.5] A complex number is any number that can be written as $a + bi$, where a and b are real numbers.

complex rational expression [6.5] A complex rational expression is a rational expression that has one or more rational expressions within its numerator and/or denominator.

composite number [1.3] A composite number is a whole number that has more than two factors.

conjugates [8.4] Expressions of the form $a\sqrt{b} + c\sqrt{d}$ and $a\sqrt{b} - c\sqrt{d}$ are called conjugates. Their product does not contain a radical term.

consistent system of equations [7.1] A system of equations that has at least one solution is said to be consistent.

constant [1.1] A constant is a known number.

constant of proportionality [7.7] In an equation of the form $y = kx$ or $y = \dfrac{k}{x}$ ($k > 0$), k is the constant of proportionality.

coordinates [3.1] Coordinates are the numbers in an ordered pair.

cube root [8.7] The number c is the cube root of a if $c^3 = a$.

D

degree of a polynomial [4.2] The degree of the term of highest degree of a polynomial is referred to as the degree of the polynomial.

degree of a term [4.2] The degree of a term is the number of variable factors in that term.

denominator [1.3] The number below the fraction line in a fraction is called the denominator.

dependent equations [7.1] If a system of equations has infinitely many solutions, the equations are dependent.

descending order [4.2] When a polynomial is written with the terms arranged according to degree, from greatest to least, it is said to be in descending order.

difference of squares [5.4] Any expression that can be written in the form $a^2 - b^2$ is called a difference of squares.

direct variation [7.7] When a situation translates to an equation described by $y = kx$, with k a constant, we say that y varies directly as x. The equation $y = kx$ is called an equation of direct variation.

distributive law [1.2] The distributive law states that multiplying a factor by the sum of two numbers gives the same result as multiplying the factor by each of the two numbers and then adding.

domain [9.7] The domain is the set of all the first coordinates of the ordered pairs in a function.

E

equation [1.1] An equation is a number sentence with the verb =.

equivalent equations [2.1] Equations with the same solutions are called equivalent equations.

equivalent expressions [1.2] Equivalent expressions are expressions that have the same value for all allowable replacements.

equivalent inequalities [2.6] Inequalities that have the same solution set are called equivalent inequalities.

exponent [1.8] An exponent is a number that indicates how many times another number is multiplied by itself.

exponential notation [1.8] Exponential notation is a representation of a number using a base raised to a power.

extrapolation [3.1] Extrapolation is the process of estimating a value that goes beyond the given data.

F

factor [1.2] To factor an expression means to write an equivalent expression that is a product. A factor can also be a number being multiplied.

FOIL [4.5] To multiply two binomials using the FOIL method, multiply the first terms, the outside terms, the inside terms, and then the last terms. Then combine like terms, if possible.

formula [2.3] A formula is an equation that uses numbers or letters to represent a relationship between two or more quantities.

fraction notation [1.3] Fraction notation is a representation of a number using a fraction (a numerator and a denominator).

function [9.7] A function is a correspondence that assigns to each member of some set called the domain exactly one member of a set called the range.

function values [9.7] Outputs from equations in function notation are called function values.

G

grade [3.5] The grade of a road is the measure of the road's steepness.

graph [3.1] A graph can be a picture or diagram of the data in a table. A graph can also be a line or a curve that represents all the solutions to an equation.

greatest common factor [5.1] The greatest or largest common factor of a polynomial is the common factor with the largest possible coefficient and the largest possible exponent.

H

higher roots [8.7] Higher roots are cube roots, fourth roots, etc.

hypotenuse [5.7] In a right triangle, the side opposite the right angle is called the hypotenuse.

I

identity property of 0 [1.5] The identity property of 0 states that the sum of a number and 0 is the original number.

identity property of 1 [1.3] The identity property of 1 states that the product of any number and 1 is that number.

imaginary number [9.5] An imaginary number is a number that can be written in the form $a + bi$, where a and b are real numbers and $b \neq 0$.

imaginary number i [9.5] The number $i = \sqrt{-1}$. That is, $i = \sqrt{-1}$ and $i^2 = -1$.

inconsistent system of equations [7.1] A system of equations for which there is no solution is said to be inconsistent.

independent equations [7.1] If a system of equations has exactly one solution or no solution, the equations are independent.

index [8.7] In the radical $\sqrt[3]{a}$, the number 3 is called the index.

inequality [1.4] An inequality is a mathematical sentence using $>$, $<$, $\geq$, or $\leq$.

integers [1.4] The integers consist of all the whole numbers and their opposites.

interpolation [3.1] Interpolation is the process of estimating a value between given values.

inverse variation [7.7] When a situation translates to an equation described by $y = k/x$, with k a constant, we say that y varies inversely as x. The equation $y = k/x$ is called an equation of inverse variation.

irrational number [1.4] An irrational number is a number that cannot be named as a ratio of two integers.

L

leading coefficient [4.2] The leading coefficient of a polynomial is the coefficient of the term of highest degree.

leading term [4.2] The leading term of a polynomial is the term of highest degree.

least common denominator [6.3] The least common denominator of two or more fractions is the least common multiple of the denominators.

least common multiple [6.3] The least common multiple of two or more numbers is the smallest nonzero number that is a multiple of each number.

legs [5.7] In a right triangle, the legs are the two sides that form the right angle.

like radicals [8.4] Like radicals are radical expressions that have a common radical factor.

like terms [1.5] Like terms are terms that have exactly the same variable factors.

line graph [3.1] A line graph is a graph in which quantities are represented as points connected by straight-line segments.

linear equation [3.2] A linear equation is any equation that can be written in the form $y = mx + b$ or $Ax + By = C$.

linear inequality [7.5] An inequality whose related equation is a linear equation is a linear inequality.

M

monomial [4.2] A monomial is a constant, a variable, or a product of a constant and one or more variables.

motion problem [6.7] Problems that deal with distance, speed, and time are called motion problems.

multiplicative inverse [1.3] Multiplicative inverses are two numbers whose product is 1.

multiplicative property of zero [1.7] The multiplicative property of zero states that for any real number a, $0 \cdot a = a \cdot 0 = 0$.

N

natural number [1.3] Natural numbers can be thought of as the counting numbers: $1, 2, 3, 4, 5 \ldots$.

nonlinear equation [3.2] A nonlinear equation is an equation that when graphed is not a line.

numerator [1.3] The number above the fraction line in a fraction is called the numerator.

O

opposite law [1.6] The law of opposites states that for any two numbers a and $-a$, $a + (-a) = 0$.

opposite of a polynomial [4.3] To find the opposite of a polynomial, replace each term with its opposite— that is, change the sign of every term.

opposites [1.6] Two numbers whose sum is 0 are opposites.

origin [3.1] The origin is the point on a coordinate plane where the two axes intersect.

P

parabolas [9.6] Graphs of quadratic equations are called parabolas.

percent [2.4] A percent is a ratio or fraction with denominator 100.

perfect square [8.1] A perfect square is a number that is the square of any rational number.

perfect-square trinomial [5.4] A trinomial that is the square of a binomial is called a perfect-square trinomial.

point–slope equation [3.7] A point–slope equation is an equation of the type $y - y_1 = m(x - x_1)$.

polynomial [4.2] A polynomial is a monomial or a sum of monomials.

prime factorization [1.3] Prime factorization is the factorization of a whole number into a product of its prime factors.

prime number [1.3] A prime number is a natural number that has exactly two different factors: the number itself and 1.

principal square root [8.1] The nonnegative square root of a number is called the principal square root of that number.

principle of zero products [5.6] The principle of zero products states that an equation $AB = 0$ is true if and only if $A = 0$ or $B = 0$, or both.

proportion [6.7] A proportion is an equation stating that two ratios are equal.

Pythagorean theorem [5.7] The Pythagorean theorem states that in any right triangle, if a and b are the lengths of the legs and c is the length of the hypotenuse, then $a^2 + b^2 = c^2$.

Q

quadrants [3.1] Quadrants are the four regions into which the axes divide a plane.

quadratic equation [5.6] A quadratic equation is an equation equivalent to one of the form $ax^2 + bx + c = 0$, where $a > 0$.

quadratic formula [9.3] The solutions of $ax^2 + bx + c = 0$ are given by $x = \dfrac{-b \pm \sqrt{b^2 - 4ac}}{2a}$.

R

radical equation [8.5] An equation in which a variable appears in a radicand is called a radical equation.

radical expression [8.1] A radical expression is an algebraic expression that contains at least one radical sign.

radical sign [8.1] A radical sign, $\sqrt{\ }$, is generally used when finding square roots and indicates the principal root.

radicand [8.1] The expression under the radical is called the radicand.

range [9.7] The range is the set of all the second coordinates of the ordered pairs in a function.

rate [3.4] A rate is a ratio that indicates how two quantities change with respect to each other.

ratio [6.7] A ratio of two quantities is their quotient.

rational equation [6.6] A rational equation is an equation containing one or more rational expressions.

rational expressions [6.1] A rational expression is a quotient of two polynomials.

rational numbers [1.4] Rational numbers are all numbers that can be named in the form $\dfrac{a}{b}$, where a and b are integers and $b \neq 0$.

rationalizing the denominator [8.3] Rationalizing the denominator is a procedure for finding an equivalent expression without a radical in the denominator.

real number [1.4] Real numbers are all the rational and irrational numbers.

reciprocal [1.3] Reciprocals are two numbers whose product is 1.

right triangle [5.7] A right triangle is a triangle with one right angle.

S

scientific notation [4.8] Scientific notation for a number is an expression of the type $N \times 10^m$, where m is an integer, $1 \leq N < 10$, and N is expressed in decimal notation.

set [1.4] A set is a collection of objects.

set-builder notation [2.6] Set-builder notation is a way of naming sets by describing basic characteristics of the elements in the set.

similar triangles [6.7] Similar triangles are triangles in which corresponding angles have the same measure and corresponding sides are proportional.

slope [3.5] Slope is the ratio of the rise to the run for any two points on a line.

slope–intercept equation [3.6] A slope–intercept equation is an equation of the type $y = mx + b$.

solution [1.1] A replacement or substitution that makes an equation or inequality true is called a solution.

solution set [2.6] A solution set is the set of all solutions for an equation or inequality.

solve [2.1] To solve an equation or inequality means to find all solutions of the equation or inequality.

square root [8.1] The number c is a square root of a if $c^2 = a$.

substitute [1.1] To substitute is to replace a variable with a number or an expression that represents a number.

system of equations [7.1] A system of equations is a set of two or more equations that are to be solved simultaneously.

system of linear inequalities [7.6] A system of linear inequalities is a set of two or more linear inequalities that are to be solved simultaneously.

T

term [1.2] A term is a number, a variable, or a product of numbers and/or variables.

trinomial [4.2] A polynomial that is composed of three terms is called a trinomial.

V

variable [1.1] A variable is a letter that represents an unknown number.

vertex [9.6] The point at which the graph of a quadratic equation crosses its axis of symmetry is called the vertex.

vertical-line test [9.7] A graph represents a function if it is impossible to draw a vertical line that intersects the graph more than once.

W

whole number [1.3] The set of whole numbers is the set of numbers: 0, 1, 2, 3, 4 . . . .

X

x-intercept [3.3] A point at which a graph crosses the x-axis is the x-intercept.

Y

y-intercept [3.3] A point at which a graph crosses the y-axis is the y-intercept.

Index

Index of Applications

In addition to the applications highlighted below, there are other applied problems and examples of problem solving in the text. An extensive list of their locations can be found under the heading "Applied problems" in the index at the back of the book.

VIDEOTAPES AND CD INDEX

Text/Video/CD Section	Exercise Numbers	Text/Video/CD Section	Exercise Numbers
Section 1.1	45, 53, 55	Section 6.1	49
Section 1.2	27, 45, 51	Section 6.2	46, 63
Section 1.3	3, 41, 53, 57, 61	Section 6.3	3, 27, 33, 53
Section 1.4	7, 13, 21, 47	Section 6.4	none
Section 1.5	9, 21, 60, 71	Section 6.5	9, 15, 19, 31
Section 1.6	17, 19, 45	Section 6.6	none
Section 1.7	27, 59, 101	Section 6.7	11, 17, 25
Section 1.8	63, 71, 91		
		Section 7.1	3, 5, 18, 22
Section 2.1	7, 29, 43	Section 7.2	23
Section 2.2	27, 71	Section 7.3	9, 15, 21
Section 2.3	2, 4, 11, 21	Section 7.4	17
Section 2.4	7, 21, 31, 57	Section 7.5	1, 25, 27
Section 2.5	none	Section 7.6	5
Section 2.6	11, 25, 33, 79	Section 7.7	9, 19, 24
Section 2.7	39		
		Section 8.1	8, 11, 27, 37, 41, 57
Section 3.1	9, 19, 43	Section 8.2	17
Section 3.2	9, 21, 39	Section 8.3	21
Section 3.3	1, 65, 67	Section 8.4	8, 35
Section 3.4	3, 13, 25	Section 8.5	12, 20, 40
Section 3.5	9, 37, 45, 51, 55	Section 8.6	7
Section 3.6	13, 19, 23, 59	Section 8.7	1, 5, 37, 51, 52
Section 3.7	41		
		Section 9.1	2, 21
Section 4.1	none	Section 9.2	31
Section 4.2	35, 57, 59	Section 9.3	11, 49
Section 4.3	29, 51(a)	Section 9.4	5, 9, 29, 31, 33, 45
Section 4.4	25	Section 9.5	5
Section 4.5	none	Section 9.6	none
Section 4.6	none	Section 9.7	5, 37, 41
Section 4.7	38		
Section 4.8	3, 21, 29, 41, 57, 103, 105, 123		
Section 5.1	13, 21, 47		
Section 5.2	9, 16, 28, 51		
Section 5.3	6, 56		
Section 5.4	1, 11, 23		
Section 5.5	25, 29, 56		
Section 5.6	49, 52, 55		
Section 5.7	8, 24, 33		

This secondary function takes the square root of number displayed.

Squares number displayed.

Activates secondary functions printed above certain keys. Also denoted INV or 2nd.

Used when entering numbers in scientific notation. Also denoted EXP.

Finds reciprocal of number displayed.

Used to raise any base to a power. Also denoted y^x, a^x, or ⌃.

Stores number displayed in memory. Also denoted MIN or M.

Recalls number stored in memory. Also denoted MR.

Clears last number displayed but not preceding operations.

Used when entering decimal notation.

Used to change sign of number displayed.

This secondary function raises 10 to any power entered.

Clears all preceding numbers and operations. Also used to turn calculator on.

Used as an approximation for pi.

Used to perform indicated operation.

Used to control order in which certain operations are performed.

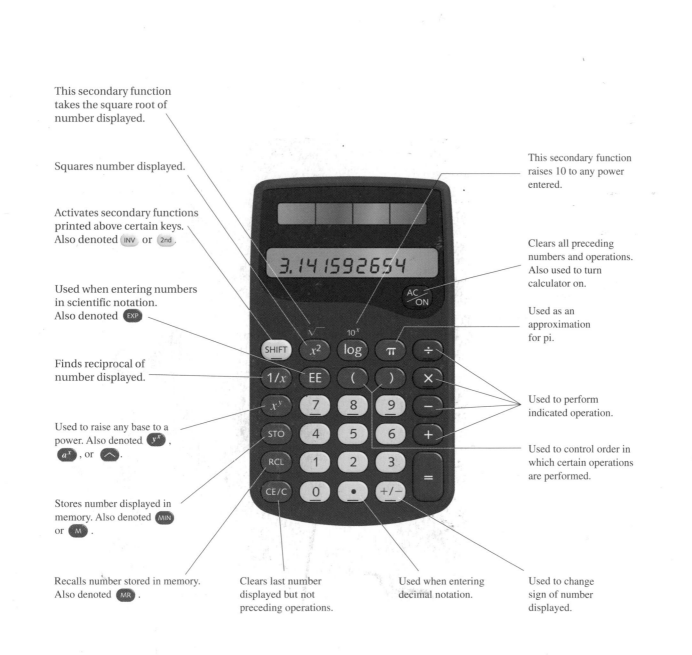